chemistry for the allied health sciences

ROGER D. BAUER
ROBERT L. LOESCHEN
Department of Chemistry, California State University, Long Beach

chemistry

Prentice-Hall, Inc., Englewood Cliffs, New Jersey 07632

for the allied health sciences

Library of Congress Cataloging in Publication Data

BAUER, ROGER D., (date)
Chemistry for the allied health sciences.

Includes index.
1. Biological chemistry. 2. Chemistry, Organic.
3. Chemistry. 4. Allied health personnel.
I. Loeschen, Robert L., (date) joint author.
II. Title.
QP514.2.B367 1980 574.1'92 79-11499
ISBN 0-13-129205-6

Chemistry for the Allied Health Sciences
by Roger D. Bauer and Robert L. Loeschen

Printed in the United States of America
10 9 8 7 6 5 4 3 2 1

Editorial/production supervision by Kim McNeily
Interior design by Mark A. Binn
Manufacturing buyer: Raymond Keating

PRENTICE-HALL INTERNATIONAL, INC., *London*
PRENTICE-HALL OF AUSTRALIA PTY. LIMITED, *Sydney*
PRENTICE-HALL OF CANADA, LTD., *Toronto*
PRENTICE-HALL OF INDIA PRIVATE LIMITED, *New Delhi*
PRENTICE-HALL OF JAPAN, INC., *Tokyo*
PRENTICE-HALL OF SOUTHEAST ASIA PTE. LTD., *Singapore*
WHITEHALL BOOKS LIMITED, *Wellington, New Zealand*

contents

CHAPTER 7

introduction to chemical reactivity 137

CHAPTER 8

acid-base chemistry 163

PART TWO

organic chemistry

CHAPTER 9

introduction to organic chemistry 195

PART THREE
Biochemistry

preface

A basic understanding of chemistry is certainly a prerequisite for the successful study and practice of the allied health sciences. The ability to think in logical terms and to approach problems in a systematic way is part of both these sciences and the natural sciences. Since living cells function because of the great variety of chemical reactions, a study of biochemical principles is important for students and practitioners in health-related fields. Many facets of these professions involve chemistry and biochemistry either directly or indirectly. The better one understands the chemistry of living systems, the more fully one can appreciate the wonder and complexity of the living process.

Most students who will use this text will do so because it is required. They may not regard chemistry as a favorite part of their preparation and training because the required attention to detail and the demands of mastering the fundamentals often tend to overshadow the beauty and order present in living systems. In this text we directed ourselves to these problems by (1) presenting systematic methods of dealing with the quantitative aspects of chemistry, and (2) relating, as much as possible, the topics to those situations encountered in daily life and in the health professions.

The study of chemistry is somewhat like learning and using a new language. One must first learn the vocabulary, grammar, and sentence construction before the literature can be appreciated and conversation is possible. We begin our study with the topics necessary as prerequisites for understanding the chemistry of life processes. The study of general and organic chemistry is presented with the emphasis on understanding the biochemical topics to be covered later in the text. In the sections on biochemistry, frequent reference is made to the preceding material.

Textbooks in introductory science courses may follow one of two modes. In the first, the text may be detailed and exhaustive in its coverage of material. The instructor usually spends the class periods explaining the text and giving the students sufficient background so they may understand the material presented in the text. The second type of text presents a limited amount of basic material which should at least set a minimum level of mastery for the student. After covering the fundamentals, the instructor still has time to develop areas of special relevance or interest to a given group of students. Our text follows the latter format.

This text provides a limited coverage of most topics of interest to students in the paramedical sciences and to others interested in an introduction to chemistry and biochemistry. The approach is toward developing chemistry in its inherently fascinating form with the desire to stimulate the student to achieve at the least a limited understanding of some of the chemical activities of living systems. The book should appeal to students in nursing, home economics, physical therapy, and health education, among other practices. Aside from these students who may spend more than a semester in the study of chemistry, this book should provide a basic treatment of the elementary aspects of chemistry and biochemistry of interest to any liberal arts student in today's universities and colleges. We hope to develop an interest in the great chemical beauty found in all of nature.

This book is divided into a number of sections which have evolved as a fairly standard approach for a book of this nature. The presentation is unique in that the material is subdivided into small portions, called *units*, which cover one general topic. These units are largely self-sufficient and are arranged in sequential fashion. Complementary topics are grouped together and are preceded by an introductory page. A number of topics can be eliminated or assigned for informational reading by the instructor without destroying the overall direction of the book. The instructor can select a number of the basic topics and then add or develop certain other areas as desired. In this way the skeleton of basic information is available in the text, and the teacher is left with the opportunity to present and expand on it in special ways.

We have included quantitative material and problems in this text. Although the students for whom this book is intended often shy away from these areas, we feel that it is essential for them to realize that living systems are a careful balance of quantitative relationships. We believe that this can be kept quite simple and still convey a scientific approach. For example, we use the mole concept almost exclusively and work with solution concentrations in terms of molarity rather than normality. Since the fundamentals remain the same, the overall thought processes are related, and the student can acquire an appreciation of scientific expression with a minimum of mathematical detail.

Many worked examples are given to help the student master the fundamentals of chemistry. Exercises found at the end of each unit cover some of the salient points of the unit and give the student an opportunity to apply and understand what was just presented. A study guide and laboratory manual are also available as part of the learning package. The manual provides exercises of both basic chemistry and applied biochemistry for beginning-level students. Exercises range from the identification of inorganic ions to the quantitative determination of constituents of blood, such as hemoglobin and glucose. As with the text, the laboratory manual offers a good choice of interesting and relevant exercises and examples.

To the Students

If this is your first experience with chemistry you may wonder just how you should study the subject. On the surface you may see many principles and seemingly unrelated topics that simply have to be memorized. As you progress in your studies, you will see that chemistry is systematic, and the principles and theories will begin to fit together in a logical fashion. This is the key to chemistry; you should constantly attempt to integrate the individual facts into the total picture. Upon finishing a particular unit or chapter, go back over it and try to see how it relates to the previous material.

Chemistry is a quantitative science, and you will be asked to learn how to apply the theories as well as understand them. One method of accomplishing both is to do as many practice problems as possible. The beginning exercises at the end of each unit are designed to help you work with a specific theory or concept, whereas the problems nearer the end are intended to help you see the relationships between specific facts and general trends. We urge you to work the problems on paper and then "talk yourself through them." Go through the problem step by step and make sure you know the reason for each step. If you experience difficulty, go back to that portion of the unit that discusses similar problems, read it again, and then redo the same or similar problems. In doing this you will learn general methods of approach to problems, not just how to work the specific problems presented in the text.

We wish you much success in your study of chemistry and biochemistry as presented in this text. In chemistry, as in anything else that is worthwhile, *steady attention*, *application*, and *plain hard work* are required. There is no easy way, and few shortcuts can be taken if you expect to arrive at a satisfactory level of understanding. Chemistry is an orderly discipline that requires an orderly approach to its mastery. By learning to study chemistry, you will learn how to learn. That's really what education is all about. As you work your way through this book, we believe that you will develop a new appreciation of the world around you and a feeling of genuine amazement at the wonder and grandeur of nature. With this will come a feeling of satisfaction at achieving a mastery and understanding of a fundamental subject.

ROGER D. BAUER

ROBERT L. LOESCHEN

Long Beach, California

PART ONE

general chemistry

CHAPTER 1

matter and energy

There are two basic entities in our world—matter and energy. Matter is anything that occupies space and has substance; energy is the source of all movement and life. In units 1 and 2 we introduce basic definitions and concepts that deal with matter and energy.

Unit 1 is concerned, primarily, with the introduction of terms and units of measurement for the metric system, the system of measurement that we shall use in this text. In the first unit we introduce the units of measurement for length, weight, volume, temperature, pressure, and density. In the latter part of the unit we present a systematic method for changing one set of units to another. This method has worked well for many of our students; we urge you to give it a serious try.

In unit 2 we introduce and define the various forms in which energy is said to exist. We also discuss transformations of energy: energy in one form is converted to another form. We also introduce the topic of chemical reactions and pave the way for the beginning of our study of chemistry.

UNIT 1 UNITS OF MEASUREMENT

The measurement of quantities such as distance, volume, weight, and temperature is a common experience. To perform these measurements, we must first adopt basic units as standards and then compare the object to be measured with these standard units. There are several different sets of standards used for measuring distance, volume, and weight, but perhaps the most common are the English system and the metric system. The metric system is used in most countries. The United States is in the process of changing from the English to the metric system. Table 1–1 lists some common units in both systems.

Table 1–1 Units of measurement in the English and metric systems

	English system	Metric system
Length	Inch, foot, mile	Centimeter, meter
Volume	Pint, quart, gallon	Milliliter, liter
Weight	Ounce, pound	Gram, kilogram

In the English system there are many different units for each type of measurement. Conversions between these units are not always easy, since we must know a separate equality for each conversion. For example, to convert inches to feet, yards, and miles, we must know that 12 inches equals 1 foot, that 36 inches equals 1 yard, and that 5280 feet equals 1 mile. If we must work with many of these different units, the problem of keeping track of the equalities often may be more time-consuming and tedious than making the measurements.

The metric system is the system of measurements preferred throughout the world. It is based on units of 10, so conversions are made simply by multiplying or dividing by powers of 10. To illustrate this process, consider the monetary system used in the United States, which is also based on units of 10. One dollar is equivalent to 10 dimes or 100 (10^2) pennies. To convert dollars into dimes, you must multiply by 10; to convert dollars into pennies, you multiply by 100. These multiplications can be done easily and quickly by moving the decimal places in the numbers.

$$5.00 \text{ dollars} = 50.0 \text{ dimes} = 500. \text{ pennies}$$

In the metric system, 1 **meter** is equivalent to 100 **centimeters** or 1000 **millimeters**. One **liter** is equivalent to 1000 **milliliters**, and a **kilogram** is equivalent to 1000 **grams**. All the conversions can be done by multiplying or dividing by some power of 10, that is, by moving the decimal points.

The metric system is also convenient to use because the units are memorized more easily than those in the English system. Each unit in the metric system consists of two parts—a prefix and a base. The **base** indicates whether we are measuring length, volume, or weight. It also gives the identity of the **standard unit**. The **prefix** (see Table 1–2) indicates what portion of the standard base is being used. The prefixes are the same for all measurements: length, volume, or weight. Thus, to use the metric system you must learn just one set of prefixes and three base units.

centi	meter	kilo	gram	milli	liter
Prefix	**Unit of length**	**Prefix**	**Unit of weight**	**Prefix**	**Unit of volume**
1/100 of the standard	meter is a standard unit	1000 of the standard	gram is a standard unit	1/1000 of the standard	liter is a standard unit

Table 1–2 Metric system prefixes

deci (d)	$\frac{1}{10}$	10^{-1}	When these prefixes are used, the unit of measure is smaller than the base unit.
centi (c)	$\frac{1}{100}$	10^{-2}	
milli (m)	$\frac{1}{1000}$	10^{-3}	
micro (μ)	$\frac{1}{1,000,000}$	10^{-6}	
nano (n)	$\frac{1}{1,000,000,000}$	10^{-9}	
deka (D)	10	10^{1}	When these prefixes are used, the unit of measure is larger than the base unit.
hecto (h)	100	10^{2}	
kilo (k)	1000	10^{3}	
mega (M)	1,000,000	10^{6}	
giga (G)	1,000,000,000	10^{9}	

Length

The basic unit of length in the metric system is the **meter** (abbreviated m). Table 1–3 gives the equivalents of the English system in the metric system. The prefix tells what multiple of a meter is present. One **centi**meter (cm) is one hundredth of a meter or 0.01 meter. There are 100 centimeters in 1 meter. One **kilo**meter (km) is 1000 meters and one thousandth of a kilometer is equal to 1 meter.

Table 1–3 Units of length

1 micrometer (μm)	= 0.001 millimeter
	= 1×10^{-6} meter
1 millimeter (mm)	= 1×10^{-3} meter
	= 0.1 centimeter
1 centimeter (cm)	= 0.01 meter
1 decimeter (dm)	= 0.1 meter
1 kilometer (km)	= 1000 meters
1 inch (in.)	= 2.54 centimeters
1 foot (ft)	= 30.5 centimeters
1 yard (yd)	= 0.914 meter
1 mile (m)	= 1.60 kilometers

Mass and Weight

Usually, there is confusion about the difference between the mass and the weight of an object. The **mass** of an object is the quantity of matter in that object. The mass is constant and does not change. The **weight** of an object on earth is a measure of the gravitational attraction between the object and the earth. The weight of that object depends upon the mass of the object, the mass of the earth, and the distance of the object from the center of the earth. On the moon, the attractive force of gravity is only about one-sixth of that on the earth, so an object that weighs 600 pounds on earth would weigh only 100 pounds on the moon. Since the weight of an object remains essentially constant everywhere on the earth's surface, the terms *mass* and *weight* are commonly used interchangeably for measurements made on the earth.

The basic unit of mass in the metric system is the **gram** (g). Table 1–4 lists units of mass. The units most often used in chemistry are grams and milligrams.

Table 1–4 Units of mass

1 microgram (μm) =	0.000001 gram =	1×10^{-6} gram
1 milligram (mg) =	0.001 gram =	1×10^{-3} gram
1 centigram (cg) =	0.01 gram =	1×10^{-2} gram
1 kilogram (kg) =	1000 grams	
1 pound (lb) =	454 grams	
1 kilogram =	2.204 pounds	

Volume

The amount of space that an object occupies is called its **volume**. In the metric system the basic unit is the **liter** (ℓ). One **milli**liter (mℓ) is one thousandth of a liter, and 1000 milliliters makes 1 liter. The milliliter is perhaps the most widely used unit of volume in the metric system and is sometimes referred to as a cubic centimeter (cm^3 or cc). We shall use mℓ exclusively to avoid confusion. Table 1–5 gives volume units and conversions, and Figure 1–1 gives a visual comparison for some units.

Table 1–5 Units of volume

1 microliter ($\mu\ell$) =	1×10^{-3} milliliter
1 microliter =	1×10^{-6} liter
1 milliliter (mℓ) =	1×10^{-3} liter
1 kiloliter (kℓ) =	1×10^{3} liters
1.057 quarts (qt) =	1 liter
1 quart =	0.96 liter

Density

The **density** of a substance is defined as the *mass per unit volume*. It gives the relationship between the mass of an object and its volume. It is expressed mathema-

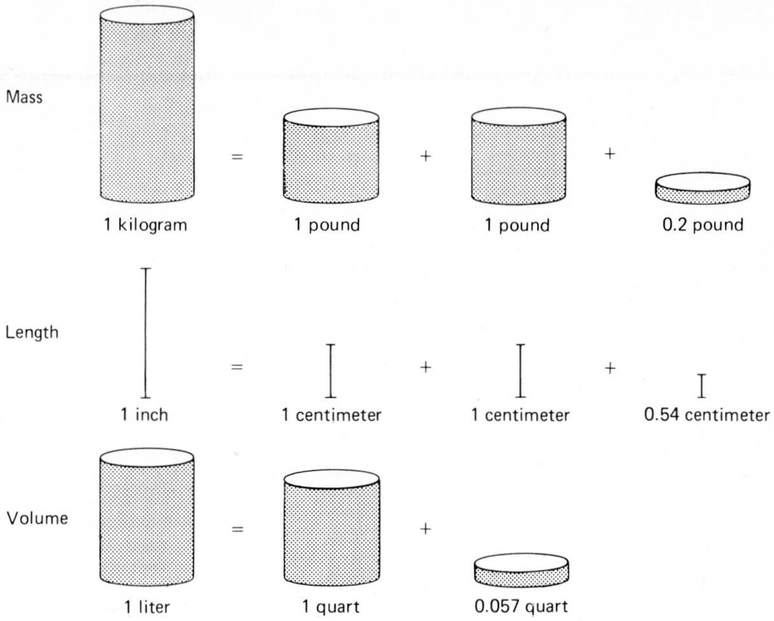
Mass
=
+
+
1 kilogram
1 pound
1 pound
0.2 pound
Length
=
+
+
1 inch
1 centimeter
1 centimeter
0.54 centimeter
Volume
=
+
1 liter
1 quart
0.057 quart

Figure 1–1 Visual comparison of the metric and English measurements.

tically as $D = M/V$, where D = density, M = mass, and V = volume. If the mass is expressed in grams and the volume is expressed in milliliters (or cubic centimeters), then the unit for density is grams per milliliter. Water has a density of approximately 1 gram per milliliter, and liquid mercury has a density of 13.5 grams per milliliter. A given volume of mercury would weigh 13.5 times as much as an equivalent volume of water.

Density values are commonly used to convert between units of weight and volume. If we know the weight and density of a substance, we can calculate its volume. Conversely, we can calculate the weight of a given object if we know its density and volume.

Pressure

When you attempt to move an object by pushing it, you are applying an amount of force to that object. When you sit on an object, you are applying a downward gravitational force equal to your weight on the object.

The term **pressure** combines the force applied to an object with the surface area of the object. Pressure is defined as the *force per unit area.* In the English system, the commonly used unit for pressure is pounds (force) per square inch (surface area), or psi. In the metric system the units for force are newtons (N; kilograms × 9.8) or dynes (dyn; grams × 980). Pressure is measured in either newtons per square meter or dynes per square centimeter. To illustrate the difference between force and pressure, we shall calculate the amount of pressure a woman weighing 51 kg (115 lb) would exert if she stood on her heels in two different types of shoes. A 51-kg woman can exert a total gravitational force of 5×10^7 dynes (5.1×10^4 g $\times$ 980), no matter what surface area she is standing on. If the size of a spike heel on dress shoes is 0.5 cm by 0.5 cm, its surface area would be 0.25 cm^2 (0.5 cm $\times$ 0.5 cm); for two heels, it would be 0.50 cm^2. The pressure the 51-kg woman would exert standing on these heels is

$$\frac{5 \times 10^7 \text{ dyn}}{0.5 \text{ cm}^2} = \frac{1 \times 10^8 \text{ dyn}}{\text{cm}^2}$$

In a more casual shoe (heel size 5 cm by 5 cm), the surface area of the heels of two shoes would be 50 cm^2, and the pressure would be

$$\frac{5 \times 10^7 \text{ dyn}}{50 \text{ cm}^2} = \frac{0.01 \times 10^8 \text{ dyn}}{\text{cm}^2}$$

The increase in surface area has the effect of greatly diminishing the pressure even though the amount of force has not changed. This is the reason that travelers wear snowshoes in heavy snow. The surface area of the snowshoe is much greater than that of common shoes, and thus the pressure of the traveler on the surface of the snow is greatly diminished. It is possible to walk over the snow rather than through it (see Figure 1–2).

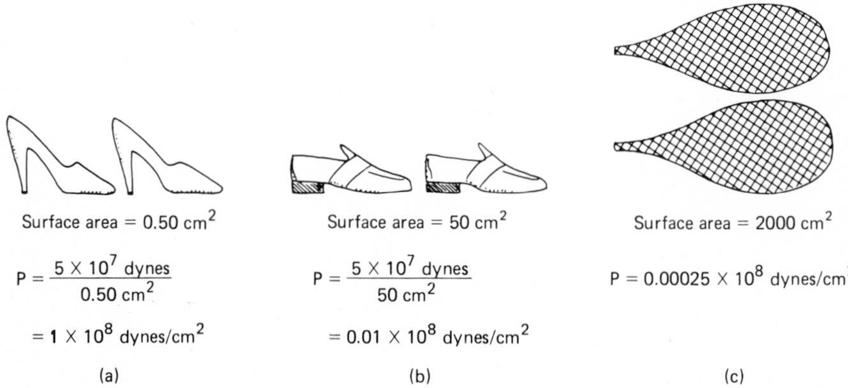

Surface area = 0.50 cm^2
$P = \frac{5 \times 10^7 \text{ dynes}}{0.50 \text{ cm}^2}$
$= 1 \times 10^8$ dynes/cm^2
(a)
Surface area = 50 cm^2
$P = \frac{5 \times 10^7 \text{ dynes}}{50 \text{ cm}^2}$
$= 0.01 \times 10^8$ dynes/cm^2
(b)
Surface area = 2000 cm^2
$P = 0.00025 \times 10^8$ dynes/cm^2
(c)

Figure 1–2 The effect of surface area on pressure. A force of 5×10^7 dynes is applied to (a) the heels of "spike" shoes, (b) the heels of loafers, and (c) snowshoes.

The pressure exerted on the surface of the earth by the gases in our atmosphere is called **atmospheric pressure**. It is commonly measured by using a **barometer**, a device that measures the height of a column of mercury whose pressure is equivalent to that of the atmosphere (see Figure 1–3). At sea level the average atmospheric pressure is equivalent to the pressure exerted by a column of mercury that is 760 mm high; thus 1 atmosphere of pressure is said to be 760 mm of mercury. At elevations higher than sea level, there is less air pushing down on the earth's surface, so atmospheric pressure is less than 760 mm. In Denver, Colorado (altitude = 1 mi), the average barometric pressure is about 620 mm.

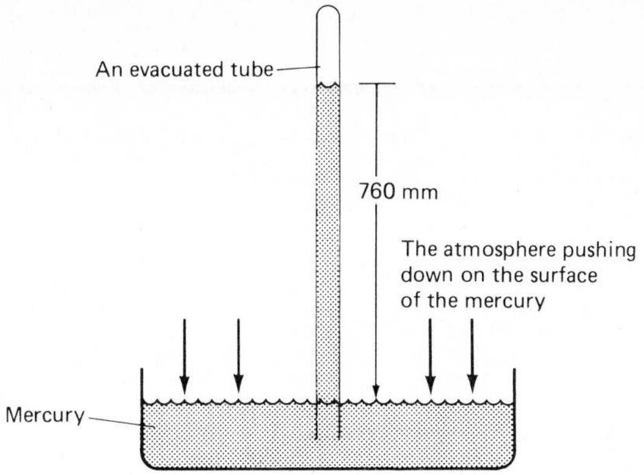

An evacuated tube
760 mm
The atmosphere pushing down on the surface of the mercury
Mercury

Figure 1–3 A barometer. As the atmosphere pushes down on the surface of the mercury, it is forced up the tube to a height of 760 mm. One atmosphere of pressure at sea level will support a column of mercury that is 760 mm high.

Temperature

The temperature of a substance is a measure of the average hotness of that substance. It is a function of the amount of thermal energy present in a substance, but it is not an absolute measurement. For example, if a straight pin and a steel rod were placed in a flame, both objects would have the same temperature. But which object would you rather hold in your hand? The steel rod would cause much greater burns since it possesses much more thermal energy than the pin. So when we want to measure how hot something is, we record its temperature. To determine how much thermal energy is present, we must perform other measurements.

As shown in Figure 1–4, three different systems are used for the measurement of temperature: the **Fahrenheit** (°F) scale, the **Celsius** scale (°C), and the **Kelvin** scale (°K). On the Fahrenheit scale, water freezes at 32°F and boils at 212°F. On the Celsius scale, water freezes at 0°C and boils at 100°C. On the Kelvin scale, water freezes at 273°K and boils at 373°K. Equations (1–1) and (1–2) show how to convert from the Fahrenheit to the Celsius scales, and vice versa.

$$°C = \frac{5}{9}(°F - 32) \qquad (1\text{–}1)$$

$$°F = \frac{9}{5}°C + 32 \qquad (1\text{–}2)$$

Neither the Celsius nor the Fahrenheit scales establishes a minimum at zero, and subzero temperatures are common. But how cold can it be? Is a minimum temperature possible? Yes, the minimum temperature is −273.16°C (we shall round it off to −273). If we wish to set zero as the lowest temperature attainable, we use the Kelvin scale. Thus 0°K = −273.16°C (−273). Water freezes at 273°K and boils at 373°K. When scientists need temperature values in calculations concerning chemical reactions, the Kelvin scale must be used. When a simple measure of temperature is all that is required, the Celsius scale is used. Conversions between the two scales

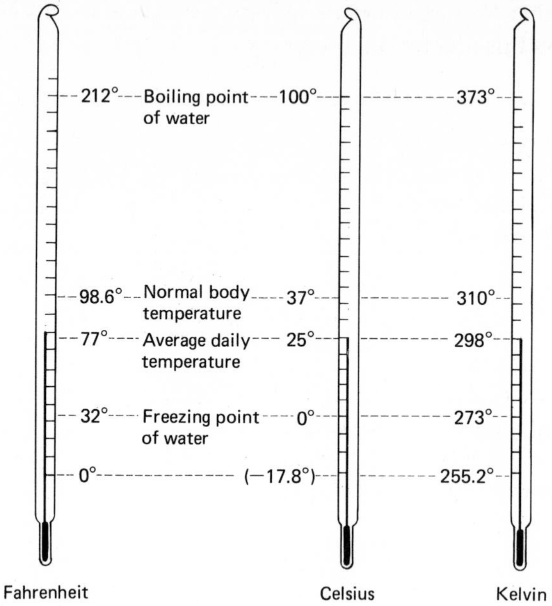
212°
Boiling point of water
100°
373°
98.6°
Normal body temperature
37°
310°
77°
Average daily temperature
25°
298°
32°
Freezing point of water
0°
273°
0°
(−17.8°)
255.2°
Fahrenheit
Celsius
Kelvin

Figure 1–4 The Fahrenheit, Celsius, and Kelvin scales.

are easy. To change °C into °K, simply add 273; to change °K into °C, simply subtract 273.

$$°K = °C + 273, \qquad °C = °K - 273 \tag{1–3}$$

Conversion of Units

At some time everyone has had to deal with changing units of numbers and with converting from one system of measurement to another. You may need to change tablespoons into cups, feet into miles, dollars into nickels, ounces to quarts, and so forth. One problem most students encounter when asked to do conversions deals with the mechanics of the change. If one has to change feet into inches, most people know intuitively that they must multiply by 12. To change inches into feet, most people know that they have to divide by 12. But when working with units that are less familiar, it is easy to become confused as to whether to multiply or divide by the conversion factor. The factor dimensional approach presents a standard technique for setting up and working these types of problems. We shall describe the procedure, give some examples, and then review.

Multiplication by Identities and Canceling

An **identity** is an algebraic expression that equates two separate units of measurement. When we write

$$12 \text{ in.} = 1 \text{ ft}$$

we mean that 12 in. and 1 ft are measures of the same distance and are equivalent. If we express this identity as a fraction

$$12 \text{ in.}/1 \text{ ft} \quad \text{or} \quad 1 \text{ ft}/12 \text{ in.}$$

either fraction is equivalent to unity. If we multiply a measurement by an identity fraction, we do not change the absolute value of that measurement; we change only the units of measure. For example, if we multiply 5 ft by the identity fraction 12 in./1 ft, we arrive at a value of 60 in.

$$5 \text{ ft} \times \frac{12 \text{ in.}}{1 \text{ ft}} = \frac{5 \text{ ft} \times 12 \text{ in.}}{1 \text{ ft}} = 60 \text{ in.}$$

The units of feet in this expression can be canceled since they occur both in the numerator and denominator. The *numbers* are not canceled, only the units. Multiplication of a unit of measurement by an identity fraction does not change the value of the size of the measurement; only the units in which that measurement is expressed are changed (see Table 1–6).

Table 1–6 Identities

Identity	Possible identity fractions
2.54 cm = 1 in.	2.54 cm/1 in. or 1 in./2.54 cm
1.06 qt = 1ℓ	1.06 qt/1 ℓ or 1 ℓ/1.06 qt
454 g = 1 lb	454 g/1 lb or 1 lb/454 g
5280 ft = 1 mi	5280 ft/1 mi or 1 mi/5280 ft
16 tbs = 1 c	16 tbs/1 c or 1 c/16 tbs

To cancel a unit, it must appear in both the numerator and denominator of an expression. Thus, when multiplying by identity fractions, you must take care to place the units in such a manner that they will cancel.

Example: Change feet into inches. We have two possible identity fractions that relate feet to inches:

$$\frac{12 \text{ in.}}{1 \text{ ft}} \quad \text{or} \quad \frac{1 \text{ ft}}{12 \text{ in.}}$$

Since we started with feet (1 ft = 1 ft/1), the first identity fraction (inches/feet) must be used so as to have feet on the bottom. Thus the units of feet can be canceled.

$$\frac{5 \text{ ft}}{1} \times \frac{12 \text{ in.}}{1 \text{ ft}} = 5 \times 12 \text{ in.} = 60 \text{ in.}$$

If we had used the other fraction, nothing would have canceled.

$$\frac{5 \text{ ft}}{1} \times \frac{1 \text{ ft}}{12 \text{ in.}} = \frac{5 \text{ ft} \times \text{ft}}{12 \text{ in.}}$$

Steps in Problem Solving by the Factor Dimensional Method

STEP 1: *Identify the units given in the problem and the units that the answer calls for.* If this is done incorrectly, the problem cannot be worked properly. Too often students speed through this step, not really knowing where to begin and where to end. Often in a word problem the answer units are preceded by the words *how many* or *into*, and the starting units are preceded by *are present in* or *are found in* or *convert.*

Example: (a) How many centimeters are present in 10 feet?

↑ Answer units (centimeters) ↑ Starting units (feet)

Problem: Change feet into centimeters.

feet ⟶ centimeters

(b) Convert 2 gallons into liters.

↑ Starting units (gallons) ↑ Answer units (liters)

Problem: Change gallons to liters.

gallons ⟶ liters

STEP 2: Set up your plan of attack. *After identifying the beginning and end units, look up identities that will help you change from the initial to the final units.* Write down a roadmap showing how the units are to be changed.

Example: Suppose that you were required to change inches into yards. In the identity table, several identities contain inches and/or yards.

$$12 \text{ in.} = 1 \text{ ft}$$
$$1 \text{ in.} = 2.54 \text{ cm}$$
$$3 \text{ ft} = 1 \text{ yd}$$
$$36 \text{ in.} = 1 \text{ yd}$$
$$5280 \text{ ft} = 1 \text{ mi}$$
$$1 \text{ mi} = 1.6 \text{ km}$$

The most direct path between inches and yards would be to choose the identity 36 in. = 1 yd. In this case you could convert directly from inches to yards simply by multiplying by a single identity fraction.

inches ⟶ yards

$$\cancel{\text{inches}} \times \frac{\text{yards}}{\cancel{\text{inches}}} = \text{yards}$$

Example: Suppose that you were required to change inches into miles. There is no single identity connecting inches and miles in the given identity table, so you may have to go through intermediate units and then cancel as you proceed to your goal. Two identity fractions are required in this case.

$$\text{inches} \longrightarrow \text{feet} \longrightarrow \text{miles}$$

$$\cancel{\text{inches}} \times \frac{\cancel{\text{feet}}}{\cancel{\text{inches}}} \times \frac{\text{miles}}{\cancel{\text{feet}}} = \text{miles}$$

We call this schematic diagram of the plan of attack a *roadmap*, as it shows you how to get from one set of units to another. This often requires going through other units first.

STEP 3: *Set up and place the identity fractions in the roadmap so that all the units but the final one cancel.*

(a) The number of arrows in a roadmap tells you how many identity fractions will be needed.

(b) The units in each identity fraction are simply those units that are connected by the arrow in a roadmap. In the preceding example, the first fraction must contain inches and feet, and the second fraction must contain feet and miles.

(c) Always start with the identity fraction for the first arrow, and then add the others in orderly fashion. This allows you to make sure that the units cancel as you proceed. For example,

$$\text{inches} \longrightarrow \text{feet} \longrightarrow \text{miles}$$

$$\cancel{\text{inches}} \times \frac{\cancel{\text{feet}}}{\cancel{\text{inches}}} \times \frac{\text{miles}}{\cancel{\text{feet}}}$$

Note that we can check for proper order and cancel before worrying about the numbers. After canceling, the only units remaining are miles, which is the unit that we wished to convert to.

STEP 4: *Insert the numbers and perform the necessary simplifying arithmetic.* The only problem here is putting the numbers where they belong.

Example: How many centimeters are present in 5 ft?

Step 1: feet → centimeters

Step 2: feet → inches → centimeters, since equalities are known to exist between feet and inches, and inches and centimeters.

Step 3: $\frac{5\,\cancel{\text{feet}}}{1} \times \frac{\cancel{\text{inches}}}{\cancel{\text{feet}}} \times \frac{\text{centimeters}}{\cancel{\text{inches}}}$. All units but centimeters (cm) cancel, so the problem must have been set up correctly.

Step 4: $\frac{5\text{ feet}}{1} \times \frac{12\text{ inches}}{1\text{ foot}} \times \frac{2.54\text{ centimeters}}{1\text{ inch}} = 5 \times 12 \times 2.54\text{ centimeters} =$ 152.3 centimeters

Example: How many tablespoons are present in 2 gallons of water?

Step 1: gallons → tablespoons

Step 2: There are a number of possible roadmaps you could use, depending upon the identities available. Obviously, you should choose the shortest route.

(a) gallons → quarts → pints → cups → tablespoons.

(b) gallons → quarts → cups → tablespoons.

(c) gallons → cups → tablespoons.

Steps 3 and 4: We will demonstrate roadmap (a):

$$\frac{2\ \cancel{\text{gallons}}}{1} \times \frac{4\ \cancel{\text{quarts}}}{1\ \cancel{\text{gallons}}} \times \frac{2\ \cancel{\text{pints}}}{1\ \cancel{\text{quarts}}} \times \frac{2\ \cancel{\text{cups}}}{1\ \cancel{\text{pints}}} \times \frac{16\ \text{tablespoons}}{1\ \cancel{\text{cup}}} = 512\ \text{tablespoons}$$

UNIT 1 Problems and Exercises

1–1. What prefixes in the metric system are indicated by the following multipliers? (Example: 1/100 = centi.)
(a) 1/1000 (b) 1000 (c) 10 (d) 1×10^{-6}
(e) 1×10^{6}

1–2. Qualitatively examine the two units in each set below and compare them by using "about equal to," "much greater than," or "much less than."
(a) A quart and a liter (b) A meter and a yard
(c) A centimeter and a foot (d) A kilometer and a mile
(e) A gallon and 4 liters (f) Two pounds and a kilogram
(g) A centimeter and an inch (h) A pound and a gram

1–3. Units in the metric system can be changed by moving the decimal point to the right or left. In the problems below, practice learning how to change units by moving the decimal point. In each case tell how many places the decimal must be moved and in which direction it must be moved. (Example: To change milliliters to liters, move the decimal three places to the left.)
(a) 40mℓ to liters (b) 0.2 ℓ to milliliters
(c) 100 g to kilograms (d) 2 kg to grams
(e) 10 cm to meters (f) 0.05 m to centimeters
(g) 10 mm to centimeters (h) 10 mm to meters

1–4. Without looking at the conversion tables, tell which of the following pairs of units is the smallest. Then write out the conversion factor between the units.
(a) Centimeter and millimeter (b) Milliliter and liter
(c) Meter and centimeter (d) Kilogram and gram
(e) Milligram and gram (f) Microliter and milliliter

1–5. Measure the following objects to get an idea of the sizes of units in the metric system.
(a) The diameter and thickness of a dime (in millimeters and centimeters)
(b) The nail of your little finger (in millimeters and centimeters)
(c) The distance from your elbow to your finger tip (in centimeters and meters)
(d) Your waist (in centimeters and meters)
(e) Your height (in centimeters and meters)

1–6. Set up and perform the following conversions. Use the examples given in the text as a format, and obtain conversion factors from the tables.
(a) How many inches are present in 5.08 cm?
(b) How many feet are present in 15.25 cm?

Table 2–1 Examples of elements, compounds, and mixtures

Elements (simplest forms of matter)	Compounds (combinations of elements)	Mixtures (groupings of elements and compounds)
[A] [B] [C]	[A B] or [B C]	[A B] + [A C] + [B C]
or or		
[A] [B] [C]	or [A B C]	[A B C] + [C]
Sodium [Na]	Sodium chloride [NaCl]	Sodium chloride mixed with sodium bromide
Chlorine [Cl]	Sodium bromide [NaBr]	
Bromine [Br]	Water [H_2O]	Water mixed with oil
Hydrogen [H]	Oil [$C_{15}H_{32}$]	
Oxygen [O]		
Carbon [C]		

Physical States of Matter

Matter exists in three different physical states, as a **solid**, **liquid**, or **gas**. The particular physical state of an element or compound is dependent on its own identity and the temperature. Water exists as a solid (ice) below temperatures of 0°C, as a liquid between 0 and 100°C, and as a gas above 100°C at atmospheric pressure. But mercury exists as a solid below −39°C, as a liquid between −39 and 357°C, and as a gas above temperatures of 357°C. Each physical state has unique physical properties associated with it. These properties shall be presented in units 11 and 12.

Energy

Energy is not something that we can see, handle, or taste; it is a property possessed by matter. If a substance has the ability to cause movement, increases in temperature, electrical currents, or chemical changes, the substance possesses energy in one form or another. For example, our muscles possess energy since they can cause movement of various parts of the body. Furthermore, since muscles receive energy from the metabolism of foods, foods themselves must possess energy. Fuels like gasoline and natural gas possess energy because they emit much heat when burned. A rock, suspended in air above the ground, has energy, because, when released, the rock will plunge to the earth and cause movement of objects there. Sunlight possesses energy because it causes plants to grow and warms our planet. Let us look now at the various forms in which energy can exist.

Kinetic Energy

A moving object is said to possess **kinetic energy**, so kinetic energy is the energy associated with motion. The amount of kinetic energy an object has depends upon the *mass* of that object and the speed at which the object is traveling (its velocity):

$$\text{Kinetic energy (KE)} = 1/2m \cdot v^2 \qquad (2\text{–}1)$$

where m = mass

v = velocity.

An automobile weighing 4000 lb (about 1800 kg) possesses twice as much kinetic energy as an automobile weighing 2000 lb (900 kg) if both are traveling at the same speed. But an automobile traveling at 60 mph possesses greater than twice the kinetic energy of an automobile traveling at 30 mph, since the velocity portion of formula (2–1) is squared. We should use automobiles that are light in weight and we should drive them at slower speeds in order to save the fuel that is used to provide kinetic energy.

Gravitational Energy

An attractive force exists between the earth and the objects on or near the surface of the earth. This attraction is called the **force of gravity**; it is what causes objects to fall toward the earth. Any unit of matter suspended at some distance above the center of the earth possesses gravitational energy. The amount of gravitational energy is proportional to the mass of the object and its height above the center of the earth.

$$\text{Gravitational energy (GE)} = m \times h$$

where m = mass of the object

h = distance above the center of the earth.

Thermal Energy

All substances possess thermal energy. The amount of thermal energy present is measured in units called **calories**:

> **Calorie**: the amount of heat required to increase the temperature of 1 g of water by 1°C (specifically from 14.5 to 15.5°C).

The process of measuring the heat content of matter is called **calorimetry**. Basically, in this process one immerses a heated object into a known amount of water, and then measures the increase in temperature of the water. The product of the temperature change (ΔT) and the mass of the water equals the number of calories of heat energy transferred to the water from the heated object.

Example: A heated iron rod weighing 10 g is plunged into a calorimeter containing 100 g of water. The temperature of the water rises from 30 to 40°C. How many calories of heat were transferred to the water?

No. of calories transferred = temperature change of water × no. of grams water × calorie conversion factor

$$\text{Temperature change} = T_{\text{final}} - T_{\text{initial}} = \Delta T = 10°\text{C}$$

$$\text{No. of grams of water} = 100 \text{ g}$$

$$\text{Calorie conversion factor for water} = \frac{1 \text{ cal}}{1°\text{C } 1 \text{ g}}$$

$$\text{No. of calories transferred} = 10°\text{C} \times 100 \text{ g} \times \frac{1 \text{ cal}}{1°\text{C } 1 \text{ g}}$$

$$= 1000 \text{ cal}$$

One important point should be mentioned before we go on to other forms of energy. The amount of thermal energy present in a substance is not determined by simply measuring its temperature. Temperature is a measure of the average *hotness* of a substance, not a measure of the amount of heat present. The average hotness times the mass is proportional to the amount of thermal energy present.

In nutrition, the basic unit of energy content for foods is the **Calorie** (big calorie). The nutritional Calorie is equivalent to 1000 cal (1 kcal) of thermal energy and is usually distinguished by using a capital C. Thus

$$1 \text{ Cal (nutrition)} = 1000 \text{ cal (thermal)}$$
$$= 1 \text{ kcal (thermal)}$$

Light Energy

A beam of light can cause many events to occur, such as an increase in temperature, a physical movement (the pupil of your eye contracts when light hits it), or a chemical change (plants grow and skin color darkens). A beam of light is also a source of energy-radiant energy.

The exact nature of light is somewhat difficult to describe; no single theory can explain all its properties. Some properties of light are best explained by considering light to be a series of waves, like the waves that travel on the surface of water. The waves of light travel through space at a speed of 3×10^{10} cm/sec. The type or color of the light wave is determined by the length of each repeating unit in the wave, the **wavelength** (see Figure 2–1). Actually, light is simply a form of electromagnetic radiation that possesses wavelengths visible to the naked eye.

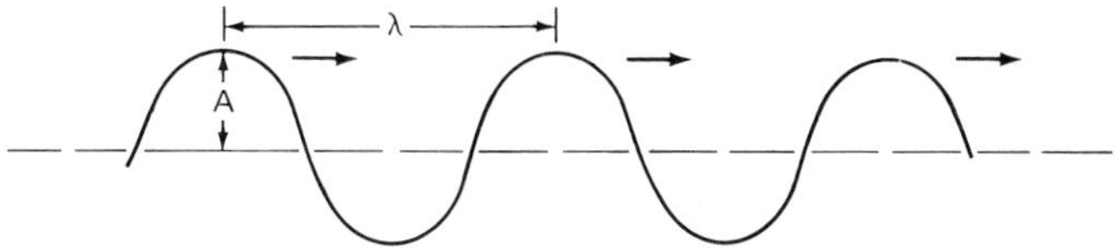
λ
A

Figure 2–1 A light wave. λ = wavelength, the length of each repeating unit, A = amplitude, the intensity of the light wave, and v = frequency, the number of wavelengths that pass a point in space each second.

The relation of energy to wavelength for electromagnetic radiation is shown in Figure 2–2. The energy associated with light (electromagnetic radiation) is best explained if we consider light to be a stream of particles called **photons**. Each particle

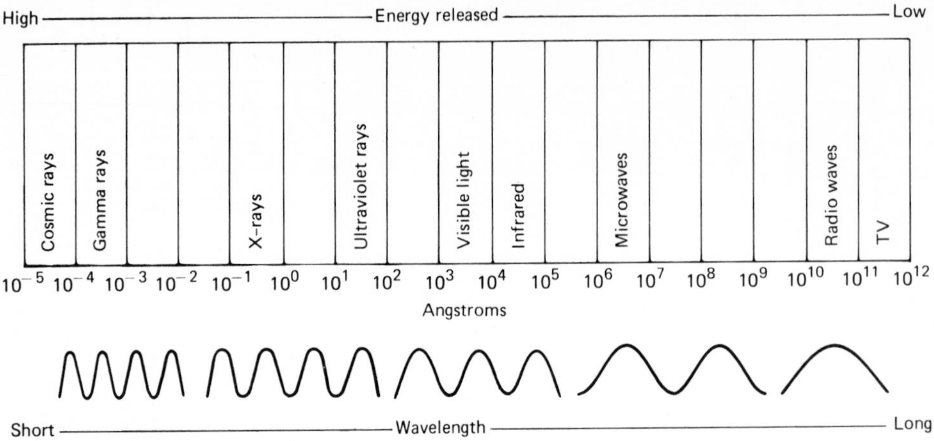
High
Energy released
Low
Cosmic rays
Gamma rays
X-rays
Ultraviolet rays
Visible light
Infrared
Microwaves
Radio waves
TV
10⁻⁵ 10⁻⁴ 10⁻³ 10⁻² 10⁻¹ 10⁰ 10¹ 10² 10³ 10⁴ 10⁵ 10⁶ 10⁷ 10⁸ 10⁹ 10¹⁰ 10¹¹ 10¹²
Angstroms
Short
Wavelength
Long

Figure 2–2 The relation of energy to wavelength for electromagnetic radiation. (1 angstrom = 1×10^{-8} cm.)

has a particular energy associated with it, and this amount of energy is a function of the wavelength or frequency of the light:

$$E = \frac{hc}{\lambda} \quad \text{or} \quad E = h \cdot v \tag{2–2}$$

where h = constant, a number

c = speed of light

λ = wavelength of light

v = frequency of the light.

Objects that absorb light energy must absorb entire photons, not just portions of them. It is a little like buying eggs. You must buy whole eggs; you cannot buy $1\frac{1}{2}$ or $2\frac{1}{4}$ eggs.

Electrical Energy

For our purposes, we can consider **electrical energy** to be energy associated with matter that bears positive and negative charges. A **repulsive** electrical force exists between two units of matter that bear the same charge. An **attractive** force exists between two units of matter that bear opposite charges (see Figure 2–3).

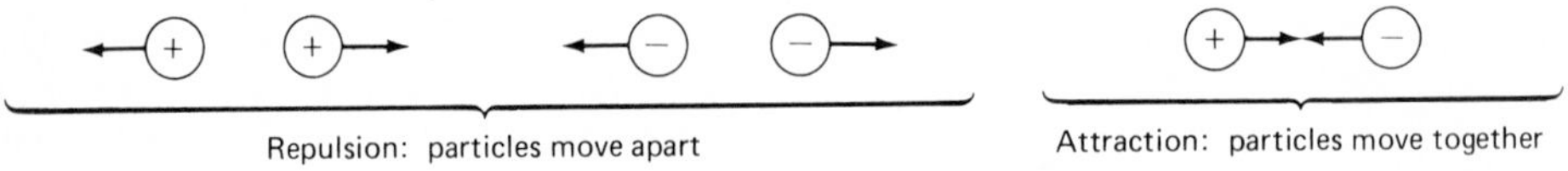
Repulsion: particles move apart
Attraction: particles move together

Figure 2–3

Chemical Energy

Every substance has a certain amount of chemical energy. Some of this energy is stored in the bonds that bind the different elements together. When a chemical reaction occurs, bonds are broken or formed, and the amount of chemical energy present in the molecule may change. When chemicals lose energy as they react, the energy is often released as thermal energy. These types of reactions are called **exothermic reactions**. When chemicals gain thermal energy as they react, the reactions are called **endothermic reactions** (see Figure 2–4).

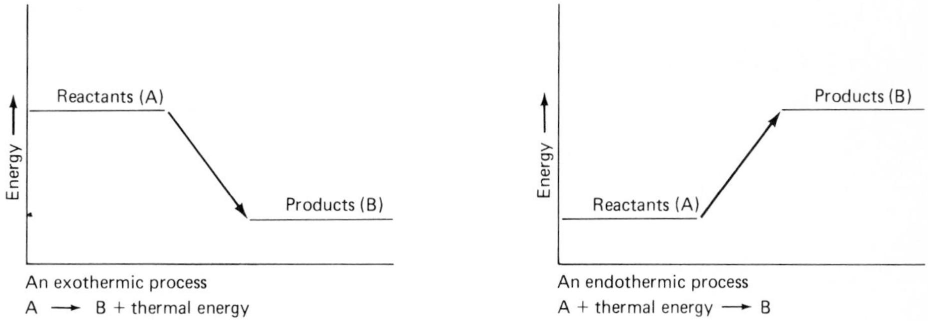
Energy
Reactants (A)
Products (B)
An exothermic process
A → B + thermal energy
Energy
Products (B)
Reactants (A)
An endothermic process
A + thermal energy → B

Figure 2–4

Chemical reactions that release energy are very important, because the energy released during these reactions is used to maintain life and run our society. During the metabolism of foods, the various chemicals present react with oxygen in the body cells and are converted mostly to carbon dioxide and water. The energy released during these reactions is used to maintain body temperature, to promote muscle movement, to cause gland secretions, and to stimulate nerve conduction, in short, to sustain life.

$$\text{Food} + \text{oxygen} \longrightarrow \text{carbon dioxide} + \text{water} + \text{energy}$$

Outside the body, most of the energy released in chemical reactions is in the form of thermal energy. We use this thermal energy to heat homes, cook food, operate automobiles, power machines, and generate electricity and light. Thus chemicals are a major storehouse of the energy used in our world.

Conservation and Transformation of Energy

In the 1840's an English scientist, James Joule, studied the transformation of heat into mechanical and kinetic energy. His experiments suggested that, although energy in one form could be converted into other forms, the total amount of energy present did not change. These observations led to the formation of the law of conservation of energy:

Law of conservation of energy: Energy can neither be created nor destroyed. It can simply be changed from one form to another.

Let's look at some examples of the transformation of energy from one form to another.

1. Solar → chemical

   Sunlight is absorbed by plants as they grow. During this process, plants convert carbon dioxide and water into glucose and other chemicals. In this process, electromagnetic energy is transformed into chemical energy.

2. Chemical → thermal → kinetic

   During the operation of an automobile, the chemicals present in gasoline release thermal energy as they are converted to water and carbon dioxide. This thermal energy is used to move the pistons in the engine, which causes movement of the automobile. The chemical energy present in gasoline is transformed into thermal energy, some of which is further transformed into kinetic energy. The excess thermal energy is emitted in the exhaust and absorbed by substances in the air, causing them to increase in temperature and movement.

3. Chemical → kinetic → gravitational

   When you pick up a book and place it on a shelf, the kinetic energy associated with that movement comes from the chemicals in the foods that you have eaten. The book has been raised to a point farther from the center of the earth, so it has gained gravitational energy. Chemical energy has been converted into kinetic energy, which is then converted into gravitational energy.

If you consider the law of conservation of energy, you may wonder about the international fuss over the pending "energy crisis." The amount of energy is constant; our problem centers on being able to use it in the forms available. It is relatively easy to extract thermal energy from substances such as coal, oil, and natural gas. Combustion of these substances releases large amounts of thermal energy, and some of this thermal energy is used to generate electricity, heat homes, run machines, and so on. But most of the thermal energy is widely dissipated into the environment, where is it absorbed by all elements and substances present in the atmosphere. At present, we do not possess the technology to retrieve very much of this thermal energy after it has been so widely dispensed. Thus we must rely on the chemicals that possess energy in large concentrations for most of our usable energy.

Another general statement concerning energy is that matter will seek to lower its energy content if possible. Many physical and chemical events occur such that systems of high energy are converted to systems of lower energy. An object will fall to the ground if allowed, thus causing a decrease in its gravitational energy. Hot objects cool as they lose thermal energy. Moving objects slow down as they lose kinetic energy. In chemical reactions, compounds may spontaneously rearrange to other compounds of lower chemical energy. Thus chemical reactions that release energy have a much greater tendency to be spontaneous than chemical reactions that absorb energy. In later portions of this text we shall use this principle as a guiding force to understand many chemical reactions.

UNIT 2 Problems and Exercises

2–1. Tell which of the following forms of matter are elements, compounds, or mixtures.
(a) Salt, NaCl (b) Water, H_2O (c) Sodium, Na
(d) Sea water (e) Sugar water

2–2. What is the kinetic energy of an object if its mass is 1 kg and its velocity is 10 m/sec?

2–3. Calculate the number of calories of heat lost when the temperature of 1 ℓ of water changes from 80 to 20°C. One liter of water weighs 1000 g.

2–4. One gram of a substance is placed in a calorimeter and burned. The calorimeter contains 5×10^3 g of water, and the temperature of the water increased by 2.00°C during combustion. How many calories of heat were transferred to the water?

2–5. How many kilocalories (kcal) are required to increase the temperature of one cup of water from 20 to 90°C? The density of water is 1 g/mℓ.

2–6. If you held a rubber ball in your hand at a height of 3 ft and dropped it,
(a) At what point would it possess maximum gravitational energy?
(b) At what point would it possess maximum kinetic energy?
(c) Would you expect the ball to return to the exact height from which it was dropped? Explain your answer.

2–7. Rank the following wavelengths of electromagnetic energy in terms of increasing energy.
(a) 10^{-2} Å (b) 100 Å (c) 10^4 Å (d) 10 Å
Which would have the highest frequency?

2–8. What types of electromagnetic energy are present in the following? Use Figure 2–2 to solve this problem.
(a) Wavelength = 10^{10} Å (b) Wavelength = 10^{-4} Å
(c) Wavelength = 10^4 Å (d) Wavelength = 10 Å
(e) Wavelength = 250 Å

2–9. Label the following events as energy releasing or energy absorbing.
(a) Gasoline burns and is converted to carbon dioxide, water, and heat.
(b) A raw egg is converted into a fried egg.
(c) A plant grows from a bud to a large stalk.
(d) An automobile travels down a highway.
(e) A bomb explodes.
(f) A building is constructed from lumber and nails.

2–10. What type of energy is being lost in the following examples?
(a) An automobile slows down.
(b) A hot liquid cools.
(c) An object rolls down a hill.
(d) Gasoline burns.
(e) You run a race.

2–11. Examine the following events and tell what forms of energy are being converted to other forms. List both the initial and final forms of energy. (Example: When you climb upstairs, you are converting kinetic energy into gravitational energy.)

(a) A raw egg is fried.
(b) A ball is thrown up in the air.
(c) A plant set outside grows.
(d) Two magnets are pulled apart.
(e) A child goes down a slide at a playground.
(f) At the beach, your skin turns red from exposure.

CHAPTER 2

atoms and elements

In units 3 through 6 we shall introduce and study the basic units of matter—atoms and elements. Unit 3 introduces the atom as a basic unit of matter and discusses its composition. In unit 3 we shall also learn that variations in the number of protons, electrons, and neutrons not only create the different elements but also can influence their chemical and physical properties.

In most chemical reactions the most important area of an atom is that region where the electrons are located. Electrons are not simply randomly located in an atom, but exist in defined volume elements that have specific energies. The shape of the volume elements and the electron energies are discussed in unit 5.

Atoms of many different elements can react in a similar manner. The creation of a periodic table (unit 6) is an attempt to bring order out of chaos: to create groups of elements with similar properties.

In unit 4, we deal with reactions of the nucleus, the center of the atom. Nuclear chemistry is a broad, exciting field just coming into its own as a useful tool to mankind.

UNIT 3 ELEMENTS AND COMPOUNDS

Suppose that you took a piece of iron wire and cut it in half. The two pieces of wire would still be iron; but if you continued this process, would you ever arrive at a piece so small that it could not be divided, even if a scissors existed that could cut it? Is there a stopping point such that the tiny particle either could not be divided or that, if it were divided, it would no longer be called iron?

Democritus, a Greek philosopher who lived during the fifth century B.C., thought about such mental experiments and decided that there must be a stopping point. There must exist small units of matter that could not be further subdivided. He called these basic units of matter *atomos*, or **atoms**, meaning *indivisible*. He reasoned that all matter was composed of these tiny particles, and that different kinds of matter contained different kinds of these atoms.

Democritus' theories on the composition of matter were opposed by Aristotle and other philosophers, who preferred the prevailing attitude that all matter was a blend of four basic components: air, earth, water, and fire. Since neither side could prove its point, Democritus' theory of matter was largely rejected for 2000 years until revived in the early 1800's by John Dalton, an English scientist. Dalton was able to do more than simply propose a theory; he backed it up with experimental observations and used his theories to predict properties of chemicals. His theories could be used to explain many of the experimental observations of his colleagues, so these theories became widely accepted and used.

Dalton reintroduced the atom as the fundamental particle of all matter. Atoms were indestructible and could not be divided. An **element** was some portion of matter that contained only one type of atom, and all the atoms of one element were identical in size, shape, and weight. Atoms of different elements did not have the same physical appearance, size, or weight, so there must exist as many different types of atoms as there were elements. Chemical compounds were simply clusters of two or more atoms of different elements, held together by some type of bond. Chemical reactions occurred when the clusters of atoms were broken up and new clusters formed. The atoms were not destroyed, but merely rearranged in their clusters.

The basic ideas behind Dalton's theories are still valid today, but many of the details have been modified in light of modern discoveries. The present postulates of atomic theory are as follows:

1. Atoms are not the smallest particles of matter that can exist, but they are the smallest units of matter that can possess the distinguishing properties of elements.
2. Atoms are the smallest units of matter that can enter into chemical reactions.
3. Atoms of different elements are different in composition and properties.
4. All atoms of the same element are not necessarily identical, but they have quite similar properties.
5. Chemical compounds are composed of combinations of atoms of different elements. The smallest units of matter that bear the properties of compounds are called **molecules**.

6. A molecule is thus a grouping of atoms of various elements. If the composition of the molecule is changed, the identity of the molecule and the compound are changed. In a chemical reaction, the compositions of molecules are changed as one chemical is converted into another.

To illustrate some of these principles, let us examine a brick wall and consider it to be some type of matter. The wall is not a single unit, but is made up of many smaller units, the bricks. Each brick is nearly identical in size, shape, and weight. If the wall were an element, each brick would be an atom (see Figure 3–1). The physical characteristics of the wall are changed by using different types of bricks, and the physical properties of elements differ because each has different atoms. So each different type of wall (element) consists of different types of bricks (atoms). To distinguish among the different types of elements, we give each a name. Every element has a different name, and there are at least 105 different elements. Furthermore, each name has a unique symbol to identify it.

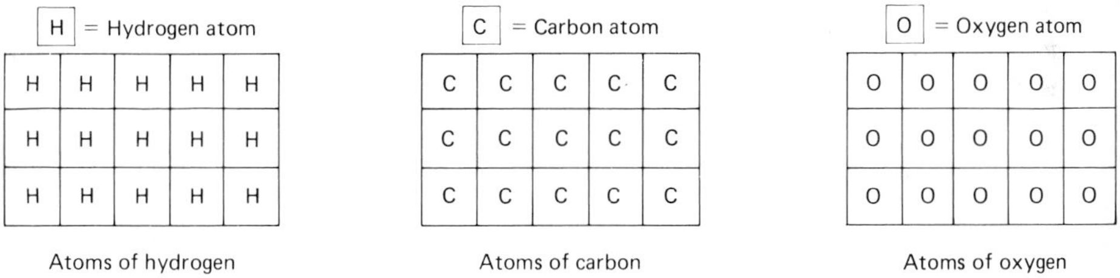
H = Hydrogen atom
C = Carbon atom
O = Oxygen atom
Atoms of hydrogen
Atoms of carbon
Atoms of oxygen

Figure 3–1

If the wall represented a compound instead of an element, each brick would be called a **molecule**. Each molecule does not consist of just one atom, but is a combination of several atoms from different elements. The atoms of different elements bond together to form new substances with properties different from those of the individual elements. If one molecule is broken apart, its chemical properties and identity change to that of some new substance. Compounds have names and properties different from those of the elements of which they consist. In the molecules of the compounds shown in Figure 3–2, the subscript numbers refer only to the elements that they immediately follow. These numbers tell how many atoms of that particular element are present in one molecule of the compound.

In a chemical reaction, the molecules of one type of compound are destroyed, and new molecules of different composition are formed. Figure 3–3 shows chemical reactions.

Atomic Structure

So far we have described matter as consisting of tiny units called atoms, but we have not discussed what atoms themselves are. Initially, it was easiest to think of atoms as tiny marble-like balls, but many experiments have shown that this is not true. During

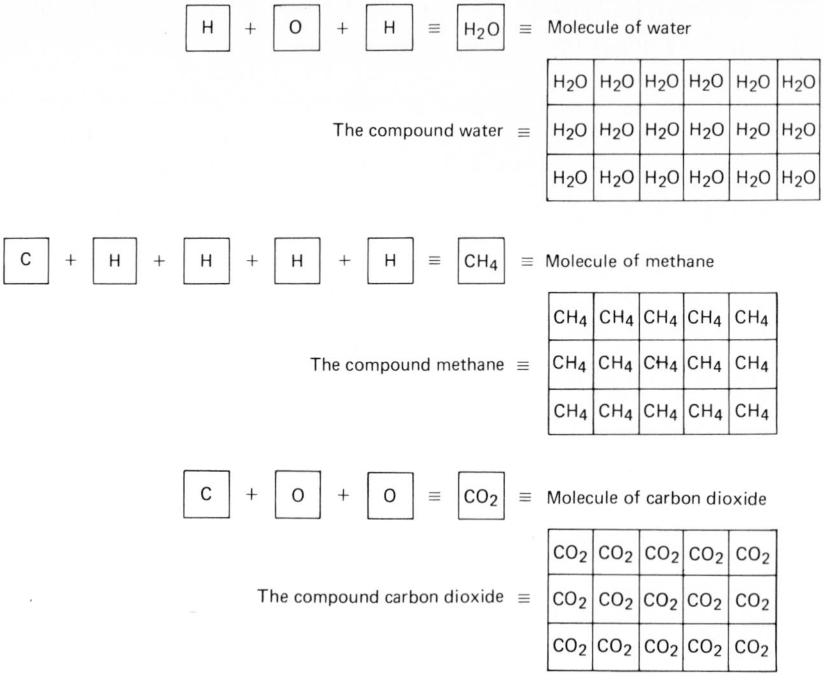
H + O + H ≡ H2O ≡ Molecule of water
The compound water ≡
H2O H2O H2O H2O H2O H2O
H2O H2O H2O H2O H2O H2O
H2O H2O H2O H2O H2O H2O
C + H + H + H + H ≡ CH4 ≡ Molecule of methane
The compound methane ≡
CH4 CH4 CH4 CH4 CH4
CH4 CH4 CH4 CH4 CH4
CH4 CH4 CH4 CH4 CH4
C + O + O ≡ CO2 ≡ Molecule of carbon dioxide
The compound carbon dioxide ≡
CO2 CO2 CO2 CO2 CO2
CO2 CO2 CO2 CO2 CO2
CO2 CO2 CO2 CO2 CO2

Figure 3–2

the late 1800's it was shown that atoms were not indivisible, but consisted of smaller units of mass. Some of these units had positive electrical charges and were called **protons**. Others had negative electrical charges and were called **electrons.** A third species bearing no charge, a **neutron**, was discovered in 1932.

Thus atoms consist of protons, electrons, and neutrons. But how are these particles arranged in atoms? During the late 1800's many theories on atomic structure were presented, but it was not until 1911 that the first real picture for atomic structure emerged. Ernest Rutherford and his coworkers, Hans Geiger and Ernest Marsden, performed an experiment which showed that most of the mass of an atom was concentrated in an extremely small portion of the atomic volume. They prepared an extremely thin sheet of gold foil and bombarded it with tiny particles

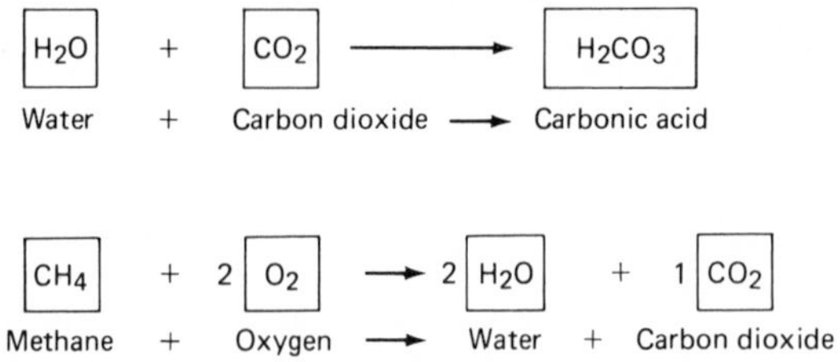
H2O + CO2 → H2CO3
Water + Carbon dioxide → Carbonic acid
CH4 + 2 O2 → 2 H2O + 1 CO2
Methane + Oxygen → Water + Carbon dioxide

Figure 3–3

called **alpha particles** (see Figure 3–4). Alpha particles are small units of matter, weighing about the same as four protons, that are emitted from some radioactive substances. The alpha particles acted like extremely small bullets being fired at the gold-foil wall. Surrounding the foil were special screens that were sensitive to the alpha particles, so Geiger and Marsden were able to detect the paths of the "bullets" after they hit the wall.

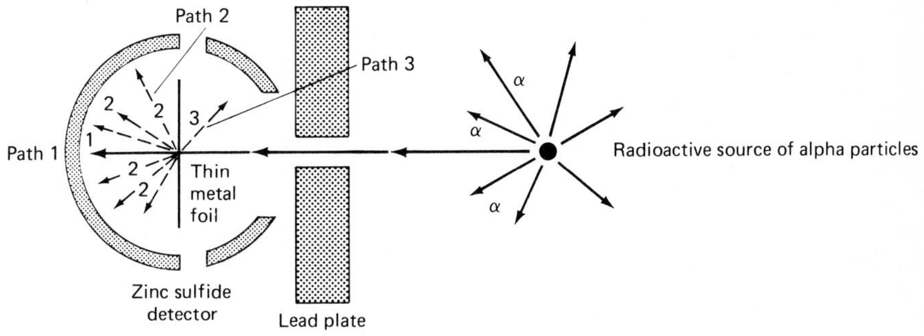
Path 2
Path 3
Path 1
2
2
3
1
2
2
Thin
metal
foil
α
α
α
Radioactive source of alpha particles
Zinc sulfide
detector
Lead plate

Figure 3–4 Ernest Rutherford's metal foil experiment. Approximately 1 in 10,000 alpha (α) particles bounded back from the metal foil, providing evidence for the existence of a heavy, compact, positively charged atomic nucleus.

The results of the gold-foil experiment were astonishing to Rutherford, as they contradicted the prevailing theories of atomic structure. It was previously thought that the mass of an atom was uniformly spread over the entire volume of the atom; all the bullets (alpha particles) should have passed through the foil (mostly path 1 of Figure 3–4). It should have been like shooting a rifle at a blanket. But some of the particles were deflected at small angles (path 2), and a smaller but significant fraction actually bounced off the foil (path 3). These particles must have hit something that was much heavier than the alpha particle. From this experiment, Rutherford proposed three theories:

1. Most of the volume of an atom must contain very little mass. (Since most of the bullets went directly through, they did not hit any heavy objects.)
2. Most of the mass of an atom must be centered in a very small portion of the atom. These centers had to be large and dense enough to cause alpha particles to bounce off.
3. Since the positively charged alpha particles bounced off the nucleus, it must also be positively charged. Negative charge must then reside outside the nucleus.

Rutherford's theories proved to be accurate and are still a part of current atomic theory. Atoms consist of three types of particles: protons, electrons, and neutrons. Protons and neutrons are relatively heavy compared to electrons. They are located in the center of the atom, which is called the **nucleus**. Surrounding the nucleus at large distances are the almost massless electrons. If the nucleus were the size of a pea, then the outer edges of the atom would be several blocks away. Notice in Table 3–1 that electrons have very little mass compared to protons and neutrons, but their presence is very real. Since electrons are located on the outer edges of the atoms, they play the most important roles in chemical reactions.

Table 3–1 Constituents of atoms

Atomic particle	Charge	Weight (in atomic mass units)*	Location
Proton	+1	1	Nucleus
Neutron	0	1	Nucleus
Electron	−1	0.005	Outside the nucleus

*An atomic mass unit (amu) is defined as the mass of one proton and is equal to 1.66×10^{-24} g.

It is an oversimplification to describe the nucleus as containing only neutrons and protons; today over 100 different nuclear particles have been identified among the products that result when atoms are bombarded by very high energy particles. They range in mass from zero (the neutrinos) to about 2400 times the mass of the electron and have very short lifetimes. However, we are primarily concerned with an understanding of the chemical reactions of atoms and molecules, so particles in the nucleus other than protons and neutrons need not concern us further.

Present Atomic Theory

Let us summarize the current ideas of atomic structure and introduce some definitions concerning atoms.

All atoms are composed of protons, electrons, and neutrons (see Table 3–1 for their properties). The heavier particles, the protons and neutrons, are located in the nucleus of the atom, and the lighter particles, the electrons, surround it. In a neutral atom the number of protons (positive electrical charges) must equal the number of electrons (negative electrical charges).

The **identity** of an atom is dependent upon the number of protons and electrons that it contains. All atoms of the same element contain the same number of protons and electrons. The number of neutrons may vary slightly. Thus any two atoms that contain the same numbers of protons and electrons are the same element. Conversely, any atoms with different numbers of protons and electrons are different elements. The atoms of the element carbon all contain six protons and six electrons; all atoms of the element oxygen contain eight protons and eight electrons.

The number of protons present in any atom is called the **atomic number**. The atomic number for the element carbon is 6, since each atom of carbon contains 6 protons. The atomic number for the element oxygen is 8, since each atom of oxygen contains 8 protons. All atoms of the same element have the same atomic number; atoms of different elements have different atomic numbers.

The atomic number of an element gives the number of protons in an atom of that element. The *weight* of an atom is the sum of the weights of all the atomic particles in the atom; since the protons and neutrons are so much heavier than the electrons, the **atomic weight** is commonly taken as the weight of the protons and neutrons in an atom. Since both protons and neutrons weigh approximately 1 atomic mass unit (1 amu), the atomic weight of an atom is the sum of the total number of protons and neutrons in an atom.

$$\text{Atomic weight} = \text{no. of protons} + \text{no. of neutrons in an atom}$$

Table 3–2 lists some of the more common elements along with some of the effects that they have on the human body. Table 3–3 lists the composition of the

Table 3–2 Biological properties of some elements

Element	Symbol	Atomic number	Atomic weight (amu)	Biological functions and/or other properties
Hydrogen	H	1	1.01	Required in most compounds; found in the body and plants
Helium	He	2	4.00	Inert; not used in the body
Lithium	Li	3	6.94	May be necessary in brain functioning
Beryllium	Be	4	9.01	Acute poison
Boron	B	5	10.80	Essential in some plants
Carbon	C	6	12.01	Essential structural element of most biochemicals
Nitrogen	N	7	14.01	Essential element in many biochemicals
Oxygen	O	8	15.99	Essential element in many biochemicals
Fluorine	F	9	18.99	Necessary for solid tooth formation and retention of calcium in bones
Neon	Ne	10	20.10	Inert; not used in the body
Sodium	Na	11	22.99	Necessary for electrochemical processes in body fluids
Magnesium	Mg	12	24.31	Necessary for energy transfer and muscle functioning
Phosphorus	P	15	30.97	Essential in biochemical synthesis and energy transfer
Sulfur	S	16	32.06	Present in proteins and other body constituents
Chlorine	Cl	17	35.45	Necessary for electrochemical processes in body fluids
Argon	Ar	18	39.95	Inert; not used in the body
Potassium	K	19	39.10	Necessary for electrochemical processes in body fluids
Calcium	Ca	20	40.08	Major component of bone
Vanadium	V	23	50.94	Essential in lower plants and certain marine animals
Chromium	Cr	24	51.99	Essential in higher animals

Table 3–2 Biological properties of some elements (cont.)

Element	Symbol	Atomic number	Atomic weight (amu)	Biological functions and/or other properties
Manganese	Mn	25	54.94	Necessary for enzyme activity, functioning of liver, kidneys, and eyes
Iron	Fe	26	55.85	Essential to oxygen transport in the body
Cobalt	Co	27	58.93	Found in vitamin B_{12}
Nickel	Ni	28	58.71	May cause lung cancer
Copper	Cu	29	63.54	Necessary for growth and blood cells
Zinc	Zn	30	65.37	Necessary for normal growth
Arsenic	As	33	74.92	Toxic? May cause cancer
Selenium	Se	34	78.96	Essential for liver fluids in low doses
Molybdenum	Mo	42	95.94	Essential in some biochemical reactions
Tin	Sn	50	118.69	Essential in rats; function unknown
Iodine	I	53	127.60	Essential to thyroid gland
Barium	Ba	56	137.34	Used in medical diagnosis
Mercury	Hg	80	200.59	Causes nerve damage and death
Lead	Pb	82	207.19	Causes brain damage and convulsions

Table 3–3 Comparison of the composition of the human body with that of seawater and the earth's crust

Element	Human body (%)	Seawater (%)	Earth's crust* (%)
H	63	66	0.22
C	9.5	0.0014	0.19
O	25.5	33	47
N	1.4	Very small, <0.001	<0.1
Ca	0.31	0.006	3.5
P	0.22	Very small, <0.001	<0.1
Cl	0.03	0.33	<0.1
K	0.06	0.006	2.5
S	0.05	0.017	<0.1
Na	0.03	0.28	2.5
Mg	0.01	0.033	2.2
All other elements	<0.01	<0.1	40.0

NOTE: Percentages are expressed as percentage of total number of atoms.

* The earth's crust also contains silicon (28%), aluminum (7.9%), iron (4.5%), and titanium (0.46%), as well as smaller amounts of other elements.

human body in comparison with that of seawater and the earth's crust. It is interesting that the percent composition of the human body more closely resembles that of seawater than that of the earth's crust. Also, the most abundant elements in the body are generally the lightest elements. Only 3 of the 24 essential elements have atomic numbers greater than 24, and they are required only in trace amounts.

Isotopes

The existence of atoms of the same element with different atomic weights was discovered in the early 1900's in the identification of elements formed from radioactive decay. In 1913, Frederick Soddy used the name **isotopes** to describe atoms of the same element that had different weights. We now know that the weight differences are due to different numbers of neutrons in atoms. Let's take the element chlorine as an example. Its atomic weight is approximately 35.45. All the chlorine

Table 3–4 Isotopes of some elements

Atomic number	Element	Isotope*	Percent natural abundance	Atomic mass of isotope (in amu)†	Atomic weight of element
1	Hydrogen	$^{1}_{1}H$	99.985	1.007825	1.00797
	Deuterium	$^{2}_{1}H$	0.015	2.01410	
	Tritium	$^{3}_{1}H$	Very small	3.01695	
6	Carbon	$^{12}_{6}C$	98.89	12.00000	12.01115
		$^{13}_{6}C$	1.11	13.00335	
7	Nitrogen	$^{14}_{7}N$	99.63	14.00307	14.0067
		$^{15}_{7}N$	0.37	15.00011	
8	Oxygen	$^{16}_{8}O$	99.759	15.99491	15.9994
		$^{17}_{8}O$	0.037	16.99914	
		$^{18}_{8}O$	0.204	17.99916	
29	Copper	$^{63}_{29}Cu$	69.09	62.9289	63.54
		$^{65}_{29}Cu$	30.91	64.9278	
34	Selenium	$^{74}_{34}Se$	0.87	73.9225	78.96
		$^{76}_{34}Se$	9.02	75.9192	
		$^{77}_{34}Se$	7.58	76.9199	
		$^{78}_{34}Se$	23.52	77.9173	
		$^{80}_{34}Se$	49.82	79.9165	
		$^{82}_{34}Se$	9.19	81.9167	
36	Krypton	$^{78}_{36}Kr$	0.35	77.9204	83.80
		$^{80}_{36}Kr$	2.27	79.9164	
		$^{82}_{36}Kr$	11.56	81.9135	
		$^{83}_{36}Kr$	11.55	82.9141	
		$^{84}_{36}Kr$	56.90	83.9115	
		$^{36}_{36}Kr$	17.37	85.9106	

* The superscript refers to atomic weight and the subscript to atomic number.

† Based on 1961 standard; C = 12.00000 amu.

atoms have an atomic number of 17, or 17 protons, but some of the atoms have 18 neutrons, and others have 20 neutrons. In actuality, about 75% of the atoms of chlorine have 18 neutrons and the other 25% have 20 neutrons.

Perhaps the most familiar set of isotopes are those of the smallest element, hydrogen. The most abundant form is called hydrogen; it has one proton and zero neutrons. One isotope, deuterium, has one proton and one neutron. Tritium, another isotope, has one proton and two neutrons. Table 3–4 lists some elements and their isotopes. When we study radioactivity, we shall learn more about the importance of the number of neutrons in an atom.

UNIT 3 Problems and Exercises

3–1. What are the main differences between Dalton's theories and the present postulates of atomic theory?

3–2. Explain the major differences among atoms of different elements.

3–3. What do protons and neutrons have in common? In what ways are they different?

3–4. How do protons and atoms differ?

3–5. What are the five most abundant elements in the earth's crust? in seawater? in the human body?

3–6. Distinguish between an element and a compound.

3–7. How is a molecule different from an atom? Is it possible to obtain one *atom* of a compound? Why or why not?

3–8. Classify each of the following as an element or compound.

(a) Iron (b) Salt (c) Sugar (d) Water
(e) Mercury (f) Gold (g) CO_2 (h) I_2
(i) NH_3

3–9. Indicate whether the following statements are true or false. If false, correct them.

(a) There are 105 different elements, so there are only 105 different types of atoms.
(b) Atoms are the smallest units of matter.
(c) An atom of copper is the same as an atom of iron.
(d) A proton in a copper atom is the same as a proton in an atom of iron.
(e) Five atoms of copper weigh as much as five atoms of iron.
(f) The nucleus is the heaviest part of the atom.
(g) Atoms of the same element may have different weights.
(h) If there are 9 protons, 10 neutrons, and 9 electrons in an atom of fluorine, its atomic weight is 18 amu.

3–10. How many atoms of each element are present in the following molecules?

(a) Water (H_2O) (b) Glucose ($C_6H_{12}O_6$)
(c) Ethyl alcohol (C_2H_6O) (d) Acetic acid ($C_2H_4O_2$)

3–11. In each of the following chemical reactions, how many molecules are on the left and right sides of the arrow?

(a) $CH_4 + 2O_2 \rightarrow 2H_2O + CO_2$
(b) $H_2SO_4 + 2NaOH \rightarrow Na_2SO_4 + 2H_2O$
(c) $H_2O + CO_2 \rightarrow H_2CO_3$

3–12. Copper and nickel atoms weigh about the same (59 amu). How can this be, if the atoms are supposed to be different?

3–13. What is the electronic charge on the following atoms?
(a) Atom 1 contains 20 protons, 20 neutrons, and 18 electrons.
(b) Atom 2 contains 18 protons, 19 neutrons, and 18 electrons.
(c) Atom 3 contains 15 protons, 17 neutrons, and 18 electrons.
(d) Atom 4 contains 16 protons, 16 neutrons, and 18 electrons.

3–14. Complete the following table. Do not use the tables unless you have to.

Element	Number of protons	Number of electrons	Number of neutrons	Atomic number	Atomic weight
Ne	10		10		
Mg		12	12		
Na		11			23
Fe				26	56
P			16		31
Cu	29				64

3–15. Give the symbols for the following elements.
(a) Calcium (b) Iron (c) Mercury (d) Potassium
(e) Silver (f) Sodium (g) Carbon (h) Nitrogen
(i) Selenium

3–16. Give the names of the elements that correspond to the following symbols.
(a) Sn (b) C (c) Fe (d) Na
(e) Ni (f) H (g) P (h) N
(i) B (j) K

3–17. How do isotopes of an element differ from one another? In what ways are they the same?

3–18. List the most abundant isotopes of the following elements.
(a) Hydrogen (b) Carbon (c) Oxygen
(d) Selenium (e) Krypton

3–19. Which of the following pairs of atoms are isotopes of the same element?

	Atoms	Atomic number	Atomic weight
(a)	X	26	54
	Y	26	58
(b)	A	30	64
	B	28	64
(c)	D	15	31
	E	13	27

3–20. Explain why the atomic weights of elements are not whole numbers.

3–21. Explain in your own words the importance of Rutherford's experiment. What would have happened if the atom had been a dense solid throughout?

UNIT 4 NUCLEAR CHEMISTRY AND RADIOACTIVITY

In most chemical reactions, the nucleus of an atom *does not change* in composition; the number of protons and neutrons present remains constant. The nucleus is said to be stable and inert—resistant to change. However, some nuclei are unstable; they spontaneously emit small particles of matter or units of energy. As a result of these emissions, the nuclei change their identities and become different, more stable nuclei. Unstable nuclei that emit particles and energy are called **radioactive** nuclei, and the study of the changes that occur in these processes is called **nuclear chemistry**.

Types of Radiation

Three types of radiation are emitted from naturally occurring radioactive atoms: **alpha** (α) rays, **beta** (β) rays, and **gamma** (γ) rays. Alpha rays consist of particles with a mass of 4 amu and a charge of +2. Each particle possesses two protons and two

Table 4–1 Characteristics of nuclear emissions

Properties	Alpha particles	Gamma rays	Beta particles (negatrons)	Beta particles (positrons)
Composition	He nuclei	Energy rays	Electrons with (−)	Electrons with (+)
Mass (amu)	4	None	$\frac{1}{1837}$	$\frac{1}{1846}$
Symbol	α, ${}_2^4$He, He − 4	${}_0^0\gamma$	${}_{-1}\beta$ or ${}_{-1}^{0}$e	${}_{+1}\beta$ or ${}_{+1}^{0}$e
Electrical charge	+ 2	0	−1	+1
Relative penetration	Few cm of air	Few miles of air, 10 mm of Pb	100 cm of air, 1 mm of Al	100 cm of air, 1 mm of Al
Relative ionizing ability	10,000	1	100	100
Velocity	Variable up to 0.1 of light	Same as light	Variable up to 0.9 of light	Variable up to 0.9 of light
Nuclear change, when particles are emitted				
Mass	Decrease by 4	No change	No change	No change
Atomic no.	Decrease by 2	No change	Increase by 1	Decrease by 1

neutrons (a helium atom nucleus). These relatively heavy particles move at speeds of about one-tenth the speed of light, but because of their larger volumes they soon collide with other particles of matter and lose their energy. They are stopped by a few centimeters of air, cannot go through the outer layer of dead skin cells, and are not dangerous in the external environment. But if a radioactive substance gets inside the body and emits an alpha particle, biological damage can occur.

Beta rays are particles, possessing the mass of electrons, that have been produced within the nucleus and then emitted. There are two types: negatrons ($_{-1}\beta$) bear a single negative charge, and positrons ($_{+1}\beta$), a single positive charge. They are much smaller than alpha rays, move much faster, and can penetrate as much as 100 cm of air or 4 mm into the skin. Radiation burns can result, but the vital internal organs are not reached unless the radioactive material is ingested by the body.

Gamma rays are not particles, but are high-energy electromagnetic radiations similar to light rays or X-rays. They can penetrate deep within the body and do great damage. Only 10% of the rays are stopped by as much as 50 cm of tissue. In nuclear disintegrations, gamma rays are often emitted along with alpha and beta rays. Table 4–1 lists the radioactive emissions with their properties.

The Unstable Nucleus: Neutron–Proton Ratios

The stability of any nucleus seems to be dependent upon the ratio of the number of neutrons to the number of protons present. There seems to be an optimum number of neutrons required for each number of protons in order that the nucleus be stable, or nonradioactive. Figure 4–1 shows that for the smaller elements there are about equal numbers of neutrons and protons in the stable isotopes. As the atomic number increases, extra neutrons are necessary to achieve stability. Table 4–2 lists some isotopes of the common elements along with their stabilities. In nuclei of some unstable isotopes, the number of neutrons is too large for the number of protons present (a high neutron to proton ratio). These nuclei tend to stabilize themselves by converting a neutron into a proton and a negatron beta particle.

$$n \longrightarrow p + {}_{-1}\beta \quad \text{(negatron is emitted)}$$

If there are not enough neutrons (too low a neutron to proton ratio), a proton may be converted to a neutron and a positron, or an alpha particle may be emitted:

$$p \longrightarrow n + {}_{+1}\beta \quad \text{(positron is emitted)}$$

$$\underset{1}{\text{Nucleus}} \longrightarrow \underset{2}{\text{Nucleus}} + \alpha \quad \text{(alpha particle is emitted)}$$

The emission of an alpha particle increases the neutron to proton ratio.

When an alpha or beta particle is emitted from the nucleus of an atom, the identity of that atom changes, since the number of protons present in the nucleus has changed. The new element may also be unstable and decay further by more emissions until finally a stable isotope of some element is attained. Uranium atoms with masses of 238 amu ($^{238}_{92}U$) decay in a series of 14 steps until a stable isotope of lead is formed ($^{206}_{82}Pb$) (see Figure 4–2).

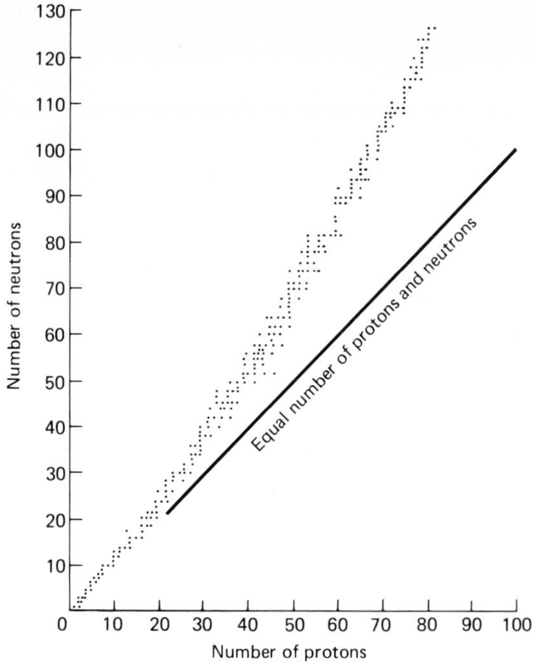
130
120
110
100
90
80
70
60
50
40
30
20
10
Number of neutrons
0
10
20
30
40
50
60
70
80
90
100
Number of protons
Equal number of protons and neutrons

Figure 4–1 Neutron–proton ratios for stable elements.

Balancing Nuclear Reactions

When writing reactions of nuclear decay, it is quite important to keep track of the number of protons and neutrons present in the nucleus. Emission of a positron increases the number of neutrons and decreases the number of protons by one. Emission of a negatron does just the reverse. Alpha particle emissions cause a loss of two neutrons and two protons. In balancing a nuclear reaction, the total sum of neutrons plus protons on one side of the arrow must equal the sum on the other side.

To keep track of neutrons and protons, the elemental symbols in a nuclear equation must always show the **mass** (left **super**script) and the **atomic number** (left **sub**script). For example, the symbol $^{238}_{92}U$ means that the element uranium has 92 protons and a mass of 238. So 238 − 92, or 146 neutrons, are present. When $^{238}_{92}U$ loses an alpha particle ($^{4}_{2}He$), it forms a new element with two less protons and two less neutrons. The new element, thorium, has an atomic number of 90 and atomic mass of 234 amu. You should be able to follow all the disintegrations shown in Figure 4–2 with a little practice.

Measuring Rates of Nuclear Reactions: Half-Lives

Each individual unstable isotope decays at its own rate or speed. Some isotopes require millions of years for any appreciable amount of decay to occur; others require only a few hours or seconds. To be able to compare rates of decay of the various radioactive isotopes, physicists have invented a measurement known as the

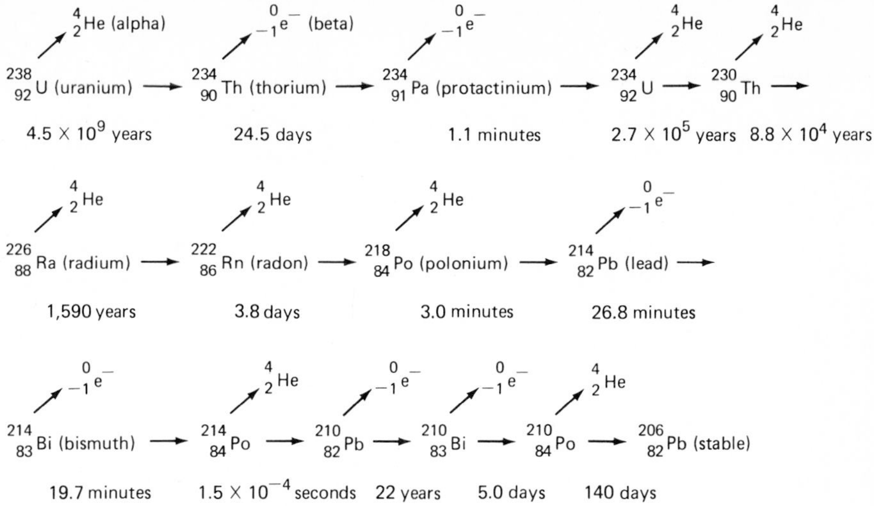

4 2 He (alpha)
0 −1 e (beta)
238 92 U (uranium)
234 90 Th (thorium)
234 91 Pa (protactinium)
234 92 U
230 90 Th
4.5 X 10⁹ years
24.5 days
1.1 minutes
2.7 X 10⁵ years
8.8 X 10⁴ years
226 88 Ra (radium)
222 86 Rn (radon)
218 84 Po (polonium)
214 82 Pb (lead)
1,590 years
3.8 days
3.0 minutes
26.8 minutes
214 83 Bi (bismuth)
214 84 Po
210 82 Pb
210 83 Bi
210 84 Po
206 82 Pb (stable)
19.7 minutes
1.5 X 10⁻⁴ seconds
22 years
5.0 days
140 days

Figure 4–2 Uranium disintegration series.

half-life. This is the time it takes for one-half of the initial number of radioactive atoms to decay. The half-life of uranium 235 is about 5 billion years. If we started with 1000 g of U-235 and waited 5 billion years, only 500 g of U-235 would remain. After another 5 billion years only one-quarter of the original mass (one-half of the 500) would remain. Obviously, scientists cannot wait 5 billion years to measure the half-life of U-235; it is possible to measure the rates of decay over much shorter periods and then calculate the half-lives of most radioactive elements. Table 4–2 shows the half-lives of some of the more important radioactive elements.

Table 4–2 Half-lives and decay modes for some important radioactive isotopes

Name	Symbol	Radiation emitted	Half-life	Medical and other uses
Tritium	$^{3}_{1}H$	Beta minus	12.3 yr	
Carbon 14	$^{14}_{6}C$	Beta minus	5700 yr	Metabolism studies
Oxygen 19	$^{19}_{8}O$	Beta minus, gamma	29.4 sec	
Sodium 24	$^{24}_{12}Na$	Beta minus, gamma	15 hr	Blood circulation studies
Phosphorus 32	$^{32}_{15}P$	Beta minus	14.3 days	Leukemia, bone, and teeth studies
Chromium 51	$^{51}_{24}Cr$	Gamma	28 days	Red blood cell studies
Cobalt 60	$^{60}_{27}Co$	Beta minus, gamma	5.3 yr	Cancer therapy
Radium 226	$^{226}_{88}Ra$	Alpha and gamma	1600 yr	Cancer therapy
Strontium 85	$^{85}_{38}Sr$	Gamma	36 yr	Bone scans
Iodine 131	$^{131}_{53}I$	Beta minus, gamma	8 days	Thyroid studies
Uranium 235	$^{235}_{92}U$	Alpha and gamma	7×10^{8} yr	Nuclear electric plants
Uranium 238	$^{233}_{92}U$	Alpha and gamma	4.5×10^{9} yr	Nuclear electric plants

Consequences and Uses of Radiation

Destruction of Body Cells: Ionizing Radiation

As alpha, beta, and gamma rays speed through matter, they knock electrons off atoms and cause the formation of charged particles (ions). These ions have a large amount of energy and generally are formed where nature never intended them to be. Their presence may impair chemical processes within the cell and break chemical bonds. These changes may lead to genetic mutations, serious injuries and sometimes death. Common symptoms of radiation poisoning include sudden baldness, formation of ulcerated lesions that do not heal, nausea, and vomiting.

Obviously, radiation penetration of healthy cells is dangerous, but the same radiation can be used beneficially to kill or slow down the growth of diseased cells. Cancer, a common name for malignant tumors, refers to the development of new tissue composed of cells that grow without the normal controls. Radiation is sometimes used to kill these cells; the success of radiation therapy depends upon the ability of the body to repair and replace the healthy cells that are also damaged. A common treatment is to irradiate the cancerous area with X-rays or with gamma rays from the isotope cobalt 60.

New Foods and Preservation of Foods

Agricultural scientists sometimes use ionizing radiation to produce mutations in food grains and vegetables. In this manner new strains possessing more desirable properties have been developed. In addition to creating new sources of food, radiation is used to destroy microorganisms and insects that cause food spoilage. An advantage of the use of radiation is that it produces no rise in the temperature of food during treatment and does not change the taste. However, there is the possibility that irradiation causes changes that may be harmful. More research and testing will have to be done before widespread use of radiation in food preservation is allowed.

Detection of Disease

The rays emitted by radioactive atoms can be detected by using **Geiger–Müller counters**. These instruments are so sensitive that very small numbers of atoms may be detected; thus small amounts of radioactive material can be injected into the human body and used as a tracer for the detection of disease. Radioactive iodine ($^{131}_{53}I$) is used in the diagnosis of thyroid gland abnormalities. By measuring the rate at which radioactive iodine is used by the thyroid gland, physicians can tell if the gland is functioning properly.

Radioactive phosphorus ($^{32}_{15}P$) has been used in research to study the formation and growth of teeth and bones, and in medicine to locate breast cancers and brain tumors. Radioactive sodium ($^{24}_{12}Ne$) in sodium chloride has been injected into the bloodstream to measure the speed and efficiency of blood circulation.

Energy Production

If you were to measure the *exact* masses of the starting radioactive isotope and the particles and isotopes generated as a result of radioactive decay, you would find that some very small amount of mass had disappeared in the process. For example, in the decay of radium 226 to radon 222 and an alpha particle, 0.0052 amu of mass has disappeared:

	$^{226}_{88}\text{Ra}$	$\longrightarrow$	$^{222}_{84}\text{Rn}$	+	$^{4}_{2}\alpha$
Exact mass:	226.0254 amu		222.0176 amu		4.0026

Mass change: 226.0254 − 226.0202 = 0.0052 amu

The small amount of mass that has disappeared in the decay has been converted into energy.

The relationship between mass and energy was derived by Albert Einstein in 1905. The equation is

$$E = mc^2 \tag{4-1}$$

where E = energy released

m = mass lost

c^2 = square of the speed of light.

This equation tells us that a tiny quantity of mass is equivalent to a fantastic quantity of energy. For example, the energy equivalent of 1 lb of uranium decaying in a nuclear reactor is approximately equal to that generated by burning 3 million lb of coal. In a nuclear electrical generating plant, the energy released by the decay of uranium is used to convert water into steam. The steam is used to derive electrical generators (see Figure 4–3).

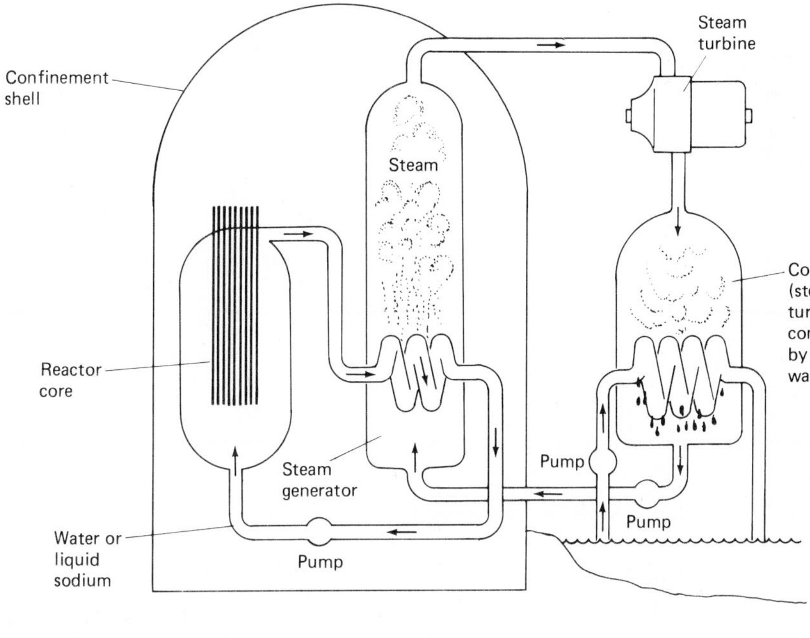
Steam turbine
Confinement shell
Steam
Condenser (steam from turbine is condensed by river water)
Reactor core
Steam generator
Pump
Pump
Water or liquid sodium
Pump

Figure 4–3 Typical design of a nuclear generating plant.

UNIT 4 Problems and Exercises

4–1. Indicate the number of protons and neutrons in each of the following nuclei.

(a) Strontium 85 (b) $^{137}_{55}Cs$ (c) Uranium 235

(d) Iodine 131 (e) $^{71}_{30}Zn$

4–2. Give the major difference among alpha, beta, and gamma rays.

4–3. Which of the three types of radioactive emissions is the most dangerous? Why?

4–4. What is meant by the *half-life* of a radioactive element?

4–5. Which of the following isotopes would you expect to accumulate in the body? Why?

(a) Strontium 83 (b) Iodine 131 (c) Carbon 14

(d) Sodium 24 (e) Cobalt 60

4–6. The half-life of hydrogen 3 (tritium) is 12.3 yr. How many milligrams of a 20-mg sample will remain after 24.6 yr?

4–7. Complete and balance the following nuclear equations by supplying the missing nuclear particle.

(a) $^{235}_{92}U \rightarrow ^{231}_{90}Th + \_\_\_\_$

(b) $^{64}_{29}Cu \rightarrow ^{64}_{30}Zn + \_\_\_\_$

(c) $^{7}_{4}Be \rightarrow ^{7}_{5}B + \_\_\_\_$

4–8. Complete and balance the following nuclear equations by supplying the missing element.

(a) $^{87}_{36}Kr \rightarrow ^{0}_{-1}e + \_\_\_\_$

(b) $^{56}_{26}Fe \rightarrow ^{0}_{-1}e + \_\_\_\_$

(c) $^{53}_{24}Cr + ^{4}_{2}He \rightarrow ^{1}_{0}n + \_\_\_\_$

(d) $\_\_\_\_ \rightarrow ^{24}_{12}Mg + ^{0}_{-1}e$

4–9. Fill in the atomic numbers and atomic weights where omitted in the following nuclear reactions. Try not to look at the tables while doing this.

(a) $^{222}_{86}Rn - \_\_He \rightarrow \_\_Po$ (b) $\_\_Po - ^{4}_{2}He \rightarrow ^{206}_{82}Pb$

(c) $\_\_Pb - ^{0}_{-1}\beta \rightarrow ^{210}_{83}Bi$ (d) $^{238}_{92}U - \_\_He \rightarrow ^{234}Th$

(e) $^{214}_{82}Pb - _{-1}\beta \rightarrow \_\_Bi$

4–10. Indicate the amount of time required for the following to occur:

(a) 1000 g of $^{238}_{92}U$ decays by 75%.

(b) 1000 g of $^{14}_{6}C$ decays by 75%.

(c) 1000 g of $^{226}_{88}Ra$ decays by 75%.

4–11. Tell what type of emission occurred in each of the following nuclear decays.

(a) $^{234}_{90}Th - \_\_\_\_ \rightarrow ^{234}_{91}Pa$ (b) $^{214}_{83}Bi - \_\_\_\_ \rightarrow ^{214}_{84}Po$

(c) $^{214}_{83}Bi - \_\_\_\_ \rightarrow ^{210}_{81}Tl$ (d) $^{234}_{90}Th - \_\_\_\_ \rightarrow ^{234}_{90}Th$

4–12. Which of the three atoms is most likely to be radioactive? Justify your choice.
$^{114}_{48}Cd$ $^{114}_{49}In$ $^{114}_{50}Sn$

4–13. Write out the nuclear reactions for the following elements. See Table 4–2 for assistance in determining the decay mode.
(a) $^{3}_{1}H \rightarrow$ (b) $^{14}_{8}C \rightarrow$ (c) $^{226}_{88}Ra \rightarrow$
(d) $^{51}_{24}Cr \rightarrow$ (e) $^{131}_{53}I \rightarrow$

4–14. Complete the following equations.

(a) $^{23}_{11}Na + ^{4}_{2}He \rightarrow ^{26}_{12}Mg +$ ______

(b) $^{64}_{29}Cu \rightarrow ^{0}_{-1}\beta +$ ______

(c) $^{106}_{47}Ag \rightarrow ^{106}_{48}Cd +$ ______

4–15. $^{18}_{9}F$ is found to undergo 88% radioactive decay in 366 min. What is its half-life?

4–16. If a nucleus has too large a neutron to proton ratio, what is likely to happen? What will happen if the ratio is too small?

UNIT 5 ELECTRON CONFIGURATIONS

The identity of an atom is determined by its atomic number, the number of protons in the nucleus. The chemical reactivity of most atoms is dictated by the number of electrons that surround the nucleus. In chemical reactions, atoms lose electrons, gain electrons, or share electrons with one another. To understand why some of these processes occur, we should first learn how electrons are situated around the nuclei of atoms.

Bohr Theory

In the Rutherford–Dalton model of atomic structure, electrons were simply located outside the nucleus and whirled about in random fashion. But in 1865 an English physicist, James Maxwell, showed mathematically that electrons moving randomly about a nucleus would gradually lose energy and finally spiral into the nucleus. If the Rutherford–Dalton model were correct, electrons would end up in the nucleus, and atoms, as we know them, would not exist. Something had to be wrong with the model.

A second observation, which also contradicted the Rutherford–Dalton model, involved the absorption and emission of light by atoms. When atoms are exposed to electromagnetic radiation of varying wavelengths, some of it is absorbed and the atoms gain energy. Often this excess energy is released as heat or by emission of light by the atoms. Calculations for the Rutherford–Dalton model indicated that atoms should absorb and emit all wavelengths of light; in actual fact atoms were quite selective. The atoms of an element absorb and emit only certain specific wavelengths. Atoms of different elements absorb different wavelengths of radiation.

By the early 1900's, scientists were seeking new models to explain the location of electrons in atoms. The model that best fit the experimental data was proposed in 1913 by Niels Bohr (see Figure 5–1). He proposed that electrons were located in specific positions at specific distances from the nucleus. He pictured the atom as a miniature solar system, with the nucleus being the center (the sun) and the electrons revolving about it (the planets). The energy possessed by an electron was dependent upon its specific position or **orbit** about the nucleus. The greater the distance between the nucleus and the orbit, the greater the energy the electron possessed. Bohr called the electron orbits **energy levels**; electrons in the energy level closest to the nucleus would possess the least energy. Electrons in the outermost level would have the highest energy in the atom.

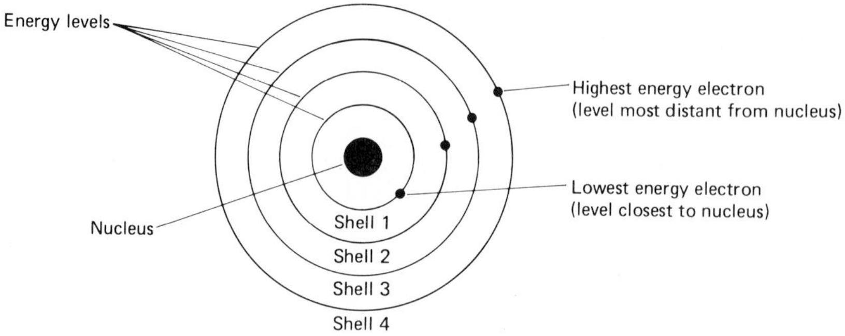
Energy levels
Highest energy electron
(level most distant from nucleus)
Lowest energy electron
(level closest to nucleus)
Nucleus
Shell 1
Shell 2
Shell 3
Shell 4

Figure 5–1 The Bohr atom.

All electrons were restricted in their movement to the specific locations of the energy levels, but they could move within those levels without losing energy. To minimize energy, electrons always occupied the lowest energy levels available to them. Thus electrons went into the first level and filled it up before occupying the second level, and so on. During the absorption of energy by an atom, electrons were promoted from low-energy levels to higher-energy levels (levels farther away from the nucleus). For an excited atom to lose its excess energy, electrons had to move from the high-energy levels at the edges of the atom to lower-energy levels closer to the nucleus (see Figure 5–2).

Since there were definite energy differences between the energy levels in any atom, an electron would move between the levels only if it absorbed exactly the amount of energy corresponding to the difference between those levels. Since the energy present in light is dependent upon the wavelength of the light wave, only certain wavelengths would be absorbed by an atom. Conversely, light energy given off by atoms as electrons move to lower-energy levels must correspond to the energy differences between those levels. So only certain wavelengths of light would be emitted. Since the spacings between the energy levels differed from element to element, each element would absorb and emit different wavelengths of light.

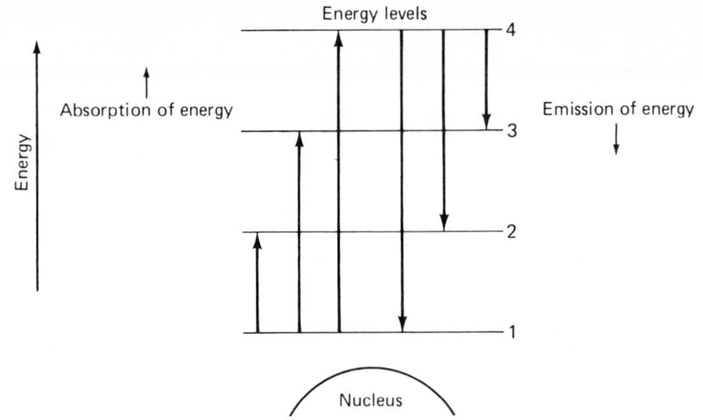

Energy levels
4
3
2
1
Absorption of energy
Emission of energy
Energy
Nucleus

Figure 5–2 Absorption and emission of energy. If an electron moves from Level 1 to Level 2, it must absorb energy equal to the difference between the two states. If an electron moves from Level 4 down to Level 2, the amount of energy released is equal to the difference in energy between Levels 2 and 4.

Quantum Mechanics and Atomic Orbitals

Bohr's theories represented a great leap forward, but they still proved insufficient to explain the experimental data. The theories worked well for hydrogen but ran into difficulties with larger atoms. These atoms absorbed and emitted many more wavelengths than were predicted by Bohr's theories. A new theory, called **quantum mechanics**, was developed in the 1920's largely as a result of the brilliant work of Albert Einstein, Erwin Schroedinger, Louis De Broglie, and Werner Heisenberg. These scientists found that electrons did not always behave like solid planet-like spheres revolving around a nucleus. Electrons could also be described as units of electrical energy and were pictured more like clouds than planets. Complex mathematical equations were developed to describe the energies and positions of these cloudlike units of matter.

Besides changing our mental picture of electrons, quantum mechanical theory also changed the method of describing the positions and energies of electrons. The major energy levels as pictured by Bohr were too limiting, so they were subdivided into more levels called **atomic orbitals,** or **subshells**. *An atomic orbital is a volume of space about a nucleus where electrons of a certain energy are most likely to be found.* All electrons in this unit of volume would have the same energy. The result is that there are many more energy levels for the electrons to be located in, and thus more wavelengths of light can be absorbed and emitted by atoms (see Figure 5–3).

The atomic orbitals represent volumes of space around the nucleus where electrons of similar energy are to be found. These orbitals can have four different shapes. In any energy level, the lowest-energy orbital is called an **s-orbital**; it is spherical (see Figure 5–4). Electrons of slightly higher energy are located in **p-orbitals**; the shapes of p-orbitals can best be pictured by taking a balloon and tying a string tightly around its middle. Two lobes are formed; the nucleus of the atom

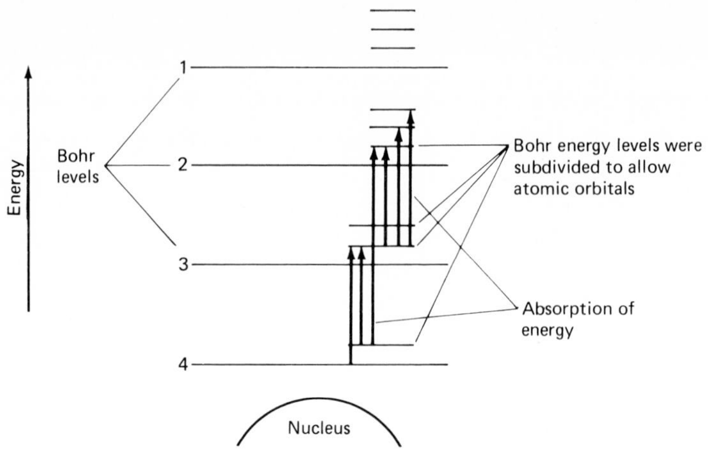
Energy
Bohr levels
1
2
3
4
Bohr energy levels were subdivided to allow atomic orbitals
Absorption of energy
Nucleus

Figure 5–3 Quantum mechanical energy levels.

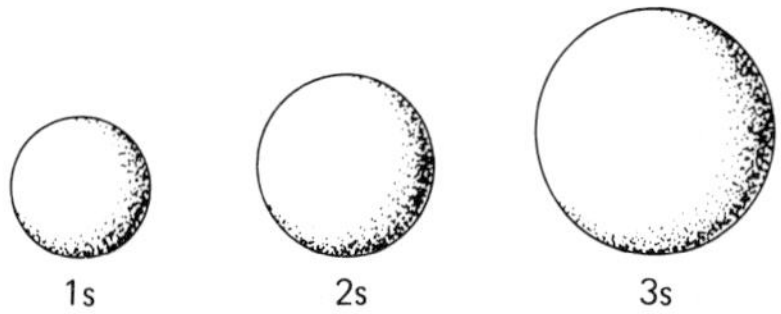
1s
2s
3s

Figure 5–4 s-orbitals in first, second, and third major energy levels.

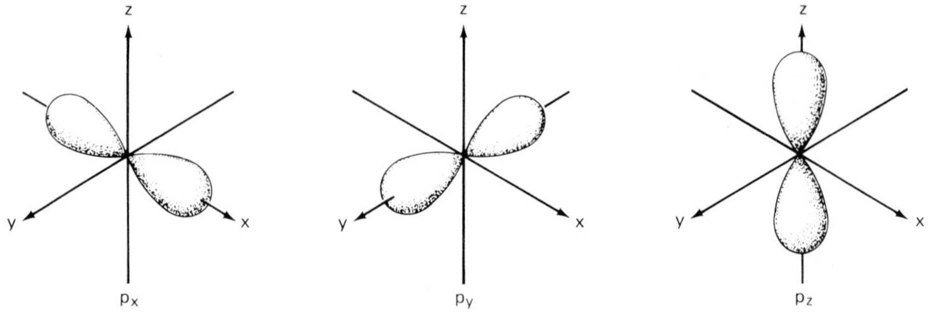
z
y
x
px
z
y
x
py
z
y
x
pz

Figure 5–5 Shapes of p-orbitals.

would be in the center where the knot is located. Unlike the s-orbitals, which point in no direction, the p-orbitals point along the x-, y-, and z-axes (see Figure 5–5). A p_x-orbital is parallel to the x-axis, a p_y-orbital is parallel to the y-axis, and a p_z-orbital is parallel to the z-axis. Where permitted, each major energy level will contain three p-orbitals of identical energy. The only differences among them are the directions in which they point. The other two major types of atomic orbitals are **d-orbitals** and **f-orbitals**. Their shapes are more difficult to picture, and it is not important to remember them. The *numbers* of each orbital per energy level are important: *there are five d-orbitals and seven f-orbitals per shell where allowed.*

Four different types of atomic orbitals can exist, but there are restrictions on the number of electrons that can be present in these orbitals and on the number of these orbitals that can be present in the major energy levels. All electrons possess a negative electrical charge and will repel each other if they get too close to one another. *Each atomic orbital can contain a maximum of two electrons,* whether it is an s-, p-, d-, or f-orbital. Furthermore, there is not room for many electrons in the energy levels closest to the nucleus, so these levels do not contain all the possible atomic orbitals.

Now we have all the facts to begin building a picture of the locations of electrons about the nucleus of an atom. Three important points must be remembered:

1. The energies of the atomic orbitals increase in the order of s $<$ p $<$ d $<$ f. To minimize energy, electrons will occupy the lowest-energy atomic orbital available. In any energy level, then, s-orbitals will always be filled before p-orbitals, and so on.
2. All atomic orbitals contain a maximum of two electrons.
3. There is one type of s-orbital, three types of p-orbitals, five types of d-orbitals, and seven types of f-orbitals. All are present in the major energy levels except for those levels closest to the nucleus.

The first major energy level (the K-shell), because of its nearness to the nucleus, contains only an s-orbital. It can possess a maximum of two electrons. The second level (the L-shell) can contain an s-orbital and three p-orbitals, so it can possess a maximum of eight electrons. The third level (the M-shell) can contain an s-orbital, three p-orbitals, and five d-orbitals so it can contain a maximum of 18 electrons. The fourth through the seventh levels contain s-, p-, d- and f-orbitals, so each may possess a maximum of 32 electrons. Figure 5–6 presents a ranking of the atomic orbitals with respect to their relative energies. Both the type and location of the atomic orbitals are designated. For example, a p-orbital in the second energy level is called a 2p-orbital, a 3p-orbital refers to the p-orbitals in the third level, and so on.

Electron Configurations of Elements

In describing the locations and energies of all electrons present in an atom, we normally make a list of all the atomic orbitals that have electrons. This list of

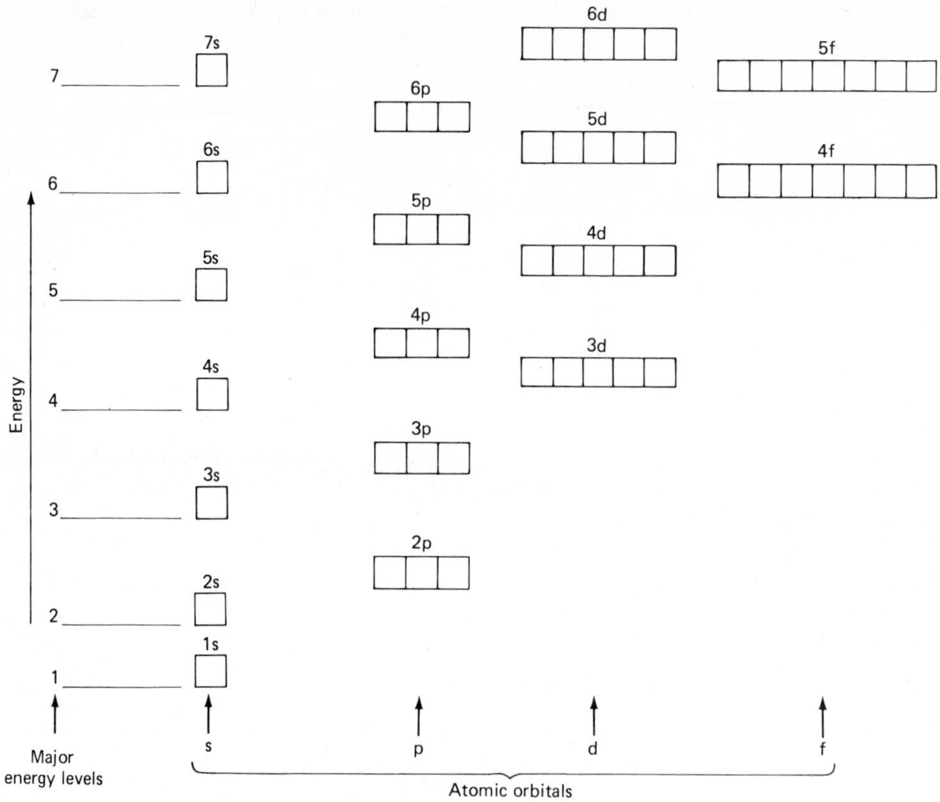
Energy
7
6
5
4
3
2
1
Major energy levels
7s
6s
5s
4s
3s
2s
1s
6p
5p
4p
3p
2p
6d
5d
4d
3d
5f
4f
s
p
d
f
Atomic orbitals

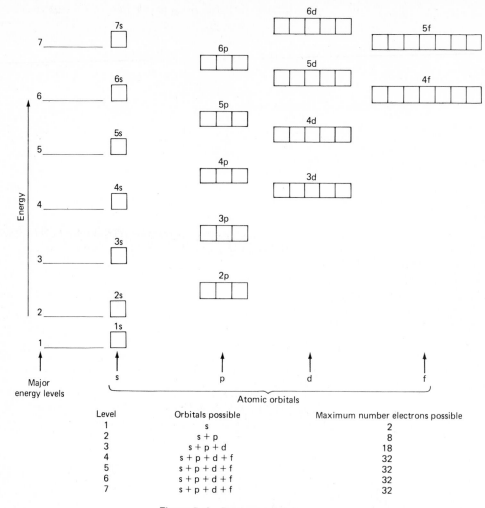

Level	Orbitals possible	Maximum number electrons possible
1	s	2
2	s + p	8
3	s + p + d	18
4	s + p + d + f	32
5	s + p + d + f	32
6	s + p + d + f	32
7	s + p + d + f	32

Figure 5–6 Electron orbitals.

occupied atomic orbitals is called an **electron configuration**. By convention, the atomic orbitals of lowest energy are listed first, followed by the higher-energy orbitals. In constructing the electron configuration for any atom the following rules must be observed:

1. First determine how many electrons an atom possesses. This is the number of electrons that must be shown in the electron configuration.
2. To minimize energy, add electrons to the lowest-energy atomic orbitals first. Normally, those atomic orbitals are filled before adding electrons to higher orbitals.

Figure 5–6 lists the atomic orbitals in order of their increasing energy. From that figure a ranking of the orbitals can be made.

1s 2s 2p 3s 3p 4s 3d 4p 5s 4d 5p 6s 4f 5d 6p 7s

Lowest-energy orbitals ⟶ Highest-energy orbitals

From this series, note that the s-orbitals of the fourth energy level are filled before the d-orbitals of the third level. Actually, the energies of 4s and 3d atomic orbitals are quite close, and it is difficult to explain why the 4s-orbitals are filled first. At this point, all we ask is that you accept it as true. As we move to energy levels farther away from the nucleus, these sorts of exceptions occur more often as the differences in the energies of the orbitals become smaller. There is no way to predict it, but there is a handy way to memorize the energy order (see Figure 5–7). We shall be concerned mostly with electrons in the lower-energy atomic orbitals (first through fourth levels), and there is only one exception in these levels.

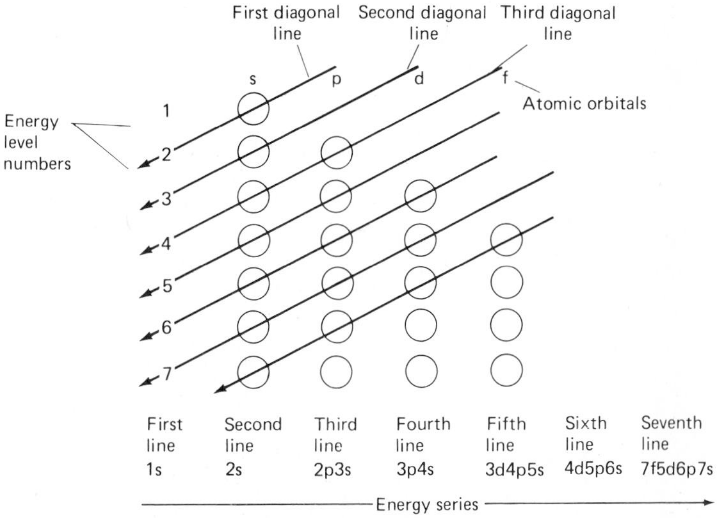
First diagonal line
Second diagonal line
Third diagonal line
s
p
d
f
Atomic orbitals
Energy level numbers
1
2
3
4
5
6
7
First line 1s
Second line 2s
Third line 2p3s
Fourth line 3p4s
Fifth line 3d4p5s
Sixth line 4d5p6s
Seventh line 7f5d6p7s
Energy series

Figure 5–7 One method of generating the atomic orbital energy sequence: construct the diagram and label it as shown. Draw diagonal lines as shown and write down the orbitals as the lines intersect them.

Now we can begin to construct the electron configurations for the elements. The simplest element, hydrogen, possesses one proton in its nucleus and one electron outside the nucleus. This electron is located in the lowest-energy atomic orbital, the 1s-orbital. The **electron configuration** for hydrogen is $1s^1$. Helium, the next element, has two electrons and both are located in the 1s-orbital ($1s^2$). Lithium has three electrons. Two of these electrons are located in the 1s-orbital; the third electron goes into the 2s-orbital since the 1s-orbital is filled. The electron configuration for lithium is $1s^22s^1$. Beryllium, atomic number 4, has an electron configuration of $1s^22s^2$. Boron, atomic number 5, has five electrons to be placed in orbitals. The first four are placed in the 1s- and 2s-orbitals, completely filling them. The fifth electron must be placed in the next lowest-energy atomic orbital, the 2p-orbital ($1s^22s^22p^1$).

Table 5–1 gives electron configurations for all the elements. Remember that the *number* of electrons present in each orbital is signified by the superscript above the orbital. The numbers present in front of the orbitals tell in which energy level the orbital is located. Remember that the 4s and 3d atomic orbitals are quite similar in energy, and that the 4s-orbital is filled before the 3d-orbitals. Thus the electron configuration for element 19, potassium, is $1s^2 2s^2 2p^6 3s^2 \mathbf{4s^1}$ and not $1s^2 2s^2 2p^6 3s^2 3p^6 \mathbf{3d^1}$. Only after the 4s-orbital is filled do we begin adding electrons to the 3d level.

The nearness in energy of the 4s- and 3d-orbitals also gives rise to the unusual electron configurations for chromium (atomic number 24) and copper (atomic number 29). Orbitals that are half-filled or completely filled have a special stability associated with them. In chromium an electron configuration of $1s^2 2s^2 2p^5 3s^2 3p^6 4s^1 3d^5$ allows both the 4s- and 3d-orbitals to be half-filled. That is a more stable arrangement than $1s^2 2s^2 2p^6 3s^2 3p^6 4s^2 3d^4$. In copper, an electron configuration of $1s^2 2s^2 2p^6 3s^2 3p^6 4s^1 3d^{10}$ has a half-filled 4s-orbital and completely filled 3d-orbitals. This arrangement is more stable than $1s^2 2s^2 2p^6 3s^2 3p^6 4s^2 3d^9$, in which only one orbital is completely filled. For these strange types of configurations to occur, the two orbitals must be very similar in energy. They will not occur between s- and p-orbitals, or between p- and d-orbitals.

Table 5–1 Electron configurations

Atomic number	Element	Electron configuration	Atomic number	Element	Electron configuration
1	H	$1s^1$	26	Fe	– $3d^6 4s^2$
2	He	$1s^2$	27	Co	– $3d^7 4s^2$
3	Li	$[He]2s(1s^2 2s^1)$	28	Ni	– $3d^8 4s^2$
4	Be	– $2s^2$	29	Cu	– $3d^{10} 4s^1$
5	B	– $2s^2 2p^1$	30	Zn	– $3d^{10} 4s^2$
6	C	– $2s^2 2p^2$	31	Ga	– $3d^{10} 4s^2 4p$
7	N	– $2s^2 2p^3$	32	Ge	– $3d^{10} 4s^2 4p^2$
8	O	– $2s^2 2p^4$	33	As	– $3d^{10} 4s^2 4p^3$
9	F	– $2s^2 2p^5$	34	Se	– $3d^{10} 4s^2 4p^4$
10	Ne	– $2s^2 2p^6$	35	Br	– $3d^{10} 4s^2 4p^5$
11	Na	$[Ne]\,3s^1(1s^2 2s^2 2p^6 3s^1)$	36	Kr	– $3d^{10} 4s^2 4p^6$
12	Mg	– $3s^2$	37	Rb	$[Kr]\,5s^1$
13	Al	– $3s^2 3p^1$	38	Sr	– $5s^2$
14	Si	– $3s^2 3p^2$	39	Y	– $4d^1 5s^2$
15	P	– $3s^2 3p^3$	40	Zr	– $4d^2 5s^2$
16	S	– $3s^2 3p^4$	41	Nb	– $4d^4 5s^1$
17	Cl	– $3s^2 3p^5$	42	Mo	– $4d^5 5s^1$
18	Ar	– $3s^2 3p^6$	43	Tc	– $4d^5 5s^2$
19	K	$[Ar]\,4s^1(1s^2 2s^2 2p^6 3s^2 3p^6 4s^1)$	44	Ru	– $4d^7 5s^1$
20	Ca	– $4s^2$	45	Rh	– $4d^8 5s^1$
21	Sc	– $3d^1 4s^2$	46	Pd	– $4d^{10}$
22	Ti	– $3d^2 4s^2$	47	Ag	– $4d^{10} 5s^1$
23	V	– $3d^3 4s^2$	48	Cd	– $4d^{10} 5s^2$
24	Cr	– $3d^5 4s^1$	49	In	– $4d^{10} 5s^2 5p^1$
25	Mn	– $3d^5 4s^2$	50	Sn	– $4d^{10} 5s^2 5p^2$

Table 5–1 Electron configurations (cont.)

Atomic number	Element	Electron configuration	Atomic number	Element	Electron configuration
51	Sb	– $4d^{10}5s^25p^3$	79	Au	– $4f^{14}5d^{10}6s^1$
52	Te	– $4d^{10}5s^25p^4$	80	Hg	– $4f^{14}5d^{10}6s^2$
53	I	– $4d^{10}5s^25p^5$	81	Tl	– $4f^{14}5d^{10}6s^26p^1$
54	Xe	– $4d^{10}5s^25p^6$	82	Pb	– $4f^{14}5d^{10}6s^26p^2$
55	Cs	[Xe] $6s^1$	83	Bi	– $4f^{14}5d^{10}6s^26p^3$
56	Ba	– $6s^2$	84	Po	– $4f^{14}5d^{10}6s^26p^4$
57	La	– $5d^16s^2$	85	At	– $4f^{14}5d^{10}6s^26p^5$
58	Ce	– $4f^26s^2$	86	Rn	– $4f^{14}5d^{10}6s^26p^6$
59	Pr	– $4f^36s^2$	87	Fr	[Rn] $7s^1$
60	Nd	– $4f^46s^2$	88	Ra	– $7s^2$
61	Pm	– $4f^56s^2$	89	Ac	– $6d^17s^2$
62	Sm	– $4f^66s^2$	90	Th	– $6d^27s^2$
63	Eu	– $4f^76s^2$	91	Pa	– $5f^26d^17s^2$
64	Gd	– $4f^75d^16s^2$	92	U	– $5f^36d^17s^2$
65	Tb	– $4f^96s^2$	93	Np	– $5f^46d^17s^2$
66	Dy	– $4f^{10}6s^2$	94	Pu	– $5f^67s^2$
67	Ho	– $4f^{11}6s^2$	95	Am	– $5f^77s^2$
68	Er	– $4f^{12}6s^2$	96	Cm	– $5f^76d^17s^2$
69	Tm	– $4f^{13}6s^2$	97	Bk	– $5f^86d^17s^2$
70	Yb	– $4f^{14}6s^2$	98	Cf	– $5f^{10}7s^2$
71	Lu	– $4f^{14}5d^16s^2$	99	Es	– $5f^{11}7s^2$
72	Hf	– $4f^{14}5d^26s^2$	100	Fm	– $5f^{12}7s^2$
73	Ta	– $4f^{14}5d^36s^2$	101	Md	– $5f^{13}7s^2$
74	W	– $4f^{14}5d^46s^2$	102	No	– $5f^{14}7s^2$
75	Re	– $4f^{14}5d^56s^2$	103	Lw	– $5f^{14}6d^17s^2$
76	Os	– $4f^{14}5d^66s^2$	104	Ku	– $5f^{14}6d^27s^2$
77	Ir	– $4f^{14}5d^76s^2$	105	Ha	– $5f^{14}6d^37s^2$
78	Pt	– $4f^{14}5d^96s^1$			

NOTE: For simplicity configurations are sometimes abbreviated; thus [He] 2s is the electron configuration for He + 2s.

UNIT 5 Problems and Exercises

5–1. How do the atomic orbitals differ from the major energy levels?

5–2. Which energy levels can possess a maximum number of electrons equal to the following?
(a) 2 (b) 8 (c) 32 (d) 18

5–3. Approximately 14 electron volts (ev) of energy is required to remove an electron from the first energy level of a hydrogen atom. Approximately 10 ev is required to cause the electron to move from the first to the second level. How much energy is required to completely remove an electron from the second level?

5–4. Complete the following table without using the text if possible.

Energy level	Number of s-orbitals	Number of p-orbitals	Number of d-orbitals	Number of f-orbitals	Total number of electrons possible
1					
2					
3					
4					
5					

5–5. What is the shape of an s-orbital? How many s-orbitals exist in each energy level? How does an s-orbital in the first level differ from an s-orbital in the second level?

5–6. What are the differences among the following?
(a) s- and p-orbitals (b) p_x-, p_y-, and p_z-orbitals

5–7. Electrons that absorb energy move to orbitals farther away from the nucleus. Loss of energy causes them to move to orbitals closer to the nucleus. For the following, tell whether the movement will involve absorption or loss of energy. Try this without looking at the tables.
(a) 3s → 3p (b) 3s → 2p (c) 2s → 3s
(d) 4f → 4d (e) 3p → 4s (f) 3d → 4s

5–8. Complete the following table by filling in all the missing data. In some cases you may need to refer to tables.

Element name	Element symbol	Number of protons present	Number of neutrons present	Atomic weight	Electron configuration
Sulphur			16	32	
	Mg	12	12		
Copper		29		64	
		7			
					$1s^2 2s^2 2p^6 3s^2 3p^6$
Chlorine					
	Na				
Fluorine					

5–9. Give the electron configurations for the following elements. Try to do it from memory.
(a) Na; atomic number 11 (b) Ca; atomic number 20
(c) P; atomic number 15 (d) Fe; atomic number 26

(e) Li; atomic number 3
(f) O; atomic number 8
(g) Br; atomic number 35
(h) Al; atomic number 13
(i) Ar; atomic number 18

5–10. Identify the elements whose neutral atoms have the following electron configurations.
(a) $1s^2 2s^2$
(b) $1s^2 2s^2 2p^3$
(c) $1s^2 2s^2 2p^6 3s^2$
(d) $1s^2 2s^2 2p^6 3s^2 3p^3$
(e) $1s^2 2s^2 2p^6 3s^2 3p^6 4s^2 3d^6$
(f) $1s^2$

5–11. What is wrong with the following electron configurations? Correct them if they are wrong.
(a) $1s^2 2s^3$
(b) $1s^2 1p^6 2s^2 2p^6$
(c) $1s^2 2s^2 3s^2 3p^6 4s^2 4p^6$
(d) $1s^2 2s^2 2p^6 3s^2 3p^{10} 4s^2$

5–12. Which of the following electron configurations represent those of excited atoms?
(a) $1s^2 2s^2 2p^6$
(b) $1s^2 2s^1 2p^6$
(c) $1s^2 2s^2 2p^5 3s^1$
(d) $1s^2 2s^2 3s^1$
(e) $1s^1 2s^2$

5–13. What is meant when we say that the electron configuration is that of an atom in its ground state?

5–14. Consider the following electron configurations: (A) $1s^2 2s^2 2p^6 3s^1$ and (B) $1s^2 2s^2 2p^6 6s^1$. Mark the following statements as true or false.
(a) Energy is required to change from level A to level B.
(b) Level A represents a sodium atom.
(c) Levels A and B represent different elements.
(d) Less energy is required to remove one electron from (A) than from (B).

5–15. Write the electron configurations for the following *ions*. Remember that they are not neutral atoms. What do they all have in common?
(a) O^{-2}
(b) F^{-1}
(c) Na^{+1}
(d) Mg^{+2}
(e) Al^{+3}

UNIT 6 THE PERIODIC TABLE

By the mid-1800's more than 50 different elements were known to exist. As their physical and chemical properties were published, it was seen that many elements had similar properties. Attempts were made to categorize the elements into groups according to their properties, for if such groupings could be done, then perhaps general theories could be advanced to explain the reactivities of the elements.

The first really successful arrangement of the elements was made in the mid-1800's by Dmitri Mendeleyev, a Russian chemist. He arranged the elements primarily in order of increasing atomic weights (see Figure 6–1). The elements with similar physical properties were placed in vertical columns or groups in the table. If the order of weights did not agree with his groupings, he changed the weights or left empty spaces. Then he boldly predicted the weights and properties of elements not yet discovered, which should fit into the empty spaces in his table (see Table 6–1).

Row	Group I	Group II	Group III	Group IV	Group V	Group VI	Group VII	Group VIII
1	H = 1							
2	Li = 7	Be = 9.4	B = 11	C = 12	N = 14	O = 16	F = 19	
3	Na = 23	Mg = 24	Al = 27.3	Si = 28	P = 31	S = 32	Cl = 35.5	
4	K = 39	Ca = 40	? = 44	Ti = 48	V = 51	Cr = 52	Mn = 55	Fe = 56, Co = 59, Ni = 59
5	Cu = 63	Zn = 63	? = 68	? = 72	As = 75	Se = 78	Br = 80	
6	Rb = 85	Sr = 87	Y = 88	Zr = 90	Nb = 94	Mo = 96	? = 100	Ru = 104, Rh = 104, Pd = 106
7	Ag = 108	Cd = 112	In = 113	Sn = 118	Sb = 122	Te = 125	T = 127	
8	Cs = 133	Ba = 137	Dy = 138	Ce = 140				
9								
10			Er = 178	La = 180	Ta = 182	W = 184		Os = 195, Ir = 197, Pt = 198
11	Au = 199	Hs = 200	Tl = 204	Pb = 207	Bi = 208			
12				Th = 231		U = 240		

Figure 6–1 Mendeleyev's periodic table.

Although he was not always correct, the accuracy of his predictions was remarkable. It was primarily the predictive value of Mendeleyev's table that led to its wide acceptance by 1890. In Figure 6–1, the elements with similar atomic weights (those in horizontal rows) do *not* have similar physical properties. But many elements with quite different atomic weights (those in vertical columns) do have similar properties. The same sorts of physical properties kept recurring as atomic weights increased. Thus the table of the elements showed *periodic* repetitions of properties of elements, and it came to be known as the **periodic table**.

Table 6–1 Verification of Mendeleyev's predicted properties for the element germanium (ekasilicon, Es)

Property	Mendeleyev's prediction 1871	Observed after discovery 1885
Atomic weight	72	72.60
Density (g/mℓ) cm^3	5.5	5.47
Gram-atomic volume (mℓ) cm^3	13	13.2
Specific heat capacity (cal/g°C)	0.073	0.076
Formula of oxide	EsO_2	GeO_2
Density of oxide (g/mℓ) cm^3	4.7	4.703
Molar volume of oxide (mℓ) cm^3	22	22.16
Formula of chloride	$EsCl_4$	$GeCl_4$
Boiling point of chloride (°C)	Under 100°	86°
Density of chloride (g/mℓ) cm^3	1.9	1.887
Molar volume of chloride (mℓ) cm^3	113	113.35

There were still problems with Mendeleyev's periodic table, for as more elements were discovered, their properties put them in the wrong place in the table. When the concept of atomic number was defined in the early 1900's, it was realized that the errors in the periodic table could be removed if it were based upon increasing atomic number (number of protons) rather than atomic weight (number of protons and neutrons). Since we know the identity of an element is dependent upon the number of protons in its nucleus and not on the number of neutrons, it is reasonable that properties of elements should also be based on atomic number.

The presently accepted version of the periodic table is shown in the inside front cover of this text and in Figure 6–2. Many of the elements in any one particular vertical column (group) have similar chemical and physical properties. In each horizontal row (period) the atomic number increases from left to right. As one goes down any group, the atomic number also increases, but by more than one unit at a time. The zigzag line is a rough dividing line for **metals** and **nonmetals**. The elements to the left of the line are called *metals.* Most metals have a luster similar to that of steel or silver. They reflect heat and light readily. They conduct heat and electricity remarkably well. Some are ductile and can be drawn into wire and some are malleable and can be hammered into thin sheets. Most are solids at room temperature.

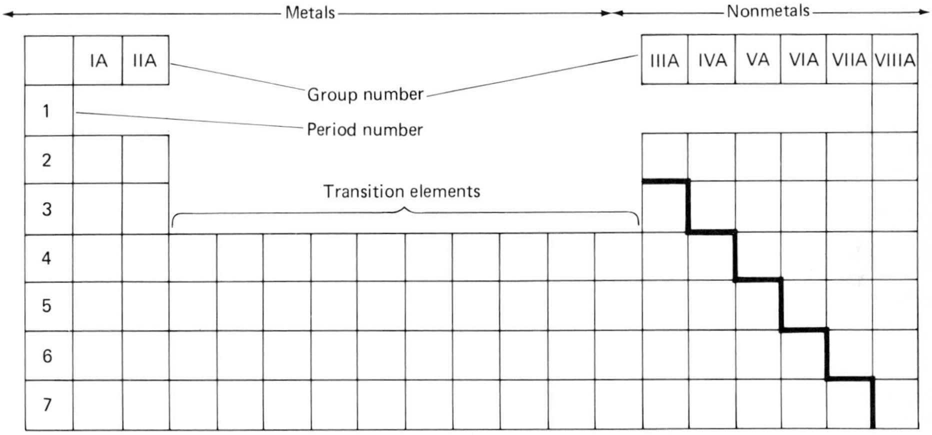

Metals
Nonmetals
IA
IIA
IIIA
IVA
VA
VIA
VIIA
VIIIA
Group number
Period number
Transition elements
1
2
3
4
5
6
7

Figure 6–2 An overall picture of the periodic table.

The elements on the right side of the line are called *nonmetals.* They usually are poor conductors of heat and electricity. They are usually too brittle to be drawn into wire or hammered into sheets. A large number of the nonmetals are gases at room temperature; others are solids. Few are liquids. The elements situated directly adjacent to the jagged line often exhibit properties characteristic of both metals and nonmetals. They do not readily fall into either class and are often called **metalloids**.

The elements in the center portion of the periodic table (groups IB through VIIIB) are called **transition metals** (Sc to Zn, Y to Cd, and La to Hg). Analysis of the electron configurations of these elements reveals that *the d-orbitals are being filled* as one moves from left to right through the periods. For example, scandium, atomic number 21, has one electron in the 3d level; the element zinc (atomic number 30) has 10 electrons in the 3d level. Most of the transition elements in the three periods have $4s^2$, $5s^2$, or $6s^2$ outer shells, and are much alike in properties. Many have the ability to lose one, two, or more electrons. They form bonds differently than the first two groups of elements and are also noted for the fact that they sometimes form colored compounds. For example, many of the compounds of chromium are green, blue, or violet; some copper compounds are blue, and some iron compounds are green or rust colored.

The two rows of elements at the bottom of the table (the lanthanide and actinide series) are placed there to make it more convenient to represent the table. In each series the **f-orbitals** are being filled as one goes from left to right. There are 14 elements in each row because there are 14 f electrons in each energy level. In the lanthanide series, the 4f-orbitals are being filled. In the actinide series, the 5f-orbitals are being filled.

Electron Configuration and Chemical Reactivity

Let's look at the periodic table and examine it more closely to see how we can use it.

1. As we proceed from left to right across any period of the table, the number of electrons present increases by one for each element. The number of electrons in the outermost occupied energy level is designated by the group numbers IA through VIIIA. This does not hold for the transition elements (groups IB through VIIIB), as d-electrons are being added to inner orbitals. You should be able to quickly determine the number of electrons present in the outer energy level of any other element simply by noting its group number. For example, there are five electrons in the outer level in an atom of nitrogen, seven in an atom of bromine, and two in an atom of calcium.
2. *All the elements in any particular group have the same number of electrons in their outermost occupied energy levels* (see Table 6–2). The difference in electronic structure of the elements in a group is in the *level being filled,* but not in the number of electrons in the outermost level. Since the physical and chemical properties of an element depend to a large extent on the number of electrons in the outer level, many of the elements in any particular group have similar properties. This is especially true for the lighter elements (periods 1, 2, and 3).
3. The energy level being filled is designated on the left side of the periodic table (the period number). Remember that there is a difference in period for the transition elements because electrons are being added to one level lower than the outermost occupied s- and p-orbital levels. For example, potassium and calcium have their last electrons in the fourth level, but scandium has its last electron in the d-orbital of the third level.

Table 6–2 Electron configurations of atoms in several groups

Group	Element	Atomic number	1	2	3	4	5	6	7
			Electron configuration (principal levels) (Energy levels)						
Group VIIIA: Noble gases	Helium	2	2						
	Neon	10	2	8					
	Argon	18	2	8	8				
	Krypton	36	2	8	18	8			
	Xenon	54	2	8	18	18	8		
	Radon	86	2	8	18	32	18	8	
Group IA: Alkali metals	Lithium	3	2	1					
	Sodium	11	2	8	1				
	Potassium	19	2	8	8	1			
	Rubidium	37	2	8	18	8	1		
	Cesium	55	2	8	18	18	8	1	
	Francium	78	2	8	18	32	18	8	1
Group IIA: Alkaline earth metals	Beryllium	4	2	2					
	Magnesium	12	2	8	2				
	Calcium	20	2	8	8	2			
	Strontium	38	2	8	18	8	2		
	Barium	56	2	8	18	18	8	2	
	Radium	88	2	8	18	32	18	8	2
Group VIIA: Halogens	Fluorine	9	2	7					
	Chlorine	17	2	8	7				
	Bromine	35	2	8	18	7			
	Iodine	53	2	8	18	18	7		
	Astatine	85	2	8	18	32	18	7	

Chemical Bonds: Losing and Gaining Electrons

Few elements are found in nature in their pure states. Generally, atoms of different elements are grouped together in clusters called compounds. The atoms are held together in these clusters by the formation of chemical bonds. In forming chemical bonds, energy is released, and the compounds are more stable than the separate, unbonded atoms.

The two major types of bonds that form are called **ionic** and **covalent** bonds. Ionic bonding occurs when two atoms bearing opposite charges (ions) get close enough together for powerful electrostatic attraction to occur. In covalent bonding, neutral, noncharged atoms join by a sharing of electrons. More will be said about each type of bonding in later units, but the important thing to remember here is that the formation of chemical bonds is an energy-releasing process.

To predict the structures of compounds, we need to know which atoms prefer to lose electrons and which tend to gain electrons as they react. We would also like to know how many electrons are gained or lost as an atom reacts. By looking at the periodic table and the electron configurations of the elements we can make pretty good predictions for the elements in the first three periods. All these elements tend to *lose or gain electrons so as to end up with completely filled or empty s- and p-orbitals in their outermost occupied energy levels.* Atoms with one, two, or three electrons in their outer level generally lose those electrons as they react. In this way their outermost occupied energy levels are emptied of electrons. Elements with four, five, six, or seven electrons in their outer occupied levels tend to gain electrons and thus fill up the s- and p-orbitals in their outermost occupied level. Elements with eight electrons in their outer energy levels are called **inert gases**; they do not readily enter into chemical reactions.

To illustrate how to predict the loss and gain of electrons, let us concentrate on the elements in the second period of the periodic table. The first element, lithium, has an atomic number of 3 and an electron configuration of $1s^2 2s^1$. To fill the s- and p-orbitals of the second energy level, lithium would have to gain seven electrons. But there is another possibility: by losing one electron, the second level is completely emptied and now the outermost *occupied* level is the first level, which already has its required electrons. It is much easier to lose one electron than to gain seven, so lithium will lose an electron and become a positive ion when it forms ionic compounds. In chemical terms, we can state this by the following equation. The plus one sign on the lithium ion signifies a charge imbalance in the atom. It means that the lithium ion has one more proton than electron.

$$Li^0 - 1e^{-1} \longrightarrow Li^{+1} \quad (\text{or } Li^0 \rightarrow Li^{+1} + 1e^{-1})$$

3 protons	=	+ 3 charges	+ 3 charges	=	3 protons
3 electrons	=	− 3 charges	− 2 charges	=	2 electrons
Net charge	=	0	+ 1		

Electron configuration:

$1s^2 2s^1$ $\qquad$ $1s^2$

Beryllium, the next element, would have to gain six electrons or lose two. It is easier to lose two, so beryllium should form positive ions. (Actually, beryllium does not form ions.) Boron would either gain five or lose three,

	$Be - 2e$	$\longrightarrow (Be^{+2})$;	$B - 3e$	$\longrightarrow B^{+3}$
No. of protons (charge):	+ 4	+ 4	+ 5	+ 5
No. of electrons (charge):	− 4	− 2	− 5	− 2
Net charge	0	+ 2	0	+ 3
Configuration:	$1s^2 2s^2$	$1s^2$	$1s^2 2s^2 2p^1$	$1s^2$

We would predict the loss of three, but it appears that both processes require high energy, because boron does not readily form ions. Carbon would have to lose or gain four electrons; loss or gain of that many electrons through ionization does not seem to be a favorable process, because most carbon atoms form covalent bonds rather than ions. Carbon thus gains four electrons through electron sharing and does not form stable ions.

Nitrogen has the choice of losing five or gaining three electrons. The smaller change is the gain of three. Again, both processes must be unfavorable since nitrogen rarely reacts to form ions. However, nitrogen can gain three or lose five electrons during the process of electron sharing while forming covalent bonds. Oxygen will gain two electrons rather than lose six, and fluorine will gain one rather than lose seven electrons.

	$N^0 + 3e^{-1}$	$\rightarrow$ N^{-3};	$0^0 + 2e^{-1}$	$\rightarrow$ 0^{-2}
No. of protons (charge):	+ 7	+ 7	+ 8	+ 8
No. of electrons (charge):	− 7	− 10	− 8	− 10
Net charge:	0	− 3	0	− 2
Configuration:	$1s^2 2s^2 2p^3$	$1s^2 2s^2 2p^6$	$1s^2 2s^2 2p^4$	$1s^2 2s^2 2p^6$

	$F^0 + 1e^{-1}$	$\rightarrow$ F^{-1}
No. of protons (charge):	+ 9	+ 9
No. of electrons (charge):	− 9	− 10
Net charge:	0	− 1
Configuration:	$1s^2 2s^2 2p^5$	$1s^2 2s^2 2p^6$

Neon, the final element in period 2, already has eight electrons in its outermost occupied shell, and so does not readily form chemical bonds. Notice that the electron configurations of all the ions formed are identical to those of the inert gases (group VIIIA). Since most elements do not have filled outermost occupied energy levels, they get them by bonding with other elements to form compounds. In the periodic table, all elements in the same group have the same number of electrons in their outermost occupied energy levels, so elements in that group tend to gain or lose the same number of electrons when forming ionic compounds.

Ionization Energy

When an atom or molecule loses an electron during a chemical reaction, it becomes positively charged; it is called a positive ion or **cation**. Energy is required to remove the electrons from the atoms of molecules, and each atom and molecule has its own particular energy requirement. *The* **ionization energy** *of an atom or molecule is the energy required to remove an electron from that atom or molecule while it is present as a gas.* Thus the ionization energy (IE) is a measure of an atom or molecule's ability to give up electrons, that is, to become positively charged. Atoms with large ionization energies do not readily lose electrons because the energy requirement for such loss is simply too large. Atoms with small ionization energies readily lose electrons during chemical reactions.

In looking at the periodic table we predicted that elements in groups IA through IIIA tend to lose electrons, whereas elements in groups IVA through VIIA tend to gain electrons during reactions. Thus the ionization energies for elements in groups IA through IIIA should be less than those in groups IVA through VIIA. Another way of saying this is that elements with few electrons in their outermost occupied energy levels will tend to lose those electrons (have low IE's) during chemical reactions. In the periodic table, ionization energy increases as we proceed from left to right across any period.

Another trend is that ionization energy decreases as we go down a group. Since all the elements in any particular group have the same number of electrons in their outer energy level, the determining factor here is the ease with which an electron can be removed from that level. Electrons in energy levels more removed from the nucleus have less attractive force for the nucleus than electrons in levels closer to the nucleus. Thus electrons in more distant levels (those for elements lower in a group) are more easily lost.

Electron Affinity

Atoms that gain electrons during a chemical reaction become negatively charged; they are called **anions**. The tendency to gain electrons is called the *electron affinity* of an atom. Specifically, *the* **electron affinity** *of an atom or molecule is the energy released when that atom or molecule in the gaseous state accepts an electron.* Atoms with large electron affinities (EA's) have the greatest tendency to accept electrons; atoms with small electron affinities have a lesser tendency to accept electrons. In the periodic table, electron affinity increases as one proceeds from left to right across any period. It also increases as one goes *up* a particular group. Elements in the upper right corner of the periodic table have both the largest electron affinities and the largest ionization energies. This is reasonable, since atoms on the right side tend to gain electrons (have large EA's) and would not want to lose electrons (would have large IE's). Elements in the bottom left of the periodic table have the greatest tendency to lose electrons (the lowest IE's) and the least tendency to gain additional electrons (the lowest EA's). Table 6–3 gives the ionization energies and the electron

Table 6–3 Ionization energy and electron affinities for some elements in the periodic table*

Group							
IA	IIA	IIIA	IVA	VA	VIA	VIIA	VIIIA
H (313) 17.3							He (567)
Li (124) 14	Be (215) 0	B (191) 7	C (260) 29	N (336) 28	O (614) 34	F (402) 80	Ne (497)
Na (119) 19	Mg (176) 0	Al (138) 12	Si (188) 32	P (254) 17	S (239) 48	Cl (300) 83	Ar (363)
K (100)	Ca (141)					Br (273) 77	Kr (323)
Rb (96)	Sr (131)					I (241) 70	
Cs (90)							

* The figure in parentheses is the ionization energy; the other number is the electron affinity. Both figures are in kcal/mol.

affinities for some of the elements. For example, to remove an electron from a hydrogen atom, 313 kcal/mol of energy must be absorbed. When one electron is added to a hydrogen atom, 17.3 kcal/mol of energy is released.

Electron Transfer Reactions

We have discussed how to predict which atoms will lose or gain electrons, but so far we have not said anything about where those electrons go when lost or come from when gained. Electrons are lost to or gained from other atoms. For an atom to lose electrons during a chemical reaction, another atom of a different element must be present that is capable of accepting electrons. This process is called **electron transfer**. In electron transfer, the number of electrons given up must always equal the number of electrons gained.

You should now be able to make some predictions on reactivity by using your knowledge of the periodic table. Would you expect lithium and beryllium to react with each other in an electron-transfer reaction? We know that both lithium and beryllium have low IE's and thus tend to give up electrons as they react. Both are electron donors and neither is an acceptor. We would predict that they would not react with each other. But lithium (a donor) should react with fluorine or oxygen, since both of these elements are electron acceptors. The balanced reactions for lithium with fluorine and oxygen are as follows. Notice that the number of electrons lost by the lithium atoms is always equal to the number of electrons gained by the fluorine or oxygen atoms.

$$Li^0 = Li^{+1} + 1e^{-1} \qquad\qquad 2(Li^0 = Li^{+1} + 1e^{-1})$$
$$F^0 + 1e^{-1} = F^{-1} \qquad\qquad 0^0 + 2e^{-1} = 0^{-2}$$

Balanced reaction:

$$Li^0 + F^0 = Li^{+1} + F^{-1} \qquad\qquad 2Li^0 + 0^0 = 2Li^{+1} + 0^{-2}$$

The rates or speeds at which the electron-transfer reactions occur can be predicted somewhat by examining the ionization energies and electron affinities of the elements involved. The fastest reactions would occur between the elements that lose electrons most easily (group IA) and the elements that have the greatest tendencies to gain electrons (group VIIA). Sodium atoms (group IA) react explosively with chlorine atoms (group VIIA), whereas the reaction between, say, magnesium atoms (group IIA) and oxygen atoms (group VIA) would occur much less vigorously. As one moves from the extremes of the periodic table, the ability of the atoms to undergo electron-transfer reactions decreases.

Some of the elements in group VIA gain electrons and become negative ions. These elements react in a different manner—by sharing electrons. We shall discuss this process in later units.

UNIT 6 Problems and Exercises

6–1. Are the following statements correct or incorrect? Explain your decision.
(a) The periodic table is based upon the atomic weights of the elements.
(b) The periodic table is based upon the number of neutrons in an element.

(c) All elements in the same group have the same electron configuration.
(d) The elements in the same group have the same number of electrons in their outermost occupied levels.
(e) The elements in the same period have the same number of electrons in their outermost occupied levels.
(f) The outermost occupied energy levels are the same for all elements in the same period.
(g) As one moves across a period (left to right), the number of electrons present increases by one each element.
(h) As one moves down a group, the number of electrons increases by one each period.
(i) Atoms tend to gain or lose electrons to look more like the electron configurations of inert gases.
(j) Transitions metals have partially filled d-orbitals.
(k) Metals are located on the left side of the periodic table.
(l) The inert gases have completely filled outermost occupied levels.
(m) Elements on the left side of the periodic table tend to lose electrons.
(n) Elements on the right side of the periodic table tend to lose electrons.

6–2. In what portion of the periodic table are the nonmetals? the metals? the transition metals?

6–3. Explain what is meant by the periodicity of the periodic table.

6–4. What is the significance of each horizontal row (period) in the periodic table? of each vertical column (group)?

6–5. Without referring to the periodic table, (1) draw the electron configurations of the elements with the following atomic numbers; (2) tell whether they are metals or nonmetals; (3) tell what group and period they are in; and (4) tell how many electrons they would lose or gain.

(a) 11	(b) 19	(c) 26	(d) 16
(e) 36	(f) 17	(g) 12	

6–6. Which of the following elements would you expect to show the greatest similarity in chemical and physical properties? Why?

Ca Li Al Ba Cl

6–7. Complete the partial periodic table shown below by filling in the atomic numbers of the elements.

	Group atomic number		
	A = 7	B =	C =
Period	D =	E = 16	F =
	G =	H =	I = 35

Are these elements metals or nonmetals?

6–8. Which portions of an atom are generally not affected by chemical changes? Which portions are altered?

6–9. In normal chemical reactions, lithium atoms form ions by losing one electron each. Fluorine atoms form ions by gaining one electron each. Explain why these two elements behave so differently.

6–10. Predict whether the following elements will gain or lose electrons when they react to form ionic compounds. You should construct the electron configuration of each element before deciding.

(a) Mg (b) O (c) Cl (d) Ca (e) N
(f) Li (g) C (h) B (i) Br

6–11. Write balanced electron-transfer reactions for the following pairs of elements.

(a) Na and Cl (b) Mg and S (c) K and O (d) Ca and O

6–12. What does ionization energy measure? When we observe that an atom of one element has a higher ionization energy than an atom of another element, what does this mean about the two atoms being compared?

6–13. Explain the general trend in ionization energy (a) as one goes down a group, and (b) as one goes across a period.

6–14. Use the periodic table to predict which of the following atoms has the greatest ionization energy. Justify your answers.

(a) Li or Ne (b) K or Cs (c) Na or Cl (d) C or N
(e) F or Br

6–15. Use the periodic table to predict which of the following atoms would possess the greatest electron affinity. Justify each choice.

(a) K or Ca (b) Li or N (c) S or Se (d) F or Cs

6–16. To which group would the following elements belong?

(a) Element A ionizes to become A^{-2}.
(b) Element B ionizes to become B^{+3}.
(c) Element C reacts with sodium to form a compound Na_2C.
(d) Element D does not lose or gain electrons.

6–17. Would you predict the electron affinity of inert gases to be high or low? Why?

6–18. On the basis of your knowledge of the periodic table, determine which of the following are reasonable reactions. Correct the reactions that do not seem reasonable.

(a) $K^0 + \text{energy} \rightarrow K^{+1} + e^{-1}$ (b) $F^0 + e^{-1} \rightarrow F^{-1} + \text{energy}$
(c) $Cl^0 \rightarrow Cl^{+1} + \text{energy} + e^{-1}$

CHAPTER 3

Bonding

We now know the basic facts about the fundamental units of matter: atoms and elements. Most of matter is not composed of individual atoms of elements. Rather, it consists of two or more atoms from different elements bound together into clusters. These clusters of atoms are called **compounds**.

To form a compound, we have to fasten together two or more atoms. This is called **bonding**, and we say that when two atoms have become fastened together, a **chemical bond** has formed between them. There are two main types of chemical bonds: **ionic** bonds and **covalent** bonds.

Ions and ionic bonds are discussed in unit 7. When a neutral atom loses or gains electrons, it becomes an electrically charged species, an **ion**. We know from unit 2 that two units of matter possessing opposite electrical charge are attracted to each other, so ions with opposite charge will be attracted to each other. This type of attraction is called **ionic bonding**.

Atoms that do not form ionic bonds can still bond together; they do it by sharing electrons between neutral atoms, thus forming covalent bonds. It is like two teams sharing a rope for a tug-of-war. The two teams are bonded together as long as each has a hold on the rope. In covalent bonding, the "rope" is a pair of electrons.

UNIT 7 IONS AND IONIC COMPOUNDS

In unit 6 we learned that atoms often lose or gain electrons as they interact with each other. After the loss or gain of electrons, an atom is no longer electrically neutral, but now bears a positive or negative electrical charge. Charged atoms or groups of atoms are called **ions**; the sign and magnitude of the electrical charge on the ion is

Table 7–1 Some important ions

		Element	Symbol for neutral atom	Symbol for its most stable ion	Name of ion
Alkali metals	IA	Lithium	Li	Li^{1+}	Lithium ion
		Sodium	Na	Na^{1+}	Sodiumion
		Potassium	K	K^{1+}	Potassium ion
Alkaline earth metals	IIA	Magnesium	Mg	Mg^{2+}	Magnesium ion
		Calcium	Ca	Ca^{2+}	Calcium ion
		Strontium	Sr	Sr^{2+}	Strontium ion
		Barium	Ba	Ba^{2+}	Barium ion
		Radium	Ra	Ra^{2+}	Radium ion
	IIIA	Aluminum	Al	Al^{3+}	Aluminum ion
	IVA	Oxygen	O	O^{2-}	Oxide ion
		Sulphur	S	S^{2-}	Sulfide ion
Halogens	VIIA	Fluorine	F	F^{1-}	Fluoride ion
		Chlorine	Cl	Cl^{1-}	Chloride ion
		Bromine	Br	Br^{1-}	Bromide ion
		Iodine	I	I^{1-}	Iodide ion
Transition elements		Chromium	Cr	Cr^{3+}	Chromium ion
		Iron	Fe	Fe^{2+}	Iron (II) ion (ferrous ion)
				Fe^{3+}	Iron (III) ion (ferric ion)
		Nickel	Ni	Ni^{2+}	Nickel ion
		Copper	Cu	Cu^{1+}	Copper (I) ion (cuprous ion)
				Cu^{2+}	Copper (II) ion (cupric ion)
		Zinc	Zn	Zn^{2+}	Zinc ion
		Silver	Ag	Ag^{1+}	Silver ion
		Mercury	Hg	Hg_2^{2+}	Mercury (I) ion (mercurous ion)
				Hg^{2+}	Mercury (II) ion (mercuric ion)

determined by the number of electrons lost or gained. Atoms that have lost electrons have positive electrical charges and are called **cations**. Conversely, atoms that have gained electrons possess negative electrical charges and are called **anions**.

Na^{1+}	sodium ion	Cations: (positively charged ions)	Cl^{1-}	chloride ion	Anions: (negatively charged ions)
Fe^{2+}	ferrous ion		O^{2-}	oxide ion	
Al^{3+}	aluminum ion		S^{2-}	sulfide ion	

Table 7–1 lists the names and charges for many of the important ions. The electronic charge on nontransition elements is determined by either emptying or completely filling the outermost occupied energy levels. All these ions have electron configurations like those of the inert gases. The electrical charges on the transition-metal ions are more difficult to predict, but many can be predicted by simply emptying the 4s-orbital or keeping both the 4s- and 3d-orbitals *half-filled.*

Names of Cations

Generally, the names of the cations are formed by adding the word *ion* to the name of the neutral element. Thus Na^{1+} is referred to as sodium ion; elemental sodium (Na) is simply called sodium. The term *ion* after a metal signifies that it is no longer neutral, but has lost electrons (ionized).

In Table 7–1, notice that several metals have more than one ionic state. Iron can be Fe^{2+} or Fe^{3+}; copper can be Cu^{1+} or Cu^{2+}. To distinguish between the various ionic states of an element, the suffixes *ous* and *ic* have been introduced. If an element has two possible ionic states, the one with the *lesser* charge is given the *ous* suffix; the one with the *larger* charge is given the *ic* suffix. (Note that the suffixes do not specify what the charge is. The magnitude of the charge must be memorized.) Another method of designating the ionic state is to place the appropriate roman numeral after the name of the element. For example, ferrous ion would be iron (II), and ferric ion would be iron (III).

Fe^{2+}, ferrous ion (lesser charge), Cu^{1+}, cuprous ion

Fe^{3+}, ferric ion (greater charge), Cu^{2+}, cupric ion

Names of Anions

When atoms gain electrons and become negatively charged, the ending of the element name is dropped and the suffix *ide* is added. Since we shall not cover any anions which have more than one possible ionic state, there is only one ending.

Cl = chlor*ine*, but Cl^{1-} = chlor*ide* ion

O = oxy*gen*, but O^{2-} = ox*ide* ion

S = sul*fur*, but S^{2-} = sul*fide* ion

Polyatomic Ions

In addition to the ions listed in Table 7–1, other more complex ions exist. The polyatomic ions are combinations of more than one element that function as groups quite separate from the individual atoms themselves. For example, the polyatomic ion consisting of one sulfur and four oxygen atoms is called sulfate, has an overall charge of 2−, and is quite different in properties from sulfide or oxide ions separately. The important thing to note is that (SO_4^{2-}) is considered as a single entity, just like Cl^- or Br^- are considered separate unique entities. Table 7–2 lists the more important polyatomic ions with their charges and formulas.

Table 7–2 Common polyatomic ions

Name of ion	Formula	Electrovalence of radical
Hydroxide ion	$(OH)^{1-}$	1−
Carbonate ion	$(CO_3)^{2-}$	2−
Bicarbonate ion	$(HCO_3)^{1-}$	1−
Sulfate ion	$(SO_4)^{2-}$	2−
Bisulfate ion (or hydrogen sulfate ion)	$(HSO_4)^{1-}$	1−
Nitrate ion	$(NO_3)^{1-}$	1−
Nitrite ion	$(NO_2)^{1-}$	1−
Phosphate ion	$(PO_4)^{3-}$	3−
Monohydrogen phosphate ion	$(HPO_4)^{2-}$	2−
Dihydrogen phosphate ion	$(H_2PO_4)^{1-}$	1−
Cyanide ion	$(CN)^{1-}$	1−
Permanganate ion	$(MnO_4)^{1-}$	1−
Chromate ion	$(CrO_4)^{2-}$	2−
Dichromate ion	$(Cr_2O_7)^{2-}$	2−
Sulfite ion	$(SO_3)^{2-}$	2−
Bisulfite ion (or hydrogen sulfite ion)	$(HSO_3)^{1-}$	1−
Ammonium ion	$(NH_4)^{1+}$	1+

There are two positions for numbers used in chemical equations and structures. *Normal-sized* numbers placed in front of ions or compounds refer to the number of the entire units that they precede. Thus $2SO_4^{2-}$ means we are discussing two of the sulfate units, not just two sulfur atoms. *Smaller numbers* placed at the bottom right of an element indicate the number of atoms of that particular element and not any other elements in that group. Thus in the dichromate ion, $Cr_2O_7^{2-}$, there are 2 chromium and 7 oxygen atoms.

There is yet another purpose for the subscript numbers when dealing with polyatomic ions. If a polyatomic ion is placed in parentheses and a subscript added *outside* the parentheses, it refers to the whole ion and not just the nearest element. For example, the ionic compound, ammonium carbonate, has the formula

$(NH_4)_2CO_3$. For every carbonate ion (CO_3^{2-}) there are 2 ammonium ions (NH_4^+). In the total formula there are 2 nitrogen atoms, 8 hydrogen atoms, 1 carbon atom, and 3 oxygen atoms. In aluminum sulfate, $Al_2(SO_4)_3$, there are 2 aluminum ions ($2Al^{3+}$) and 3 sulfate ions $3(SO_4^{2-})$. So the formula contains 2 aluminum, 3 sulfur, and 12 oxygen atoms.

Ionic Bonding: Ionic Compounds

In nature, electrical neutrality is the preferred condition. In any solution or compound there must exist the same number of positive and negative charges. Since we know that particles of opposite charge are attracted to each other, it is reasonable to picture anions and cations being brought together through these attractive forces. When anions and cations combine through electrostatic attraction to form units that are electrically neutral, the compounds formed are called **ionic compounds**. This type of bonding, the attractive forces between ions of opposite charge, is called **ionic bonding**.

Ionic compounds, then, are formed by the attraction between ions of opposite charge. Since not all the ions have the same number of charges, ionic compounds are not always composed of equal numbers of anions and cations. The overall guiding force that determines the number of each ion that will appear in the compound is electrical neutrality. *The total number of positive charges must equal the total number of negative charges.* Ferric ion has a charge of 3+ and chloride is 1−. If we make the salt, ferric chloride, we have to have 3 chloride ions for every ferric ion.

$$
\begin{aligned}
Fe^{3+} &= 1 \times (3+) = 3+ = \text{no. of positive charges} \\
3Cl^{-} &= 3 \times (1-) = \underline{3-} = \text{no. of negative charges} \\
& \qquad\qquad\qquad\; 0 = \text{overall charge}
\end{aligned}
$$

The formula for ferric chloride is $FeCl_3$. Sodium chloride has the formula NaCl, since each ion has the same charge. And calcium chloride must have the formula $CaCl_2$ since calcium has 2 positive charges per atom.

Here are two helpful hints to aid you in constructing the correct formulas for the salts:

1. If the anion and cation in the salt have the *same charge*, then the structure of the salt is simply a one-to-one combination.

   Sodium chloride (Na^{1+}, Cl^{1-}) = NaCl
   Calcium sulfate (Ca^{2+}, SO_4^{2-}) = $CaSO_4$
   Aluminium phosphate (Al^{3+}, PO_4^{3-}) = $AlPO_4$

   Note that the name of the salt is simply the combination of the names of each ion, with the name of the cation given first. The word *ion* is *omitted.*

2. If the charges on the ions in a salt differ, the correct formula may be derived by cross-multiplying. For example, let us use a cation $\mathbf{X}^{3+}$ and an anion $\mathbf{Y}^{2-}$. The charge on **X** is 3+, so by cross-multiplying we need 3 atoms of **Y** in the

formula. Since the charge on **Y** is 2−, we need 2 atoms of **X** in the formula (X_2Y_3):

$$2X^{3+} + 3Y^{2-} \longrightarrow X_2Y_3$$

We have a total of 6 positive charges and 6 negative charges. The net charge is 0. By cross-multiplying in this manner, you ensure that the total number of positive charges is equal to the number of negative charges. More examples are given below.

Ferric hydroxide = $Fe^{3+} + (OH)^{1-} = Fe(OH)_3$
Ammonium carbonate = $(NH_4)^{1+} + (CO_3)^{2-} = (NH_4)_2(CO_3)$
Mercuric phosphate = $Hg^{2+} + (PO_4)^{3-} = Hg_3(PO_4)_2$

To avoid confusion, it is important to include the parentheses when complex ions are used. If calcium cyanide were written without parentheses, $CaCN_2$, the formula would be incorrect since it states that each unit contains one calcium atom, *one* carbon atom, and *two* nitrogen atoms. The correct formula is $Ca(CN)_2$, which says that each unit contains 1Ca and 2C and 2N. Remember that subscripts outside the parentheses multiply the numbers of every atom within the parentheses.

When to Expect Ionic Bonding

Ionic bonding occurs between ions of opposite electrical charge. In any reaction where anions and cations are formed, we expect ionic compounds to result. Elements in groups IA and IIA tend to lose electrons and become cations as they react; elements in groups VIA and VIIA tend to gain electrons and become anions as they react. Thus we can predict that the reaction of elements from groups IA or IIA with elements from groups VIA or VIIA should form ionic compounds. The ionic compound sodium chloride results when elemental sodium and chlorine are mixed. The reaction of sodium and oxygen forms sodium oxide.

$$\begin{array}{lcl}
Na - 1e^{1-} & \longrightarrow & Na^{1+} \\
Cl + 1e^{1-} & \longrightarrow & Cl^{1-} \\
\hline
Na^0 + Cl^0 & \longrightarrow & NaCl
\end{array}
\qquad
\begin{array}{lcl}
2(Na - 1e^{1-} & \longrightarrow & Na^{1+}) \\
0 + 2e^{1-} & \longrightarrow & O^{2-} \\
\hline
2Na^0 + 0^0 & \longrightarrow & Na_2O
\end{array}$$

Conversely, we would not expect ionic compounds to form if elements of similar ionization energy or electron affinity are mixed. Both sodium and calcium have low ionization energies and are electron donors. Neither would show a tendency to become an anion and no ionic compound would form between sodium and calcium. Likewise, the elements in groups VA through VIIA have large electron affinities and tend to accept electrons as they react. Mixing of these elements together should not form ionic compounds.

In general, then, we expect ionic compounds to form when elements on the far left side of the periodic table interact with elements on the far right side (excluding the inert gases). The closer elements are in the periodic table, the less the tendency for electron transfer and ionic bonding to occur.

Table 7–3 Some ions important to the body

Ion	Location	Function
Na^{1+}	Body fluids outside cells	Regulates and controls body fluid neutrality
K^{1+}	Body fluids inside cells	Regulates and controls body fluid neutrality
Ca^{2+}	90% of calcium in body is in bone as $Ca_3(PO_4)_2$ or $CaCO_3$	Major component of bone and teeth; necessary for maintenance of proper rhythm of heartbeat
Mg^{2+}	Body fluids outside cells (70% of Mg^{2+} in the body is present in bone structure)	Essential for certain enzymes and nerve control
Fe^{3+}	Blood	Important in oxygen transport
Cl^{1-}	Principal anion outside cells	Regulates and controls body fluid neutrality
HCO_3^{1-}	Body fluids outside cells	Controls the acid–base balance in body fluids
HPO_4^{2-}	Body fluids inside cells	Controls the acid–base balance in body fluids; important in metabolic processes

Table 7–4 Some important ionic compounds

Formula and name	Uses
$(NH_4)_2CO_3$, ammonium carbonate	Constituent of smelling salts
$BaSO_4$, barium sulfate	White pigment for paper and linoleum; used for X-raying gastrointestinal tract
$CaSO_4$, calcium sulfate (plaster of Paris)	Plaster casts, wall stucco
$MgSO_4$, magnesium sulfate (Epsom salts)	Purgative
$HgCl_2$, mercuric chloride	Disinfectant (dilute solutions) for hands and instruments that cannot be boiled
$KMnO_4$, potassium permanganate	In dilute solutions, a disinfectant for irrigation of wounds; also used for treatment of athlete's foot and poison ivy
$AgNO_3$, silver nitrate	Antiseptic and germicide (used in eyes of infants to prevent gonorrhea)
$NaHCO_3$, sodium bicarbonate (baking soda)	Baking powders, stomach antacid, fire extinguishers
Na_2CO_3, sodium carbonate (soda ash)	Water softener; used in glass and soap manufacture
$NaNO_2$, sodium nitrite	Major food additive to prevent spoilage
SnF_2, tin (II) fluoride	Toothpaste additive to combat dental cavities

Some Biological and Physical Properties of Ions and Ionic Compounds

Table 7–3 lists some of the principal ions in the body, as well as their purposes. It is interesting that electrical neutrality in the body fluids outside the cells is maintained mainly by sodium and chloride ions; but inside the body cells, potassium and monohydrogen phosphate ions perform this function. Table 7–4 lists some ionic compounds and their functions.

The melting temperature of a compound is defined as the temperature at which a compound changes from solid to liquid. The boiling temperature is defined as the temperature at which a compound changes from liquid to gas. Generally, ionic compounds have much higher melting and boiling temperatures than compounds of similar molecular weight that are not ionic. For example, sodium chloride (molecular weight of 58) melts at 801°C and boils at 1413°C, whereas ethyl alcohol (molecular weight of 46; a nonionic compound) melts at −117°C and boils at 78°C. Because of their high melting points, most ionic compounds are solids at room temperature.

A second important property of ionic compounds is that they are generally soluble in water, much more so than nonionic compounds. We shall learn how the dissolving process works in later units.

UNIT 7 Problems and Exercises

7–1. What is the difference between an atom and an ion?

7–2. Write the electron configuration for each of the following ions.
(a) Na^{1+} (b) Cu^{1+} (c) O^{2-} (d) Ca^{2+}

7–3. Give the symbols and charges for each of the following ions. You should try to do this by memory. (Example: sodium = Na^{1+}; cyanide = $(CN)^{1-}$.)

(a) Potassium ion	(b) Magnesium ion	(c) Strontium ion
(d) Radium ion	(e) Aluminum ion	(f) Sulfide ion
(g) Ferrous ion	(h) Cupric ion	(i) Mercuric ion
(j) Carbonate ion	(k) Hydroxide ion	(l) Nitrate ion
(m) Phosphate ion	(n) Permanganate ion	(o) Bicarbonate ion
(p) Ammonium ion	(q) Bisulfate ion	(r) Sulfate ion

7–4. Give the correct names for the following ions.

(a) Li^{1+}	(b) Ba^{2+}	(c) Al^{3+}	(d) Cl^{1-}
(e) Br^{1-}	(f) Fe^{3+}	(g) Ni^{2+}	(h) Cu^{1+}
(i) Ag^{1+}	(j) $(Hg_2)^{2+}$	(k) CO_3^{2-}	(l) HSO_4^{1-}
(m) NH_4^{1+}	(n) NO_3^{1-}	(o) HPO_4^{2-}	(p) $Cr_2O_7^{2-}$
(q) SO_4^{2-}	(r) OH^{1-}		

7–5. How many atoms are present in each of the following formulas?

(a) H_2SO_4	(b) $LiNO_3$	(c) $K_2Cr_2O_7$
(d) $Pd(CH_3CO_2)_4$	(e) $Mg(NO_3)_2$	(f) $Al(CN)_3$
(g) $Al_2(SO_4)_3$		

7–6. Fill in the table by giving the balanced formulas for each of the ionic compounds formed by the two ions that intersect in the rows and columns. In addition, write the name of each ionic compound. (Example: Ca^{2+} and Br^{1-} yield $CaBr_2$: calcium bromide.)

		Anions					
		O^{2-}	Cl^{1-}	HCO_3^{1-}	CO_3^{2-}	PO_4^{3-}	OH^{1-}
Cations	Li^{1+}						
	Al^{3+}						
	Cu^{1+}						
	Ca^{2+}						
	Fe^{2+}						
	NH_4^{1+}						

7–7. Give the structures for the following ionic compounds.

(a) Mercuric sulfate
(b) Mercurous chloride
(c) Potassium carbonate
(d) Ferric hydroxide
(e) Aluminum bisulfate
(f) Ammonium bromide
(g) Cupric cyanide
(h) Ammonium chloride
(i) Calcium bicarbonate
(j) Silver oxide
(k) Sodium hydroxide
(l) Aluminum oxide

7–8. Draw the chemical formulas for a compound formed between elements X and Z, where:

(a) X has 1 and Z has 7 electrons in the outer level.
(b) X has 2 and Z has 6 electrons in the outer level.
(c) X has 5 and Z has 1 electron in the outer level.
(d) X has 5 and Z has 2 electrons in the outer level.

7–9. Why is the charge on a sodium ion never 2+?

UNIT 8 COVALENT BONDING

Ionic bonding occurs whenever atoms with widely differing electron affinities interact. One atom is an electron donor and the second atom is an electron acceptor. But when two atoms with similar electron affinities (or ionization energies) interact, it is rare that complete electron transfer occurs. Actually, there are two possibilities: (1) the elements will not interact with each other, or (2) they will interact to form **covalent** bonds (share electrons).

The elements on the left side of the periodic table have low ionization energies and thus are electron donors. Since none of these atoms readily accepts electrons, they do not interact with each other to form ionic compounds. Since elements must be electron acceptors to form covalent bonds, these atoms do not form covalent compounds either. In general, we can predict that no reaction will occur between elements on the left side of the periodic table.

Elements on the right side of the periodic table have large ionization energies and electron affinities so they tend to accept electrons. Again, ionic compounds would not be expected to form when two of these elements are mixed, since neither is a good electron donor. But interaction is still possible by the **sharing** of electrons. *This process is called* **covalent bond** *formation.* In this process each atom contributes one or more electrons to a volume shared by both atoms. As long as the atoms stay together, both atoms have gained electrons:

Ionization: A· + B ⟶ A^+ + $B\cdot^-$

Na· + :C̤̈l· ⟶ Na^{1+} + :C̤̈l:$^{1-}$

Ionization involves complete loss and gain of electrons. Sodium has lost an electron; chlorine has gained an electron.

Covalent bond formation: A· + B· ⟶ A:B (both electrons are shared by the two atoms; no ions are formed)

H· + Cl· ⟶ H:Cl (H and Cl share two electrons)

Single, Double, and Triple Covalent Bonds

A **single covalent bond** is defined as the sharing of two electrons by two atoms. These types of bonds are generally represented as two dots located between the two atoms or as a solid line connecting the two atoms. Both representations indicate the presence of two shared electrons.

X· + Y· ⟶ X:Y or X—Y (a single covalent bond; two shared electrons)

It is also possible for two atoms to share more than two electrons. The most common numbers of electrons shared are two, four, and six. When four electrons are shared between two atoms, the arrangement is similar to two single covalent bonds between two atoms and is often called a **double** covalent bond. A **triple** covalent bond is formed by the sharing of six electrons between the two atoms involved. Double bonds are represented by the presence of four dots or two solid lines between the two atoms. Triple bonds have six dots or three solid lines between the two atoms.

X: + :Y ⟶ X::Y or X=Y (a double covalent bond; four shared electrons)

X⁝ + ⁝Y ⟶ X⁝⁝Y or X≡Y (a triple covalent bond; six shared electrons)

Electron Dot Structures and the Octet Rule

The electrons present in the outermost occupied energy levels of atoms are often referred to as the **valence electrons**, and these outermost occupied levels are often called the **valence levels**. Valence electrons are the electrons that we have to keep close track of, since they are involved in forming covalent bonds.

When we write structures for ions and molecules that show all the electrons present in the valence levels (the outermost occupied energy levels), we are drawing **electron dot representations** of the molecules and ions. The object of drawing electron dot structures is to keep track of the valence electrons and to ensure that each atom has *eight* valence electrons. This requirement of eight electrons is known as the **octet rule**. Hydrogen is an exception; it requires only two electrons.

In units 5 and 6 we discussed how atoms tend to lose or gain electrons in order to completely empty or fill up the s- and p-orbitals in their valence levels. By doing this, atoms attain electron configurations that resemble those of the inert gases, the elements that possess the greatest stability. Since there is a maximum of eight electrons in the s- and p-orbitals of any level, we can see the reasons behind the octet rule: atoms gain electrons to have valence levels containing eight electrons just like those of the inert gases.

In using the octet rule to draw electron dot structures for polyatomic ions and molecules, we must remember that each pair of shared electrons is *owned* by both atoms. Each atom can use those two shared electrons to count toward its goal of eight valence electrons. Two atoms joined by a double bond share four electons, and both atoms can count those four. Similarly, atoms forming a triple bond can each count the six shared electrons. Thus the total numbers of valence electrons in an ion or molecule need not be eight times the number of atoms present; the sharing allows for smaller numbers of electrons to be present to obey the octet rule. For example, nitrogen gas exists as a diatomic molecule, N_2, and its electron dot structure is

Each nitrogen atom has eight electrons in its valence level: two that it possesses outright and six that it shares with the other atom. The total number of valence electrons present is 10, not 16, but through sharing, the octet rule has been satisfied.

Drawing Electron Dot Structures

In drawing electron dot structures for molecules and polyatomic ions, the first goal is to try to satisfy the octet rule by making single covalent bonds. If there are not enough valence electrons present to allow this, we have to make double or triple bonds. The following guidelines should help you in drawing electron dot structures for many polyatomic ions and molecules. We will illustrate each guideline by using sulfate ion, SO_4^{2-}, and then present more examples.

1. Add up the total number of valence electrons present in the atoms that form the ion or molecule under consideration. If the species possesses negative charges, add one more electron for each negative charge. Subtract one electron for each positive charge. For SO_4^{2-}, the total number of valence electrons is 32 (6 from sulfur, 6 from each of 4 oxygens, and 2 from the negative charges).
2. Arrange the atoms in as symmetrical a fashion as possible, and connect the atoms with single covalent bonds.

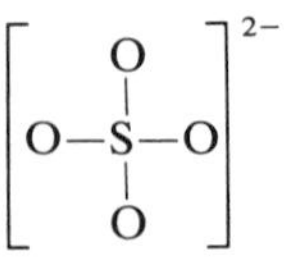

Since each single covalent bond contains two electrons, we have used $4 \times 2 = 8$ valence electrons and have $32 - 8 = 24$ valence electrons left to add to the structure.

3. Add the remaining valence electrons to the structure, attempting to provide each atom with eight electrons (hydrogen should have only two).

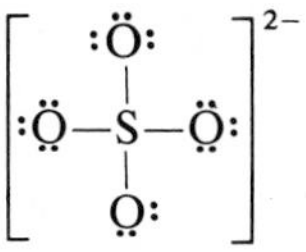

4. If all the valence electrons have been added and all the atoms have eight valence electrons, the structure is complete. If some atoms have less than eight electrons, make double or triple bonds until the octet rule is satisfied. The sulfate ion is complete as shown above.

Now let us look at some examples of using the rules.

Example 1:

F_2 — There are 14 valence electrons (7 from each atom). (rule 1)

F—F — The atoms are joined in a single bond. Two electrons have been used; 12 remain to be added. (rule 2)

:F̤̈—F̤̈: — All valence electrons have been added and each atom has 8 electrons. The structure is complete. (rule 3)

Example 2:

CO_2 — There are 16 valence electrons (4 from carbon and 6 from each oxygen).

O—C—O — A symmetrical structure has been drawn. Four electrons have been used; 12 remain to be added.

Ö̤—C̤̈—Ö̤ — All electrons have been added, but the structure is incomplete since oxygen atoms have only 6 electrons.

:Ö̤—C—Ö̤: — All electrons have already been added, but the structure is still incomplete since the carbon atom has only 4 electrons. We need to make double bonds.

Ö̤::C::Ö̤ or Ö̤=C=Ö̤ — The structure is complete.

Example 3:

CO_3^{2-} — There are 24 valence electrons (4 from carbon, 6 from each oxygen, and 2 from the −2 charge).

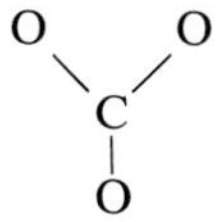

Symmetrical structure (18 electrons remain to be added).

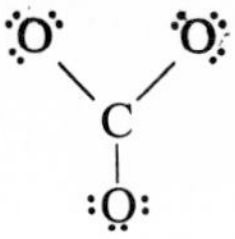

All electrons have been added. The structure is incomplete and we need to make one double bond.

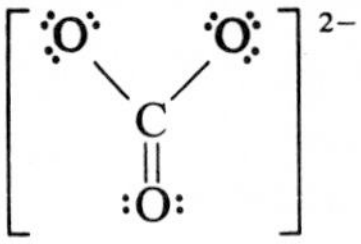

The structure is complete.

Example 4: HCN There are 10 valence electrons.

H—C—N Six valence electrons remain to be added.

H—C̤̈—N̈ or H—C—N̤̈: The structure is incomplete.

H—C:::N: or (H—C≡N:) The structure is complete.

Drawing covalent structures for molecules that contain large numbers of atoms can sometimes be difficult. One reason is that many different structures can often be drawn that represent the same molecular formula. For example, there are nine different compounds that have the molecular formula of C_3H_6O:

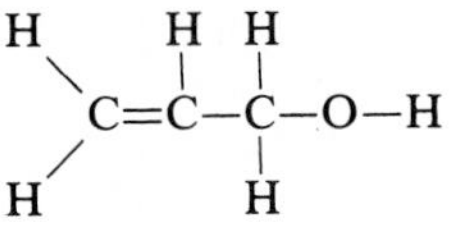

allyl alcohol

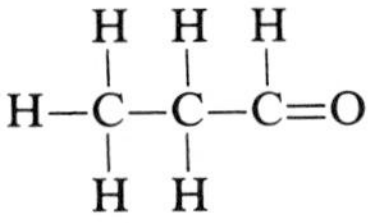

propanal

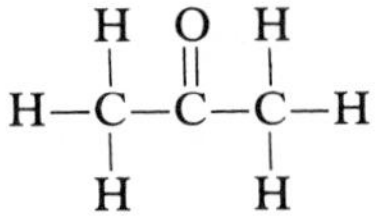

acetone

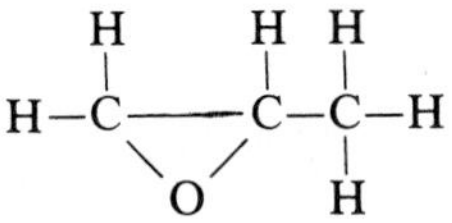

propene epoxide

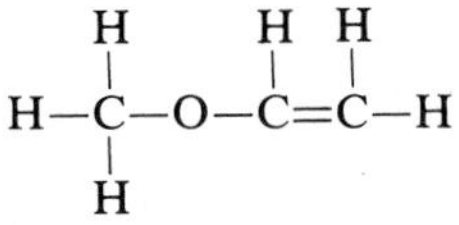

methyl vinyl ether

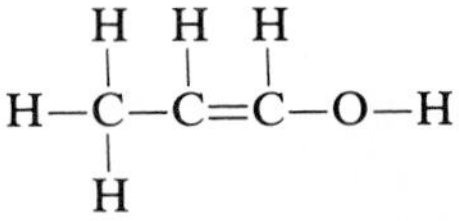

1-hydroxypropene

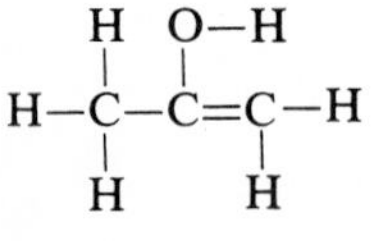

2-hydroxypropene

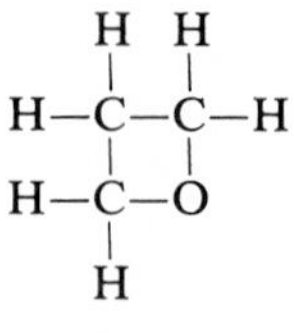

oxetane

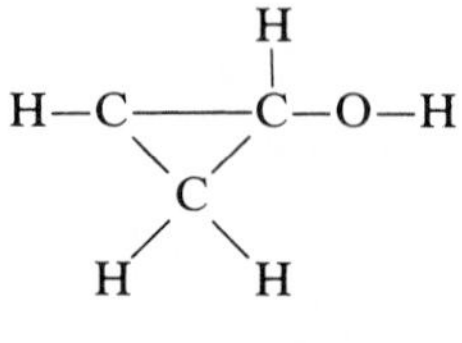

cyclopropyl alcohol

We shall work with these types of compounds later in the text and develop theories that will help you to understand their structures. There are some similarities among all the structures, however. Every carbon atom has four covalent bonds (each atom has gained four electrons), each hydrogen atom has one covalent bond (gain of one electron), and each oxygen atom has two covalent bonds (gain of two electrons).

Electronegativity

In unit 6 we discussed electron affinity, the ability of atoms to gain electrons and form anions. In covalent bonding we do not have complete charges and electron transfer, but only electron sharing. The two atoms in a covalent bond both have a hold or *pull* on the pair of shared electrons. The strength of this pull is defined as the electronegativity of the atom. Thus **electronegativity** *of an atom is the pulling power or attracting ability of that atom for the pair of shared electrons.* As with electron affinity, electronegativity increases as you go from left to right and from bottom to top in the periodic table. Table 8–1 lists the electronegativities for some of the elements.

Table 8–1 Electronegativities and covalent bonds

H (2.1)				
B (2.0)	C (2.5)	N (3.0)	O (3.5)	F (4.0)
Al (1.5)	S (1.8)	P (2.0)	S (2.5)	Cl (3.0)
				Br (2.8)

Very polar covalent bonds
H—F, H—O, H—Cl, C—O, N—H

Slightly polar covalent bonds
N—C, H—S, H—C, C—Cl, N—O, P—O

Nonpolar covalent bonds
Cl—Cl, O—O, C—C, H—H

Polar Covalent Bonds

A covalent bond involves the sharing of two electrons by two atoms, but this need not be equal sharing. If one of the atoms is more electronegative than the other, it will have a greater hold on the pair of shared electrons, and a **polar** covalent bond will result. Let us illustrate by examining the covalent bond between hydrogen and chlorine. Since chlorine is much more electronegative than hydrogen, the shared electrons are pulled closer to chlorine than to hydrogen. Chlorine has gained electron density at the expense of the hydrogen atom, which has lost some. The chlorine atom then possesses a **partial negative charge** (symbolized by δ^-), the hydrogen atom possesses a **partial positive charge** (δ^+). Neither atom has a full charge; that would be the case if they were ions. Thus within the covalent bond we have set up a positive and a negative end. This is called a **polar covalent bond**.

$$\mathrm{H{:}\ddot{\underset{\cdot\cdot}{Cl}}{:}} \longrightarrow \overset{\delta+}{\mathrm{H}}{:}\overset{\delta-}{\mathrm{\ddot{\underset{\cdot\cdot}{Cl}}}}{:}$$

In theory, any covalent bond between two atoms that possess different electronegativities will be a polar covalent bond; it will involve unequal sharing of the electron pair. The greater the difference in electronegativity between the two atoms, the greater the polarity. To simplify this concept, we will identify covalent bonds as very polar, slightly polar, and nonpolar. We define very polar covalent bonds as occurring between atoms whose difference in electronegativity is between 0.9 and

2 units. Slightly polar covalent bonds occur when the electronegativity differences are less than 0.9 units. Nonpolar bonds occur between the same elements. Table 8–1 lists some of the covalent bonds that fit our definitions.

Hydrogen Bonding

Molecules containing polar covalent bonds have partial electrical charges owing to unequal sharing of the bonding electrons. The presence of these weak charges allows two or more of these polar molecules to be attracted toward each other with a weak electrostatic attraction. When the polar covalent bond is H—O, H—N or H—F, the attraction is called **hydrogen bonding**. Thus the presence of H—F, H—O, and N—H bonds in covalent molecules allows them to be attracted to each other to form hydrogen bonds (see Figure 8–1).

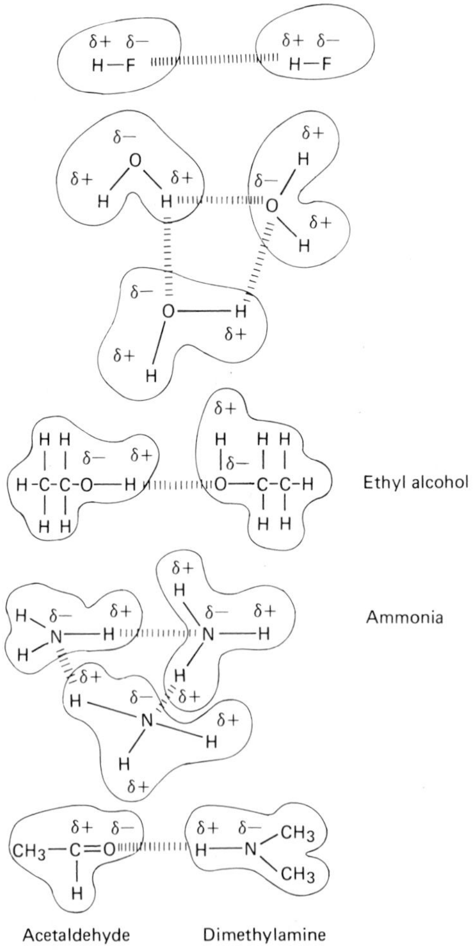
Ethyl alcohol
Ammonia
Acetaldehyde
Dimethylamine

Figure 8–1 Hydrogen bonding. Hatched lines represent hydrogen bonds and solid lines represent covalent bonds.

The presence of hydrogen bonding in polar molecules causes many physical changes. It increases the boiling and freezing points of compounds, increases their water solubility, and is responsible for determining the overall structures of many large biochemical molecules. We shall study some of these effects in later units.

Comparing the Properties of Ionic and Covalent Compounds

Ionic compounds are formed when atoms from opposite sides of the periodic table interact with each other and transfer electrons. In general, ionic compounds have very high melting and boiling temperatures. They are usually fairly soluble in water, a very polar solvent. Most ionic compounds are solids at room temperature.

Covalent compounds are formed when atoms on the right side of the periodic table interact with each other and share electrons. They tend to have much lower melting and boiling points than ionic compounds, are generally much less soluble in water, and may exist as solids, liquids, or gases at room temperature. For example, the covalent compound propane, C_3H_8, is a gas at room temperature, melts at −190°C, boils at −44.5°C, and is only slightly soluble in water. Contrast this with the ionic compound, sodium chloride, which is a solid at room temperature, melts at 801°C, boils at 1413°C, and is quite soluble in water.

UNIT 8 Problems and Exercises

8–1. Energetically, which is more stable, a hydrogen atom or a hydrogen molecule? Why?

8–2. Which would be more stable, a flask full of hydrogen atoms or a flask full of helium atoms? Why?

8–3. How does the number of electrons gained by an atom during covalent bonding relate to the electron configuration of the inert gases? Explain using N, O, and Cl as examples.

8–4. Determine how many covalent bonds each element will need if it is covalently bonded.

(a) F (b) O (c) N (d) C
(e) P (f) Br (g) S (h) H
(i) Cl (j) Si

8–5. Give reasons to explain why a triple covalent bond might be shorter than a single covalent bond.

8–6. Predict whether the following pairs of elements will bond covalently, ionically, or not at all.

(a) S and O (b) Li and O (c) Li and Na (d) C and Na
(e) C and H (f) C and Br (g) S and Cl (h) C and S

8–7. Except for hydrogen, every atom in a covalent bond must have eight electrons surrounding it. Draw the structures for the following compounds, given the following molecular formulas (see the examples on page 75).

(a) Br_2 (b) H_2 (c) H_2O (d) N_2
(e) CH_4 (f) HCl (g) H_2O

8–8. The structures of some of the following compounds will have double or triple bonds. Give possible structures. Work them step by step.

(a) C_2H_4 (b) CO_2 (c) C_2H_2 (d) C_2H_4O
(e) C_3H_4 (f) CN^{1-} (g) HCO_3^{1-} (h) NO_3^{1-}
(i) H_2CO (j) NH_4^{1+}

8–9. The will-o'-the'wisp, a faint light sometimes seen over marshland at night, is attributed by some people to the burning of a compound of phosphorus and hydrogen. What would the formula for this compound be?

8–10. Persons who clean the outsides of buildings by sandblasting sometimes contract a disease known as silicosis from inhaling dust formed during the blasting. Assuming that the dust consists solely of extremely fine particles of a compound formed from silicon and oxygen, what would its formula be?

8–11. Can two molecules exist with the same composition yet be different compounds? Explain your answer.

8–12. Which of the bonds in each pair is most polar?

(a) H—O or S—O (b) H—O or H—S (c) B—O or C—O
(d) N—Cl or N—F (e) P—Cl or B—Cl

8–13. Which of the following is the least polar covalent bond? Why?

(a) S—O (b) Cl—Cl (c) Cl—O

8–14. In each of the following polar covalent bonds, indicate which end of the bond is partially negative ($\delta-$) and which end is partially positive ($\delta+$).

(a) H:F (b) O:H (c) N:H (d) C:O
(e) C:N (f) C:Cl (g) H:S (h) H:Br

8–15. Use Table 8–1 to arrange the following bonded atoms in order of increasing polarity.

(a) N—O (b) H—Cl (c) H—Br (d) C—O
(e) C—Br (f) N—C (g) B—F

8–16. Which would you expect to have stronger covalent bonds, H_2O or H_2S? Explain on the basis of electronegativity.

8–17. Indicate which of the following molecules can be involved in hydrogen bonding.

(a) $CH_3—CH_2—O—H$ (b) $H_3C—O—CH_3$ (c) Br—Br
(d) NaCl (e) H—Br (f) $H_3C—S—H$
(g) $H_3C-\underset{\substack{|\\ H}}{N}-CH_3$ (h) $CH_3—CH_2—NH_2$ (i) $H_3C-\underset{\substack{|\\ CH_3}}{N}-CH_3$

8–18. Show by means of dotted lines the hydrogen bonding between the following pairs of compounds.

(a) $H_3C—S—H$ and H_2O (b) H_2O and NH_3 (c) $H_3C—O—H$ and H—Cl

8–19. Predict if the following should be covalent, polar covalent, or ionic.

(a) NO (b) LiF (c) PH_3 (d) H_2O
(e) Cl_2Ca

8–20. What tests might you perform to determine whether a solid substance is ionic or covalent?

CHAPTER 4

Atomic weights and chemical formulas

In units 9 and 10 we shall begin to work with some of the quantitative aspects of chemistry: we shall begin to work with numbers and weights of atoms and molecules. Atoms and molecules are units of matter so small that we cannot accurately work with them on that scale. To be able to weigh and measure matter, arbitrary units of measurement called **gram-atomic weights** and **moles** were introduced. These units allow us to concentrate on measurements at the gram level rather than at the 1×10^{-22} gram level.

The two most important definitions in unit 9, gram-atomic weight and mole, are tied together in a unique way. The mole is a term used to describe a definite number of atoms or molecules (6×10^{23}), and the gram-atomic weight is the weight in grams of that many atoms or molecules.

In unit 10 we shall learn how chemists actually work out the chemical formulas for many of the compounds. You will learn how to convert percentages into empirical and molecular formulas.

UNIT 9 ATOMIC WEIGHTS, MOLES, AND AVOGADRO'S NUMBER

When two or more elements react with each other to form a chemical compound, the atoms of those elements must combine in *whole-number ratios*, not in decimal ratios. For example, a molecule may contain 1, 2, 3, . . . , 30, atoms of one element but it cannot contain 1.5, 2.3, 4.6, . . . atoms of that element. The chemical formula for water is H_2O so each molecule contains 2 atoms of hydrogen and 1 atom of oxygen. The ratio of hydrogen atoms to oxygen atoms in water is always 2:1, no matter how much water is present. In glucose ($C_6H_{12}O_6$) the atom ratios are C:H:O = 6:12:6. Notice in both of these examples that *numbers* of atoms, not *weights* of atoms, are being used. This is an essential point. We cannot interchange numbers of atoms and weights, since all atoms do not weigh the same. The combination of 2 *atoms* of hydrogen with 1 *atom* of oxygen will yield 1 *molecule* of water, but combining 2 *grams* of hydrogen with 1 *gram* of oxygen will not yield 3 *grams* of water. Thus any unit of measure that we adopt for working with elements must be related to the numbers of atoms present.

Atomic Weights

The **atomic weight** of any element is defined as the weight of one atom of that element. Since the weight of one proton or neutron is about 1.66×10^{-24} g (1 amu), we can calculate an approximate weight for any atom simply by summing the numbers of protons and neutrons in one atom and then multiplying by 1.66×10^{-24} g (1 amu). One atom of hydrogen (one proton, zero neutrons) would weigh 1.66×10^{-24} g (1 amu). One atom of carbon (six protons, six neutrons) would weigh $12 \times (1.66 \times 10^{-24}$ g) or 20×10^{-24} g (12 amu). Actually, atomic weights calculated in this manner are only approximate because (1) neutrons and protons do not weigh exactly the same, (2) the weight of electrons has not been included, and (3) we have not made allowances for the presence of isotopes. In order to standardize the atomic weights of the elements, the atomic weights of the most abundant isotope of carbon, ${}^{12}_{6}C$, is defined as weighing exactly 12.000 amu. The weights of all other elements are then calculated by comparison to carbon. An atom of hydrogen is slightly greater than one twelfth the weight of carbon, so its atomic weight is 1.008 amu. The average weight of a magnesium atom is slightly greater than twice that of the carbon standard, so its weight is 24.3 amu. In this manner, all the atomic weights given in the periodic table were calculated by comparison to ${}^{12}_{6}C$. Rounded off, the atomic weight of each element is equal to the average number of protons and neutrons in an atom of that element.

Gram-Atomic Weights

It's fine to know the absolute weights of single atoms, but those numbers are too small to be used. No balance can weigh just one atom of an element. What we really need is a larger unit of measure that can be used easily on standard balances. This unit must also be proportional to the numbers of atoms present. Such a unit is called the **gram-atomic weight** (GAW):

> **Gram-atomic weight:** The mass of an element in grams that is numerically equal to its atomic weight.

It is a convenient definition because we do not have to worry about more than one number for each element. The units listed in the periodic table refer both to the atomic weights of atoms (the sum of the numbers of protons and neutrons) and to the gram-atomic weights of elements (an arbitrary unit expressed in grams). Table 9–1 gives weights for a few elements.

Table 9–1 Atomic weights and gram-atomic weights

Element	Atomic weight (amu)	Gram-atomic weight (g)
H	1.008	1.008
C	12.011	12.011
O	15.999	15.999
N	14.007	14.007

Now we have a unit of measure that we can work with, but how does it relate to the numbers of atoms present? To see the relationship, let us calculate the numbers of atoms present in 1 gram-atomic weight of the elements hydrogen, carbon, and oxygen.

Example: How many atoms are present in 1 gram-atom of (a) H, (b) C, and (c) O?

(a) Hydrogen:

$$\text{atomic wt} = 1.008\ \text{amu}$$

$$1\ \text{gram-atomic wt} = 1.008\ \text{g}$$

$$\text{g} \longrightarrow \text{amu} \longrightarrow \text{atoms}$$

$$1.008\ \text{g H} \times \frac{1\ \text{amu}}{1.66 \times 10^{-24}\ \text{g}} \times \frac{1\ \text{atom H}}{1.008\ \text{amu}} = 6.023 \times 10^{23}\ \text{atoms H}$$

(b) Carbon:

$$\text{atomic wt} = 12.011\ \text{amu}$$

$$1\ \text{gram-atomic wt} = 12.011\ \text{g}$$

$$12.011\ \text{g C} \times \frac{1\ \text{amu}}{1.66 \times 10^{-24}\ \text{g}} \times \frac{1\ \text{atom C}}{12.011\ \text{amu}} = 6.023 \times 10^{23}\ \text{atoms C}$$

(c) Oxygen:

$$\text{atomic wt} = 15.999\ \text{amu}$$

$$1\ \text{gram-atomic wt} = 15.999\ \text{g}$$

$$15.999\ \text{g O} \times \frac{1\ \text{amu}}{1.66 \times 10^{-24}\ \text{g}} \times \frac{1\ \text{atom O}}{15.999\ \text{amu}} = 6.023 \times 10^{23}\ \text{atoms O}$$

Notice that 1 gram-atomic weight of all three elements contains the same number of atoms: 6.023×10^{23} atoms. This number is called **Avogadro's number** in honor of Amadeo Avogadro, an Italian chemist who first derived it in 1811. For our purposes, it is more convenient to round it off to 6×10^{23}. Now let's put this all together and summarize.

1. A gram-atomic weight of any element is defined as a mass in grams equal to the atomic weight of that element.
2. A gram-atomic weight of any element contains 6×10^{23} atoms of that element.
3. The **gram-atomic weights** for different elements are not the same, but they all contain the same number of atoms. Since individual atoms of different elements do not weigh the same, obviously 6×10^{23} atoms of different elements will not weigh the same.

To illustrate, let us use as an example a number of different types of athletic balls (see Table 9–2).

Table 9–2 Different types of athletic balls

Type	Weight of 1 ball	Weight of 12 balls
Ping-pong ball	0.16 oz	1.92 oz
Tennis ball	2 oz	24 oz (1.5 lb)
Golf ball	2.4 oz	28.8 oz (1.8 lb)
Baseball	7 oz	84 oz (5.25 lb)
Soccer ball	1 lb	12 lb
Football	2 lb	24 lb
Basketball	2 lb	24 lb

All the balls are different and have different weights, just as atoms of different elements are different and do not weigh the same. If we take a dozen of each type of ball, we have the same *number* of balls but not the same *weights* of the balls. If we take 6×10^{23} atoms of different elements, we have the same *number* of atoms of each element, but not the same *weights*. We can also calculate the number of balls present if we know the total weight present. Thus, if we have 1.5 lb of tennis balls, we have 12 tennis balls. The same can be done for atoms. If we have 1.008 g of hydrogen, we have 6×10^{23} atoms of hydrogen. If we have 2.016 g of hydrogen, we have 12×10^{23} atoms of hydrogen. Avogadro's number is not some strange number; it's just a number used to keep track of the amounts of elements present.

Moles

It is cumbersome to write 6×10^{23} all the time: we need a term that will mean 6×10^{23} of anything. This term is the **mole** (abbreviated, mol).

A **mole** *of anything is* 6×10^{23} *units.*

A mole of golf balls is 6×10^{23} golf balls. A mole of hydrogen atoms is 6×10^{23} atoms of hydrogen. The mole then is just a grouping unit used to simplify calculations and allow for the use of smaller numbers, just like a dozen, meaning 12, or a pair, meaning two (see Table 9–3).

Table 9–3 Terms that signify numbers

Term	Number of units	Portions	
Pair	2	1/2 pair = 1	2 pair = 4
Dozen	12	1/2 dozen = 6	2 dozen = 24
Score	20	1/2 score = 10	2 score = 40
Gross	144	1/2 gross = 72	2 gross = 288
Mole	6×10^{23}	1/2 mole = 3×10^{23}	2 moles = 12×10^{23}

Weight Units for Compounds

The terms *atoms*, *atomic weight*, and *gram-atomic weight* are used when referring to individual elements. Other units are necessary when dealing with compounds, since compounds are not made up of single elements, but rather groups of elements bound together in some fashion. **Atoms** are the basic unit in elements, and **molecules** are the basic unit in compounds. The **molecular formula** of a compound gives the composition in atoms of one molecule of that compound. The molecular formula of glucose is $C_6H_{12}O_6$, so one molecule of glucose contains 6 atoms of carbon, 12 atoms of hydrogen, and 6 atoms of oxygen. To calculate the weight of one molecule of glucose, we simply sum up the total atomic weights of all the atoms present in the molecule. The molecular weight of one molecule of glucose is 180.159 amu. By analogy with gram-atomic weights, the **gram-molecular weight** is 180.159 g. One mole of glucose is 6×10^{23} molecules of glucose.

A **mole** *of an element is* 6×10^{23} **atoms**,

a **mole** *of a compound is* 6×10^{23} **molecules**.

$$
\begin{aligned}
\text{6 atoms of carbon} = 6 \times 12.011 &= 72.067 \text{ amu} \\
\text{12 atoms of hydrogen} = 12 \times 1.008 &= 12.096 \text{ amu} \\
\text{6 atoms of oxygen} = 6 \times 15.999 &= \underline{95.996 \text{ amu}} \\
\text{Molecular weight of glucose} &= 180.159 \text{ amu} \\
\text{Gram-molecular weight} &= 180.159 \text{ g}
\end{aligned}
$$

The term *molecule* really applies only to covalent compounds and not to ionic compounds. In an ionic compound the ions are not permanently attached to other specific ions, so the molecules really do not exist. In calculating the weights of ionic compounds we use a formula as the base unit. The **formula weight** is the weight of one group of atoms shown in the chemical formula. It is calculated in the same manner as molecular weights. A **gram formula weight** is the formula weight expressed in grams. For sodium chloride (NaCl), the formula weight is 58.44 amu, the gram formula weight is 58.44 g, and 1 mol of sodium chloride contains 6×10^{23}

formula units (NaCl). For all practical purposes, however, the formula unit and molecule can be thought of as the same thing.

All the important new terms we need to use when calculating weights of compounds and elements have now been introduced. A comparison of the units for elements and compounds is shown in Table 9–4. Study it for a while to make sure that you understand the similarities and the differences. Remember, the formula of a compound gives both the number of atoms of each element present in one molecule or formula unit and the number of moles of each element in 1 mol of the compound. For glucose ($C_6H_{12}O_6$),

1 molecule = 6 atoms of carbon + 12 atoms of hydrogen + 6 atoms of oxygen

1 mol of glucose = 6 moles of carbon + 12 moles of hydrogen + 6 moles of oxygen

Table 9–4 Comparison of weight units for elements and compounds

	Representation	Basic unit	Weight unit	Mole unit
Elements:	Element symbol	Atom	Atomic wt.	1 mole = 6×10^{23} atoms GAW given in periodic table
Compounds:				
Covalent:	Molecular formula defines a molecule	Molecule	Molecular wt.	1 mole = 6×10^{23} molecules GMW = sum of GAW's
Ionic:	Ionic formula	Formula unit	Formula wt.	1 mol = 6×10^{23} formula units GFW = sum of GAW's

Unit Conversions With Elements and Compounds

Often it is necessary to convert from molecules or formula units to grams, from grams to atoms, and so on. The use of the factor dimensional approach provides the easiest mechanism of doing this, so the necessary identities and a roadmap are presented here to aid you in the conversions. Table 9–5 gives the necessary identities and Figure 9–1 shows the roadmap. Notice in Figure 9–1 how moles occupy a central

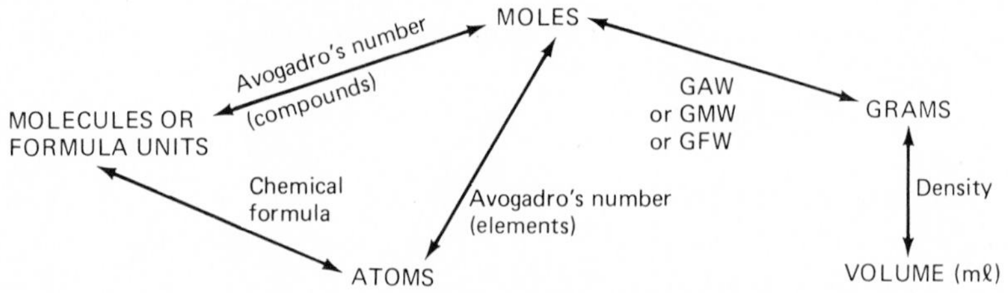
MOLES
Avogadro's number
(compounds)
MOLECULES OR
FORMULA UNITS
GAW
or GMW
or GFW
GRAMS
Chemical
formula
Avogadro's number
(elements)
Density
ATOMS
VOLUME (mℓ)

Figure 9–1 Roadmap for conversions.

Table 9–5 Identities of conversion roadmap

Name	Units	Uses
Avogadro's number	6×10^{23} atoms $= 1$ mol (element)	Atoms ↔ moles
	6×10^{23} molecules $= 1$ mol (covalent compound)	Molecules ↔ moles
	6×10^{23} formula units $= 1$ mol (ionic compound)	Formula units ↔ moles
Molecular formula	Atoms/molecule	Atoms ↔ molecules
Ionic formula	Atoms/formula unit	Atoms ↔ formula unit
Gram-atomic wt.	Grams/mole element	Grams ↔ moles
Gram-molecular wt.	Grams/mole compound	Grams ↔ moles
Gram formula wt.	Grams/mole compound	Grams ↔ moles
Density	Grams/milliliter	Grams ↔ volume

position in the map and virtually every conversion passes through them. In the following problems we show how to use the roadmap and the conversion identities.

Problem 1: The density of water is 1 g/mℓ. How many molecules are present in 36 mℓ of water?

$$\overset{\text{(Start)}}{\text{m}\ell} \rightarrow \text{g} \rightarrow \text{mol} \rightarrow \overset{\text{(End)}}{\text{molecules } (H_2O)}$$

$$\frac{36\ \text{m}\ell}{1} \times \underbrace{\frac{1\ \text{g}}{1\ \text{m}\ell}}_{\substack{\text{Changing}\\ \text{m}\ell \rightarrow \text{g}\\ \text{using}\\ \text{density}}} \times \underbrace{\frac{1\ \text{mol}}{18\ \text{g}}}_{\substack{\text{Changing}\\ \text{g} \rightarrow \text{mol}\\ \text{using GMW}\\ \text{of } H_2O =\\ 18\ \text{g/mol}}} \times \underbrace{\frac{6 \times 10^{23}\ \text{molecules}}{1\ \text{mol}}}_{\substack{\text{Changing}\\ \text{mol} \rightarrow \text{molecules}\\ \text{using Avogadro's}\\ \text{number}}} = 12 \times 10^{23}\ \text{molecules}$$

On the roadmap if you start at volume and want to end at molecules you must proceed along the line to grams, then to moles, and finally to molecules.

Problem 2: How many atoms of carbon are present in 10 mℓ of benzene? The formula for benzene is C_6H_6 and its density is 0.9 g/mℓ. From the formula, the GMW = 78 g = 1 mol:

$$\overset{\text{(Start)}}{\text{m}\ell} \rightarrow \text{g} \rightarrow \text{mol} \rightarrow \text{molecules} \rightarrow \overset{\text{(End)}}{\text{atoms C}}$$

$$\frac{10\ \text{m}\ell}{1} \times \frac{0.9\ \text{g}}{1\text{m}\ell} \times \frac{1\ \text{mol}}{78\ \text{g}} \times \frac{6 \times 10^{23}\ \text{molecules}}{1\ \text{mol}} \times \frac{6\ \text{atoms C}}{1\ \text{molecule}} = 4.2 \times 10^{23}\ \text{atoms of carbon}$$

In this example we are interested only in the number of atoms of carbon present, not the total number of atoms. If we wanted the total number of atoms, the last term would be $\frac{12\ \text{atoms}}{1\ \text{molecule}}$.

Problem 3: How many grams are present in 2.4×10^{24} molecules of hexane (C_6H_{12})?

$$\text{GMW hexane} = \overset{(C+H)}{72 + 12} = 84 \text{ g/mol}$$

$$\overset{(\text{Start})}{\text{molecule}} \rightarrow \text{mol} \rightarrow \overset{(\text{End})}{\text{g}}$$

$$\frac{2.4 \times 10^{24} \text{ molecules}}{1} \times \frac{1 \text{ mol}}{6 \times 10^{23} \text{ molecules}} \times \frac{84 \text{ g}}{1 \text{ mol}} = 336 \text{ g}$$

The roadmap directs traffic in both directions.

UNIT 9 Problems and Exercises

9–1. Complete the following expressions.
(a) One mole of oxygen atoms (O) contains ______________ atoms.
(b) One mole of oxygen molecules (O_2) contains ______________ molecules.
(c) One mole of oxygen molecules contains ______________ atoms.
(d) One mole of oxygen atoms weighs ______________ grams.
(e) One mole of oxygen molecules weighs ______________ grams.

9–2. Calculate the gram-molecular weights for the following compounds.
(a) Hydrogen sulfide, H_2S
(b) Ethyl alcohol, C_2H_6O
(c) Acetic acid, $C_2H_4O_2$
(d) Freon 12, CCl_2F_2
(e) Nicotine, $C_{10}H_{14}N_2$
(f) Aspirin, $C_9H_8O_4$
(g) TNT, $C_7H_5N_3O_6$
(h) DDT, $C_{14}H_9Cl_5$

9–3. Calculate the gram formula weights for the following ionic compounds.
(a) Calcium oxide
(b) Sodium bicarbonate
(c) Potassium sulfate
(d) Aluminum bromide
(e) Sodium carbonate
(f) Aluminum sulfide
(g) Ammonium sulfate
(h) Potassium dichromate

9–4. Epinephrine (adrenaline) is a hormone secreted in the bloodstream in times of danger or stress. Its chemical formula is $C_9H_{13}NO_3$.
(a) What is its molecular weight?
(b) How many moles are present in 1 g of this substance?
(c) How many carbon atoms are present in 1 mol of this substance?
(d) What is the weight of 6×10^{23} molecules of this substance?

9–5. How many molecules are present in each of the following amounts?
(a) 2 mol
(b) 0.1 mol
(c) 1×10^{-23} mol
(d) 5×10^2 mol
(e) 0.008 mol

9–6. How many moles are present in the following?
(a) 10 g of H_2SO_4
(b) 5 g of O_2
(c) 3.0 g of Ne
(d) 1.3×10^{20} atoms of Ne
(e) 20 g of ferric carbonate
(f) 5 mg of water

9–7. How many moles are present in 1 ton of sand (SiO_2)?

9–8. How many grams are present in each of the following units (moles → grams)?
(a) 0.004 mol of sodium chloride
(b) 2×10^{-4} mol of potassium dichromate
(c) 1×10^{21} formula units of potassium sulfate

9–9. Calculate the number of grams present in the following.
(a) 0.1 mol of aspirin ($C_9H_8O_4$)
(b) 2 mol of sodium carbonate
(c) 0.001 mol of acetic acid ($C_2H_4O_2$)
(d) 2×10^1 mol of ethyl alcohol (C_2H_6O)
(e) 0.007 mol of ammonium sulfate

9–10. Acetylsalicylic acid, aspirin, has the molecular formula $C_9H_8O_4$. An average oral dose is 0.65 g. How many moles is this?

9–11. Calculate the total weights of the following.
(a) $2Fe(NO_3)_2$ (b) $6HgCl_2$ (c) $3Cu(HCO_3)_2$ (d) $4C_3H_8$

9–12. How many molecules are present in the following weights of compounds (grams → moles → molecules)? (See problem 9–2.)
(a) 4.6 g of ethyl alcohol
(b) 0.0180 g of aspirin
(c) 600 g of acetic acid
(d) 1.7×10^{-10} g of hydrogen sulfide

9–13. How many grams do each of the following numbers of molecules weigh (molecules → moles → grams)? (See problem 9–2.)
(a) 6×10^{20} molecules of Freon 12
(b) 2.4×10^{27} molecules of water
(c) 4×10^{18} molecules of DDT

9–14. What will 1.0×10^{23} formula units of sodium cyanide weigh?

9–15. Carbon tetrachloride (CCl_4) has a density of 1.6 g/mℓ. What is the volume of 1 mol of CCl_4?

9–16. One-half mole of a compound contains 3×10^{23} atoms of carbon and 6×10^{23} atoms of oxygen. What is the molecular formula for the compound? What is its GMW?

9–17. A compound has a molecular formula of $C_2H_6O_3$ and a density of 0.8 g/mℓ.
(a) How many molecules exist in 10 mℓ?
(b) How many atoms of oxygen are present in 10mℓ?

9–18. A liquid has a GMW of 20 g and a density of 2 g/mℓ. How many moles of the liquid are present in 40 mℓ?

9–19. How many molecules of benzene (C_6H_6) are present in 14.4 g of benzene? How many atoms of carbon are present in 14.4 g of benzene?

9–20. Hemoglobin is a complex protein molecule that has the power to transfer oxygen to the body cells and to remove carbon dioxide from them. About 0.33% of the weight of hemoglobin is iron ions. If the molecular weight of hemoglobin is 68,000 g, how many moles of hemoglobin would need to be present to carry 900 mg of iron?

9–21. If the U.S. government is spending in excess of \$200 billion (2×10^{11}) a year, how many years would it take to spend 1 mol of dollars?

9–22. How many water molecules are present in one drop (0.05 g) of water? How many years would it take for that drop to evaporate if 1 billion molecules leave the drop every second?

UNIT 10 EMPIRICAL AND MOLECULAR FORMULAS

Molecular Formulas

A chemical formula is called a **molecular formula** when it indicates the actual numbers of atoms of each element in one molecule of a particular compound. It also indicates the numbers of moles of each element present in 1 mol of the compound. For example, the molecular formula for glucose is $C_6H_{12}O_6$. This means that *one molecule* of glucose contains 6 atoms of carbon, 12 atoms of hydrogen, and 6 atoms of oxygen, no more and no less. On the mole scale it means that 1 *mol* of glucose contains 6 mol of carbon, 12 mol of hydrogen, and 6 mol of oxygen. Since the molecular formula represents the actual composition of a molecule and mole, each chemical compound can have only one molecular formula. The formula $C_3H_6O_3$ has the same ratios of elements as $C_6H_{12}O_6$, but it is not the molecular formula for glucose, since it gives the wrong composition of a molecule and mole.

Empirical Formulas

The **empirical formula** of a compound differs from a molecular formula in that it does not necessarily give the exact numbers of atoms of each element present in one molecule. It gives only the smallest whole-number ratio of the different atoms that are present in 1 mol or one molecule. For example, the smallest whole-number ratio of atoms in a molecule of glucose is $C_1H_2O_1$. Now $C_1H_2O_1$ is not the formula for the complete composition of one molecule of glucose; it is simply the lowest ratio of the atoms of each element in one molecule. The true molecular formula of a compound is some multiple of its empirical formula, and one empirical formula may refer to several different molecular formulas and compounds. The empirical formula $C_1H_2O_1$ can refer to several different compounds, as shown in Table 10–1. Note that even the molecular formula is not sufficient to identify completely the individual compound, as many compounds may also have the same molecular formula.

Table 10–1 Possible molecular formulas for the empirical formula $C_1H_2O_1$

Empirical formula		Molecular formula	Possible compounds
$C_1H_2O_1$	× 1 =	$C_1H_2O_1$	Formaldehyde
	× 2 =	$C_2H_4O_2$	Acetic acid
	× 3 =	$C_3H_6O_3$	Glyceraldehyde, 2-hydroxypropanoic acid
	× 4 =	$C_4H_8O_4$	Erythrose, threose
	× 5 =	$C_5H_{10}O_5$	Ribose, xylose, arabinose
	× 6 =	$C_6H_{12}O_6$	Glucose, fructose, or 14 other monosaccharides

When scientists isolate a new compound, they have to identify it and determine its chemical structure. Chemical analysis determines only the identity and amount of each element present in a given sample of the compound. From these data, it is possible to calculate only the simple ratios of the moles of the elements present (the **empirical formula**). Other experiments have to be done to find the molecular weight of the substance; then the true molecular formula can be calculated and the substance identified.

Calculation of the Empirical Formula of a Compound

The empirical formula gives the atom or mole ratios of the elements present, *not* the gram ratios. Analysis gives gram ratios, so the first thing we must do is convert the grams of each element present into moles of each element. Then, to obtain the empirical formula, we must find the lowest whole-number ratios of the moles present.

The data from analysis are usually presented in one of two ways. The actual grams of each element or the percent by weight of each element present may be given. In problems (1) and (2) we show methods of working with both types of data.

Problem 1: When 8.4 g of an unknown compound was analyzed, it was found to contain 7.2 g of carbon and 1.2 g of hydrogen. Determine the empirical formula for the compound.

STEP 1: Convert grams of each element to moles.

$$7.2 \text{ g of C} \times \frac{1 \text{ mol of C}}{12 \text{ g}} = 0.6 \text{ mol of C}$$

$$1.2 \text{ g of H} \times \frac{1 \text{ mol of H}}{1 \text{ g}} = 1.2 \text{ mol of H}$$

STEP 2: Find the lowest whole-number ratio of the elements.

One method of doing this is to divide both of the above numbers by the smaller number.

$$\frac{0.6 \text{ mol of C}}{0.6 \text{ mol}} = 1 \text{ mol of C}$$

$$\frac{1.2 \text{ mol of H}}{0.6 \text{ mol}} = 2 \text{ mol of H}$$

$$\text{Empirical formula} = C_1H_2$$

We have calculated only the empirical formula; the actual molecular formula may be any multiple of C_1H_2. It may be CH_2, C_2H_4, C_3H_6, C_4H_8, and so forth. Until we have the molecular weight, we cannot determine the actual molecular formula.

Problem 2: Problem (1) could have been represented in percents of composition instead of grams. A sample was analyzed and found to contain 85.7% carbon and 14.3% hydrogen. Determine its empirical formula.

STEP 1: Determine the number of moles of each element present.

In 100 g of the sample there are 85.7 g of carbon and 14.3 g of hydrogen. Using these weights, we can calculate the number of moles present.

$$\text{No. mol of C} = 85.7\text{ g} \times \frac{1\text{ mol of C}}{12\text{ g}} = 7.15\text{ mol of C}$$

$$\text{No. mol of H} = 14.3\text{ g} \times \frac{1\text{ mol of H}}{1\text{ g}} = 14.3\text{ mol of H}$$

STEP 2: Determine the smallest whole-number ratios: divide both of the calculated numbers of moles present by the smaller number.

$$\frac{7.15\text{ mol of C}}{7.15} = 1\text{ mol of C}$$

$$\frac{14.3\text{ mol of H}}{7.15} = 2\text{ mol of H}$$

$$\text{Empirical formula} = CH_2$$

Note that the steps are the same for treating both types of data after the percentages are converted into grams of each element.

Problem 3: Often the experimental data do not allow calculation of mole ratios that are exactly whole numbers. Then you have to round off the numbers to get the best fit. An analysis of vitamin C shows that it contains 40.9% carbon, 4.6% hydrogen, and 54.5% oxygen. Find the empirical formula.

STEP 1: Conversion to moles: in a 100-g sample there are 40.9 g of C, 4.6 g of H, and 54.5 g of 0.

$$\text{No. mol of C} = 40.9\text{ g} \times \frac{1\text{ mol of C}}{12\text{ g}} = \frac{3.4\text{ mol of C}}{3.4} = 1\text{ C}$$

$$\text{No. mol of H} = 4.6\text{ g} \times \frac{1\text{ mol of H}}{1\text{ g}} = \frac{4.6\text{ mol of H}}{3.4} = 1.35\text{ H}$$

$$\text{No. mol of O} = 54.5\text{ g} \times \frac{1\text{ mol of O}}{16\text{ g}} = \frac{3.4\text{ mol of O}}{3.4} = 1\text{ O}$$

After dividing all three values by the smallest number, we do not arrive at whole numbers. We have to try to make them whole numbers by multiplying each number by 2, 3, 4, and so on.

No. mol of C = 1	2C	3C	Best fit: all are either whole numbers or close to whole numbers
No. mol of H = 1.35; × 2 =	2.7H; × 3 =	4.05H ~ 4 H	
No. mol of O = 1	2O	3O	

$$\text{Empirical formula} = C_3H_4O_3$$

Calculation of Molecular Formulas

With practice you should be able to calculate the empirical formulas for any compound. When given the molecular weight of the sample, you can use the empirical formula to get the true molecular formula for the compound.

Problem 4: The gram-molecular weight of the compound in problem (1) was found to be 56 g. What is the molecular formula for the compound? The calculated empirical formula is CH_2, so the molecular formula is some whole-number multiple of that. Let's make a table of the possibilities and calculate the gram-molecular weight of each.

Formula:	CH_2	C_2H_4	C_3H_6	C_4H_8	C_5H_{10}
GMW:	14	28	42	56	70

Note that only C_4H_8 has a gram-molecular weight that is equal to the one given in the problem. Therefore, the molecular formula for the compound is C_4H_8.

Problem 5: The gram-molecular weight of vitamin C is 176 g. What is its molecular formula? From problem (3) we know that the empirical formula is $C_3H_4O_3$.

Possible formulas:	$C_3H_4O_3$	$C_6H_8O_6$	$C_9H_{12}O_9$
Gram-molecular weights:	88	176	264

Only the formula $C_6H_8O_6$ has a gram-molecular weight that agrees with the calculated value.

Calculation of Molecular Weights of Samples

When given the chemical analysis and the molecular weight of a sample, you should be able to determine its molecular formula. The determination of the molecular weight of a sample can be difficult. Some of the more common methods are given next.

Mass Spectroscopy

In a sophisticated instrument called a mass spectrograph, chemical samples are bombarded with high-energy electrons. This bombardment causes the chemicals to ionize to positive and negative ions. These ions are then swept into a magnetic field where the positively charged ions are sped up and sent into a detector. The speeds of the positive ions depend upon their charges and their masses. These ions are categorized in the detector by measurement of their mass-to-charge ratios. The ion with the largest mass-to-charge ratio will generally be that ion formed by loss of just one electron from the parent chemical. The mass of that ion is then the molecular weight of the compound.

Gas-Volume Measurements at Standard Temperature and Pressure

Under standard conditions (STP; 25°C and 1 atm pressure), 1 mol of any gas will occupy a volume of 22.4 ℓ.

By measuring the volume of a known weight of a gas under standard conditions (or by correcting to STP), we can calculate the molecular weight of the gas.

Problem 6: Twenty-four grams of a gas has a volume at STP of 11.2 ℓ. What is its gram-molecular weight?

$$\text{At STP } 1 \text{ mol} = 22.4\ \ell$$

$$11.2\ \ell \times \frac{1 \text{ mol}}{22.4\ \ell} = 0.5 \text{ mol}$$

We have 0.5 mol of the gas, and it weighs 24 g:

$$0.5 \text{ mol} \times \left(\text{gram-molecular wt} \frac{\text{g}}{\text{mol}}\right) = 24 \text{ g}$$

$$\text{Gram-molecular wt} = \frac{24}{0.5} = 48 \text{ g}$$

Chromatography and High-Speed Centrifugation

In the study of the large molecules often found in the body, mass spectroscopy and gas-volume calculations often cannot be used. Other methods must be used to calculate the approximate molecular weights of these molecules. In liquid chromatography, chemical samples are placed on a solid support (usually some form of silica), and a variety of liquid solvents are allowed to pass through the support. The chemical sample is pulled along with the moving solvent at a rate depending upon the polarity and size of the chemical. By measuring the distance a chemical sample travels and comparing that distance to distances traveled by other compounds of known molecular weight, we can approximate the unknown compound's molecular weight.

In high-speed centrifugation, mixtures of compounds are placed in tubes and spun at very high speed. The heavier the sample, the farther it will move to the bottom of the tube. Again, by placing the unknown in a tube with a mixture of compounds of known molecular weights, we can obtain an approximate molecular weight by comparison.

UNIT 10 Problems and Exercises

10–1. Determine the empirical formulas of the following from their formulas.

(a) Acetone, C_3H_6O
(b) Methyl benzoate, $C_8H_8O_2$
(c) Benzene, C_6H_6
(d) Nicotine, $C_{10}H_{14}N_2$
(e) Sucrose, $C_{12}H_{22}O_{11}$
(f) Naphthalene, $C_{10}H_8$
(g) Aspirin, $C_9H_8O_4$

10–2. What are the percentage compositions of (a) water, and (b) hydrogen peroxide, H_2O_2? Which element constitutes the largest percentage composition in aspirin ($C_9H_8O_4$)?

10–3. Find the percentage composition of the following.

(a) K_2CO_3
(b) Aluminum bicarbonate
(c) Calcium cyanide
(d) Potassium dichromate

10–4. Determine the empirical formulas for the compounds having the following composition.

(a) 82.3% nitrogen, 17.7% hydrogen
(b) 85.7% carbon, 14.3% hydrogen
(c) 32% carbon, 6.7% hydrogen, 18.7% nitrogen, 42.6% oxygen

10–5. Vinegar contains acetic acid, which is a compound containing carbon, oxygen, and hydrogen. Analysis yields the following percent compositions: carbon, 40.0%; hydrogen, 6.7%; and oxygen, 53.3%. Calculate the empirical formula of acetic acid.

10–6. Fool's gold is actually a compound of iron and sulfur called iron pyrite. A 0.64-g sample contains 0.46 g of iron and 0.18 g of sulfur. Compute the empirical formula for iron pyrite.

10–7. Ethylene glycol, the substance used as a primary component of antifreeze solutions, has an empirical formula of CH_3O and a gram-molecular weight of 62 g. What is its molecular formula?

10–8. Nicotine has an empirical formula of C_5H_7N and gram-molecular weight of 162 amu. What is its molecular formula?

10–9. Some of the major ingredients of smog that result from combustion are the oxides of nitrogen.

(a) Initially, oxygen reacts with nitrogen at high temperatures to produce nitric oxide, whose composition is 47% nitrogen and 53% oxygen. Its gram-molecular weight is 30 g/mol. Calculate its molecular formula.
(b) The second gas, nitrogen dioxide, is yellow brown and gives smog its color. It is formed by reaction of nitric oxide with oxygen, and contains 30% nitrogen and 70% oxygen. If its gram-molecular weight is 46 g/mol, what is its molecular formula?
(c) Can you write the two balanced reactions for the formation of both gases?

10–10. The insecticide, lindane, is 24.7% carbon, 2.1% hydrogen, and 73.2% chlorine. If its gram-molecular weight is 291, what is its molecular formula?

10–11. Mannose is a sugar containing only carbon, hydrogen, and oxygen. It has a gram-molecular weight of 180. A 2.36 g sample of mannose was found to contain 0.944 g of carbon and 0.158 g of hydrogen. Calculate its molecular formula.

10–12. One of the most effective remedies developed in primitive cultures was an antimalarial brew made from the bark of the Peruvian cinchona tree. The principal medicinal ingredient was shown to be quinine. Analysis showed quinine to be 74% carbon, 7.4% hydrogen, 8.6% nitrogen, and 9.9% oxygen. Its gram-molecular weight was 324 g. Calculate the molecular formula for quinine.

10–13. What are the gram-molecular weights of the following?

(a) 39 g benzene at STP with a volume of 11.2 ℓ
(b) 0.22 g of propane with a volume of 112 mℓ at STP

10–14. One component of natural gas is butane. It is 82.7% carbon and 17.3% hydrogen. Under standard conditions, 5.8 g of the gas occupies a volume of 2.24 ℓ. Calculate the molecular formula for butane.

10–15. A sample of potato starch was ground in a ball mill to give a starch-like molecule of low molecular weight. This product was analyzed and found to contain 0.086% phosphorus. If each molecule of the compound is assumed to contain one atom of phosphorus, what is the average gram-molecular weight of the compound?

CHAPTER 5

solids, liquids, and gases

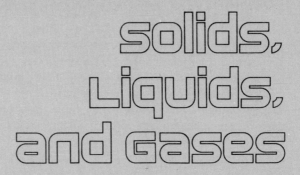

So far we have discussed matter from a molecular view and have concentrated on the composition of matter. In units 11 and 12 we shall deal with matter as we see it in the real world: as solids, liquids, and gases.

All compounds can exist as solids, liquids, or gases. The factors that determine whether a compound is a solid, liquid, or gas are temperature, pressure, and chemical bonding. In unit 11 we define the properties of solids, liquids, and gases, and then see how their volumes are affected by changes in pressure and temperature. Finally, we shall present postulates (a kinetic theory) to help explain on the molecular level some of the properties of solids, liquids, and gases.

In unit 12 we shall discuss changes of state, the factors involved when solids melt to become liquids, when liquids boil to become gases, and when the reverse processes occur. Our discussions will center on the energy requirements for the processes.

UNIT 11 THE PHYSICAL STATES OF MATTER

There are three physical states in which all matter can exist: **solid**, **liquid**, and **gas**. Most substances can exist in any of these states under the proper conditions. The factors that determine whether a given substance is a solid, liquid, or gas include the physical characteristics of that substance, the external atmospheric pressure, and temperature. In this unit we shall examine some of the principal characteristics of each state and see how they are affected by changes in temperature and pressure. We shall also develop theoretical models for gases, liquids, and solids in an attempt to explain some of these characteristics.

Solids

The basic particles (ions, atoms, or molecules) in a solid are quite close to each other; strong forces of attraction exist between these particles and more or less hold them in fixed positions within the solid. The particles may vibrate back and forth at these fixed positions, but rarely do they move past one another. In many solids the particles are packed in very orderly arrangements, which are called **crystal lattices**. Figure 11–1 shows the lattice structures for sodium chloride and the diamond and graphite forms for carbon.

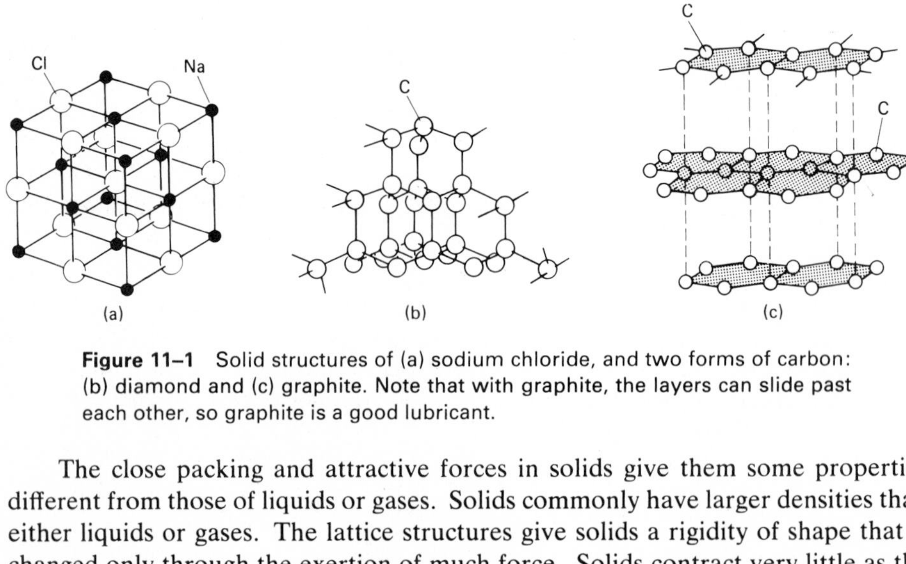

Cl
Na
C
C
C
(a)
(b)
(c)

Figure 11–1 Solid structures of (a) sodium chloride, and two forms of carbon: (b) diamond and (c) graphite. Note that with graphite, the layers can slide past each other, so graphite is a good lubricant.

The close packing and attractive forces in solids give them some properties different from those of liquids or gases. Solids commonly have larger densities than either liquids or gases. The lattice structures give solids a rigidity of shape that is changed only through the exertion of much force. Solids contract very little as the external pressure is increased. They make good building materials because of their rigidity and resistance to contraction.

The volumes of solids increase only slightly as their temperatures increase. The increase per degree Celsius in the length of a solid rod is given by a unit called the **coefficient of thermal expansion.** For a steel rod, the coefficient of expansion is

2×10^{-6} °C, and the length of the rod will increase by 0.0002% for every degree Celsius of increase in temperature. Obviously, this expansion is small, but it must be taken into account during the construction of buildings, bridges, and the like. If space is not allowed for this expansion, tremendous pressures can build up and serious structural damage may result.

The coefficients of thermal expansion of most metals differ from each other; one use of these differences is in the construction of bimetal thermometers and thermostats used in furnaces, stoves, and a variety of engines. If two thin flat strips of different metals are welded together to form a thin bar, it will bend as its temperature changes because one side expands more than the other. The amount of bending is measured and used to measure temperature changes. When used as a thermostat (see Figure 11–2), one end of the bimetal strip is fixed and the other is positioned next to an electrical contact switch. As temperature increases, the strip bends and makes contact with the switch, turning off the heating element.

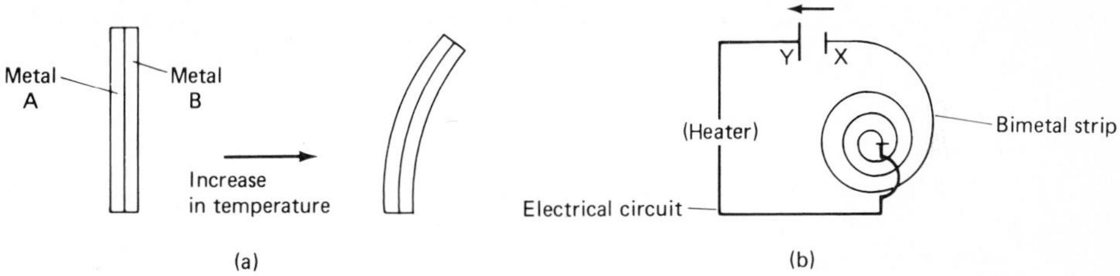

Metal A
Metal B
Increase in temperature
(a)
Y
X
(Heater)
Bimetal strip
Electrical circuit
(b)

Figure 11–2 Thermostats made of bimetal strips. (a) Metal A expands more than metal B, causing a curving to the right as temperature increases. (b) As temperature increases, the coil bends such that end X makes contact with the switch Y and shuts off the heater.

Liquids

Like solids, the atoms and molecules in a liquid are closely packed. Unlike solids, the molecules of a liquid are not rigidly held in place so they have more freedom of movement. Because of this molecular motion, liquids do not possess rigid shapes, but rather will assume the shape of the container in which they are placed. Since the particles are still tightly packed, liquids have densities in the same ranges as solids.

As with solids, the volumes of liquids change only slightly as temperature changes. In general, the volume of a liquid increases with increasing temperature. For example, 100 g of water has a volume of 100.0028 mℓ at 4°C and a volume of 100.0078 at 40°C. A tenfold increase in temperature has caused only a 0.005% increase in volume. This expansion is small but fairly regular with temperature increases. Thus the length of a small column of a liquid in a thin glass tube can be used as a measure of temperature. In mercury thermometers, the height of the column of mercury has been calibrated against temperature and provides a quick method of measuring the temperature of objects or people. Again, as with solids, these expansions are small, but they will occur and allowances must be made for them. In the canning industry care must be taken to leave air spaces in the containers before they are sealed to prevent explosion of the cans when they are heated.

The expansion and contraction of water differs somewhat from most liquids because its volume does not continually decrease with decreasing temperature. The volume of a given weight of water will decrease until its temperature reaches 4°C. Below this temperature the volume increases, until water freezes at 0°C. Since the weight remains constant and the volume has increased, the density of ice is less than that of liquid water at 4°C. When lakes freeze at the surface, the ice stays at the top and does not sink. This prevents the entire lake from becoming one solid ice cube, thus permitting marine and plant life under the surface to survive.

As with solids, liquid volumes are not greatly affected by pressure changes. Increasing the pressure by 200 atm causes only a 1% decrease in the volume of water. Because liquids do not compress to any extent, they are used in hydraulic jacks. Pressure exerted on a liquid reservoir at one end of the jack is transmitted to the other end, where a movable piston is forced upward, as shown in Figure 11–3.

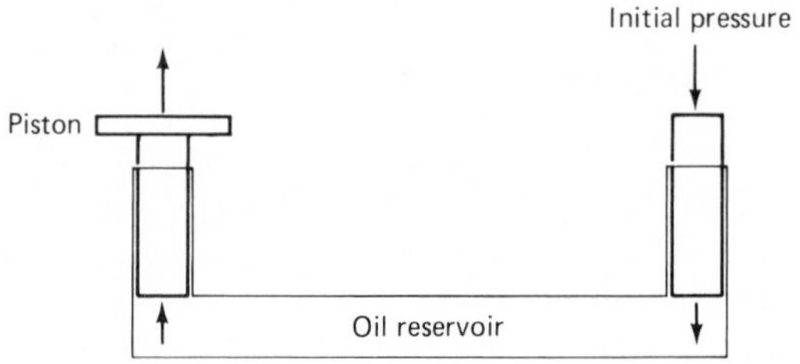
Initial pressure
Piston
Oil reservoir

Figure 11–3

The same sort of hydraulic effect is responsible for blood circulation in our bodies. The pressure exerted on the blood in the heart as the heart muscles contract forces it out of the heart and into the vessels. Repeated contractions of the heart push the blood throughout the body, where it supplies oxygen and food to the living cells. When we measure blood pressure, we measure the amount of external pressure that must be exerted on an artery to close it and thus prevent circulation of the blood. This pressure is related to the ability of the heart to force blood through the vessels. Low blood pressure may indicate weakened heart muscles and blood vessels; high blood pressure may be caused by a shrinking of the vessels.

Gases

Unlike solids and liquids, the atoms and molecules in gases are far apart. Because of the large amount of open space, the particles in gases move about rapidly and mix freely with each other. Gases have no defined shapes like solids and cannot be contained in an open flask like liquids. Gases must be stored in closed containers because they will expand to fill all available space. Gas densities are much lower than those of solids or liquids and are commonly expressed in grams per *liter* rather than grams per *milliliter*. Table 11–1 lists the densities of some common solids, liquids, and gases.

Table 11–1 Approximate densities of some common substances

Solids		Liquids		Gases	
Substance	Density (g/mℓ)	Substance	Density (g/mℓ)	Substance	Density (g/ℓ)
Lead	11.4	Mercury	13.6	Air	1.3
Iron	7.9	Water	0.999	Oxygen	1.43
Cement	3	Butter	0.9	Nitrogen	1.25
Sand	2.3	Oil	0.8	Helium	0.18
Ice	0.917	Gasoline	0.7	Hydrogen	0.09
Balsa wood	0.11				

NOTE: All measurements are at STP.

Boyle's Law

The volumes of liquids and solids change little with changes in pressure and temperature, but the volume occupied by a given amount of a gas is dramatically affected by pressure and temperature. In 1662, Sir Robert Boyle discovered that at constant temperature an **inverse relationship** existed between the volume of a gas and the external pressure. In an inverse relationship one variable increases as the other decreases proportionately; so the volume of a gas *increases* as the applied pressure *decreases.* Conversely, the volume of a gas decreases as pressure increases. Mathematically, this can be expressed as

$$P \cdot V = \text{constant} \qquad (11\text{–}1)$$

The change in the volume of a gas caused by a change in pressure can be calculated using

$$P_1V_1 = P_2V_2 \qquad (11\text{–}2)$$

where P_1 and V_1 are the initial volume and pressure, and V_2 is the new volume of the gas at new pressure P_2. The effects are illustrated in Figure 11–4.

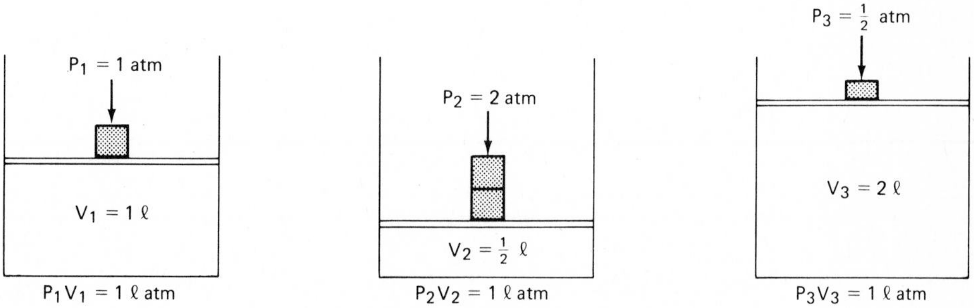

P1 = 1 atm
V1 = 1 ℓ
P1V1 = 1 ℓ atm
P2 = 2 atm
V2 = 1/2 ℓ
P2V2 = 1 ℓ atm
P3 = 1/2 atm
V3 = 2 ℓ
P3V3 = 1 ℓ atm

Figure 11–4 Boyle's law; $P \times V$ = a constant.

Charles's Law

In 1787, a French physicist, J. A. C. Charles, discovered a **direct relationship** between gas volumes and temperature changes. Here we see an opposite effect as compared to pressure changes. In a direct relationship both variables increase or decrease together. The volume of a gas increases proportionately as the **absolute temperature** increases, and vice versa. Mathematically, we write this as

$$\frac{V}{T} = \text{constant (some number)} \qquad (11\text{–}3)$$

Calculations of volume changes with temperature changes are made using equation (11–4):

$$\frac{V_1}{T_1} = \frac{V_2}{T_2} \qquad (11\text{–}4)$$

Again, V_1 is the volume of the gas at temperature T_1, and V_2 is the new volume at new temperature T_2. The changes are illustrated in Figure 11–5. Note that absolute temperatures (degrees Kelvin) must be used.

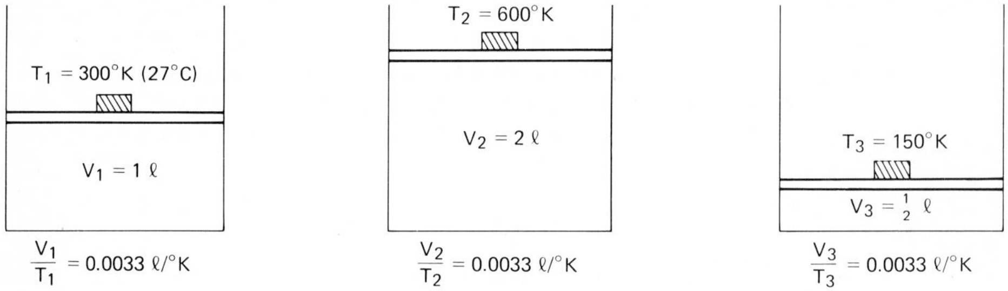

T1 = 300°K (27°C)
V1 = 1 ℓ
V1/T1 = 0.0033 ℓ/°K
T2 = 600°K
V2 = 2 ℓ
V2/T2 = 0.0033 ℓ/°K
T3 = 150°K
V3 = 1/2 ℓ
V3/T3 = 0.0033 ℓ/°K

Figure 11–5 Charles's law; V/T = a constant.

Ideal Gas Equation

When Charles's and Boyle's laws are combined into one equation, we get

$$\frac{PV}{T} = \text{constant } (K)$$

The constant K consists of two factors, the number of moles of the gas present (n) and an ideal gas constant (R). The final form of the ideal gas equation is then

$$\frac{PV}{T} = nR \quad \text{or} \quad PV = nRT \qquad (11\text{–}5)$$

When pressure is expressed in units of atmospheres, volume in liters, and temperature in degrees Kelvin, the ideal gas constant R is equal to $0.082\ \ell \cdot \text{atm}/°\text{K (mol)}$. Problems (1) and (2) illustrate the use of this equation.

Problem 1: How many moles of an ideal gas are present in 2 ℓ if the pressure is 2 atm and the temperature is 300°K?

$$\frac{PV}{T} = nR$$

where

$$P = 2 \text{ atm}$$
$$T = 300°\text{K}$$
$$V = 2\ \ell$$
$$R = 0.082 \frac{\ell \cdot \text{atm}}{°\text{K} \cdot \text{mol}}.$$

$$n = \frac{P \cdot V}{T \cdot R} = \frac{(2 \text{ atm})(2\ \ell)}{(300°\text{K})(0.082\ \ell \cdot \text{atm}/°\text{K} \cdot \text{mol})} = 0.164 \text{ mol}$$

Problem 2: How many liters are present in 1 mol of an ideal gas at a pressure of 1 atm and a temperature of 0°C.

$$V = \frac{nRT}{P} = \frac{(1 \text{ mol})(0.082\ell \cdot \text{atm}/°\text{K} \cdot \text{mol})(273°\text{K})}{1 \text{ atm}} = 22.4\ell$$

The conditions listed in problem (2) are called **standard conditions** ($P = 1$ atm, $T = 0$°C). *Under standard conditions,* 1 *mol of an ideal gas will occupy a volume of* 22.4 ℓ. Remember that this relationship exists only under standard conditions. Corrections to standard conditions are made by using equation (11–6).

$$\frac{P_1 V_1}{T_1} = \frac{P_S V_S}{T_S} \qquad (11\text{–}6)$$

where

V_1 = initial volume
P_1 = initial pressure (atm)
T_1 = initial temperature (°K)
P_S = standard pressure
= 1 atm
T_S = standard temperature
= 273°K
V_S = volume of gas under standard conditions.

Kinetic Theory

Table 11–2 lists most of the properties of gases, liquids, and solids that we have discussed so far. Note that liquids and solids behave in similar manners and are quite different from gases. Now let us generalize from these data and formulate a set of theoretical rules to explain them. Such rules have been set forth in the development of a **kinetic theory**, a working model for the behavior of gases, liquids, and solids. Some of the main postulates are given on the next page.

Table 11–2 Comparison of the properties of solids, liquids, and gases

	Solid	Liquid	Gas
Structure	Close-packed, orderly arrangements; little motion except vibration; rigid shapes	Close-packed but molecules move slowly past one another; nonrigid shapes	Wide separation of molecules; much freedom of movement; no definite shape
	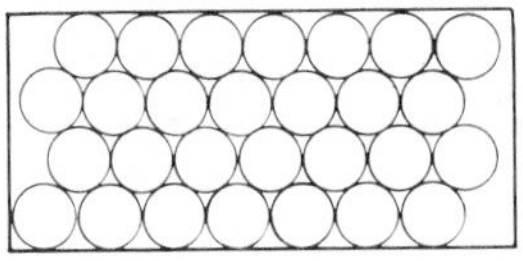	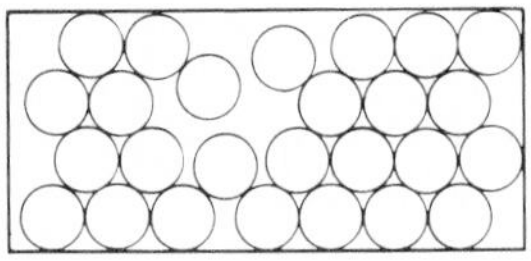	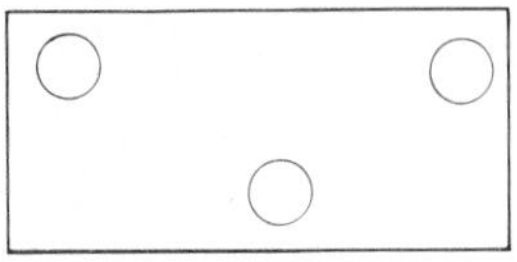
Density	Large 0.1–20 g/mℓ	Large 0.5–10 g/mℓ	Small 0.01–12 g/ℓ
Effects of temperature changes	Volume changes little	Volume changes little	Dramatic volume change: $\frac{V_1}{T_1} = \frac{V_2}{T_2}$
Effects of pressure change	Little change in volume	Little change in volume	Dramatic volume change: $P_1V_1 = P_2V_2$

1. Gases consist of tiny discrete particles that are very small in comparison to the total volume that they occupy. The volume of a gas is mostly empty space, and the atoms of a gas are very far apart.
2. There are no attractive forces between the atoms of a gas.
3. Gas molecules are constantly in rapid, random motion, colliding with themselves and the walls of the container. Collisions are elastic; no energy is lost during a collision.
4. In liquids, the molecules or atoms are close together, essentially in contact with each other. The attractive forces between the molecules are strong enough to keep them together but still allow some freedom of motion within the liquid.
5. In solids, the molecules are also in close contact with each other. There are many strong attractive forces between the molecules that hold them in rigid positions.
6. The kinetic energy of molecules and atoms is increased by absorption of heat. Likewise, kinetic energy is lost when temperature decreases.

Postulates 1 through 3 help to explain the changes of gas volumes with changes in pressure (Boyle's law). Since the molecules are far apart and in constant motion, they strike the walls of the container, pushing it outward (exerting pressure). If the walls of the container are brought in (the volume decreased), there is less space for the molecules to travel and they strike the walls with greater frequency (the pressure increases). If the volume is increased, the molecules strike the walls *less* frequently since they have more free space (the pressure decreases).

Postulates 1 through 3 and 6 can be used to explain Charles's law. Increasing the temperature of a gas causes an increase in the kinetic energy (and speed) of the molecules. They will then strike the walls of the container with greater force and more often. If the external pressure on the walls remains constant, the walls will move out (the volume increases). Cooling the gas will cause a loss of kinetic energy and molecular speed. There will be fewer, less forceful collisions with the walls, and the walls will move inward (the volume decreases).

Postulates 4 and 5 describe how liquids and solids differ from gases. The molecules, ions, and atoms of liquids and solids are close together, held there by forces of attraction. Decreasing temperature and increasing pressure will slow down motion to some extent, but volumes will decrease only slightly since the particles are already closely packed. But if the pressure is decreased and the temperature increased enough, the molecules can fly apart and boil or sublime.

Departures From Ideal Behavior

The equations and kinetic theory presented so far work quite well for an ideal gas (a theoretical picture of a gas). Real gases, however, sometimes behave differently from ideal gases. Real gases are not composed of perfectly round particles and collisions are not perfectly elastic. If the molecules of a real gas get close enough together, the forces of attraction will apply just as they do with liquids and solids. If we keep lowering the temperature and increasing the pressure on a gas, theory predicts that the volume will keep getting smaller. For a real gas, a temperature and pressure will be reached at which the gas will condense and become liquid, and large decreases in volume will no longer occur. Liquid nitrogen can be made by cooling nitrogen gas to a temperature of 77°K, and liquid oxygen is formed when the temperature is lowered to 90°K.

UNIT 11 Problems and Exercises

11–1. The basic units of a solid and a liquid are both closely packed. What then is the major difference between them?

11–2. Give the effects on the volumes of solids, liquids, and gases resulting from the following changes. Tell why the change occurred.

(a) The temperature of a solid is doubled.
(b) The temperature of a liquid is decreased by one-half.
(c) The temperature of a gas is doubled.
(d) The pressure exerted on a solid is increased.
(e) The pressure exerted on a liquid is increased.
(f) The pressure exerted on a gas is decreased.

11–3. An object floats on a liquid because its density is less than that of the liquid.

(a) Which of the solids and liquids in Table 11–1 would float on water?
(b) Which would float on mercury?

11–4. Which gas laws relate most closely to the following effects? Explain your conclusions.
(a) A soft drink fizzes when the lid is removed.
(b) A pressure cooker explodes if it is heated too hot and the steam valve is clogged.
(c) As a balloon rises, its volume increases.

11–5. Why is the phrase *a cubic foot of air* an unsatisfactory unit of measure by itself?

11–6. What will happen to the volume of a balloon under the following conditions?
(a) It is placed in ice water.
(b) It is placed in an oven.
(c) Atmospheric pressure drops.
(d) Atmospheric pressure increases.

11–7. At constant volume, how is the pressure exerted by a gas related to the Kelvin temperature?

11–8. Explain how kinetic theory accounts for the following.
(a) The compressibility of gases
(b) The pressure of gases
(c) The ability of gases to mix with each other

11–9. Use kinetic theory to explain how an odor moves across a room.

11–10. From the viewpoint of a molecule, explain how increasing temperature causes an increase in the pressure of a confined gas.

11–11. A set of automobile tires is charged with 32 $lb/in.^2$ of air pressure on a day when the tires are at a temperature of 24°C. Later the car is driven and the temperature of the tires climbs to 44°C. Assuming no increase in volume, what is the pressure in the tires?

11–12. Can the pressure of a gas in a container be kept constant if the temperature is raised? How?

11–13. A gas at 25°C and 1 atm of pressure has a volume of 2 ℓ. If the pressure is kept constant and the temperature is raised to 100°C, what is the new volume?

11–14. Use Boyle's and Charles's laws to calculate the new volumes for a gas under the following conditions. Initial conditions: 1 ℓ of gas at 0°C and 1 atm of pressure. Final conditions:
(a) T = 27°C, P = 1 atm
(b) T = −53°C, P = 1 atm
(c) T = 0°C, P = 2 atm
(d) T = 0°C, P = 0.2 atm
(e) T = 100°C, P = 4 atm

11–15. Many aerosol cans will explode if the internal pressure reaches 3 atm. If at 27°C the pressure is 2 atm, to what temperature must a can be heated before it will explode? (Remember that the volume is constant.)

11–16. If the number of molecules in a container of fixed volume is increased, what will happen to the pressure? Why?

11–17. How many moles of a gas are present if 10 ℓ has a pressure of 4 atm at a temperature of 80°C?

11–18. Determine the volume occupied by 4.0 g of oxygen (O_2) at STP.

11–19. At 18°C and 765 mm, 1.29 ℓ of a gas weighs 2.71 g. Calculate the approximate molecular weight of the gas.

11–20. Determine the molecular weight of a gas if 560 mℓ weighs 1.55 g at STP.

11–21. Compute the density of a gas at STP if 970 mℓ of the gas at 64°C and 723 mm has a mass of 1.64 g.

11–22. Flask A contains 20.0 ℓ of CH_4 gas. Flask B contains 30.0 ℓ of CO gas. Each volume is measured at the same temperature and pressure. If flask A contains 6.18 mol of CH_4, how many moles of CO are in flask B?

11–23. An organic compound has the following analysis: C = 55.8%, H = 7.03%, O = 37.2%. A 1.500-g sample was vaporized and found to occupy 530 mℓ at 100°C and 740 mm. What is the molecular formula for the compound?

11–24. One gram of a gaseous compound of boron and hydrogen occupies 0.820 ℓ at 1.0 atm and +3.0°C.

(a) The number of moles present is ____________.

(b) The molecular weight is ________________.

(c) The molecular formula is (BH_3, B_4H_{10}, B_2H_6, B_3H_{12}, or B_5H_{14})____________.

UNIT 12 CHANGES OF STATE

Solids, liquids, and gases are not independent states of matter totally unrelated to each other; they coexist. By changing the temperature and pressure of a system, it is possible to convert one state into another. In this unit we shall examine some of the factors that determine whether a compound exists as a solid, liquid, or gas.

Melting Points

As the temperature of a solid is increased, the particles (ions or molecules) in that solid gain energy and begin to vibrate more vigorously. As more energy is added, a temperature is reached at which the particles shake free of their rigid lattice structures and begin to move past one another. In other words, the compound *melts* and becomes a liquid. Each pure compound has a specific temperature at which melting occurs; this temperature is called the **melting point** of the compound.

Sometimes the identity of a solid can be determined by measuring its melting temperature, but this method is not perfect as many solids have similar melting points. It is a good measure of the purity of a compound, however, as mixtures generally melt over much wider temperature ranges than pure solids.

Sometimes when a solid is heated it does not melt and become liquid, but rather changes directly into a gas. This process is known as **sublimation**, and is quite dependent upon the temperature and pressure of the solid. Solid carbon dioxide (dry ice) is the most common example of a solid that sublimes. At atmospheric pressure solid dry ice changes directly to gas at −78.5°C; if the pressure is increased to 5.2 atm, dry ice will melt and become liquid at −56.6°C.

If thermal energy is removed from a liquid, its temperature decreases until the liquid freezes and becomes solid. The process is just the reverse of melting, and the **freezing temperature** for a liquid is the same as the melting point of the solid state of the same substance. It's just that the same temperature is being approached from different directions. So ice melts at 0°C as liquid water is formed, and liquid water freezes to become ice at 0°C.

Boiling Points

If thermal energy is added to a liquid, its temperature will rise until the **boiling point** is reached and the liquid is converted into a gas. At this temperature, the molecules in the liquid state have sufficient energy to separate from each other and move about freely in space. Each liquid also has its own specific boiling point, but this temperature is quite dependent upon the external pressure. A decrease in atmospheric pressure causes a corresponding decrease in the boiling point of the compound. For example, at 1 atm water boils at 100°C; but at 0.5 atm (an elevation of 3.8 mi) water boils at 78°C. Thus the pressure of the system must always be noted when discussing boiling points. The reverse of boiling is **condensation**, the conversion of gas back into liquid. The **condensation point** of a gas occurs at the same temperature and pressure as the boiling point of the liquid.

Factors That Govern Physical States

The melting and boiling points for the compounds given in Table 12–1 show wide variations, even among compounds with similar molecular weights. Although all the factors that govern these variations cannot be explained, some general theories can

Table 12–1 Melting and boiling points of some common compounds

Name	Structure	Molecular weight	Melting point (°C)	Boiling point (°C)
Ionic substances				
Sodium chloride	$NaCl$	58	801	1413
Potàssium fluoride	KF	58	846	1505
Polar covalent compounds				
Water	H_2O	18	0	100
Propyl alcohol	$CH_3—CH_2—CH_2—OH$	60	−127	97
Nonpolar covalent compounds				
Methyl ethyl ether	$CH_3—CH_2—O—CH_3$	60	< −200	7
Butane	$CH_3—CH_2CH_2CH_3$	58	−138	−0.5

account for the trends. One important factor is the molecular weight of a compound. *Other factors being equal, melting and boiling points generally increase with increasing molecular weight.* This generalization is difficult to use, however, because the other factors often take precedence.

A second more important factor that governs boiling and melting temperatures involves the types of attractive forces that exist between the ions, atoms, or molecules in a substance. We shall consider four main types: **ionic attractions**, **hydrogen bonding**, **polar attractions**, and **van der Waals forces**.

Of the four types, **ionic attractions** are the strongest. In the crystal lattice of an ionic substance, each positive ion is surrounded by many negative ions, and vice versa, so each ion has electrostatic attractions to many other ions. These attractions create strong forces that hold the ions in tight formations in the lattices. Disruption of these formations requires the input of much energy, so ionic substances generally have much higher melting and boiling points than covalent compounds. In Table 12–1 we see that the salts sodium chloride and potassium fluoride melt and boil at much higher temperatures than the covalent compounds listed, even though their molecular weights are similar.

When the pair of electrons in a covalent bond is not shared equally by the two atoms, partial positive and negative charges are created, and the molecule is called a **polar molecule**. Two or more polar molecules can interact with each other by attraction between the negative end of one molecule and the positive end of another. These **polar attractions** are short-range interactions; they occur mostly in the solid and liquid states. The greater the polarity of a molecule, the greater are the polar attractions between molecules, and the greater the tendency of the compound to be a solid or liquid rather than a gas.

Hydrogen bonding is the strongest type of polar bonding. Hydrogen is much less electronegative than oxygen or nitrogen, so when it is covalently bonded to these atoms, strong dipoles occur. Hydrogen bonding occurs when the partially positive hydrogen atom of one molecule is attracted to a partially negatively charged oxygen or nitrogen atom in a second molecule. Compounds that can hydrogen bond usually possess higher boiling points than other polar covalent compounds of similar weight. In Table 12–1, water and propyl alcohol can hydrogen bond, whereas ether and butane cannot. Both butane and ether boil at lower temperatures than water and propyl alcohol.

In the absence of ionic and polar attractions, a third type of attraction exists between atoms of different molecules: **van der Waals attractions**. Nuclei of atoms, being positively charged, attract the negatively charged electrons. Strong attractions exist between a nucleus and the electrons surrounding it, but weaker attractions can also exist between the *nucleus of one atom* and the electrons of another. *Van der Waals attraction is defined as the attraction between the nucleus of one atom and the electrons of another atom.* The strength of the attraction falls off rapidly as the two centers of charge (the atoms) are separated, so van der Waals forces of attraction are important only when atoms or molecules are close together. The strength of this attraction is also a function of the magnitude of the nuclear charge: larger atoms with greater numbers of protons have stronger van der Waals forces than smaller atoms. Figure 12–1 summarizes the types of attractive forces.

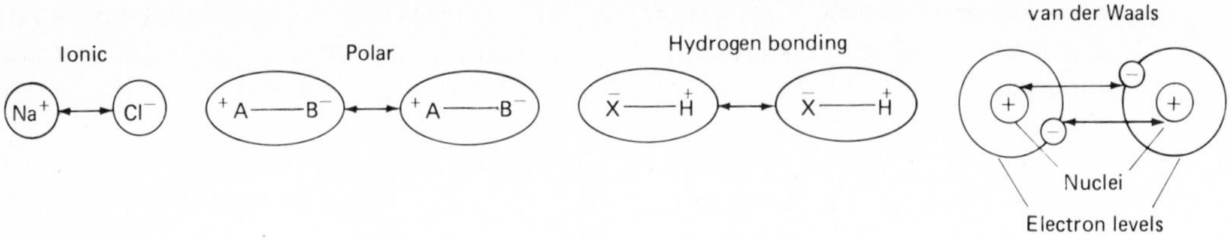
Ionic
Na+
Cl−
Polar
+A—B−
+A—B−
Hydrogen bonding
X—H
X—H
van der Waals
+
+
Nuclei
Electron levels

Figure 12–1 Summary of the types of attractive forces.

Equilibrium and Changes of State

A condition of **equilibrium** is said to exist when two opposing actions occur at the same rate such that no net change occurs. To illustrate a condition of equilibrium, let's consider the number of autos on a superhighway during a busy rush hour. Large numbers of autos may be entering and exiting the highway, but if those numbers are equal, a constant number of autos will always be present on the highway. The opposing forces (entering and exiting) cancel each other out and **equilibrium** exists.

The same type of equilibrium conditions can exist at the boiling and melting temperatures of substances. At the melting temperature of a substance, the number of molecules leaving the solid phase and becoming liquid is equal to the number of molecules leaving the liquid phase and becoming solid. If the temperature is raised, equilibrium no longer exists, since more of the solid molecules enter the liquid phase than are returning, causing complete melting. A lowering of temperature below the melting point causes just the reverse.

At the boiling point of a liquid, equilibrium exists when the number of molecules leaving the liquid equals the number of molecules returning to the liquid from the gas phase. Again, as the temperature is lowered, more molecules will enter the liquid phase than leave it, so condensation occurs. In boiling, more molecules leave the liquid than enter it. Figure 12–2 shows conditions of equilibrium and nonequilibrium.

Nonequilibrium	*Equilibrium*	*Nonequilibrium*
Melting: Solid ⟶ Liquid	Solid ⇄ Liquid	Freezing: Solid ⟵ Liquid
Boiling: Liquid ⟶ Gas	Liquid ⇄ Gas	Condensation: Liquid ⟵ Gas

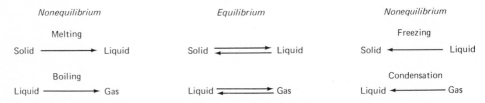

Figure 12–2 Conditions of equilibrium and nonequilibrium. The arrows point in the direction of net change. At equilibrium there is no net change.

Vapor Pressure and Boiling

The boiling point is not the only temperature at which molecules of a substance enter and leave the surface of a liquid. Actually, these processes occur all the time at temperatures much lower than the boiling point. The water in a glass set out in the open will completely evaporate because the molecules that leave the liquid phase are swept away and prevented from returning. But if that glass is placed in a closed container, only part of the water will evaporate; an equilibrium will be established between the liquid water and the water vapor. It is then possible to measure the pressure exerted by the molecules in the gas phase. This pressure at equilibrium is called the **vapor pressure** of the liquid.

All substances in the liquid state exert vapor pressures but these pressures vary considerably from liquid to liquid. The vapor pressure is a measure of the ability of molecules to escape the surface of a liquid, and it depends to a great extent upon the physical properties of the particular liquid.

The vapor pressure of a liquid also depends upon the temperature and pressure. As the temperature of a liquid increases, vapor pressure increases since the molecules have absorbed more energy and move about much more rapidly. When the vapor pressure just exceeds the opposing *atmospheric pressure*, bubbles of vapor appear throughout the liquid and rise quickly to the top, producing the effect we call boiling.

Thus *the boiling point of a liquid is defined as that temperature at which the vapor pressure just equals atmospheric pressure.* Standard boiling points are measured at 1 atm of pressure. At pressures of less than 1 atm, liquids boil at temperatures that are less than their standard temperatures. In a pressure cooker the atmospheric pressure is greater than 1 atm, so water boils at temperatures that are higher than 100°C and food can be cooked more rapidly. Table 12–2 lists the boiling point of water at various pressures.

Table 12–2 Boiling point of water at different pressures

Altitude above sea level (ft)	Atmospheric pressure (mm)	Boiling point of water (°C)
0	760 (1 atm)	100
5,280 (Denver, Colo.)	630	92
14,500 (Mt. Whitney, Calif.)	450	86
29,000 (Mt. Everest)	253	71

Heats of Fusion and Vaporization; Heat Capacity

Before going on to other subjects, the energetics of the heating and cooling of states of matter should be considered. In the beginning of this text, we defined a calorie as the amount of thermal energy required to raise the temperature of 1 g of water by 1°C. This relationship only holds for water; other substances require different amounts of thermal energy to change temperature. *The* **heat capacity** *of any pure substance is defined as the quantity of heat required to increase the temperature of* 1 g *of that substance by* 1°C*.* Thus for water the heat capacity is 1 cal/g°C. The heat capacities for some other substances are listed in Table 12–3. Using heat capacities, we can calculate the amount of thermal energy required to increase the temperature of a substance.

Table 12–3 Heats of fusion and vaporization, and heat capacities

Substance	Melting point (°C)	Boiling point (°C)	Heat of fusion (cal/g)	Heat of vaporization (cal/g)	Heat capacity (cal/g°C)
Solids					
Sodium chloride	801	1413	124		0.204
Sodium fluoride	992	1695	186		0.258
Copper	1083	2336	42	1138	0.09
Lead	327	1620	5.9	205	0.03
Aluminum	659	2450	76	2515	0.21
Liquids					
Acetic acid	16.7	118.3	43.2	97	0.47
Benzene	5.4	80.1	30.2	94.3	0.41
Ethyl alcohol	−114.4	78.3	25	204	0.58
Water	0	100	80	540	1
Gases					
Oxygen	−219	−183	3.3	50.9	0.218

Heat capacities are used only when no change of phase occurs. To calculate the thermal energies required during phase changes, we need to know **heats of fusion** and **vaporization**. To help explain heats of fusion and vaporization, consider the following example. If a solid sample is placed in an oven and thermal energy is added at a constant rate, the temperature of the sample does not increase in a smooth fashion. As shown in Figure 12–3, the temperature of the solid will rise smoothly until the melting point is reached. At this point, the temperature remains constant *as long as both solid and liquid are present*, even though more thermal energy is being added. This extra energy is needed to change the ordered structure of the solid into the more disordered molecular arrangement present in the liquid. It causes some of the intermolecular bonds to be broken. The amount of thermal energy needed is a characteristic of each individual compound and is called the **heat of fusion**. *The heat of fusion is thus defined as the number of calories of heat required to change* 1 *g of a solid into liquid.* There are no temperature units involved in the heat

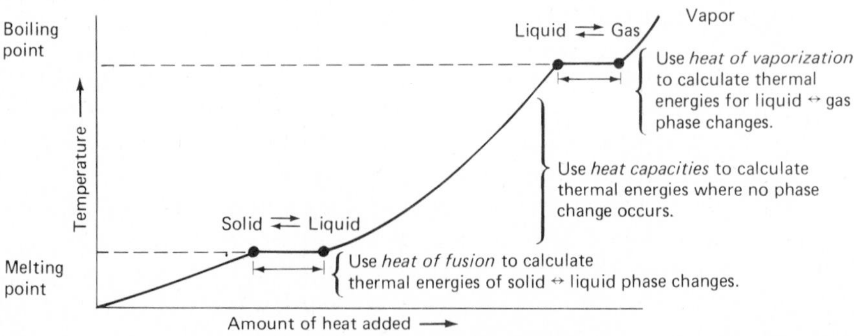
Vapor
Boiling point
Liquid ⇄ Gas
Use *heat of vaporization* to calculate thermal energies for liquid ↔ gas phase changes.
Temperature
Use *heat capacities* to calculate thermal energies where no phase change occurs.
Solid ⇄ Liquid
Melting point
Use *heat of fusion* to calculate thermal energies of solid ↔ liquid phase changes.
Amount of heat added

Figure 12–3 Change-of-state diagram.

of fusion since temperature is constant until all the solid has changed into liquid. Conversely, if we remove heat and cool a liquid, the heat of fusion is a measure of the number of calories of thermal energy that will be released when freezing occurs.

A second break in the curve in Figure 12–3 occurs at the boiling point of the liquid. A great deal of energy is required to break all the intermolecular forces and completely free the molecules from each other. *The* **heat of vaporization** *of a liquid is defined as the number of calories required to change* 1 g *of a liquid* (*at its boiling point*) *to a gas.* Table 12–3 lists the heats of vaporization and fusion for several compounds. Notice that the heats of vaporization are by far the larger values.

Using heats of fusion and vaporization along with heat capacities, scientists are able to calculate the amount of thermal energy required or released as the temperatures of many substances change. Examples 1 and 2 illustrate the processes.

Example 1: How many calories are required to convert 10 g of ice at 0°C to gas at 100°C? There are three separate processes here. Calculate the number of calories for each process and then sum them.

1. To melt the ice:

$$\text{Heat of fusion} = 80 \text{ cal/g}$$

$$10 \text{ g} \times \frac{80 \text{ cal}}{\text{g}} = 800 \text{ cal}$$

2. To warm from 0° to 100°:

$$\text{Specific heat} = \frac{1 \text{ cal}}{\text{g}\cdot\text{deg}}$$

To warm 1 g from 0° to 100°:

$$1 \text{ g} \times \frac{1 \text{ cal}}{1 \text{ g deg}} \times 100^\circ = 100 \text{ cal}$$

$$10 \text{ g} \times \frac{1 \text{ cal}}{1 \text{ g deg}} \times 100^\circ = 1000 \text{ cal}$$

3. To convert water to steam:

$$\text{Heat of vaporization} = 540 \text{ cal/g}$$

$$10 \text{ g} \times \frac{540 \text{ cal}}{\text{g}} = 5400 \text{ cal}$$

4. Total number of calories = 800 + 1000 + 5400 = 7200 cal.

Example 2: How many calories of heat are given off when 10 g of ethyl alcohol is changed from vapor at 78.3°C to solid at −114.3°C? Since the temperature is decreasing, we use the heats of fusion and vaporization and the heat capacity to determine the number of calories of heat *given off*, not added, to the system.

1. To condense the vapor: the heat of vaporization for the alcohol is 204 cal/g.

$$10 \text{ g} \times 204 \text{ cal/g} = 2040 \text{ cal released during condensation}$$

2. To cool the liquid from 78.3 to $-114.3°C$: we must use the heat capacity of the alcohol (0.58 cal/g deg) to calculate the calorie change. The total change in temperature is 192.6°.

$$10 \text{ g} \times 192.6° \times \frac{0.58 \text{ cal}}{\text{g deg}} = 1117 \text{ cal released}$$

3. To freeze the liquid: the heat of fusion for the alcohol is 25 cal/g.

$$10 \text{ g} \times 25 \text{ cal/g} = 250 \text{ cal released}$$

4. Total number of calories of heat released:

$$2040 + 1117 + 250 = 3407 \text{ cal.}$$

UNIT 12 Problems and Exercises

12–1. Define the following.
(a) Sublimation
(b) Evaporation
(c) Boiling
(d) Melting

12–2. Are the following true or false? Explain your reason.
(a) The melting point and freezing point of a compound are identical.
(b) To become a gas, a solid must first be melted to a liquid.
(c) When a substance changes from gas to liquid, much heat is released to the environment.
(d) At equilibrium, no action is occurring.
(e) The vapor pressure of a liquid is always equal to atmospheric pressure.
(f) At the melting point, the temperature remains constant as long as liquid and solid are present.

12–3. Two pans of boiling water are heated side by side on a stove. One is boiling rapidly; the other is boiling slowly. Is the water in the two pans at different temperatures? Explain.

12–4. Would it be possible to boil water in a paper cup without burning the cup? Explain.

12–5. In each of the following pairs, which substance would you expect to have a higher boiling point? Justify your answer.
(a) O_2 or N_2 (b) HF or HCl

12–6. Why do NH_3, H_2O, and HF have abnormally high boiling points when compared to PH_3, H_2S, and HCl?

12–7. List the following substances in order of increasing boiling points.
(a) CH_4 (b) CCl_4
(c) CBr_4 (d) CF_4

12–8. Predict which of the following pairs of substances is more likely to be a solid. Explain your decision.
(a) Oxygen or polonium
(b) Sodium oxide or water
(c) H_2S or HCl

12–9. Arrange the following compounds in order of increasing predicted boiling points. All the compounds have similar molecular weights. Tell why you arranged them as you did:

$CH_3CH_2CH_2CH_3$ NaCl $HO{-}CH_2CH_2{-}OH$ $CH_3CH_2CH_2{-}OH$

12–10. Liquid molecules cannot escape from the surface into the gas phase unless we supply energy equal to the ________________ for the liquid.

12–11. One gram of a solid will melt if an amount of heat equal to its ________________ is added.

12–12. What happens to the boiling temperature of a liquid if external pressure is decreased? Why?

12–13. If 10 mℓ of water is poured into a flask and the air in the flask is pumped out, the water will boil. Why?

12–14. Can two containers of water exist at the same temperature but possess different quantities of thermal energy? Explain.

12–15. Two beakers, A and B, are both filled with a mixture of ice cubes and water. Extra ice cubes are added to beaker A and extra water is added to beaker B, yet both solutions still remain at the same temperature.
(a) What is the temperature?
(b) Why does the temperature remain constant?

12–16. Why is the heat of vaporization of benzene, C_6H_6, so much less than that of ethyl alcohol?

12–17. Why does your arm feel cool after it is rubbed with alcohol?

12–18. Steam sterilizers (autoclaves) used in hospitals attain temperatures of 250°F with boiling water. Explain how this is possible.

12–19. How many calories of heat are given off when the following substances freeze? The temperature is at the freezing point.
(a) 10 g of water   (b) 2 g of copper

12–20. How many calories are required to convert the following liquids (at their boiling point) into gases?
(a) 10 g of H_2O (b) 5 g of benzene   (c) 5 g of ethyl alcohol

12–21. How many calories of heat are required to change the temperature of the following substances from 20 to 100°C?
(a) 1 g of lead   (b) 1 g of acetic acid
(c) 2 g of water   (d) 1 lb of NaCl

12–22. Assuming that the specific heat is the same in both liquid and solid states, calculate the number of calories of heat required to perform the following.
(a) Convert 20 g of ice (−10°C) to water vapor at 100°.
(b) Convert 10 g of benzene (0°) to vapor at 80.1°.

CHAPTER 6

solutions

Elements and compounds are seldom found in their pure states. Usually they are mixed together in solutions, colloids, and mixtures. In unit 13 we define each of these combinations, and then concentrate on **aqueous** solutions. In discussing aqueous solutions, we shall describe how they form, their special properties, and methods by which they can be separated.

In unit 14 we cover the quantitative aspects of solution chemistry: units of concentration. Hospitals and industrial laboratories often use percentage units of concentration, so we shall describe weight-to-volume, weight-to-weight, and volume-to-volume percentages in concentration. Chemical laboratories and research centers often use **molarity** and **normality** concentration units. Both of these units deal with the number of moles of solutes dissolved in the solvent, not the number of grams of solute dissolved.

UNIT 13 SOLUTIONS, COLLOIDS, AND MIXTURES

So far we have dealt mainly with pure compounds, but in the real world mixtures are much more prevalent. The combination of two or more elements or compounds can result in the formation of a mixture, a colloid, or a true solution. In this unit we shall define each type of combination and then dwell on the characteristics and properties of solutions.

A **solution** exists when the separate compounds are mixed so thoroughly that no clusters of molecules of each compound are present. The mixing is complete, thorough, and **homogeneous**, that is, uniform in composition throughout. If you removed small samples of the solution and analyzed them, each sample would contain the same amounts of each compound present. The individual components in a solution will not settle out and cannot be separated by filtration. The mixing of vinegar with water produces a true solution, as does the dissolving of sugar or salt in water.

In a **mixture**, large clusters of molecules of each compound exist, and the separate components can often be seen. A mixture is **heterogeneous** (not uniform throughout) and can often be separated into its components by filtration. In a solid–liquid mixture the solid particles do not dissolve in the liquid and will settle out if the mixture is not continually shaken. Another very common mixture is rock, in which many different types of ionic compounds are randomly mixed together. Vinegar and oil salad dressing is a mixture since the separate components are quite visible and separation occurs if the liquids are not continually shaken.

In between true solutions and mixtures are **colloids**. Generally, a colloid is homogeneous and the individual components do not settle out and cannot be separated by filtration. In a colloid, one component is often present as large molecular clusters, not as individual molecules, as in a solution. Tea, coffee, fog, whipping cream, and some inks are common colloids.

A colloid then is more similar to a solution than to a mixture. A solution and colloid may appear identical, but colloids can often be identified because they exhibit a **Tyndall effect**: a beam of light is visible as it passes through the colloid since the

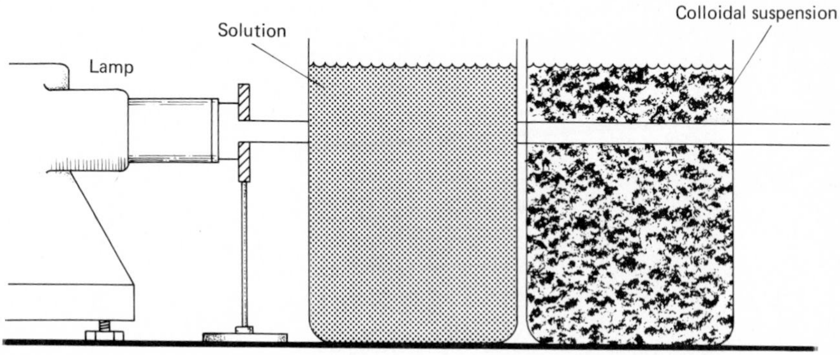
Colloidal suspension
Solution
Lamp

Figure 13–1 The Tyndall effect. A light beam is visible in a colloid.

clusters of molecules are large enough to reflect some of the light (see Figure 13–1). A true solution contains no clusters large enough to reflect the light beam, and we do not see the beam as it passes through the solution. Table 13–1 summarizes the various properties of the three combinations.

Table 13–1 Properties of solutions, colloids, and mixtures

	Solution	Colloid	Mixture
Consistency	Homogeneous	Homogeneous	Heterogeneous
Particle size	Molecular size: less than 10 Å	Molecular clusters of 100–1000 Å	Greater than 1000 Å
Separation by gravity or filtration	Cannot be separated	Some partial separation	Fairly easily separated
Appearance	Clear, may be colored	May be clear or cloudy, some particles may be seen in a high-power microscope	Many times light rays will not pass through at all
Behavior toward light	No Tyndall effect; light rays not reflected	Positive Tyndall effect; light beam is seen as it passes through solution	Many times light rays will not pass through at all

Table 13–2 gives some common examples of solutions and colloids. The dispersing medium, that component present in the largest amount, is called the **solvent**. The other components suspended or dissolved in the solvent are called **solutes**. The distinction is not always clear-cut as the proportions of the two components mixed will determine which is solvent and which solute. Since solids, liquids, and gases can all be either solvents or solutes, there can be nine different types of solutions and nine different types of colloids.

Table 13–2 Examples of solutions and colloids

Solute	Solvent	Solution	Colloid
Gas	Gas	Air (80% N_2, 19% O_2, 1% other)	None; gas particles are not large enough
Gas	Liquid	Oxygen in water; CO_2 in soda	Whipped cream, soap bubbles
Gas	Solid	Rare; hydrogen in platinum	Marshmallow, activated charcoal
Liquid	Gas	Humid air (not clouds)	Fog, clouds, mist
Liquid	Liquid	Many; alcohol in water, vinegar in water	Milk, cream
Liquid	Solid	Water dissolved in salts	Cheese
Solid	Gas	Some types of smoke	Some types of smoke
Solid	Liquid	Numerous; salt or sugar in water	Paint, coffee, tea
Solid	Solid	Some rock, metal alloys	Some gems, black diamonds

Water as a Solvent: Aqueous Solutions

In the most common types of solutions water is the solvent and ionic compounds are the solutes. To understand how these solutions are created, we need to look more closely at the properties of water. Each molecule of water contains two O—H covalent bonds. These bonds are quite polar, so the hydrogen atoms bear partial positive charges and the oxygen atom bears a partial negative charge. Water is one of the most strongly hydrogen-bonding liquids known, since all three atoms can be involved. The molecules of water are attracted to each other and also to any other centers of electronic charge.

The beading of water on surfaces like wax paper and the *meniscus* effect of water on glassware are caused by the high polarity of water. Wax, like grease and oil, is composed of nonpolar molecules that do not hydrogen bond to water. Since water molecules are not attracted to this nonpolar surface, the molecules on the surface of the drop of water hydrogen bond to the molecules in the interior of the drop. The water does not spread out on the nonpolar surface, but rather forms beads as if in an attempt to remove itself from the nonpolar substance. The surface of the drop is resistant to change; it possesses **surface tension** (see Figure 13–2).

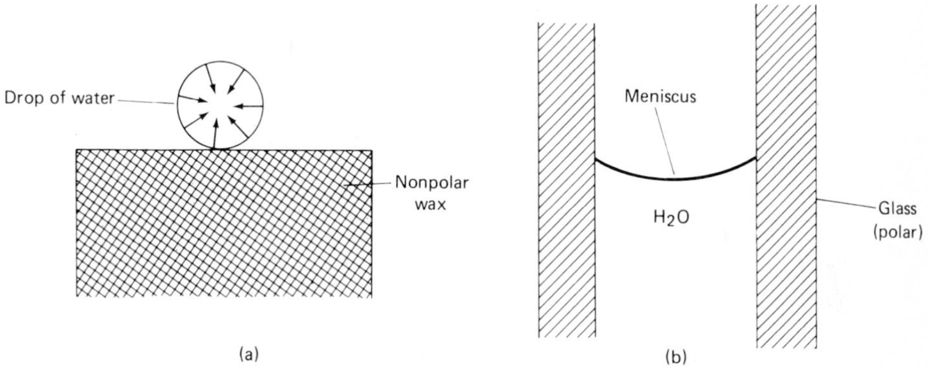
Drop of water
Nonpolar
wax
(a)
Meniscus
H_2O
Glass
(polar)
(b)

Figure 13–2 Water on nonpolar and polar surfaces. (a) Water molecules are attracted to the center of the drop and pull away from a nonpolar surface. (b) Water clings to the glass surface and a meniscus is formed.

Glass and metals contain charged ions, so the molecules of water are attracted to their surfaces. Water does not bead upon glass but spreads out on it and seems to cling to it. Because of these attractions, the level of water at the sides of a glass is slightly higher than that in the middle, and a meniscus is formed. We can also tell when a car needs a new waxing by observing if beading occurs. If the layer of wax is sufficient, water beads up rather than spreading out on the metal surface.

Hydration

Since ionic compounds consist of charged ions, they can form weak polar bonds with water. In this process, called **hydration**, water molecules become loosely bound to the surfaces of the salts. A salt is called **anhydrous** if no water molecules are bound to its surface; it is **hydrated** if water molecules do exist on its surface. The attached waters of hydration are often shown in chemical formulas at the end of the formula. Thus $CaCl_2 \cdot 2H_2O$ means that two molecules of water are loosely bound to every formula unit of calcium chloride. Heating often drives off these molecules; anhydrous salts are formed by heating hydrated ones.

Hydrated forms of ionic compounds often possess slightly different physical properties than the anhydrous forms. Many anhydrous copper salts are light-green crystalline solids, but the hydrated forms are blue. Nickel salts turn green when exposed to water, and calcium chloride gradually changes from a flaky powder to a thick syrup. Plaster of Paris is partially hydrated calcium sulfate. When exposed to water, it hydrates and sets into a hard solid called gypsum. This is how plaster casts are made.

$$\underset{\text{(light green)}}{\underset{\text{anhydrous}}{CuSO_4}} + 5H_2O \longrightarrow \underset{\text{(blue)}}{\underset{\text{hydrated}}{CuSO_4 \cdot 5H_2O}}$$

$$\underset{\text{(powder)}}{\underset{\text{plaster of Paris}}{(CaSO_4)_2 \cdot H_2O}} + 3H_2O \longrightarrow \underset{\text{(hard solid)}}{\underset{\text{gypsum}}{2CaSO_4 \cdot 2H_2O}}$$

Cement is a mixture of impure calcium carbonate (called *limestone*) and clay. It is ground to a fine powder and heated to 1500°C to drive off all the water. When remixed with water, it hydrates and sets to form a hard, rigid solid.

The Dissolving Process

Ionic compounds dissolve in water in much the same way that hydration occurs. Dipole–ion attractions that form between water molecules and the ions cause each ion to become surrounded by water molecules as it separates from the solid crystal. These envelopes of water keep the ions apart and distributed throughout the aqueous solution. Figure 13–3 illustrates the action of dissolving, using sodium chloride and water. In the solid crystal, each ion is rigidly held in place by ionic attractions to other ions of opposite charge. As water molecules form ion–dipole bonds with the ions on the surfaces of the crystal, the ionic bonds within the crystal are gradually weakened until the ion is removed and completely surrounded by water. This process continues until the solid crystal is completely dissolved.

In the dissolving process there is a limit to the amount of solute that can be dispersed in a given amount of solvent. A solution is said to be *saturated* when it has dissolved as much solute as it can. At this point a condition of equilibrium exists between dissolved and undissolved solute. Every time one ion is removed from the solid surface and dissolved, another ion comes out of solution and precipitates. The maximum amounts of the various ionic compounds that will dissolve in water vary considerably. Table 13–3 lists the solubilities of some common ionic compounds in water. Note that there are vast differences in solubilities.

Factors That Influence Solubility

The solubility of liquids and solids in liquid solvents is not dramatically affected by changes in atmospheric pressure, but the solubility of gases in liquids is quite dependent upon pressure. Generally, the solubility of a gas in a liquid increases as

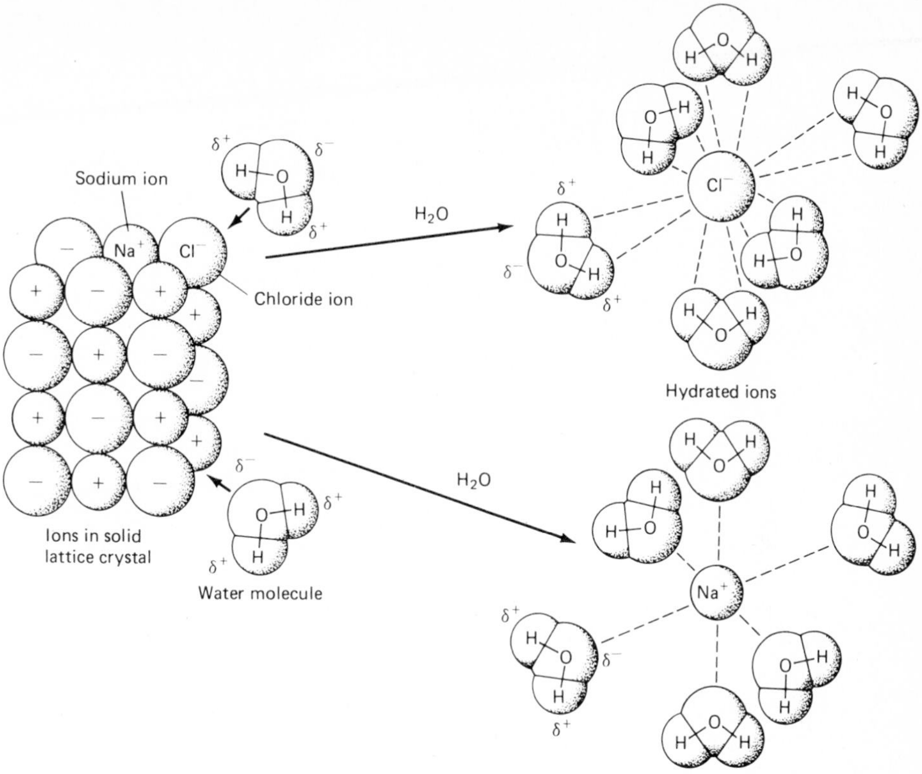
Sodium ion
Na+
Cl−
Chloride ion
Ions in solid lattice crystal
Water molecule
H₂O
H₂O
Hydrated ions
Cl−
Na+

Figure 13–3 The dissolving of sodium chloride. The water molecules become loosely attached (hydrated) to the sodium and chloride ions on the edges of the solid crystal and gradually remove the ions from the solid lattice. Eventually the crystal dissolves.

Table 13–3 Solubilities of ionic compounds in water

Ionic compound	Solubility in cold water (g/100 mℓ)	Solubility in hot water (g/100 mℓ)
NaCl	35.7	39.2
AgCl	9×10^{-5}	7×10^{-3}
$CaCl_2$	74.5	159
NaBr	116	121
AgBr	8.4×10^{-6}	4×10^{-4}
$CaBr_2$	142	312
$NaNO_3$	92.1	180
$AgNO_3$	122	952

the pressure of the gas increases. A decrease in pressure decreases the solubility. Carbonated beverages are made when carbon dioxide is forced into a liquid by increasing the pressure. When the lid from a bottle of carbonated soda is removed, the pressure lowers to atmospheric pressure, and the solution fizzes as some of the dissolved carbon dioxide escapes.

Increasing the temperature of a solution affects different types of solutions in different ways. *Generally, an increase in temperature will increase the solubility of a solute in the solvent,* but there are also cases when increasing the temperature decreases solubility. As a general rule, solids and liquids are more soluble in hot liquid solvents than in cold solvents, and gases are less soluble in hot liquids than in cold liquids. For example, as ocean, lake, and stream waters are warmed by industrial input, the amount of dissolved oxygen and carbon dioxide in the waters decreases. The result is that the waters are less able to support marine life.

A general rule often used when attempting to predict solubility is stated as *like dissolves like.* This means that substances of similar polarity are more likely to dissolve in each other than substances of widely differing polarity. Highly polar liquids will dissolve in each other because they can form polar bonds with each other. If a polar liquid and a nonpolar liquid are mixed, few polar attractions can occur between the two substances; the molecules of the polar liquid will form polar bonds with themselves and squeeze the nonpolar molecules out of the way. Two layers will form and little dissolving occurs. Polar solvents include water, acetic acid, methyl alcohol, and ethyl alcohol; some nonpolar solvents are hexane, carbon tetrachloride, and benzene. Note in Figure 13–4 that all the polar solvents can hydrogen bond, whereas nonpolar solvents cannot hydrogen bond to any appreciable degree. The medium-polarity solvents have electronegative atoms but cannot hydrogen bond.

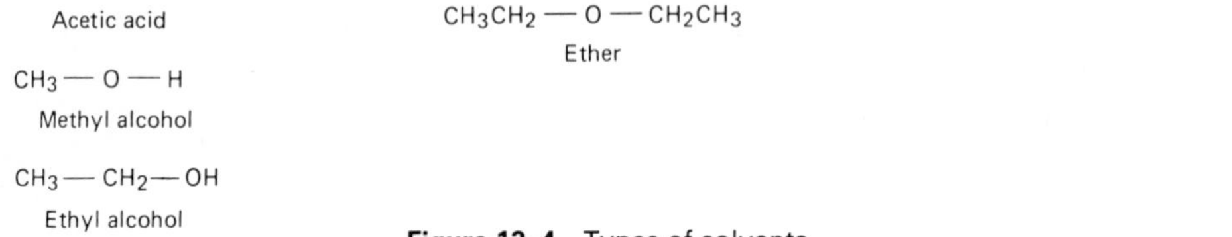
Highly polar solvents
Water
Acetic acid
Methyl alcohol
Ethyl alcohol
Medium polarity solvents
Acetone
HCCl3
Chloroform
Ether
Nonpolar solvents
CH3CH2CH2CH2CH2CH3
Hexane
C6H6
Benzene

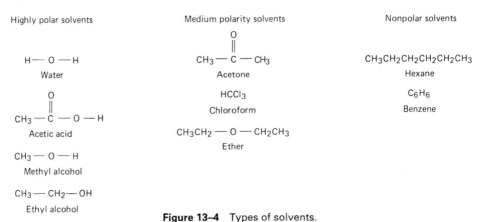

Figure 13–4 Types of solvents.

Physical Properties of Solutions

The boiling point of a solution is higher than that of the pure solvent. Thus salt water will boil at higher temperatures than pure water. Qualitatively, the increase in boiling temperature is a function of the absolute number of separate particles in solution. Thus 1 mol of calcium chloride dissolved in water will cause a greater increase in the boiling point than 1 mol of sodium chloride. There are three ions in calcium chloride but only two in sodium chloride. Quantitatively, the increase in boiling temperature (ΔT) is given by the following equation:

$$\Delta T = K_b \cdot m$$

where ΔT = degrees Celsius increase in boiling point
K_b = boiling point elevation constant
m = molality of the solution
= moles of solute/1000 g of solvent.

For water, $K_b = 0.52°\text{C/cal molal}$, so a 1-molal solution would have a boiling point of $100.52°\text{C}(\Delta T = 0.52 \times 1 = 0.52°\text{C})$.

The melting points of solutions change in the opposite direction from boiling points. A solution melts at a *lower* temperature than the pure solvent. Again, the amount of decrease is a function of the number of separate particles in solution; calcium chloride (three ions) lowers the melting point of water to a greater extent than sodium chloride (two ions). Quantitatively, the equation for the temperature decrease is similar to that for temperature increases in the boiling point:

$$\Delta T = K_f \cdot m$$

where ΔT = degrees Celsius decrease
K_f = freezing point depression constant
m = molality of solution.

The freezing point depression constant for water is $-1.86°\text{C/molal}$, so the addition of 1 mol of sodium chloride lowers the freezing point by 3.72°C.

$$\Delta T = -1.86(2 \text{ molal}) = -3.72°\text{C}$$

These effects have many uses in everyday life. In cooking, salt is sometimes added so that food will boil at a somewhat higher temperature. If salt is poured on icy streets, some of the ice melts, since the salt solution will still be liquid at temperatures that cause pure water to freeze. Antifreeze is added to the radiator of an automobile to lower the temperature at which the cooling water will freeze. The external temperature would have to get down to, say, −20°C instead of 0°C before the radiator water would freeze and ruin the engine. These additives also help protect a car during the hot summer by raising the temperature necessary for the water to boil. Thus the engine would have to get much hotter before it would boil over.

Diffusion of Gases and Respiration

The movement of gas molecules through space is called **diffusion**. Different gases will diffuse, or mix, together until the pressure of each separate gas (the **partial pressure** of the gas) is constant throughout the entire volume of the enclosure. Gases can also diffuse in and out of liquids and solids; the gases dissolved in liquids also exert a partial pressure.

Diffusion always takes place from regions of *high* concentration to regions of *low* concentration. A gas will diffuse into a liquid solvent if the partial pressure of the gas above the liquid is greater than the partial pressure of the gas dissolved in the liquid. A gas will diffuse out of a liquid if the reverse is true: the pressure of the gas dissolved in the liquid must exceed the partial pressure of the gas above the liquid.

These rules for the direction of diffusion can be illustrated by considering our respiratory processes. When you inhale, you fill your lungs with air composed mostly of nitrogen and oxygen. The partial pressure of the oxygen in the lungs is greater than the partial pressure of the oxygen dissolved in your blood, so oxygen diffuses into the blood and is carried through the body to the cells. Oxygen then diffuses out of the blood into the cells, because the pressure of dissolved oxygen in the blood is greater than that in the cells.

Carbon dioxide is formed in the cells of the body as a by-product of their function. When the pressure of carbon dioxide dissolved in cellular fluids exceeds the pressure of carbon dioxide dissolved in the blood, it diffuses out of the cells into the blood and is carried to the lungs. Here it diffuses out of the blood into the air and is removed from the body when you exhale.

O_2 diffusion →				
P_{O_2}	$>$	P_{O_2}	$>$	P_{O_2}
Lungs		Blood		Cellular fluid
P_{CO_2}	$<$	P_{CO_2}	$<$	P_{CO_2}
←				CO_2 diffusion

Osmosis

Osmosis is the diffusion of solvent molecules through a membrane, such as a piece of animal tissue. A membrane of this type can be thought of as similar to a piece of filter paper, but with holes so small that only very small molecules can pass through.

As with diffusion of gases in and out of liquids, the diffusion of solvents through osmotic membranes occurs in a direction that equalizes the concentration of solute on both sides of the membrane. Since it is the solvent and not the solute that moves, flow occurs from the *dilute* side to the *concentrated* side of the membrane. In this

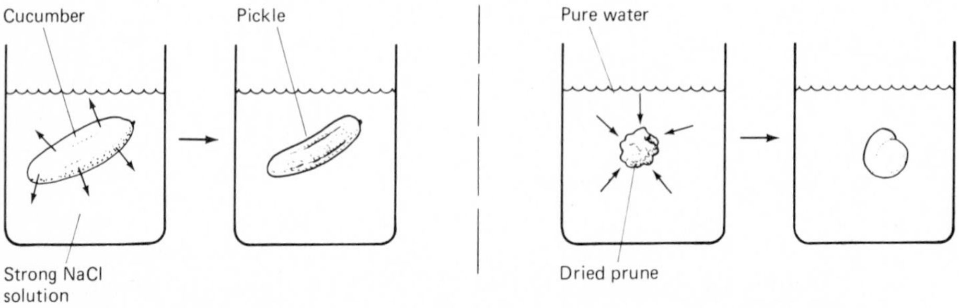
Cucumber
Pickle
Pure water
Strong NaCl solution
Dried prune

Figure 13–5 Osmosis related to cucumbers and prunes. (a) Water moves out of the cucumber, causing it to shrink. (b) Water moves into the prune, causing it to swell.

manner, the concentrated side becomes more dilute. If the osmotic membrane is made in the form of a sack, the net direction of the water flow can be seen readily. If the contents of the sack are more concentrated than the outside, water will enter and the sack will swell. If the contents outside the sack are more concentrated, water will leave the sack and it will shrink. When making a cucumber pickle, you soak the cucumber in a concentrated salt solution. Water leaves the cucumber, since the salt solution is more concentrated, and the cucumber shrinks to form a pickle. Often prunes are made more palatable by soaking them in water. Water enters the prune and it swells, becoming juicier (see Figure 13–5).

Osmotic Pressure

If a sugar solution is placed in a thistle tube separated from pure water by a semipermeable membrane, water will pass into the tube and increase the volume of the sugar water. The sugar water will rise in the thistle tube to a given level and then stop. The difference in the levels of the sugar solution and the pure water solution is a measure of the tendency of the water to pass through the membrane. The column of sugar water exerts a pressure on the membrane; this pressure is called the **osmotic pressure** (see Figure 13–6). If an additional external pressure is brought to bear on

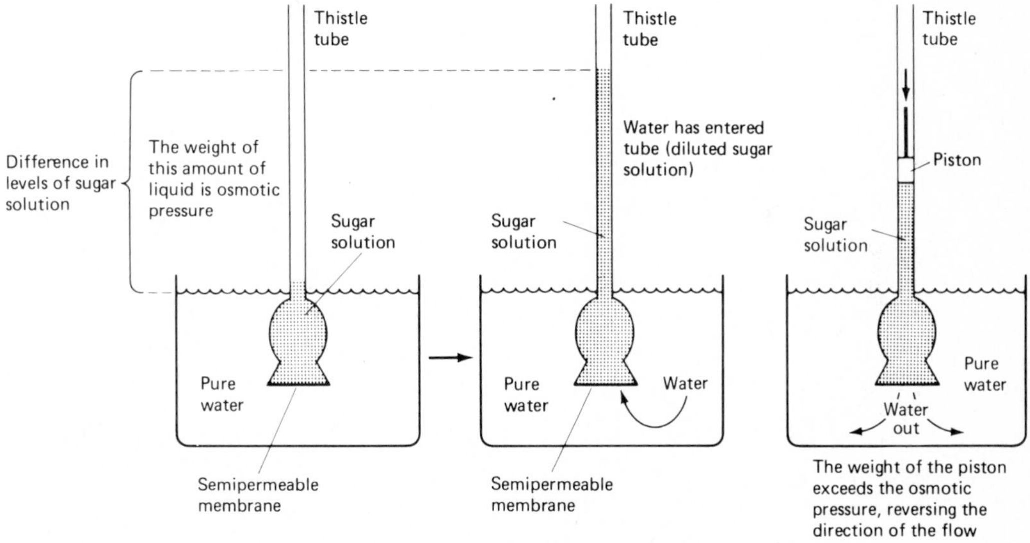

Thistle tube
Difference in levels of sugar solution
The weight of this amount of liquid is osmotic pressure
Sugar solution
Pure water
Semipermeable membrane
Thistle tube
Water has entered tube (diluted sugar solution)
Sugar solution
Pure water
Water
Semipermeable membrane
Thistle tube
Piston
Sugar solution
Pure water
Water out
The weight of the piston exceeds the osmotic pressure, reversing the direction of the flow

Figure 13–6 Measurement of osmotic pressure.

the sugar-solution side of the membrane, water can be made to flow in the **opposite** direction, that is, from the sugar solution into the pure water. It is possible to reverse the net direction of solvent flow in osmosis if we exert a pressure greater than the osmotic pressure on the more concentrated side of the membrane. One very important use of this phenomenon is in the desalination of water.

Water and some dissolved solutes flow in and out of the bloodstream by osmosis because of differences between the blood pressure and the osmotic pressure on the outside of the blood vessels. As blood moves from the heart and is pumped through the arteries and capillaries, it may have a greater pressure than the outside osmotic pressure, so water and dissolved nutrients diffuse into the body cells. Blood returning to the heart through the veins has a lower pressure than the osmotic pressure so water and waste molecules diffuse back into the bloodstream. This process is one mechanism by which the cells are provided with nutrients and cleansed of wastes.

The red blood cells themselves are also affected by differences in osmotic pressure. If the cells are placed in a 0.9% solution of sodium chloride (an **isotonic** solution), the osmotic pressure on the inside and outside of the cells is the same, and no osmosis occurs. If the cells are placed in pure water or in a water solution containing less than 0.9% sodium chloride (a **hypotonic** solution), water diffuses into the cells, and they will swell and burst. If the cells are placed in a water solution containing more than 0.9% sodium chloride (a **hypertonic** solution), water will diffuse out of the cells, and they will shrink. Medical personnel must be very careful to select the correct solutions for intravenous injections into patients.

Dialysis

Osmosis involves the movement of *solvent* molecules through a membrane. **Dialysis** involves the diffusion of *solute*, the dissolved molecules and ions, through a membrane. The membrane has holes large enough to pass ions and small molecules, but not the larger colloidal clusters. Ions and small molecules move from the more concentrated side of the membrane to the more dilute side, which tends to balance the concentrations.

One important use of dialysis is in purifying the blood of patients with kidney failure. In this process, known as *hemodialysis*, a patient's blood is circulated through dialysis tubes of an artificial kidney machine. The harmful substances that build up in the bloodstream diffuse out into the surrounding solution, but the albumin, other proteins, blood cells, and other colloidal particles remain and are returned to the patient.

UNIT 13 Problems and Exercises

13–1. How does a solution differ from a compound?

13–2. What is the main difference between a solution and a colloid?

13–3. Which of the following would exhibit a positive Tyndall effect?

(a) Air
(b) Milk in water
(c) Tea
(d) Salt water
(e) Ocean water
(f) Opals
(g) Vinegar water
(h) Fog
(i) Smoke
(j) Clouds

13–4. Is hydration necessary for dissolving ionic compounds? Why?

13–5. Draw a picture of the hydration of silver and oxide ions by water. Place dotted lines between the ions and the portion of the water molecule to which they are attracted.

13–6. Explain how anhydrous cupric sulfate could be used as a measure of the amount of moisture present.

13–7. Is the bonding between the water molecules and ions in hydration as strong as a covalent bond? What evidence can you provide to justify your conclusion?

13–8. How would you remove trace amounts of moisture from gasoline?

13–9. What will happen to the solubility of a gas in a liquid under the following conditions?
(a) The temperature of the liquid is raised.
(b) The pressure of the gas over the liquid is lowered.
(c) The amount of liquid is increased.
(d) The solution is cooled.

13–10. What will happen to the solubility of sodium chloride in water under the following conditions?
(a) The external pressure is increased.
(b) The temperature is increased.
(c) More salt is dumped into the solution.
(d) Benzene is used as solvent instead of water.

13–11. Which of the following solutions are saturated?
(a) 20 g of NaCl in 100 mℓ of H_2O
(b) 1.42 g of $CaBr_2$ in 1 mℓ of hot water
(c) 1.22 g of $AgNO_3$ in 1 mℓ of cold water

13–12. How many grams of silver chloride will dissolve in 500 mℓ of cold water? in 500 mℓ of hot water?

13–13. How many grams of calcium chloride must you add to make a saturated solution using 25 mℓ of boiling water? If this 25 mℓ were then cooled to 25°C, how many grams of $CaCl_2$ would remain in solution? How many would precipitate?

13–14. Sugar is added to 200 mℓ of water until a saturated solution exists. Excess sugar is located on the bottom of the flask. What will happen to the amount of undissolved sugar if the following occur?
(a) The solution is stirred.
(b) The solution is cooled.
(c) The solution is heated.
(d) More water is added.

13–15. If you wished to rapidly dissolve a lump of sugar in a given amount of water, which of the following processes would help? Explain your reasoning.
(a) Put water in the refrigerator.
(b) Stir the water solution.
(c) Warm the water solution.
(d) Break up the lump of sugar.
(e) Increase the atmospheric pressure.

13–16. Why does some of the dissolved carbon dioxide bubble out of a solution of soda when the lid is removed?

13–17. Why does freshwater evaporate faster than seawater under the same conditions?

13-18. Predict if the pairs of compounds given below will be soluble in each other. Explain your predictions.

(a) $CH_3{-}CH_3$ and $CH_3{-}CH_2{-}CH_2{-}CH_3$
(b) H_2O and $CH_3CH_2CH_3$
(c) H_2O and NaCl
(d) $CH_3{-}O{-}CH_3$ and H_2O
(e) $CH_3{-}NH_2$ and H_2O

13-19. Two bottles have lost their labels. One is pure water; the other is concentrated sodium cyanide (toxic) in water. All you have available is a thermometer, test tubes, a refrigerator, and a heat source. How would you decide which is which?

13-20. What type of membrane (*filter paper, dialysis, or osmotic*) would you need to effect a separation of the following?

(a) Cream from water
(b) Sugar from water
(c) Crushed glass and water
(d) Ink from its solvent
(e) Ocean water from its components

13-21. Two aqueous solutions, a 2% sucrose solution and an 8% sucrose solution, are separated by a semipermeable membrane.

(a) In which direction will osmosis occur?
(b) Which solution will increase in volume?

13-22. In the following examples, what is the net direction of solvent movement?

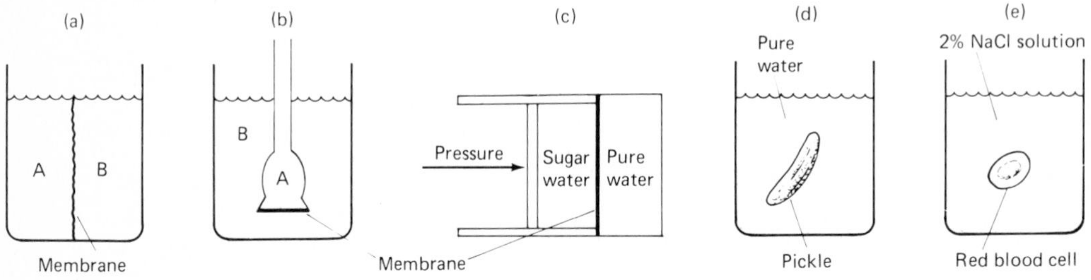

(a)
A
B
Membrane
(b)
B
A
(c)
Pressure
Sugar water
Pure water
Membrane
(d)
Pure water
Pickle
(e)
2% NaCl solution
Red blood cell

Solution A consists of 10 g of NaCl in 100 mℓ of H_2O. Solution B consists of 100 g of NaCl in 100 mℓ of H_2O. There is distilled water in A and sugar water in B.

UNIT 14 UNITS OF CONCENTRATION

When working with solutions, it is often important to know the concentrations of the dissolved solute. There are several methods of expressing the concentration; each has its own advantages and disadvantages. In this unit we shall introduce some of the more important methods.

Percentage Concentrations

In hospitals and industrial laboratories where many repetitive measurements are made, it is often convenient to measure concentrations of solutes as a percent of the total solution. Then, if a certain number of grams or milliliters of solute is required, all we have to do is pour out a given volume of the solution or obtain a given weight of solution. There are three types of percentage concentrations: weight to volume, weight to weight, and volume to volume.

A **weight to volume** (w/v) concentration unit is defined as the number of grams of solute dissolved in 100 mℓ of the solution. A 10% (w/v) solution contains 10 g of solute in every 100-mℓ solution. If you needed 5 g of solute, you would need 50 mℓ of the 10% (w/v) solution. If 25 g were needed, then 250 mℓ of the solution would be required.

A **weight to weight** (w/w) concentration unit is defined as the number of grams of solute present in 100 **grams** of solution. To obtain a specific amount of solute, we need to weigh a given amount of solution, not measure its volume. Thus in a 10% (w/w) solution 50 g of solution contains 5 g of solute, and 250 g of solution contains 25 g of solute.

A third type of unit for percentage concentration is **volume to volume** (v/v). This unit is often used when the solution consists of two liquids. A volume-to-volume percent solution (v/v) is defined as the number of milliliters of solute present in a 100-mℓ solution. A 10% (v/v) solution thus contains 10 mℓ of solute for every 100 mℓ of solution.

$$\%\,(\text{w/v}) = \frac{\text{no. g of solute}}{100\text{ m}\ell\text{ of solution}}$$

$$\%\,(\text{w/w}) = \frac{\text{no. g of solute}}{100\text{ g of solution}}$$

$$\%\,(\text{v/v}) = \frac{\text{no. m}\ell\text{ of solute}}{100\text{ m}\ell\text{ of solution}}$$

Notice that the amount of *solution* (not solvent) is used in each type of solution. To make up any of these concentrations, we normally place the proper amount of solute (grams or milliliters) in a flask and then add solvent until the total volume is 100 mℓ or the total weight is 100 g. You do not begin with 100 g or 100 mℓ of solvent and then add solute. For example, to prepare 500 mℓ of a 3% (w/v) sodium chloride solution, we weigh 15 g of NaCl (3 g/100 mℓ × 500 mℓ) into a flask, and then add water until the total volume is 500 mℓ.

Parts per Million, Parts per Billion

When analyzing for the presence of trace amounts of impurities in solutions, concentration units of weight and volume percent are often too large. The units of **parts per million** (ppm) or **parts per billion** (ppb) are more convenient. The term *parts* may refer to any unit of measure (grams, milliliters, liters, grains, and so on). A 1-ppm solution would contain one part of solute for every 1 million parts of solution. A 1-ppm solution of sodium chloride in water contains 1 g of NaCl for every 1 million g of water. Hard water usually contains about 100 ppm dissolved calcium ion. Thus about 100 g of calcium ion is present in every 1 million g of water, and there are about 0.01 g of calcium ion in every 100 g of water. In Los Angeles, California, health alerts are called when the concentrations of ozone, nitrogen oxides, sulfur oxides, and other smog ingredients in the atmosphere maintain an hourly average of 0.2 ppm.

When very small concentrations are measured, the units of parts per billion may be used. A 1-ppb solution contains one part solute for every 1 billion (10^9) parts of solvent.

Molarity

In a chemical reaction the molecules of one substance interact in some fashion with molecules of another substance to form new compounds. The exact ratios of molecules or formula units of each compound needed in the reaction are given in balanced chemical equations. The balanced chemical equation for the reaction of sodium carbonate with hydrochloric acid is written as follows:

$$\begin{aligned} Na_2CO_3 + 2HCl &\longrightarrow CO_2 + 2NaCl + H_2O \\ \text{1 unit} + \text{2 molecules} &\longrightarrow \text{1 molecule} + \text{2 units} + \text{1 molecule} \\ \text{1 mol} + \text{2 mol} &\longrightarrow \text{1 mol} + \text{2 mol} + \text{1 mol} \\ 106\text{ g} + 73\text{ g} &\longrightarrow 44\text{ g} + 117\text{ g} + 18\text{ g} \end{aligned}$$

In this reaction, two molecules of hydrochloric acid (HCl) are required for every formula unit of sodium carbonate that reacts. On the mole scale, 2 mol ($2 \times 6 \times 10^{23}$ molecules) of HCl reacts with 1 mol (6×10^{23} formula units) of Na_2CO_3. Since chemists are interested in the combining ratios of *moles* of each compound, a unit of concentration based on moles (not grams) is most convenient.

The **molarity** of a solution is defined as the number of moles of solute dissolved in 1 ℓ of solution.

$$\text{molarity } (M) = \frac{\text{no. of moles of solute}}{1 \ell \text{ of solution}}$$

A 1 *M* solution of sodium chloride contains 1 mol (58.5 g) of sodium chloride dissolved in a total of 1ℓ of solution, or 0.1 mol of NaCl in 100 mℓ of solution, or 0.5 mol of NaCl in 500 mℓ of solution, and so on. *Molarity simply expresses the ratio of moles of solute to volume of solution.* To determine the actual amount of solute present in any given solution, we have to know both the molarity and the volume of that solution. Equation (14–1) is a convenient tool when working with molarity units. It is simply a conversion problem with the separate units given names.

$$\begin{aligned} \text{Molarity} \times \text{volume (liters)} &= \text{no. of moles present} \\ M \times V &= \text{no. of moles} \qquad (14\text{–}1) \\ \frac{\text{moles}}{\text{liter}} \times \text{liters} &= \text{moles} \end{aligned}$$

In this equation there are three variables: molarity, volume, and number of moles. To calculate one of the variables, you need to know the other two. The following problems show typical examples of working with molarity.

Problem 1: Portions of sodium bromide weighing 10.3 g are dissolved in water to make up the following solutions: (a) 100 mℓ, (b) 250 mℓ, (c) 500 mℓ, and (d) 1000 mℓ. What are the molarities of each solution?

To calculate the molarity of any solution, we need to know the number of moles of solute present, not the number of grams. So first we must convert grams of solute into moles of solute (see Figure 14–1).

$$\frac{10.3 \text{ g NaBr}}{1} \times \frac{1 \text{ mol}}{103 \text{ g}} = 0.1 \text{ mol NaBr present in each solution}$$

(a) Molarity × volume = number of moles

$$\text{Molarity} = \frac{\text{no. of mol present}}{\text{no. of liters solution present}} = \frac{0.1 \text{ mol}}{0.1\ \ell} = 1\ M$$

(b) $$M = \frac{0.1 \text{ mol of NaBr}}{0.25\ \ell} = 0.4\ M$$

(c) $$M = \frac{0.1 \text{ mol of NaBr}}{0.5\ \ell} = 0.2\ M$$

(d) $$M = \frac{0.1 \text{ mol of NaBr}}{1.0\ \ell} = 0.1\ M$$

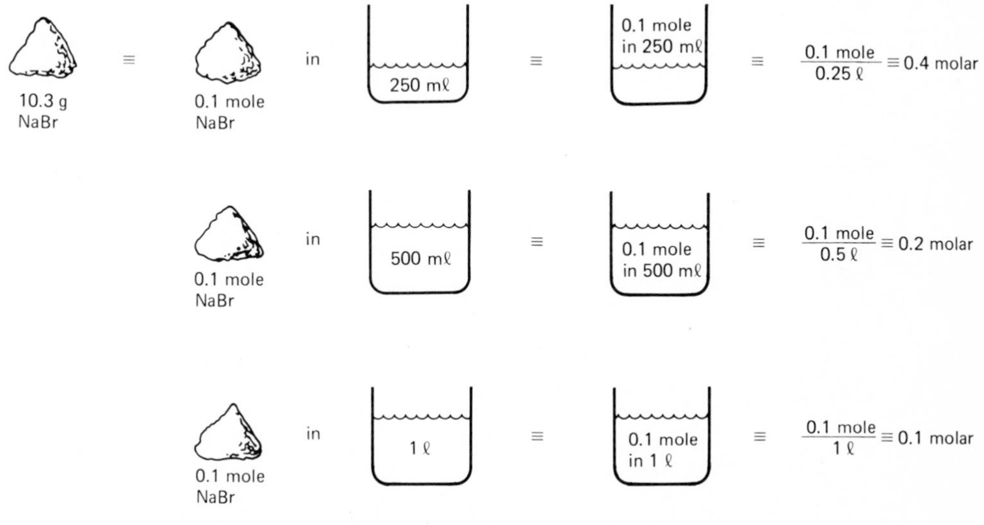
10.3 g NaBr
≡
0.1 mole NaBr
in
250 mℓ
≡
0.1 mole in 250 mℓ
≡
0.1 mole / 0.25 ℓ ≡ 0.4 molar
0.1 mole NaBr
in
500 mℓ
≡
0.1 mole in 500 mℓ
≡
0.1 mole / 0.5 ℓ ≡ 0.2 molar
0.1 mole NaBr
in
1 ℓ
≡
0.1 mole in 1 ℓ
≡
0.1 mole / 1 ℓ ≡ 0.1 molar

Figure 14–1 Molarities of solutions.

Problem (1) illustrates a point that we wish to emphasize. To calculate the actual amount of solute present, we have to know both the molarity and the volume. In all four cases the same number of grams of NaBr is present, but the molarities of each solution are different because the solution volumes are different.

Problem 2: How many grams of solute are present in each of the following solutions?

(a) 100 mℓ of a 0.6 M potassium chloride solution. Here we know the volume and the molarity. Equation (14–1) can be used to determine the number of moles present. Once this number is known, the number of grams of solute present can be calculated.

$$M \times V = \text{no. of mol present}$$

$$0.6\,M \times 0.1\,\ell = 0.06 \text{ mol of KCl}$$

$$0.06 \text{ mol KCl} \times \frac{74.5 \text{ g}}{1 \text{ mol}} = 4.47 \text{ g KCl}$$

(b) 200 mℓ of a 0.20 M KCl solution.

$$M \times V = \text{no. of mol present}$$

$$0.20 \times 0.2\,\ell = 0.040 \text{ mol of KCl}$$

$$0.040 \text{ mol KCl} \times \frac{74.5 \text{ g}}{1 \text{ mol}} = 2.98 \text{ g}$$

In this problem note that solution (b) with the larger volume does not contain the largest amount of solute.

Problem 3: What volume of solution is needed if 117 g of sodium chloride is used to make up (a) a 1 M solution, and (b) a 0.4 M solution?

(a) In a 1 M solution there is 1 mol of solute per liter. In this problem we know the number of grams present and the molarity needed. We need to know what volume of solution to make up. We need to know moles present, so first we must convert grams to moles.

$$117 \text{ g NaCl} \times \frac{1 \text{ mol}}{58.5 \text{ g}} = 2 \text{ mol of NaCl to be used}$$

Then we can calculate the volume required.

$$M \times V = \text{no. of mol}$$

$$1\,M \times \text{V} = 2 \text{ mol}$$

$$V = \frac{2 \text{ mol}}{1 \text{ mol}/\ell} = 2\,\ell \text{ of solution needed}$$

(b) Number of moles present is still 2 mol. Molarity necessary is 0.4 M.

$$V = \frac{\text{no. of mol}}{\text{molarity}} = \frac{2 \text{ mol}}{0.4 \text{ mol}/\ell} = 5\,\ell \text{ of solution}$$

Now let us examine problems (1), (2), and (3) once more to reemphasize the important points. All three problems deal with three variables: molarity, volume, and number of moles. In problem (1), moles and volume were given, so molarity could be calculated. In problem (2), molarity and volume were given, so the number of moles could be calculated. In problem (3), the molarity and number of moles were given, so volume could be calculated. In all three problems, two of the three variables were given; the third then could be calculated.

Problem 4: Four liters of pure water are added to 2 ℓ of a 0.4 M solution of sodium chloride. What is the new molarity of the solution?

$$\text{Molarity} = \frac{\text{total mol of solute present}}{\text{total vol. of solution present}}$$

The new total volume will be 6 ℓ. Since sodium chloride is the solute, we get it from only one of the initial solutions, the 2 ℓ of 0.4 M NaCl. From this solution we know that 0.4 mol/ℓ $\times$ 2 ℓ = 0.8 mol of NaCl

So

$$\text{Molarity} = \frac{0.8 \text{ mol of NaCl}}{6\ \ell \text{ of solution}} = \frac{0.133 \text{ mol}}{1\ \ell} = 0.133\ M$$

Molality

In some instances a different concentration term including moles is used. In unit 13 the term molality was used when calculating the amount of boiling-point elevation or the amount of melting-point lowering in a solution. Molality gives the number of moles of solute dissolved in every **1000 g of solvent**.

$$\text{Molality } (m) = \frac{\text{no. of mol of solute}}{1000 \text{ g of solvent}}$$

This is one of the few concentration expressions in which the weight of the *solvent* is used instead of the weight of the solution.

Normality

Another unit of concentration often used is called **normality**. The normality of a solution is defined as the number of **equivalent weights** of a solute present in 1 ℓ of solution:

$$\text{Normality } (N) = \frac{\text{no. of equivalent weights of solute}}{\text{liter of solution}}$$

Equivalent weights can be defined in several ways; one convenient definition deals with acids and bases. *The equivalent weight of an acid is that weight of acid which supplies* 1 *mol of hydrogen ions, and the equivalent weight of a base is that weight of base which will react with* 1 *mol of hydrogen ions or supply* 1 *mol of hydroxide ions.* In hydrochloric acid, HCl, there is 1 mol of hydrogen atoms per mole of HCl, so 1 gram-equivalent weight of HCl is simply 1 gram-molecular weight (see Table 14–1). Sulfuric acid, H_2SO_4, contains 2 mol of H per mole of acid, so the gram-equivalent weight of sulfuric acid is **one-half** its gram-molecular weight. In the base sodium hydroxide, NaOH, there is 1 mol of hydroxide ions per mole of sodium hydroxide. The gram-equivalent weight for sodium hydroxide is equal to its gram-molecular weight. The normality of a solution that contains 1 mol (1 equivalent) of hydrochloric acid per liter of solution is then 1 N. For sulfuric acid (1 mol = 2 equivalents), the normality of a solution containing 1 mol/ℓ is 2 N (2 eq./1 ℓ).

Table 14–1 Equivalent weights*

Substance	No. of mol of H^+ or OH per mol of compound	GMW	GEW	Normality of 1 *M* solution
HCl, hydrochloric acid	1	36.5	36.5	1 *N*
H_2SO_4, sulfuric acid	2	98.0	49.0	2 *N*
H_3PO_4, phosphoric acid	3	98.0	32.7	3 *N*
NaOH, sodium hydroxide	1	40	40	1 *N*
$Ca(OH)_2$, calcium hydroxide	2	74	37	2 *N*
$Al(OH)_3$, aluminum hydroxide	3	78	26	3 *N*

$$\text{* Gram-equivalent weight} = \frac{\text{gram-molecular weight}}{\text{no. of mol of H}^+ \text{ or OH per mol}}$$

Finally, when measuring very small amounts of substances in solution, the unit **milliequivalents** can be used. One milliequivalent (mEq) is equivalent to 0.001 equivalents. Sometimes the units milliequivalents per liter (mEq/ℓ) are used in hospitals to express the concentrations of certain substances present in our bodies. The concentration of calcium ion in the bloodstream of a healthy adult ranges from 4.5 to 5.6 mEq/ℓ. For magnesium ion, the range is 1.6 to 2.2 mEq/ℓ.

UNIT 14 Problems and Exercises

14–1. Calculate the number of moles of solute present in each of the following solutions.

(a) 100 mℓ of 0.25 *M* H_2SO_4
(b) 100 g of 10% (w/w) NaCl
(c) 100 mℓ of 10% (w/v) NaCN
(d) 100 mℓ of 10% (v/v) ethyl alcohol (C_2H_6O); density of pure alcohol = 0.8 g/mℓ
(e) 100 g of a 200-ppm solution of NO_2

14–2. Which of the following pairs of solutions contains the greater amount of solute?

(a) 100 mℓ of 1% (w/v) NaCl or 100 mℓ of 0.1 *M* NaCl
(b) 1 ℓ of 10% (w/v) $NaCO_3$ or 500 mℓ of 1 *M* Na_2CO_3

14–3. Complete the following chart.

Solute	GMW	Weight of solute present	Moles of solute	Milliliters of solution	Liters of solution	Molarity of solution
NaOH	____	400 g	____	____	____	1.0
NaCl	____	11.7	____	____	0.02	____
$NaHCO_3$	____	____	0.25	____	2.5	____
$Ca(OH)_2$	____	____	____	250	____	0.1
$C_6H_{12}O_6$	____	____	6.0	____	____	12

14–4. Calculate the molarities of the following solutions.
(a) 0.98 g of sulfuric acid in 1 ℓ of solution
(b) 5.85 g of sodium chloride in 100 mℓ of solution
(c) 5.8×10^{-3} g of potassium fluoride in 10 mℓ of solution
(d) 3.24 g of calcium bicarbonate in 250 mℓ of solution

14–5. The average concentration of calcium ion in human blood serum is about 20 mg/100 mℓ. What is the molarity of calcium ion in this solution?

14–6. How you would prepare the following solutions? In each case, give the number of grams of solute and the amount of solvent or solution necessary. All solvents are water.
(a) 1 ℓ of 0.5 *M* sodium hydroxide
(b) 1 ℓ of 0.5 *N* sodium hydroxide
(c) 100 mℓ of 0.03 *M* sodium chloride
(d) 6000 g of 5% by weight silver nitrate
(e) 2000 g of a 10-ppm ferric chloride solution
(f) 200 mℓ of a 0.2 *M* silver chloride solution
(g) 1 ℓ of a 4.0 *N* sulfuric acid solution

14–7. Determine the number of grams of solute in each of the following solutions.
(a) 1ℓ of a 5% NaCl (w/v) solution
(b) 2000 g of a 6% NaI (w/w) solution
(c) 10,000 g of a 3-ppm HCl solution
(d) 250 mℓ of a 0.02 *M* H_2SO_4 solution
(e) 5 ℓ of a 3×10^{-2} *M* sodium sulfate solution
(f) 200 mℓ of a 2 *N* sodium hydroxide solution
(g) 500 mℓ of a 1 *N* sulfuric acid solution
(h) 2000 g of a 0.015 molal solution of acetic acid ($C_2H_4O_2$)

14–8. What volumes of solution are necessary if the following weights of solute are to be used to make up 1 *M* solutions?
(a) 9.8 g of sulfuric acid
(b) 5.85×10^{-3} g of sodium chloride
(c) 0.180 g of glucose ($C_2H_{12}O_6$)

14–9. How many grams of bromide ion are present in 200 mℓ of a 0.1 *M* calcium bromide solution?

14–10. How many grams of sodium carbonate would you have to add to 500 mℓ of water (a) to make the molarity of sodium ion equal to 0.3 *M*? (b) to make the molarity of carbonate ion equal to 0.3 *M*?

14–11. Three solutions bear the labels of 10% (w/v) NaCl, 10% (w/v) NaBr, and 10% (w/v) NaI. Translate these labels into molarity.

14–12. A concentration of 0.065 *M* alcohol (C_2H_5OH) in the blood causes obvious signs of intoxication in most humans. A concentration of 0.17 *M* is usually lethal. If an average adult has a total blood volume of 7.0 ℓ, what is the weight in grams of C_2H_5OH that corresponds to the difference between intoxicating and lethal doses?

14–13. Carbon monoxide reaches a danger level when its concentration in a room is about 0.005% (v/v).
(a) What is this concentration in ppm? in ppb?

(b) If a room capacity is 3.24×10^5 ℓ, how many milliliters (at STP) of carbon monoxide are present?

14–14. If 200 mℓ of 0.1 *M* sodium chloride is diluted to 1 ℓ with distilled water, what is the molarity of the final solution?

14–15. What will be the molarity of the resultant solution when the following solutions are mixed?
(a) 100 mℓ of 6 *M* HCl + 100 mℓ of H_2O
(b) 20 mℓ of 0.2 *M* HCl + 60 mℓ of H_2O

14–16. How many milliliters of 12 *M* HCl would you have to add to 100 mℓ of H_2O to make the molarity equal to the following?
(a) 1 *M*
(b) 2×10^{-4} *M*
(c) 0.01 *M*

14–17. Two solutions were mixed. Calculate the molarity of the resultant solution.
Solution 1: 2 ℓ of 0.1 *M* NaCl
Solution 2: 1 ℓ of 0.5 *M* NaCl

14–18. A solution is made by dissolving 8.7 g of lithium bromide, 10.3 g of sodium bromide, and 1.5 g of sodium iodide in 1 ℓ of solution. Calculate the molarities of each ion present in solution.

14–19. How many gram equivalents of solute are present in the following?
(a) 1 ℓ of 2 *N* solution
(b) 1 ℓ of 0.5 *N* solution
(c) 0.5 ℓ of 0.2 *N* solution

CHAPTER 7

introduction to chemical reactivity

In any chemical reaction, the starting materials (the **reactants**) interact in some fashion and are changed into other compounds (the **products**). Thousands of chemical reactions have been discovered, so it is next to impossible to attempt to describe each one. It is possible, however, to categorize and classify the major reactions into general types. In unit 15 we introduce some of the major types of chemical reactions, and then discuss the creation and use of balanced chemical reaction equations.

Any further study of chemical reactions beyond simply identifying the reactants and products involves research into the hows and whys of the reactions. In unit 16 we introduce some of the theoretical principles behind the energetics of reactions. We shall learn how **entropy**, **enthalpy**, and **Gibbs free energy** influence the spontaneity of chemical reactions. In unit 17 we discuss reaction rate theory and learn why one reaction proceeds faster than another. We shall also discuss some of the factors that influence the rates of chemical reactions.

UNIT 15 CHEMICAL REACTIONS

In any chemical reaction the first thing we have to determine is what changes occurred. We must know the identities and amounts of the reactants used and then determine the identities and amounts of products formed. In writing a chemical equation, the reactants are usually written on the left and the products on the right. An equals sign or an arrow separates the reactants and products; the arrow shows the direction in which the reaction occurs. Sometimes two arrows ($\leftrightarrows$) pointing in opposite directions are used. This indicates that the reaction is reversible; that is, it may occur in either direction.

When it is important to know the specific conditions under which the reaction occurs, the conditions are placed above or below the arrow. The symbol Δ is used to indicate that thermal energy must be added; $h\nu$ indicates that solar energy must be added. Actual reaction temperatures or wavelengths of light used may also be given. Solvents and catalysts required are also placed above or below the arrows. In the following example, palladium is the catalyst, the reaction temperature is 60°C, and the solvent is hexane.

$$\underset{\text{butene}}{C_4H_8} + \underset{\text{hydrogen}}{H_2} \xrightarrow[\text{Hexane}]{\text{Pd,60°C}} \underset{\text{butane}}{C_4H_{10}}$$

Types of Chemical Reactions

It is impossible in this text to describe all the chemical reactions that are known to occur. We can, however, define the general types of reactions. If you are familiar with these, new reactions can often be better understood and categorized. Table 15–1 lists some of the common types of reactions that occur. Table 15–2 lists some specific reactions.

Additions, Decompositions, and Substitutions

In an **addition** reaction, two or more chemicals are joined together in some fashion to form a single larger chemical. **Decompositions** are just the opposite: a single large chemical breaks apart into two or more smaller chemicals. In many decomposition reactions, small stable molecules like water, carbon dioxide, carbon monoxide, or sulfur dioxide are formed. Finally, in a **substitution** reaction, portions of chemicals are exchanged with different groups of atoms.

Oxidation and Reduction

Examples of oxidation and reduction reactions were introduced when we studied the loss and gain of electrons in unit 6. A chemical is said to be **oxidized** when it (1) loses electrons, (2) loses hydrogen atoms, or (3) gains oxygen atoms. **Reduction** is just the opposite: (1) gain of electrons, (2) gain of hydrogen atoms, or (3) loss of oxygen atoms. Oxidation and reduction always occur together: for one species to gain electrons (be reduced), another species must be present to lose electrons (be oxidized). The chemical species that is oxidized is the donor of electrons, and is often

Table 15–1 Types of chemical reactions

Class of reaction	Type	Examples
Additions	$A + B \rightarrow C$ (combinations of molecules)	$Fe + S \rightarrow FeS$ $C_2H_4 + HBr \rightarrow C_2H_5Br$
Decompositions	$A \rightarrow B + C$ (breakup of a molecule)	$CaCO_3 \rightarrow CaO + CO_2$ $Ca(OH)_2 \rightarrow CaO + H_2O$
Substitutions (displacements)		
Single displacement	$AB + C \rightarrow AC + B$ (replacement of one portion of a compound)	$CuSO_4 + Zn \rightarrow ZnSO_4 + Cu$ $CH_3Br + OH^- \rightarrow CH_3OH + Br^-$
Double displacement	$AB + CD \rightarrow AD + BC$	$AgNO_3 + NaCl \rightarrow AgCl + NaNO_3$

called the **reducing agent**; the species that is reduced is the acceptor of electrons, and is called the **oxidizing agent**. In later portions of this text we shall see many examples of oxidation and reduction reactions.

Acid–Base Reactions

The most general definitions of acid–base reactions deal with the transfer of hydrogen ions (H^+) and hydroxide ions (OH^-). An **acid** is a chemical that donates hydrogen ions during a chemical reaction; a **base** is a chemical that either accepts hydrogen ions or donates hydroxide ions during the reaction. Like oxidation-reduction reactions, acids and bases always occur together in the same reaction. They are quite important to body chemistry, and we shall spend more time studying them in later units.

Table 15–2 Specific names for some reactions

Name	Description	Examples
Oxidation-reduction:		
A substance is **oxidized** if it:	Loses electrons	$Fe \rightarrow Fe^{2+} + 2e^{1-}$
	Loses hydrogen	$CH_3{-}OH \rightarrow CH_2O$
	Gains oxygen	$H{-}\overset{\overset{\displaystyle O}{\|}}{C}{-}H \rightarrow H{-}\overset{\overset{\displaystyle O}{\|}}{C}{-}OH$
A substance is **reduced** if it:	Gains electrons	$Cl + 1e^{1-} \rightarrow Cl^{1-}$
	Gains hydrogen	$C_2H_4 + H_2 \rightarrow C_2H_6$
	Loses oxygen	$H_2CO_2 \rightarrow H_3COH$
Acid–base:		
A substance is an **acid** if it:	Can lose H^+ ions	$H_2SO_4 \rightarrow HSO_4^- + H^+$
A substance is a **base** if it:	Can gain H^+ ions	$CN^- + H^+ \rightarrow HCN$
	Can lose OH^- ions	$NaOH \rightarrow Na^+ + OH^-$

Balancing Chemical Reactions

Written correctly, a complete chemical equation tells us the formulas of all reactants and products. It also tells us the amounts of each element and compound consumed or formed during the reaction. In writing a chemical equation, we must accurately represent the experimental facts. The correct chemical formulas for all reactants and products must be shown. Then the chemical equation must satisfy the laws of conservation. This means that the same number of atoms of each element must be present on both sides of the equation. Problem (1) illustrates the process of producing balanced chemical reactions.

Problem 1: When sodium bicarbonate reacts with sulfuric acid, the products formed are carbon dioxide, water, and sodium sulfate. Write a balanced chemical equation for this reaction.

STEP 1: Write down the correct formulas for all reactants and products.

$$NaHCO_3 + H_2SO_4 \longrightarrow Na_2SO_4 + CO_2 + H_2O$$

STEP 2: Balance the equation so that each side contains the same number of atoms of each element. To do this, you balance each element separately by adjusting the coefficients in front of each chemical formula. *You must not change the composition of any formula.* The process of balancing is one of trial and error, but balancing the elements present in the fewest compounds first is often the easiest method.

(a) Balance Na: There are two Na on the right, and one on the left. To increase the number of sodium atoms on the left, the number of $NaHCO_3$ units must be increased.

$$2(NaHCO_3) + H_2SO_4 \longrightarrow CO_2 + H_2O + Na_2SO_4$$

(b) Balance C: Two on left, one on right; so increase the number of CO_2 units by one.

$$2(NaHCO_3) + H_2SO_4 \longrightarrow 2CO_2 + H_2O + Na_2SO_4$$

(c) Balance S: One S one each side; already balanced.

(d) Balance O: Ten on the left, nine on the right. One more oxygen is needed on the right. If the number of CO_2 or Na_2SO_4 units were increased, then C and Na and S would no longer be balanced. But if the amount of H_2O is increased, only O and H are changed.

$$2(NaHCO_3) + H_2SO_4 \longrightarrow 2CO_2 + 2H_2O + Na_2SO_4$$

(e) Now all but H have been balanced. If this has been done correctly, then hydrogen will automatically be balanced.

Thus balancing the last element is a check on the earlier steps. In this example there are four H on each side; the balancing was correct.

Using Balanced Equations

A balanced chemical equation gives you the exact ratio of reactants used to products formed. It is like a recipe from a cookbook in that it gives the amounts of reactants that must be used to yield a certain amount of product. Let's look at the balanced equation from problem (1):

	$2NaHCO_3 + H_2SO_4$	$\rightarrow$	$2CO_2 + 2H_2O + Na_2SO_4$
Moles:	2 mol + 1 mol	$\rightarrow$	2 mol + 2 mol + 1 mol
Grams:	168 g + 98 g	$\rightarrow$	88 g + 36 g + 142 g

This equation tells us how many moles of each compound are used and formed. Thus if we combine exactly 2 mol of $NaHCO_3$ with 1 mol of H_2SO_4, we will make 2 mol of CO_2, 2 mol of H_2O, and 1 mol of Na_2SO_4. If we double the moles of each reactant used, we double the moles of each product formed.

Notice that moles may not be conserved in a balanced reaction, but that grams always are. This is because moles of different compounds usually do not weigh the same. Problem (2) illustrates how a balanced chemical reaction is used.

Problem 2: In the reaction of sodium bicarbonate with sulfuric acid in problem (1), 16.8 g of $NaHCO_3$ were used.

(a) How many grams of H_2SO_4 must be used? Chemical equations use moles, not grams, so the clearest method of solving these types of problems is to use conversion processes working through moles.

$$\text{GMW}(NaHCO_3) = 84\text{ g},\qquad \text{GMW}(H_2SO_4) = 98\text{ g}$$

1. Grams of $NaHCO_3 \rightarrow$ moles of $NaHCO_3$

$$16.8\text{ g of }NaHCO_3 \times \frac{1\text{ mol of }NaHCO_3}{84\text{ g of }NaHCO_3} = 0.2\text{ mol of }NaHCO_3\text{ used}$$

2. Moles of $NaHCO_3 \rightarrow$ moles of H_2SO_4

The conversion factor here is obtained from the **balanced chemical equation**, which states that 2 mol of $NaHCO_3$ react with 1 mol of H_2SO_4.

$$0.2\text{ mol of }NaHCO_3 \times \frac{1\text{ mol of }H_2SO_4}{2\text{ mol of }NaHCO_3} = 0.1\text{ mol of }H_2SO_4\text{ needed}$$

3. Moles of $H_2SO_4 \rightarrow$ grams of H_2SO_4

$$0.1\text{ mol of }H_2SO_4 \times \frac{98\text{ g of }H_2SO_4}{1\text{ mol of }H_2SO_4} = 9.8\text{ g of }H_2SO_4\text{ needed}$$

(b) How many grams of CO_2, H_2O, and Na_2SO_4 are formed? These problems are worked just as in part (a).

$$\text{GMW}(CO_2) = 44\text{ g},\qquad \text{GMW}(H_2O) = 18\text{ g},\qquad \text{GMW}(Na_2SO_4) = 142\text{ g}$$

1. Grams of CO_2: moles of $NaHCO_3 \rightarrow$ moles of $CO_2 \rightarrow$ grams of CO_2

$$0.2\text{ mol of }NaHCO_3 \times \frac{2\text{ mol of }CO_2}{2\text{ mol of }NaHCO_3} \times \frac{44\text{ g of }CO_2}{1\text{ mol of }CO_2} = 8.8\text{ g of }CO_2$$

2. Grams of H_2O: moles of $NaHCO_3$ → moles of H_2O → grams of H_2O

$$0.2 \text{ mol of } NaHCO_3 \times \frac{2 \text{ mol of } H_2O}{2 \text{ mol of } NaHCO_3} \times \frac{18 \text{ g of } H_2O}{1 \text{ mol of } H_2O} = 3.6 \text{ g of } H_2O$$

3. Grams of Na_2SO_4: moles of $NaHCO_3$ → moles of Na_2SO_4 → grams of Na_2SO_4

$$0.2 \text{ mol of } NaHCO_3 \times \frac{1 \text{ mol of } Na_2SO_4}{2 \text{ mol of } NaHCO_3} \times \frac{142 \text{ g}}{1 \text{ mol of } Na_2SO_4} = 14.2 \text{ g of } Na_2SO_4$$

4. Total grams of reactants = Total grams of products

   16.8 + 9.8 = 26.6 g   8.8 + 3.6 + 14.2 = 26.6 g

Limiting Reactants

The balanced chemical reaction tells us the exact ratio of reactants to use, and it allows us to calculate the amounts of products that should form if the reaction proceeds to 100% completion. There are many instances, however, when one or more of the reactants will be present in excess, and one reactant will be the **limiting reactant**. To calculate the amounts of products that will form, we must first determine which reactant is limiting.

Let's use a common recipe from cooking to illustrate the limiting reactant concept. Suppose that a pancake recipe calls for the mixing of 1 egg, 2 cups of mix, and 1 cup of milk to make 6 pancakes:

$$1 \text{ egg} + 2 \text{ cups mix} + 1 \text{ cup milk} = 6 \text{ pancakes}$$

If we had 2 eggs, 3 cups of mix, and 2 cups of milk, how many pancakes could we make? By looking at the ratios given in the recipe (the balanced chemical reaction), we can see that there is not enough mix for us to use all the eggs and milk. The mix is the limiting reactant. If we wanted to use all the mix, we should use $1\frac{1}{2}$ eggs, $1\frac{1}{2}$ cups of milk, and 3 cups of mix. We would then be able to make 9 pancakes. If we had only 1 egg, 3 cups of mix, and 2 cups of milk, the eggs would be the limiting reactant, and only 6 pancakes could be made.

One point must be emphasized: the limiting reactant is *not always* the reactant present in the *smallest quantity*. We must first look at the balanced reaction and the ratios therein to determine the limiting reactant. In the first example, using the pancake recipe, the mix was the limiting reactant even though it was present in the largest numerical amount, because the recipe called for more mix than eggs or milk. Problems (3) and (4) illustrate the process for determining and using the limiting reactant in chemical reactions.

Problem 3: If 4.9 g of H_2SO_4 were mixed with 16.8 g of $NaHCO_3$, which is the limiting reagent?

$$\frac{4.9 \text{ g of } H_2SO_4}{1} \times \frac{1 \text{ mol}}{98 \text{ g}} = 0.05 \text{ mol of } H_2SO_4$$

There are 0.2 mol of $NaHCO_3$ present. The mole ratio in the balanced equation of problem (1) is

$$\frac{2 \text{ mol of } NaHCO_3}{1 \text{ mol of } H_2SO_4} = \frac{2}{1}$$

We have an experimental ratio of

$$\frac{0.2 \text{ mol of } NaHCO_3}{0.05 \text{ mol of } H_2SO_4} = \frac{4}{1}$$

so there is an excess of $NaHCO_3$ and not all of it will react. The 0.05 mol of H_2SO_4 is the limiting reactant. Only 0.1 mol of $NaHCO_3$ would react.

Problem 4: How many grams of CO_2 are formed? Use the amounts listed in problem (3). Since the limiting reagent is H_2SO_4, we use it to calculate product yields.

$$\text{Mol of } H_2SO_4 \longrightarrow \text{mol } CO_2 \longrightarrow \text{g } CO_2$$

$$\frac{0.05 \text{ mol of } H_2SO_4}{1} \times \frac{2 \text{ mol of } CO_2}{1 \text{ mol of } H_2SO_4} \times \frac{44 \text{ g of } CO_2}{1 \text{ mol of } CO_2} = 4.4 \text{ g of } CO_2 \text{ formed}$$

Balancing Oxidation-Reduction Reactions

Reactions that involve oxidation and reduction are more difficult to balance than other reactions because we must conserve electrons as well as atoms. The number of electrons given up by the substance oxidized must equal the number of electrons accepted by the substance that was reduced. To balance oxidation-reduction reactions, use the following steps:

1. Separate the reaction into two half-reactions, one showing the oxidation and the other showing the reduction portion.
2. Balance each half-reaction with respect to atoms present.
3. Balance each half-reaction with respect to *charges* present by adding the appropriate numbers of electrons (negative charge) to the side needing them.
4. Couple the two half-reactions by making the number of electrons present in one reaction equal to the number of electrons present in the other. Do this by multiplying one or both of the entire half-reactions by the appropriate numbers.
5. Add both half-reactions together and cancel to get the final balanced oxidation–reduction reaction.

Problems (5) and (6) illustrate the processes.

Problem 5: Balance the following oxidation–reduction reaction:

$$Na + Cl_2 \longrightarrow Na^{1+} + Cl^{1-}$$

STEP 1: Divide into half-reactions:

$$\text{Oxidation: loss of electrons} \quad Na \longrightarrow Na^{1+}$$
$$\text{Reduction: gain of electrons} \quad Cl_2 \longrightarrow Cl^{1-}$$

STEP 2: Balance the numbers of atoms in each separate half-reaction:

$$\text{Oxidation:} \quad Na \longrightarrow Na^{1+}$$
$$\text{Reduction:} \quad Cl_2 \longrightarrow 2Cl^{1-}$$

STEP 3: Add electrons to balance the charges:

$$\text{Oxidation:} \quad Na \longrightarrow Na^{1+} + 1e^{1-}$$
$$\text{Reduction:} \quad Cl_2 + 2e^{1-} \longrightarrow 2Cl^{1-}$$

STEP 4: Make the number of electrons present in each half-reaction equal. In this example, multiply the oxidation half-reaction by 2.

$$\text{Oxidation:} \quad 2Na \longrightarrow 2Na^{1+} + 2e^{1-}$$
$$\text{Reduction:} \quad Cl_2 + 2e^{1-} \longrightarrow 2Cl^{1-}$$

STEP 5: Sum both half-reactions, canceling where appropriate:

$$2Na + Cl_2 \longrightarrow 2Na^{1+} + 2Cl^{1-}$$

If we had not worried about conserving the number of electrons exchanged, the balanced reaction would have been

$$Na + Cl_2 \longrightarrow Na^{1+} + 2Cl^{1-}$$

an incorrect reaction.

Problem 6: Balance the following oxidation-reduction reaction:

$$H^{+} + NO_3^{1-} + S^{2-} \longrightarrow NO + S^{0} + H_2O$$

STEPS 1 and 2: By looking at the reactants and products, we can tell that sulfide ion has been oxidized to elemental sulfur. Therefore, the nitrate ion must have been reduced.

$$\text{Oxidation:} \quad S^{2-} \longrightarrow S^{0}$$
$$\text{Reduction:} \quad 4H^{+} + NO_3^{1-} \longrightarrow NO + 2H_2O$$

STEP 3: Balance charges by adding electrons:

$$\text{Oxidation:} \quad S^{2-} \longrightarrow S^{0} + 2e^{1-}$$
$$\text{Reduction:} \quad 4H^{+} + NO_3^{1-} + 3e^{1-} \longrightarrow NO + 2H_2O$$

STEP 4: Conserve electrons: multiply the oxidation half-reaction by 3 and the reduction half-reaction by 2.

$$\text{Oxidation:} \quad 3S^{2-} \longrightarrow 3S^{0} + 6e^{1-}$$
$$\text{Reduction:} \quad 8H^{+} + 2NO_3^{1-} + 6e^{1-} \longrightarrow 2NO + 4H_2O$$

STEP 5: Sum the half-reactions:

$$8H^{+} + 2NO_3^{1-} + 3S^{2-} \longrightarrow 3S^{0} + 2NO + 4H_2O$$

Common Oxidation-Reduction Reactions

Bleaches and Cleaning Agents

Oxidizing agents are often used to remove stains and to bleach material. Large molecules possessing color are often readily oxidized to materials which can be removed from the fabric. Sodium hypochlorite (NaOCl) is used as a laundry bleach for cotton fabric but will damage wool and silk. Hydrogen peroxide (H_2O_2) is used to remove blood from cotton fabric. Potassium permanganate is used to remove stains from white fabric, but it will discolor many dyed fabrics.

Combustion

Oxygen gas is the most widely used oxidizing agent. The burning of a substance in air involves oxidation of the materials by oxygen. In the combustion of methane, the carbon atom is changed from an oxidation number of -4 to $+4$ by oxygen. The process is highly exothermic.

$$CH_4 + 2H_2O \longrightarrow CO_2 + 8H^+ + 8e^- \quad \text{(oxidation)}$$
$$2(O_2 + 4H^+ + 4e^- \longrightarrow 2H_2O) \quad \text{(reduction)}$$
$$\text{Net reaction: } CH_4 + 2O_2 \longrightarrow CO_2 + 2H_2O + \text{energy}$$

Corrosion

The conversion of metals to metal oxides is a redox process called **corrosion**. Oxygen is generally the oxidizing agent. Some metals react so rapidly that enough heat is evolved to cause flames. Magnesium is so reactive with oxygen that it is used in flares.

$$2Mg + 2H_2O \longrightarrow 2MgO + 4H^+ + 4e^- \quad \text{(oxidation)}$$
$$O_2 + 4H^+ + 4e^- \longrightarrow 2H_2O \quad \text{(reduction)}$$
$$\text{Net eq.: } 2Mg + O_2 \longrightarrow 2MgO$$

Iron is oxidized to ferric oxide (rust), but more slowly.

$$4Fe + 6H_2O \longrightarrow 2Fe_2O_3 + 12H^+ + 12e^- \quad \text{(oxidation)}$$
$$3O_2 + 12H^+ + 12e^- \longrightarrow 6H_2O \quad \text{(reduction)}$$
$$\text{Net eq.: } 4Fe + 3O_2 \longrightarrow 2Fe_2O_3$$

Biological Redox Reactions

In living systems many redox reactions are involved in energy production and food metabolism. In the metabolic process, food is oxidized, giving up energy and electrons. The electrons are passed along an electron-transport system, finally being accepted by oxygen.

$$\text{Food} \longrightarrow \text{waste products} + \text{energy} + \text{electrons} \quad \text{(oxidation)}$$
$$\text{Electrons} + \text{oxidized transporters} \longrightarrow \text{reduced transporters} + H^+ \quad \text{(transport)}$$
$$\text{Reduced transporters} \longrightarrow \text{oxidized transporters} + e^-$$
$$4e^- + 4H^+ + O_2 \longrightarrow 2H_2O \quad \text{(reduction)}$$
$$\text{Net eq.: Food} + O_2 \longrightarrow \text{waste products} + H_2O + \textit{energy}$$

UNIT 15 Problems and Exercises

15–1. Suppose that you had three different solids in open vessels and heated them to 200°C. After cooling you weighed them and discovered that the first had gained weight, the second had lost weight, and the third had no change in weight. How can you reconcile these facts with the law of conservation of mass: the total weight of products must equal that of reactants? Suggest types of reactions that may have occurred.

15–2. Place the following reactions into one or more of the categories listed in Tables 15–1 and 15–2.

(a) $AgNO_3 + NaI \rightarrow AgI + NaNO_3$
(b) $NaHCO_3 + HCl \rightarrow H_2O + CO_2 + NaCl$
(c) $Na - 1e^- \rightarrow Na^{1+}$
(d) $AgNO_3 + I^- \rightarrow AgI + NO_3^-$
(e) $NH_4^+ \rightarrow NH_3 + H^+$
(f) $NaOH \rightarrow Na^+ + OH^-$
(g) $Br^- + CH_3{-}I \rightarrow CH_3Br + I^-$
(h) $C_2H_4 + H_2 \rightarrow C_2H_6$
(i) $C_2H_6O + H^+ \rightarrow C_2H_4 + H_2O$

15–3. When coal burns, the ashes weigh less than the starting coal. When iron filings are allowed to stand in the open, they rust, and the rust weighs more than the original iron filings. Reconcile these facts with the law of conservation of mass.

15–4. Balance the following.

(a) Ammonium hydroxide + sulfuric acid → ammonium sulfate + water
(b) Sodium nitrate → sodium nitrite + oxygen
(c) Sodium nitrate + sulfuric acid → nitric acid + sodium bisulfate
(d) Barium chloride + sodium sulfate → sodium chloride + barium sulfate
(e) Zinc chloride + ammonium sulfide → zinc sulfide + ammonium chloride
(f) Ferric oxide + carbon monoxide → iron + carbon dioxide
(g) Calcium carbonate + hydrochloric acid → calcium chloride + water + carbon dioxide
(h) Calcium hydroxide + ammonium sulfate → calcium sulfate + ammonia + water
(i) Sodium hydroxide + sulfuric acid → sodium sulfate + water
(j) Ferric oxide + hydrochloric acid → ferric chloride + water
(k) Ammonium sulfate + sodium hydroxide → ammonia + water + sodium sulfate
(l) Aluminum bicarbonate + sulfuric acid → aluminum sulfate + carbon dioxide + water

15–5. Write balanced equations for the following reactions.

(a) Magnesium bromide + chlorine (Cl_2) → magnesium chloride + bromine (Br_2)
(b) Zinc + hydrochloric acid (HCl) → zinc chloride + hydrogen gas
(c) Sodium hydroxide + carbon dioxide → sodium bicarbonate + water
(d) Cuprous chloride + iron → ferrous chloride + copper
(e) Sodium + water → sodium hydroxide + hydrogen gas (H_2)
(f) Magnesium + ferric ion → magnesium ion + ferrous ion

15–6. Balance the following equations.

(a) Cupric carbonate decomposes to cupric oxide and carbon dioxide.

(b) Calcium carbonate reacts with hydrochloric acid to yield carbon dioxide, water, and calcium chloride.

15–7. The following are addition reactions. Predict the products and balance the reactions.
(a) $Na + O_2 \rightarrow$
(b) $CaO + SO_3 \rightarrow$
(c) $Al + Br_2 \rightarrow$
(d) $Fe + S \rightarrow$

15–8. The following are decomposition reactions. Predict the products and balance the reactions.
(a) $H_2SO_3 \rightarrow H_2O +$
(b) $ZnCO_3 \rightarrow CO_2 +$
(c) $KNO_3 \rightarrow KNO_2 +$

15–9. The following are substitution reactions. Write down the products and balance the reactions.
(a) $AgNO_3 + H_2S \rightarrow$
(b) $Cu_2O + HCl \rightarrow$
(c) $Zn(OH)_2 + HCl \rightarrow$
(d) $FeCl_2 + NaOH \rightarrow$

15–10. The reaction between nitric oxide, NO, and oxygen gas, O_2, is written $2NO + O_2 \rightarrow 2NO_2$.
(a) Two molecules of nitric oxide give how many molecules of nitrogen dioxide, NO_2?
(b) How many moles of oxygen atoms are present in 2 mol of NO?
(c) If 1 mol of nitric oxide is consumed, how many moles of O_2 are consumed? How many moles of nitrogen dioxide are formed?
(d) Verify that conservation of mass exists by calculating the weights of all compounds present in the ratios given in the equation.

15–11. A paraffin candle burns in air to form water and carbon dioxide.

$$\underset{\text{paraffin}}{C_{25}H_{52}} + O_2 \longrightarrow CO_2 + H_2O$$

(a) Balance the above equation.
(b) How many moles of O_2 are needed to burn 5 mol of paraffin?
(c) How many moles of water would be formed?

15–12. Consider the combustion of pentyl alcohol, $C_5H_{11}OH$.

$$2C_5H_{11}OH + 15O_2 \longrightarrow 10CO_2 + 12H_2O$$

(a) How many moles of O_2 are needed for the combustion of 1 mol of alcohol?
(b) How many moles of H_2O are formed for each mole of O_2 consumed?
(c) How many grams of CO_2 are produced for each mole of alcohol burned?
(d) How many grams of CO_2 are produced for each gram of alcohol burned?

15–13. One type of bacteria has the ability to oxidize sulfur to bisulfate according to the following equation:

$$3S + 3O_2 + 2H_2O \longrightarrow 2H^+ + 2HSO_4^{1-} + \text{energy} + \text{other products}$$

What weight of sulfur would be required to synthesize 6.1 g of bisulfate ion?

15–14. One hundred and thirty grams of zinc was dropped into a solution containing 100 g of HCl.

$$Zn + 2HCl \longrightarrow ZnCl_2 + H_2$$

(a) Which compound is present in excess?
(b) How many grams of hydrogen gas were produced?

15–15. Rust (Fe_2O_3) is removed by reacting with acids.

$$\text{(Not balanced)} \quad Fe_2O_3 + HCl \longrightarrow FeCl_3 + H_2O$$

How many grams of rust will be removed if a solution containing 7.3 g of HCl is added?

15–16. In each of the following problems, balance the reaction and calculate the number of grams of product that will be formed.
(a) Twenty-five grams of mercuric oxide are converted by heat to oxygen gas and mercury. What weights of O_2 + Hg are formed?
(b) How many grams of H_2O, CO_2, and NaCl are formed when 10.6 g of sodium carbonate are treated with 3.65 g of hydrochloric acid?
(c) How many grams of cupric oxide are formed when 1 kg of copper is oxidized by oxygen gas?
(d) How many grams of calcium chloride are formed when 148 g of calcium hydroxide is reacted with 73 g of hydrochloric acid?
(e) How many grams of methyl bromide (CH_3Br) are formed when 7.1 g of methyl iodide (CH_3I) is reacted with 5.15 g of sodium bromide?

15–17. What is the formula for the oxide of mercury formed by the combination of 100.3 g of mercury and 8.00 g of oxygen?

15–18. Ethyl alcohol (C_2H_5OH) is made by the fermentation of glucose ($C_6H_{12}O_6$) as indicated by the balanced equation:

$$C_6H_{12}O_6 \longrightarrow 2C_2H_5OH + 2CO_2$$

How many pounds of alcohol can be made from 2000 lb of glucose?

15–19. Caustic soda, NaOH, is prepared commercially by the reaction of Na_2CO_3 with slaked lime, $Ca(OH)_2$. How many grams of NaOH can be prepared by treating 1 kg of Na_2CO_3 with excess $Ca(OH)_2$?

15–20. Plants use carbon dioxide and water to make glucose:

$$6CO_2 + 6H_2O \longrightarrow C_6H_{12}O_6 + 6O_2$$

How many liters of CO_2 at STP are consumed by a plant in the production of 454 g of glucose?

15–21. One formula for gasoline is C_8H_{18}. Burning it in the presence of oxygen gas yields carbon dioxide and water. If an automobile gets 10 mi/gal of gas, how many grams of oxygen are necessary to drive 100 mi? The density of gasoline is about 0.70 g/mℓ.

UNIT 16 ENERGETICS OF REACTIONS

Many physical and chemical processes occur spontaneously. By spontaneously, we mean that the processes occur naturally without our help. In physical events, gravitational potential energy, kinetic energy, or thermal energy is often lost to the surroundings during spontaneous events. For example, the water in a river spontaneously flows downhill, losing gravitational potential energy as it does so. Hot objects lose thermal energy as they cool, and moving objects lose kinetic energy as they slow down.

In chemical processes, most spontaneous events occur such that there is a net release of energy as the reactants are converted to products. Thermal energy is released as wood and paper burn, forming carbon dioxide and water. Therefore, wood and paper must possess more chemical energy than carbon dioxide and water.

However, many chemical reactions do not occur spontaneously. For these reactions to occur, energy must be added. For example, plants will not grow unless solar energy is absorbed. The chemical reactions that take place during the preparation of many foods do not occur unless thermal energy is added. Obviously, there must be guidelines to help us understand the differences between spontaneous and nonspontaneous processes. In this unit we shall introduce some of the principles.

Enthalpy

Since most spontaneous reactions occur such that the products possess less energy than the reactants, there must have occurred a net transfer of energy from the chemicals to the surroundings. Conversely, in nonspontaneous reactions, the products possess greater energy than the reactants, and energy must have been absorbed from the surroundings. **The energy liberated or absorbed during chemical or physical processes is called the enthalpy change for the process.** Enthalpy is pictured by the symbol H (for heat) and the enthalpy change for a reaction (the heat of reaction) is noted as ΔH. Thus ΔH is defined as the difference between the enthalpy of the products and that of the reactants:

$$\Delta H(\text{reaction}) = H_{\text{products}} - H_{\text{reactants}}$$

The enthalpy change will be negative ($\Delta H = -$) for reactions in which energy is released; it will be positive ($\Delta H = +$) for reactions in which energy is absorbed. When ΔH is negative, the energy released is often in the form of thermal energy, so the reactions are called **exothermic reactions**. A reaction that absorbs thermal energy (ΔH = positive) is often called an **endothermic reaction**. These reactions are shown in Figure 16–1.

Enthalpy changes for chemical reactions can often be measured by use of a *calorimeter*. In this device, a flask containing the reactants is suspended in a known amount of water at a known temperature. If thermal energy is released during the chemical reaction, it will be absorbed by the water, and can be calculated by accurately measuring the temperature increase of the water. If thermal energy is

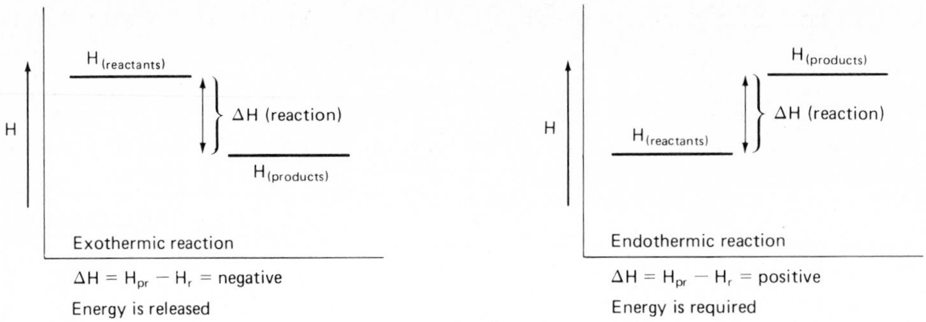
H
H (reactants)
ΔH (reaction)
H (products)
Exothermic reaction
ΔH = H_pr − H_r = negative
Energy is released
H
H (products)
ΔH (reaction)
H (reactants)
Endothermic reaction
ΔH = H_pr − H_r = positive
Energy is required

Figure 16–1

absorbed by the chemicals during a reaction, the temperature of the water will decrease.

Calculation of Enthalpy Changes From Bond Energies

The formation of covalent bonds from individual atoms is a favorable process, so molecules are generally more stable (possess less energy) than the individual atoms that compose them. If we measure the difference in energy between the covalent molecules and the individual atoms, we can calculate the enthalpy changes for formation of covalent bonds. We call these enthalpy changes **bond formation energies**. They are negative when the process is favorable.

$$\underset{\text{individual atoms}}{X\cdot + \cdot Y} \longrightarrow \underset{\text{a covalent bond}}{X{-}Y}, \qquad \Delta H = \text{usually negative}$$

Bond formation energy values for many types of covalent bonds are given in Table 16–1. The processes for measuring these energies can be quite complex, and we shall not concern ourselves with them; but we can still use the values to predict enthalpy changes for chemical reactions. Let's illustrate the process by examining

Table 16–1 Bond formation energies (ΔH) in kilocalories/mole

H—H	−103.4	C—H	−87.3	C=C	−100
C—C	−58.6	N—H	−83.7	C≡C	−123
N—N	−20.0	O—H	−110.2	C=O	−149
O—O	−34.9	H—Cl	−102.7	N≡N	−170
S—S	−63.8	H—Br	−87.3		
Cl—Cl	−57.8	H—F	−71.4		
Br—Br	−46.1	H—I	−71.0		
I—I	−36.2	C—Cl	−66.5		
		C—Br	−54.5		
		C—I	−45.5		

the reaction between chlorine gas (Cl_2) and methane (CH_4). The products that form are methyl chloride (CH_3Cl) and hydrogen chloride (HCl).

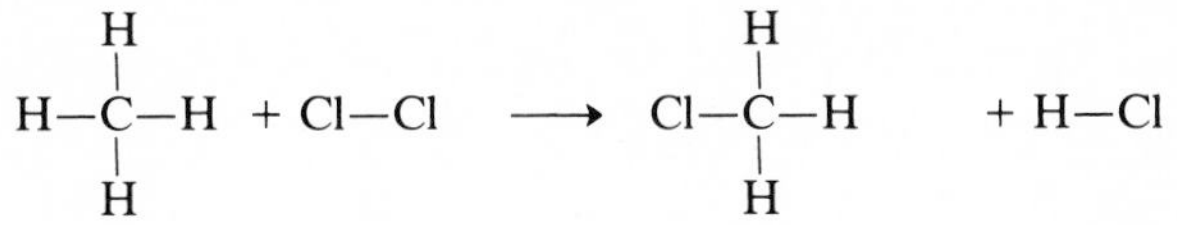

methane + chlorine ⟶ chloromethane + hydrogen chloride

The first goal is to determine how much energy has been released during the formation of each of the molecules when they were formed from the individual atoms. Then we can compare the differences between the reactants and products, and calculate an enthalpy change for the reaction.

1. Formation of methane: In methane there are 4 C—H bonds, so the formation of methane from one carbon and 4 hydrogen atoms should liberate four times the energy released when one carbon–hydrogen bond formed:

$$\text{C}\cdot + \cdot\text{H} \longrightarrow \text{C—H}, \quad \Delta H = -87.3 \text{ kcal/mol}$$

So

$$\cdot\dot{\text{C}}\cdot + 4\text{H}\cdot \longrightarrow CH_4 \quad \Delta H = 4 \times (-87.3) = -349.2 \text{ kcal/mol}$$

2. Formation of chlorine: There is only one covalent bond present in chlorine gas:

$$\text{Cl}\cdot + \text{Cl}\cdot \longrightarrow \text{Cl—Cl}, \quad \Delta H = -57.8 \text{ kcal/mol}$$

3. Formation of hydrogen chloride:

$$\text{H}\cdot + \cdot\text{Cl} \longrightarrow \text{H—Cl}, \quad \Delta H = -102.7 \text{ kcal/mol}$$

4. Formation of methyl chloride: Here we need to sum the values for formation of 3 C—H bonds and one C—Cl bond:

$$\cdot\dot{\text{C}}\cdot + 3\text{H}\cdot \longrightarrow CH_3, \quad \Delta H = 3 \times (-87.3) = -261.9 \text{ kcal/mol}$$

$$\text{C}\cdot + \text{Cl}\cdot \longrightarrow \text{C—Cl}, \quad \Delta H = -66.5 \text{ kcal/mol}$$

Summing,

$$\cdot\dot{\text{C}}\cdot + 3\text{H}\cdot + \text{Cl}\cdot \longrightarrow CH_3Cl, \quad \Delta H = -328.4 \text{ kcal/mol}$$

Now let's look at the complete reaction again and add the H values.

$$\begin{array}{lcc} & CH_4 + Cl_2 \longrightarrow & CH_3Cl + HCl \\ \Delta H \text{ formation} = & \underbrace{-349.2 \quad -57.8}_{-407} & \underbrace{-328.4 \quad -102.7}_{-431} \end{array}$$

$$\Delta H(\text{reaction}) = H_{pr} - H_r = (-431) - (-407) = -24.1 \text{ kcal/mol}$$

Formation of the products (CH_3Cl and HCl) from the individual atoms has released more energy than formation of the reactants (CH_4 and Cl_2) from the same individual atoms. Thus the products are 24.1 kcal/mol more stable than the reactants, and the enthalpy change for the reaction is − 24.1 kcal/mol.

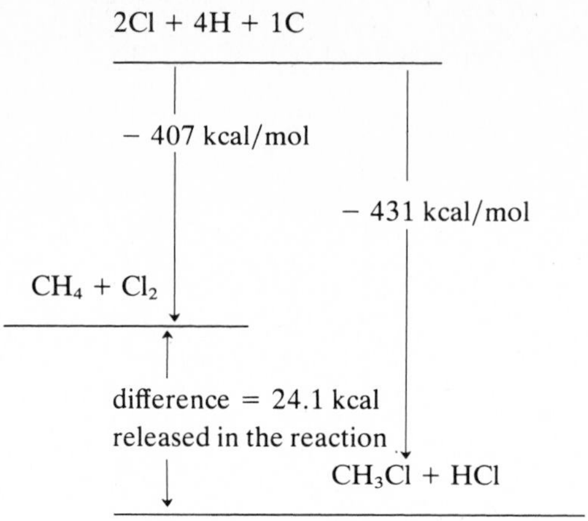
2Cl + 4H + 1C
− 407 kcal/mol
− 431 kcal/mol
CH4 + Cl2
difference = 24.1 kcal
released in the reaction
CH3Cl + HCl

$$\Delta H = (-431 \text{ kcal/mol}) - (-407) = -24.1 \text{ kcal/mol}.$$

Entropy

A second basic principle that governs physical events is the **maximum randomness principle**. In nature there seems to be a tendency to go from highly ordered systems to more disordered, random states. Changes tend to occur such that the end result is more dilute, mixed, or spread out than the initial state. For example, when ice melts the rigid orderly crystal lattice breaks apart and the liquid molecules spread out over a surface as the disorder of the system increases. A gas sample will not stay in one portion of a room but will spread throughout the entire room, mixing with all the other gases. Obviously, the mixing increases the disorder of the system. In many spontaneous chemical reactions, large molecules break apart into smaller ones that move away from each other, decreasing the order in the system and increasing the mixing and freedom of motion. Thus spontaneity seems to be associated with decrease in internal energy (energy release) and increase in disorder.

The degree of disorder of a system is represented by the term **entropy**, *denoted by the symbol S. The greater the disorder or freedom of motion of a system, the greater is the entropy of the system.* The **entropy change** in an event is the difference between the entropy of the products and that of the starting materials.

$$\Delta S = S_{\text{products}} - S_{\text{reactants}}$$

A positive ΔS indicates that disorder and freedom of motion have increased, a *negative* ΔS indicates an increase in order or a decrease in freedom of motion. In chemical reactions the entropy change will be positive (1) if the reactants change from solid to liquid or from liquid to gas, (2) if the reactant molecules break apart to form more molecules of products, or (3) if the reactant molecule somehow changes such that there is more movement of the individual atoms possible in the product. A negative entropy would result from condensation, precipitation, or the addition of two reactants to make one molecule of product (see Table 16–2).

Table 16–2 Entropy changes

$\Delta S = S_{pr} - S_{sm}$ = positive	$\Delta S = S_{pr} - S_{sm}$ = negative
1. Change of state:	
Solid → liquid	Liquid → solid
Liquid → gas	Gas → liquid
2. Increase in no. of molecules:	Decrease in no. of molecules:
$H_2CO_3 \rightarrow H_2O + CO_2$	$2NO + O_2 \rightarrow 2NO_2$
3. Increase in freedom of motion:	Decrease in freedom of motion:
Pressure decreases	Pressure increases
Bond reorganization:	
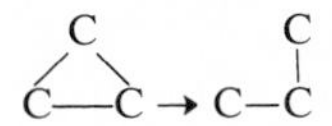	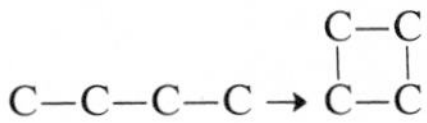
4. Increase in disorder:	Increase in order:
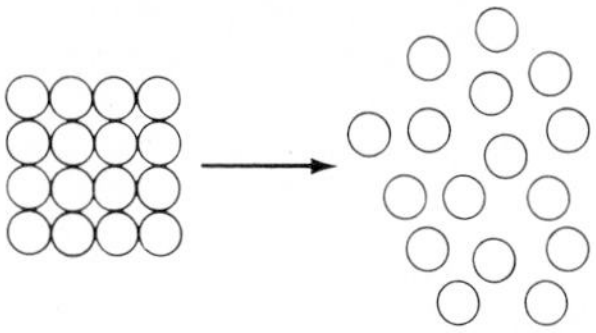	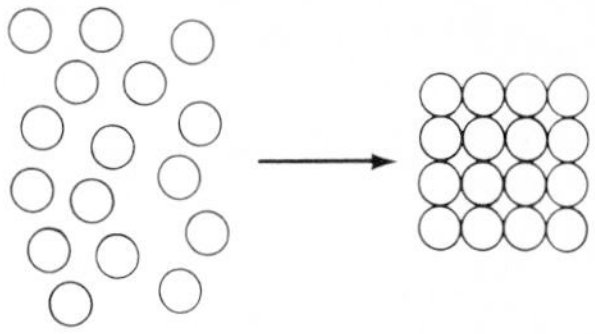

Gibbs Free Energy

Enthalpy and entropy both govern the spontaneity of physical and chemical events. They have been combined into one term called the **Gibbs free energy**, denoted as G. *The change in the Gibbs free energy for a process is the difference between the free energy of the products and that of the reactants.* This change represents the amount of usable energy that is available or that must be put into any process. A negative free energy change indicates that usable energy has been liberated by the process. *All spontaneous processes have negative free energy changes.* Processes that do not occur spontaneously have positive free energy changes. This indicates that work must be done *on* the system in order for it to occur.

$$\Delta G = G_{\text{products}} - G_{\text{reactants}}$$

$$\Delta G = \text{negative} \quad \text{(spontaneous process; usable energy released)}$$

$$\Delta G = \text{positive} \quad \text{(unfavorable process; energy must be added)}$$

At constant temperature and pressure, the relationship between entropy, enthalpy, and Gibbs free energy is given by

$$\Delta G = \Delta H - T\,\Delta S$$

If ΔH is negative (exothermic process) and ΔS is positive (increase in disorder), then ΔG will always be negative and the process will always be spontaneous. Conversely, if ΔH is positive (endothermic process) and ΔS is negative (increase in order), then ΔG will always be positive and the process will never occur spontaneously. When ΔH

and ΔS are both negative or both positive, the sign of ΔG will depend upon the temperature of the system. Table 16–3 gives some examples.

Table 16–3 Gibbs free energy changes

ΔG always negative; *combustion*:	
C_3H_8(propane) + $5O_2$ = $3CO_2$ + $4H_2O$ + heat	ΔH = negative (heat is released) ΔS = positive (increase in the number of molecules) ΔG = negative (spontaneous)
ΔG always positive; *photosynthesis*:	
$6CO_2$ + $6H_2O$ + solar energy = $C_6H_{12}O_6$ (glucose) + $3O_2$	ΔH = positive (energy added) ΔS = negative (decrease in number of molecules) ΔG = positive (not spontaneous)
ΔG positive or negative:	
Melting of ice: Solid + heat energy = liquid	ΔH = positive (endothermic) ΔS = positive (increase in disorder) ΔG = positive or negative, depending on temperature
Condensation of gases: Gas = liquid + energy	ΔH = negative (energy released) ΔS = negative (increase in order of system) ΔG = positive or negative, depending on temperature

The Gibbs free energy is dependent on the temperature and the concentration of a compound in solution. Standard free energies, denoted as G^0, have been defined as the free energy of a compound whose concentration is 1 molar at a temperature of 25°C. At other concentrations and temperatures, the free energy can be calculated from equation (16–1).

$$G = G^0 + 2.3RT \log C \qquad (16\text{–}1)$$

where G = free energy at other than standard conditions
G^0 = free energy of compound at 25°C and 1 M concentration
R = ideal gas constant
T = temperature (°K)
C = concentration (molarity).

The Relationship Between Free Energy and Equilibrium

A reversible reaction may occur in either direction. At equilibrium there is no net change in the concentrations of either the starting materials or the products. The ratio of the concentrations of the products and starting materials at equilibrium is called the **equilibrium constant** (K).

$$\text{A} \rightleftharpoons \text{B}, \qquad \text{equilibrium constant } (K) = \frac{\text{concentration B (molarity)}}{\text{concentration A (molarity)}}$$

The free energy change of any reversible reaction is given by

$$\Delta G = \Delta G^0 + 2.3RT \log K \qquad (16\text{–}2)$$

At equilibrium, there is no spontaneous change in either direction, so the free energy change, ΔG, is zero and equation (16–2) reduces to

$$\Delta G^0 = -2.3RT \log K \qquad (16\text{–}3)$$

This equation enables us to calculate the standard free energy change for any reversible reaction by measurement of the temperature and the concentrations of the reactants and products at equilibrium. We shall use this equation later when discussing the energetics of reactions that occur in the body.

UNIT 16 Problems and Exercises

16–1. Identify the following changes as endothermic or exothermic.
(a) Burning of coal
(b) Using food as fuel for our bodies
(c) Green plants making starch out of carbon dioxide, water, and sunlight
(d) Changing liquid water to vapor
(e) Shooting off a flashbulb
(f) Freezing water

16–2. Give three examples of exothermic processes and three examples of endothermic processes.

16–3. Write down three spontaneous changes common to everyday life, and two nonspontaneous changes.

16–4. When two substances are mixed, does entropy increase or decrease? Why?

16–5. Predict the sign of the entropy change in each of the following examples.
(a) A house is blown up.
(b) A gas condenses to a liquid.
(c) A cup of water is spilled on the floor.
(d) An aerosol can is emptied.
(e) An automobile is built from its parts.
(f) $CH_2{=}CH_2 + H_2 \rightarrow CH_3CH_3$
(g) $2KClO_3 \rightarrow 2KCl + 3O_2$
(h) $NaHCO_3$ (solid) + HCl (liquid) $\rightarrow$ NaCl + CO_2 (gas) + H_2O

16–6. Consider the following combinations of entropy and enthalpy changes. Which reactions will definitely occur? Which will not occur? Which might occur, depending on the temperature?
(a) Negative enthalpy change + positive entropy
(b) Negative enthalpy change + negative entropy
(c) Positive enthalpy change + positive entropy
(d) Positive enthalpy change + negative entropy

16–7. Consider the process of converting ice to liquid and then to water vapor.
(a) Is each step endothermic or exothermic?
(b) Does the enthalpy increase or decrease during the process?
(c) Does entropy increase or decrease during the process?

16–8. Using Table 16–1, calculate the bond formation energies for each of the products and reactants in the following reactions. Determine whether the reaction is endothermic or exothermic as written.

(a) $CH_4 + Br_2 \rightarrow CH_3Br + HBr$
(b) $CH_4 + I_2 \rightarrow CH_3I + HI$
(c) $H_2C{=}CH_2 + H_2 \rightarrow H_3C{-}CH_3$
(d) $H_2 + Cl_2 \rightarrow 2HCl$

16–9. Predict the sign of ΔG for each of the following. Calculate the sign of ΔH from the graphs and that of ΔS from the given reaction. Tell whether the reaction is spontaneous, would not occur, or whether it is possible to tell.

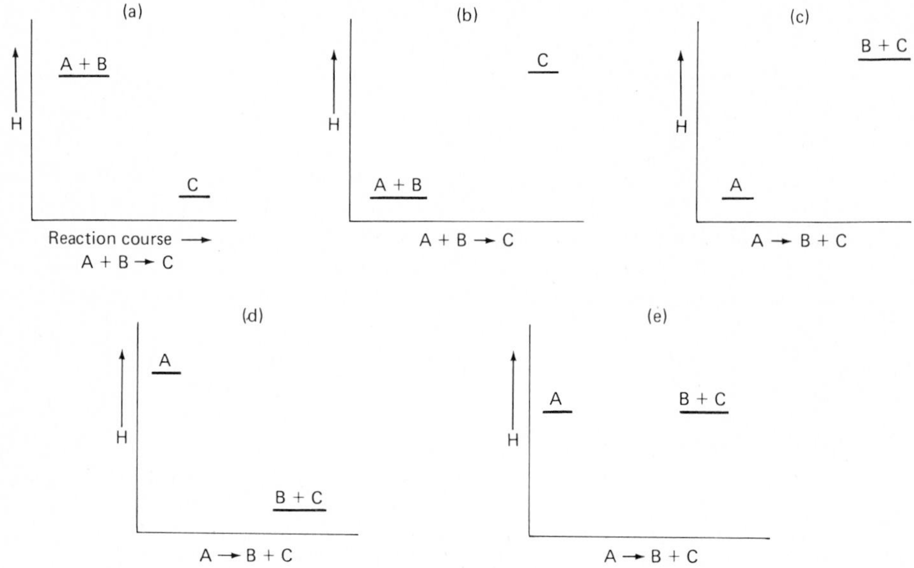
(a)
A + B
H
C
Reaction course →
A + B → C
(b)
C
H
A + B
A + B → C
(c)
B + C
H
A
A → B + C
(d)
A
H
B + C
A → B + C
(e)
A
B + C
H
A → B + C

UNIT 17 REACTION RATES AND EQUILIBRIUM

The calculation of free energy changes allows us to predict whether or not chemical reactions will be spontaneous, but the free energy changes tell us little about how *fast* these reactions occur. We know that a reaction with a negative free energy change should occur, but it may occur in seconds or it may take years for significant reaction to occur.

The reason that free energy changes tell us nothing about the rate of reactions is that they compare only the energy differences between the starting and ending points of the reactions; they do not deal with the reaction paths. For example, refer to Figure 17–1; suppose that you set out to bicycle between cities A and B, and a mountain existed between those two cities. Suppose that two paths were available to you: path X proceeded over the mountain and path Y proceeded through a tunnel in

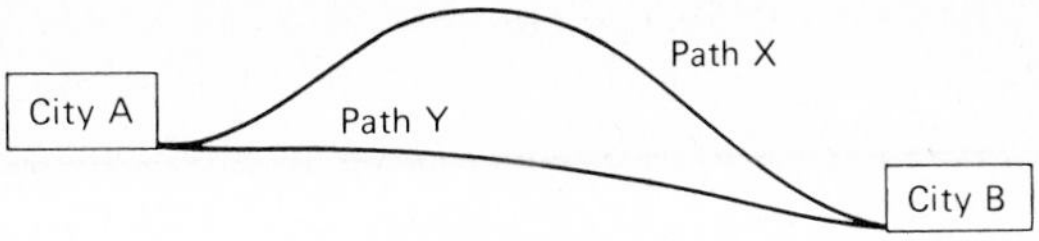

Path X
City A
Path Y
City B

Figure 17–1

the mountain. Which path would allow you to reach your destination sooner? Even though the two cities are a specific distance apart, you would arrive at city B much sooner if path Y were taken.

We can look at the rates of chemical reactions in much the same fashion. Each chemical reaction has a specific energy path that it must travel before significant reaction can occur. In some reactions, the path includes a large *energy barrier* that must be overcome. Those types of reactions proceed at a slower rate than reactions with lower energy barriers, because a smaller percentage of the reactant molecules possess the necessary energy to pass over the larger barrier. Look at the two reactions depicted in Figure 17–2. Both have the same ΔG value, the free energy change, and we predict both would occur spontaneously. But reaction X → Y has a much larger energy barrier in its path than reaction X → Z, so the rate of reaction X → Y should be much slower than reaction X → Z. In this unit we shall discuss some of the factors that help determine the rates of chemical reactions.

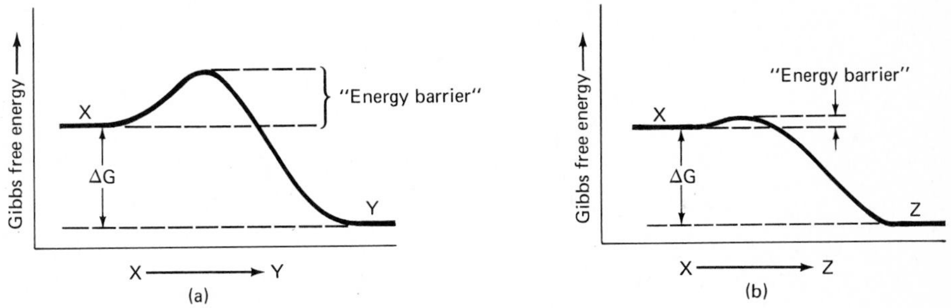

Gibbs free energy
X
"Energy barrier"
ΔG
Y
X → Y
(a)
Gibbs free energy
"Energy barrier"
X
ΔG
Z
X → Z
(b)

Figure 17–2 Paths for two chemical reactions.

Theory of Reaction Rates; Activation Energies

In most chemical reactions, the reactants need some sort of push to begin reacting. They need to attain an *activated state* before significant amounts of reaction can begin. For example, in a decomposition reaction (A → B + C), the reactant A must absorb sufficient energy to begin the bond-breaking process. In addition reactions (A + B → C), the reactants A and B must possess sufficient speeds to bring them close enough together to allow bond formation to begin. In substitution reactions (A—B + C → A—C + B), the reactants not only must have sufficient energy in the collision, but they must also collide in the proper orientation. Reactant C must interact with the A end of A—B in order to form the A—C bond.

For reactants to reach the activated state, they must absorb some form of energy. Once in the activated state, they can either react to form products, or lose the energy and revert to stable reactants. The amount of energy required for a reactant to reach the activated state is called the **activation energy** (*Ea*), and it differs for every reaction. The activation energies of reactions then are one factor that determines the rates at which reactions will occur. Reactions that occur rapidly have low activation energies, so many of the reactant molecules easily gain the energy required to reach the activated state. A reaction that has a large activation energy will usually proceed much more slowly, since larger amounts of energy must be absorbed by the reactant molecules in order for them to be activated. Fewer of the reactant molecules will possess the necessary activation energy at any time, so fewer molecules will react per unit time. In Figure 17–3, compound A will react to form B more rapidly than the reverse (B → A), since the activation energy for the A → B reaction is smaller than the activation energy for the B → A reaction.

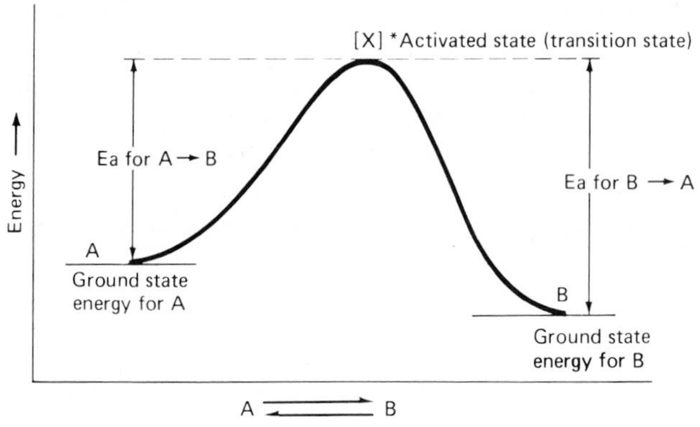
[X] *Activated state (transition state)
Energy
Ea for A → B
Ea for B → A
A
Ground state energy for A
B
Ground state energy for B
A ⇄ B

Figure 17–3 Activation energy of $A \rightleftharpoons B$.

Factors That Influence Rate of Reaction

We have already discussed the importance of activation energies on the rate of chemical reactions. Other factors that influence reaction rate include the concentrations of reactants, the temperature at which reactions are run, and the presence of catalysts.

Concentration

In many reactions, two or more reactants must somehow interact for reaction to occur. Any process that increases the probability of contact between the reactants may then increase the rate of the reaction. In cooking, we stir mixtures thoroughly to increase the contact between the separate ingredients and thus increase the rate of reaction. We often grind or pulverize solids to increase the surface area and thus the reactivity. For example, a log that has been chopped into kindling will burn much faster because there is a greater surface area of wood to come into contact with oxygen from the air. When working with solutions, we must increase the concentrations of the reactants in order to increase the probability of contact. In gas solutions, this is done by increasing the pressure; in liquid solutions, this is done by adding more reactants to the solution.

The dependence of reaction rates on concentrations of reactants is often expressed as a mathematical equation, which is called the **rate law** for that particular reaction. For a reaction between reactants A and B, ($mA + nB \rightarrow$ products), the rate law expression can be written as

$$\text{rate} = k[\text{A}]^x[\text{B}]^y$$

The brackets [ ] represent the molar concentrations of the reactants and the superscripts x and y represent the powers to which those concentrations must be raised. The proportionality constant, k, is called the **rate constant** for the reaction; it contains the activation energy and temperature parameters.

Superscripts x and y often are related to the numbers m and n present in the balanced equation, but they are not always identical to them. Superscripts x and y are determined experimentally by varying the concentrations of the reactants and observing how the rate of the reaction changes. The numbers m and n are simply those numbers necessary to satisfy the law of conservation of mass. Let's look at the rate laws for some reactions to illustrate the dependence of rate on concentration.

Example 1:

$$Ca(HCO_3)_2 \xrightarrow{\text{Heat}} CaCO_3 + H_2O + CO_2 \quad \text{Rate} = k[Ca(HCO_3)_2]$$

The rate of this reaction is proportional to the concentration of the starting material, calcium bicarbonate. A doubling of the concentration of calcium bicarbonate will cause the rate of the reaction to double.

Example 2:

$$N_2 + O_2 \longrightarrow 2NO \quad \text{Rate} = k[N_2][O_2]$$

Here the rate is proportional to the concentrations of both reactants. A doubling of the concentration of either gas will cause the rate to double. A doubling of the concentrations of both gases will increase the rate by a factor of 4.

Example 3:

$$2N_2O_5 \rightarrow 4NO_2 + O_2 \quad \text{Rate} = k[N_2O_5]$$

Notice here that the rate is directly proportional to the concentration of N_2O_5, not to the square of the concentration of N_2O_5. Remember that the numbers used in balancing the reactions need not be the superscripts in the rate law.

Example 4:

$$C_4H_9Br + OH^{1-} \longrightarrow C_4H_9OH + Br^{1-} \quad \text{Rate} = k[C_4H_9Br]$$

In some reactions the concentration of one or more of the reactants is not important to the rate of the reaction.

Temperature

The temperature of a system determines the amount of thermal and kinetic energy present in the reactants. At a given temperature, all the reactant molecules do not possess the same energy; there is a wide variation in the energies possessed. Figure 17–4 shows the distribution of kinetic energies in a sample of gas molecules at two

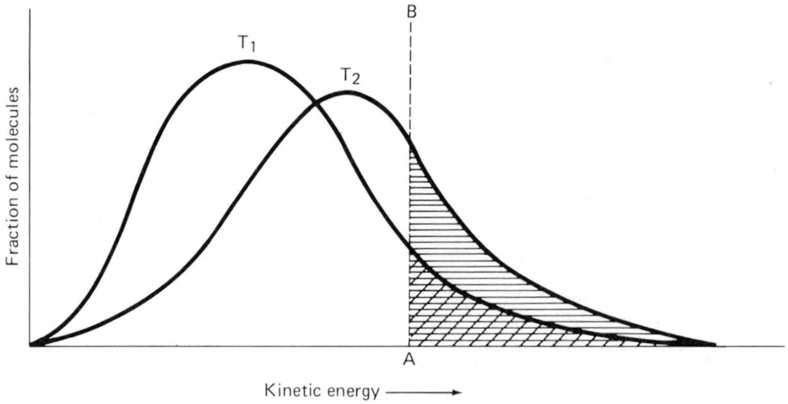

B
T_1
T_2
Fraction of molecules
A
Kinetic energy

Figure 17–4 The distribution of kinetic energies in a sample of gas molecules at two temperatures. Shaded portions indicate the numbers of molecules with energy greater than *AB*.

temperatures, and the areas under the curves represent the percentages of the molecules that possess a particular energy. Suppose that line *AB* represents the minimum activation energy for the reaction. Then all molecules with energy less than *AB* will not react, and all molecules with energy greater than *AB* (the shaded portion) will react. The rate of the reaction will depend to some extent on the proportion of the molecules with energies greater than *AB*. If the temperature is increased from T_1 to T_2, all the molecules possess greater energy, and a larger proportion of them will possess the minimum energy necessary to react; the rate of the reaction will increase.

The rate of a reaction normally increases with increasing temperature because the reactant molecules possess greater kinetic energy at higher temperatures. This means that the reactant molecules move about more rapidly. Not only is the number of collisions increased at these higher temperatures, but the collisions also occur with greater impact. Thus more of the collisions possess the necessary activation energy for reaction. One general rule of thumb is that an increase of 10°C can often increase the rate of a reaction by a factor of 2 or 3.

Catalysts

A **catalyst** is a substance that speeds up the rate of a reaction, but is itself not consumed during the reaction. Usually a catalyst does this by somehow interacting with the reactants to provide a new, lower-energy reaction pathway. Thus a catalyst acts to lower the activation energy for a reaction, and this allows the reaction to

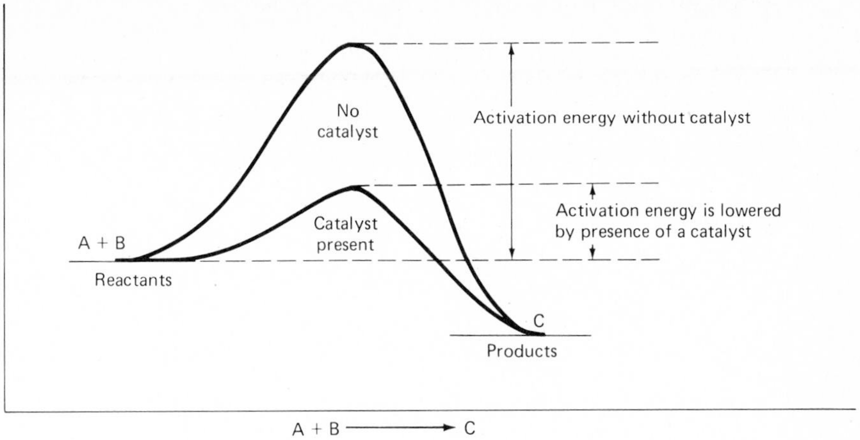
No
catalyst
Activation energy without catalyst
Activation energy is lowered
by presence of a catalyst
Catalyst
present
A + B
Reactants
C
Products
A + B ⟶ C

Figure 17–5 Influence of catalysts.

proceed at a faster rate and to occur at lower temperatures (see Figure 17–5). A construction foreman could be called a catalyst because, although he does not perform the physical labor himself, his presence generally increases the rate of activity at the construction site.

In our bodies most of the reactions involved in metabolism and general body function are catalyzed. These catalysts are called *enzymes*, and each enzyme catalyzes a different type of reaction. Most of these reactions could not occur fast enough at body temperature without the enzymes, so their presence is essential to life. Some genetic diseases are caused because an individual does not possess the ability to produce certain enzymes, so certain types of chemicals cannot be metabolized in the body. Sometimes these diseases can be controlled through proper diet.

UNIT 17 Problems and Exercises

17–1. Draw a representative reaction curve for an exothermic reaction, and label the activation energies for both the forward and reverse reactions. Do the same for an endothermic reaction.

17–2. Which of the following theoretical reactions would be likely to have the fastest rate? Why?

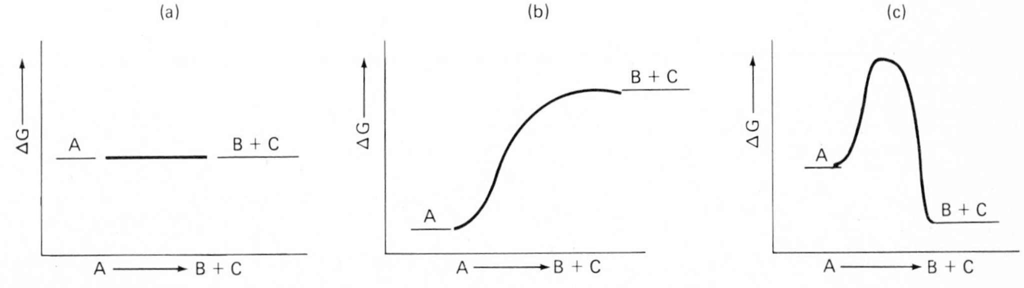
(a)
(b)
(c)
ΔG
A
B + C
A ⟶ B + C

17–3. List three conditions that must be satisfied before two molecules can react with each other.

17–4. List the factors that influence reaction rates, and describe how these features are interpreted in terms of kinetic molecular theory.

17–5. Milk will sour in two days at room temperature but remain unspoiled for two weeks in a refrigerator. Why?

17–6. Based on what you have learned about reactions, collision theory, and reaction rates, label each of the following as true or false. Explain your choice.
- (a) When two gases that react with each other are placed in a sealed container, decreasing the volume of the container generally decreases the reaction rate.
- (b) An increase in the concentration of only one of the gases will increase the reaction rate in part (a).
- (c) An increase in the number of collisions between the molecules automatically increases the reaction rate.

17–7. Explain why a mixture of H_2 and Cl_2 is stable if kept in the dark but explodes to form HCl if exposed to intense light.

17–8. How would each of the following changes affect the activation energy of a particular reaction?
- (a) Increasing the concentrations of the reactants
- (b) Increasing the temperature
- (c) Adding more solvent
- (d) Adding a catalyst

17–9. Predict the effect of the following changes on the rate of the forward reaction for the actual reaction written below. Explain your prediction.

$$A_{solid} + B_{solid} \rightleftharpoons C_{solid}$$

- (a) Atmospheric pressure is increased.
- (b) [A] is increased.
- (c) [B] is increased.
- (d) [C] is increased.
- (e) Temperature is increased.
- (f) A catalyst is added.
- (g) Activation energy is raised.

17–10. In the laboratory, sugar is converted into CO_2 and H_2O only at high temperature. In the body this reaction occurs at 37°C. Why?

CHAPTER 8

Acid-Base chemistry

Acids and bases play important roles in our everyday lives. Acids formed as a result of air pollution are dangerous to our health and slowly destroy the surfaces of buildings and statues. Lake and river water is being contaminated by acids that are dumped from industrial plants and leached out of strip-mining operations. Radio and television advertisements constantly bombard us with reasons why we should buy products to relieve acid indigestion. Labels on soaps and cleaners warn of the presence of strong bases or alkali. Gardeners and farmers worry about the soil being too acidic or basic.

In chemistry a great number of reactions involve acids and bases. Many of the reactions that occur in living systems are acid-base reactions. For our bodies to function properly, the concentrations of acid and base in the body fluids must be maintained at certain critical levels. An understanding of acid-base chemistry is crucial to those involved in the study of body functions.

In unit 18 we define and introduce acids and bases. Methods of measuring the concentrations of acids and bases are presented in unit 19, and many of the important acid-base reactions are given in unit 20. Finally, in unit 21 we deal with the mechanisms used by the body to control the concentrations of acid and base.

UNIT 18 INTRODUCTION TO ACIDS AND BASES

There are three major definitions of acids and bases: the **Arrhenius concept**, the **Bronsted-Lowry concept**, and the **Lewis concept**. Each definition focuses on a slightly different aspect of what is termed acidic or basic behavior.

Arrhenius Acids and Bases

In 1884, Svante Arrhenius, a Swedish chemist, proposed a definition for acids and bases that depended upon their interactions in aqueous solutions.

1. An **acid** is any substance that can supply hydrogen ions (H^{1+}) when dissolved in water.

$$\text{HA (acid)} \xrightarrow{H_2O} H^{1+} + A^{1-}$$

Acidic solutions (a) have a sour taste, (b) turn litmus paper (an acid–base indicator) from blue to red, (c) conduct electricity, and (d) react readily with basic solutions to yield water.

2. A **base** is any substance that can supply hydroxide ions (OH^{1-}) when dissolved in water.

$$\text{BOH (base)} \xrightarrow{H_2O} B^{1+} + OH^{1-}$$

Basic solutions (a) have a bitter taste, (b) turn litmus paper from red to blue, (c) feel soapy or slippery to the touch, (d) react with acidic solutions to yield water, and (e) conduct electricity.

A *strong acid* is a good provider of H^{1+} ions when it dissolves in water; a *weak acid* does not ionize much in water or provide many hydrogen ions. For example, hydrochloric acid (HCl) is a strong acid because virtually 100% of the HCl molecules ionize to H^{1+} and Cl^{1-} when dissolved in water. Acetic acid (CH_3CO_2H) is considered a weak acid because only about 1% of the molecules ionize in water to yield hydrogen ions. Conversely, strong bases are good providers of hydroxide ions, and weak bases do not ionize readily in water to yield hydroxide ions. Table 18–1 lists some of the more common acids and bases.

Bronsted-Lowry Acids and Bases

The Arrhenius definition of acids and bases works only in aqueous solutions and for bases that actually contain hydroxide ions. It cannot explain why compounds like Na_2S, Na_2O, and NH_3 act like bases when dissolved in water. In 1923, J. N. Bronsted and J. M. Lowry proposed new definitions of acids and bases to account for those compounds that did not fit the Arrhenius definitions.

Table 18–1 Arrhenius acids and bases

Acid		Base	
Name	Structure	Name	Structure
Strong acids:		Strong bases:	
Hydrochloric acid:	$HCl \rightarrow H^{1+} + Cl^{1-}$	Sodium hydroxide:	$NaOH \rightarrow Na^{1+} + OH^{1-}$
Sulfuric acid:	$H_2SO_4 \rightarrow H^{1+} + HSO_4^{1-}$	Potassium hydroxide:	$KOH \rightarrow K^{1+} + OH^{1-}$
Nitric acid:	$HNO_3 \rightarrow H^{1+} + NO_3^{1-}$	Barium hydroxide:	$Ba(OH)_2 \rightarrow BaOH^{1+} + OH^{1-}$
Moderate acids:		Weak bases:	
Phosphoric acid:	$H_3PO_4 \rightarrow H^{1+} + H_2PO_4^{1-}$	Magnesium hydroxide:	$Mg(OH)_2 \rightarrow MgOH^{1+} + OH^{1-}$
Sulfurous acid:	$H_2SO_3 \rightarrow H^{1+} + HSO_3^{1-}$	Aluminium hydroxide:	$Al(OH)_3 \rightarrow Al(OH)_2^{1+} + OH^{1-}$
Weak acids:		Ammonia water:	$NH_3 \cdot H_2O \rightarrow NH_4^{1+} + OH^{1-}$
Acetic acid:	$CH_3CO_2H \rightarrow H^{1+} + CH_3CO_2^{1-}$		
Carbonic acid:	$H_2CO_3 \rightarrow H^{1+} + HCO_3^{1-}$		
Citric acid:	$C_6H_8O_7 \rightarrow H^{1+} + C_6H_7O_7^{1-}$		

1. An **acid** is any molecule or ion that can *donate* hydrogen ions to solution.

$$HA\ (\text{acid}) = H^{1+} + A^{1-}$$

2. A **base** is any molecule or ion that can *accept* hydrogen ions from solution.

$$B\ (\text{base}) + H^{1+} = BH^{1+}$$

Bronsted-Lowry thus expanded the definitions of basic compounds. To identify whether a particular ion or molecule is an acid or base, we must examine the ions in the particular reaction under study. The same molecule or ion may act either as an acid or a base, depending upon the particular reaction of interest. In Table 18–2, note that water is an acid in reactions 3, 4, and 5, whereas it is a base in reaction 2. Ammonia is an acid in reaction 6 and a base in reactions 1 and 3.

Table 18–2 Bronsted-Lowry acids and bases

	Acid (proton donor)		Base (proton acceptor)		Acid		Base
1.	HCl	+	NH_3	$\longrightarrow$	NH_4^{1+}	+	Cl^{1-}
2.	HCl	+	H_2O	$\longrightarrow$	H_3O^{1+}	+	Cl^{1-}
3.	H_2O	+	NH_3	$\longrightarrow$	NH_4^{1+}	+	OH^{1-}
4.	H_2O	+	CO_3^{2-}	$\longrightarrow$	HCO_3^{1-}	+	OH^{1-}
5.	H_2O	+	S^{2-}	$\longrightarrow$	HS^{1-}	+	OH^{1-}
6.	NH_3	+	CH_3^{1-}	$\longrightarrow$	CH_4	+	NH_2^{1-}

Conjugate Acid-Base Pairs

Bronsted-Lowry definitions also created the concept of acid-base pairs. If the reactions shown in Table 18–2 are reversible, then the compounds on the right are also acids and bases. For example, HCl is an acid in reaction 1 since it gives up a hydrogen ion to yield chloride ion. But if the reverse reaction occurred, chloride ion would accept a hydrogen ion and become HCl. Thus Cl^{1-} is a base and we have the HCl/Cl^{1-} acid-base pair. For every acid there is a base. The acid has the greater number of hydrogen atoms; the base is what is left after one hydrogen has been

Table 18–3 Bronsted-Lowry conjugate acids and bases*

Increasing acid strength (arrow pointing up)

Increasing base strength (arrow pointing down)

Conjugate acid		Conjugate base	
Name	Formula	Formula	Name
Hydrogen iodide	HI	I^{1-}	Iodide ion
Hydrogen bromide	HBr	Br^{1-}	Bromide ion
Sulfuric acid	H_2SO_4	HSO_4^{1-}	Hydrogen sulfate ion
Hydrogen chloride	HCl	Cl^{1-}	Chloride ion
Nitric acid	HNO_3	NO_3^{1-}	Nitrate ion
Hydronium ion	H_3O^{1+}	H_2O	Water
Hydrogen sulfate ion	HSO_4^{1-}	SO_4^{2-}	Sulfate ion
Phosphoric acid	H_3PO_4	$H_2PO_4^{1-}$	Dihydrogen phosphate ion
Acetic acid	$CH_3\overset{O}{\overset{\|}{C}}-OH$	$CH_3\overset{O}{\overset{\|}{C}}-O^{1-}$	Acetate ion
Carbonic acid	H_2CO_3	HCO_3^{1-}	Bicarbonate ion
Hydrogen sulfide	H_2S	HS^{1-}	Hydrogen sulfide ion
Ammonium ion	NH_4^{1+}	NH_3	Ammonia
Hydrogen cyanide	HCN	CN^{1-}	Cyanide ion
Bicarbonate ion	HCO_3^{1-}	CO_3^{2-}	Carbonate ion
Water	H_2O	OH^{1-}	Hydroxide ion
Methyl alcohol	CH_3-O-H	CH_3-O^{1-}	Methoxide ion
Ammonia	NH_3	NH_2^{1-}	Amide ion
Methane	CH_4	CH_3^{1-}	Methyl carbanion

* Reciprocal relations:
If acid is strong, conjugate base is weak.
If acid is weak, conjugate base is strong.
If base is strong, conjugate acid is weak.
If base is weak, conjugate acid is strong.

removed from the acid. Many texts use the term **conjugate acid-base pairs** when speaking of these groupings. Thus Cl^{1-} is the **conjugate base** of the acid HCl, and NH_4^+ is the **conjugate acid** of the base NH_3. Table 18–3 lists several conjugate acid-base pairs.

A strong Bronsted-Lowry acid is a good provider of hydrogen ions. If the acid HA is a strong acid, the reaction $HA \rightarrow H^{1+} + A^{1-}$ is favored over the reverse reaction, $A^{1-} + H^{1+} \rightarrow HA$. This means that A^{1-}, the conjugate base of HA, is a *weak* Bronsted-Lowry base and is not a good acceptor of hydrogen ions. The pairings strong acid–weak conjugate base and weak acid–strong conjugate base always exist. Notice in Table 18–3 that, as the strength of the acid increases, the strength of the conjugate base decreases.

Direction of Reversible Acid-Base Reactions

The reactions given in Table 18–2 all have the form

$$\text{Acid} + \text{base} \rightleftharpoons \text{acid} + \text{base}$$

If they are reversible, we would like to be able to predict which direction will be the prominent one. Normally, the direction is dictated by the relative strengths of the acids and bases present and goes from strength to weakness:

$$\text{Stronger acid} + \text{stronger base} \longrightarrow \text{weaker acid} + \text{weaker base}$$

For the reaction

$$HCl + NH_3 \rightleftharpoons NH_4^{1+} + Cl^{1-}$$

HCl is a stronger acid than NH_4^{1+}, and NH_3 is a stronger base than Cl^{1-} (see Table 18–3). The major direction of the reversible reaction is from left to right, from strength to weakness.

Lewis Theory of Acids and Bases

The Arrhenius theory is limited to aqueous solutions and compounds that contain H^+ and OH^- ions. Bronsted-Lowry theory can be used in any solvent, but is involved only with transfer of protons. In 1916, G. N. Lewis introduced a more general theory of acids and bases that could be used even when no proton transfer occurred:

1. An **acid** is an electron pair acceptor ($A + {:}X \rightarrow A{:}X$). (electron pair, pointing to ":")
2. A **base** is an electron pair donor ($B{:} + X \rightarrow B{:}X$).

This definition allows for a broader classification that is especially useful in organic and biochemistry. In this type of covalent bond formation, one of the partners contributes both electrons to make the bond:

$$\underset{\text{base}}{X{:}} + \underset{\text{acid}}{Y} \longrightarrow \underset{\substack{\uparrow \\ \text{new covalent bond}}}{X{:}Y}$$

Here X: is a base since it contributes the electron pair to the bond, and Y is an acid since it accepts the electron pair. Therefore, any molecule or ion with a pair of unshared electrons can act as a Lewis base. Any molecule or ion in need of electrons can act as a Lewis acid. Often, we have to write out the electron dot configurations of ions and molecules in order to determine which is the Lewis acid and base.

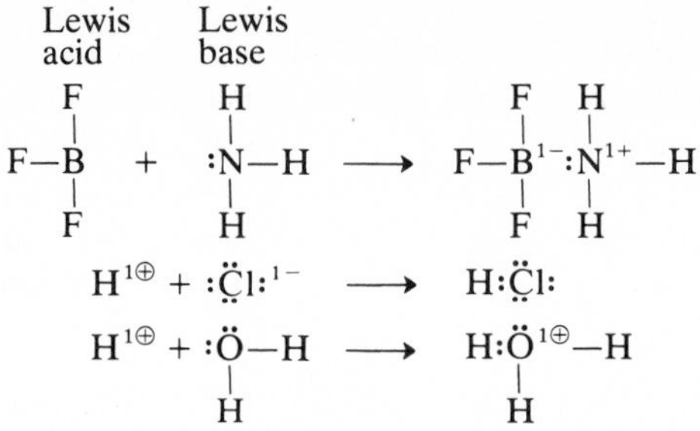

$$
\begin{array}{cc}
\text{Lewis} & \text{Lewis} \\
\text{acid} & \text{base}
\end{array}
$$

$$
\underset{\underset{\displaystyle F}{|}}{\overset{\overset{\displaystyle F}{|}}{F\!-\!B}} \;+\; \underset{\underset{\displaystyle H}{|}}{\overset{\overset{\displaystyle H}{|}}{:N\!-\!H}} \longrightarrow \underset{\underset{\displaystyle F}{|}\;\underset{\displaystyle H}{|}}{\overset{\overset{\displaystyle F}{|}\;\overset{\displaystyle H}{|}}{F\!-\!B^{1-}\!:\!N^{1+}\!-\!H}}
$$

$$H^{1\oplus} + :\ddot{\underset{..}{C}}l:^{1-} \longrightarrow H\!:\!\ddot{\underset{..}{C}}l:$$

$$H^{1\oplus} + :\ddot{O}\!-\!\underset{\displaystyle H}{\overset{|}{}} \longrightarrow H\!:\!\ddot{O}^{1\oplus}\!-\!\underset{\displaystyle H}{\overset{|}{}}$$

In conclusion, the three major definitions of acids and bases are similar in many respects but still deal with different circumstances. Most of the time we shall consider an acid as a hydrogen ion donor and a base as either a hydroxide donor or a hydrogen ion acceptor. Occasionally, we may use the Lewis concept. Table 18–4 compares the three definitions.

Table 18–4 Comparison of definitions of acids and bases

	Acid	Base
Property	Sour taste	Bitter taste
	Turns litmus red	Turns litmus blue
	Reacts with bases	Reacts with acids
		Soapy to the touch
Arrhenius theory	Donates H^{1+} ions	Donates OH^{1-} ions
	$(HA \rightarrow H^{1+} + A^{1-})$	$(BOH \rightarrow B^{1+} + OH^{1-})$
Bronsted-Lowry theory	Donates H^{1+} ions	Accepts H^{1+} ions
	$(HA \rightarrow H^{1+} + A^{1-})$	$(A^{1-} + H^{1+} \rightarrow HA)$
Lewis theory	Accepts electron pair	Donates electron pair
	$(A + :X \rightarrow A:X)$	$(B: + X \rightarrow B:X)$

One final point should be mentioned before going on to reactions of acids and bases. In water there is no such thing as a hydrogen ion. It always reacts with water to form a **hydronium ion** (H_3O^{1+}). When we speak of the concentration of hydrogen ion in solution, we actually mean the concentration of hydronium ions since that is the carrier for H^{1+} ions. For accuracy, many texts always use the form H_3O^{1+} in acid-base reactions, but we shall use H^{1+} almost exclusively because it tends to be less confusing for the beginning student.

$$H^{1+}(\text{hydrogen ion}) + H_2O \rightarrow H_3O^{1+}(\text{hydronium ion})$$

UNIT 18 Problems and Exercises

18–1. Vinegar, lemon juice, and curdled milk all taste sour. What other properties would you expect them to have in common?

18–2. In the following reactions, identify the compounds underlined as acids or bases.

(a) $\underline{HCl} + H_2O \rightarrow Cl^{1-} + H_3O^{1+}$

(b) $\underline{NH_3} + Na \rightarrow NH_2^{1-} + NaH + \frac{1}{2}H_2$

(c) $\underline{BF_3} + NH_3 \rightarrow F_3B^{1-}\text{—}NH_3^{1+}$

(d) $\underline{CH_4} + H^{1-} \rightarrow CH_3^{1-} + H_2$

(e) $\underline{CH_3^{1-}} + NH_4^{1+} \rightarrow CH_4 + NH_3$

(f) $\underline{HNO_3} \rightarrow H^{1+} + NO_3^{1-}$

18–3. In the following reactions, label each compound underlined as either acid or base.

(a) $\underline{NaOH} + \underline{HCl} \leftrightharpoons Na^{1+} + \underline{Cl^{1-}} + \underline{H_2O}$

(b) $\underline{HNO_3} + \underline{KOH} \leftrightharpoons \underline{H_2O} + \underline{NO_3^{1-}} + K^{1+}$

(c) $\underline{H_2O} + \underline{NH_3} \leftrightharpoons NH_4^{1+} + \underline{OH^{1-}}$

(d) $\underline{H_2O} + \underline{HCl} \leftrightharpoons \underline{H_3O^{1+}} + \underline{Cl^{1-}}$

18–4. Label the underlined compounds as Arrhenius, Bronsted-Lowry, or Lewis bases. (Some may be more than one.)

(a) $\underline{:NH_3} + H^{1+} \rightarrow NH_4^{1+}$

(b) $\underline{:CH_3^{1-}} + H_2O \rightarrow CH_4 + OH^{1-}$

(c) $\underline{NaOH} \rightarrow Na^{1+} + OH^{1-}$

(d) $\underline{Br^{1-}} + CH_3^{1+} \rightarrow Br\text{—}CH_3$

18–5. According to the Bronsted-Lowry definition, determine which of the ions in each pair is the stronger base. Use Table 18–3.

(a) Br^{1-} or HCO_3^{1-} (b) $H_2PO_4^{1-}$ or HSO_4^{1-}
(c) NO_3^{1-} or I^{1-} (d) NH_3 or NH_2^{1-}
(e) H_2O or OH^{1-}

18–6. Give the conjugate acids for the following bases.

(a) H_2O (b) NH_2^{1-} (c) NH_3 (d) Cl^{1-}
(e) HS^{1-} (f) CH_3^{1-} (g) HSO_4^{1-} (h) $H_2PO_4^{1-}$
(i) CN^{1-}

18–7. Give the conjugate bases for the following acids.

(a) CH_4 (b) H_2O (c) NH_3 (d) HCl
(e) HNO_3 (f) HS^{1-} (g) HSO_4^{1-} (h) $H_2PO_4^{1-}$

18–8. What is the major direction of reaction for the following acid-base reactions?

(a) $CH_3\overset{\overset{O}{\|}}{C}\text{—}OH + HSO_4^{1-} \leftrightharpoons CH_3\text{—}\overset{\overset{O}{\|}}{C}\text{—}O^{1-} + H_2SO_4$

(b) $NH_4^{1+} + H_2O \leftrightharpoons NH_3 + H_3O^{1+}$

(c) $CH_3^{1-} + H_2O \leftrightharpoons CH_4 + OH^{1-}$

(d) $Cl^{1-} + HSO_4^{1-} \leftrightharpoons HCl + SO_4^{2-}$

(e) $H_2PO_4^{1-} + CH_3OH \leftrightharpoons CH_3O^{1-} + H_3PO_4$

(f) $H_2S + H_2O \leftrightharpoons HS^{1-} + H_3O^{1+}$

18–9. In the following reactions, label the conjugate acid-base pairs and identify all acids and bases. Which acid is stronger?

(a) $NH_3 + H_3O^{1+} \rightarrow NH_4^{1+} + H_2O$
(b) $CH_3OH + NH_2^{1-} \rightarrow CH_3O^{1-} + NH_3$
(c) $NH_2^{1-} + H_2O \rightarrow NH_3 + OH^{1-}$

UNIT 19 CONCENTRATIONS OF HYDROGEN AND HYDROXIDE IONS AND pH

Water ionizes to a small extent to yield hydrogen ions and hydroxide ions, so any aqueous solution always contains some hydrogen and hydroxide ions. The amount of ionization is small, but the concentrations of hydrogen ions and hydroxide ions present can be calculated from the equilibrium constant expression.

$$H_2O \rightleftharpoons H^{1+} + OH^{1-}, \qquad K_{eq} = \frac{[H^{1+}][OH^{1-}]}{[H_2O]} \tag{19–1}$$

The equilibrium constant, K_{eq}, has been calculated to be 1.8×10^{-16} at room temperature, so only about one molecule out of every 550 million water molecules is converted to hydrogen and hydroxide ions. For purposes of calculation, we can assume that the concentration of water is constant and does not change, even during ionization; it has been calculated to be 55.6 *M*. If we insert the values for K_{eq} and $[H_2O]$ into equation (19–1) and simplify, we can derive an expression relating the concentrations of hydrogen ion and hydroxide ion in any aqueous solution:

$$1.8 \times 10^{-16} = \frac{[H^{1+}][OH^{1-}]}{55.6\,M}$$

$$(1.8 \times 10^{-16})(55.6\,M) = [H^{1+}][OH^{1-}]$$

$$1.0 \times 10^{-14} = [H^{1+}][OH^{1-}] \tag{19–2}$$

The net result is equation (19–2), which tells us that the product of the concentration of hydrogen ion and the concentration of hydroxide ion in aqueous solution (at room temperature) must always equal 1×10^{-14}. If the $[H^{1+}]$ is 1×10^{-2} *M*, then the $[OH^{1-}]$ must be 1×10^{-12} *M*. As the concentration of one ion increases, the concentration of the other must decrease. Table 19–1 lists some examples.

Acidic and Basic Solutions

Equation (19–2) also allows us to calculate the concentrations of hydrogen and hydroxide ions in acidic, neutral, and basic solutions. In a **neutral solution** the concentration of hydrogen ion equals the concentration of hydroxide ion, so both must be equal to 1×10^{-7} *M*. In an **acidic solution** the concentration of hydrogen ion is greater than that of hydroxide ion, so the hydrogen ion concentration must be greater than 1×10^{-7} *M*. In a **basic solution** the reverse exists; the concentration of hydroxide is greater than that of hydrogen ion. The concentration of hydroxide ion must be greater than 1×10^{-7} *M*.

1. Acidic solution: $[H^{1+}]$ greater than $[OH^{1-}]$; $[H^{1+}]$ greater than 1×10^{-7} *M*; $[OH^{1-}]$ less than 1×10^{-7} *M*.

2. Neutral solution: $[H^{1+}]$ equal to $[OH^{1-}]$; both are 1×10^{-7} *M*.

3. Basic solution: $[OH^{1-}]$ greater than $[H^{1+}]$; $[OH^{1-}]$ greater than 1×10^{-7} *M*; $[H^{1+}]$ less than 1×10^{-7} *M*.

pH Units

The pH and pOH scales are convenient methods of expressing concentrations of acids and bases, since the use of exponential notation is not required. Hydrogen ion and hydroxide ion concentration are expressed in simple numbers by this method.

The mathematical symbol p in front of a term means that we must take the logarithm of that term and then multiply the result by negative one. This means that any number containing a negative exponential term will be converted to a positive simple number:

$$\log 10^x = x, \quad \text{so} \log(10^{-7}) = -7$$

$$p(10^{-7}) = -1 \log(10^{-7}) = -1 \times (-7) = 7$$

The term pH refers to the *negative logarithm of the concentration of hydrogen ion*, and the term pOH refers to the *negative logarithm of the concentration of hydroxide ion in solution*:

$$\text{pH} = -1 \log(H^{1+}), \qquad \text{pOH} = -1 \log(OH^{1-})$$

If the concentration of hydrogen ion in solution is 1×10^{-1} *M*, the pH of the solution is

$$-1 \log(1 \times 10^{-1}) = 1$$

The concentration of hydroxide in this same solution is 1×10^{-13}, so the pOH is

$$-1 \log(1 \times 10^{-13}) = 13$$

pH values normally range from 0 to 14, but negative pH values are also possible. If the concentration of acid is greater than 1 *M*, then pH will be negative. The same holds for pOH values. The following problems show some sample calculations of pH and pOH. You should refer to the student supplement for a review of logarithms.

Problem 1: What are the pH and pOH of a neutral solution?

$$\text{Neutral:}\quad [H^{1+}] = 1 \times 10^{-7}\,M \qquad [OH^{1-}] = 1 \times 10^{-7}\,M$$

$$\begin{aligned} pH &= -1\log(1 \times 10^{-7}) \\ &= -1(\log 1 + \log 10^{-7}) \\ &= -1(0 - 7) \\ &= -1(-7) \\ &= 7 \end{aligned} \qquad \begin{aligned} pOH &= -1\log(1 \times 10^{-7}) \\ &= -1(\log 1 + \log 10^{-7}) \\ &= -1(0 - 7) \\ &= -1(-7) \\ &= 7 \end{aligned}$$

Problem 2: From equation (19–2), calculate the pH and pOH when $[H^{1+}] = 5 \times 10^{-4}\,M$.

$$[OH^{1-}] = \frac{1 \times 10^{-14}}{5 \times 10^{-4}} = 2 \times 10^{-11}$$

$$[H^{1+}] = 5 \times 10^{-4}\,M \qquad [OH^{1-}] = 2 \times 10^{-11}\,M$$

$$\begin{aligned} pH &= -1\log(5 \times 10^{-4}) \\ &= -1(\log 5 + \log 10^{-4}) \\ &= -1(0.7 - 4.0) \\ &= -1(-3.3) \\ &= +3.3 \end{aligned} \qquad \begin{aligned} pOH &= -1\log(2 \times 10^{-11}) \\ &= -1(\log 2 + \log 10^{-11}) \\ &= -1(0.3 - 11.0) \\ &= -1(-10.7) \\ &= +10.7 \end{aligned}$$

From equation (19–2) we know that $[H^{1+}] \times [OH^{1-}]$ must always be equal to 1×10^{-14}. In pH and pOH terms we see from Problems (1) and (2) that pH + pOH always equals 14.

Reverse calculations are more difficult. Problem (3) illustrates how one calculates $[H^{1+}]$ and $[OH^{1-}]$ from pH and pOH.

Problem 3: Calculate the $[H^{1+}]$ and $[OH^{1-}]$ if the pH is 5.4.

$$\begin{aligned} pH = -1\log[H^{1+}] &= 5.4 \\ \log[H^{1+}] &= -5.4 \\ &= (-6 + 0.6) \\ [H^{1+}] &= 10^{-6} \times 4 \\ &= 4 \times 10^{-6}\,M \end{aligned} \qquad \begin{aligned} pOH = -1\log[OH^{1-}] &= 8.6 \\ \log[OH^{1-}] &= -8.6 \\ &= (-9 + 0.4) \\ [OH^{1-}] &= 10^{-9} \times 2.5 \\ &= 2.5 \times 10^{-9}\,M \end{aligned}$$

Remember that we have to rewrite the log in order that the decimal portion be positive. To do this, change the integer to one larger negative number and add a positive decimal to it:

$$-5.4 = -6 + 0.6$$

We did not have to use both pH and pOH to calculate $[H^{1+}]$ and $[OH^{1-}]$. Once having calculated $[H^{1+}]$ from the pH, we could have used equation (19–2):

$$[OH^{1-}] = \frac{1 \times 10^{-14}}{4 \times 10^{-6}} = 2.5 \times 10^{-9}\,M$$

Table 19–1 lists pH values along with $[H^{1+}]$ and $[OH^{1-}]$. A pH of 7 indicates a neutral solution. Below pH values of 7, the solution is acidic, and above 7 it is basic. Remember also that a solution of pH = 3 is *more* acidic than a solution at pH = 5.

Table 19–1 pH values for some body fluids and consumer goods

$[OH^{1-}]$	$[H^{1+}]$		pH	Body fluids	Foods and goods
1×10^{-14}	1×10^{0}	Strongly acidic	0		1 *M* HCl (0)
1×10^{-13}	1×10^{-1}	↑	1		0.1 *M* HCl (1)
1×10^{-12}	1×10^{-2}		2	Gastric juices (1.6–1.8)	Lemon (2.2–2.4)
1×10^{-11}	1×10^{-3}		3		Vinegar (3.5)
1×10^{-10}	1×10^{-4}	Increasingly acidic	4		Orange (3.0–4.0) Coca Cola (3.6) Tomato (4.0–4.4)
1×10^{-9}	1×10^{-5}		5	Urine (5.5–7.0)	Tap water (5–6) Black coffee (5)
1×10^{-8}	1×10^{-6}		6	Saliva (6.2–7.8)	Potato (5.6–6.0) Milk (6.3–6.6)
1×10^{-7}	1×10^{-7}	Neutral	7	Blood (7.35–7.45) Bile (7.8–8.6)	Fresh egg (7.7)
1×10^{-6}	1×10^{-8}		8	Pancreatic juice (7.8–8.0)	Sodium bicarbonate solution (8–9)
1×10^{-5}	1×10^{-9}		9		
1×10^{-4}	1×10^{-10}	Increasingly basic	10		Milk of magnesia (10)
1×10^{-3}	1×10^{-11}		11		Aqueous ammonia (11)
1×10^{-2}	1×10^{-12}		12		Na_2CO_3; washing soda (12)
1×10^{-1}	1×10^{-13}	↓	13		Lye (13)
1×10^{0}	1×10^{-14}	Strongly basic	14		1 *M* NaOH (14)

Acid-Base Indicators

Acid-base indicators are compounds that exhibit one color at or below a certain pH and a different color above that pH. They are commonly used to determine the approximate pH of a solution. One common indicator called *litmus* is impregnated in strips of paper, and it is used by wetting it with the test solution. The paper turns red if the pH is less than 4.5 and blue if the pH is above 7.0. Another common indicator, phenolphthalein, is colorless at pH values less than 8 and red in color above pH values of 9. Table 19–2 lists some of the more common indicators and their color changes.

pH Meters

The use of acid-base indicators is generally limited to colorless solutions, and allows us to determine only the approximate pH of a solution. When a more exact measure of the pH is required, pH meters are generally used. These instruments permit the

Table 19–2 Common acid-base indicators

Name	pH of color change	pH 1 2 3 4 5 6 7 8 9 10 11 12
Methyl orange	3.1–4.4	... red .....(pink).... yellow ....
Bromothymol blue	6.0–7.6	.... yellow...........(green)..... blue ....
Litmus	4.5–7.0	... red ....(indistinct) .... blue ....
Phenolphthalein	8.0–9.2	... colorless ....(pink)........ red...
Universal indicator (a composite of several indicators)		.. red ....(orange) .. yellow...(green) .. blue ....

instantaneous determination of the pH of any aqueous solution, and can also be used to monitor the changes in pH of a solution during a chemical reaction.

UNIT 19 Problems and Exercises

19–1. When lemon juice is added to tea, the color of the solution changes. Suggest an explanation for this occurrence.

19–2. When would a pH meter be used in place of indicators? When would acid-base indicators be preferred?

19–3. Mark the following as true or false. Explain why you marked some statements false.
(a) If water is added to a hydrochloric acid solution, the pH will increase.
(b) A solution whose pOH is 8 is more acidic than a solution whose pOH is 10.
(c) pH values can never become negative numbers.
(d) A pH meter allows you to calculate the $[H^{1+}]$ in a solution.

19–4. Calculate the hydroxide ion concentration of a solution if the hydrogen ion concentration is (a) 3×10^{-10} *M*, (b) 1.5×10^{-5} *M*, (c) 4×10^{-3} *M*.

19–5. Calculate the pH values from the following hydrogen ion concentrations.
(a) 3.5×10^{-5} *M* (b) 8×10^{-13} *M* (c) 6×10^{-4} *M*
(d) 1.0 *M* (e) 6×10^{-2} *M* (f) 0.5×10^{-1} *M*
(g) 5.0 *M*

19–6. Calculate the hydroxide ion concentrations in the following solutions.
(a) pOH = 6 (b) pOH = 3.8 (c) pH = 8
(d) pH = 4.5 (e) pH = 3.2

19–7. Calculate the hydrogen ion concentration for the following pH values.
(a) 6.0 (b) 5.5 (c) 7.0 (d) 8.6
(e) 10.8 (f) −1.0 (g) 0 (h) 7.2

19–8. Complete the following table by filling in the empty spaces.

$[H^{1+}]$	$[OH^{1-}]$	pH	Solution is (acidic/basic)
	10^{-3}		
		8	
5×10^{-9}			
	3×10^{-7}		
		4.5	
		8.6	
		0.0	

19–9. Urine normally has a pH of 6. If a patient eliminates 1300 mℓ of urine per day, how many grams of HCl have been eliminated? (Assume all acid is HCl.)

19–10. Calculate the pH of a solution resulting from mixing 500 mℓ of HCl (pH = 2) with 100 mℓ of HNO_3 (pH = 1).

19–11. Which acid-base indicator would you choose to distinguish between the following pairs of solutions? For each case give the color of the indicator used in each solution.

	Solution 1	Solution 2
(a)	pH = 0	pH = 3
(b)	pH = 5	pH = 7
(c)	$[H^{1+}] = 10^{-8}$	$[H^{1+}] = 10^{-10}$
(d)	$[OH^{1-}] = 10^{-8}$	$[OH^{1-}] = 10^{-6}$

19–12. How many grams of NaOH must be added to 1 ℓ of pure water to make the pH = 11?

19–13. How many grams of HCl must be added to pure water (1 ℓ) to make the pH = 2?

19–14. Estimate the pH values of the following solutions, given the acid-base indicator colors.

	Indicator	Color
(a)	Solution 1	Methyl orange = yellow
		Bromothymol blue = yellow
(b)	Solution 2	Universal = greenish blue
(c)	Solution 3	Phenolphthalein = red
		Universal = blue

19–15. Separate solutions of two acids, A and B, have the same pH. Does this mean that the molarities of the acids, A and B, are the same?

UNIT 20 REACTIONS OF ACIDS AND BASES, AND TITRATIONS

Acids and bases react with many different types of compounds. In this unit we shall introduce the reactions of acids with metals, carbonates, and bicarbonates. Then we shall discuss titrations, the reactions of acids with bases. In later units we shall learn of acid and base reactions with the chemicals present in our bodies.

Reactions of Acids With Metals

When pure metals react with acids, the products are usually metal salts and hydrogen gas. These reactions are really oxidation-reduction reactions, since the metal is oxidized and the hydrogen ions are reduced.

$$\text{Acid} + \text{metal} \longrightarrow \text{metal salt} + \text{hydrogen gas}$$

$$2H_2O + 2Na \longrightarrow 2NaOH + H_2$$

$$\text{Oxidation:}\quad Na - 1e^- = Na^{1+}$$

$$\text{Reduction:}\quad 2H^{1+} + 2e^- = H_2$$

Table 20–1 Reactivities of metals with acids

Group no.	Metal			
IA	K, Na, Li	React violently with water (a weak acid)	All react with acids, liberating hydrogen gas (K through Pb)	↑ Reactivity with acids increases (from Au up to K)
IIA	Ca, Mg	React more slowly with water		
Transition Metals	Al, Mn, Zn, Cr, Fe, Ni, Sn, Pb	React with steam		
	Cu, Bi, Sb, Hg, Ag, Pt, Au	React very slowly, if at all, with acids		

As is shown in Table 20–1, not all metals react with acids. Metals from group IA of the periodic table react quite vigorously with acids, even a weak acid like water. The reactions are so rapid that elements like sodium, lithium, and potassium must be stored in airtight containers to prevent their reactions with water vapor from the air. Group IIA metals also react with water, but not as vigorously as the group IA metals. Transition metals react only slowly with weak acids, and some do not react even with strong acids.

There are many industrial uses for acid-metal reactions, but there are also many instances where great effort is expended in trying to prevent such reactions. When exposed for long periods of time to water vapor, metals like iron, aluminum, and tin are converted to their oxides and hydroxides. These processes are called **corrosion** and cause millions of dollars of damage annually to cars, electrical instruments, and many other types of machines.

Reactions of Acids With Carbonates and Bicarbonates

When carbon dioxide dissolves in water, carbonic acid is formed. Carbonic acid can then ionize to form bicarbonate and carbonate ions:

$$CO_2 + H_2O \rightleftharpoons H_2CO_3 \text{ (carbonic acid)}$$
$$H_2CO_3 \rightleftharpoons H^{1+} + HCO_3^{1-}$$
$$HCO_3^{1-} \rightleftharpoons H^{1+} + CO_3^{2-}$$

The reactions shown are reversible, so the treatment of bicarbonates and carbonates with acids yields carbon dioxide and water:

$$\text{Acids} + \text{carbonates} \longrightarrow \text{salts} + \text{carbon dioxide} + \text{water}$$
$$2HCl + Na_2CO_3 \longrightarrow 2NaCl + CO_2 + H_2O$$
$$H_2SO_4 + MgCO_3 \longrightarrow MgSO_4 + CO_2 + H_2O$$

$$\text{Acids} + \text{bicarbonates} \longrightarrow \text{salts} + \text{carbon dioxide} + \text{water}$$
$$HCl + NaHCO_3 \longrightarrow NaCl + CO_2 + H_2O$$
$$2HNO_3 + Ca(HCO_3)_2 \longrightarrow Ca(NO_3)_2 + 2CO_2 + H_2O$$

Marble and limestone are composed principally of calcium carbonate, so structures built of these rocks are susceptible to any acids that might be present in the air. Sulfur dioxide and sulfur trioxide present in smoggy, polluted air react with water vapor to produce sulfurous acid and sulfuric acid. These acids in turn react with calcium carbonate surfaces of many statues, monuments, and buildings, gradually ruining their appearance.

Air pollutants:

$$SO_2 \text{ (sulfur dioxide)} + H_2O = H_2SO_3 \text{ (sulfurous acid)}$$
$$SO_3 \text{ (sulfur trioxide)} + H_2O = H_2SO_4 \text{ (sulfuric acid)}$$
$$H_2SO_3 + CaCO_3 \longrightarrow CaSO_3 + CO_2 + H_2O$$

Baking powder is essentially a mixture of sodium bicarbonate and some weak acid (usually tartaric acid or calcium monohydrogen phosphate). When in contact with water in cake or pastry dough, the acid reacts with the bicarbonate to liberate CO_2. The CO_2 causes the cake or pastry to rise during baking.

A soda-acid fire extinguisher works on the same bicarbonate–acid reaction (see Figure 20–1). The larger main tank of the extinguisher is filled with H_2CO_3, HCO_3^{1-}, and water. A small stoppered bottle containing acid is placed inside the top of the tank. Inversion of the extinguisher causes rapid mixing of acid with HCO_3^{1-} and H_2CO_3. This causes the formation of large amounts of carbon dioxide, which builds up pressure and forces the water out of the hose onto the fire.

Another major use of carbonates and bicarbonates is as stomach acid neutralizing agents, or antacids. In the stomach, hydrochloric acid is secreted to aid in the digestion of foods. Under conditions of stress or anxiety, the glands sometimes secrete too much HCl and a condition of hyperacidity exists. Most of the antacids sold in drugstores contain carbonates or bicarbonates to neutralize this excess HCl. One important caution should be noted. If too much antacid is ingested, all the HCl can be neutralized and a condition of *alkalosis* can result. Hyperacidity is uncomfortable and may eventually cause ulcers, but alkalosis can be lethal. So follow the directions on the antacid preparation carefully.

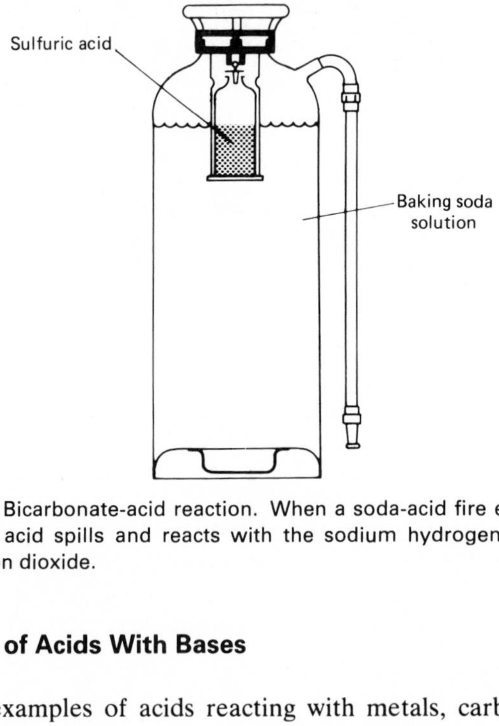

Sulfuric acid
Baking soda solution

Figure 20–1 Bicarbonate-acid reaction. When a soda-acid fire extinguisher is inverted, the acid spills and reacts with the sodium hydrogen carbonate to liberate carbon dioxide.

Neutralization: Reaction of Acids With Bases

There are many examples of acids reacting with metals, carbonates, and bicarbonates, but the most commonly encountered reaction is the reaction of acids with bases:

$$\text{Acids} + \text{bases} \longrightarrow \text{salts} + \text{water}$$

$$HCl + NaOH \longrightarrow NaCl + H_2O$$

$$H_2SO_4 + 2KOH \longrightarrow K_2SO_4 + 2H_2O$$

Actually, the net reaction that occurs here is simply the combination of a hydrogen ion (from the acid) with a hydroxide ion (from the base) to form water. The salt is made up of the other ions remaining in solution:

$$\left.\begin{array}{l} HCl \longrightarrow H^{1+} + Cl^{1-} \\ NaOH \longrightarrow OH^{1-} + Na^{1+} \end{array}\right\} \text{These ions make up the salt}$$

$$\text{Neutralization:} \quad H^{1+} + OH^{1-} \longrightarrow H_2O$$

To counteract the presence of an acid, we add base; conversely, to counteract or *neutralize* a base, we add acid. *Neutralization* means "to destroy the effect of."

Titrations

Acid-base neutralization reactions are commonly used to quantitatively determine the concentration of an acid or a base in some aqueous solution. For example, suppose that we have an HCl solution and need to know the exact molarity of HCl in that solution. If we take a known volume of that solution and add just enough base to it, the solution will be neutralized. By knowing the molarity of the base and by measuring the exact volume of base used, we can calculate the number of moles of base that were added. Then we can use this value to calculate the number of moles of HCl that were present in the volume of acid that we used. If we know both the volume of the acid solution used and the number of moles of HCl present in that solution, we can calculate the molarity of the solution.

Experimentally, acid-base indicators are added to the acid solution to aid in determining the exact volume of base that must be added to just neutralize the solution. See Figure 20–2 for an illustration of this technique. The most commonly

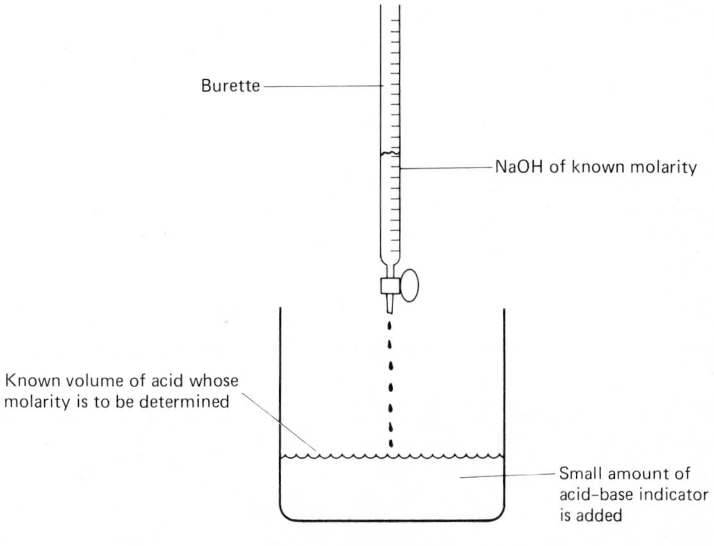
Burette
NaOH of known molarity
Known volume of acid whose molarity is to be determined
Small amount of acid–base indicator is added

Figure 20–2

used indicator, phenolphthalein, is colorless in acid solution and red in basic solution. In a titration, the indicator is placed in the acid sample, and base is slowly added from a burette until the solution just barely remains pink in color.

Working Acid-Base Titration Problems

Acid-base titration problems are easy to work if you remember to use the balanced chemical reaction, to change volumes to liters, and to keep track of which volumes are for base and which are for acids. Problem (1) illustrates a step-by-step method that has worked well for many students. Problems (2) and (3) illustrate variations in the procedure.

Problem 1: What is the molarity of a solution of sulfuric acid if 200 mℓ is neutralized by the addition of 300 mℓ of 0.2 M NaOH?

STEP 1: *Write down the balanced acid-base reaction* and note the mole ratios with which the acid and base react.

$$H_2SO_4 + 2NaOH \longrightarrow Na_2SO_4 + 2H_2O$$

To neutralize every mole of sulfuric acid present, 2 mol of NaOH are required. This is because sulfuric acid has two acid protons per molecule.

STEP 2: Below the balanced equation, *make a table with spaces for the molarity, volume (in liters), and number of moles* for both the acid and the base. Fill in as much data from the problem as is given.

H_2SO_4	+	$2NaOH$	$\longrightarrow Na_2SO_4 + 2H_2O$
M ?		M 0.2 M	
V 0.02 ℓ		V 0.03 ℓ	
No. of moles ____		No. of moles ____	

STEP 3: *Calculate number of moles of the acid or base.* In the above table the required answer is M H_2SO_4. Notice that the three units in each column are related by

$$M \cdot V = \text{no. of mol}$$

On the base side we know M NaOH and V NaOH, so we can calculate the number of moles of NaOH used.

$$M \cdot V = \text{no. of mol}$$

$$0.2 \times 0.03 = 0.006 \text{ mol of NaOH}$$

were consumed as it neutralized all the H_2SO_4.

STEP 4: *Calculate the number of moles of H_2SO_4 present.* Since we know the ratio in which H_2SO_4 and NaOH react (step 1), we can calculate how many moles of H_2SO_4 were present:

$$0.006 \text{ mol of NaOH used} \times \frac{1 \text{ mol of } H_2SO_4}{2 \text{ mol of NaOH}} = 0.003 \text{ mol of } H_2SO_4 \text{ present}$$

STEP 5: Calculate final answer.

$$M(H_2SO_4) = \frac{\text{no. mol of } H_2SO_4}{V\ H_2SO_4} = \frac{0.003 \text{ mol}}{0.02\ \ell} = 0.15\ M$$

Problem 2: If 20 g of sulfuric acid is present in a flask, how many milliliters of 0.5 M sodium hydroxide must be added to neutralize it?

STEP 1: $H_2SO_4 + 2NaOH \rightarrow 2H_2O + Na_2SO_4$

STEPS 2 and 3:

	H_2SO_4	+		2NaOH	$\longrightarrow$
M	×		M	0.5	
V	×		V	?	
20 g →	No. of moles ______			No. of moles ______	

Here M (acid) and V (acid) are not required or given.

We need to know the number of moles of NaOH in order to calculate the volume required. If we know the number of moles of acid present, we can calculate the number of moles of NaOH.

$$20 \text{ g of } H_2SO_4 \times \frac{1 \text{ mol of } H_2SO_4}{98 \text{ g}} = 0.22 \text{ mol of } H_2SO_4$$

STEP 4:

$$0.22 \text{ mol of } H_2SO_4 \times \frac{2 \text{ mol of NaOH}}{1 \text{ mol of } H_2SO_4} = 0.44 \text{ mol of NaOH required}$$

STEP 5:

$$M \cdot V = \text{no. of mol}$$

$$V = \frac{\text{no. of mol}}{M} = \frac{0.44}{0.5} = 0.88\ \ell = 880\ m\ell \text{ of the NaOH solution must be added}$$

Problem 3: How many grams of sodium hydroxide must be added to neutralize 40 mℓ of 0.2 M acetic acid?

STEP 1:

$$CH_3CO_2H + 1NaOH \longrightarrow CH_3CO_2^{1-}Na^{1+} + H_2O$$

STEP 2:

	CH_3CO_2H	+		NaOH	$\longrightarrow$
M	0.2		M	×	Not required
V	0.04 ℓ		V	×	Not required
No. of moles	0.008		No. of moles	______	$\longrightarrow$ g

STEPS 3 and 4:

$$0.008 \text{ mol of } CH_3CO_2H \times \frac{1 \text{ mol of NaOH}}{1 \text{ mol of } CH_3CO_2} = 0.008 \text{ mol of NaOH}$$

STEP 5: We know that 0.008 mol of NaOH are needed. To get grams required,

$$0.008 \text{ mol of NaOH} \times \frac{40 \text{ g}}{1 \text{ mol}} = 0.320 \text{ g of NaOH}$$

must be added.

Acid-base neutralization problems can also be worked using normality instead of molarity as the concentration units. If normalities are used, titration calculations are relatively easy since the *number of equivalents of acid present must always equal the number of equivalents of base present.* Since normality × volume equals the number of equivalents.

$$N_{\text{acid}} \times V_{\text{acid}} = N_{\text{base}} \times V_{\text{base}}$$

Problem 4: What volume of 0.1 N sodium hydroxide is required to neutralize 20 mℓ of 0.3 M sulfuric acid? First change the molarity of sulfuric acid to normality:

$$\frac{0.3 \text{ mol of acid}}{1\ \ell} \times \frac{2 \text{ equiv.}}{1 \text{ mol of } H_2SO_4} = \frac{0.6 \text{ equiv. of acid}}{\ell} = 0.6\ N\ H_2SO_4$$

$$N_{\text{acid}} \times V_{\text{acid}} = N_{\text{base}} \times V_{\text{base}}$$

$$0.6 \times 0.02 = 0.1 \times V_{\text{base}}$$

$$\frac{0.6 \times 0.02}{0.1} = V_{\text{base}} = 0.12\ \ell$$

In these problems we do not have to write the balanced equation; we do, however, have to be able to determine the normalities of the solutions.

UNIT 20 Problems and Exercises

20–1. Vinegar, lemon juice, and orange juice are commonly listed as antidotes in first-aid treatment for someone who has swallowed lye (NaOH). What do you think is the purpose of these compounds?

20–2. Complete and balance the following reactions of acids.

(a) $HCl + NaOH \rightarrow$
(b) $HCl + NaHCO_3 \rightarrow$
(c) $HCl + Na_2O \rightarrow$
(d) $HCl + Na_2CO_3 \rightarrow$
(e) $H_2SO_4 + KOH \rightarrow$
(f) H_2SO_4 + calcium bicarbonate →
(g) H_2SO_4 + ferric oxide →
(h) HNO_3 + aluminum hydroxide →
(i) HNO_3 + calcium bicarbonate →
(j) Sodium + $H_2O \rightarrow$
(k) Potassium + $H_2O \rightarrow$
(l) $Zn + HCl \rightarrow$

20–3. How many grams of calcium carbonate were present if 200 mℓ of 0.1 M hydrochloric acid solution was required to entirely destroy it?

20–4. The main acid in the gastric juice of the stomach is hydrochloric acid. If a 75-mℓ sample of gastric juice has a hydrochloric acid concentration of 0.17 M, how many grams of sodium bicarbonate must be added to exactly neutralize this acid?

20–5. How many milliliters of stomach acid will be neutralized by ingestion of 2 g of sodium bicarbonate? The pH of stomach acid (HCl) is 1.

20–6. What is the molarity of a hydrochloric acid solution if 100 mℓ of it are neutralized by the addition of 200 mℓ of 0.3 *M* sodium hydroxide?

20–7. What is the molarity of a sodium hydroxide solution if 50 mℓ is neutralized by the addition of 80 mℓ of 0.1 *M* hydrochloric acid?

20–8. What is the molarity of a sulfuric acid solution if 400 mℓ is neutralized by the addition of 40 g of sodium hydroxide?

20–9. What is the molarity of a phosphoric acid solution if 100 mℓ is neutralized by the addition of 200 mℓ of 0.3 *M* sodium hydroxide?

20–10. How many grams of sodium hydroxide must you add to neutralize 500 mℓ of 0.01 *M* sulfuric acid?

20–11. One hundred milliliters of 0.2 *M* sulfuric acid was mixed with 50 mℓ of 0.4 *M* sodium hydroxide. Is the resulting solution acidic, basic, or neutral?

20–12. The addition of 1.06 g of sodium carbonate neutralized 500 mℓ of a hydrochloric acid solution. What was the pH of the original solution?

20–13. If 200 mℓ of 0.1 *M* HCl is mixed with 400 mℓ of 0.02 *M* sodium hydroxide, is the resulting solution acidic or basic? Calculate the pH of the resultant solution.

20–14. A sample containing an unknown amount of sodium oxide was reacted with HCl. Only the sodium oxide reacted. If the reaction required 500 mℓ of 0.5 *M* HCl, how many grams of sodium oxide were present?

20–15. A 10-g sample of rock was analyzed for calcium carbonate by treating the ground-up rock with acid. If 50 mℓ of 0.1 *M* HCl was required, what percent by weight of the rock was calcium carbonate?

20–16. One hundred grams of a solid containing aluminum hydroxide is dissolved and neutralized by adding it to 200 mℓ of 0.1 *M* HCl. The HCl remaining in solution is titrated to neutrality by adding 50 mℓ of 0.05 *M* sodium hydroxide. How many grams of aluminum hydroxide were present in the original sample?

UNIT 21 WEAK ACIDS AND BASES: HYDROLYSIS AND BUFFERS

So far we have dealt mostly with solutions of strong acids and bases. In these solutions we assume that the acid or base is almost completely ionized, and thus we do not worry about the reverse reaction, the recombination of the ions to form the original acid or base. Since weak acids and bases do not ionize 100%, the reverse reaction is important. In this unit we shall study some of the effects caused by reversible acid-base reactions.

Writing Equilibrium Constant Expressions

In any reversible chemical reaction, the reactants and products are both present at the same time. The relative concentrations of these materials are related through an expression known as the equilibrium constant (K_{eq}). For a reaction A $\leftrightarrows$ B, the equilibrium constant expression is

$$K_{eq} = \frac{\text{concentrations of products}}{\text{concentrations of reactants}} = \frac{[B]}{[A]}$$

The value of K_{eq} simply tells us which chemical is present in greatest concentration at equilibrium. Large values of K_{eq} (10 to 10,000) indicate that at equilibrium the products are present in much larger concentrations than the reactants. This means that the reaction occurs primarily toward the right, the product side. A small value for K_{eq} (10^{-8} to 10^{-1}) indicates the reverse, that the concentrations of reactants are much greater than the concentrations of products at equilibrium. These reactions occur primarily in the reverse direction, to the left, in favor of the reagents. Strong acids that ionize almost completely have large K_{eq} values, whereas weak acids have small K_{eq} values.

When a reversible reaction contains more than one reactant or product, the equilibrium constant expression is somewhat more complex. The following examples illustrate:

$$2A \leftrightarrows B + C \qquad A + B \leftrightarrows C \qquad A \leftrightarrows B + C \qquad A + B \leftrightarrows C + D$$

$$K_{eq} = \frac{[B][C]}{[A]^2} \qquad K_{eq} = \frac{[C]}{[A][B]} \qquad K_{eq} = \frac{[B][C]}{[A]} \qquad K_{eq} = \frac{[C][D]}{[A][B]}$$

Several points should be emphasized about the above expressions:

1. The symbols [ ] are used to indicate the molarity of the substances inside them. Thus [HCl] means "molarity of hydrochloric acid." When working with an equilibrium constant expression, simply replace the different molarity symbols with the actual molarities of the compounds present.
2. Notice that when two or more reactants or products are present their molarities are *multiplied*, not added, in the equilibrium constant expression. Thus [A][B] means the "molarity of A multiplied by the molarity of B."
3. When one reactant or product is present in the balanced equation in greater than 1 mol amounts, the concentration unit for that compound must reflect this. For example, if 2A are present, then $(A)^2$ is used in the equilibrium expression. If 3B are present, then $(B)^3$ is used in the equilibrium expression.
4. The balanced reaction must be written before you write the equilibrium constant expression.
5. The value of K_{eq} is a constant for a given set of reactants and products. It changes only if the temperature of the system is changed.

Weak Acid Equilibrium Expressions

When a weak acid is added to water, the water acts as a base by accepting the hydrogen ion. The equilibrium expression can be written as

$$HA + H_2O \rightleftharpoons H_3O^{1+} + A^{1-}$$

$$K_{eq} = \frac{[H_3O^{1+}][A^{1-}]}{[HA][H_2O]} \tag{21-1}$$

Actually, the amount of water converted to hydronium ion is very small compared to the actual amount of water present, so the concentration of water in the system remains essentially constant. It can be included in the value for K_{eq} to yield another constant, K_a:

$$K_a = K_{eq}[H_2O] = \frac{[H_3O^{1+}][A^{1-}]}{[HA]} \tag{21-2}$$

K_a values are more convenient to use than K_{eq} expressions since we do not have to use the concentration of water each time. Since $[H_3O^{1+}] = [H^{1+}]$, we can also write the expression as

$$K_a = \frac{[H^{1+}][A^{1-}]}{[HA]} \tag{21-3}$$

Table 21–1 lists K_a values for several compounds.

Table 21–1 K_a values for several compounds

Compound	Reaction	Dissociation constant K_a (at 25°C)	pK_a
Acids			
Sulfuric acid	$H_2SO_4 \rightleftharpoons H^{1+} + HSO_4^{1-}$	K_1 = very large	
	$HSO_4^{1-} \rightleftharpoons H^{1+} + SO_4^{2-}$	$K_2 = 1.2 \times 10^{-2}$	1.98
Phosphoric acid	$H_3PO_4 \rightleftharpoons H^{1+} + H_2PO_4^{1-}$	$K_1 = 7.5 \times 10^{-3}$	2.12
	$H_2PO_4^{1-} \rightleftharpoons H^{1+} + HPO_4^{2-}$	$K_2 = 6.2 \times 10^{-8}$	7.21
	$HPO_4^{2-} \rightleftharpoons H^{1+} + PO_4^{3-}$	$K_3 = 2.2 \times 10^{-13}$	12.66
Hydrofluoric acid	$HF \rightleftharpoons H^{1+} + F^{1-}$	$K = 6.8 \times 10^{-4}$	3.17
Carbonic acid	$H_2CO_3 \rightleftharpoons H^{1+} + HCO_3^{1-}$	$K_1 = 4.3 \times 10^{-7}$	6.37
	$HCO_3^{1-} \rightleftharpoons H^{1+} + CO_3^{2-}$	$K_2 = 5.6 \times 10^{-11}$	10.25
Hydrocyanic acid	$HCN \rightleftharpoons H^{1+} + CN^{1-}$	$K = 4.9 \times 10^{-10}$	9.31
Acetic acid	$CH_3C(=O){-}O{-}H \rightleftharpoons H^{1+} + CH_3C(=O){-}O^{1-}$	$K = 1.8 \times 10^{-5}$	4.75
Formic acid	$H{-}C(=O){-}O{-}H \rightleftharpoons H^{1+} + H{-}C(=O){-}O^{1-}$	$K = 1.7 \times 10^{-4}$	3.77
Ammonia	$NH_4^{1+} \rightleftharpoons H^{1+} + NH_3$	$K = 5.76 \times 10^{-10}$	9.24
Ethylamine	$CH_3CH_2NH_3^{1+} \rightleftharpoons H^{1+} + CH_3CH_2NH_2$	$K = 2.3 \times 10^{-11}$	10.64

Calculation of the pH of Solutions of Weak Acids

In aqueous solutions, weak acids partially ionize to provide hydrogen ions. The $[H^{1+}]$ and pH of the solutions can be calculated using the equilibrium equations if some simple approximations are made. Problem (1) illustrates the technique.

Problem 1: Calculate the $[H^{1+}]$ and pH of a 0.1 M solution of acetic acid. $K_a = 2 \times 10^{-5}$.

$$CH_3CO_2H \rightleftharpoons CH_3CO_2^{1-} + H^{1+}$$

$$K_a = \frac{[CH_3CO_2^{1-}][H^{1+}]}{[CH_3CO_2H]}$$

Initially, the $[CH_3CO_2H]$ is 0.1 M, but as ionization occurs, its concentration decreases as it is converted to $CH_3CO_2^{1-}$ and H^{1+}. If some portion, say X, ionized, then at equilibrium

$$[CH_3CO_2H] = 0.1 - X$$
$$[H^{1+}] = X$$
$$[CH_3CO_2^{1-}] = X$$

By inserting these values into the equilibrium equation, we get

$$2 \times 10^{-5} = \frac{(X)(X)}{0.1 - X}$$

The object is to solve for X, but it is difficult to do so from this equation unless we simplify. Let's make an assumption—that *the amount of ionization is small compared to 0.1 M* ($X \ll 0.1$). Then $0.1 - X$ is similar and not too different from 0.1 ($0.1 - X \sim 0.1$). So

$$2 \times 10^{-5} = \frac{X \cdot X}{0.1 - X} = \frac{X^2}{0.1}$$

(This X [in the denominator $0.1 - X$] is dropped out)

Solving for X,

$$2 \times 10^{-6} = X^2$$
$$1.4 \times 10^{-3} = X = [H^{1+}]$$
$$pH = 2.85$$

We have to go through this process because weak acids do not ionize completely. For a strong acid that is completely ionized, a 0.1 M solution would possess pH = 1.

One important point should be made regarding acidity. In solution, weak acids can exist in the protonated form (HA) and the nonprotonated form (A^-). The pH of the solution determines which predominates.

1. If $[H^{1+}] = K_a$, then $[HA] = [A^{1-}]$ at equilibrium.
2. If $[H^{1+}]$ is much greater than K_a, the protonated form is present in largest concentration.
3. If $[H^{1+}]$ is much less than K_a, the nonprotonated form predominates.

Hydrolysis

Acid-base reactions involve the ionization of acids and bases when dissolved in water. **Hydrolysis** involves somewhat the opposite, the formation of acids and bases by combinations of ions with water. Actually, these types of reactions are simply the actions of Bronsted-Lowry bases accepting hydrogen ions from water, a weak acid. Let's examine what happens when we dissolve sodium acetate in water.

$$\underset{\text{sodium acetate}}{NaCH_3CO_2} + H_2O \longrightarrow \underset{\text{sodium ion}}{Na^{1+}} + \underset{\text{acetate ion}}{CH_3CO_3^{1-}}$$

After ionization we have two ions in the solution. Sodium ion has little if any tendency to combine with OH^{1-} ions, since the reverse reaction ($NaOH \rightarrow Na^{1+} + OH^-$) is a favored reaction. But acetate ion is the conjugate base of a relative weak acid, acetic acid, and thus can accept hydrogen ions from water.

$$CH_3CO_2^{1-} + H_2O \rightleftharpoons CH_3CO_2H + OH^{1-} \tag{21–4}$$

The net effect is that the concentration of hydroxide ion in solution increases, and the pH of the system increases. The actual pH of the system will depend upon the equilibrium constant expression for equation (21–4). Since this is hydrolysis, we call the equilibrium constant K_{hydr}.

$$K_{\text{hydr}} = K_{\text{eq}}[H_2O] = \frac{[CH_3CO_2H][OH^{1-}]}{[CH_3CO_2^{1-}]} \tag{21–5}$$

The larger the K_{hydr}, the more basic the solution. K_{hydr} values can be calculated from the K_a values of the conjugate acids by using equation (21–6).

$$K_{\text{hydr}} = \frac{1 \times 10^{-14}}{K_a} \tag{21–6}$$

If the acid is very weak, then the K_a value is very small, and K_{hydr} for its conjugate base will be large; the solution composed of the conjugate base dissolved in water will be basic.

Problem 2: Which solution is more basic?
Solution 1: 1 *M* sodium fluoride
Solution 2: 1 *M* sodium formate

In solution 1, fluoride ion is the conjugate base of the acid HF ($K_a = 3.5 \times 10^{-4}$). Fluoride ion will accept hydrogen ions from water:

$$F^{1-} + H_2O \rightleftharpoons HF + OH^{1-}$$

$$K_{\text{hydr}} = \frac{[HF][OH^{1-}]}{[F^{1-}][H_2O]} = \frac{1 \times 10^{-14}}{K_a} = 2.8 \times 10^{-11}$$

In solution 2, formate ion is the conjugate base of formic acid ($K_a = 1.8 \times 10^{-4}$). It will also accept hydrogen ions from water:

$$HCO_2^{1-} + H_2O \rightleftharpoons HCO_2H + OH^{1-}$$

$$K_{\text{hydr}} = \frac{[HCO_2H][OH^{1-}]}{[HCO_2^{1-}][H_2O]} = \frac{1 \times 10^{-14}}{K_a} = 5.5 \times 10^{-11}$$

Since K_{hydr} (formate) is greater than K_{hydr} (fluoride), solution 2 will be more basic than solution 1 if both are present at the same concentration.

Buffers

Many chemical and biological reactions are pH dependent; that is, they occur readily only if the pH of the solution is at a certain level. For these reactions to occur, some mechanism must be present in solution to keep the pH at the desired level even though extra acid or base may be added. The solution must be **buffered**; it must be kept at a constant pH level.

A **buffer** is a substance that helps maintain the pH of a solution at a nearly constant level. It can absorb excess hydrogen ions if some are added to solution, and it can also supply hydrogen ions if the $[H^{1+}]$ in solution begins to decrease. *The most common type of buffer system consists of a mixture of a weak acid and its conjugate base.* If a weak acid (HA) and its conjugate base (A^{1-}) are added to water, an equilibrium is established:

$$HA \rightleftharpoons H^{1+} + A^{1-} \qquad (21\text{–}7)$$

If the acid HA ionizes, hydrogen ions are added to the solution and the $[H^{1+}]$ would increase. If the base A^{1-} reacts with hydrogen ions, then the $[H^{1+}]$ would decrease. It's as if we have two large reservoirs present. One reservoir furnishes hydrogen ion if the $[H^{1+}]$ in solution starts to decrease; the second reservoir removes hydrogen ion if the $[H^{1+}]$ gets too large.

Acid				Conjugate base
HA (furnishes H^{1+}) ($HA \longrightarrow H^{1+} + A^{1-}$)	$\rightleftharpoons$	H^{1+} $[H^{1+}]$ is maintained at a constant level	+	A^{1-} (removes H^{1+}) ($A^{1-} + H^{1+} \longrightarrow HA$)

Suppose that the pH of an acetic acid–acetate buffer system is 4.76. If 1 mℓ of 1 *M* HCl is added to 1 ℓ of this solution, the pH becomes 4.75 (a change of only 0.01 units). In contrast, if 1 mℓ of 1 *M* HCl is added to 1 ℓ of pure water (pH = 7), the pH becomes 3 (a change of 4.0 units).

A buffer, then, is made by adding both a weak acid and its conjugate base to solution. These compounds are in equilibrium and establish the $[H^{1+}]$ at a certain level. This level is maintained even though extra hydrogen ions or hydroxide ions are added to solution. The actual $[H^{1+}]$ established by the buffer is determined by the K_a value of the weak acid and the concentrations of acid and conjugate base present. For example, the equilibrium constant expression for equation (21–7) is

$$K_a = \frac{[H^{1+}](\text{concentration of conjugate base})}{\text{concentration of acid}} = \frac{[H^{1+}][A^{1-}]}{[HA]} \qquad (21\text{–}8)$$

If we know the K_a (Table 21–1) and the molarities of the acid and its conjugate base, then the $[H^{1+}]$ can be calculated using equation (21–8). Problem (3) illustrates.

Problem 3: Calculate the $[H^{1+}]$ and the pH of a buffer solution consisting of acetic acid and acetate ion if

(a) [Acetic acid] = 0.1 *M*, [acetate] = 0.1 *M*
(b) [Acetic acid] = 0.1 *M*, [acetate] = 0.2 *M*
(c) [Acetic acid] = 0.2 *M*, [acetate] = 0.1 *M*

$$\underset{\text{acetic acid}}{CH_3CO_2H} \rightleftharpoons H^{1+} + \underset{\substack{\text{acetate ion}\\\text{(conjugate base)}}}{CH_3CO_2^{1-}}$$

$$K_a = 1.75 \times 10^{-5} = \frac{[H^{1+}][\text{acetate ion}]}{[\text{acetic acid}]}$$

Substitute the actual concentrations of acetic acid and acetate into the equilibrium expression.

(a) $$1.75 \times 10^{-5} = \frac{[H^{1+}][0.1\,M]}{(0.1\,M)}. \quad \text{Solve for } [H^{1+}].$$

$$[H^{1+}] = 1.75 \times 10^{-5}, \qquad pH = 4.76$$

(b) $$1.75 \times 10^{-5} = \frac{[H^{1+}](0.2\,M)}{(0.1\,M)}, \quad [H^{1+}] = 8.75 \times 10^{-6}, \quad pH = 5.06$$

(c) $$1.75 \times 10^{-5} = \frac{[H^{1+}](0.1\,M)}{(0.2\,M)}, \quad [H^{1+}] = 3.5 \times 10^{-5}, \quad pH = 4.46$$

Now let's look at problem (3) and see what it can tell us.

1. If the concentrations of an acid and its conjugate base are equal, as in problem (2a), then the $[H^{1+}]$ in solution is *equal to the* K_a for the acid.
2. If the concentration of acid in solution is greater than the concentration of the conjugate base, as in problem (2c), then the $[H^{1+}]$ in solution will be *greater than the* K_a value.
3. If the concentration of the acid in solution is less than the concentration of the conjugate base, as in problem (2b), then the $[H^{1+}]$ in solution will be *less than the* K_a value.

In general, then, to prepare a buffer system, you must first decide what pH or $[H^{1+}]$ you wish to maintain. Then select a weak acid whose K_a value is close to the desired $[H^{1+}]$. To get the $[H^{1+}]$ exactly at the desired level, you must adjust slightly the concentrations of the weak acid and conjugate base added to the solution.

Problem 4: Make up a buffer system of pH = 4. What concentrations of acid and conjugate base would you use? If the pH is to be 4, then $[H^{1+}]$ is 1×10^{-4}, and we need a weak acid whose K_a value is close to 10^{-4}. From Table 21–1, the K_a for formic acid is closest to 10^{-4} (1.8×10^{-4}).

$$\underset{\substack{\text{formic}\\\text{acid}}}{HCO_2H} \rightleftharpoons H^{1+} + \underset{\substack{\text{formate}\\\text{ion}}}{HCO_2^{1-}}$$

$$K_a = 1.8 \times 10^{-4} = \frac{[H^{1+}][HCO_2^{1-}]}{[HCO_2H]}$$

Substitute $[H^{1+}] = 1 \times 10^{-4}$ into this expression and solve for $[HCO_2^{1-}]/[HCO_2H]$.

$$\frac{[HCO_2^{1-}]}{[HCO_2H]} = \frac{1.8 \times 10^{-4}}{1.0 \times 10^{-4}} = \frac{1.8}{1}$$

The concentration of formate ion must be 1.8 times greater than the concentration of formic acid if $[H^{1+}]$ is to be 1×10^{-4} *M*. If [formate ion] is 0.18 *M*, then [formic acid] must be 0.1 *M*.

The Henderson-Hasselbach Equation

It is often convenient to express the equilibrium constant expression for a buffer [equation (21–8)] in a slightly different form. If equation (21–8) is solved for $[H^+]$, it becomes

$$[H^{1+}] = K_a \frac{[\text{weak acid}]}{[\text{conjugate base}]} \qquad (21\text{–}9)$$

Now if the negative logarithm of each factor is taken, equation (21–9) can be changed into a form that contains pH units.

$$-1 \log [H^{1+}] = -1 \log K_a - \log \frac{[\text{acid}]}{[\text{conjugate base}]}$$

It can also be written as

$$-1 \log [H^{1+}] = -1 \log K_a + \log \frac{[\text{conjugate base}]}{[\text{weak acid}]}$$

$$\text{pH} = \text{p}K_a + \log \frac{[\text{conjugate base}]}{[\text{weak acid}]} \qquad (21\text{–}10)$$

Equation (21–10) is called the **Henderson-Hasselbach equation**; it is often used by biochemists because it relates directly to pH. To pick an appropriate buffer system, we simply choose a weak acid whose pK_a is close to the desired pH of the system and then calculate the necessary base-to-acid ratio.

Buffers in the Body

All body fluids are buffered, but not all exist at the same pH levels (see Table 19–1). Therefore, many buffers must exist in the body to maintain the different pH levels of the different body fluids. The most common buffers are proteins, the carbonic acid–bicarbonate system, and the $H_2PO_4^{1-}$–HPO_4^{2-} system.

It is interesting to examine the carbonic acid–bicarbonate buffer system in more detail to see just how the body uses it to maintain pH.

$$\underset{\text{carbonic acid}}{H_2CO_3} \rightleftharpoons H^{1+} + \underset{\text{bicarbonate}}{HCO_3^{1-}} \qquad K_a = 4.3 \times 10^{-7}$$

$$4.3 \times 10^{-7} = \frac{[H^{1+}][HCO_3^{1-}]}{[H_2CO_3]}$$

Carbonic acid is formed in body fluids when carbon dioxide dissolves in water:

$$CO_2 + H_2O \rightleftharpoons H_2CO_3$$

As the cells in the body function, they use oxygen and release carbon dioxide to the bloodstream. The lungs pick up carbon dioxide from the bloodstream and release it out of the body, so the lungs actually control the amount of dissolved carbon dioxide and thus the concentration of carbonic acid in the bloodstream. The concentration of bicarbonate in the bloodstream is controlled by the kidney, which can convert carbonic acid to bicarbonate. In the bloodstream the pH is about 7.4, so the ratio of carbonic acid to bicarbonate is maintained at about 1:10.

UNIT 21 Problems and Exercises

21–1. Suppose that you have two solutions: one is 1 *M* HCl and the other is 1 *M* acetic acid + 1 *M* acetate. How could you distinguish between them using only a pH meter and solutions of HCl and NaOH?

21–2. Explain how a carbonic acid buffer solution acts to limit changes in pH on addition of acid or base.

21–3. Write equilibrium constant expressions for the following reversible reactions. Balance the reactions where necessary.

(a) $CH_3OH + CH_3CO_2H \rightleftharpoons CH_3CO_2CH_3 + H_2O$
(b) $HCl + O_2 \rightleftharpoons H_2O + Cl_2$
(c) $HBr \rightleftharpoons Br_2 + H_2$
(d) $N_2O_4 \rightleftharpoons 2NO_2$
(e) $C + O_2 \rightleftharpoons CO$

21–4. The equilibrium concentrations for reactants and products are given for the following reactions. Calculate the value of the equilibrium constant in each case.

(a) $\underset{0.324\,M}{N_2O_4} \rightleftharpoons \underset{1.85 \times 10^{-2}\,M}{2NO_2}$

(b) $\underset{0.172\,M}{H_2} + \underset{0.406\,M}{I_2} \rightleftharpoons \underset{1.78\,M}{2HI}$

(c) $\underset{0.009\,M}{CH_3CH_2OH} + \underset{0.829\,M}{CH_3CO_2H} \rightleftharpoons \underset{0.171\,M}{CH_3CO_2CH_2CH_3} + \underset{0.171\,M}{H_2O}$

21–5. List the following acids in increasing acid strength.
HCO_2H, H_2CO_3, HF, H_2SO_4, CH_3CO_2H.

21–6. Calculate the pK_a from the following K_a values.

(a) 2×10^{-6} (b) 4×10^{-8}
(c) 1×10^{-10} (d) 10

21–7. Rank the following acids in terms of decreasing acid strength.

Acid A: $K_a = 1 \times 10^{-5}$ Acid B: $K_a = 2 \times 10^{-6}$
Acid C: $K_a = 5 \times 10^{-2}$ Acid D: $K_a = 8 \times 10^{-4}$

21–8. Calculate the K_{hydr} values for the conjugate bases of the following acids.

(a) Formic acid (b) Ethylammonium ion $(CH_3CH_2\overset{1+}{N}H_3)$
(c) Hydrofluoric acid

21–9. Which solution is more basic?

(a) 1 *M* sodium acetate or 1 *M* sodium formate
(b) 1 *M* sodium sulfate or 1 *M* sodium carbonate

21–10. If you were titrating acetic acid with sodium hydroxide, what would the pH be when exactly half of the acetic acid had been converted to acetate ion?

21–11. Which of the following buffer solutions has the highest value of pH? Which has the lowest pH?
(a) 0.1 *M* $NaHCO_3$ + 0.1 *M* Na_2CO_3
(b) 0.1 *M* NaF + 0.1 *M* HF
(c) 0.1 *M* H_3PO_4 + 0.1 *M* NaH_2PO_4

21–12. The HPO_4^{2-} and the $H_2PO_4^{1-}$ pair form an important component of the buffer system that controls blood pH. At the pH of blood, say 7.4, what is the $HPO_4^{2-}:H_2PO_4^{1-}$ ratio?

21–13. Calculate the pH of a solution composed of 0.05 *M* ammonia and 0.1 *M* ammonium chloride.

21–14. Calculate the pK_a of a buffer if the pH of the solution is 4.2, the concentration of acid is 0.2 *M*, and the concentration of its conjugate base is 0.4 *M*.

21–15. Calculate the $[H^{1+}]$, $[OH^{1-}]$, and pH of the following buffer solutions.
(a) 0.2 *M* acetic acid and 0.1 *M* sodium acetate
(b) 0.2 *M* sodium formate and 0.1 *M* formic acid
(c) 500 mℓ of a solution containing 2 g of HF and 4.2 g of sodium fluoride

21–16. What must the pH of a solution be in order that the concentration of acetic acid be twice that of its conjugate base, acetate ion?

21-17. Using Table 21–1, determine what buffer system would be the best for the following solutions.
(a) pH = 5 (b) pH = 10 (c) pH = 2
Calculate the ratios of acid to *conjugate* base necessary to maintain the pH values. Use the buffer system you chose for each part.

21–18. The addition of 40 mℓ of 0.1 *M* NaOH partially neutralizes 100 mℓ of a 0.1 *M* solution of acetic acid. Calculate the concentrations of acetic acid and acetate ion present. Calculate the pH of the resultant buffer solution.

21–19. Calculate the $[H^{1+}]$ and the pH of a 0.1 *M* solution of formic acid ($K_a = 1.8 \times 10^{-4}$).

21–20. What would be the pH of a solution containing 1 *M* ammonium ion?

$$NH_4^{1+} \rightleftharpoons H^{1+} + NH_3, \qquad K_a = 5.76 \times 10^{-10}$$

What would the pH be if $NH_4^{1+} = 0.01\,M$?

21–21. The K_a for ethylamine hydrochloride

$$CH_3CH_2NH_3^{1+}Cl^{1-} \rightleftharpoons CH_3CH_2NH_2 + H^{1+} + Cl^{1-}$$

is 2.3×10^{-11}. Which form ($CH_3CH_2NH_3^{1+}$ or $CH_3CH_2NH_2$) will be present in larger concentration if the pH is equal to (a) 8? (b) 10.64? (c) 12?

21–22. At what pH will the ratio $[CH_3CO_2^{1-}]:[CH_3CO_2H]$ be equal to 3:1?

PART TWO

organic chemistry

CHAPTER 9

introduction to organic chemistry

For centuries scientists have isolated and investigated the chemical compounds found in nature. Until the middle of the nineteenth century, all known compounds were divided into two groups: **organic** compounds and **inorganic** compounds. Inorganic compounds were those chemicals obtained from rocks, minerals, and any substances that had never been *alive.* Organic chemicals came from animal and vegetable sources, from living or once living systems. It was also known that organic compounds were composed of relatively few elements: carbon, hydrogen, oxygen, nitrogen, and sulfur; inorganic compounds could consist of all the elements.

The major difference between organic and inorganic chemicals was that organic chemicals were thought to possess a special *vital force* that was necessary for life. Since this vital force could not be defined or isolated, it could not be given to inorganic chemicals in the laboratory. It was thus impossible to convert inorganic chemicals into organic chemicals.

The vital force theory held until the middle of the nineteenth century when two chemists, Friedrich Wohler and Adolph Kolbe, conclusively demonstrated the synthesis of two organic chemicals from inorganic reactants. Wohler synthesized urea from ammonium cyanate, and Kolbe synthesized acetic acid from various inorganic materials. The vital force theory was disproved, and the distinction between compounds in nonliving and living systems was no longer valid.

Wohler synthesis:

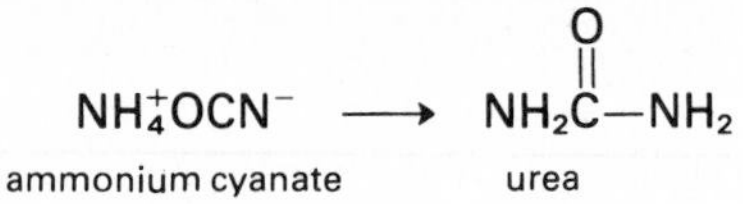
NH$_4^+$OCN$^-$
NH$_2$C—NH$_2$
O
ammonium cyanate
urea

Kolbe synthesis:

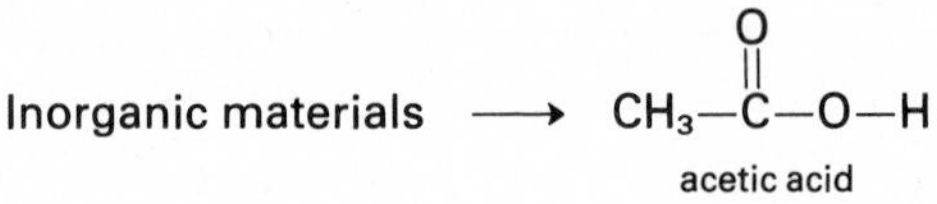
Inorganic materials
CH$_3$—C—O—H
O
acetic acid

The terms organic and inorganic no longer mean what they used to, but they are used to classify chemicals. All organic chemicals contain carbon, whereas very few inorganic chemicals contain carbon. Most organic chemicals are covalent compounds; most inorganic chemicals are ionic or become ions when dissolved in water. The molecular weights of organic compounds range from 16 g to the millions, but it is rare for molecular weights of inorganic chemicals to exceed 2000 g. In general, then, when we study organic chemistry, we study the properties and reactions of carbon-containing compounds.

As we study organic chemistry, we shall concentrate on learning the names, structures, properties, and reactions of the different types of organic chemicals. Unit 22 is a preview of the entire field. Besides introducing theory on the structure of organic molecules, we also introduce the various types of reactive centers present in organic molecules and review some of the basic types of reactions that we shall be covering.

A wide variety of different structures is possible for organic chemicals, even though many of them may possess the same molecular formula. For this reason, the names of organic molecules are somewhat more complex than the names of the inorganic chemicals that we have discussed so far. Fortunately, there is a systematic method, the IUPAC system, for naming all organic chemicals. This system is based on the names of the simplest organic molecules, the alkanes. In unit 23 we introduce alkanes and the IUPAC method of naming alkanes.

UNIT 22 STRUCTURES, TYPES, AND REACTIONS OF ORGANIC MOLECULES

Clearly, carbon must be quite unique to have a complete branch of chemistry devoted to it. A single carbon atom can covalently bond to 1, 2, 3, or 4 other atoms at the same time. Carbon atoms can bond to each other to form long chains, a feat very rare in the chemical world. Carbon atoms can form single, double, and triple bonds with themselves and with other atoms. No other element is as versatile as carbon. In fact, there exist more compounds that contain carbon than compounds that do not contain carbon. This is quite surprising, considering that there is a total of 104 elements.

Structures of Organic Molecules

All carbon atoms possess four electrons in their valence levels and must gain four more electrons through covalent bonding in order to fill the valence level. Thus, *all carbon atoms seek to form four covalent bonds with other atoms.* They do this by forming combinations of single, double, and triple bonds at the same time; the overall geometry and structure of an organic molecule is determined by the types of bonds formed. For example, a carbon atom possessing four single covalent bonds has a *tetrahedral geometry.* A carbon atom bonded to three other atoms and possessing a double covalent bond has a *trigonal planar geometry*, and a carbon atom possessing a triple bond has a *linear geometry.* See Figure 22–1.

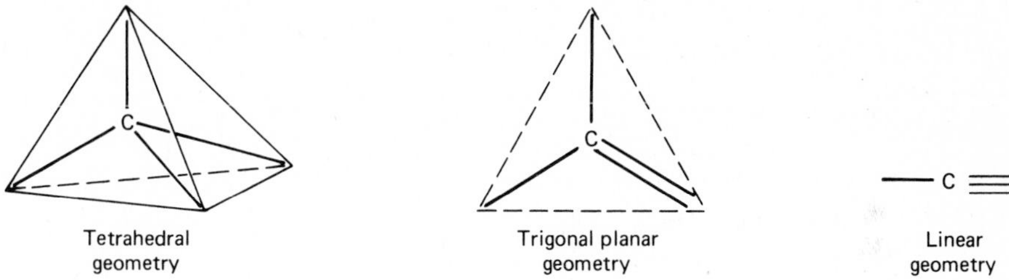
C
C
C
Tetrahedral geometry
Trigonal planar geometry
Linear geometry

Figure 22–1

Hybrid Orbitals

To form a covalent bond between two atoms, one half-filled atomic orbital from one atom must overlap (mix) with a half-filled atomic orbital from the other atom. After overlap, the atomic orbitals no longer exist—a new *molecular orbital* has been created in their place. See Figure 22–2.

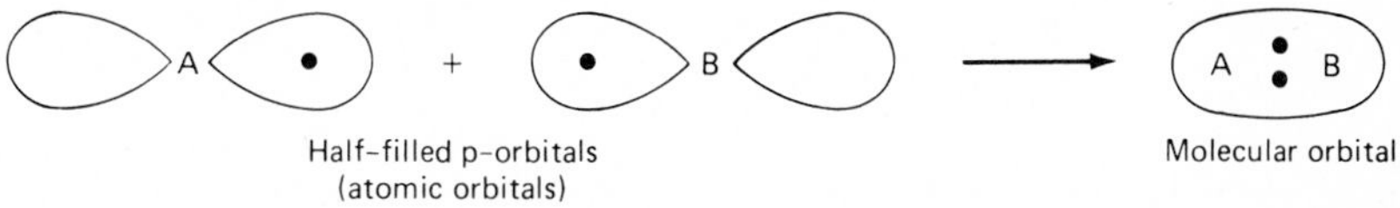
A
+
B
A
B
Half-filled p-orbitals (atomic orbitals)
Molecular orbital

Figure 22–2

To form the four covalent bonds required for carbon, each atom must possess four half-filled atomic orbitals. But examination of the electron configuration of carbon shows only two half-filled atomic orbitals. The solution to this dilemma is not to throw out the previous bonding theory, but to expand on it and create new atomic orbitals different from the s-, p-, d-, and f-orbitals previously introduced. Mathematically we can mix s- and p-orbitals to create new atomic orbitals that are called *hybrid orbitals.* These hybrid orbitals possess some of the properties of both the s- and p-orbitals, yet are different from both. There are three types of hybrid orbitals associated with carbon atoms: sp^3 hybrid orbitals, sp^2 hybrid orbitals, and sp hybrid orbitals.

sp^3 hybrid orbitals: *Any carbon atom possessing four single bonds is said to be sp^3 hybridized,* and it will have four sp^3 hybrid orbitals. The four sp^3-orbitals arise from mathematical mixing of one s-orbital and three p-orbitals (thus the name s^1p^3). The optimum geometrical arrangement for the four sp^3-orbitals about the carbon center is in the shape of a tetrahedron. As a rule, then, whenever a carbon atom is bonded to *four other atoms*, that carbon will be sp^3 hybridized and *the four atoms will be arranged about the central carbon atom in a tetrahedral geometry.* This is illustrated in Figure 22–3.

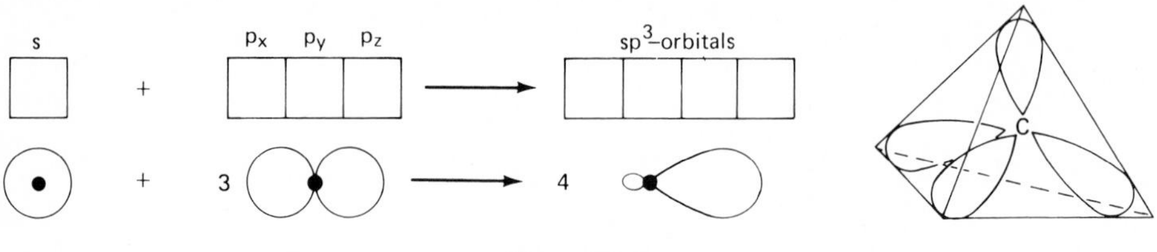
s
px py pz
sp3-orbitals
+
3
4
C

Figure 22–3

sp^2 hybrid orbitals: When a carbon atom forms a double covalent bond to one atom, it can bond to only two other atoms at the same time. *A carbon atom bonded to only three atoms is said to be sp^2 hybridized*; it will possess three sp^2 hybrid orbitals and one p_z-orbital in its valence level. The three sp^2-orbitals arise from a mixing of the s-, p_x-, and p_y-orbitals of the valence level; the geometry of the three hybrid orbitals is such that the three sp^2-orbitals point to the corners of a triangle (trigonal planar). The unused p_z-orbital is perpendicular to them. Therefore, whenever a carbon atom is covalently bonded to *three* other atoms, *those three atoms will be situated about the central atom in a trigonal planar geometry*, as shown in Figure 22–4.

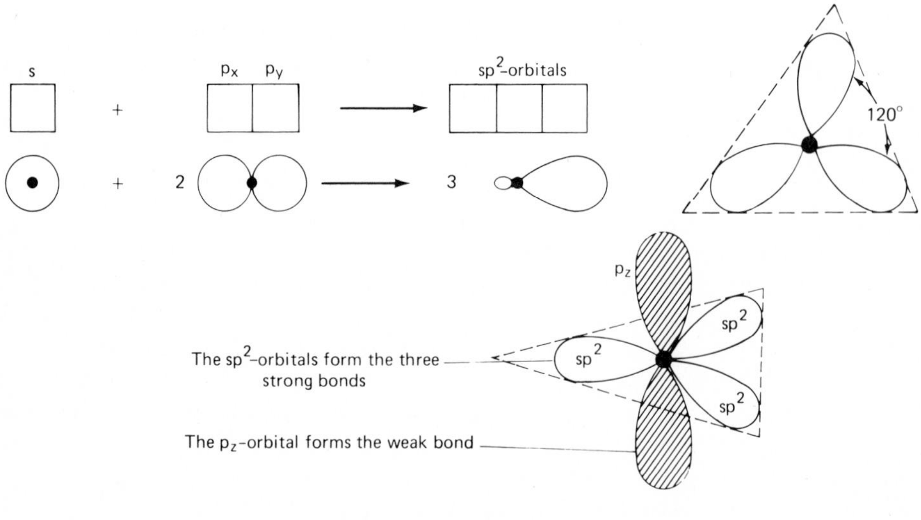
s
px py
sp2-orbitals
+
2
3
120°
pz
sp2
sp2
sp2
The sp2-orbitals form the three strong bonds
The pz-orbital forms the weak bond

Figure 22–4

sp hybrid orbitals: If a carbon atom forms a triple bond to another atom, it will bond with but one more atom. *Carbon atoms forming triple bonds are said to be sp hybridized*; they possess two sp hybrid orbitals and one p_y- and one p_z-orbital in the valence level. The sp-orbitals are arranged in a linear geometry: they point in opposite directions. The two remaining p-orbitals are perpendicular to them. *Any carbon atom bonded to butt two other atoms will have a linear geometry.* See Figure 22–5.

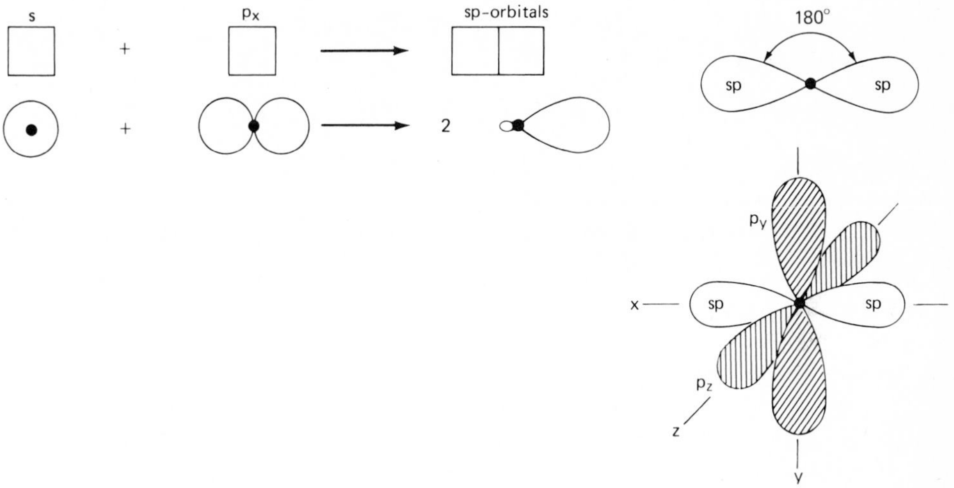

s
px
sp-orbitals
+
2
180°
sp
sp
py
x
sp
sp
pz
z
y

Figure 22–5

Types of Organic Molecules: Functional Groups

When an organic molecule undergoes a chemical reaction, most of the time only specific portions of that molecule undergo change. For example, in the reaction shown below, the —HC=CH— portion of the molecule has been changed to $-\overset{\overset{\text{HO}}{|}}{\text{HC}}-\overset{\overset{\text{H}}{|}}{\text{CH}}-$. The rest of the molecule has remained unchanged.

$$CH_3-CH_2-\overset{\overset{O}{\|}}{C}-CH_2-\boxed{HC=CH}-CH_3 + H_2O \longrightarrow CH_3-CH_2-\overset{\overset{O}{\|}}{C}-CH_2-\boxed{\overset{\overset{HO}{|}}{HC}-\overset{\overset{H}{|}}{CH}}-CH_3$$

The portion of the molecule that changes in a chemical reaction is called the *reactive site* or *functional group.* When we look at organic molecules, we look for the presence or absence of these functional groups. Actually, the names of organic molecules are derived from the functional groups they contain. Table 22–1 lists some of the common functional groups and also gives their names. We shall study the chemistry of each specific type of functional group in the following units of this text.

Table 22–1 Types and names of functional groups

Type of functional group	Name	Type of functional group	Name
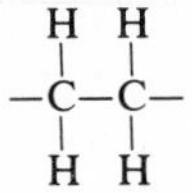 All single bonds; contain only carbon and hydrogen	Alkane	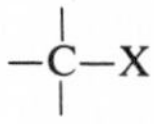 X = F, Cl, Br, I	Alkyl halide
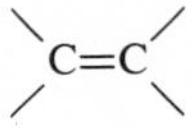	Alkene	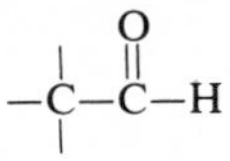	Aldehyde
—C≡C—	Alkyne	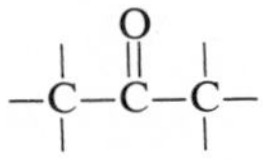	Ketone
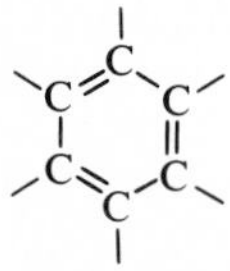 A six-membered ring with three double bonds	Benzene	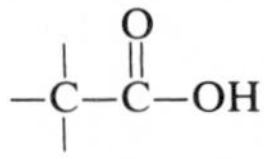	Carboxylic acid
		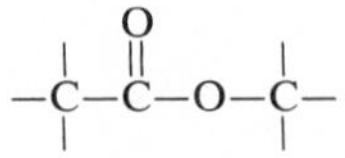	Ester
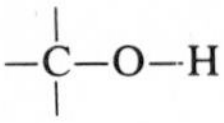	Alcohol	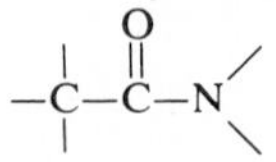	Amide
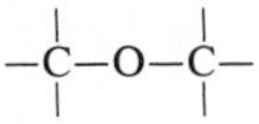	Ether		
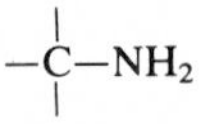	Amine	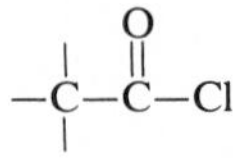	Acid chloride

Reactions of Organic Molecules

Most of the reactions of organic molecules that we shall study involve the conversion of one type of functional group into another. In working the problems, you will become accustomed to looking for the functional group in a molecule and then following it through to the product to see what changes have occurred. Later on you will be able to predict which reagents to add to cause the conversions to occur.

Most of the reactions we shall discuss can be grouped into one of the following categories: additions, eliminations, substitutions on sp^3 centers, and substitutions on sp^2 centers. We shall also cover reduction (a type of addition) and oxidation (a type of elimination). Table 22–2 lists the common types of reactions and gives examples. Remember that only the functional center will change: the remainder of the molecule generally remains as it was in the reactant.

Table 22–2 Types of reactions of organic molecules

Name-description	Examples

ADDITIONS: Small molecules are added to double and triple bonds.

$$\ce{>C=C< + A-B ->} \overset{A\quad B}{>C-C<}$$

$$CH_3-CH=CH_2 + HO-H \rightarrow CH_3-\overset{\overset{OH}{|}}{C}H-\overset{\overset{H}{|}}{C}H_2$$

$$CH_3-\underset{\underset{H}{|}}{C}=O + CH_3O-H \rightarrow CH_3-\overset{\overset{CH_3-O}{|}}{\underset{\underset{H}{|}}{C}}-\overset{H}{O}$$

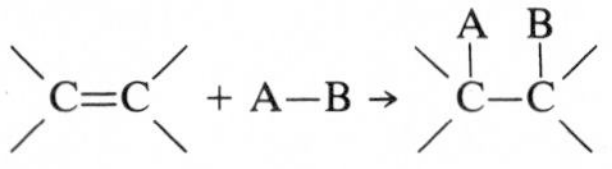

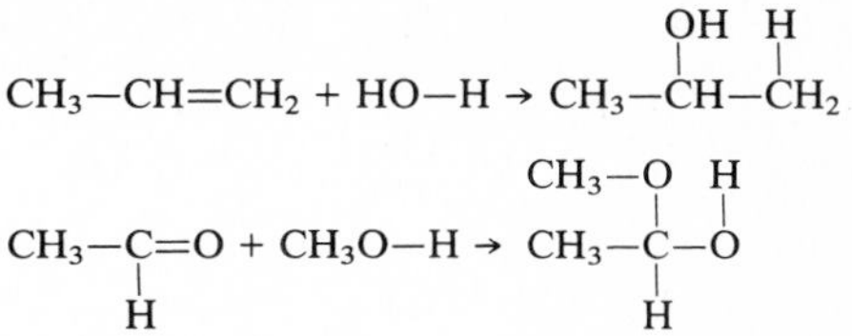

ELIMINATIONS: The reverse of additions; small molecules are removed to form double bonds.

$$-\overset{\overset{H}{|}}{C}-\overset{\overset{X}{|}}{C}- \rightarrow -C=C- + HX$$

$$\overset{\overset{H}{|}}{C}H_2-\overset{\overset{Br}{|}}{C}H_2 \rightarrow CH_2=CH_2 + HBr$$

$$\overset{\overset{H}{|}}{C}H_2-\overset{\overset{OH}{|}}{C}H_2 \rightarrow CH_2=CH_2 + HOH$$

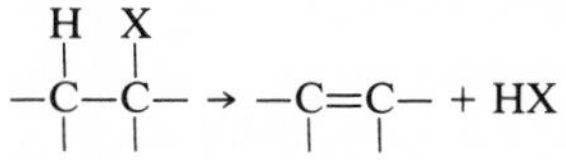

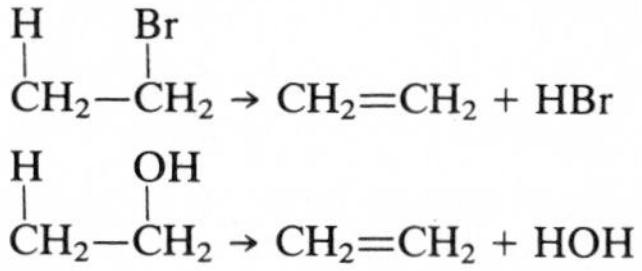

SUBSTITUTIONS ON sp^3 CENTERS: One portion of a molecule is removed and another group takes its place.

$$-\overset{|}{\underset{|}{C}}-X + Z \rightarrow -\overset{|}{\underset{|}{C}}-Z + X$$

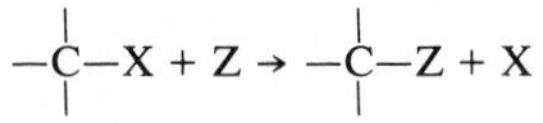

$$CH_3-Br + OH^- \rightarrow CH_3-OH + Br^-$$

$$CH_3-CH_2-Cl + NH_3 \rightarrow CH_3CH_2-\overset{+}{N}H_3 + Cl^-$$

SUBSTITUTIONS ON sp^2 CENTERS: Similar to other substitutions in effect, but the actual pathway is one of addition-elimination.

$$-\overset{\overset{O}{||}}{C}-X + Z \rightarrow -\overset{\overset{O^-}{|}}{\underset{\underset{Z}{|}}{C}}-X \rightarrow -\overset{\overset{O}{||}}{\underset{\underset{Z}{|}}{C}} + X$$

$$CH_3-\overset{\overset{O}{||}}{C}-Cl + HO^- \rightarrow CH_3\overset{\overset{O}{||}}{C}-OH + Cl^-$$

$$CH_3\overset{\overset{O}{||}}{C}-OH + CH_3OH \rightarrow CH_3\overset{\overset{O}{||}}{C}-OCH_3 + H_2O$$

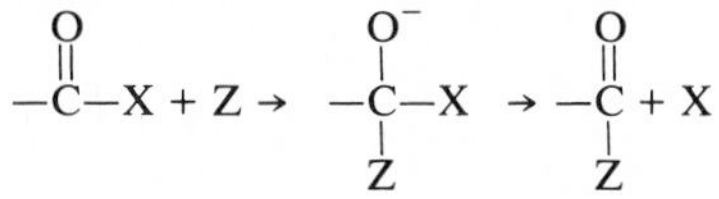

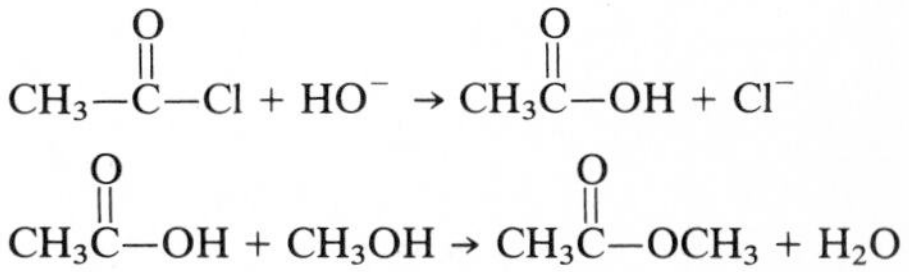

REDUCTION: Addition of H_2 to a double bond. Reduction will sometimes involve loss of oxygen.

$$>C=C< + H_2 \rightarrow \overset{H\quad H}{>C-C<}$$

$$CH_2=CH_2 + H_2 \rightarrow \overset{\overset{H}{|}}{C}H_2-\overset{\overset{H}{|}}{C}H_2$$

$$CH_3-\overset{\overset{O}{||}}{C}-OH + H_2 \rightarrow \rightarrow CH_3-CH_2-OH$$

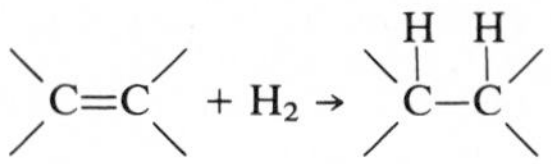

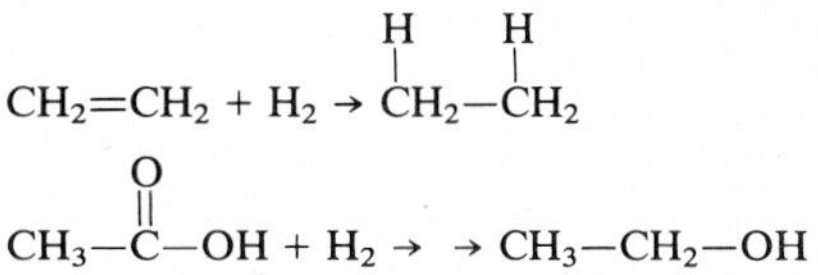

OXIDATION: Removal of hydrogen from a molecule. Oxidation will sometimes involve addition of oxygen.

$$-\overset{\overset{O-H}{|}}{\underset{|}{C}}-\overset{|}{\underset{|}{C}}-H \rightarrow -\overset{|}{\underset{|}{C}}-\overset{\overset{O}{||}}{C}$$

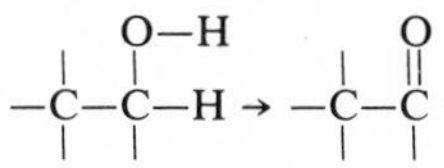

$$CH_3-OH \rightarrow H-\overset{\overset{O}{||}}{C}H \rightarrow H-\overset{\overset{O}{||}}{C}-OH$$

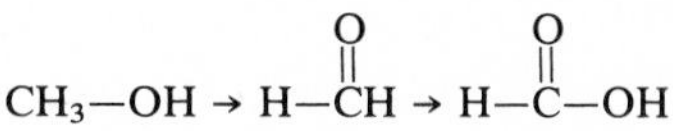

UNIT 22 Problems and Exercises

22–1. What is the difference between an atomic orbital and a molecular orbital?

22–2. Fill out the following chart concerning the hybrid atomic orbitals.

Orbital	Number	Geometry
sp^3	______	______
sp^2	______	______
sp^1	______	______

22–3. In the compounds below tell what type of hybridization is present at the carbon atoms bearing an asterisk (*).

(a) 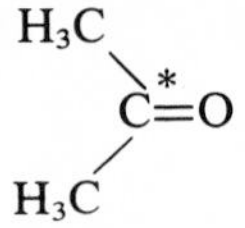

(b) $H_3C-\overset{*}{C}H_2-CH_3$

(c) $H_3C-\overset{*}{C}H=CH_2$

(d) $CH_3-\overset{*}{C}\equiv C-CH_3$

(e) $H-\overset{*}{C}\equiv N$

(f) 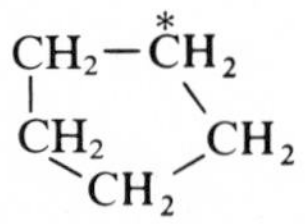

22–4. Examine the molecules shown below. Circle and name the functional groups present.

(a) $CH_3-CH_2-CH=CH-CH_3$

(b) $CH_3-\underset{NH_2}{\underset{|}{C}H}-CH_3$

(c) $CH_3-\overset{O}{\overset{\|}{C}}-O-CH_3$

(d) $CH_3-CH_2-\overset{O}{\overset{\|}{C}}-H$

(e) $CH_3-CH_2-\overset{OH}{\overset{|}{C}H}-CH_3$

(f) 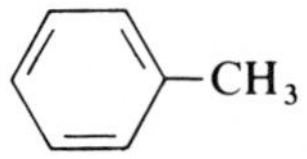

(g) $CH_3-CH_2-CH_2-CH_3$

(h) $CH_3-O-CH_2-CH_2-\overset{O}{\overset{\|}{C}}-CH_3$

(i) $CH_3CH_2-\overset{O}{\overset{\|}{C}}-OH$

(j) $\underset{Br}{\underset{|}{C}H_2}-CH_2-CH_2-\overset{O}{\overset{\|}{C}}-NH_2$

22–5. Examine the reactions shown below and (1) circle the portion of the molecule which underwent change; (2) use Table 22–2 to indicate the type of reaction which occurred.

(a) $CH_3CH_2-OH \rightarrow CH_3-CH_2-Br$

(b) $CH_3-\overset{OH}{\overset{|}{C}H}-CH_3 \rightarrow CH_2=CH-CH_3$

(c) $CH_3-\overset{OH}{\overset{|}{C}H}-CH_3 \rightarrow CH_3-\overset{O}{\overset{\|}{C}}-CH_3$

(d) $CH_3-\underset{Br}{\underset{|}{\overset{CH_3}{\overset{|}{C}}}}-CH_2-CH=CH_2 \rightarrow CH_3-\underset{OH}{\underset{|}{\overset{CH_3}{\overset{|}{C}}}}-CH_2-CH=CH_2$

(e) $CH_3-\overset{\overset{O}{\|}}{C}-OCH_3 \rightarrow CH_3-\overset{\overset{O}{\|}}{C}-NH_2$ 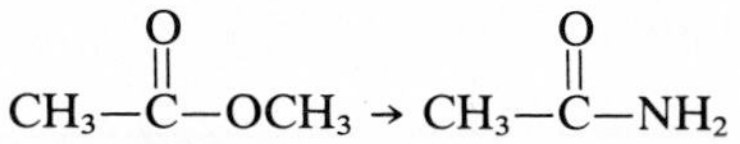

(f) $CH_3-\overset{\overset{CH_3}{|}}{\underset{\underset{Br}{|}}{C}}-CH_2-CH=CH_2 \rightarrow CH_3-\overset{\overset{CH_3}{|}}{\underset{\underset{Br}{|}}{C}}-CH_2-CH_2-CH_3$ 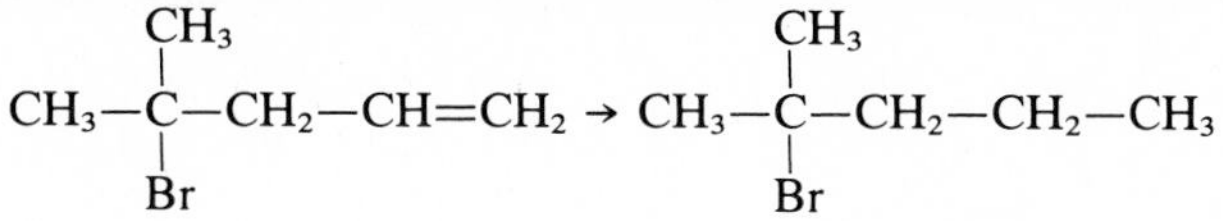

(g) $CH_3-\underset{\underset{OH}{|}}{CH}-CH_2-\overset{\overset{O}{\|}}{C}-OH \rightarrow CH_3-\overset{\overset{O}{\|}}{C}-CH_2-\overset{\overset{O}{\|}}{C}-OH$ 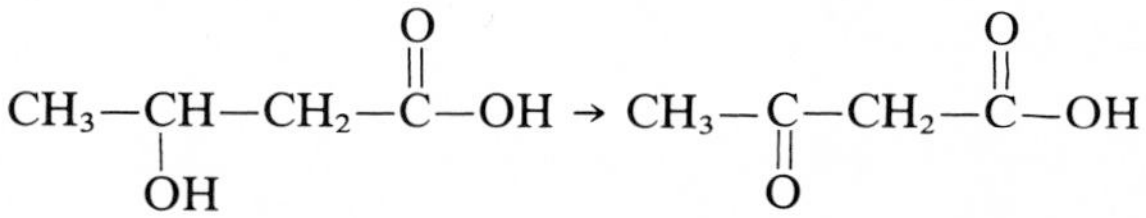

(h) $CH_3-\underset{\underset{O}{\|}}{C}-CH_2-CH=CH_2 \rightarrow CH_3\overset{\overset{O}{\|}}{C}-CH_2-\underset{\underset{OH}{|}}{CH}-CH_3$ 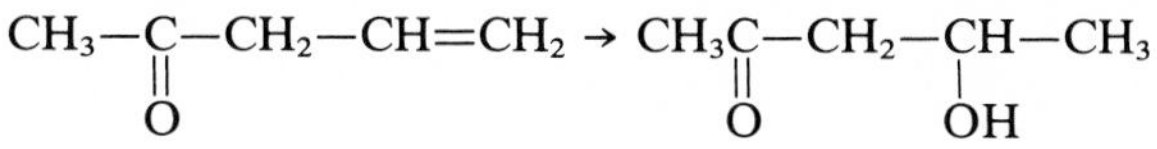

UNIT 23   ALKANES

The simplest members of the class of organic chemicals called hydrocarbons are the *alkanes*. They are also called *saturated* hydrocarbons since they contain as much hydrogen as is theoretically possible. The alkenes and alkynes are called *unsaturated* hydrocarbons since they can react to absorb more hydrogen. All the carbon atoms in an alkane have sp^3 hybridization, and each is bonded covalently to four other atoms. No double or triple bonds exist in alkanes.

Structures of Alkanes

Each carbon atom in an alkane has sp^3 hybridization, so the four atoms attached to it are arranged in a tetrahedral geometry. When the geometry of molecules is important, textbooks often show pictures of orbital models, ball and stick models, or van der Waals models (see Figure 23–1). Actually, the best way for you to become familiar with the structures is to construct them from molecular models yourself.

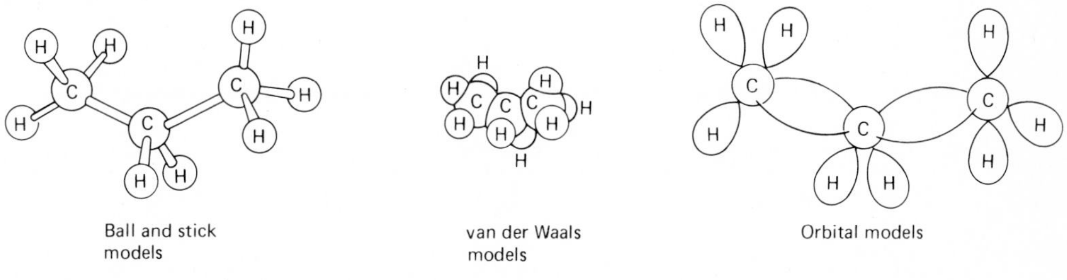

Ball and stick models   van der Waals models   Orbital models

Figure 23–1   Molecular model representations of propane (C_3H_8).

When it is not necessary to use three-dimensional models, a variety of other methods for picturing alkanes are used. These methods are convenient since they take less time to draw and occupy less space, but none of them gives accurate representations of the actual geometries. Notice in Figure 23–2 that the shorthand representations of *n*-hexane can be drawn in a straight line or in a variety of other ways. Since all the atoms are linked by single bonds, the atoms are free to rotate and twist. All the structures pictured in Figure 23–2 are representations of the same molecule.

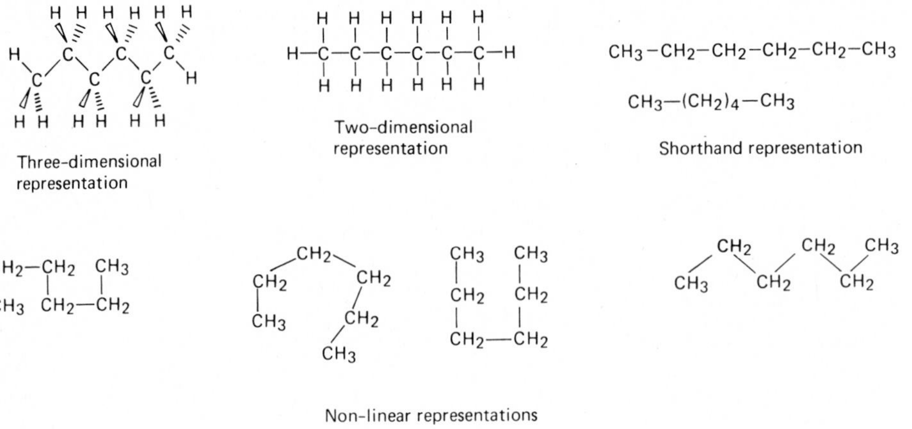

Three-dimensional representation
Two-dimensional representation
CH3−CH2−CH2−CH2−CH2−CH3
CH3—(CH2)4—CH3
Shorthand representation
Non-linear representations

Figure 23–2 Different representations of *n*-hexane (C_6H_{14}).

Isomers

A molecular formula is often not sufficient to identify the actual structure of organic molecules, as many different structures can generally be drawn from one molecular formula. Figure 23–3 shows that five different molecules can be drawn from the molecular formula C_6H_{14}. Each different molecule is an **isomer** of the other, a molecule with the same molecular formula but a different structure. Two molecules are isomers if they have the same molecular formulas but different structures. The

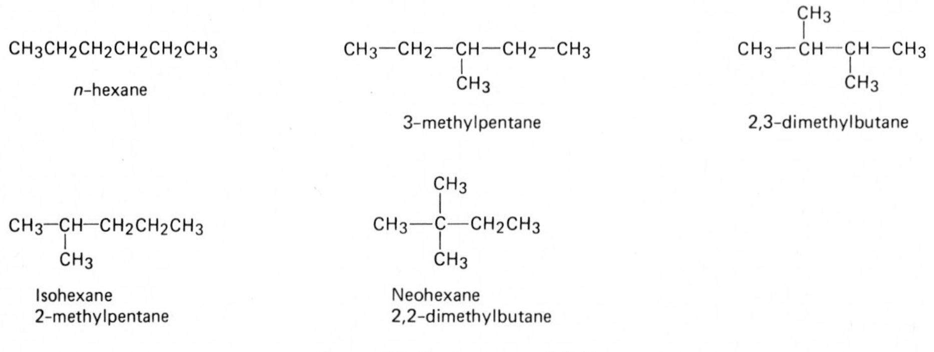

CH3CH2CH2CH2CH2CH3
n-hexane
3-methylpentane
2,3-dimethylbutane
Isohexane
2-methylpentane
Neohexane
2,2-dimethylbutane

Figure 23–3 Isomers of C_6H_{14}.

structures given in Figure 23–2 are not isomers but simply different *representations* of the same molecule. In Figure 23–3, different *molecules* are pictured.

Names of Alkanes

The IUPAC Method

There are several methods of naming organic molecules. Historically, the names of organic molecules were derived either from their source or from some particular property that they possessed. Formic acid (HCO_2H) was first isolated by heating crushed ants (*Latin*, *formica*, ants) and is responsible for the sting of an ant bite. Methane (CH_4) came from methyl alcohol. At one time, methyl alcohol was made by distilling wood chips, and the name comes from the Greek *metky* (wine) and *hyl* (wood). Ethane (C_2H_6) was derived from the Greek *aithein*, to blaze. Propane came from propanoic acid, a three-carbon compound whose name is derived from the Greek *protos* (first)and *pion* (fat). The name butane, came from butyric acid, which is found in rancid butter.

As more and more compounds were discovered, it was quickly realized that the historic method of naming organic molecules would lead to chaos, as soon it would be impossible to keep all the names straight. A systematic method was established in 1892 by the International Chemical Congress. The rules and their subsequent modifications are now known as the **IUPAC system** (International Union of Pure and Applied Chemistry).

In the IUPAC system, the name for an organic chemical is composed of three major portions. The **suffix** is used to indicate the major class of the compound. Thus alk**anes** have no double bonds, alk**enes** have a carbon-carbon double bond, alk**ynes** have a triple bond, alcoh**ols** have an OH group, and ket**ones** a $>C=O$ group. The **center portion** of the name indicates the number of carbon atoms and the structure of the main parent skeleton of the molecule. The word *cyclohexane* means a six-carbon alkane that forms a ring. The initial portion, the **prefix**, of a name tells how a particular compound differs from the parent skeleton. It gives the names and locations of groups that have been attached to the parent skeleton.

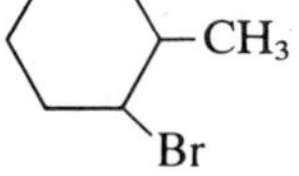

CH3
Br

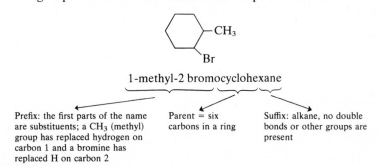

1-methyl-2 bromocyclohexane

Prefix: the first parts of the name are substituents; a CH_3 (methyl) group has replaced hydrogen on carbon 1 and a bromine has replaced H on carbon 2

Parent = six carbons in a ring

Suffix: alkane, no double bonds or other groups are present

It is extremely important to learn how to name alkanes correctly, as the names of all other organic compounds are derived from these names. The following rules present the method for naming alkanes.

The ending of all alkanes is *ane*. Table 23–1 shows the ending for the other classes of organic molecules.

Table 23–1 Endings for the various classes of organic compounds

Group	Name	Ending	Group	Name	Ending
>C=C<	Alkene	-ene	—C—N̈<	Amine	-amine
—C≡C—	Alkyne	-yne	—C(=O)—OH	Carboxylic acid	-oic acid
—C—OH	Alcohol	-ol	—C(=O)—Cl	Acid chloride	-oyl chloride
—C—C(=O)—H	Aldehyde	-al	—C(=O)—O—C—	Ester	-ate
—C—C(=O)—C—	Ketone	-one	—C(=O)—N̈<	Amide	-amide

Base or parent skeleton: *The longest continuous chain of carbon atoms in any molecule is the parent skeleton.*

To name any alkane, find the longest carbon chain and give it the name shown in Table 23–2. The first 10 are the most important.

Table 23–2 Names of continuous alkane chains

No. of carbons	Molecular formula	Name	Structural formula
1	CH_4	*Meth*ane	CH_4
2	C_2H_6	*Eth*ane	$CH_3—CH_3$
3	C_3H_8	*Prop*ane	$CH_3CH_2CH_3$
4	C_4H_{10}	*But*ane	$CH_3—(CH_2)_2CH_3$
5	C_5H_{12}	*Pent*ane	$CH_3—(CH_2)_3CH_3$
6	C_6H_{14}	*Hex*ane	$CH_3—(CH_2)_4CH_3$
7	C_7H_{16}	*Hept*ane	$CH_3—(CH_2)_5CH_3$
8	C_8H_{18}	*Oct*ane	$CH_3—(CH_2)_6CH_3$
9	C_9H_{20}	*Non*ane	$CH_3—(CH_2)_7CH_3$
10	$C_{10}H_{22}$	*Dec*ane	$CH_3—(CH_2)_8CH_3$
11	$C_{11}H_{24}$	*Undec*ane	$CH_3—(CH_2)_9CH_3$
12	$C_{12}H_{26}$	*Dodec*ane	$CH_3—(CH_2)_{10}CH_3$
13	$C_{13}H_{28}$	*Tridec*ane	$CH_3—(CH_2)_{11}CH_3$
15	$C_{15}H_{32}$	*Pentadec*ane	$CH_3—(CH_2)_{13}CH_3$
20	$C_{20}H_{42}$	*Eicos*ane	$CH_3—(CH_2)_{18}CH_3$

Note that the word *straight* was not used. Remember that alkanes have freedom of rotation about their bonds. You must find the longest chain, *not* the longest *straight* chain. The word *continuous* means that no other types of atoms (O, S, N, and so on) are present in the chain, and there are no breaks in the chain.

If the parent has a cyclic structure, indicate this by placing *cyclo* in front of the parent name.

Note the type of substituents attached to the parent chain (Table 23–3). Any groups of atoms attached to the parent skeleton are called **substituents**, as they have

Table 23–3 Names and structures of common substituents*

Br—bromo Cl—chloro I—iodo	HO—hydroxy CH_3—methyl CH_3CH_2—ethyl
Normal group (*n*-): CH_3—CH_2— $CH_3CH_2CH_2$—*n*-propyl $CH_3CH_2CH_2CH_2$—*n*-butyl	Secondary group (sec): CH_3 | ∿CH— CH_3 | CH_3CH_2CH—secbutyl $CH_3CH_2CH_2CH$—secpentyl | CH_3
Iso group (iso): CH_3 \\ CH∿ / CH_3 CH_3 \\ CH—isopropyl / CH_3 CH_3 \\ CH—CH_2—isobutyl / CH_3 CH_3 \\ $CHCH_2CH_2$—isopentyl / CH_3	Tertiary group (tert.- or *t*-): CH_3 | ∿C— | CH_3 CH_3 | CH_3—C—*t*-butyl | CH_3 CH_3 | CH_3CH_2C—*t*-pentyl | CH_3 CH_3 | $CH_3CH_2CH_2$—C—*t*-hexyl | CH_3
Neo group (neo): CH_3 | CH_3CCH_2∿ | CH_3 CH_3 | CH_3CCH_2—neopentyl | CH_3 CH_3 | $CH_3CCH_2CH_2$—neohexyl | CH_3 CH_3 | $CH_3CCH_2CH_2CH_2$—neoheptyl | CH_3	

*The symbol (—) refers to the point of attachment to the parent chain. The wavy line refers to a series of CH_2 and CH_3 groups (—CH_2—CH_2...CH_3).

been substituted for hydrogen atoms. The names of the substituents that contain only carbon and hydrogen are similar to those for the alkanes, with two important changes: (1) a carbon substituent has the ending *yl* instead of ane; (2) the substituent is often named *in toto* by using prefixes that were derived from the common names.

Table 23–3 lists the names of the most common substituents. Often the only difference in substituents comes from the specific carbon atom that is the point of attachment to the parent skeleton. An ***n*-propyl** group is fastened by the first carbon, whereas an **isopropyl** group is fastened by the middle carbon.

Locate the substituents on the present chain by numbering the carbon atoms in the parent. *Number the carbon atoms of the parent chain starting at the end that is closest to a substituent.* When the first substituents nearest each end are present on carbon atoms of the same number, go to the next substituent.

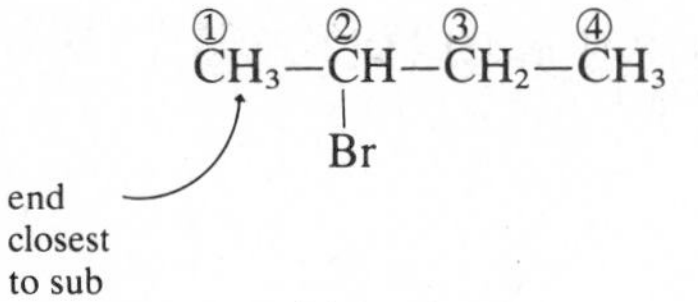
end
closest
to sub

2-bromobutane

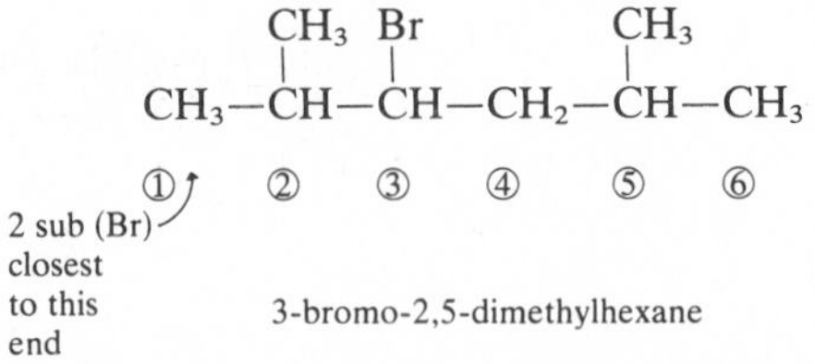
2 sub (Br)
closest
to this
end

3-bromo-2,5-dimethylhexane

The *names of the substituents are preceded by the number of the carbon atom in the parent chain to which they are attached.* Every substituent (other than hydrogen) must have a number.

If there are more than one of a particular substituent, indicate this by the prefixes di, tri, tetra, and so on.

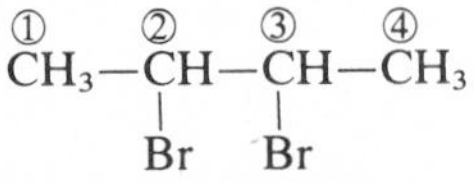

2,3-*di*bromobutane

The substituents are arranged in front of the parent name in either of two methods: (1) *Increasing size*: CH_3 is placed before CH_3CH_2—, which appears before $CH_3CH_2CH_2$—, and so on; (2) *alphabetically*: ethyl > methyl > propyl, and so on. *They are not placed in the name in the order of appearance on the carbon chain.*

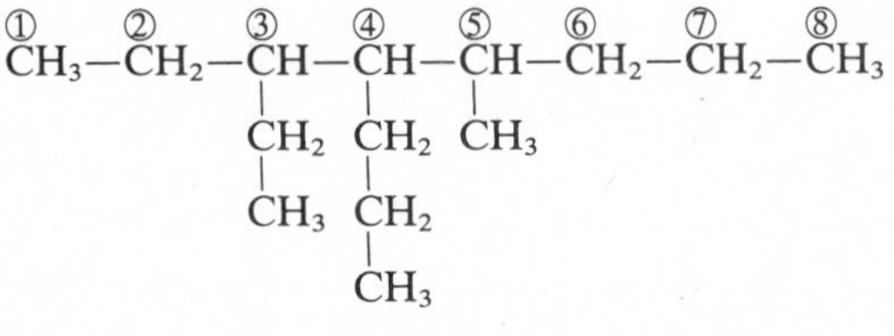

5-methyl-3-ethyl-4-*n*-propyloctane, NOT 3-ethyl-4-*n*-propyl-5-methyloctane

Punctuation: The entire name is one word; commas separate numbers, and hyphens separate numbers and letters (see Figure 23–4).

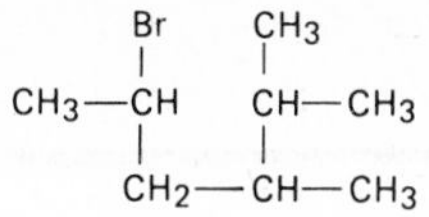

5-bromo-2,3-dimethylhexane

(not 4-bromo-2-isopropylpentane)

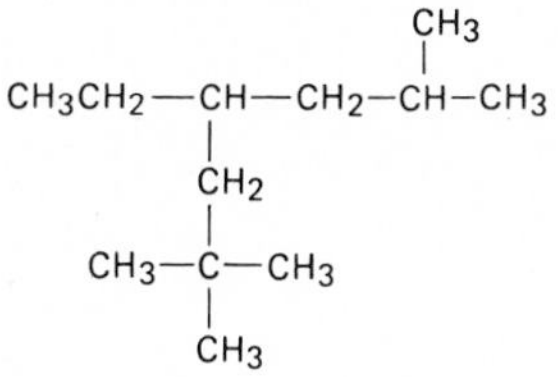

2,2,6-trimethyl-4-ethylheptane

(not 5-methyl-3-neopentylpentane)

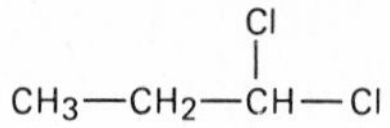

1,1-dichloropropane

(not 3,3-dichloropropane)

(not 1-dichloropropane)

(not 1,1-chloropropane)

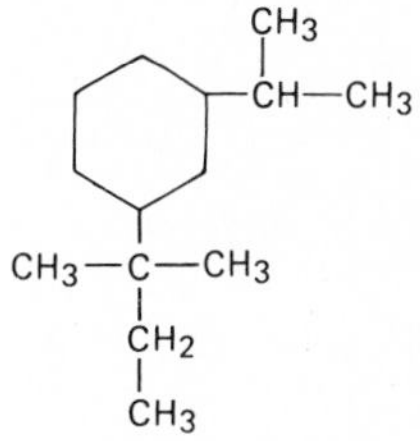

1-isopropyl-3-*t*-pentylcyclohexane

(not 1-isopropyl-5-*t*-pentylcyclohexane)

Figure 23–4 Names of alkanes.

When the substituents on a parent chain cannot be named by using the common names, we have to number the substituent chain also, and then place the name of the substituent in brackets when we put it in the entire compound name.

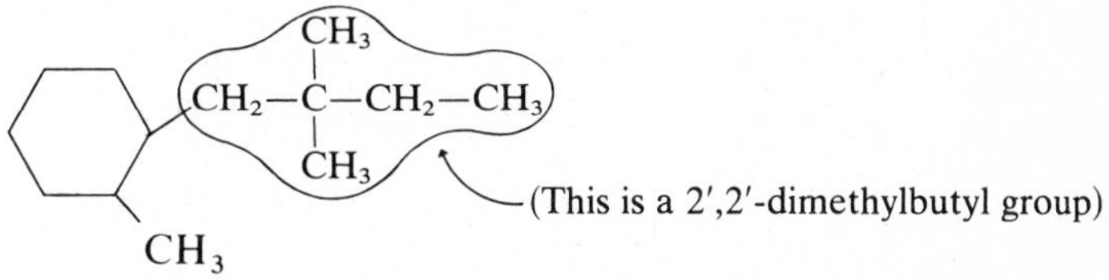
(This is a 2′,2′-dimethylbutyl group)

1-methyl-2-(2′,2′-dimethylbutyl)cyclohexane

(In order to avoid confusion in the numbering of the parent and side chain, we use primed numbers for the side chain.)

Sources and Uses of Alkanes

There are two major sources of alkanes: natural gas and crude oil. Both were formed millions of years ago as decaying animal and vegetable matter was subjected to high pressures over long periods of time. Both sources are limited. The alkanes are isolated from natural gas and crude oil by careful distillation; Table 23–4 lists the components in each distillation fraction.

Various alkanes can also be extracted from plants. Heptane is found in several kinds of pine, tetradecane (14 carbons) in chrysanthemums, and hexadecane (16 carbons) in roses. Many waxy coatings on plants are composed in part of alkanes.

Table 23–4 Typical crude oil distillation

Fraction	Range of alkanes	Approximate boiling range (°C)	Uses
Gas	CH_4 to C_4H_{10}	Less than 40	Fuels, starting materials for plastics
Gasoline	C_5H_{12} to $C_{12}H_{26}$	40–200	Solvents, fuels
Kerosene	$C_{12}H_{26}$ to $C_{16}H_{34}$	175–275	Diesel, jet fuel, home heating; may be further refined with gasoline
Heating oil	$C_{15}H_{32}$ to $C_{18}H_{38}$	250–400	Industrial heating; may be cracked to provide gasoline
Lubricating oil	$C_{17}H_{36}$ and up	Above 300	Lubricants in machines
Residual	$C_{20}H_{42}$ and up	Above 350	Tar, asphalt, paraffin

Physical Properties of Alkanes

The boiling temperatures of alkanes increase regularly as the length of the alkane increases. As a rule alkanes boil at lower temperatures than the other types of organic compounds because they are quite nonpolar and cannot participate in hydrogen bonding. The first four alkanes are gases at room temperature. The liquid alkanes increase in viscosity as the size increases, and it is difficult to distinguish the liquid–solid breakpoint, since many of the so-called solids are still soft and pliable. Alkanes are relatively nontoxic, but can cause death by asphyxiation since those which are gases can prevent oxygen from entering the lungs.

Chemical Properties of Alkanes

In comparison to the other types of organic compounds, alkanes are relatively inert. They are quite stable and are good solvents for nonpolar solutes. They undergo a relatively small number of reactions.

Combustion: As can be seen from Table 23–4, the major use of alkanes is as fuels. Alkanes react with oxygen in the presence of a flame to yield water, carbon dioxide, and energy. The energy released in the combustion process is used to heat homes, operate factories, power automobiles, generate electricity, and cook food.

$$\text{Alkanes} + \text{oxygen} \xrightarrow{\text{Flame}} CO_2 + H_2O + \text{heat}$$

If there is not enough oxygen present when alkanes are burned, carbon monoxide and various partially oxidized alkanes result. We should never operate internal combustion engines in closed buildings, for these products of combustion are toxic.

Halogenation: Chlorine and bromine atoms can replace hydrogen atoms on an alkane when mixtures of bromine or chlorine and alkanes are exposed to light or high temperatures. These reactions are merely substitution reactions in which the chlorine or bromine atom takes the place of a hydrogen atom. The hydrogen atom removed from the alkane reacts with the other chlorine or bromine atom to form HCl and HBr, so only one chlorine atom becomes attached to the alkane for every Cl_2 molecule that reacts:

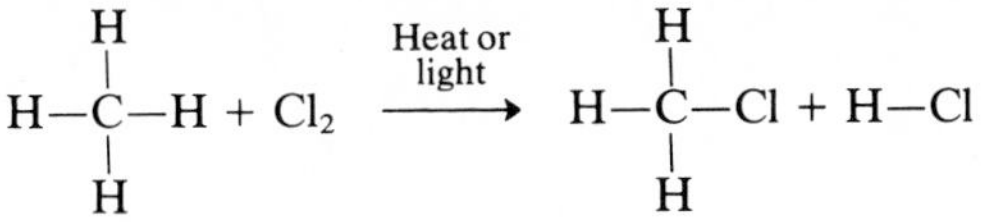

$$\underset{\underset{H}{|}}{\overset{\overset{H}{|}}{H-C-H}} + Cl_2 \xrightarrow{\substack{Heat\ or \\ light}} \underset{\underset{H}{|}}{\overset{\overset{H}{|}}{H-C-Cl}} + H-Cl$$

When a larger alkane is halogenated, a variety of products can be formed. Three different monobromo compounds are formed by bromination of *n*-pentane. Dibrominated and tribrominated products may also result.

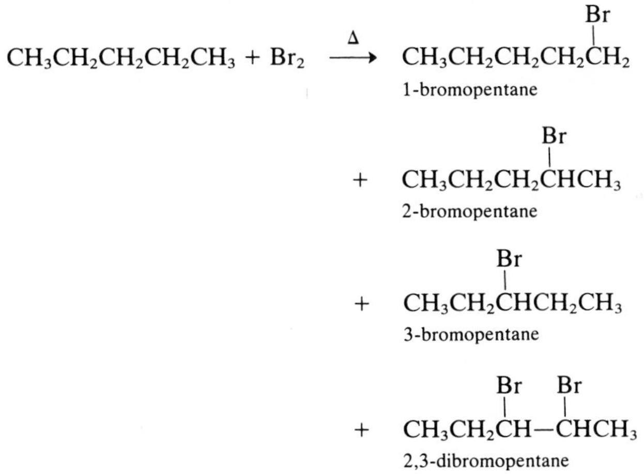

$$CH_3CH_2CH_2CH_2CH_3 + Br_2 \xrightarrow{\Delta} \underset{\text{1-bromopentane}}{CH_3CH_2CH_2CH_2\overset{\overset{Br}{|}}{C}H_2}$$

$$+ \quad \underset{\text{2-bromopentane}}{CH_3CH_2CH_2\overset{\overset{Br}{|}}{C}HCH_3}$$

$$+ \quad \underset{\text{3-bromopentane}}{CH_3CH_2\overset{\overset{Br}{|}}{C}HCH_2CH_3}$$

$$+ \quad \underset{\text{2,3-dibromopentane}}{CH_3CH_2\overset{\overset{Br}{|}}{C}H-\overset{\overset{Br}{|}}{C}HCH_3}$$

and so on

Industrially, chloroform ($CHCl_3$) and carbon tetrachloride (CCl_4) are produced by combining chlorine and methane under high temperatures.

$$CH_4 + Cl_2 \xrightarrow{\Delta} CH_3Cl \xrightarrow[\Delta]{Cl_2} CH_2Cl_2 \xrightarrow[\Delta]{Cl_2} CHCl_3 \xrightarrow[\Delta]{Cl_2} CCl_4$$
$$\quad\quad\quad\quad + \quad\quad\quad\quad + \quad\quad\quad\quad + \quad\quad\quad\quad +$$
$$\quad\quad\quad\quad HCl \quad\quad\quad HCl \quad\quad\quad HCl \quad\quad\quad HCl$$

UNIT 23   Problems and Exercises

23–1. Draw a line through the parent chains in each of the following, and name the parent.

(a) $\quad \underset{CH_3\overset{\overset{|}{CH_2CH_3}}{CHCH_2CH_2CH_3}}{}$ 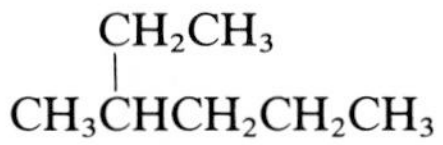

(b) $\underset{CH_3}{\overset{CH_3}{\diagdown}}CH-CH_2-\underset{}{\overset{CH_3}{\overset{|}{CH_2}}}$ 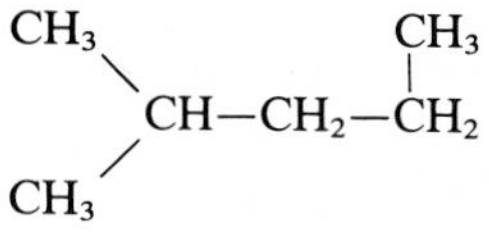

(c)

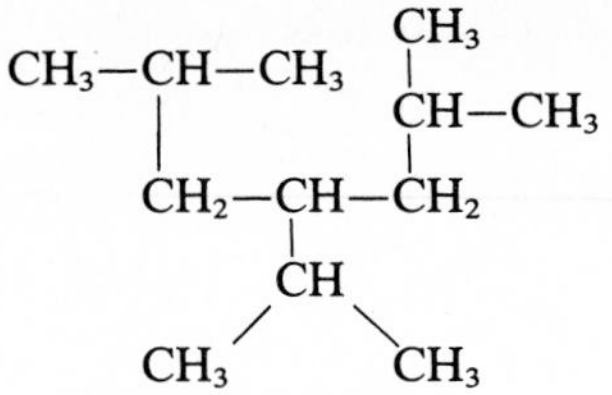

(d)

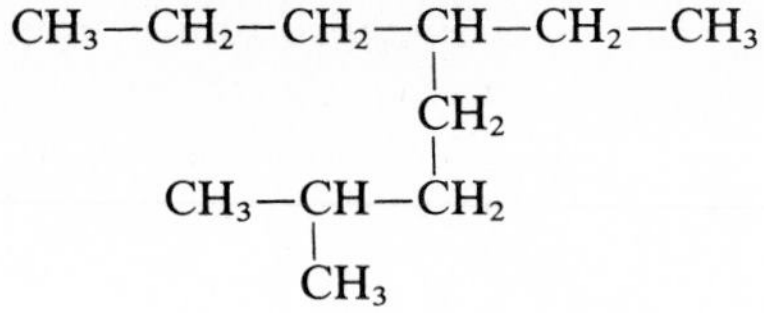

(e)

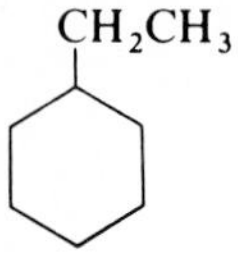

(f)

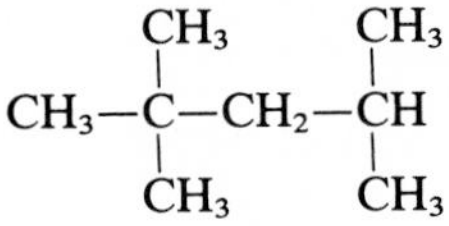

23–2. The thick line represents a parent chain that has substituents attached to it. Name each substituent.

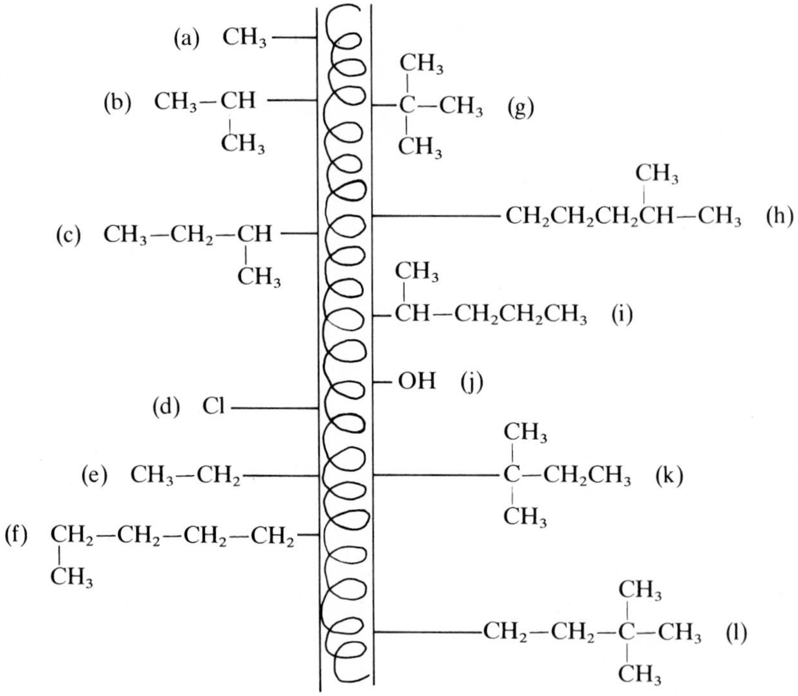
(a) CH3
(b) CH3—CH—CH3
(c) CH3—CH2—CH—CH3
(d) Cl
(e) CH3—CH2
(f) CH2—CH2—CH2—CH2—CH3
(g) C(CH3)2—CH3
(h) CH2CH2CH2CH(CH3)—CH3
(i) CH(CH3)—CH2CH2CH3
(j) OH
(k) C(CH3)2—CH2CH3
(l) CH2—CH2—C(CH3)2—CH3

23–3. Name the following.

(a)

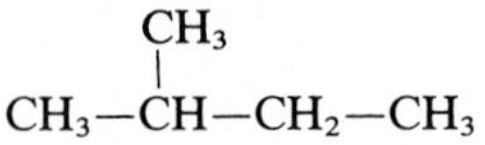

(b)

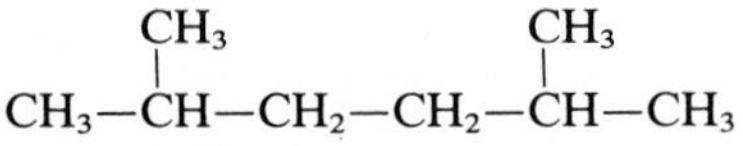

(c)

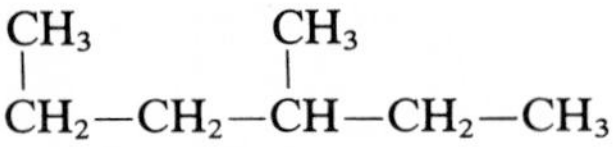

(d)

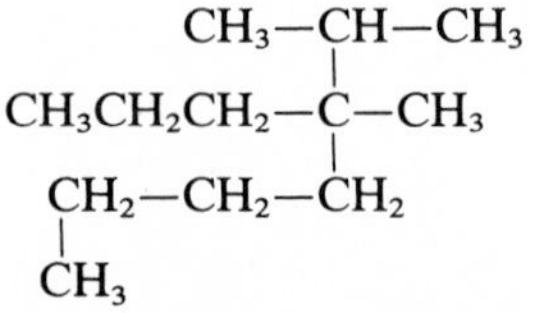

(e)

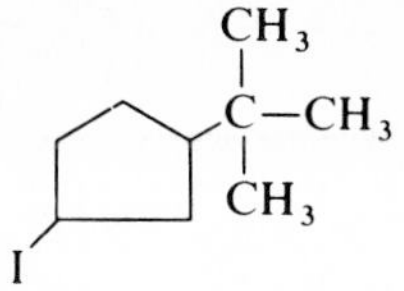

(f)

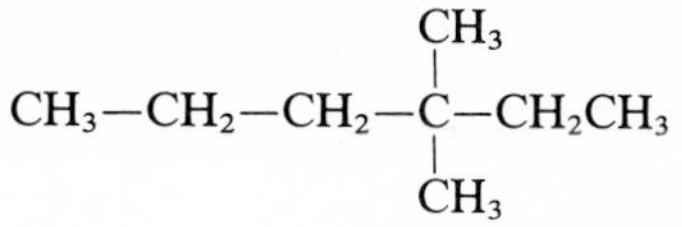

(g)

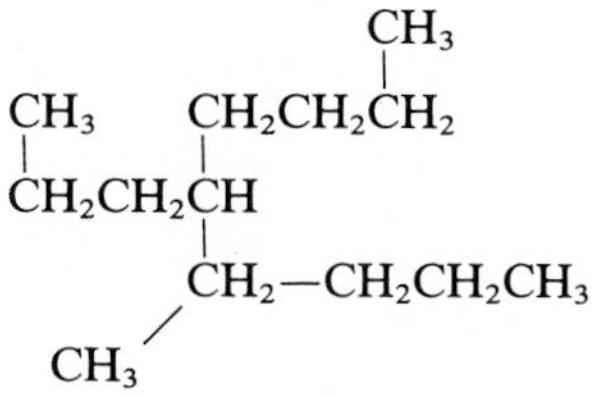

(h)

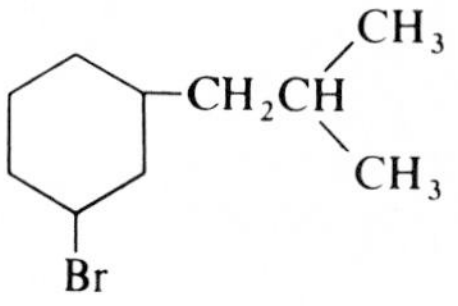

(i) $(CH_3)_2CH(CH_2)_4CH_3$

(j) $(CH_3CH_2)_2CHCH_2CH(CH_2CH_3)_2$

23–4. Draw structures for the following.

(a) 2,3-dimethylpentane
(b) 3-bromo-2-methylheptane
(c) 2,2-dimethyloctane
(d) 3,4-dimethyl-4-ethyloctane
(e) 3-isopropyloctane
(f) 4-methyl-3,3-dimethyl-5-isobutyldecane
(g) 1-methyl-3-secbutylcyclobutane
(h) 5,5-dichloro-3-bromo-2-methylheptane

23–5. Check each of the following names against the structure to see if it is correct. If not, tell what is wrong and correct it.

(a)

1-methylbutane

(b)

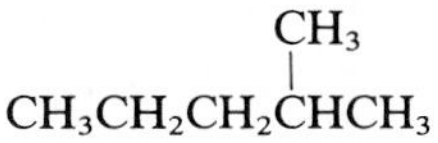

4-methylpentane

(c)

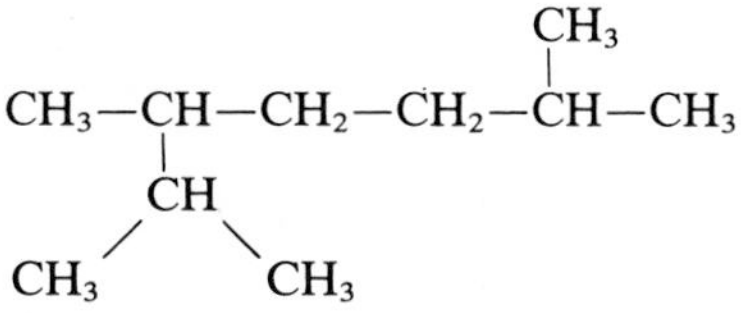

2-methyl-5-isopropylhexane

(d)

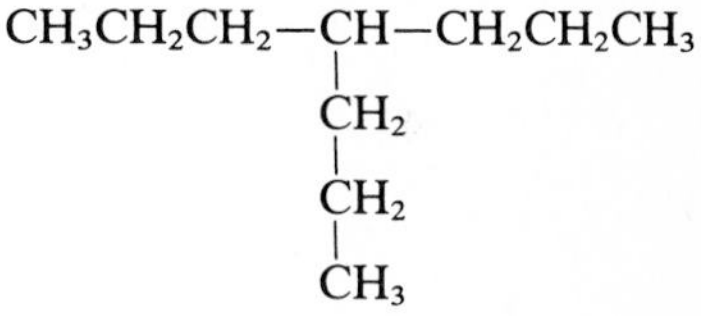

4-isopropylheptane

(e)

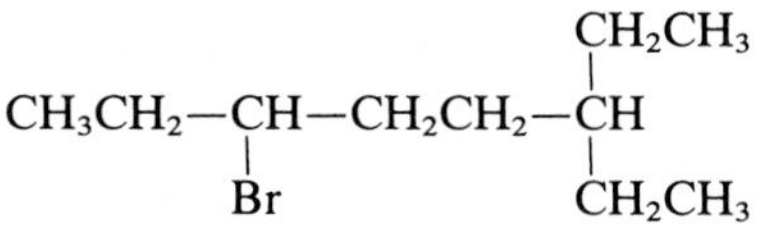

3-bromo-6,6-diethylhexane

(f)

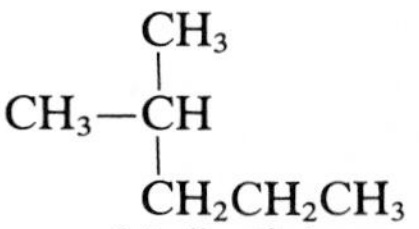

1,1-dimethylpentane

(g)

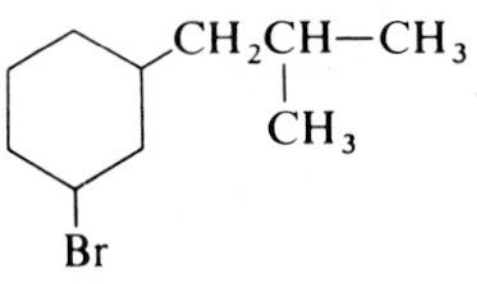

1-bromo-3-secbutyl-
cyclohexane

(h)

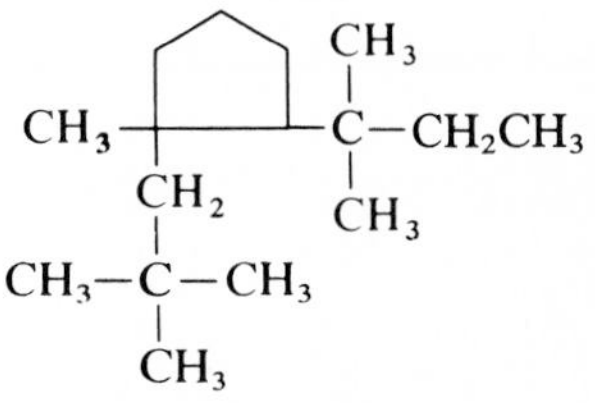

1-methyl-1-neopentyl-
2-*t*-pentylcyclopentane

(i)

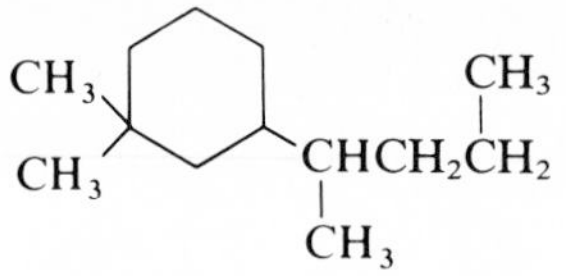

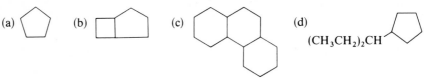

1,1-dimethyl-3-isopentyl-
cyclohexane

23–6. Give the molecular formulas for the following compounds. Remember that many of the hydrogen atoms are not shown in the abbreviated structures.

(a) 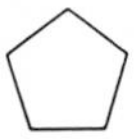 (b) 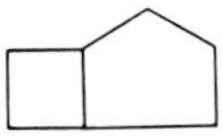 (c) 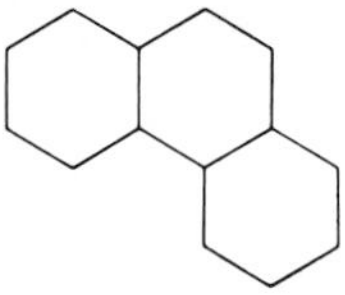 (d) 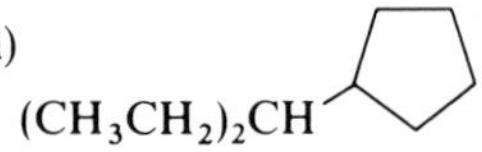

23–7. Draw the structural formula and give the IUPAC name for the following.
(a) $(CH_3)_2CHCH_2CH_2CH_3$ (b) $CH_3CH_2CH_2C(CH_3)_2CH_2CH_3$
(c) $(CH_3CH_2)_2CH{-}CH_2CH(CH_2CH_3)_2$ (d) $(CH_3)_3C(CH_2)_4CH_2CH_3$
(e) $CH_3(CH_2)_7CH_3$

23–8. Write the structural formulas for all the isomers of each of the following formulas and name them (the correct number of isomers is indicated in parentheses).
(a) C_6H_{14} (5) (b) C_4H_9Br (4) (c) $C_3H_6Cl_2$ (4)
(d) C_7H_{16} (9) (e) $C_2H_3Cl_3$ (2)

23–9. Are the following pairs of compounds the same or different compounds?

(a) 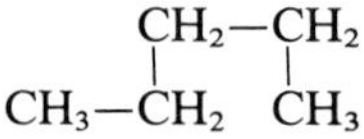 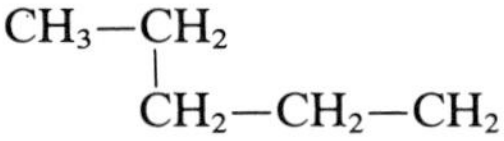

(b) 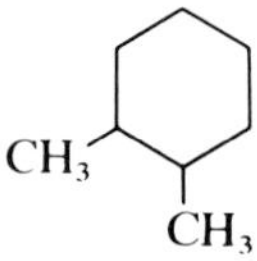 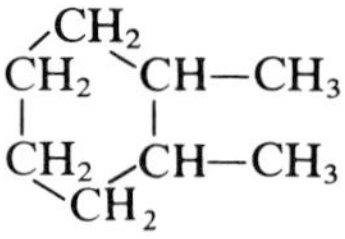

(c)  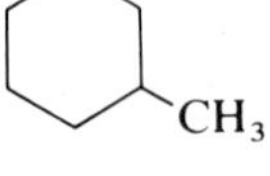

(d) 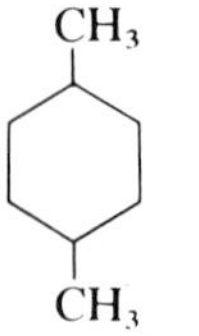 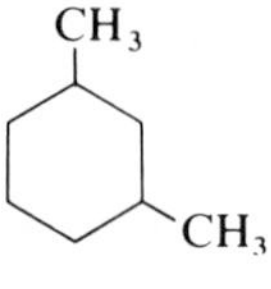

(e) 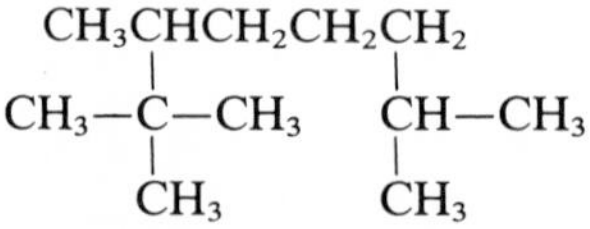 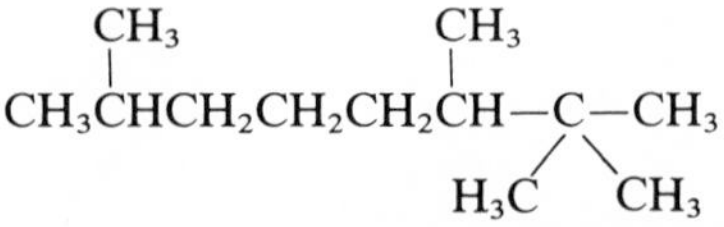

23–10. Write balanced equations for the following.

(a) The preparation of 2-chloropropane from the appropriate alkane and other reagents

(b) The preparation of 1,2-dibromoethane from ethane and other reagents

23–11. Complete the following.

(a) Butane + O_2 $\xrightarrow{\text{Heat}}$ (balance this reaction)

(b) Cyclohexane + O_2 $\xrightarrow{\text{Heat}}$ (balance this reaction)

(c) Butane + Cl_2 $\xrightarrow{\Delta}$ (two products; react only one Cl_2 per molecule butane)

(d) Ethane + Br_2 $\longrightarrow$ (all possible products)

23–12. What is the major difference between gasoline and heating oil?

23–13. An alkane with the formula C_6H_{14} is treated with Br_2 + light. The products are isolated and separated.

(a) What is the structure and name of the alkane if only two monobromo products are possible?

(b) What is a possible structure and name if only three monobromo products are possible?

(c) What is the structure and name if five monobromo products are possible?

CHAPTER 10

the chemistry of unsaturated hydrocarbons

One of the most reactive types of functional groups is the carbon-carbon double bond. The compounds that possess these double bonds, the **alkenes**, are introduced in unit 24. In that unit we discuss some of the basic properties, uses, and reactions of compounds containing the carbon-carbon double bond.

In unit 25 we begin what many chemists believe to be the chief study of organic chemistry, the mechanisms of the reactions. It is not sufficient to understand what happened in a chemical reaction; we must also know how and why it happened so that we can develop theories to explain the events. One means of learning how a reaction occurs is to attempt to outline each separate step involved in the entire reaction process—to develop the **mechanism** for the reaction. In unit 25 we shall develop mechanisms for addition and elimination reactions.

In unit 26 we introduce companion compounds to alkenes, the **alkynes**. Their names and chemistry are similar to those for alkenes.

Finally, in unit 27 we introduce a more complicated type of unsaturated molecule, an **aromatic** compound. These compounds possess multiple bonds like alkenes and alkynes but are not nearly as reactive. They are much more stable, and enter into chemical reactions only under vigorous conditions. Much of this inertness is explained through the concept of resonance, the movement of electrons from atom to atom.

UNIT 24 ALKENES

The second members of the hydrocarbon class contain carbon-carbon double bonds. These compounds are known as the **alkenes**, **olefins**, or **unsaturated hydrocarbons**. They are much more reactive than alkanes, and have structures and bonding different from alkanes.

Structure and Bonding in Alkenes

In alkanes the four covalent bonds of each carbon atom all possess similar strength, but this is not the case for the four bonds at the $>C=C<$ center in alkenes. Three of the covalent bonds at the carbon atoms in this center are similar in strength to the alkane bonds, but the fourth bond (one of the bonds in the double bond) is much weaker. The three strong covalent bonds are called **sigma** (σ) **bonds** and arise from sp^2 hybrid orbitals of the carbon atoms. These orbitals possess a planar triangular geometry (see Figure 24–1).

The fourth covalent bond in the alkene functional group is formed by side-to-side overlap of the $2p_z$-orbitals remaining on the carbon atoms, which were not used in making the sp^2 hybrid orbitals. This side overlap is not nearly as effective as the direct head-to-head overlap of the other orbitals, so this fourth bond is not nearly as strong as the other three. It is called a **pi bond** (π) to distinguish it from the other three (see Figure 24–2).

Effects of the Pi Bond in Alkenes

Cis-trans isomers: The presence of a double bond between two carbon atoms in an alkene restricts the freedom of rotation of the carbon atoms. This restricted rotation gives rise to **geometrical isomers** (compounds that have the same formula

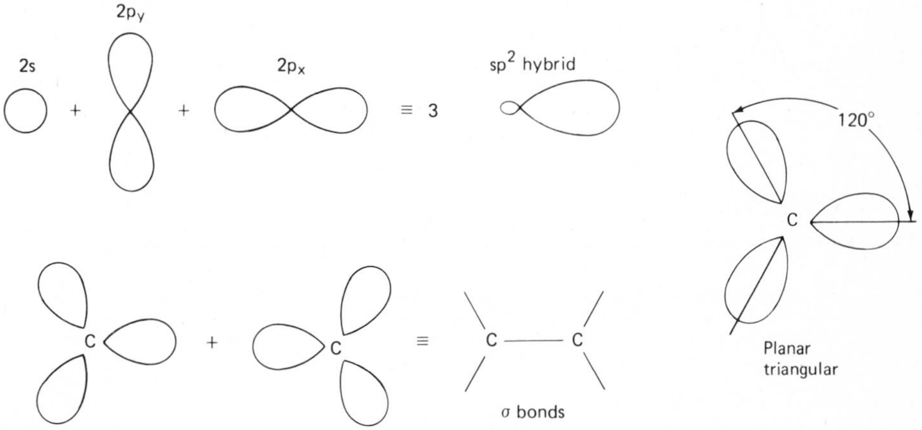

2py
2s
2px
sp2 hybrid
+
+
≡ 3
120°
C
Planar
triangular
C
+
C
≡
C
C
σ bonds

Figure 24–1

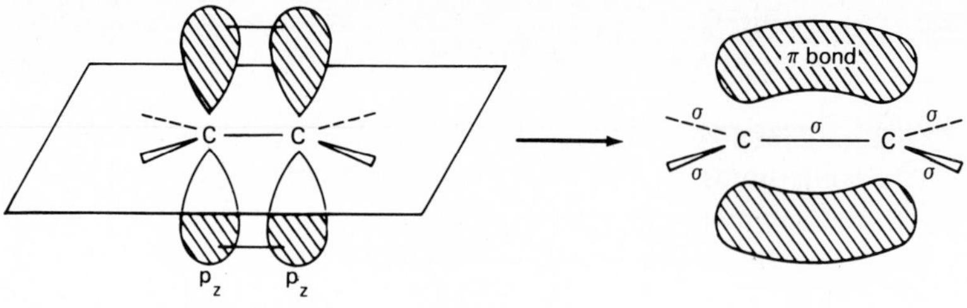
π bond
σ
σ
σ
σ
σ
C
C
C
C
p_z
p_z

Figure 24–2

but slightly different structures). Thus *cis*-2-butene is not the same compound as *trans*-2-butene (see Table 24–1). In order to determine and identify cis and trans geometrical isomers you should follow the steps outlined below:

Table 24–1

	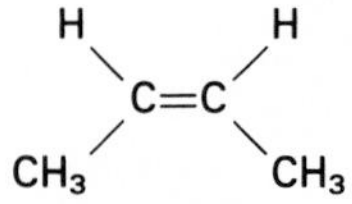 *cis*-2-butene	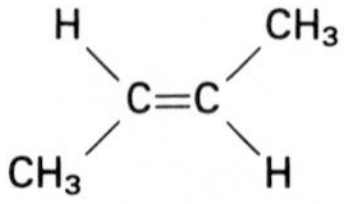 *trans*-2-butene
Melting point	−108°C	−105°C
Boiling point	3.7°C	0.9°C
Density	0.65 g/mℓ	0.68 g/mℓ

1. Divide the double-bond portion of the compound into four quadrants (see Figure 24–3).

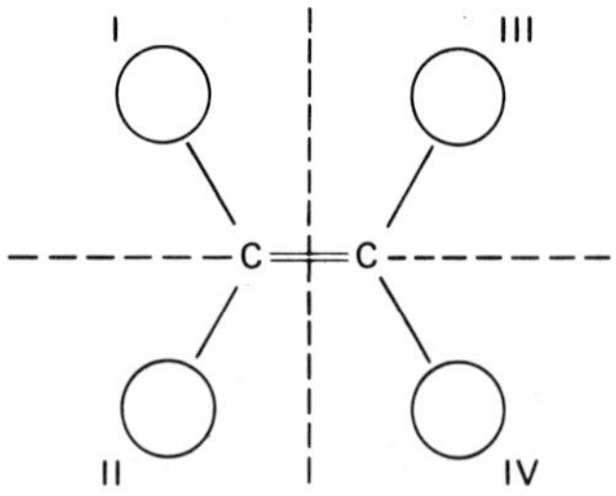
I
III
C
C
II
IV

Figure 24–3

2. Cis and trans isomers can exist if the groups in quadrants I and II differ and the groups in quadrants III and IV differ. It does not matter if similar groups are present in I and III, I and IV, and so on. If the groups in I and II (or III and IV) are identical, no cis and trans isomer can exist.
3. Locate the larger of the groups in quadrants I and II. Do the same for quadrants III and IV. *If these two larger groups are both above or below the*

double bond (I and II) or (III and IV), *the geometry is* **cis**. *If they exist on opposite sides* (I and IV or II and III), *the geometry is* **trans**.

Examples:

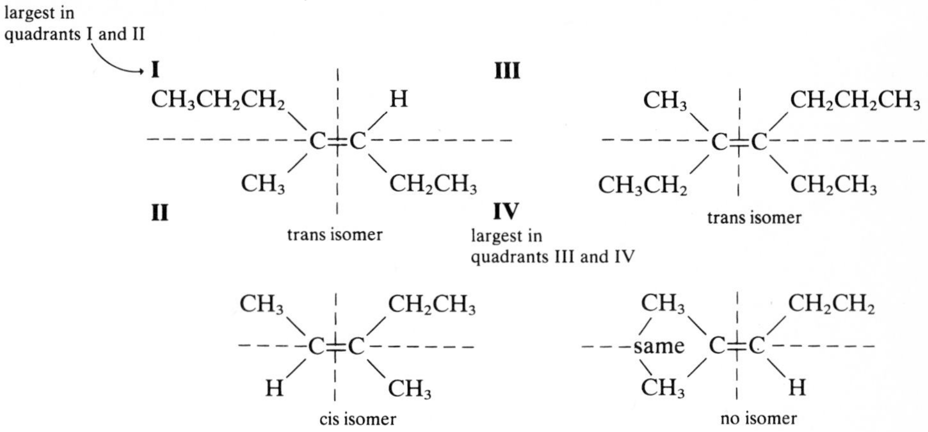
largest in
quadrants I and II
I
II
III
IV
largest in
quadrants III and IV
trans isomer
trans isomer
cis isomer
same
no isomer

It is important to note that the restricted rotation is present only in the double-bond portion of the molecule. The single bonds in the rest of the molecule still have freedom of rotation.

Increased chemical reactivity: The weak pi bond in alkenes is easily broken, so the vast majority of all the reactions of alkenes are additions in which some reagent adds to the π bond of the double bond. The carbon atoms of the alkene rehybridize and become sp^3 as the weak pi bond disappears and is replaced by two strong sigma bonds. In this process the reactive alkene has been converted to a more stable alkane or substituted alkane.

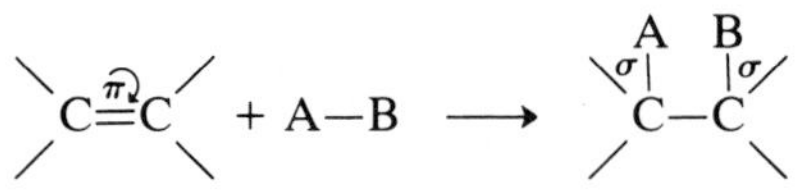

Nomenclature of Alkenes

The rules for naming alkenes are similar to those for the alkanes, but there are some important changes. The parent chain is now the longest continuous chain *containing the double bond.* The name of the parent is similar to that for alkanes, except the suffix *ane* is replaced by *ene.* Remember, there may be longer chains in the compound, but if they do not contain the double bond, they are not the parent chain.

The positions of the double bond and the positions of the substituents are located by numbering the parent chain. But you must *start numbering from the end nearest to the double bond.* In naming the alkene, *place the number of the first carbon of the double bond directly before the name of the parent alkene.* As with alkanes, the names and positions of the substituents are placed in front of the parent compound (see Figure 24–4).

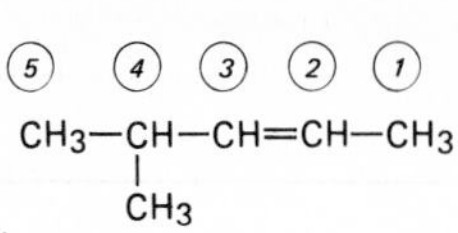

4-methyl-2-pentene

(not 2-methyl-3-pentene)

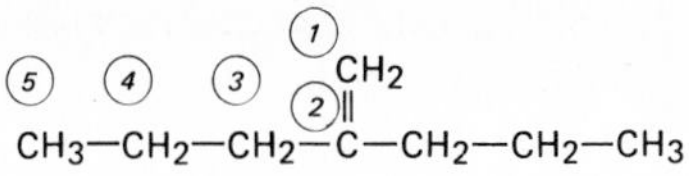

2-*n*-propyl-1-pentene

(here the longest chain is 7, but the longest chain containing the double bond is 5)

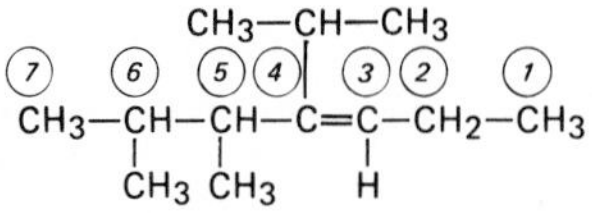

5,6-dimethyl-4-isopropyl-3-heptene

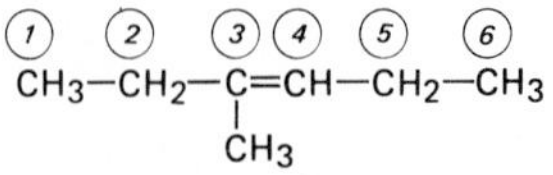

3-methyl-3-hexene
(not 4-methyl-3-hexene)

Figure 24–4 Examples of names of alkenes.

Uses of Alkenes

The compounds found in our world are rarely simple molecules containing only one type of group, so it is difficult to describe all the types of reactions and uses for compounds that contain carbon-carbon double bonds. Instead we shall illustrate some of the areas where the alkene bond plays a central role.

Polymer formation: The word polymer means *many parts*, so a polymer is made up of many smaller, identical parts (called **monomers**) as a chain is made up of many links. The polymers formed by joining alkene molecules together are called *plastics*, and are long chains of carbon atoms with substituents attached at regular intervals:

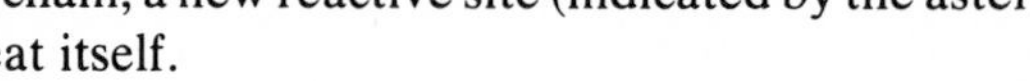

(a chain made up of links)

$$C{=}C + C{=}C + C{=}C + C{=}C \longrightarrow -(C{-}C){-}(C{-}C){-}(C{-}C){-}(C{-}C)-$$

(a polymer made up of alkanes)

During the polymerization process the long chains are formed by successive additions to the double bonds of the alkene. Each time an alkene molecule is added to the growing chain, a new reactive site (indicated by the asterisk) is created, so the process can repeat itself.

$$\text{Initiator} + C{=}C \longrightarrow I{-}C{-}C^*$$

$$I{-}C{-}C^* + C{=}C \longrightarrow I{-}C{-}C{-}C{-}C^*$$

$$I{-}C{-}C{-}C{-}C^* + C{=}C \longrightarrow I{-}C{-}C{-}C{-}C{-}C{-}C^* \text{ and so on}$$

The properties of a polymer or plastic depend upon the alkenes used as monomers and the length of the polymer chain formed. Table 24–2 gives some common plastics and their uses.

Table 24–2 Types of polymers

Alkene	Polymer formula	Name	Uses
$CH_2{=}CH_2$	$\{CH_2-CH_2\}_n$	Polyethylene	Packaging, bottles, toys
$CH_2{=}CH(CH_3)$	$\{CH_2-CH(CH_3)\}_n$	Polypropylene	Carpet fibers, pipes, valves, bottles
$CH_2{=}CH(Cl)$ Vinyl chloride (chloroethene)	$\{CH_2-CH(Cl)\}_n$	Polyvinyl chloride PVC	Floor tile, pipes, packaging, phonograph records
$CF_2{=}CF_2$	$\{CF_2-CF_2\}_n$	Teflon	Cooking utensil coating, gaskets, electrical insulation
$CH_2{=}CH(CN)$ Acrylonite (cyanoethene)	$\{CH_2-CH(CN)\}_n$	Polyacrylonitryls Orlon, Acrilan	Textile fibers
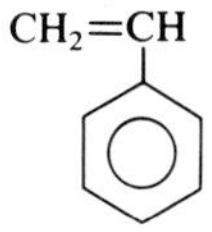 Styrene (phenylethene)	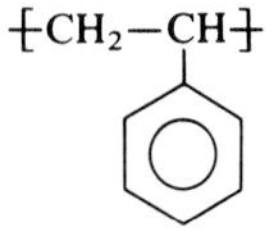	Polystyrene	Toys, styrofoam
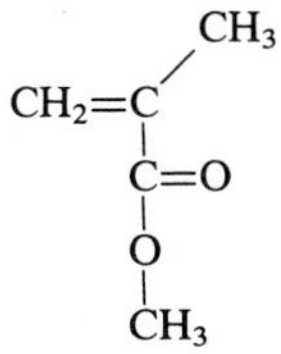 Methylmethacrylate	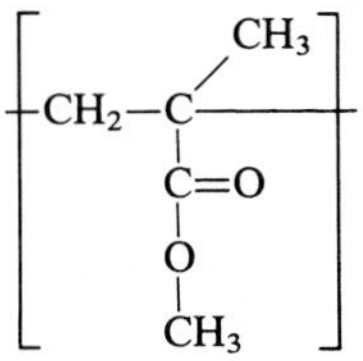	Acrylic Lucite, Plexiglas	Windows, decorative objects, costume jewelry

Many of the compounds present in nature are similar to polymers in that they are formed by linking together several smaller molecules. Steroids are formed by joining together several molecules of a di-alkene called isoprene (2-methyl-1,3-butadiene). Isoprene is also the base molecule for natural rubber. The isoprene units in each of the compounds shown in Figure 24–5 are circled.

The Chemistry of Vision

The photosensitive pigment of the eye is called *visual purple* or *rhodopsin.* It consists of a protein opsin and an unsaturated compound called neoretinene B. During the visual process, light bleaches the visual purple, causing rearrangement of *cis*-neoretinene to a trans form. During this conversion a nerve impulse is also sent to the brain. The entire process is vastly more complicated than presented here, but it is

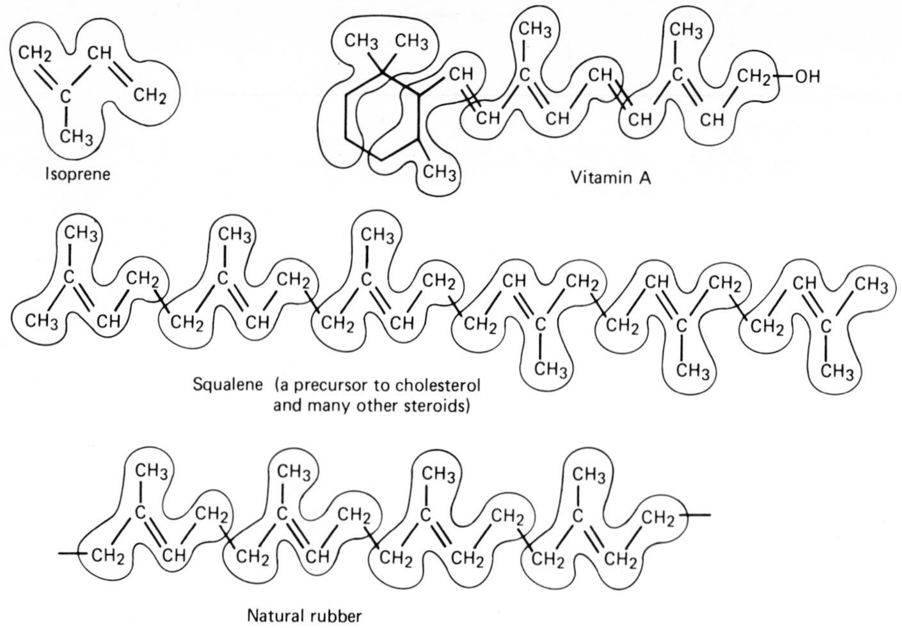
Isoprene
Vitamin A
Squalene (a precursor to cholesterol and many other steroids)
Natural rubber

Figure 24–5 Molecules synthesized from isoprene units.

interesting to note that a major reaction in the vision process is simply the rearrangement of one cis double bond to a trans double bond.

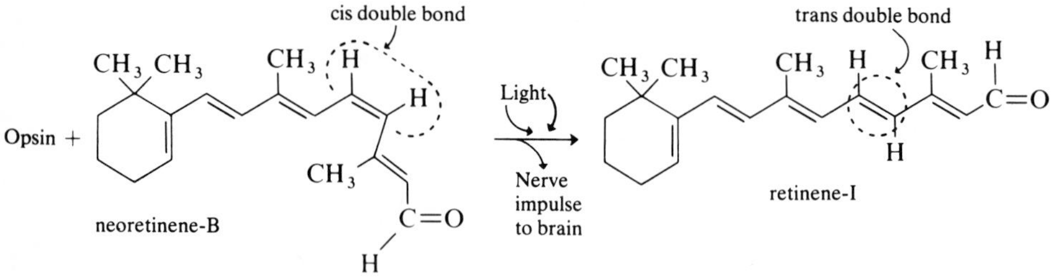
cis double bond
trans double bond
Opsin +
neoretinene-B
Light
Nerve impulse to brain
retinene-I

Synthesis and Reactions of Alkenes

As has been mentioned previously, the weak pi bond in alkenes makes them particularly susceptible to attack by a variety of reagents. The overall processes and some important reactions are discussed next.

Hydrogenation: The addition of hydrogen to a double bond is called **hydrogenation**. Industrially, the addition of hydrogen gas to alkenes is brought about with the use of a metal catalyst, such as platinum or palladium. The ingredients for many margarines are produced by hydrogenation of inexpensive, highly unsaturated liquid vegetable oils.

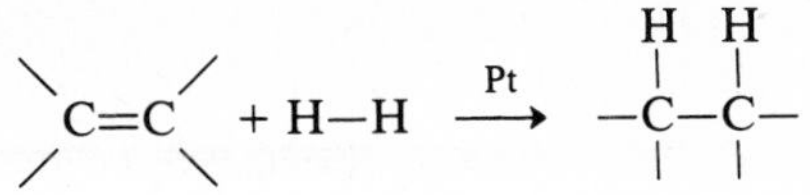

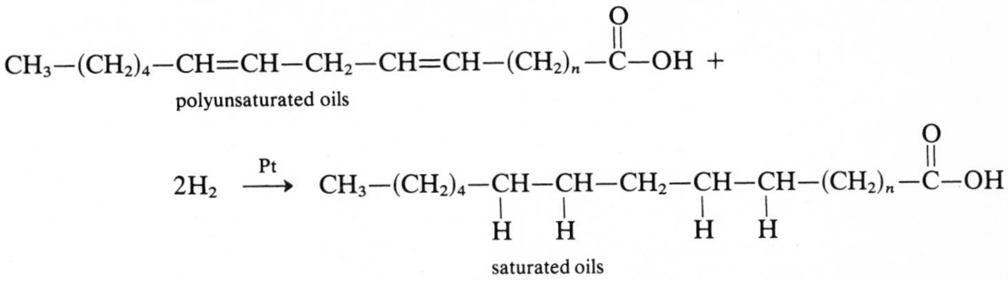
polyunsaturated oils
saturated oils

Hydrogen gas does not exist in our bodies, but reduction of alkenes to alkanes goes on nonetheless. Instead of using hydrogen gas, our bodies use several compounds that act as sources of hydrogen atoms and perform the same type of reaction. For example, unsaturated fatty acids are reduced to saturated acids by compounds called **coenzymes**.

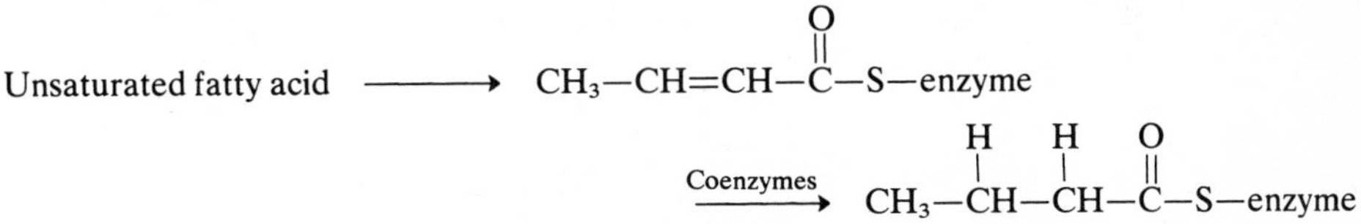
Unsaturated fatty acid
Coenzymes

Hydration: Addition of water to alkenes is called **hydration**. Alcohols are formed by the addition of water to a carbon-carbon double bond in the presence of an acid catalyst or an enzyme. The general reaction sequence is

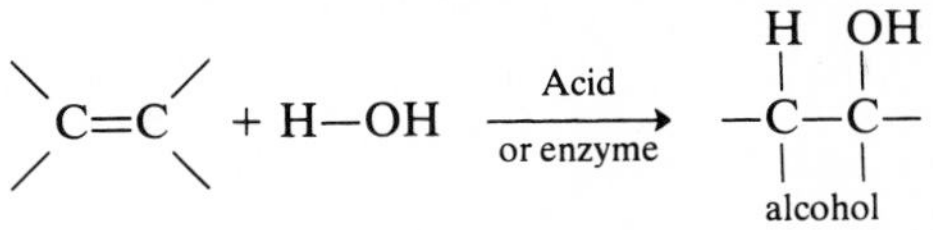
Acid
or enzyme
alcohol

Biologically, hydration of alkenes is found in many metabolic cycles.

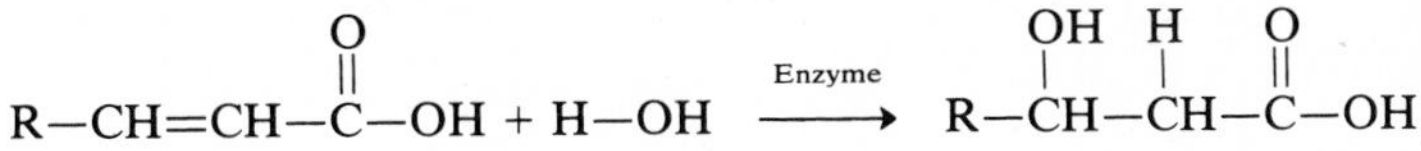
Enzyme

(part of the metabolism of fatty acids)

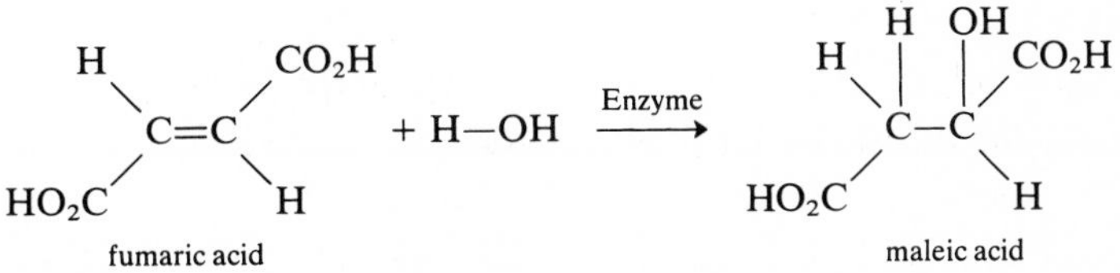
Enzyme
fumaric acid
maleic acid

(part of the citric acid cycle)

Halogenation: Addition of bromine or chlorine to double bonds is called **halogenation**. A useful method of determining the number of double bonds in an unsaturated compound is by titration with bromine. Bromine, a deep red-brown molecule, adds rapidly and quantitatively to alkenes to yield colorless dibromoalk-ane products. No indicator is needed in these titrations as the color change of the red elemental bromine to colorless products works just as well as indicators.

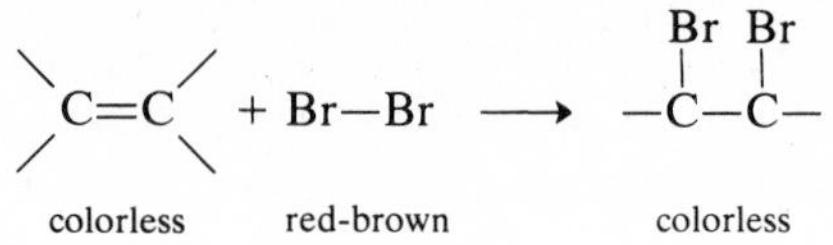

$$\underset{\text{colorless}}{\overset{\diagdown}{\underset{\diagup}{C}}=\overset{\diagup}{\underset{\diagdown}{C}}} + \underset{\text{red-brown}}{Br-Br} \longrightarrow \underset{\text{colorless}}{-\overset{\overset{\displaystyle Br}{|}}{C}-\overset{\overset{\displaystyle Br}{|}}{C}-}$$

Oxidation: The reaction of oxidizing agents with double bonds is called **oxida-tion**. Oxidizing agents like permanganate ion and ozone (O_3) readily react with alkenes to form diols and to cleave the $C=C$ group:

$$-\overset{|}{\underset{|}{C}}-\overset{|}{\underset{|}{C}}=\overset{|}{\underset{|}{C}}-\overset{|}{\underset{|}{C}}- + MnO_4^- \rightarrow \rightarrow \quad -\overset{|}{\underset{|}{C}}-\overset{\overset{\displaystyle OH}{|}}{\underset{|}{C}}-\overset{\overset{\displaystyle OH}{|}}{\underset{|}{C}}-\overset{|}{\underset{|}{C}}-$$

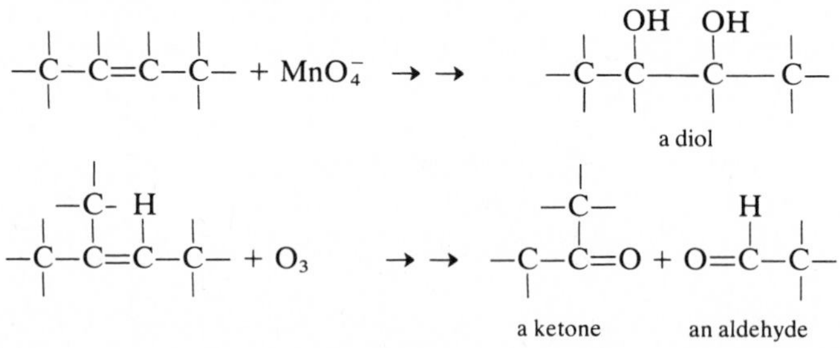

a diol

$$-\overset{\overset{\displaystyle -\overset{|}{C}-\ H}{|}}{\underset{|}{C}}-\overset{|}{\underset{|}{C}}=\overset{|}{\underset{|}{C}}-\overset{|}{\underset{|}{C}}- + O_3 \quad \rightarrow \rightarrow \quad -\overset{\overset{\displaystyle -\overset{|}{C}-}{|}}{\underset{|}{C}}-\overset{}{C}=O + O=\overset{}{\underset{|}{C}}-\overset{\overset{\displaystyle H}{|}}{\underset{|}{C}}-$$

a ketone   an aldehyde

Direct oxidation of compounds by the oxygen in the air is the chief factor in destroying fats and fatty portions of food. Chemically, oxygen reacts with the alkene portions of the fats to form hydroperoxides $(R-O-O-H)$, which then decompose to other products:

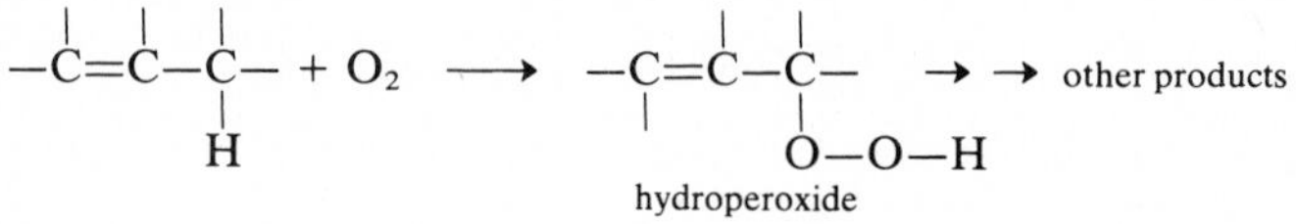

$$-\overset{|}{\underset{\overset{\displaystyle |}{H}}{C}}=\overset{|}{\underset{|}{C}}-\overset{|}{\underset{|}{C}}- + O_2 \longrightarrow -\overset{|}{\underset{|}{C}}=\overset{|}{\underset{|}{C}}-\overset{\overset{\displaystyle |}{}}{\underset{\overset{\displaystyle O-O-H}{}}{C}}- \rightarrow \rightarrow \text{ other products}$$

hydroperoxide

Dehydration: The removal of water from alcohols to form alkenes is called **dehydration**. If an alcohol is heated in the presence of an acid catalyst, water is driven off and an alkene is formed. This process is just the reverse of hydration.

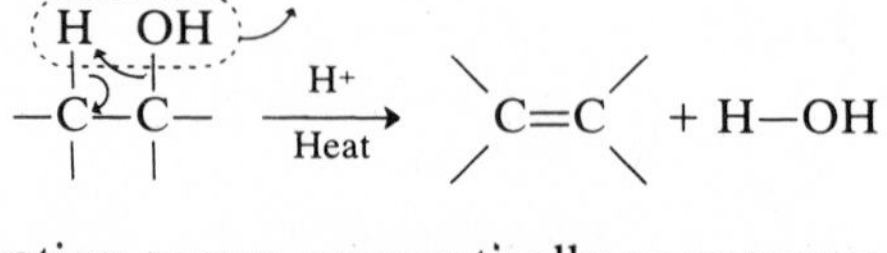

$$-\overset{\overset{\displaystyle \boxed{H \ \ OH}}{|}}{\underset{|}{C}}-\overset{|}{\underset{|}{C}}- \xrightarrow[\text{Heat}]{H^+} \overset{\diagdown}{\underset{\diagup}{C}}=\overset{\diagup}{\underset{\diagdown}{C}} + H-OH$$

In our bodies, dehydration occurs enzymatically as enzymes act as catalysts. One example is the metabolism of the amino acid serine.

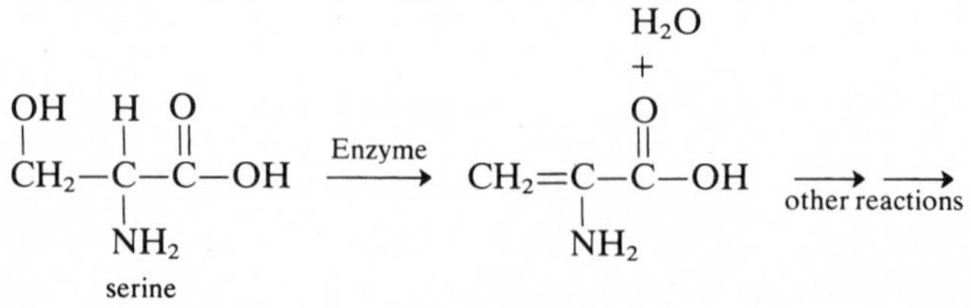

$$\underset{\text{serine}}{\overset{\overset{\displaystyle OH \ \ H \ \ O}{|\ \ \ \ |\ \ \ ||}}{\underset{\overset{\displaystyle |}{NH_2}}{CH_2-C-C-OH}}} \xrightarrow{\text{Enzyme}} \underset{\overset{\displaystyle |}{NH_2}}{\overset{\overset{\displaystyle H_2O}{+}}{\overset{\overset{\displaystyle O}{||}}{CH_2=C-C-OH}}} \xrightarrow{\text{other reactions}} \rightarrow$$

UNIT 24 Problems and Exercises

24–1. Name the following.

(a) $CH_3CH_2CH{=}CH{-}CH_3$

(b) $CH_3CH_2CH_2CH{=}CH_2$

(c) 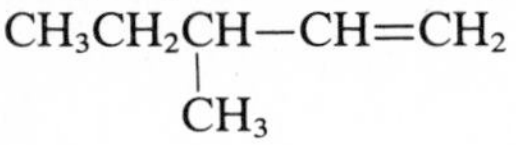

(d) 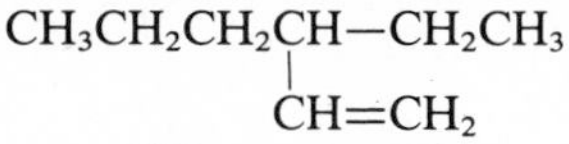

(e) 

(f) 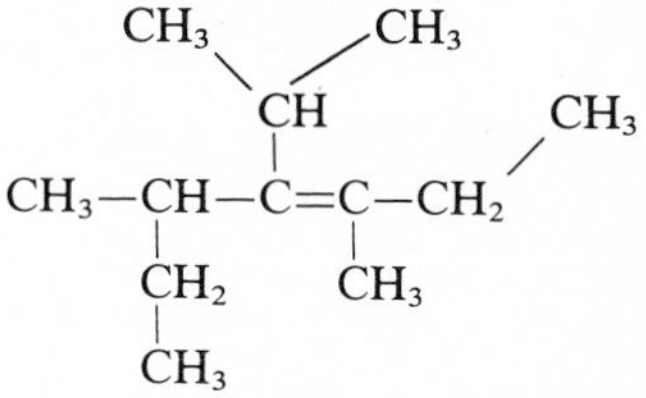

(g) 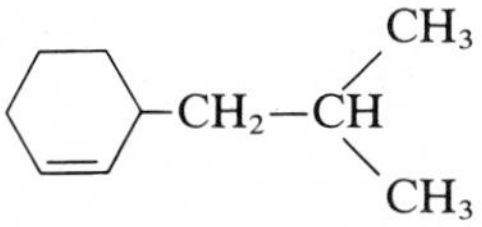

(h) 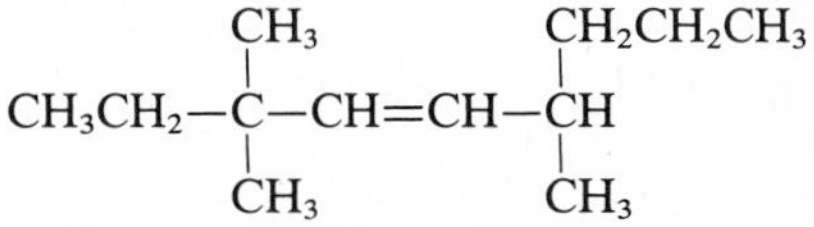

(i) 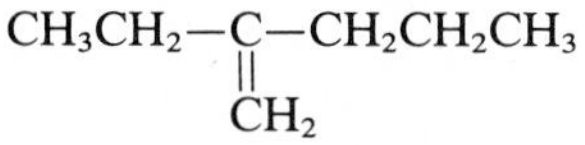

(j) 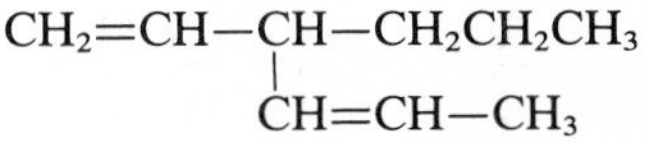

24–2. Draw structures for the following.

(a) 3-methyl-2-pentene

(b) 4-isopropyl-1-heptene

(c) 2,3-dimethyl-2-butene

(d) 2-chloro-2-pentene

(e) 3,6-diethyl-1-octene

(f) 1-methylcyclopentene

(g) 3-secbutyl-1-cyclohexene

(h) 4-isobutyl-3,5-dimethyl-4-nonene

24–3. Can the following compounds exist as cis or trans isomers? If they can, draw and name both.

(a) $CH_3CH{=}CHCH_3$

(b) 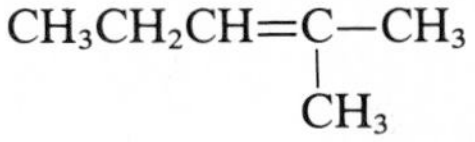

(c) 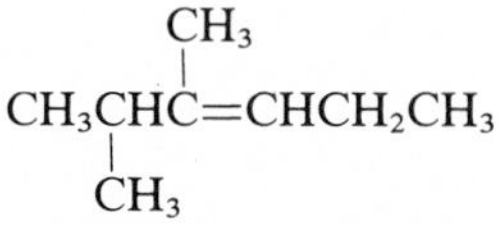

(d) 3,4-dimethyl-3-heptene

(e) 2,3-dimethyl-1-butene

(f) 3-ethyl-3-hexene

24–4. Examine the names of the following compounds. Are they correct or incorrect? Correct them if they are wrong.

(a) $CH_3CH_2CH{=}CHCH_3$

3-pentene

(b) 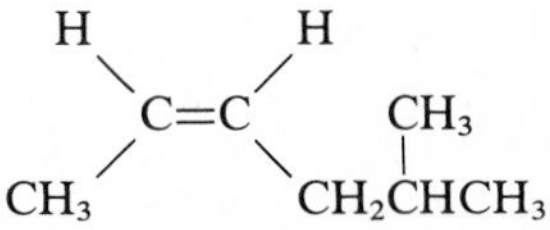

cis-1,4-dimethyl-1-butene

(c) 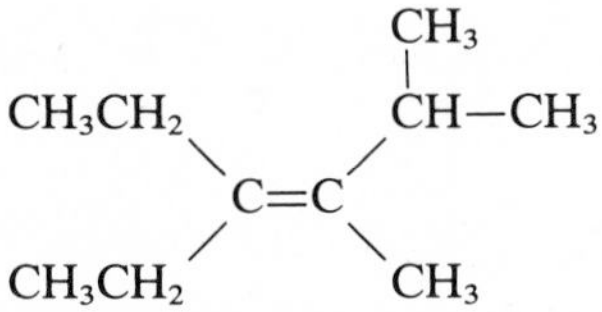

cis-4-ethyl-2-isopropyl-2-pentene

(d) 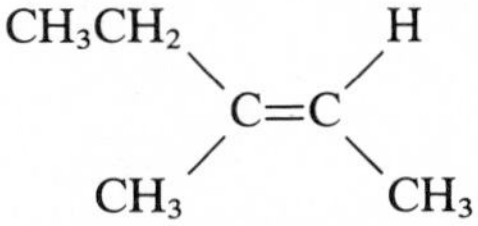

cis-3-methyl-2-pentene

(e) 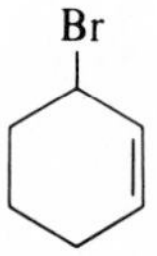

1-bromo-2-cyclohexene

(f) 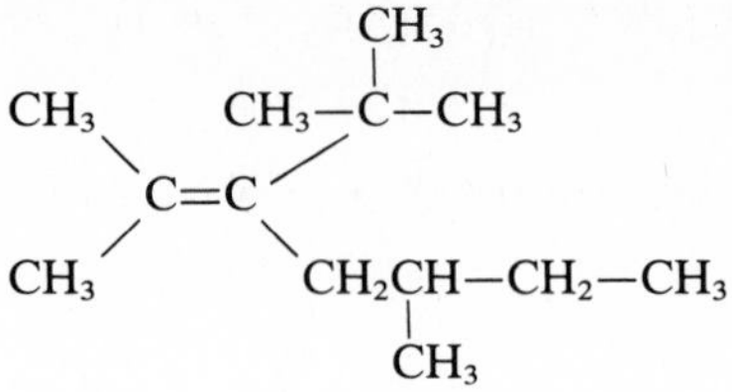

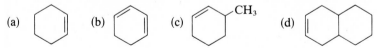

cis-2,5-dimethyl-3-isopentyl-3-heptene

24–5. Write out the molecular formulas for each of the molecules given below.

(a) 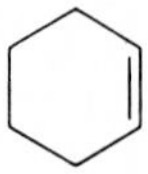 (b) 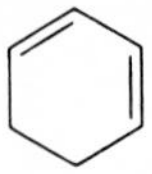 (c) 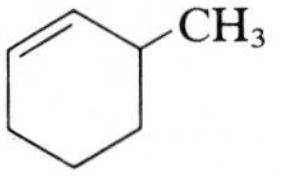 (d) 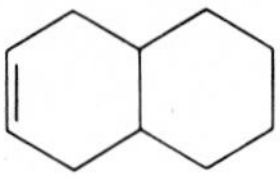

24–6. Draw and name all the alkenes with the following formulas:

(a) C_6H_{12} (11) (b) C_5H_8(6) cyclic alkenes only

24–7. Examine the following and tell what type of reaction occurred (bromination, hydration, and so on).

(a) 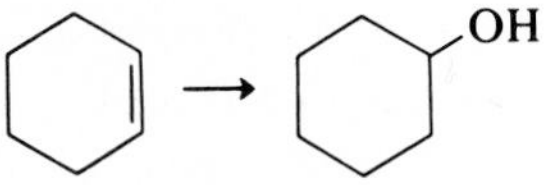

(b) $CH_3CH{=}CHCH_2CH_3 \longrightarrow CH_3\overset{O}{\overset{\|}{C}}H + H\overset{O}{\overset{\|}{C}}CH_2CH_3$

(c) $CH_3CH{=}CH{-}CH_2\overset{OH}{\overset{|}{C}}H_2 \longrightarrow CH_3CH{=}CH{-}CH{=}CH_2$

(d)  (e) 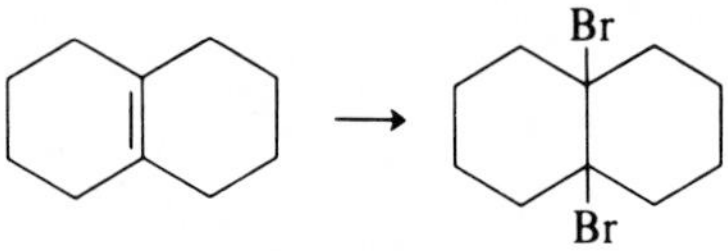

24–8. Complete the following reactions.

(a) 2-butene + $H_2O \xrightarrow{H^{\oplus}}$ (b) 2-butene + $Br_2 \longrightarrow$

(c) 2-butene + $H_2 \xrightarrow{Pt}$ (d) 3-hexene + $Br_2 \longrightarrow$

(e) Cyclohexene + $H_2O \xrightarrow{H^{\oplus}}$

(f) $CH_3{-}CH{=}\overset{CH_3}{\overset{|}{C}}{-}CH_3 + H_2O \xrightarrow{H^{\oplus}}$ how many products?

(g) $CH_3CH{=}\overset{CH_3}{\overset{|}{C}}{-}CH_3 + Br_2 \longrightarrow$ how many products?

(h) $CH_3\underset{OH}{\underset{|}{C}}H{-}CH_3 \xrightarrow[\Delta]{H^{\oplus}}$ how many products?

(i) $CH_3\underset{OH}{\underset{|}{C}}HCH_2CH_3 \xrightarrow[\Delta]{H^{\oplus}}$ how many products?

UNIT 25 MECHANISMS OF ADDITION AND ELIMINATION

If we look only at the reactants and products, many chemical reactions can appear to be quite similar. For example, when wood or paper is burned, the carbohydrates present are converted to carbon dioxide, water, and thermal energy. In our bodies other carbohydrates are metabolized, the products also being carbon dioxide, water, and chemical energy:

$$\text{Combustion:}\quad \underset{\text{(a carbohydrate)}}{C_6H_{12}O_6} + O_2 \xrightarrow{\text{Flame}} CO_2 + H_2O + \text{thermal energy}$$

$$\text{Metabolism:}\quad C_6H_{12}O_6 + O_2 \xrightarrow{\text{Enzymes}} CO_2 + H_2O + \text{chemical energy}$$

The two reactions appear to be identical but do not occur in the same manner. Our body furnaces do not reach the temperatures required for combustion.

In investigating chemical reactions, then, we must do more than simply analyze the structures of the reactants and products. We must also investigate the *mechanisms* of the reactions to identify the separate steps that occur between the beginning and end points. In this unit we shall concentrate on addition and elimination reactions and seek to explain how they occur.

Carbonium Ions

When discussing the mechanism of a reaction, chemists try to break down the reaction into small discrete steps and then look at each separate step to see if it is logically sound. Often some of the products of these smaller steps are unstable species. They are formed and react quite quickly; their lifetimes are very short. One of the most common types of unstable species formed during reactions are **carbonium ions**, carbon atoms bearing positive charges. Carbonium ions are formed when a C—X bond is broken in such a way that the pair of electrons shared by the C and X atoms leaves with the X group. The carbon atom has thus lost an electron; it now bears a positive charge. Carbonium ions may also be formed when alcohols lose water to form alkenes (dehydration), when organic halides lose HX to form alkenes, and when reagents add to alkenes.

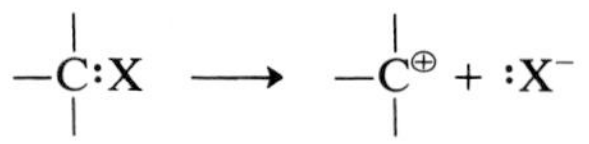

—C:X
—C⊕ + :X⁻

Dehydration (X = OH):

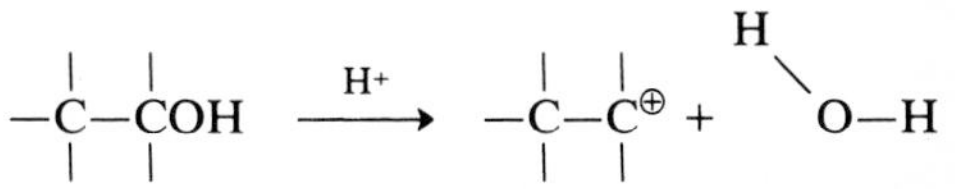

—C—COH
H⁺
—C—C⊕ +
H
O—H

Dehydrohalogenation (X = Br, Cl):

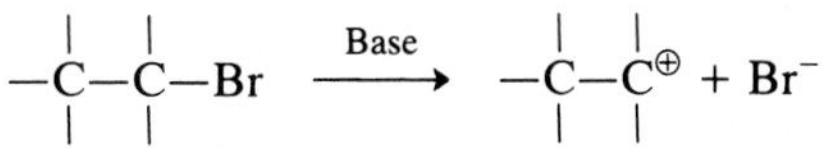

—C—C—Br
Base
—C—C⊕ + Br⁻

Addition: 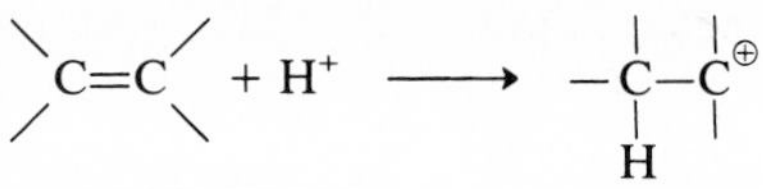

Carbonium ions are generally classed into three types, depending upon the number of other carbon atoms directly attached to the carbon atom bearing the charge. A **primary carbonium ion** has one carbon and two hydrogen atoms attached. A **secondary carbonium ion** has two carbons attached (one hydrogen), and a **tertiary carbonium ion** has three carbon atoms attached.

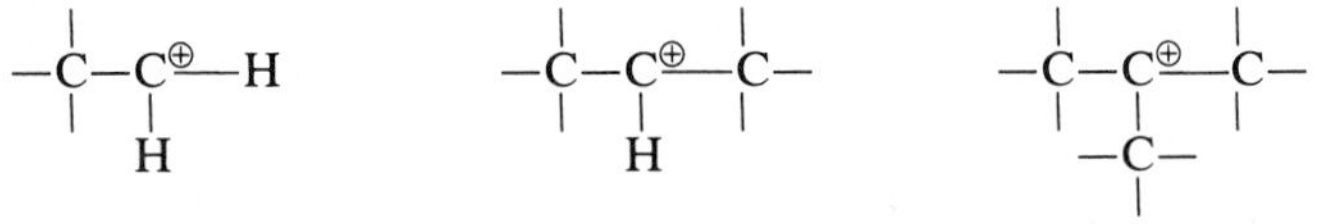

primary (1°) carbonium ion　　secondary (2°) carbonium ion　　tertiary (3°) carbonium ion

The three types of carbonium ions are all unstable, very reactive intermediates in reactions, but they do not all have the same stabilities. The stability of the ions depends to a large extent upon the abilities of the attached groups to lend electron density and help decrease the positive charge on the carbonium ion. Since carbon atoms have more electrons than hydrogen atoms, they are better able to lend a "helping hand" than are hydrogen atoms. Tertiary carbonium ions, with three carbon atoms attached, are more stable than secondary carbonium ions, which have only two carbon atoms and one hydrogen attached. Likewise, secondary carbonium ions are more stable than primary carbonium ions.

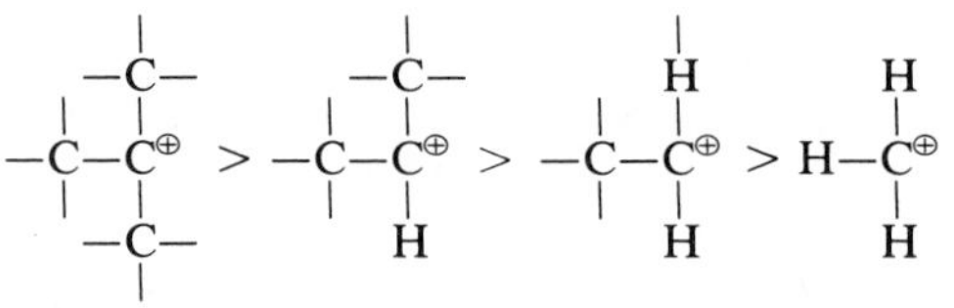

3° most stable, least energy　　2°　　1°　　methyl, least stable, highest energy

The stability series of the carbonium ions is quite important when discussing mechanisms. Since carbonium ions are often the highest-energy species present in many reaction pathways, the energies required to form them often determine the speed or rate at which a reaction proceeds. The activation energy barrier is often the difference in energy between the starting materials and the intermediate carbonium ions. Since 3° carbonium ions are more stable than 2° and 1° carbonium ions, they have the smallest activation energies and are formed the fastest. Consequently, reactions that form 3° ions will usually occur at faster rates than reactions that form 2° or 1° ions (see Figure 25–1).

$$\text{Rate} = k \cdot e^{\frac{-Ea}{Rt}}$$

$$Ea(3°) < Ea(2°) < Ea(1°)$$

$$\text{Rate} = \text{ease of formation} = 3° > 2° > 1°$$

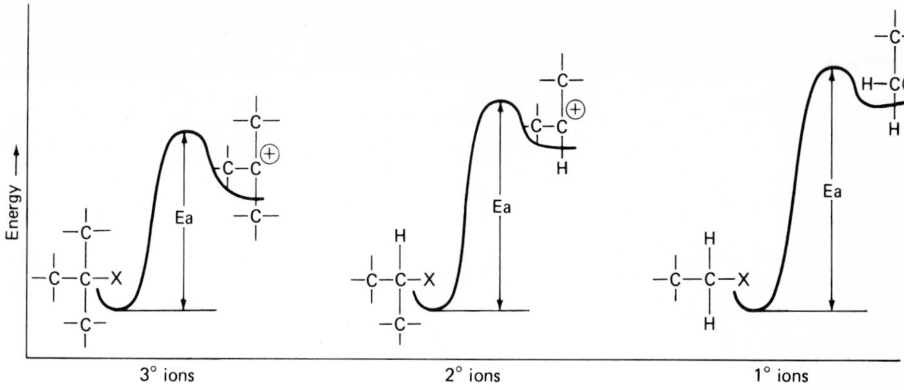
Energy
Ea
Ea
Ea
3° ions
2° ions
1° ions

Figure 25–1

Elimination Reactions: Synthesis of Alkenes

In unit 24 we introduced two methods of synthesizing alkenes. Both were elimination reactions; one involved loss of water from alcohols, and the other involved loss of HCl and/or HBr from organic halides. We shall discuss each separately.

Dehydration of Alcohols

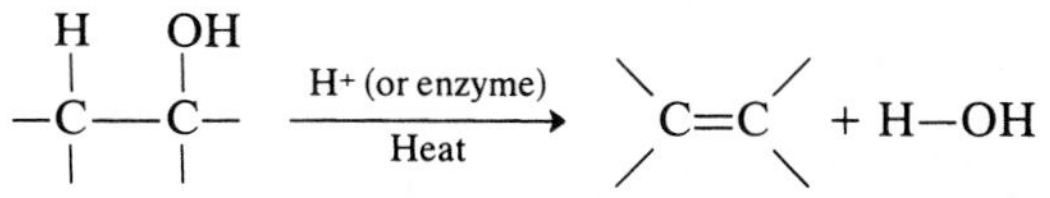
H+ (or enzyme)
Heat
+ H—OH

In most of the dehydrations of alcohols, the carbon bearing the OH becomes part of the alkene group. The alkene double bond can occur between this carbon and *any adjacent carbon atom that has a hydrogen atom on it.* For our purposes, then, a dehydration will not occur if there are no hydrogen atoms on the carbon atoms adjacent to the —C—OH group. The molecule shown below would actually react, but in a more complicated manner which we shall not discuss.

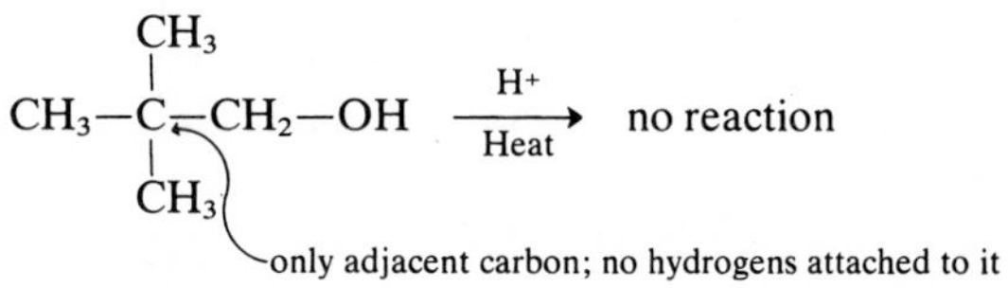
H+
Heat
no reaction
only adjacent carbon; no hydrogens attached to it

If there is more than one adjacent carbon atom, more than one alkene may be found. Usually, when two or more alkenes are found, the *major product is the alkene that is most substituted*, that is, has the greater number of carbon atoms directly attached to the (>C=C<) center.

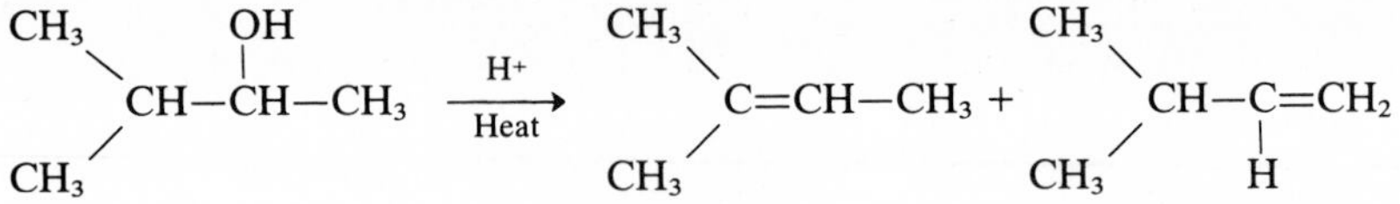

$$\underset{\text{CH}_3}{\overset{\text{CH}_3}{\diagdown}}\underset{\qquad}{\overset{\text{OH}}{\underset{|}{\text{CH}-\text{CH}-\text{CH}_3}}} \xrightarrow[\text{Heat}]{\text{H}^+} \underset{\text{CH}_3}{\overset{\text{CH}_3}{\diagdown}}\text{C}=\text{CH}-\text{CH}_3 + \underset{\text{CH}_3}{\overset{\text{CH}_3}{\diagdown}}\underset{\text{H}}{\overset{|}{\text{CH}-\text{C}=\text{CH}_2}}$$

major product:
3 C, 1 H attached
to $>$C$=$C$<$

minor product:
3 H, 1 C attached
to $>$C$=$C$<$

To make an alkene from an alcohol, it is necessary to remove an OH from one of the carbon atoms, to break a C—O bond. To break the C—O bond, we must make it easy for the OH to leave the molecule. The addition of an acid as a catalyst accomplishes this purpose. When H$^+$ is added, it is attracted to the oxygen atom rather than carbon, because oxygen atoms are more electronegative than carbon atoms. Oxygen atoms also have unshared pairs of electrons that can be used in bonding to H$^+$. The first step in dehydration, then, is addition of H$^+$ to the alcohol, resulting in the formation of an **oxonium ion**.

$$-\overset{|}{\underset{\underset{H}{|}}{\text{C}}}-\overset{|}{\underset{|}{\text{C}}}-\ddot{\text{O}}-\text{H} + \text{H}^+ \;\rightleftharpoons\; -\overset{|}{\underset{\underset{H}{|}}{\text{C}}}-\overset{|}{\underset{|}{\text{C}}}-\overset{\overset{\displaystyle H}{|}}{\ddot{\text{O}}^{\oplus}}-\text{H}$$

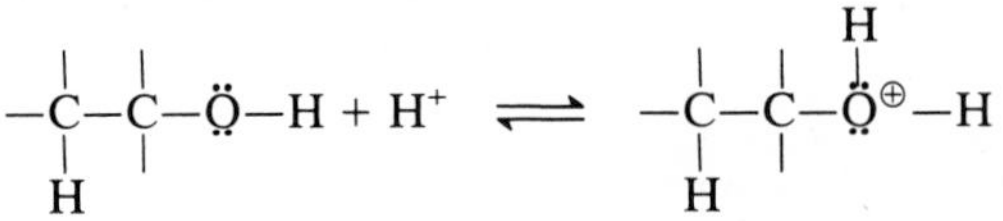

oxonium ion

The oxygen atom now has a positive charge (it has given up an electron to H$^+$) and is not stable. To get back its electron density, it must acquire an electron from either the C—O bond or the O—H bond. Cleavage of the O—H bond would result in formation of the alcohol, but cleavage of the C—O bond would result in formation of a **carbonium ion**.

$$-\overset{|}{\underset{\underset{H}{|}}{\text{C}}}-\overset{|}{\underset{|}{\text{C}}} : \overset{\oplus}{\ddot{\text{O}}}\diagup^{\text{H}} \;\rightleftharpoons\; \overset{|}{\underset{\underset{H}{|}}{\text{C}}}-\overset{|}{\text{C}}^{\oplus} + :\ddot{\text{O}}\diagdown_{\text{H}}$$

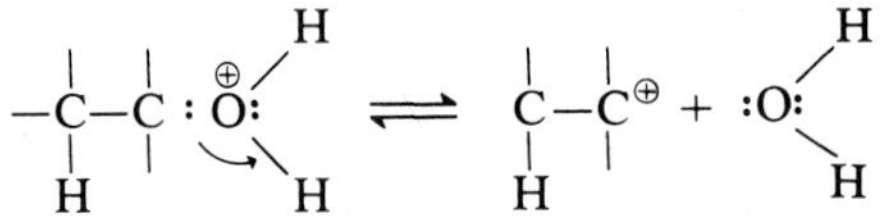

Once formed, the carbonium ion is also quite reactive and needs an electron. It will form a covalent bond with any negative ions that happen to be present. But if few anions are present, it can take electrons from the $-\overset{|}{\underset{|}{\text{C}}}-$H bond of an adjacent carbon atom. The result is expulsion of H$^+$ from the molecule and formation of the alkene.

$$\underset{\diagup}{\overset{\diagdown}{\underset{\underset{\text{H}}{\overset{\displaystyle |}{\cdot\cdot}}}{\text{C}}}}\text{-}\overset{|}{\text{C}}^{\oplus} \longrightarrow \underset{\diagup}{\overset{\diagdown}{\text{C}}}\text{--}\overset{\diagup}{\underset{\diagdown}{\text{C}}} + \text{H}^+$$

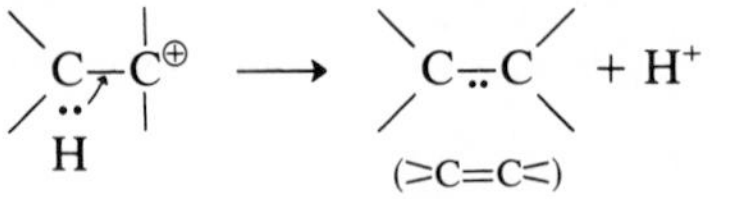

($>$C$=$C$<$)

The overall reaction mechanism for the dehydration of alcohols can be written as

$$-\overset{|}{\underset{\underset{H}{|}}{\text{C}}}-\overset{|}{\underset{|}{\text{C}}}-\ddot{\text{O}}\text{H} + \text{H}^+ \rightleftharpoons -\overset{|}{\underset{\underset{H}{|}}{\text{C}}}-\overset{|}{\underset{\underset{H}{|}}{\text{C}}}-\overset{\overset{\displaystyle H}{\diagup}}{\overset{\oplus}{\ddot{\text{O}}}} \rightleftharpoons -\overset{|}{\underset{\underset{H}{|}}{\text{C}}}-\overset{|}{\text{C}}^{\oplus} \rightleftharpoons \underset{\diagup}{\overset{\diagdown}{\text{C}}}=\overset{\diagup}{\underset{\diagdown}{\text{C}}} + \text{H}^+$$

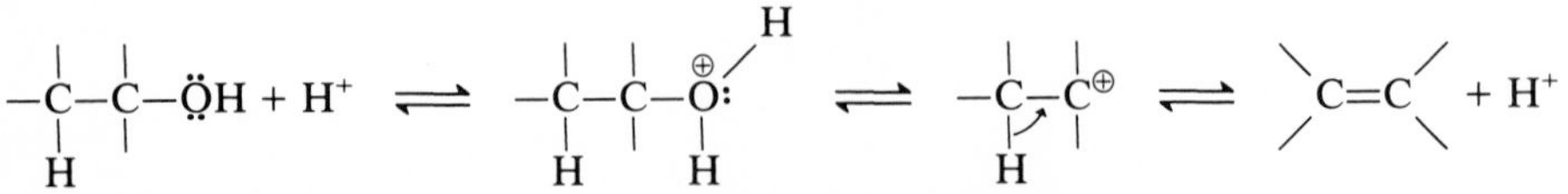

Alcohol $\longrightarrow$ oxonium ion $\longrightarrow$ carbonium ion $\longrightarrow$ alkene

Notice that H$^+$ is a catalyst and is regenerated at the end. Only a small amount of acid needs to be added to cause the dehydration to occur.

The rate of reaction of alcohols depends upon the stabilities of the carbonium ions formed in the intermediate steps. Tertiary alcohols yield 3° carbonium ions and so react faster than 2° or 1° alcohols. The differences in the rates can also be seen from the reaction coordinate diagrams (Figure 25–2). The activation energy leading to the formation of the 3° carbonium ion is smaller than that for formation of the other ions, so that the rate of formation of the tertiary carbonium ion is the fastest.

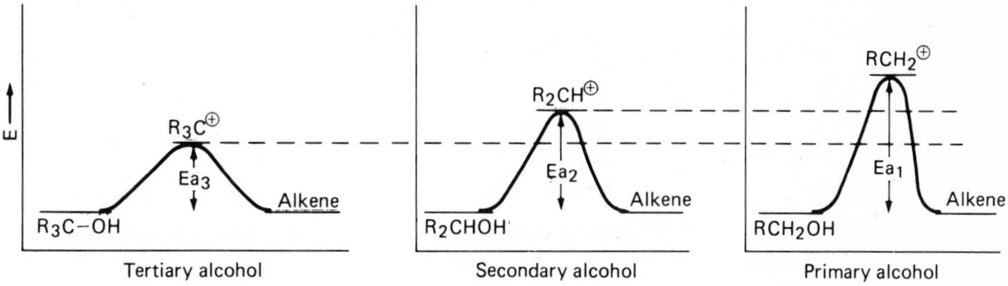
E
R3C⊕
Ea3
Alkene
R3C–OH
Tertiary alcohol
R2CH⊕
Ea2
Alkene
R2CHOH
Secondary alcohol
RCH2⊕
Ea1
Alkene
RCH2OH
Primary alcohol

Figure 25–2 Formation of carbonium ions from alcohols. The activation energy (*Ea*) for loss of water from the tertiary alcohol is lower than for other alcohols. Therefore it occurs more rapidly.

Dehydrohalogenation of Alkyl Halides

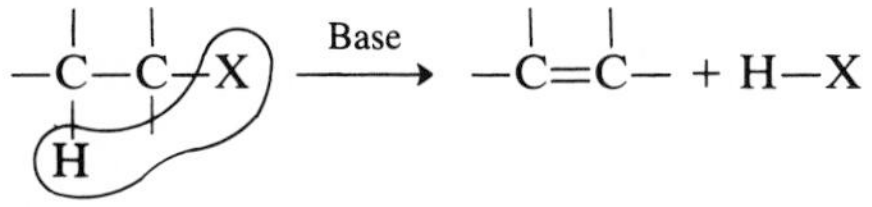
Base
+ H–X

X = Br, Cl

The base-catalyzed loss of H and X from an organic halide also leads to formation of alkenes. As with dehydration of alcohols, the $-\overset{|}{\underset{|}{C}}-X$ carbon is part of the alkene group. The other carbon in the alkene group is *any adjacent carbon bearing a hydrogen.* The *major product again is the more carbon substituted alkene.*

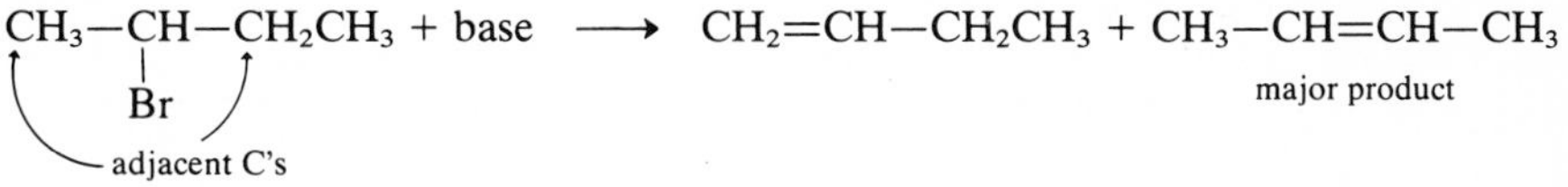
CH3–CH–CH2CH3 + base
Br
adjacent C's
CH2=CH–CH2CH3 + CH3–CH=CH–CH3
major product

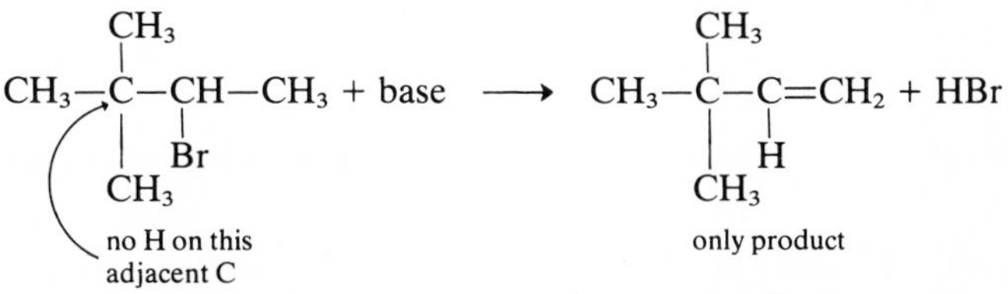
+ base
Br
no H on this
adjacent C
+ HBr
only product

There are two general mechanisms by which organic halides can lose HX to form alkenes. In the **E-1 mechanism**, the halogen is lost from the molecule to form a carbonium ion. The carbonium ion then abstracts an electron pair from an adjacent C—H bond to form an alkene. In this mechanism the order of reactivity is identical to that for alcohol dehydration: 3° halides react faster than 2° halides, which react faster than 1° halides. The reasoning is the same: the more stable the carbonium ion intermediate, the faster it will form.

In the **E-2 mechanism** the bromide atom does not leave of its own accord; its departure must be assisted. The base pulls off a hydrogen atom at an adjacent carbon at the same time that the bromide ion is ejected. The reorganization of electron density in the molecule provides some of the impetus for ejection of Br⁻. The reaction products are the same in either mechanism, but the E-2 mechanism requires a stronger base, and no carbonium ion intermediates are formed.

E-1 $\quad -\overset{|}{\underset{\underset{H}{|}}{C}}-\overset{|}{\underset{|}{C}}-X + \text{solvent} \quad \rightleftharpoons \quad -\overset{|}{\underset{\underset{H}{|}}{C}}-\overset{|}{C}{}^{\oplus} \quad \rightleftharpoons \quad \diagdown C{=}C\diagup + H^+$ 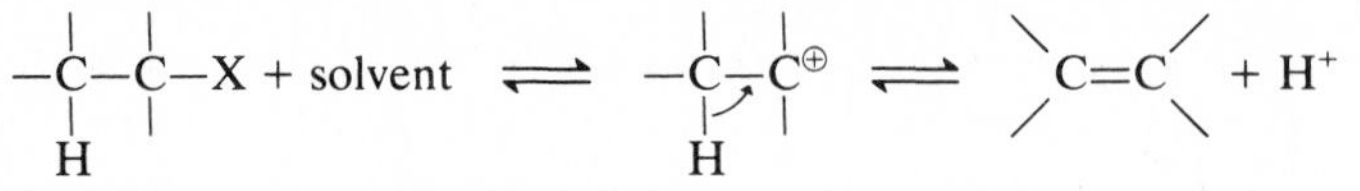

E-2 $\quad -\overset{|}{\underset{\underset{H}{\cdot}}{C}}-\overset{|}{C}{:}X \longrightarrow \left[-\overset{|}{\underset{\underset{H\,\vdots\,B}{\vdots}}{C}}{=}\overset{|}{\underset{\vdots}{C}}{:}\overset{\delta^-}{Br} \right] \longrightarrow \diagdown C{=}C\diagup + :Br^-$ 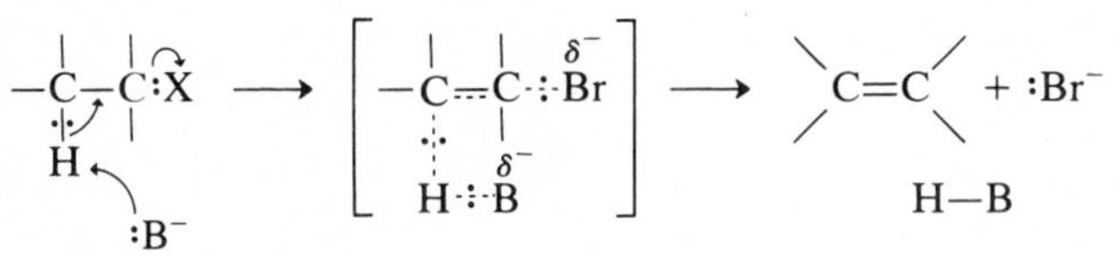

:B⁻ $\qquad\qquad\qquad\qquad$ H—B

Additions to Alkenes

The major reactions of alkenes are additions of some reagent AB across the double bond to yield stable sp³ hybridized carbon atoms. The major reagents used in the additions are the following:

$$\diagdown C{=}C\diagup + A{-}B \longrightarrow -\overset{\overset{\displaystyle B}{|}}{C}-\overset{\overset{\displaystyle A}{|}}{C}-$$

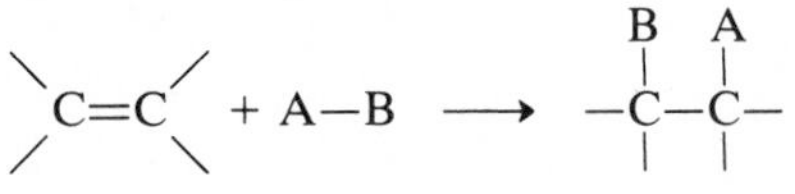

Hydrogenation:

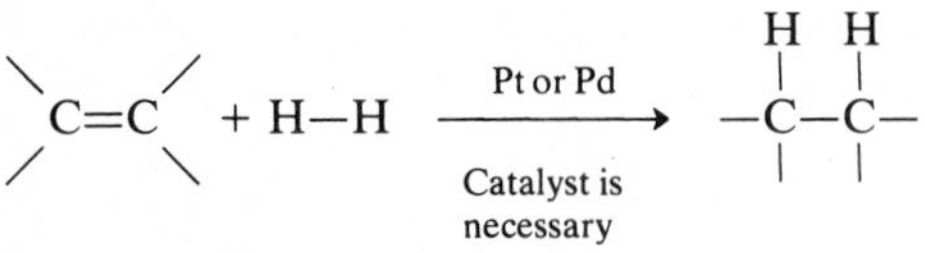

$$\diagdown C{=}C\diagup + H{-}H \xrightarrow[\substack{\text{Catalyst is}\\\text{necessary}}]{\text{Pt or Pd}} -\overset{\overset{\displaystyle H}{|}}{C}-\overset{\overset{\displaystyle H}{|}}{C}- \qquad \text{synthesis of alkanes}$$

Hydrohalogenation:

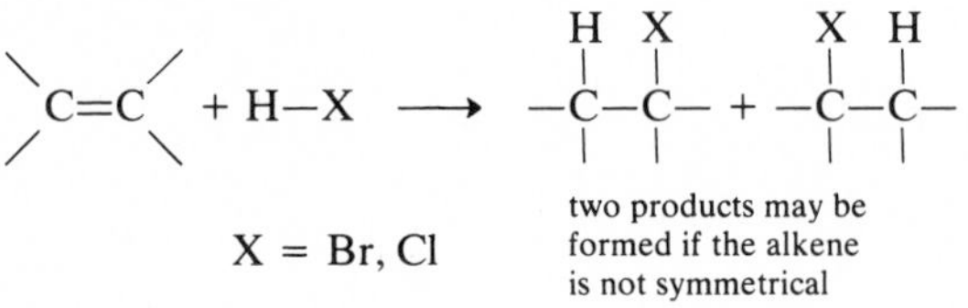

$$\diagdown C{=}C\diagup + H{-}X \longrightarrow -\overset{\overset{\displaystyle H}{|}}{C}-\overset{\overset{\displaystyle X}{|}}{C}- + -\overset{\overset{\displaystyle X}{|}}{C}-\overset{\overset{\displaystyle H}{|}}{C}- \qquad \text{synthesis of organic halides}$$

X = Br, Cl $\qquad$ two products may be formed if the alkene is not symmetrical

Hydration:

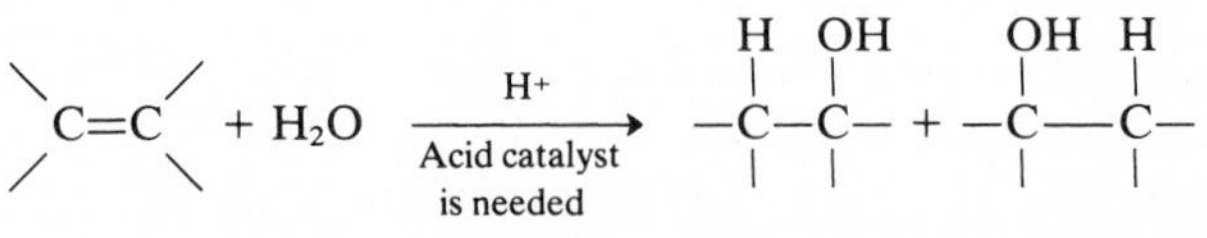

$$\diagdown C{=}C\diagup + H_2O \xrightarrow[\substack{\text{Acid catalyst}\\\text{is needed}}]{H^+} -\overset{\overset{\displaystyle H}{|}}{C}-\overset{\overset{\displaystyle OH}{|}}{C}- + -\overset{\overset{\displaystyle OH}{|}}{C}-\overset{\overset{\displaystyle H}{|}}{C}- \qquad \text{synthesis of alcohols}$$

Bromination:

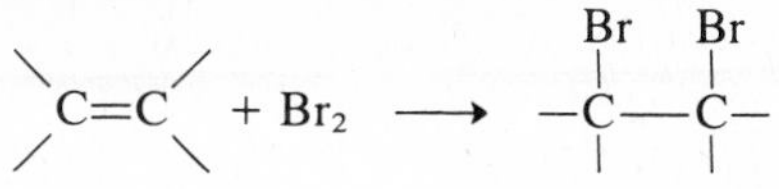

C=C + Br2 ⟶
Br Br
—C—C—

General Mechanism of Additions

The alkene group has four electrons shared between two carbon atoms, so there is a lot of electron density in a relatively small area. It seems reasonable that these electron-dense centers should be attracted to positively charged centers, atoms that need electrons. In other words, the carbon atoms of alkene centers can act as Lewis bases. In the additions of HX and H_2O to alkenes, the initial step involves interaction of the alkene with H^+ to form a carbonium ion. Once the carbonium ion is formed, it can react with the other half of the reagent Br^-, H_2O, Cl^-).

Step 1:

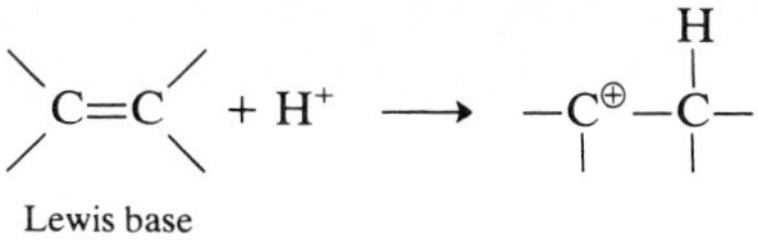

C=C + H⁺ ⟶
Lewis base

Step 2:

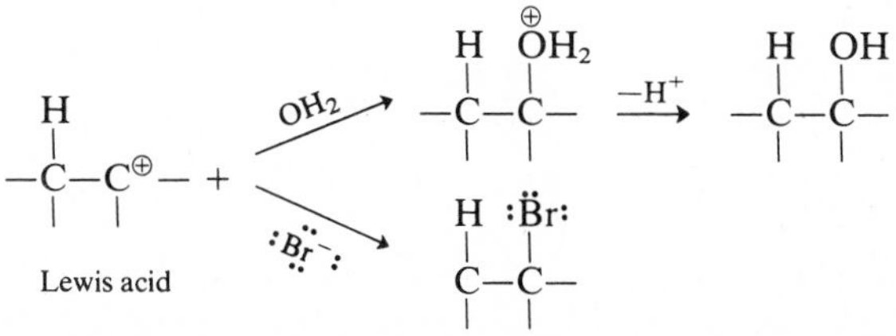

Lewis acid
OH2
−H⁺
H OH2
H OH
H :Br:

If we start with an unsymmetrical alkene like propene and add HCl, two different carbonium ions can result; but they are not formed in equal amounts because they do not have the same energies (see Figure 25–3). The secondary carbonium ion, being more stable than the primary ion, is formed much more rapidly

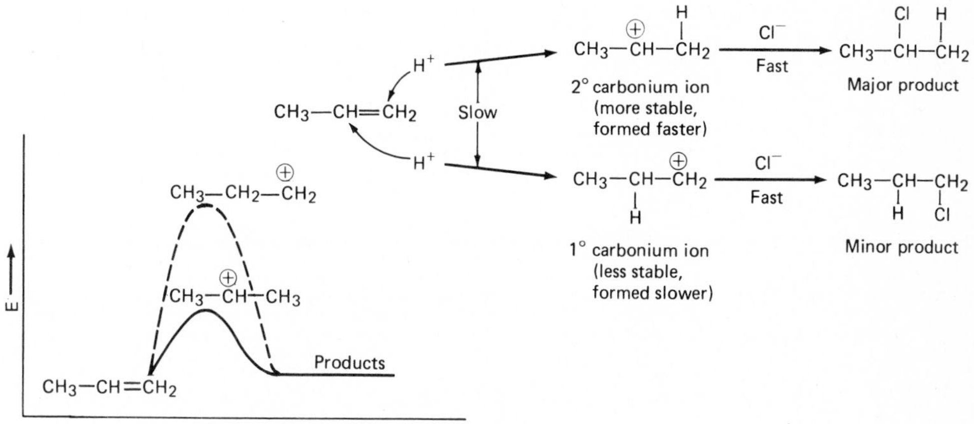

CH3—CH=CH2
H⁺
Slow
CH3—CH—CH2
2° carbonium ion
(more stable,
formed faster)
Cl⁻
Fast
CH3—CH—CH2
Major product
CH3—CH—CH2
1° carbonium ion
(less stable,
formed slower)
Minor product
E
CH3—CH2—CH2
CH3—CH—CH3
Products

Figure 25–3.

and thus in larger amounts. So the product, 2-chloropropane, which is derived from the secondary carbonium ion, is formed in larger amounts than 1-chloropropane, which is derived from the primary carbonium ion.

In the additions of HX and H_2O to alkenes, two products are possible if the alkene is not symmetrical. The product formed in the largest amount, the major product, will be that which is derived from the more stable of the two carbonium ions formed in the initial addition steps. The OH or Br or Cl ends up on the end of the alkene that has the most carbon atoms directly attached. *The hydrogen atom ends up preferentially on the end of the double bond that has the most hydrogen atoms directly attached.*

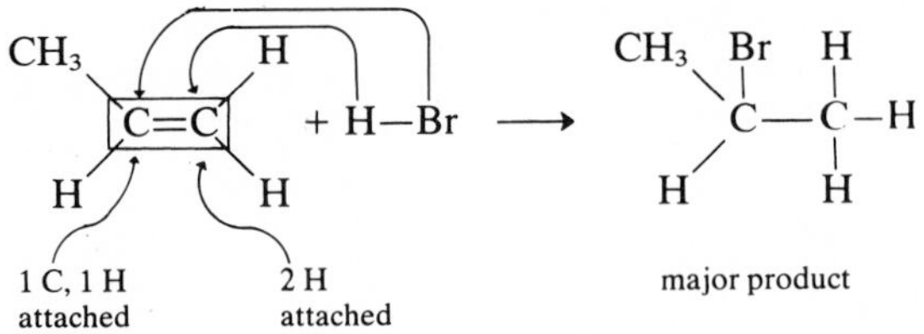
1 C, 1 H attached
2 H attached
major product

A Russian chemist, Markovnikoff, said the same thing in 1870 when he was investigating this reaction.

Markovnikoff's Rule: In the addition of unsymmetrical reagents to alkenes, the positive half of the reagent (usually H^+) adds preferentially to that end of the alkene containing the greater number of hydrogen atoms.

The reactions we have discussed in this unit are quite interrelated. An examination of the mechanisms will show a large degree of similarity. It is as though a two-lane highway exists between alkenes and alcohols or alkyl halides; you use essentially the same mechanistic path to go in either direction.

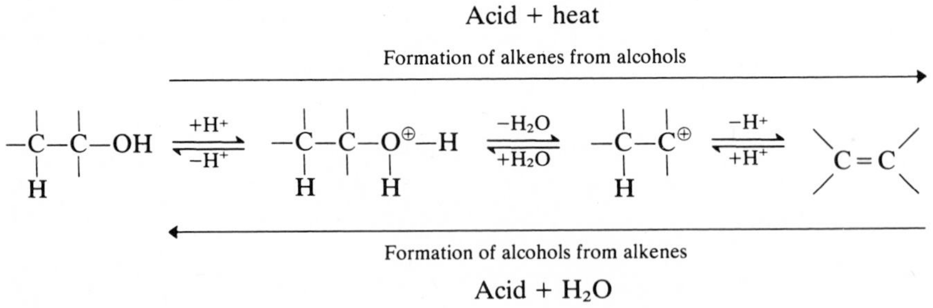
Acid + heat
Formation of alkenes from alcohols
Formation of alcohols from alkenes
Acid + H2O

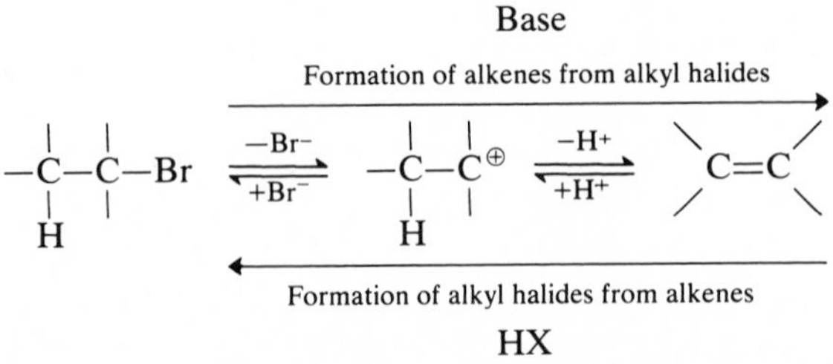
Base
Formation of alkenes from alkyl halides
Formation of alkyl halides from alkenes
HX

Alkenes seem to occupy a central role in conversion, much as moles occupied a central position in earlier conversions.

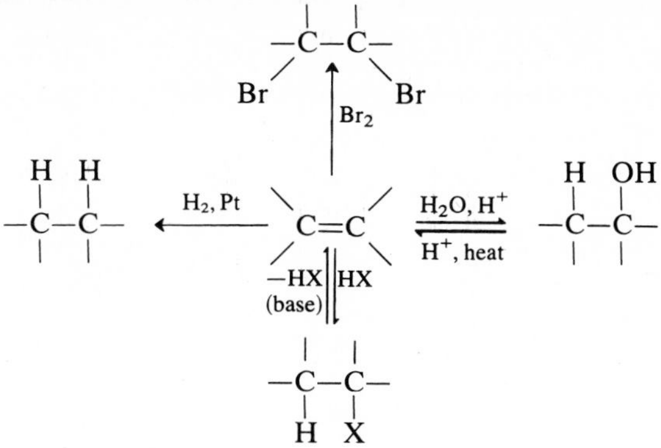

Br
Br
Br_2
H H
H_2, Pt
H_2O, H^+
H^+, heat
H OH
−HX
(base)
HX
H X

UNIT 25 Problems and Exercises

25–1. Complete the following reactions giving the expected products. If more than one product is formed, indicate which is the major product.

(a) $CH_3-\underset{\displaystyle CH_3}{\underset{|}{C}}=CH_2 + H_2O \xrightarrow{H^+}$

(b) $CH_3-\underset{\displaystyle CH_3}{\underset{|}{C}}=CH_2 + Br_2 \longrightarrow$

(c) $CH_3-\underset{\displaystyle CH_3}{\underset{|}{C}}=CH_2 + H_2 \xrightarrow{Pt}$

(d) $CH_3-\underset{\displaystyle CH_3}{\underset{|}{C}}=CH_2 + HCl \longrightarrow$

(e) 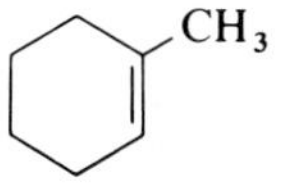 $+ H_2 \xrightarrow{Pt}$

(f) 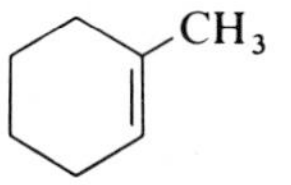 $+ H_2O \xrightarrow{H^+}$

(g) $CH_3-\underset{\displaystyle OH}{\underset{|}{CH}}-CH_2CH_3 + H^+ \xrightarrow{Heat}$

(h) $CH_3-\underset{\displaystyle Br}{\underset{|}{CH}}-CH_2CH_3 + OH^- \longrightarrow$

(i) $CH_3-\overset{\displaystyle CH_3}{\overset{|}{\underset{\displaystyle CH_3}{\underset{|}{C}}}}-CH_2Br + OH^- \longrightarrow$

(j) $CH_3-CH(CH_3)-C(CH_3)(OH)-CH_2CH_3 + H^+ \xrightarrow{\text{Heat}}$

25–2. Draw the structures of the starting material necessary for the synthesis of the products shown. Use only one starting material for each reaction.

(a) ______________ $+ H_2O \xrightarrow{H+} CH_3-C(H)(OH)-CH(CH_3)-CH_3 + CH_3-CH_2-C(CH_3)(OH)-CH_3$
major product

(b) ______________ $+ Br_2 \longrightarrow$ 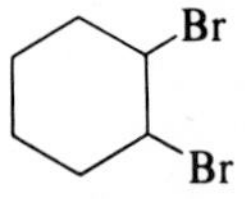

(c) ______________ $+ HCl \longrightarrow$ 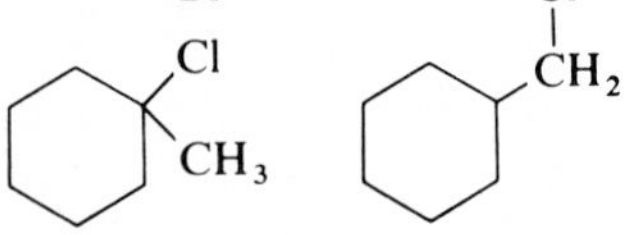

(d) ______________ $+ H^+ \xrightarrow{\text{Heat}} (CH_3)_2C=CH-CH_3 + CH_3-CH(CH_3)-CH=CH_2$

(e) ______________ $+ OH^- \longrightarrow CH_3-CH=CH-CH_3$
$+ CH_2=CH-CH_2CH_3$

(f) ______________ $+ H_2 \xrightarrow{Pt}$ 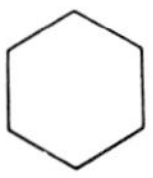

(g) ______________ $+ HBr \longrightarrow$ 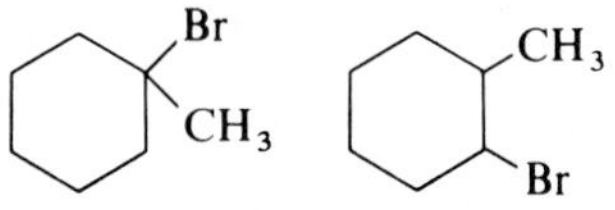

25–3. Give the mechanism steps in the dehydration of $CH_3-C(OH)(CH_3)-CH_2-CH_3$ to form 2-methyl-2-butene and 2-methyl-1-butene. Which is the major product, and why?

25–4. Give the steps in the reaction of 2-methyl-2-butene with H_2O to form $CH_3-C(OH)(CH_3)-CH_2CH_3$ and $CH_3-CH(CH_3)-CH(OH)-CH_3$. Which is the major product, and why?

25–5. Using mechanisms and/or reaction coordinate diagrams, rank the following alcohols in decreasing order of reaction with H^+ and heat. Explain your rankings.

$$CH_3OH \qquad CH_3CH_2OH \qquad CH_3\underset{CH_3}{\underset{|}{C}}HOH \qquad CH_3\overset{CH_3}{\overset{|}{\underset{CH_3}{\underset{|}{C}}}}-OH$$

25–6. Rank the following alkenes in order of decreasing rate of reaction with HBr. Use mechanisms and/or reaction diagrams to explain your rankings.

$$H_2C{=}CH_2 \qquad CH_3-CH{=}CHCH_3 \qquad CH_3-\overset{CH_3}{\overset{|}{C}}{=}\underset{CH_3}{\underset{|}{C}}-CH_3$$

25–7. By using two or three of the reactions discussed in this unit, show how it would be possible to synthesize the following products from the starting materials given.

(a) Starting with $CH_3CH_2CH_2{-}OH$, make $CH_3\overset{OH}{\overset{|}{C}}HCH_3$.
(b) Starting with $CH_3CH_2CH_2OH$, make propane.
(c) Starting with $CH_3CH_2CH_2OH$, make 2-bromopropane.
(d) Starting with 1-bromopropane, make 2-bromopropane.
(e) Starting with 1-bromopropane, make $CH_3\overset{OH}{\overset{|}{C}}HCH_3$.
(f) Starting with 2-bromopropane, make propane.
(g) Starting with 2-bromopropane, make 2-chloropropane.

UNIT 26 ALKYNES

The third member of the hydrocarbon family is the **alkynes**, compounds containing a carbon-carbon triple bond. Each carbon atom involved in the triple bond has sp hybridization, so each carbon atom has two sp-orbitals and two p-orbitals. Each carbon then has two strong sigma bonds and two weaker pi bonds. The molecular orbital picture of ethyne (acetylene) shows the two carbon atoms almost enveloped in a tube of electron density (see Figure 26–1). As with alkenes, we should expect similar reactions to occur.

Nomenclature of Alkynes

The names of alkynes are quite similar to those of alkenes.

1. The suffix for the parent chain is **yne**.
2. The parent chain is again the longest continuous carbon chain containing the carbon-carbon triple bond.

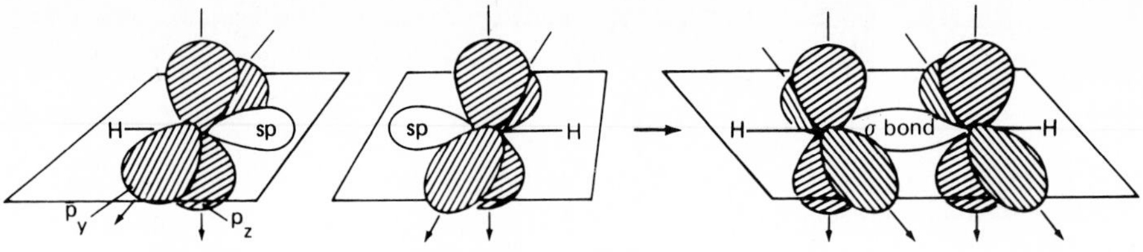
H
sp
p_y
p_z
sp
H
H
σ bond
H

Formation of the σ bond in acetylene

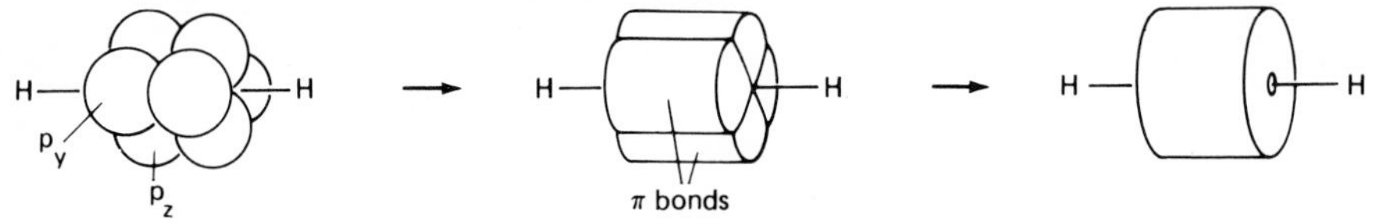
H
H
p_y
p_z
H
H
π bonds
H
H

Formation of the π bonds of acetylene

Figure 26–1

3. The parent chain is numbered, starting at the end closest to the triple bond, and the alkyne triple bond is located by placing the number of the first carbon atom in front of the parent name.

	IUPAC	Common
H—C≡C—H	Ethyne	Acetylene
CH_3—C≡C—H	1-propyne	
CH_3—C≡C—CH_3	2-butyne	
H—C≡C—CH_2CH_3	1-butyne	

4. Substituents are named as before.
5. Since the triple bond is linear, there are no cis or trans isomers of alkynes.

Sources and Uses of Alkynes

The alkyne of chief industrial importance is ethyne, better known as acetylene. Acetylene is synthesized by the reaction of water with calcium carbide, CaC_2, which itself is prepared by the reaction of calcium oxide and coke at very high temperatures. Calcium oxide and coke are obtained from limestone and coal, respectively, so acetylene is obtained relatively cheaply from three abundant raw materials: coal, water, and limestone.

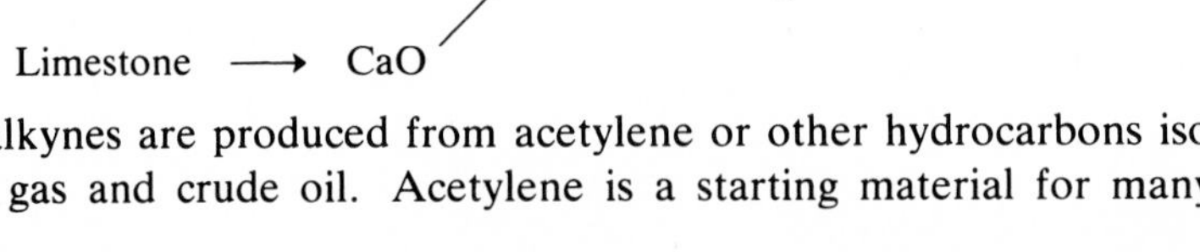
Coal → coke
2000°C
CaC_2
H_2O
HC≡CH
Limestone → CaO

Other alkynes are produced from acetylene or other hydrocarbons isolated from natural gas and crude oil. Acetylene is a starting material for many industrial processes, leading to the synthesis of acetic acid, alkenes, alcohols, aldehydes, ketones, acids, and many polymers. Biologically, alkynes are used in some drugs, but are not prevalent as metabolic intermediates in the body.

Reactions of Alkynes

The carbon-carbon triple bond has a high electron density and thus is susceptible to many of the reactions characteristic of alkenes. Alkynes react with H_2, water, acids, and bromine in a fashion similar to alkenes. The products sometimes differ because the presence of the triple bond allows for double additions.

Hydrogenation:

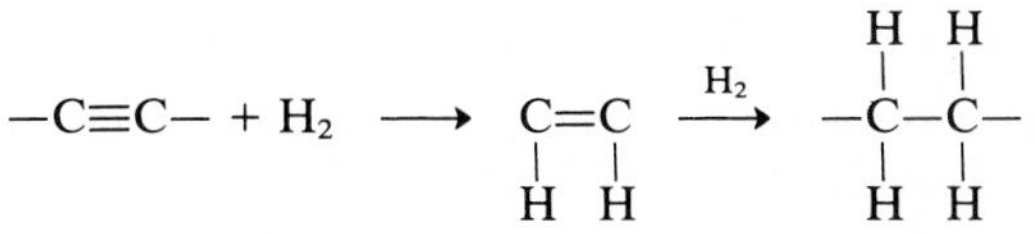

$$-C\equiv C- + H_2 \longrightarrow \underset{\underset{H\ H}{|\ \ |}}{C=C} \overset{H_2}{\longrightarrow} \underset{\underset{H\ H}{|\ \ |}}{\overset{\overset{H\ H}{|\ \ |}}{-C-C-}}$$

alkanes

Hydration:

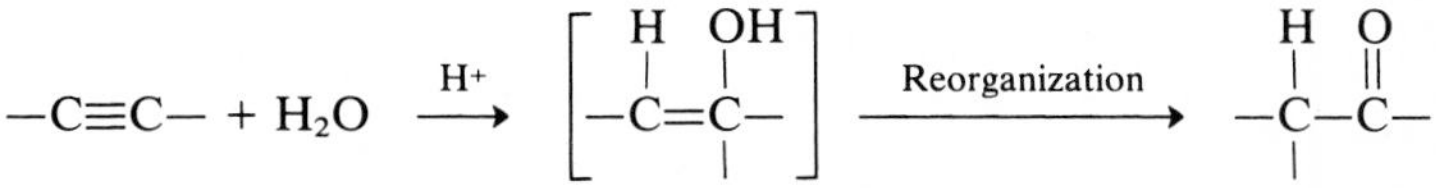

$$-C\equiv C- + H_2O \overset{H^+}{\longrightarrow} \left[\underset{}{\overset{\overset{H\ OH}{|\ \ |}}{-C=C-}} \right] \overset{Reorganization}{\longrightarrow} \underset{}{\overset{\overset{H\ O}{|\ \ ||}}{-C-C-}}$$

aldehydes and ketones

Hydrobromination:

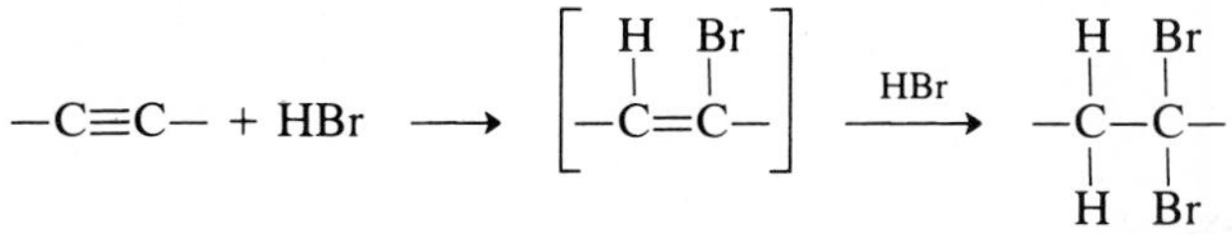

$$-C\equiv C- + HBr \longrightarrow \left[\underset{}{\overset{\overset{H\ Br}{|\ \ |}}{-C=C-}} \right] \overset{HBr}{\longrightarrow} \underset{\underset{H\ Br}{|\ \ |}}{\overset{\overset{H\ Br}{|\ \ |}}{-C-C-}}$$

dibromo alkanes

Bromination:

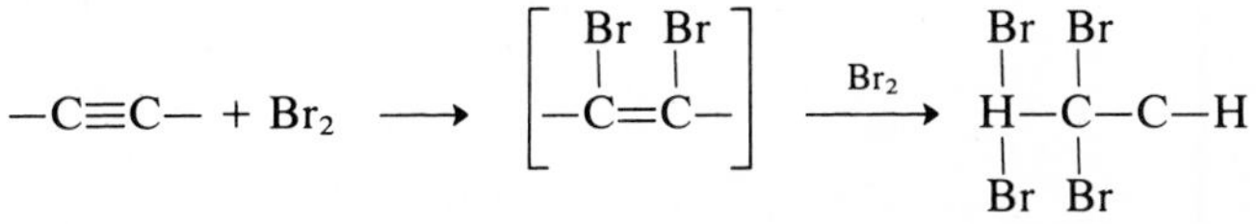

$$-C\equiv C- + Br_2 \longrightarrow \left[\underset{}{\overset{\overset{Br\ Br}{|\ \ |}}{-C=C-}} \right] \overset{Br_2}{\longrightarrow} \underset{\underset{Br\ Br}{|\ \ |}}{\overset{\overset{Br\ Br}{|\ \ |}}{H-C-C-H}}$$

tetrabromo alkanes

The direction of additions of reagents like H_2O or HBr to alkynes is governed by the same rules that govern additions to alkenes. The hydrogen ion adds preferentially to the end of the triple bond that possesses the greater number of hydrogen atoms.

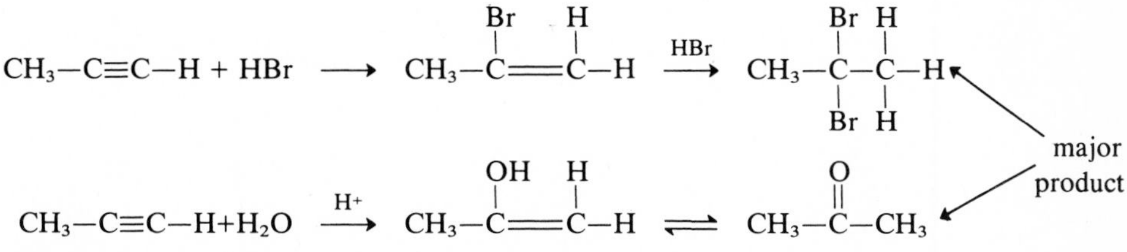

$$CH_3-C\equiv C-H + HBr \longrightarrow CH_3-\underset{\underset{H}{|}}{\overset{\overset{Br}{|}}{C}}=\underset{}{\overset{}{C}}-H \overset{HBr}{\longrightarrow} CH_3-\underset{\underset{Br\ H}{|\ \ |}}{\overset{\overset{Br\ H}{|\ \ |}}{C-C}}-H$$

major product

$$CH_3-C\equiv C-H+H_2O \overset{H^+}{\longrightarrow} CH_3-\underset{}{\overset{\overset{OH\ H}{|\ \ |}}{C}}=\underset{}{\overset{}{C}}-H \rightleftharpoons CH_3-\overset{\overset{O}{||}}{C}-CH_3$$

If the triple bond is not located at the end of the chain, then neither end of the triple bond possesses a hydrogen, and a mixture of products results:

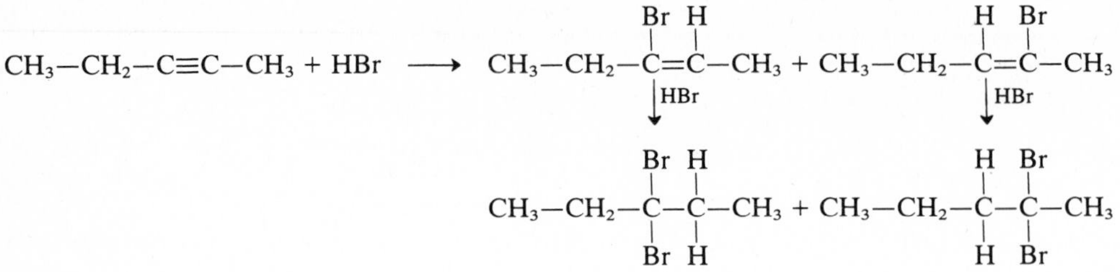

Acidity of Alkynes

Alkynes in which the triple bond is located on the end of the chain are called **terminal alkynes** and possess a property different from alkenes, alkanes, and other alkynes. They are much more acidic than other hydrocarbons. The hydrogen atom of a terminal acetylene can be removed using a treatment with sodium metal or soda amide $NaNH_2$:

$$R{-}C{\equiv}C{-}H \xrightleftharpoons{NaNH_2} R{-}C{\equiv}C^- + H^+$$

Neither of these reagents abstracts hydrogen atoms from alkenes or alkanes. The theoretical explanation for this phenomenon deals with the amount of s-character in the hybrid sp-orbital. Electrons in an sp-orbital are thought to be held closer to the carbon nucleus than electrons in sp^2- or sp^3-orbitals. This creates a more polar C—H bond in alkynes, and allows the H^+ to leave more readily.

UNIT 26 Problems and Exercises

26–1. Name the following compounds.

(a) $CH_3{-}C{\equiv}CH$

(b) $CH_3{-}CH_2{-}C{\equiv}C{-}CH_3$

(c) 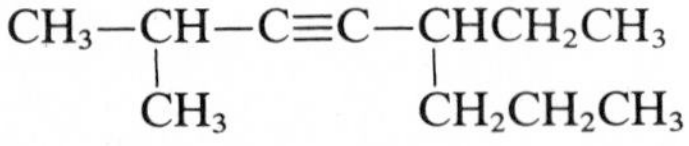

(d) $Br{-}CH_2{-}C{\equiv}CH$

(e) 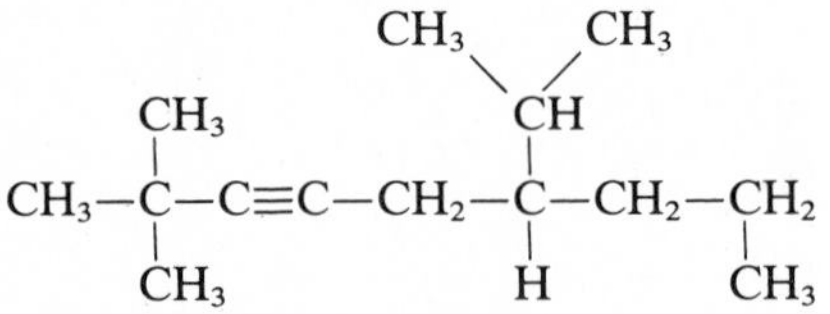

(f) 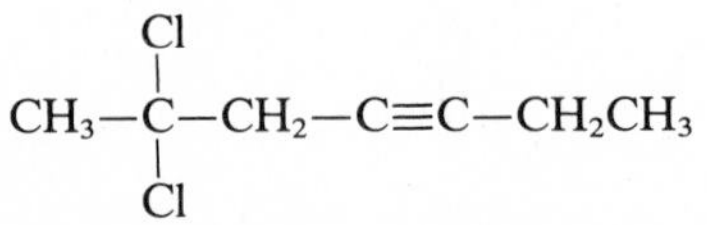

26–2. Draw structures for the following.

(a) 1-butyne
(b) 2-hexyne
(c) 4,5-dimethyl-2-hexyne
(d) 2,2-dimethyl-5-*t*-butyl-3-octyne
(e) 1,5-dichloro-2-pentyne
(f) 3-methyl-1,4-pentadiyne

26–3. Complete the following reactions (circle major products).

(a) $CH_3—C{\equiv}C—CH_3$ + excess Br_2 $\longrightarrow$

(b) $CH_3—C{\equiv}C—CH_3$ + excess H_2 $\xrightarrow{Pt}$

(c) $CH_3—C{\equiv}C—CH_3 + H_2O$ $\xrightarrow{H^+}$

(d) $CH_3—C{\equiv}CH$ + 2(HBr) $\longrightarrow$

(e) $CH_3—C{\equiv}CH + H_2O$ $\xrightarrow{H^+}$

(f) $CH_3—C{\equiv}CH$ + excess Br_2 $\longrightarrow$

(g) $CH_3—C{\equiv}C—CH_3$ + Na $\longrightarrow$

(h) $CH_3—C{\equiv}C—H + NH_2^-$ $\longrightarrow$

26–4. Give the structure of the alkyne reagent that must be used to produce the following products.

(a) ______________ + excess H_2 $\longrightarrow$ pentane

(b) ______________ + excess Br_2 $\longrightarrow$ 1,1,2,2-tetrabromohexane

(c) ______________ + H_2O $\longrightarrow$ $CH_3CH_2\overset{\overset{\displaystyle O}{\|}}{C}CH_2CH_2CH_3$

(d) ______________ + 2(HBr) $\longrightarrow$ 2,2-dibromo-3-methylbutane

(e) ______________ + Na $\longrightarrow$ $CH_3CH_2CH_2—C{\equiv}C^-Na^+$

26–5. Give all the possible structures of alkynes with the formula of C_5H_8.

26–6. An unknown compound reacts readily with Br_2 in a ratio of 2 mol of Br_2/1 mol of compound. It also reacts with $NaNH_2$. If its formula is C_5H_8, what is its structure and name?

26–7. Describe the experiments you might perform to distinguish the following.

(a) 1-butyne from 2-butyne
(b) 2-butyne from 2-butene

UNIT 27 AROMATIC HYDROCARBONS

Discovery of the Structure of Benzene

In the early 1800's a group of compounds was isolated from coal tar that possessed properties quite different from other classes of compounds. Most of these compounds were very stable, resistant to chemical reactions, and all had agreeable, pleasant odors. They were given the name **aromatic compounds**, principally because of their odors. Today we characterize aromatic compounds by properties other than aroma, but the name is still used for this class of compounds.

The simplest of all aromatic hydrocarbons is the compound **benzene**, which has the molecular formula C_6H_6. Compared to hexane (C_6H_{14}), it appears that benzene is quite unsaturated and thus must possess more than one double bond. It should then undergo addition reactions in the same fashion as alkenes. But benzene does not act that way. It is very inert, and does not react with most of the reagents that add so easily to alkenes.

$$\text{Benzene} + Br_2 \longrightarrow \text{no reaction;} \qquad \text{Alkenes} + Br_2 \longrightarrow \begin{array}{c} \text{Br}\ \ \text{Br} \\ |\quad | \\ -\text{C}-\text{C}- \\ |\quad | \end{array}$$

$$\text{Benzene} + H_2O \xrightarrow{H^+} \text{no reaction;} \qquad \text{Alkenes} + H_2O \longrightarrow \begin{array}{c} \text{H}\ \ \text{OH} \\ |\quad | \\ -\text{C}-\text{C}- \\ |\quad | \end{array}$$

$$\text{Benzene} + \text{HBr, HCl} \longrightarrow \text{no reaction;} \qquad \text{Alkenes} + \text{HBr} \longrightarrow \begin{array}{c} \text{H}\ \ \text{Br} \\ |\quad | \\ -\text{C}-\text{C}- \\ |\quad | \end{array}$$

Benzene does react with hydrogen gas, but only under conditions of high pressure. The product of hydrogenation is cyclohexane, and this indicates that benzene must have three double bonds in a ring; but for some reason these double bonds do not act like typical alkene double bonds.

$$C_6H_6 + 3H_2 \xrightarrow[\text{High pressure}]{\text{Pt}} C_6H_{12}\ \text{(cyclohexane)}$$

The Structure of Benzene: Resonance

In 1865, August Kekule proposed a structure for benzene that helped to explain its inertness. He reasoned that the actual structure of benzene was not structures I or II shown in Figure 27–1, but some hybrid or combination of them. Both I and II have complete double bonds and should react like alkenes. In a **hybrid** of the two structures, each carbon atom would have one and a half bonds to each of its adjacent neighbors rather than one or two. We would expect the reactivity of such a hybrid to differ from the reactivity of alkenes.

Such a hybrid structure makes more sense when we examine the hybridization of the electron orbitals in benzene. Each carbon atom has three sp^2-orbitals and a p-orbital that is perpendicular to the plane of the ring. Notice in Figure 27–2 that the six p-orbitals are symmetrically arranged, and each p-orbital is equidistant from two other p-orbitals. Instead of confining the p-electrons to a molecular orbital covering just two carbon atoms, as in an alkene bond, let us envisage a larger molecular orbital encompassing all six carbon atoms. This molecular orbital will look like two doughnuts located above and below the ring formed by the carbon atoms. In this new **molecular orbital** the electrons no longer belong to any one carbon atom but are delocalized into a super-orbital covering all six carbon atoms.

The freeing or delocalization of electrons greatly stabilizes the benzene molecule. As long as there are six electrons and each carbon atom is sp^2 hybridized, the

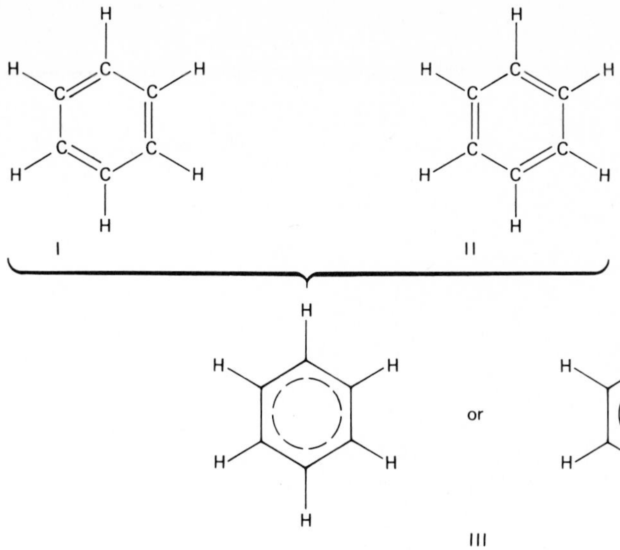
I
II
or
III

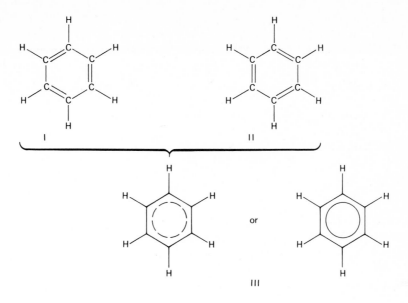

Figure 27–1 Hybrid orbitals of benzene.

resonating picture of benzene can exist. Any change in hybridization at one or more of the carbon atoms will destroy the electron cloud and greatly decrease the stability of benzene.

The amount of stabilization attained by benzene has been estimated by comparing the amounts of energy released during hydrogenation of similar alkenes and dienes. As shown in Figure 27–3, addition of H_2 to cyclohexene releases 28.6 kcal/mol of energy. Reduction of cyclohexadiene releases about double that amount; so reduction of three double bonds should release 3×28.6 or about 85.8 kcal/mol. But hydrogenation of benzene releases only 49.8 kcal/mol. The difference between the predicted (85.8) and actual (49.8) amount of energy release is called the **resonance stabilization energy** of benzene. The delocalized type of molecular orbital allows benzene to be 36 kcal more stable than theoretically predicted.

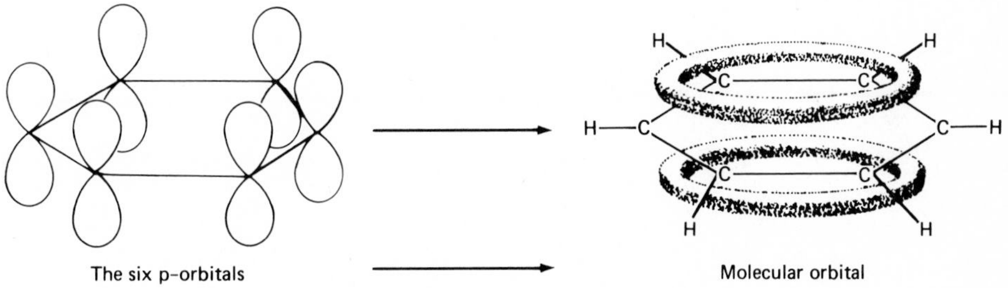
The six p-orbitals
Molecular orbital

Figure 27–2 The molecular orbital in benzene.

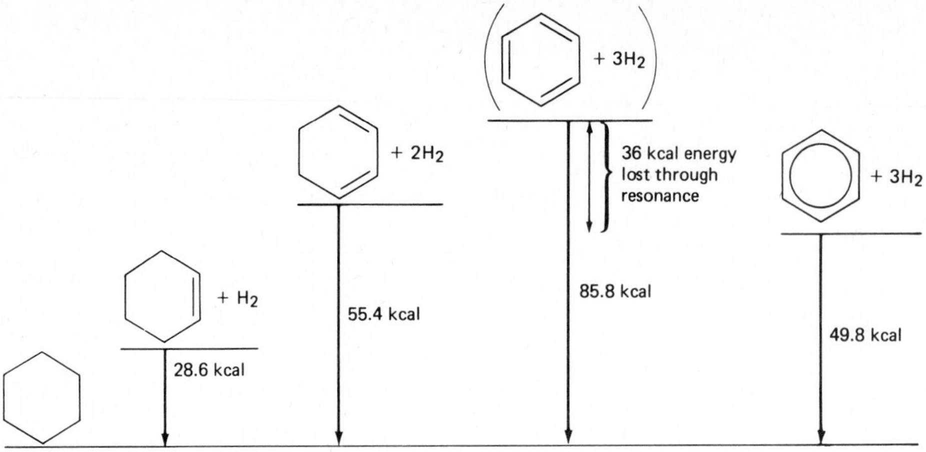
+ 3H2
+ 2H2
36 kcal energy lost through resonance
+ 3H2
+ H2
55.4 kcal
85.8 kcal
49.8 kcal
28.6 kcal

Figure 27–3 Hydrogenation energies for cyclic hydrocarbons.

Nomenclature of Aromatic Compounds

Aromatic compounds are named differently from the alkanes because the six-membered benzene ring is usually the parent. In addition, many common names for aromatic compounds have persisted and are the preferred names.

Compounds containing a single benzene ring with simple groups attached are named as derivatives of benzene itself. The carbon atoms of benzene are numbered in order to designate the locations of substituents.

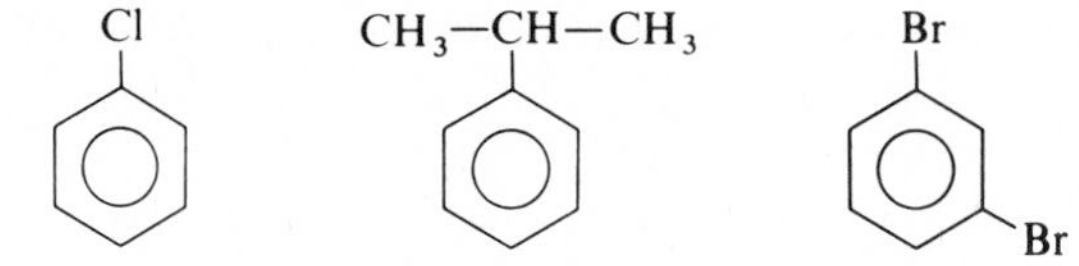
Cl
$CH_3-CH-CH_3$
Br
Br

chlorobenzene isopropylbenzene 1,3-dibromobenzene

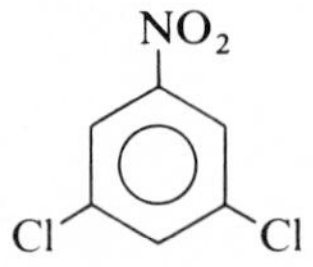
NO_2
Cl
Cl

3,5-dichloro-1-nitrobenzene

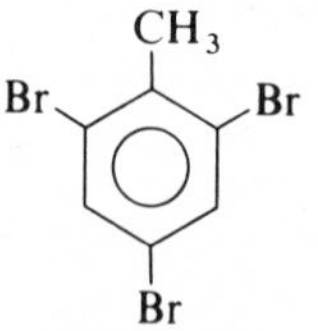
CH_3
Br
Br
Br

2,4,6-tribromomethylbenzene

There are several monosubstituted benzenes whose common names are preferred and should be memorized. Often compounds containing these particular structures are named as derivatives of their names rather than benzene.

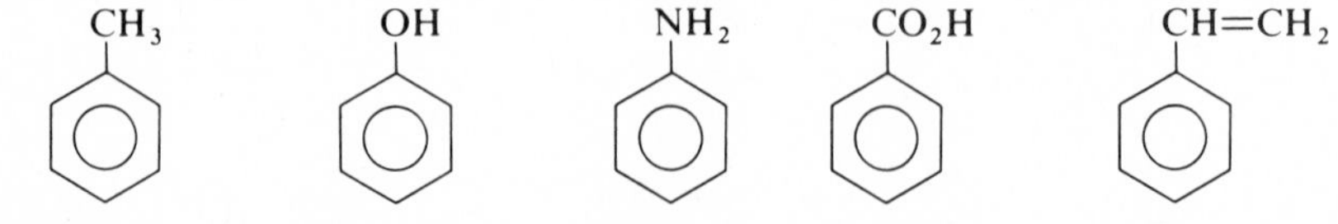
CH_3
OH
NH_2
CO_2H
$CH=CH_2$

toluene (methylbenzene) phenol (hydroxybenzene) aniline (aminobenzene) benzoic acid styrene

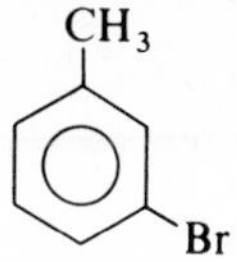
CH3
Br

3-bromotoluene
m-bromotoluene

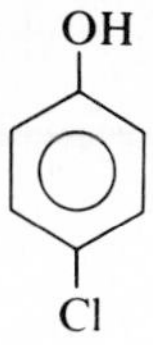
OH
Cl

4-chlorophenol
p-chlorophenol

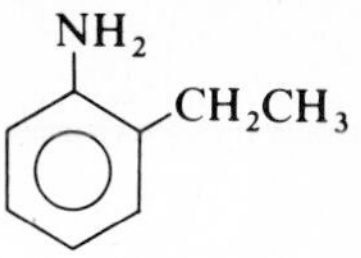
NH2
CH2CH3

2-ethylaniline
o-ethylaniline

The parent molecules here are toluene, phenol, and aniline, not benzene.

A second method of locating substituents on disubstituted benzenes is often used in preference to numbers. The prefixes *ortho* (*o*-), *meta* (*m*-) and *para* (*p*-) are used to designate the relative positions of the *two substituents* with respect to each other.

ortho (*o*)
substituents are on
adjacent carbons:

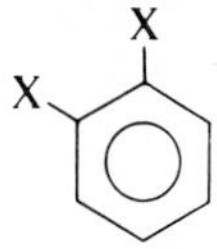

meta (*m*)
substituents are on carbons
separated by one carbon:

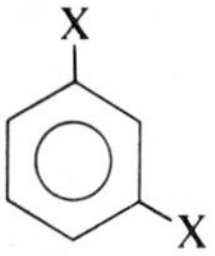

para (*p*)
substituents are opposite
each other on ring:

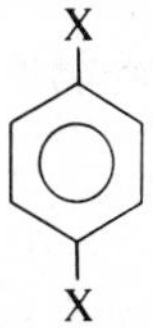

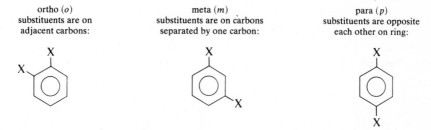

The use of either *o*, *p*, *m*, or numbers is correct, but the student should be familiar with both designations. The use of *o*, *p*, and *m* as designations of position is done only on disubstituted benzenes. No other types of compounds use this designation.

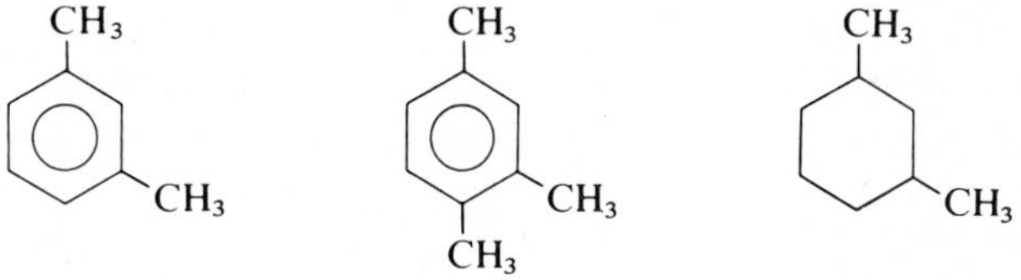

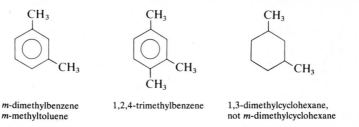

m-dimethylbenzene
m-methyltoluene

1,2,4-trimethylbenzene

1,3-dimethylcyclohexane,
not *m*-dimethylcyclohexane

Benzene is not always used as the parent unit. In some instances it is more convenient to consider benzene as a substituent like methyl or ethyl groups. When it is a substituent, benzene is called a **phenyl** group (not a benzyl group).

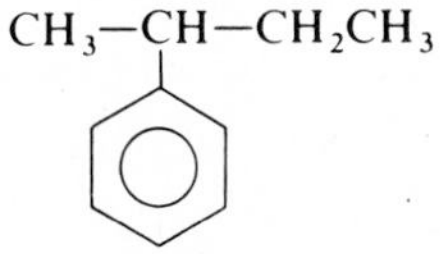
CH3—CH—CH2CH3

2-phenylbutane

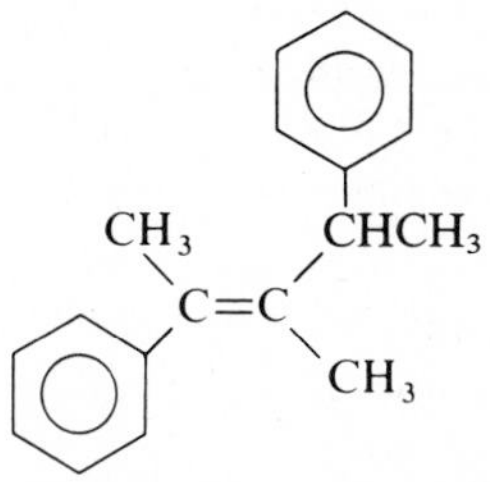
CH3
CHCH3
C=C
CH3

trans-3-methyl-2,4-diphenyl-2-pentene

Aromatic compounds containing two or more benzene rings fused together are known by their common names.

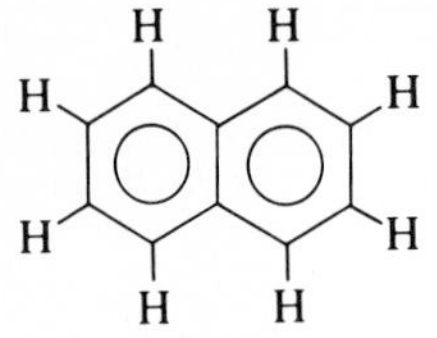

H
H
H
H
H
H
H
H

naphthalene

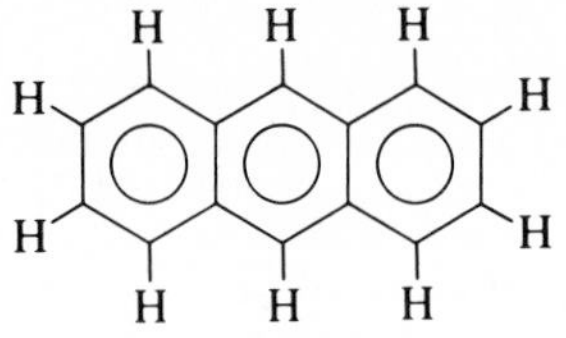
anthracene

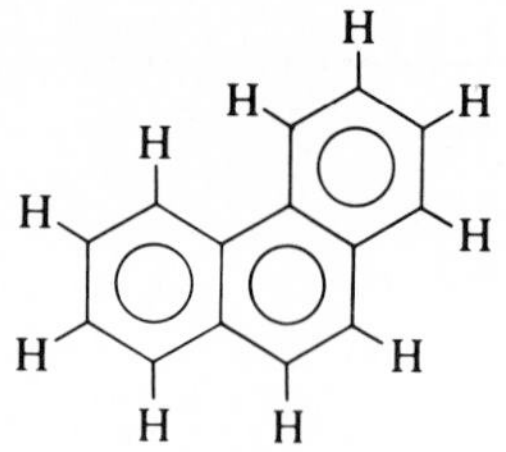
phenanthrene

Sources and Uses of Aromatic Compounds

Coal heated in the absence of air produces coke, coal gas, and coal tar. Coal gas consists mainly of hydrogen, hydrogen sulfide, methane, and other low-molecular-weight hydrocarbons. Coal tar is a complex mixture that can be further refined to yield a number of aromatic hydrocarbons, including benzene, toluene, dimethylbenzenes, phenols, naphthalenes, and anthracenes. Industrially, some important aromatic compounds are made by dehydrogenating (removing H_2) from alkanes and alkenes. Toluene can also be synthesized in large amounts by passing *n-heptane* over catalysts at 500°C.

Aromatic compounds are used in the synthesis of a wide variety of compounds, such as pharmaceuticals, explosives, plastics, perfumes, dyes, insecticides, lacquers, and solvents. The simpler aromatic hydrocarbons, benzene, toluene, and dimethyl benzenes, are good industrial solvents. Toluene is often used as the solvent in model airplane glue. Phenol in dilute solutions is a common disinfectant; substituted phenols are impregnated into wood to protect it from decay and insects. Aniline derivatives are widely used in the dye industry. Naphthalene is a major constituent of moth balls; anthracenes can be used to synthesize steroids.

Physical and Biological Properties of Aromatic Hydrocarbons

As a class, aromatic hydrocarbons are not very polar. The smaller members are liquids at room temperature and are used as solvents because of their inertness. Most polyaromatic hydrocarbons are solids, and molecules larger than anthracene are often highly colored.

Biologically, aromatic hydrocarbons are generally dangerous. Benzene and toluene are toxic if inhaled in large amounts; in smaller amounts they can damage the liver. The large polyaromatic hydrocarbons are dangerous because many are carcinogenic (cancer producing). A large number of aromatic compounds containing four or more fused rings have been shown to be highly carcinogenic. These compounds are often emitted to the environment by combustion of coal and other fuels. Figure 27-4 lists some of the more dangerous examples.

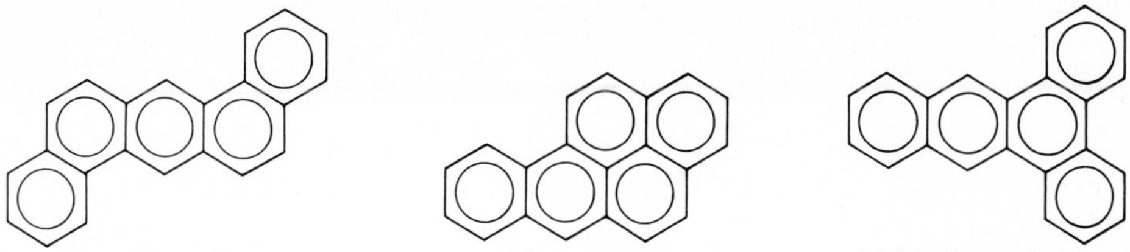

1,2,5,6-dibenzanthracene (a) 3,4-benzpyrene (b) 1,2,3,4-dibenzanthracene (c)

Figure 27–4 Carcenogenic aromatic compounds. (a) and (c) These compounds occur in cigarette smoke, mineral oils, smoked meat, and many other places. (b) This is the most deadly cancer-producing hydrocarbon in polluted air. It is produced from incomplete combustion of tobacco and coal, and is also found in many foods.

Reactions of Aromatic Compounds

Generally, substituted benzenes are regarded as fairly inert compounds. The stability formed by the delocalization of the electrons makes them resistant to the addition reactions so characteristic of alkenes and alkynes. When benzenes do undergo reactions, *substitution* instead of addition usually occurs. In these substitution reactions a hydrogen atom is removed and some group takes its place. Several types of substitution reactions can be made to occur, but only if harsh conditions are used.

Halogenation: High-temperature treatment of benzene with halogens in the presence of iron catalysts causes halogenation of the benzene ring.

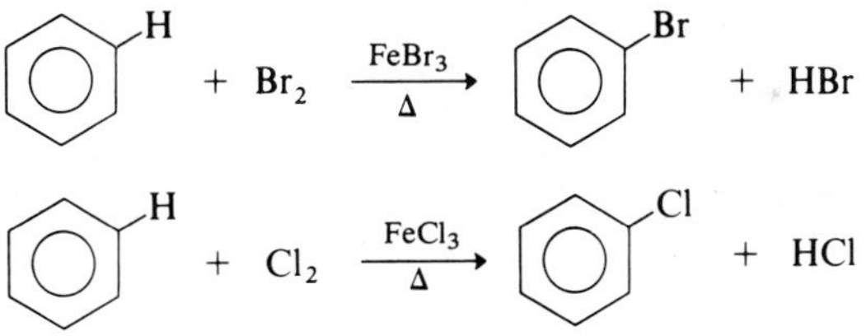
H
+ Br2
FeBr3
Δ
Br
+ HBr
H
+ Cl2
FeCl3
Δ
Cl
+ HCl

Nitration: A nitro group can be introduced into benzene rings by treatment with concentrated nitric acid, using sulfuric acid as a catalyst.

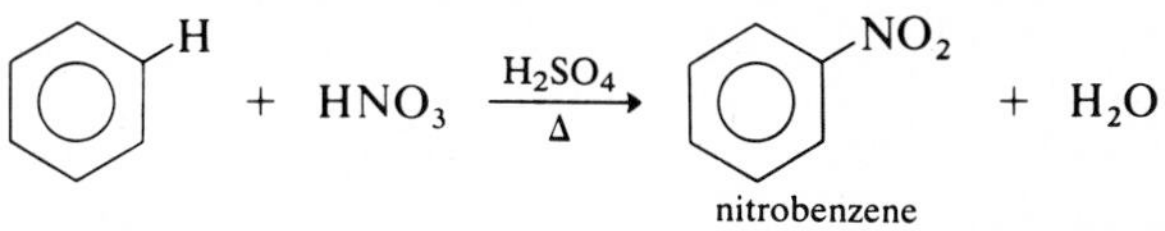
H
+ HNO3
H2SO4
Δ
NO2
+ H2O

nitrobenzene

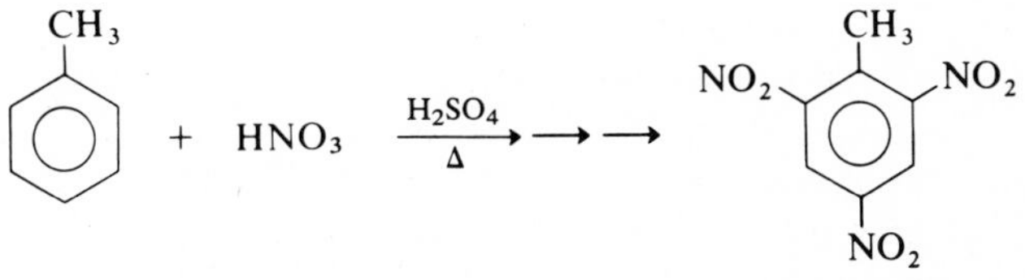
CH3
+ HNO3
H2SO4
Δ
CH3
NO2
NO2
NO2

2,4,6-trinitrotoluene (TNT)

Sulfonation: Benzene sulfonic acids are synthesized commercially by treatment of benzene with SO_3 and H_2SO_4.

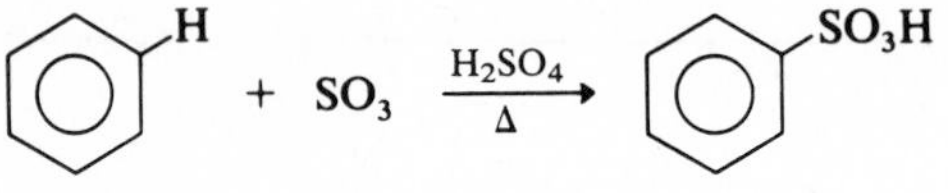

Oxidation of carbon side chains: Treatment of aromatic hydrocarbons that possess alkane substituents with strong oxidizing agents results in the conversion of those side chains to carboxylic acids and CO_2. *Only the carbon atoms directly attached to the benzene rings remain.* All other carbon atoms are converted to carbon dioxide. Other groups are, for the most part, unaffected.

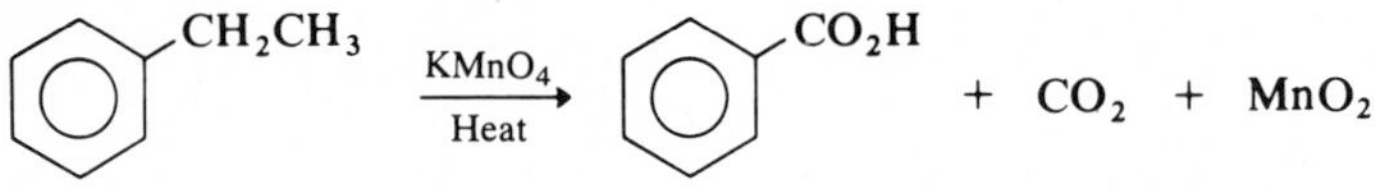

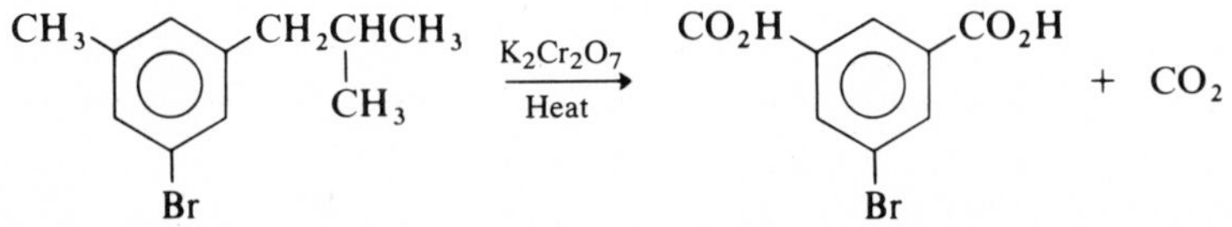

The metabolism of aromatic hydrocarbons in the body follows a similar course. We do not have enzymes that can break up the benzene rings, but alkyl side chains are oxidized to yield either benzoic acid or phenyl acetic acid. Both of these are excreted in the urine.

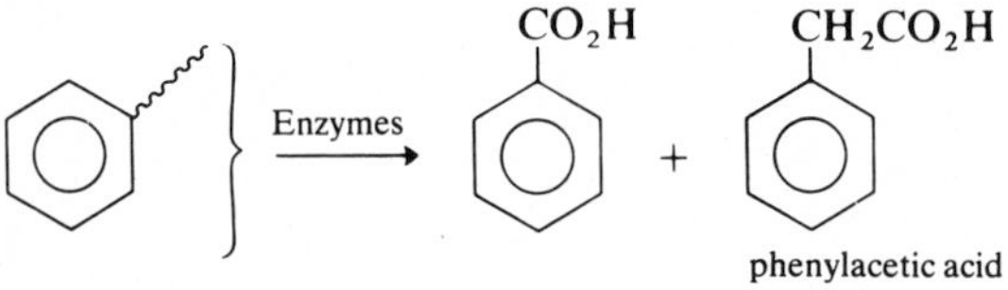

phenylacetic acid

Mechanisms of Aromatic Substitution: Directive Effects

If monosubstituted benzenes are halogenated or nitrated, theoretically three products should be formed in almost equal amounts:

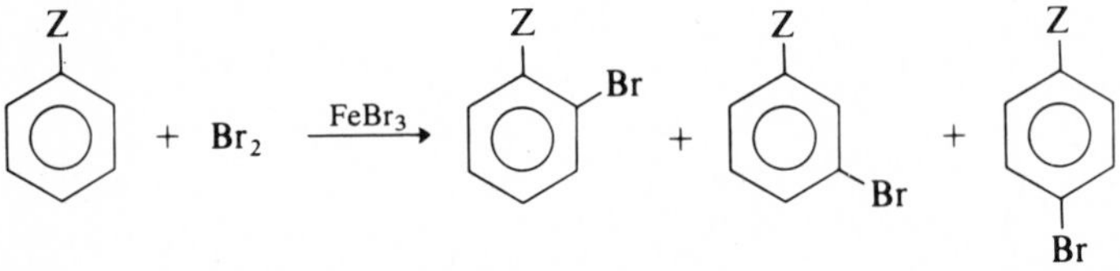

In actual practice, some of the products are formed in preference to others. *The Z group already present on the benzene ring directs where the incoming group will preferentially attack.* Some Z groups allow attack at the ortho and para positions; others favor meta substitution. Table 27–1 lists the possible Z-group directing

Table 27–1 Ortho-para and meta directors

Ortho-para directors	Meta directors
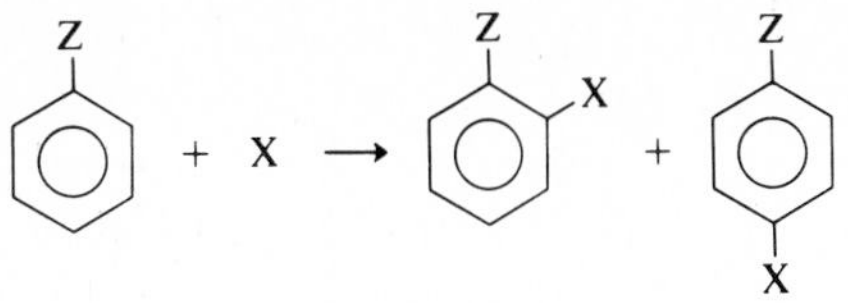	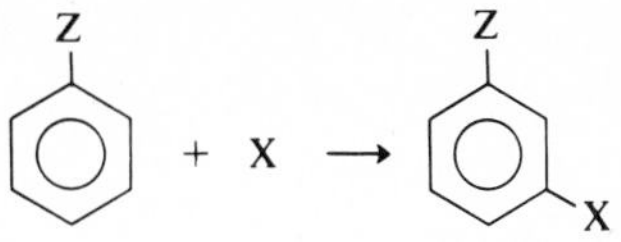
Z = —N(—C—)(—C—) or $-NH_2$	Z = $-NO_2$
= —O—H	= —C(=O)—C—
= —O—C—	= —C(=O)—H
= —C—	= —C(=O)—OH
= —Cl, —Br, —I	

activities. The directing effects can be explained somewhat if the reactions of aromatic hydrocarbons are seen as a composite of the addition–elimination reactions of alkenes.

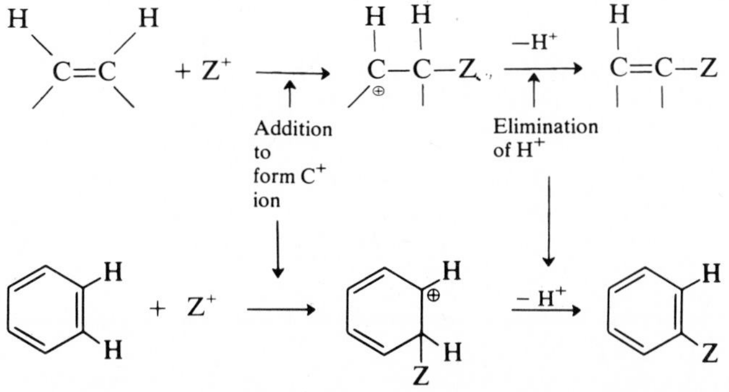

+ Z^+
Addition to form C^+ ion
$-H^+$
Elimination of H^+

All the ortho–para directors are electron-donating groups, which help to stabilize the carbonium ion. The carbonium ion is stabilized to the largest extent if the attacking group goes ortho or para; it is least affected if X attacks at the meta position. Meta directors are electron-withdrawing groups.

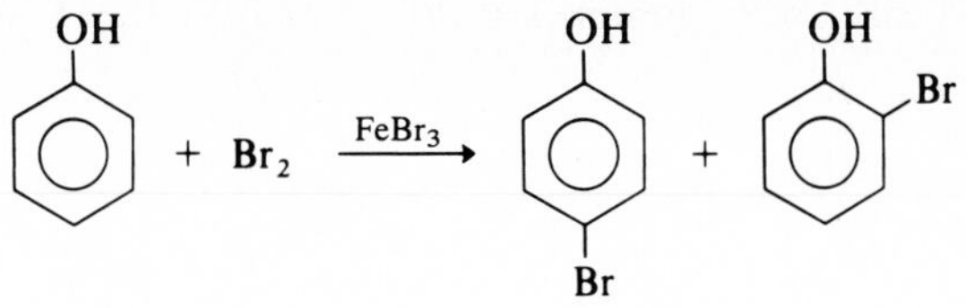

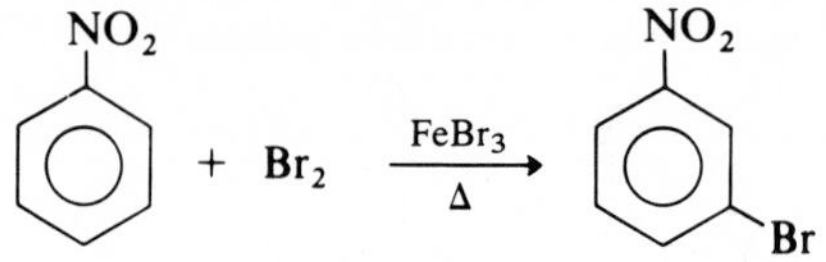

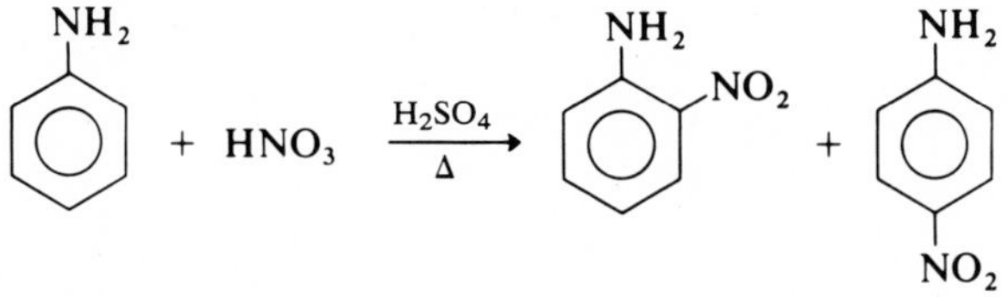

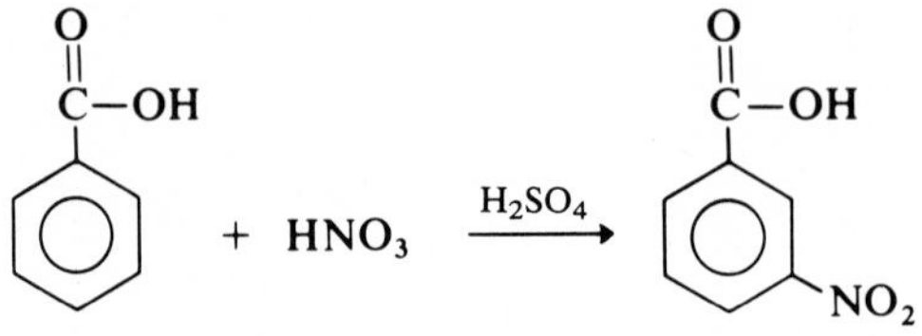

UNIT 27 Problems and Exercises

27–1. What characteristic features of aromatic hydrocarbons distinguish them from the other hydrocarbons?

27–2. Name the following.

(a) 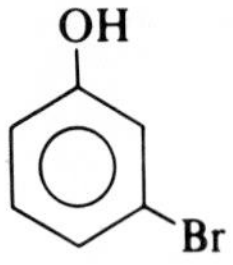 (b) 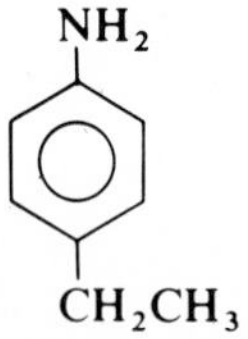 (c) 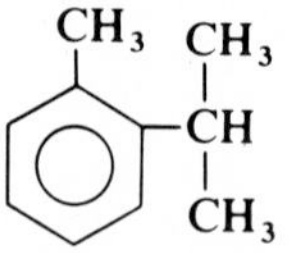 (d) 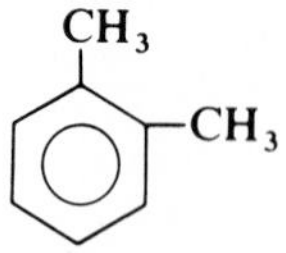

(e) 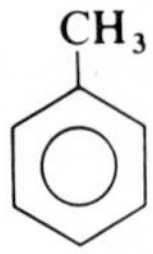 (f) 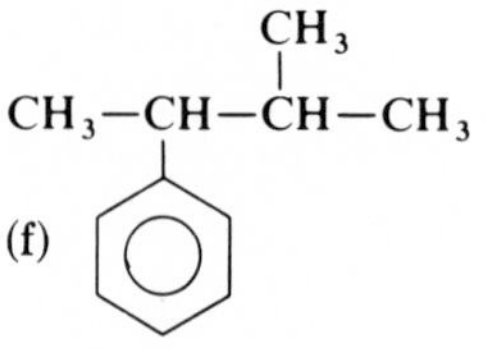 (g) 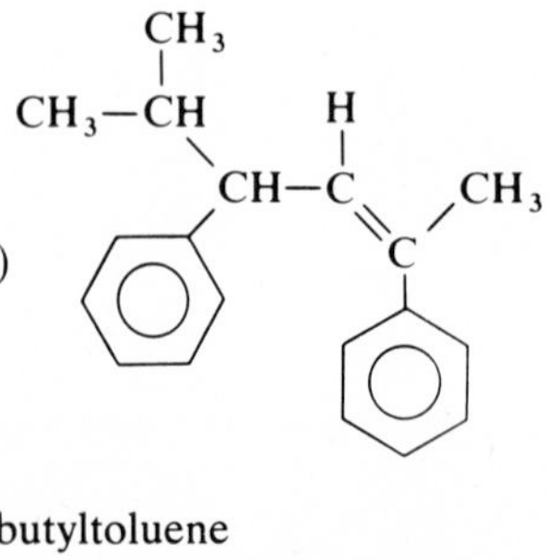

27–3. Draw the structures for the following.

(a) *m-sec*-butylaniline
(b) *p*-isobutyltoluene
(c) *o*-isopropylphenol
(d) 1,3-dimethyl-5-ethylbenzene
(e) *cis*-2-phenyl-2-butene
(f) 3,4-dimethyl-1,5-diphenylheptane

27–4. Correct the names for the following if they are incorrect.

(a) 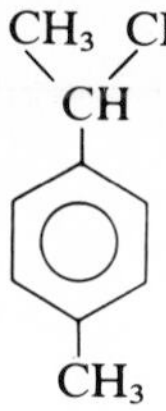

m-isopropyltoluene

(b) 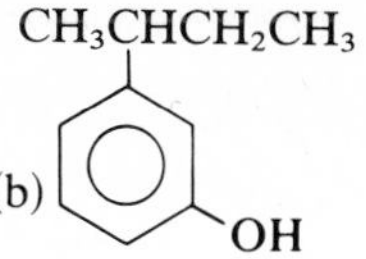

o-isobutylaniline

(c) 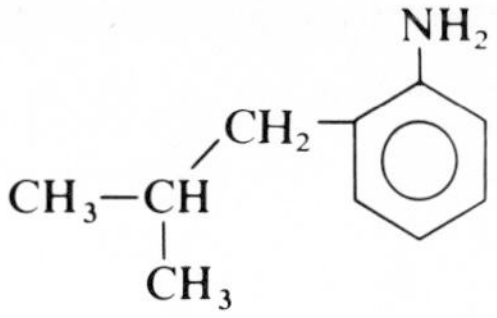

p-*sec*-butylphenol

(d) 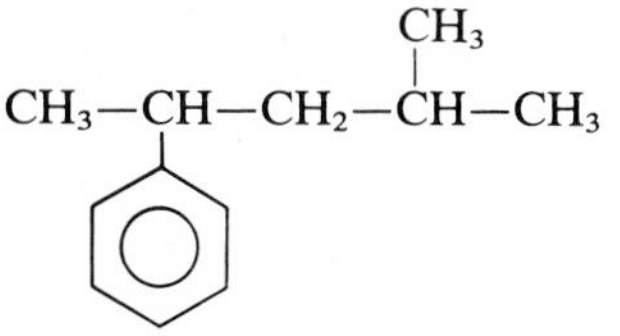

2-methyl-4-benzylpentane

(e) 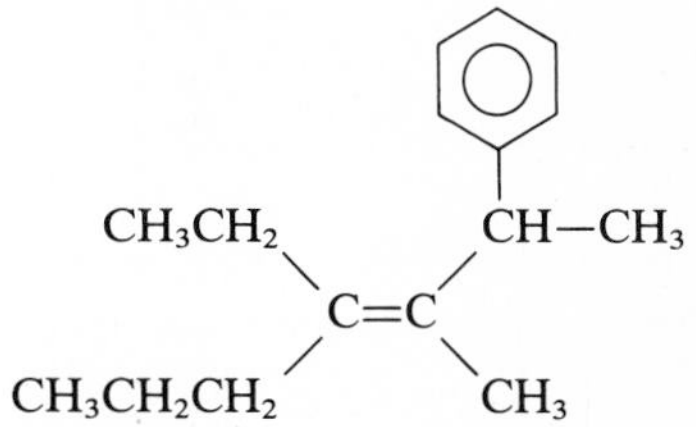

trans-3-methyl-4-ethyl-2-phenyl-3-heptene

27–5. Write the molecular formulas for the following compounds.

(a) 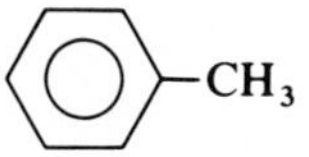

(b) 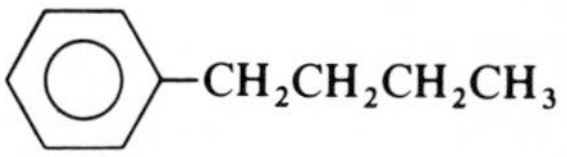

(c) 2-phenyl-2-butene
(d) 2-phenyl-1-butyne
(e) Naphthalene
(f) Anthracene
(g) 1-phenylcyclohexane

27–6. Draw and name the eight benzene-containing compounds that possess a molecular formula of C_9H_{12}.

27–7. How many monochloro derivatives of naphthalene are possible?

27–8. Complete the following reactions.

(a) 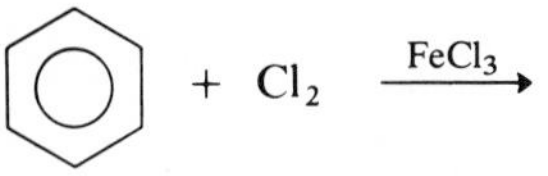

(b) 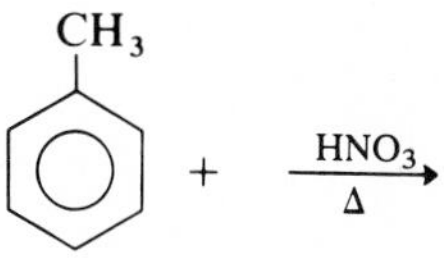

(c) 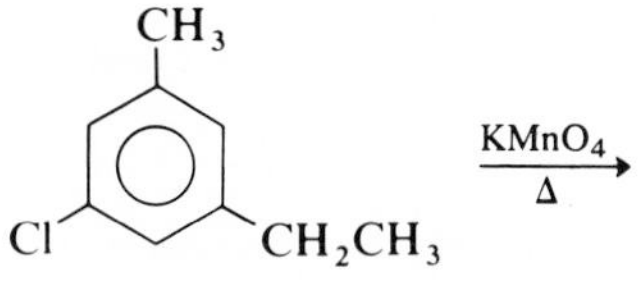

(d) 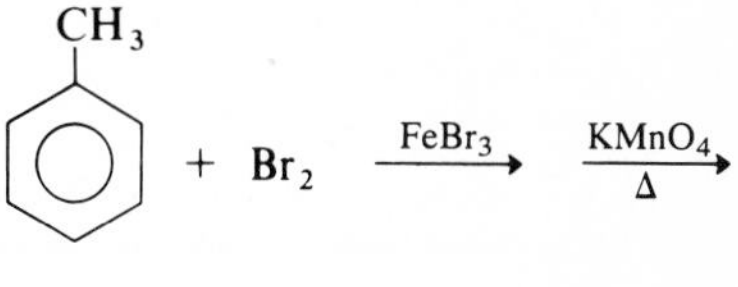

(e) 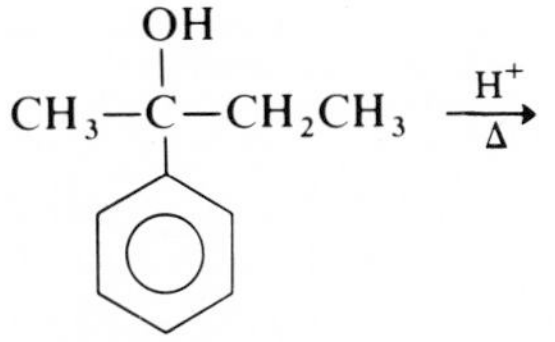

27–9. Use Table 27–1 to predict the principal products in the following reactions.

(a) 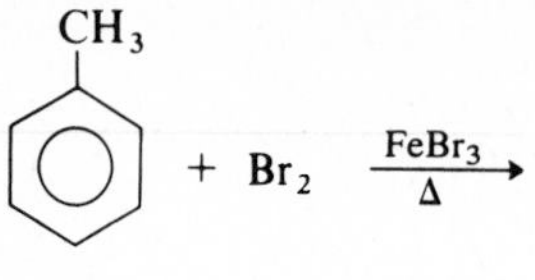

CH3
+ Br2
FeBr3
Δ

(b) 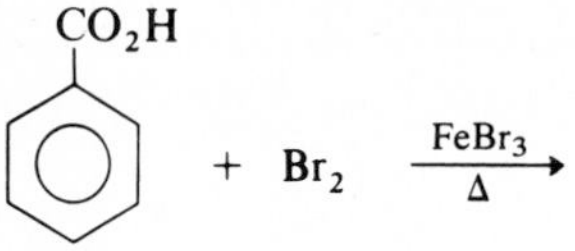

CO2H
+ Br2
FeBr3
Δ

(c) 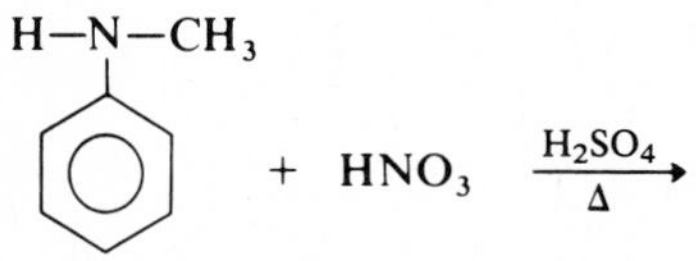

H—N—CH3
+ HNO3
H2SO4
Δ

(d) 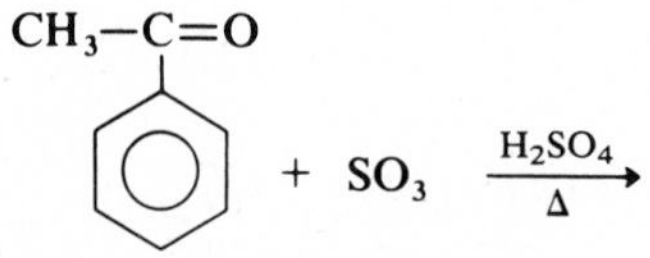

CH3—C=O
+ SO3
H2SO4
Δ

27–10. Write equations to show how each of the following conversions might be accomplished. Make sure that you prepare the correct isomer in the largest yield. Several steps may be required for each compound.

(a) Make *p*-chloronitrobenzene from benzene.

(b) Make 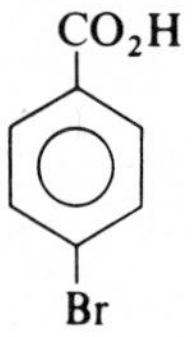 from toluene.

CO2H
Br

(c) Make *m*-chloronitrobenzene from benzene.

CHAPTER 11

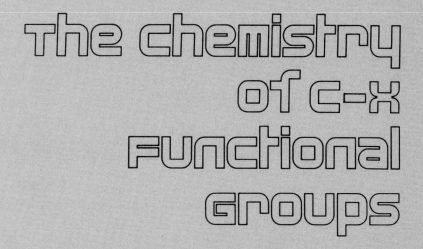

the chemistry of C-X functional groups

Besides carbon and hydrogen, elements such as oxygen, sulfur, the halogens, and phosphorus are also found in organic molecules. Oxygen is the most common with nitrogen being next. In units 28 through 31 we introduce and describe the properties of many of the compounds that contain the elements listed above.

Saturated organic molecules containing oxygen are introduced in units 28 and 30. **Alcohols** are compounds that possess a C—O—H center, and **ethers** possess a C—O—C center. Unit 30 also introduces some of the sulfur-containing compounds found in nature.

There are not many molecules in our bodies that contain halogens, but this is not to say that organic halogen compounds are not important to study. As will be seen in unit 29, the organohalides are important as reactants for many reactions and are also widely used industrially.

In unit 31 we introduce **amines**, the first of several types of nitrogen-containing organic compounds. In unit 32 we review the reactions covered in units 28 through 31 and discuss the mechanisms of substitution reactions.

UNIT 28 ALCOHOLS

After carbon and hydrogen, the third most abundant element in organic compounds is oxygen. Oxygen is present in the $-\overset{|}{\underset{|}{C}}-O-H$ group in alcohols, as $-\overset{|}{\underset{|}{C}}-O-\overset{|}{\underset{|}{C}}-$ in ethers, and as $\overset{O}{\overset{\|}{\underset{|}{C}}}-O-\overset{|}{\underset{|}{C}}-$ in esters. In aldehydes, ketones, and carboxylic acid groups it is present as $\rangle C=O$. In this unit we shall discuss the —OH group in alcohols.

Nomenclature of Alcohols (IUPAC)

1. The suffix for the alcohol class is *-ol.*
2. The parent chain must contain the carbon atom to which the OH group is bound. The parent then is *the longest continuous carbon chain containing the C—OH group.* There may be longer chains present, but they are not to be considered as the parent chain.
3. The parent chain is numbered from one end so as to give the carbon bearing the OH group the lower number. This number is used in the name of the compound to indicate the location of the OH group. It may precede the parent name or the *-ol* suffix.

$$\underset{④}{CH_3}\underset{③}{CH_2}\underset{②}{\overset{\overset{OH}{|}}{C}H}-\underset{①}{CH_3}$$

2-butanol or butan-2-ol

4. All substituents are named and placed in the name as previously done for alkanes and alkenes.
5. Compounds containing more than one alcohol are called diols, triols, and so on.

Common Names of Alcohols

The common names of many alcohols are often used in preference to the IUPAC names. In many common names the alkane portion of the alcohol is named like a substituent, and then the suffix *-ol* or the word *alcohol* is added. For example $CH_3-\overset{\overset{CH_3}{|}}{C}H-OH$ is named 2-propanol in the IUPAC system, but commonly it is called isopropanol or isopropyl alcohol. Other common names are listed in Figure 28–1.

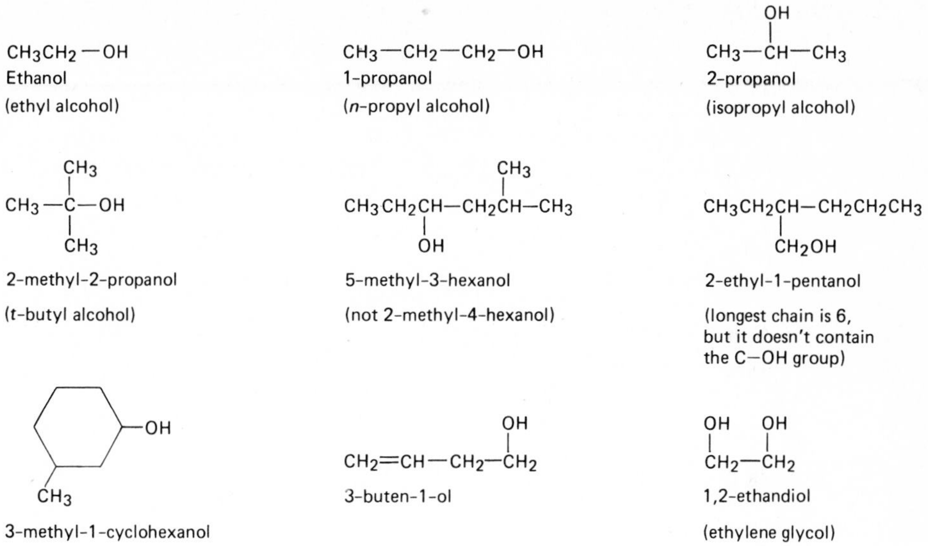
Ethanol
(ethyl alcohol)
1-propanol
(n-propyl alcohol)
2-propanol
(isopropyl alcohol)
2-methyl-2-propanol
(t-butyl alcohol)
5-methyl-3-hexanol
(not 2-methyl-4-hexanol)
2-ethyl-1-pentanol
(longest chain is 6, but it doesn't contain the C—OH group)
3-methyl-1-cyclohexanol
3-buten-1-ol
1,2-ethandiol
(ethylene glycol)

Figure 28–1 Names and structures of alcohols.

Types of Alcohols

Alcohols are divided into four major subclasses. A **primary** alcohol (1°) is defined as a $-\overset{|}{\underset{|}{C}}-OH$ group that has two hydrogens and one carbon bonded to it. A **secondary** alcohol (2°) has two carbons and one hydrogen, whereas a **tertiary** alcohol (3°) has three carbons and no hydrogens. An **aromatic** alcohol (phenol) has an OH group on the benzene ring.

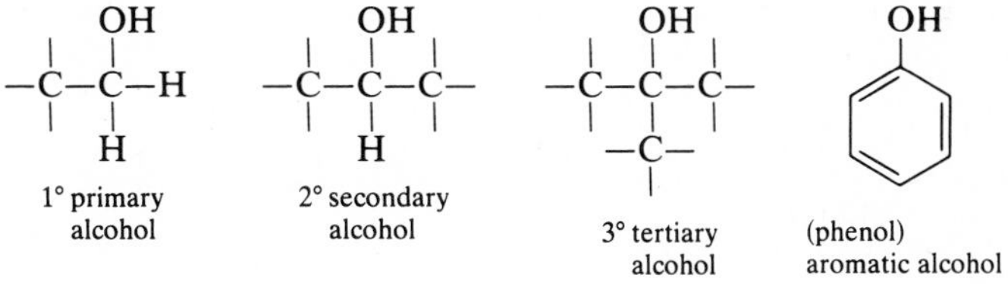
1° primary alcohol
2° secondary alcohol
3° tertiary alcohol
(phenol) aromatic alcohol

Sources and Uses of Alcohols

The alcohols most widely used in industry are methanol, ethanol, and the propanols. Methanol was isolated by distilling wood chips in the absence of oxygen, but now it is synthesized by reduction of carbon monoxide (CO). It is widely used as a precursor to formaldehyde ($H-\overset{\overset{O}{\|}}{C}-H$, used in polymers) and formic acid ($H-\overset{\overset{O}{\|}}{C}-OH$), and is also widely used in the synthesis of many drugs, perfumes, and dyes. Methanol is a good solvent for polar compounds, but care must be exercised in its use since it is toxic. Ingestion or inhalation of methanol vapors can lead to blindness and death, so it should never be used in unventilated areas. Methanol burns with almost no flame color, so care must be taken when using it around flames.

Ethanol or ethyl alcohol is perhaps the most widely known and used alcohol. It is the only small simple alcohol that is not toxic in small quantities. It can be synthesized by yeast fermentation of the sugars present in almost any grain, vegetable, or fruit. This fermentation will produce a maximum concentration of about 13% ethanol. Alcoholic beverages with higher alcohol content are manufactured by additional refinement of the initial fermentation process. Industrially, ethanol is synthesized by hydration of ethene. It is used in great quantities as solvents and as a starting point for further synthesis. To prevent the ingestion of ethyl alcohol not sold for that purpose, chemicals are sometimes added to make it toxic; then ethyl alcohol is called **denatured alcohol**.

2-propanol, commonly known as isopropanol or rubbing alcohol, is made by reduction of acetone ($CH_3—\overset{\overset{\large O}{\|}}{C}—CH_3$). It is widely used in hand lotions, after-shave lotions, cosmetics, inks, and paints.

Alcohols with more than four carbon atoms are generally synthesized by hydration of alkenes or reduction of acids and ketones.

Physical Properties of Alcohols

Since alcohols contain the highly polar O—H bond, they hydrogen-bond readily. Consequently, their boiling points are much higher than those for other organic compounds of similar molecular weight. Note in Table 28–1 that the differences in boiling points between alkanes and alcohols decrease as the molecular weights increase. As the alkane (nonpolar) portion of the molecule increases in size relative to the OH (polar end), the molecules take on more of the properties of alkanes and

Table 28–1 Comparison of boiling points of alkanes and alcohols of approximately the same molecular weight

Approx. MW	Alkane*	BP (°C)	Alcohol	BP (°C)	Difference in BP (°C)	Alcohol solubility in H_2O
30–32	CH_3CH_3	−89	CH_3OH	65	+154	Soluble
44	$CH_3CH_2CH_3$	−44	CH_3CH_2OH	78	+132	Soluble
58–60	$CH_3CH_2CH_2CH_3$	−0.5	$CH_3CH_2CH_2OH$	97	+98	Soluble
72–74	$CH_3(CH_2)_3CH_3$	36	$CH_3CH_2CH_2CH_2OH$	117	+81	
100–102	$CH_3(CH_2)_5CH_3$	98	$CH_3(CH_2)_4CH_2OH$	158	+60	
128–130	$CH_3(CH_2)_7CH_3$	151	$CH_3(CH_2)_6CH_2OH$	194	+43	Insoluble
156–158	$CH_3(CH_2)_9CH_3$	195	$CH_3(CH_2)_8CH_2OH$	229	+34	Insoluble

* Alkanes are insoluble in water.

less of the properties of water. The smaller alcohols are quite polar and can even dissolve ionic compounds to some extent. The first three are soluble in water in all proportions, but the solubility of larger alcohols decreases as the alkane portion increases. Methanol is so polar that nonpolar compounds such as hexane are not soluble in it. Ethanol and propanol occupy a middle ground; they will dissolve in both water and hexane.

Chemical Reactions and Synthesis of Alcohols

Hydration–Dehydration

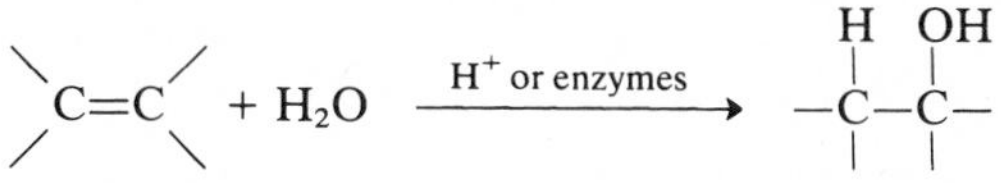

$$\underset{/}{\overset{\backslash}{C}}=\underset{\backslash}{\overset{/}{C}} + H_2O \xrightarrow{H^+ \text{ or enzymes}} \overset{H \quad OH}{\underset{|\quad\;\;|}{-C-C-}} \quad \text{(hydration)}$$

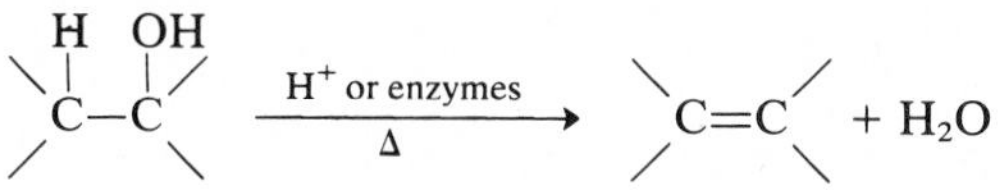

$$\overset{H \quad OH}{\underset{|\quad\;\;|}{-C-C-}} \xrightarrow[\Delta]{H^+ \text{ or enzymes}} \underset{/}{\overset{\backslash}{C}}=\underset{\backslash}{\overset{/}{C}} + H_2O \quad \text{(dehydration)}$$

In the presence of acid catalysts or enzymes, water may add to alkenes to form alcohols (hydration) or be eliminated from alcohols to form alkenes (dehydration). The processes are essentially the reverse of one another. Hydration occurs readily at room temperature, but heat is often added in dehydrations to drive off either the alkene or water as it is formed. This prevents the alkene from reacting with water to re-form the alcohol.

Both processes, hydration and dehydration, occur readily because of several factors. Alkenes are susceptible to attack by acids and form carbonium ions. These ions are attracted to the unshared electron pairs on the oxygen atom of the water molecule. The water molecule can then lose H^+ to complete the hydration.

$$\underset{/}{\overset{\backslash}{C}}=\underset{\backslash}{\overset{/}{C}} \;\underset{\rightleftharpoons}{\overset{H^+}{}}\; \overset{H}{\underset{|\;\;\;\;|}{-C-C^{\oplus}}} \xrightarrow{H_2O:} \overset{\overset{\oplus}{H-O-H}}{\underset{|\;\;\;\;|}{-C-C-}} \xrightarrow{-H^+} \overset{H \quad :\ddot{O}-H}{\underset{|\;\;\;\;\;|}{-C-C-}}$$

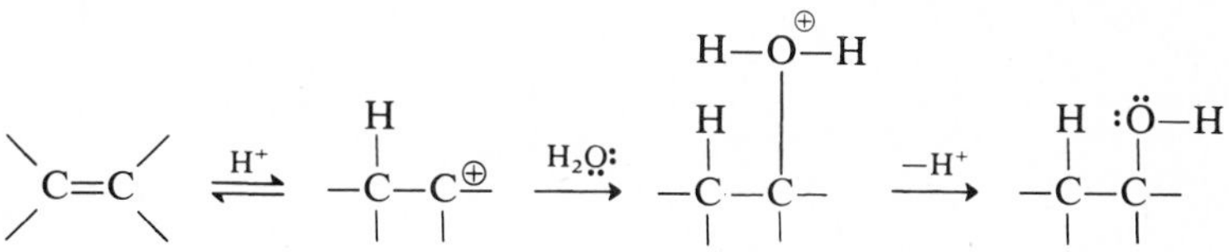

In dehydration the acid attacks the unshared electron pairs on the oxygen and removes it as water.

$$\overset{H \quad :\ddot{O}H}{\underset{|\;\;\;\;\;|}{-C-C-}} \underset{\rightleftharpoons}{\overset{H^+}{}} \overset{H \quad :\overset{\oplus}{O}-H}{\underset{|\;\;\;\;\;|}{-C-C-}} \underset{\rightleftharpoons}{\overset{-H_2\ddot{O}:}{}} \overset{H}{\underset{|\;\;\;\;\;|}{-C-C^{\oplus}-}} \underset{\rightleftharpoons}{\overset{-H^+}{}} \underset{/}{\overset{\backslash}{C}}=\underset{\backslash}{\overset{/}{C}}$$

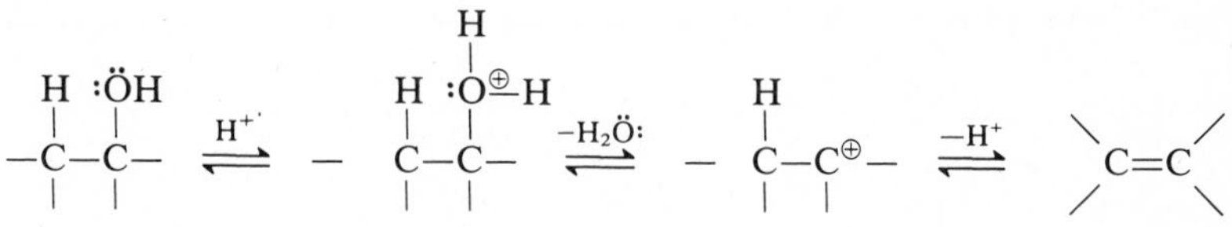

These mechanisms were discussed in more detail earlier. Students should refer back to unit 25 if review is necessary.

Biologically, hydration and dehydration are both important processes during food metabolism. The sequence

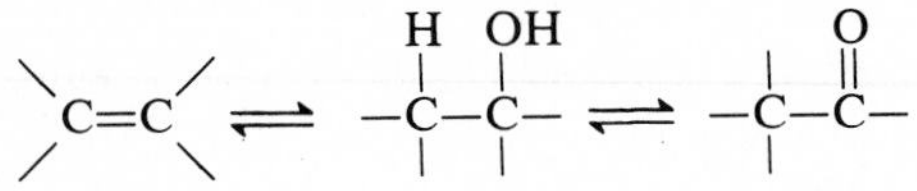

seems to be quite prevalent in our bodies. In the citric acid cycle (food metabolism), hydration and dehydration occur several times. They also occur in many other cycles.

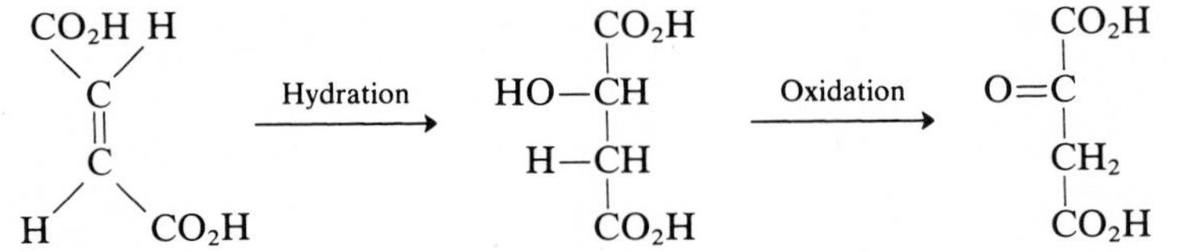
Hydration
Oxidation

fumaric acid — malic acid — oxaloacetic acid

Oxidation–Reduction

In organic chemistry, the term *oxidation* usually refers to loss of hydrogen or gain of oxygen. *Reduction* refers to gain of hydrogen or loss of oxygen. Alcohols are oxidized to >C=O groups by enzymes or by oxidizing agents like $KMnO_4$, O_2, and $K_2Cr_2O_7$. Conversely, >C=O groups are reduced to alcohols by the use of reducing agents such as $LiAlH_4$, H_2 + Pt, $NaBH_4$, or enzymes.

Oxidation:

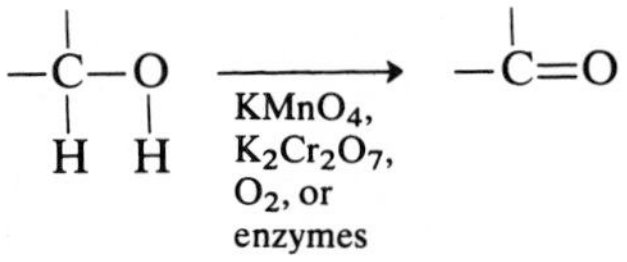
KMnO4,
K2Cr2O7,
O2, or
enzymes

Reduction:

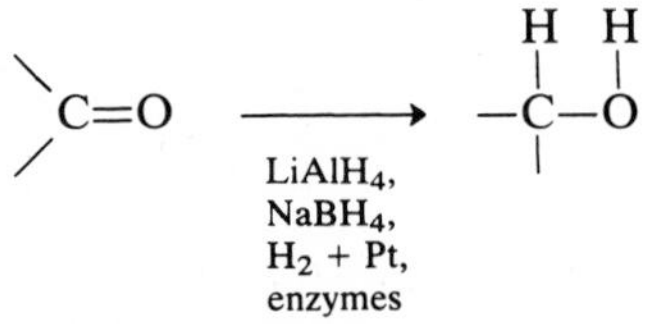
LiAlH4,
NaBH4,
H2 + Pt,
enzymes

Primary alcohols (—CH_2OH) are oxidized to aldehydes, but aldehydes are quite susceptible to further oxidation and yield carboxylic acids. For example, when exposed to air, wines can become vinegary as the ethanol is air-oxidized to acetic acid.

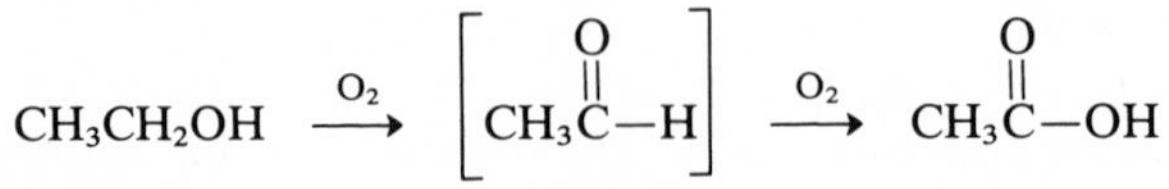
O2
O2

ethanol — acetaldehyde — acetic acid

Secondary alcohols are oxidized to form ketones, which are resistant to further oxidation.

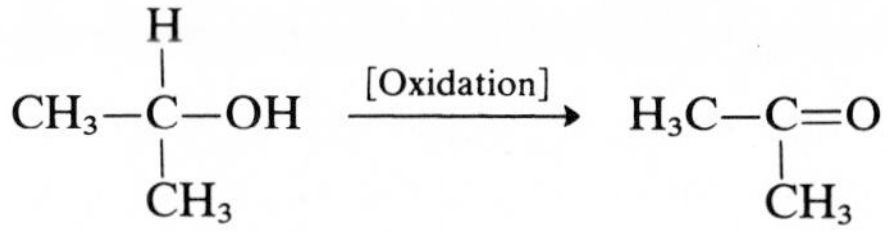

isopropyl alcohol — acetone

Tertiary alcohols are not subject to oxidation since they do not have hydrogen on the $-\overset{|}{\underset{|}{C}}-OH$ carbon.

If we reverse the above processes and add hydrogen to the $>C=O$ groups, reduction occurs. Thus reduction of carboxylic acids and aldehydes leads to primary alcohols, whereas reduction of ketones leads to secondary alcohols:

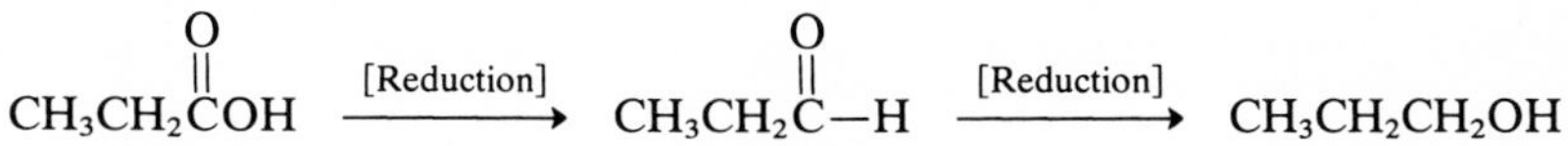

propanoic acid — propanaldehyde — 1-propanol

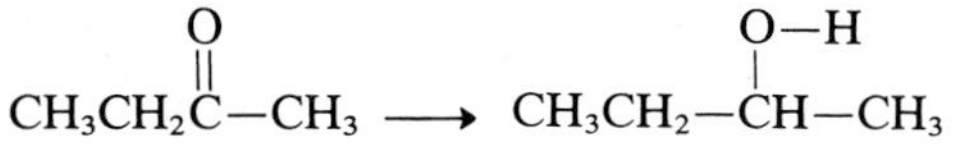

2-butanone — 2-butanol

Both oxidation and reduction also occur readily in biochemical systems.

In lipid metabolism,

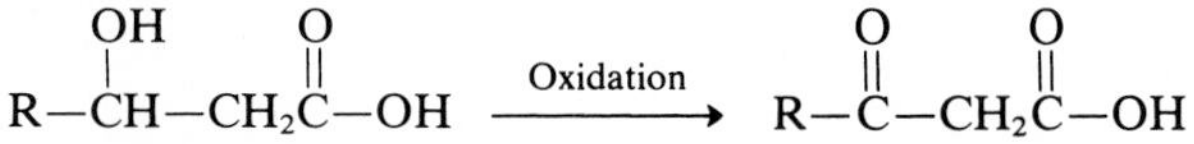

a β-hydroxy acid; a secondary alcohol — a β-keto acid; a ketone

In carbohydrate metabolism,

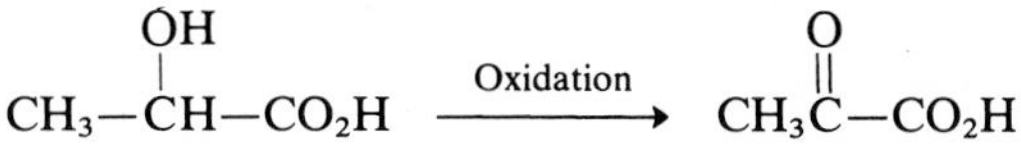

lactic acid — pyruvic acid

Alcohols as Acids

Since alcohols possess an OH group, you may think that they can act as bases. If alcohols readily ionized to give up hydroxide ions, they would be bases, but this is not the case. Since O—H bonds are weaker than C—O bonds, alcohols can act as acids, yielding hydrogen ions when treated with strong bases. Typical K_a values for alcohols are in the range of 10^{-16} to 10^{-18}. Compared to acetic acid ($K_a = 10^{-5}$), alcohols are quite weak, but they are still much stronger acids than alkanes ($K_a = 10^{-50}$). Of the alcohols, the aromatic alcohols, the phenols, are most acidic. They have K_a values of approximately 10^{-10}.

Treatment of alcohols with sodium or potassium metal is normally required to cleave the O—H bond. In this oxidation–reduction reaction, sodium or potassium salts of the alcohol are produced along with hydrogen gas.

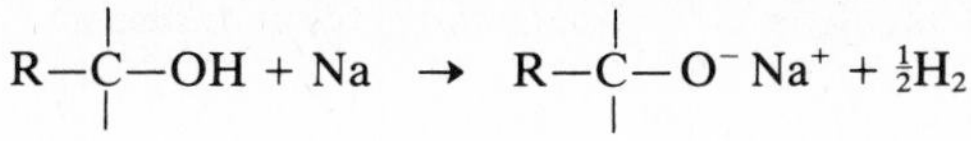

This reaction is sometimes used as a quick chemical test to determine the presence or absence of alcohol centers within a molecule. If bubbling occurs (liberation of hydrogen gas) when elemental sodium is added to a compound, some form of acidic hydrogen atom is probably present. These acidic hydrogen atoms are probably present on an oxygen atom either in an alcohol center (—C—O—H) or in a carboxylic acid center $\left(-\overset{\overset{\large O}{\|}}{C}-O-H\right)$.

UNIT 28 Problems and Exercises

28–1. Name the following alcohols.

(a) 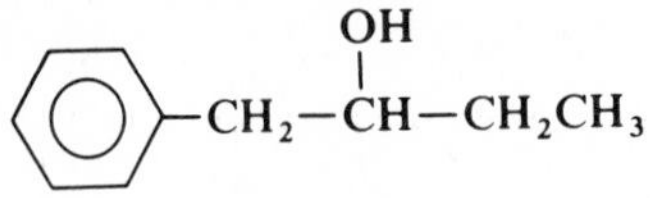

(b) 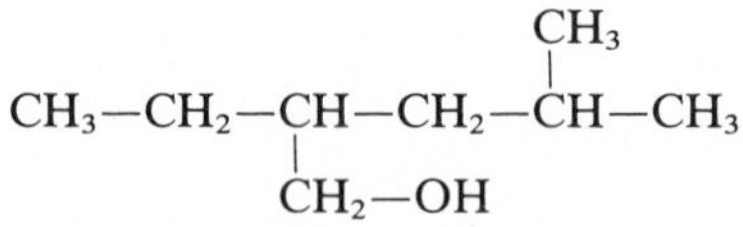

(c) 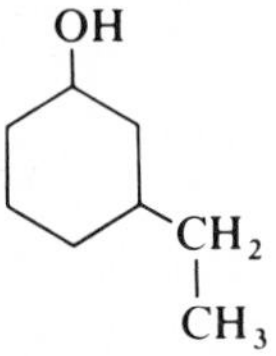

(d) 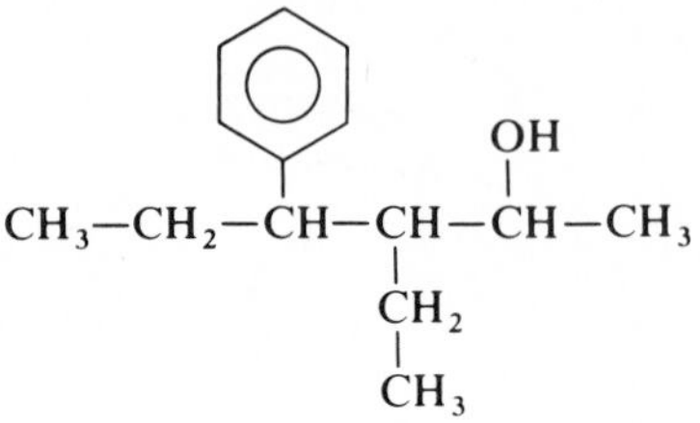

(e) 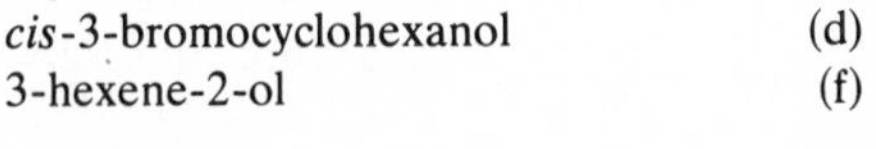

28–2. Draw the eight possible structures for alcohols with the formula $C_5H_{11}OH$. Name each and tell whether it is a primary, secondary, or tertiary alcohol.

28–3. Draw structures for the following.

(a) 3-phenylcyclopentanol
(b) 4,4-diethyl-2-heptanol
(c) *cis*-3-bromocyclohexanol
(d) 4-isopropyl-2,2-dimethyl-1-octanol
(e) 3-hexene-2-ol
(f) 2,4-dimethyl-1,5-hexandiol

28–4. Give the common and IUPAC names for the following.

(a)

(b) 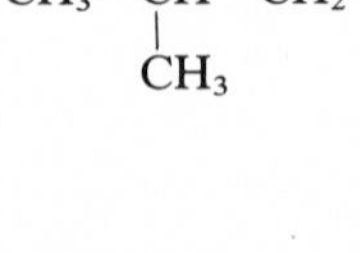

(c)

(d) 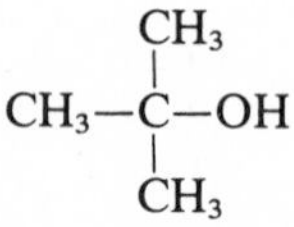

28–5. Write out the structural formulas for the following compounds.

(a) 2-butanol (b) 1-butanol
(c) 2-methyl-1-propanol (d) 2-phenyl-2-butanol
(e) 3-buten-2-ol (f) 3-pentyn-1-ol

28–6. Write out the reactions between 2-pentanol and the following.

(a) H^+, Δ (b) $KMnO_4$ (c) Na

28–7. Complete the following and circle the major product.

(a) $CH_3{-}\underset{\displaystyle OH}{\underset{|}{C}H}{-}\underset{\displaystyle CH_3}{\underset{|}{C}H}{-}CH_3 + \xrightarrow[\Delta]{H^+}$ 2 products

(b) 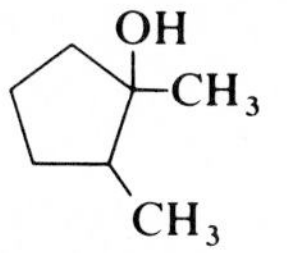 $\xrightarrow{H^+}$ 3 products

(c) $CH_3{-}\overset{\displaystyle CH_3}{\overset{|}{\underset{\displaystyle CH_3}{\underset{|}{C}}}}{-}CH_2OH \xrightarrow{H^+}$

(d) 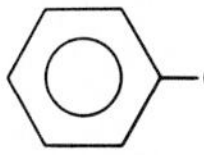 $-CH_2OH \xrightarrow{KMnO_4}$

(e) $CH_3{-}\overset{\displaystyle OH}{\overset{|}{C}H}{-}CH_3 \xrightarrow{KMnO_4}$

(f) $CH_3CH_2OH + Na \longrightarrow$

(g) 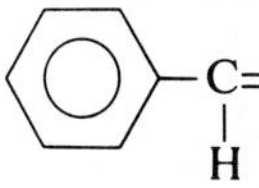 $+ H_2 \xrightarrow[Pt]{}$

(h) $CH_3{-}\overset{\displaystyle CH_3}{\overset{|}{\underset{\displaystyle CH_3}{\underset{|}{C}}}}{-}OH + KMnO_4 \longrightarrow$

28–8. Synthesize the following compounds from the given starting material. More than one step is needed for each synthesis.

(a) Make $CH_3{-}\overset{\displaystyle Br}{\overset{|}{C}H}{-}CH_3$ from $CH_3CH_2\overset{\displaystyle O}{\overset{\|}{C}}{-}H$.

(b) Make 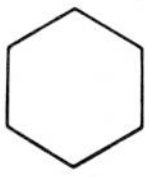 from 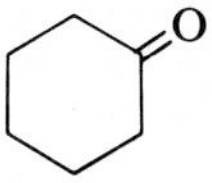

(c) Make $CH_3\underset{\displaystyle CH_3}{\underset{|}{C}H}{-}\overset{\displaystyle O}{\overset{\|}{C}}{-}CH_3$ from $CH_3\overset{\displaystyle OH}{\overset{|}{\underset{\displaystyle CH_3}{\underset{|}{C}}}}{-}CH_2CH_3$.

(d) Make 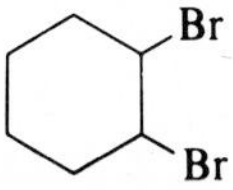 from 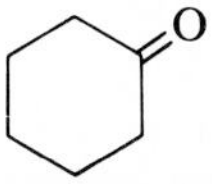

28–9. Arrange the following compounds in order of increasing boiling points and justify your choice.

(a) $CH_3(CH_2)_5CH_3$

(b) $CH_3\underset{\displaystyle OH}{\underset{|}{C}H}{-}\underset{\displaystyle OH}{\underset{|}{C}H}{-}\underset{\displaystyle OH}{\underset{|}{C}H_2}$

(c) $CH_3CH_2CH_2CH_2\underset{\displaystyle \underset{\displaystyle \underset{|}{OH}}{CH_2}}{\underset{|}{C}H_2}$

(d) $CH_3CH_2CH_2\underset{\displaystyle OH}{\underset{|}{C}H}{-}\underset{\displaystyle OH}{\underset{|}{C}H}$

28–10. Outline a procedure of chemical tests that would enable you to distinguish among hexane, 1-hexene, 1-hexyne, and 1-hexanol. What would happen to each compound in each chemical reaction?

28–11. A compound with the formula C_4H_8O reacts with sodium to give off hydrogen gas. It reacts rapidly with bromine and reacts with $KMnO_4$ to yield a carboxylic acid. Give a possible structure for the compound.

28–12. A compound with the molecular formula of $C_4H_{10}O$ reacts rapidly with sodium metal to give off hydrogen gas. It reacts with H^+, Δ to yield only *one* alkene, but does not react with $KMnO_4$ to yield an aldehyde or ketone. Give a structure for the compound.

UNIT 29 ORGANIC HALOGEN COMPOUNDS

The elements present in group VIIA (F, Cl, Br, and I) of the periodic table are often called **halogens**. When these elements replace a hydrogen atom on an alkane or aromatic hydrocarbon, the resulting compound becomes an **alkyl halide** or an **aryl halide**. The symbol R—X refers to an alkyl halide (R = alkyl portion and X = halogen) whereas Ar—X refers to an aryl halide. In this unit we shall discuss the properties of some important alkyl halides.

Nomenclature of Organohalogen Compounds

In the IUPAC method, the halogens are regarded as substituents and thus do not have suffix endings like alcohols and alkenes. When halogens are covalently bound to carbon, the suffix *o* is added to the halogens. In naming organic compounds containing halogens, the halogens are generally grouped together at the beginning in front of the alkyl substituents.

Br = bromo	I = iodo
Cl = chloro	F = fluoro

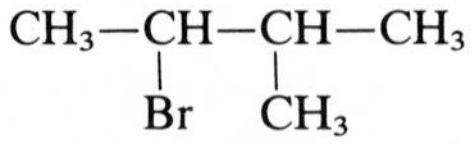

CH3—CH—CH—CH3
Br
CH3

2-bromo-3-methylbutane

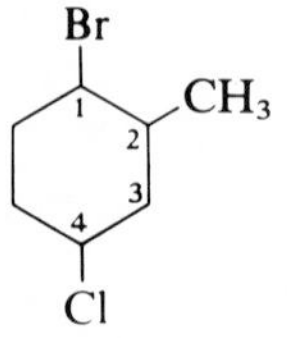

Br
CH3
1
2
3
4
Cl

1-bromo-4-chloro-2-methylcyclohexane

In the common method of nomenclature, alkyl halides are named in a manner similar to the common names of alcohols. The alkyl group name precedes the word bromide, chloride, iodide, or fluoride.

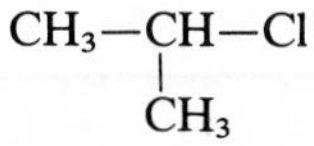

isopropyl chloride
(2-chloropropane)

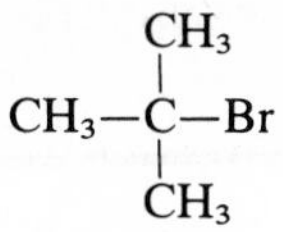

t-butyl bromide
(2-bromo-2-methyl propane)

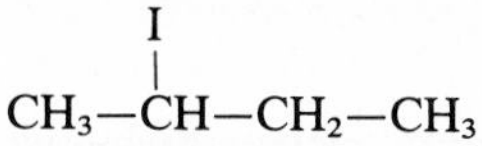

sec-butyl iodide
(2-iodobutane)

Like alcohols, organohalides are divided into four subclasses: primary, secondary, tertiary, and aromatic.

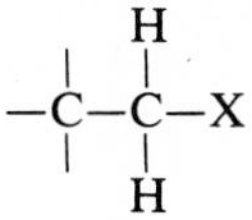

1° primary
halide

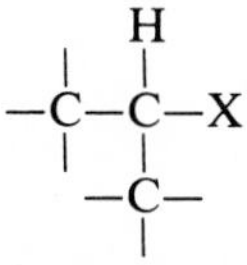

2° secondary
halide

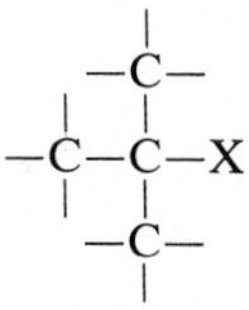

3° tertiary
halide

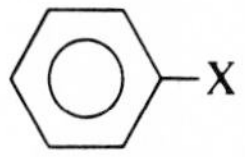

aromatic
halide

Sources and Uses of Organohalides

Organic halogen compounds are rare in nature but widely used in industry as solvents and in the synthesis of other compounds. Generally, they are synthesized by halogenation of alkanes and aromatics or by addition of HX to alkenes. Table 29–1 lists some of the more common organohalides and their uses.

Table 29–1 Common organic halogen compounds

Structure	Name	BP (°C)	Use
CH_3Cl	Chloromethane	−24	Synthesis and low-temperature solvent
CH_2Cl_2	Dichloromethane (methylene chloride)	40	Solvent
$CHCl_3$	Trichloromethane (chloroform)	62	Solvent; formerly used as inhalation anesthetic
CCl_4	Tetrachloromethane (carbon tetrachloride)	77	Dry cleaning fluid; solvent for oils, fats, waxes; some use as insecticide and medicine
CF_2Cl_2	Difluorochloromethane (Freon-12)	−29	Fluid used in refrigerators; propellant for some aerosols
$CH_3CH_2{-}Cl$	Chloroethane (ethyl chloride)	13.1	Local anesthetic for minor surgery (freeze technique)
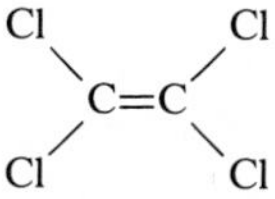	Tetrachloroethene	121	Dry cleaning fluid, degreasing solvent for automobile parts
$CH_2{=}CHCl$	Chloroethene (vinyl chloride)	14.0	Used in polymer synthesis (polyvinyl chloride)

Physical and Biological Properties of Organohalides

Since the halogens are heavy compared to hydrogen atoms, halogenated compounds have higher densities and boiling points than their hydrocarbon analogues. Like alkanes, they are not polar enough to dissolve in water, but they are good solvents for waxes and oils. Industrially, they make good solvents because they are readily available, stable, and are liquids at room temperature.

Biologically, the chlorinated hydrocarbons have received a great deal of attention since they are widely used as insecticides. The most notable of this group is DDT; others are dieldrin, aldrin, chlordane, and lindane (see Table 29–2). Several properties of chlorinated hydrocarbons make them threats to our environment. They have a wide range of biological activity, and thus are broad-spectrum poisons that affect many different organisms. They have great stability, so they remain in our environment for long periods of time and are not biodegraded to more harmless chemicals. They are quite mobile and have spread over the entire world, causing environmental problems. Finally, they become concentrated in the fatty tissue of organisms and are thus passed up the food chain to animals, birds, and humans. People with large concentrations of DDT in their bodies have developed darkened skin, eye discharge, and acne, but the longer-range effects of these chlorinated hydrocarbons are not presently known.

Recently, worldwide attention has been focused on the effects of organic halogens in the upper atmosphere. The fluorochloroalkanes are widely used as refrigerants and propellants in aerosol cans. When released, they rise into the upper

Table 29–2 Chlorinated hydrocarbons as insecticides

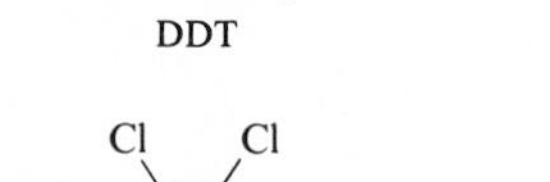

Cl
CH
CCl3
Cl

DDT

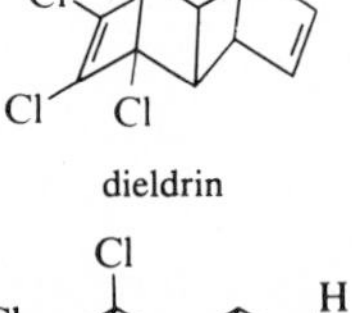

Cl
Cl
CH2
Cl
Cl
Cl
Cl

dieldrin

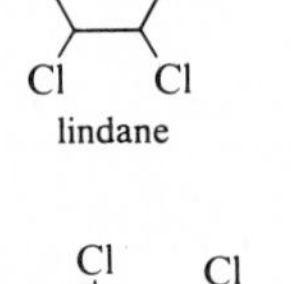

Cl
Cl
Cl
Cl
Cl
Cl

lindane

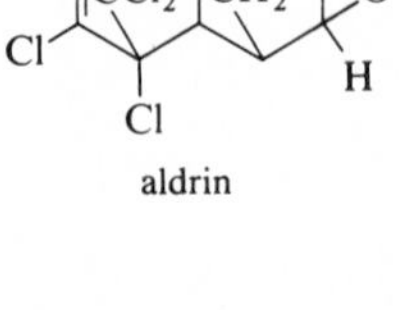

Cl
Cl
H
CCl2
CH2
O
Cl
H
Cl

aldrin

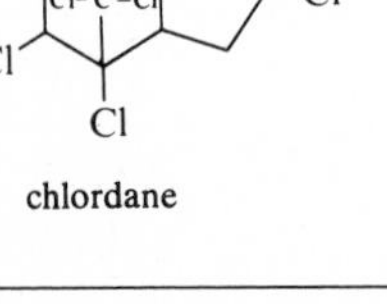

Cl
Cl
Cl
Cl-C-Cl
Cl
Cl
Cl

chlordane

atmosphere, and break down into fluorine and chlorine radicals (F·, Cl·). These radicals are capable of destroying the valuable ozone layer that protects the earth from the dangerous ultraviolet rays given off by the sun.

Chemical Reactions and Synthesis of Organohalides

The major methods of synthesis have already been introduced in earlier units but will be reviewed briefly at this time.

1. Halogenation of alkanes (substitution of X for H):

$$R{-}H + X_2 \xrightarrow[\text{heat}]{\text{Light or}} R{-}X + HX \quad (X = Br, Cl)$$

$$CH_3CH_3 + Cl_2 \xrightarrow{\Delta} CH_3CH_2{-}Cl + HCl$$

2. Addition of HX or X_2 to alkenes:

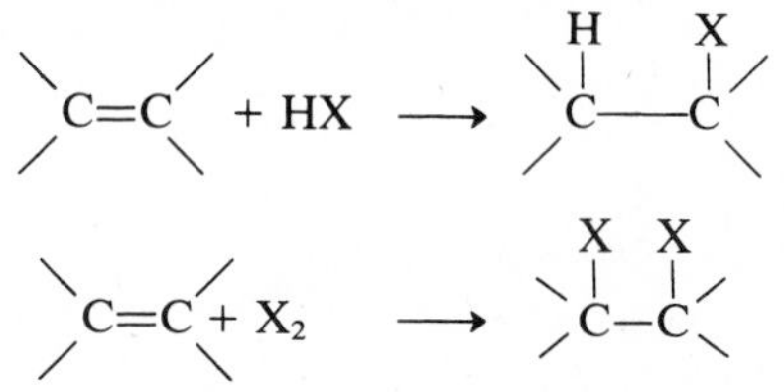

H
X
C=C
+ HX
C
C
X
X
C=C
+ X2
C
C

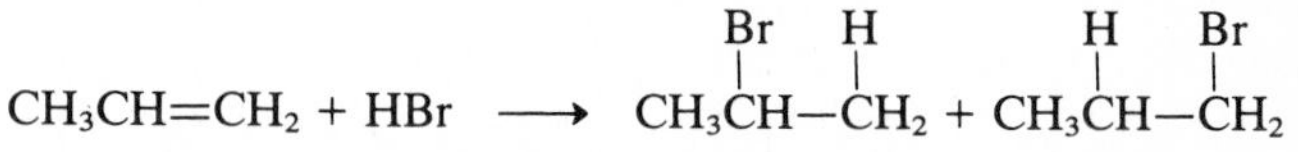

CH3CH=CH2 + HBr
Br
H
H
Br
CH3CH—CH2 + CH3CH—CH2

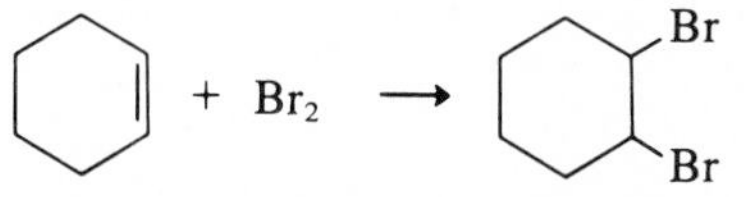

+ Br2
Br
Br

3. Halogenation of benzenes (substitution of X for H):

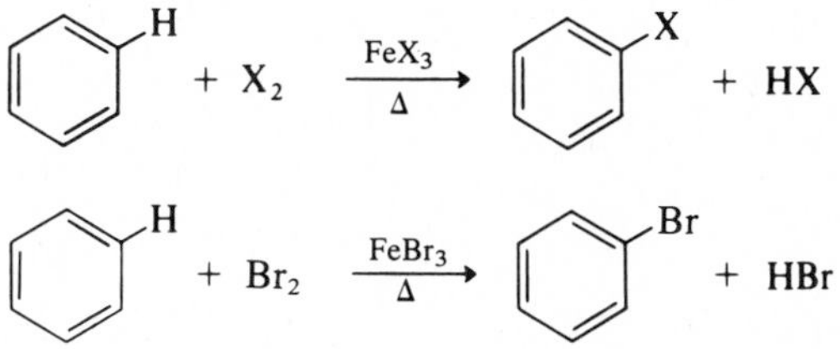

H
+ X2
FeX3
Δ
X
+ HX
H
+ Br2
FeBr3
Δ
Br
+ HBr

There are two main reactions that organohalides undergo: substitution and elimination.

Elimination

Treatment of halo-alkanes with strong bases can lead to elimination of HX and the formation of alkenes. The HX lost from the halo-alkane reacts with the base, so we do not see HX but rather water and halide ions.

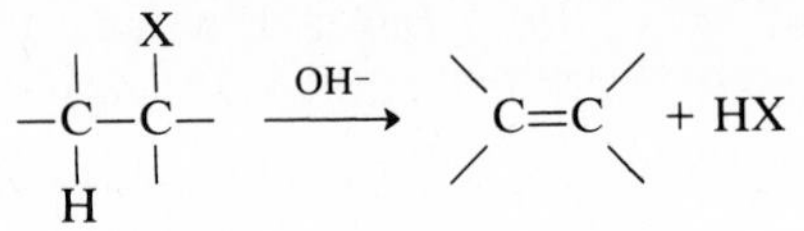

$$HX + OH^- \longrightarrow H_2O + X^-$$

The carbon bearing the X and an adjacent carbon become the alkene carbons. More than one alkene may be formed, depending on the structure of the organic reactant. For example, both 1-butene and 2-butene can be formed by reaction of 2-bromobutane, but they will not be formed in equal amounts. The major product is generally the alkene that has the greatest number of carbon atoms directly attached to the >C=C< group.

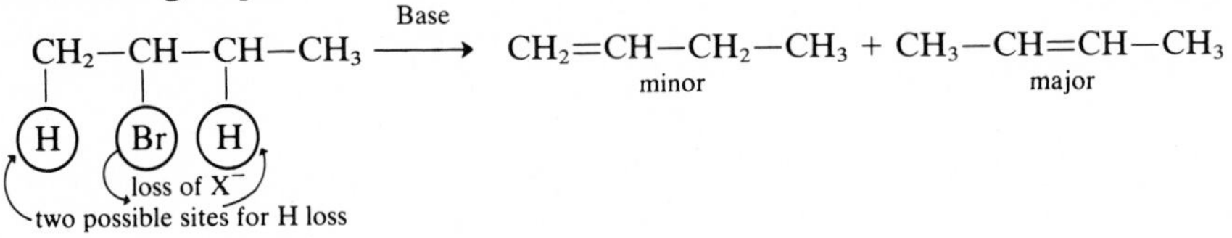
Base
minor
major
loss of X⁻
two possible sites for H loss

Table 29–3 Substitution reactions

Reaction	Synthesis of
$-C-X + OH^- \longrightarrow -C-OH$	Alcohols
$CH_3CH(Br)CH_3 + OH^- \longrightarrow CH_3CH(OH)CH_3$	
$-C-X + {}^-OR \longrightarrow -C-O-R$	Ethers
$CH_3CH_2Br + {}^-O-CH_3 \longrightarrow CH_3CH_2-O-CH_3$	
$-C-X + {}^-SH \longrightarrow -C-SH$	Thiols
$CH_3CH_2CH(Cl)-CH_3 + {}^-SH \longrightarrow CH_3CH_2CH(SH)CH_3$	
$-C-X + CN^- \longrightarrow -C-CN$	Cyanides

$-C-X + :N(H)- \longrightarrow -C-N-$	Amines
$CH_3CH(Br)-CH_3 + :NH_2CH_3 \longrightarrow CH_3-CH(CH_3)-NHCH_3$	

Substitution

In substitution reactions of organohalides, a reagent (Z) takes the place of Br, Cl, or I on the molecule. The Z reagents are called *nucleophiles* (nucleus lovers) and generally attack centers of positive charge. They generally possess a negative charge or at least a readily available pair of unshared electrons. Common nucleophiles include CN^-, ^-O-R, OH^-, SH^-, and $:\overset{|}{\underset{|}{N}}-$. Table 29–3 lists examples of the reactions. In most cases *Z becomes attached to the carbon atom originally bearing the halogen atom.* Since a large number of other classes of organic compounds can be synthesized by substitution reactions on alkyl halides, they are quite useful chemicals to synthetic chemists.

UNIT 29 Problems and Exercises

29–1. Give both the IUPAC name and the common name for the following.

(a) CH_3CH_2-I

(b) $CH_3CH(Br)-CH_3$

(c) 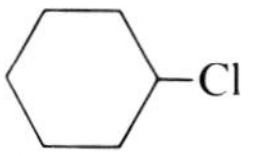

(d) $CH_3CH(Br)-CH_2CH_3$

(e) $CH_3-CH(CH_3)-CH_2-Cl$

(f) $CH_3-C(CH_3)_2-I$

(g) 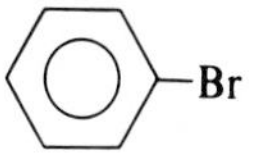

29–2. Draw and name all the isomers of $C_5H_{11}Br$.

29–3. Complete the following reactions. If there is more than one product, indicate which is the major product.

(a) $CH_4 + Cl_2 \xrightarrow{\Delta}$

(b) 2-methyl-2-butene + HBr $\longrightarrow$

(c) 1,3-cyclohexadiene + $2Br_2$ $\longrightarrow$

(d) Benzene + Br_2 $\xrightarrow[\Delta]{FeBr_3}$

(e) 2-bromo-2-methylbutane $\xrightarrow{\text{Base}}$

(f) 1-chloro-1-methylcyclohexane + base $\longrightarrow$

29–4. Use the appropriate reagents to synthesize each of the following compounds. Synthesis may involve more than one step.

(a) 1-bromobutane from butane

(b) 2-bromobutane from 1-butene

(c) 2-bromobutane from 1-butanol

(d) 2,3-dichlorobutane from 2-butanol

(e) *p*-chlorotoluene from toluene

(f) Bromophenylmethane from toluene

(g) 2-butanol from 1-bromobutane

29–5. Give the products resulting from reaction of isopropyl bromide with the following.

(a) OH^-

(b) CH_3O^-

(c) $CH_3-CH(CH_3)-O^-$

(d) CN^-

(e) $^-SCH_3$

(f) NH_3

(g) CH_3-NH_2

(h) 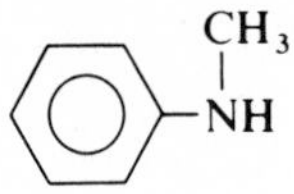

29–6. Complete the following substitution reactions.

(a) 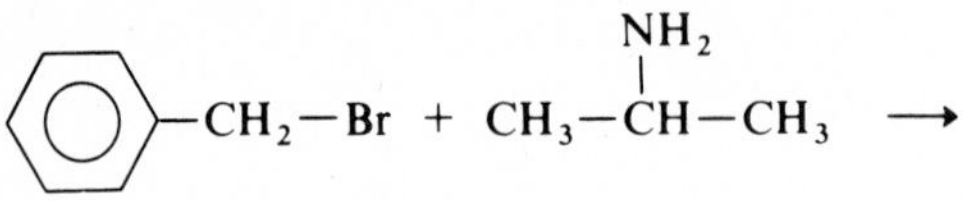

(b) $CH_3CH_2-CH(I)-CH_3 + CH_3CH_2-O^- \longrightarrow$

(c) 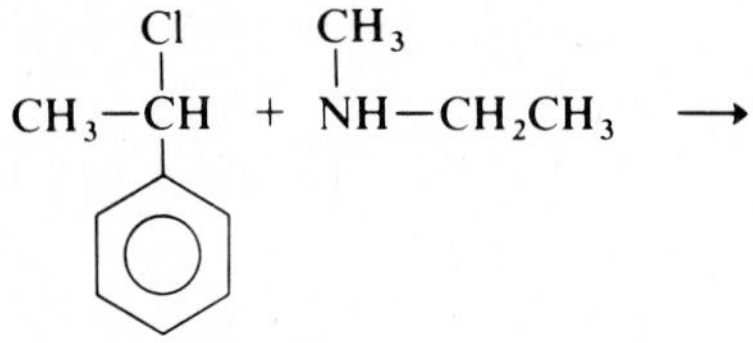

(d) 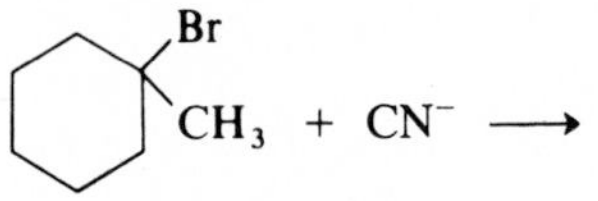

(e) $CH_3CH_2-Br + N(CH_3)_2-CH_3 \longrightarrow$

29–7. Use the appropriate reagents to synthesize the following compounds, starting with the reactant given. Each synthesis may require more than one step.

(a) Make $CH_3CH(CN)-CH_3$ from $CH_3CH{=}CH_2$.

(b) Make $CH_3CH(OH)-CH_3$ from $CH_3CH_2-CH_2-Br$.

(c) Make 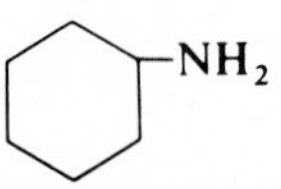 from 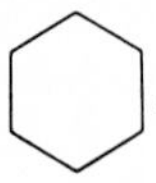.

(d) Make $CH_3-CH(CN)-CH_3$ from $CH_3CH_2C(H){=}O$.

29–8. A compound with the formula of C_4H_9Br reacts with a strong base to yield only one alkene (C_4H_8). Give a possible structure for both compounds.

29–9. Give the structure of a compound with the formula of C_8H_{18} (an alkane) that would yield only one monobromo compound upon treatment with Br_2 and light.

UNIT 30 ETHERS, THIOLS, SULFIDES, AND DISULFIDES

Water is the parent structure for alcohols and ethers. In alcohols one of the hydrogens of the water molecule has been replaced by an alkyl group. In ethers, both hydrogens have been replaced. **Ethers** are characterized by the presence of a C—O—C bond.

H—O—H	R—O—H	R—O—R′
	alcohol	ether

If we use H_2S as the base structure instead of water and replace the hydrogens with alkyl groups, **thiols** and **sulfides** are formed.

H—S—H	R—S—H	R—S—R′
	thiols (mercaptans)	sulfides (thioethers)

In the chemistry of oxygen-containing compounds a $-\overset{|}{\underset{|}{C}}-O-O-\overset{|}{\underset{|}{C}}-$ group is called a **peroxide**. In sulfur chemistry, a $-\overset{|}{\underset{|}{C}}-S-S-\overset{|}{\underset{|}{C}}-$ group is called a **disulfide**. Actually, there exist many more sulfur compounds that are analogues of the common oxygen-containing organic compounds, but in this unit we shall discuss only ethers and the sulfur analogues of alcohols and ethers.

Nomenclature of Ethers

The common names for simple ethers are derived by naming the two alkyl groups attached to the oxygen and adding the word ether. If the groups are identical, the prefix *di-* may be used.

CH_3-O-CH_3	$CH_3-O-CH_2CH_3$	$CH_3-\underset{}{\overset{CH_3}{\overset{\vert}{C}}}H-O-CH_2CH_3$
dimethyl ether or methyl ether	methyl ethyl ether	ethyl isopropyl ether

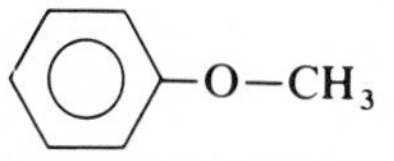

methyl phenyl ether

In more complex ethers where an alkyl group has no simple name, the R—O group is often regarded as a substituent of the other portion of the molecule. These R—O— groups are called **alkoxy** groups, and their names are placed at the beginning of the name along with other substituents.

$CH_3{-}O{-}$
methoxy

$CH_3CH_2{-}O{-}$
ethoxy

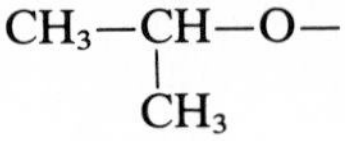
isopropoxy

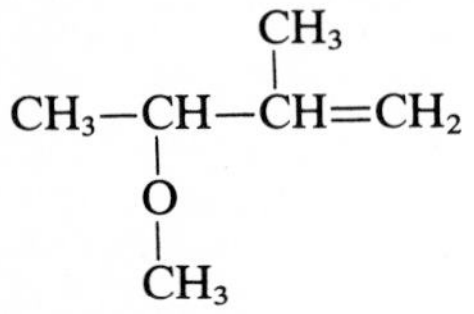
2-methyl-3-methoxy-1-butene

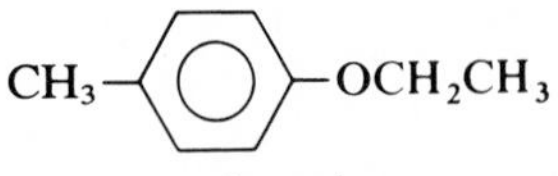
p-ethoxytoluene

$CH_3O{-}CH_2CH_2CH_2{-}OCH_3$
1,3-dimethoxypropane

Nomenclature of Sulfur Analogues

The term **thio** in a name always denotes replacement of an oxygen by a sulfur atom. A common method of naming sulfur-containing compounds is to name them as you would the corresponding oxygen compound, and then to add thio before the unit that denotes the oxygen linkage.

$CH_3CH_2{-}O{-}H$
ethanol

$CH_3{-}O{-}CH_2CH_3$
methyl ethyl ether

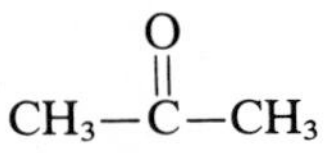
2-propanone

$CH_3{-}CH_2{-}S{-}H$
ethanethiol
(ethyl mercaptan)

$CH_3{-}S{-}CH_2{-}CH_3$
methyl ethyl thioether
(methyl ethyl sulfide)

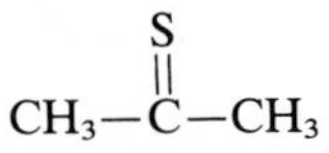
2-propanthione

In naming sulfur analogues of alcohols, the term **mercaptan** historically was used, but the suffix **thiol** is now more prevalent. It is combined with the parent just like the suffix *-ol* was added to alcohols. The only difference is that the *e* from the alkane name remains in thiols.

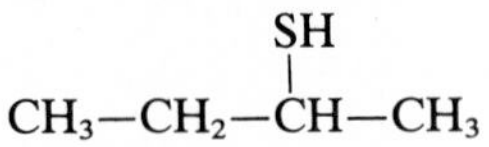
2-butanethiol
(2-butyl mercaptan)

4-methyl-1-cyclohexanethiol

In naming sulfur analogues of ethers, the word **sulfide** is added after naming the alkyl groups attached to the sulfur:

$CH_3{-}S{-}CH_3$
dimethyl sulfide

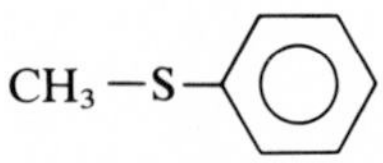
methyl phenyl sulfide

As with more complex ethers, the R—S group is often considered a substituent. It is then called an alkylthio group and is often placed in parentheses for clarity.

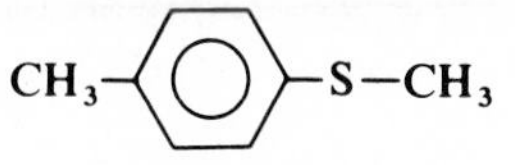

p-(methylthio)toluene

$$CH_3{-}\underset{\displaystyle CH_3}{\underset{|}{CH}}{-}CH_2{-}CH_2{-}S{-}CH_2CH_3$$

3-methyl-1-(ethylthio)butane

$$CH_3{-}\underset{\displaystyle CH_3}{\underset{|}{CH}}{-}S{-}CH_2CH_2CH_2CH_3$$

1-isopropylthiobutane

Compounds with —S—S— linkages are called **disulfides**. They are named by adding the word **disulfide** to the names of the alkyl groups attached to the sulfur atoms.

$$CH_3{-}CH_2{-}S{-}S{-}CH_3$$

methyl ethyl disulfide

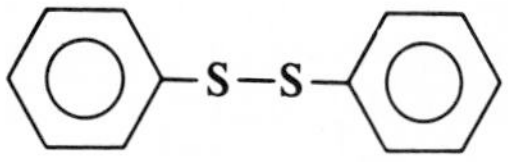

diphenyl disulfide

Physical Properties of Ethers

The physical properties of some ethers are listed in Table 30–1 along with alcohols for comparison. One important fact stands out immediately. Ethers have much lower boiling points than alcohols with the same molecular weights. This fact is easily explained by the absence of hydrogen bonding in them. Without this extra attraction between molecules, boiling occurs at much lower temperatures.

Ethers have a polarity somewhere between nonpolar alkanes and highly polar alcohols. They are not as soluble as alcohols in water, but generally are more soluble

Table 30–1 Comparison of some physical properties of ethers with alcohols

Compound	Name	MW	FP (°C)	BP (°C)	Soluble in H_2O	Soluble in hexane
$CH_3{-}O{-}CH_3$	Methyl ether	46	−138	−23	Completely	Completely
$CH_3CH_2{-}OH$	Ethanol	46	−114	78	Completely	Insoluble
$CH_3{-}CH_2{-}O{-}CH_3$	Methyl ethyl ether	60	—	10.8	Completely	Soluble
$CH_3CH_2CH_2{-}OH$	1-propanol	60	−126	97	Completely	Slightly soluble
$CH_3CH_2{-}O{-}CH_2CH_3$	Ethyl ether	74	−116	34.5	10 g/100 mℓ	Soluble
$CH_3CH_2CH_2CH_2{-}OH$	1-butanol	74	−89	117	8 g/100 mℓ	Soluble
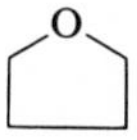	Tetrahydrofuran	72	−65	64	Soluble	Soluble
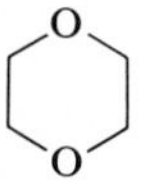	1,4-dioxane	88	120	101	Soluble	Soluble

in alkanes. Because of this intermediate polarity, they are often used as solvents. Another reason that they make good solvents is that ethers are chemically relatively inert. They undergo few reactions and can be easily removed from the reaction mixture. One danger is that they are highly flammable. Diethyl ether boils at only 10°C above room temperature and can burn explosively if ignited.

Biological Properties and Uses of Ethers

In the past, one of the more common medical uses of small ethers was as anesthetics. Diethyl ether has been used routinely in operating rooms since 1900, but is being replaced by safer chemicals today. Divinyl ether ($CH_2{=}CH{-}O{-}CH{=}CH_2$) is still widely used by doctors for office use, as it has less serious after-effects and is a rapid-acting, short-term anesthetic. Methyl propyl ether is claimed to be less irritating and more potent than ethyl ether, but it is not widely used.

Because of its chemical inertness, the ether group is not a significant functional group in the metabolic processes that occur in our bodies. This is not to say that ether linkages are not present. Many drugs and other organic compounds found in the body do possess ether linkages, but usually these sites remain unchanged. Often these linkages involve ethoxy or methoxy groups.

Preparations of Ethers

There are several methods of preparing ethers, but we shall discuss only two. Treatment of alcohols with acid catalysts results in the formation of symmetrical ethers along with alkenes. In this reaction, water is eliminated as the $-\overset{|}{\underset{|}{C}}-O-\overset{|}{\underset{|}{C}}-$ bond is formed. It is of limited synthetic usefulness since only symmetrical ethers are easily formed, and there are often side products.

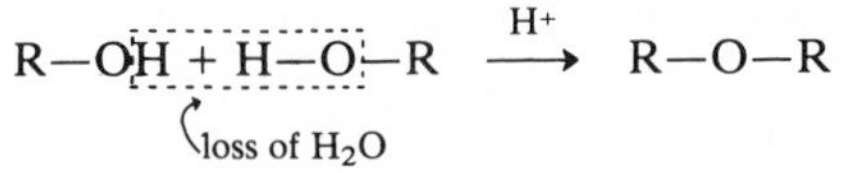

$$2CH_3CH_2OH \xrightarrow{H^+} CH_3CH_2{-}O{-}CH_2CH_3$$

A second more useful method of ether preparation is called the *Williamson ether synthesis.* It involves a substitution reaction between the salt of an alcohol and an alkylbromide. This reaction is an example of a substitution reaction, which was introduced in unit 29.

$$R{-}OH \xrightarrow{Na} R{-}O^- \ Na^+ \quad \text{(salt of alcohol)}$$

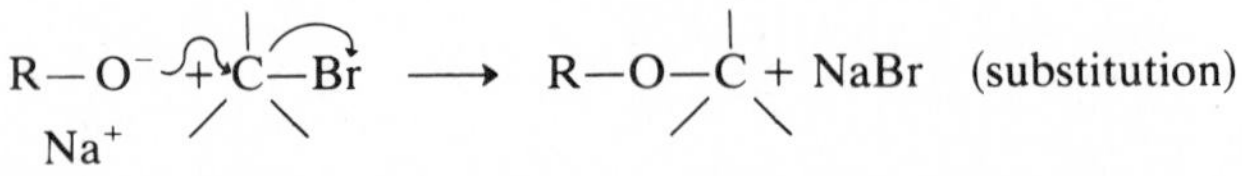

There are two ways of making any unsymmetrical ether by this method. Either alkyl group may come from the alcohol or the alkyl halide.

Example 1: Synthesize methyl ethyl ether. Either method should work.

$$CH_3OH \xrightarrow{Na} \underset{Na^+}{CH_3O^-} \xrightarrow{CH_3CH_2Br} CH_3{-}O{-}CH_2CH_3$$

or

$$CH_3CH_2OH \xrightarrow{Na} \underset{Na^+}{CH_3CH_2O^-} \xrightarrow{CH_3Br} CH_3{-}O{-}CH_2CH_3$$

Example 2: Synthesize methyl isopropyl ether.

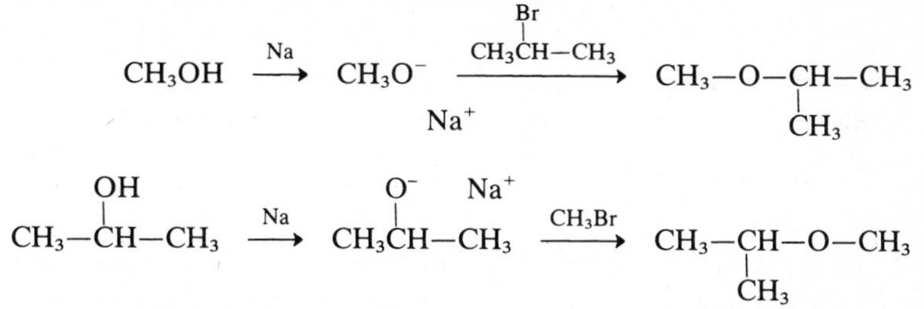

Substitution occurs most smoothly at primary halogen centers so the second method in Example 2 is preferred.

Properties and Reactions of Sulfur Analogues

Thiols have much lower boiling points than alcohols. They are comparable to other compounds of similar molecular weight that hydrogen bond weakly. Sulfur is much less electronegative than oxygen (about the same as carbon), so there is little hydrogen bonding to increase the boiling point. The sulfides act in similar fashion. Ethyl sulfide (GMW = 90 g) boils at 92°C, about the same as an ether of similar weight (propyl ether = 91°C). Some examples are as follows:

$CH_3{-}S{-}H$, 6°C

$CH_3CH_2{-}S{-}H$, 37°C

$CH_3{-}S{-}CH_3$, 37.3°C

$CH_3CH_2{-}S{-}CH_3$, 66°C

$CH_3CH_2{-}S{-}CH_2CH_3$, 92°C

$CH_3{-}S{-}S{-}CH_3$, 116°C

$CH_3CH_2{-}S{-}S{-}CH_2CH_3$, 153°C •

Since H_2S is more acidic than water

$$[K_a(H_2S) = 1.1 \times 10^{-7};\ K_a(H_2O) = 1 \times 10^{-14}]$$

we would expect thiols to be more acidic than alcohols. Ethanethiol is completely ionized by hydroxide, but ethanol is unaffected.

$$(K_a = 10^{-20})\quad CH_3CH_2{-}OH + OH^- \not\rightarrow CH_3CH_2{-}O^-$$
$$(K_a = 10^{-10})\quad CH_3CH_2{-}SH + OH^- \longrightarrow CH_3CH_2{-}S^- + H_2O$$

A most impressive property of thiols is their intensely disagreeable odor, which makes their use in laboratories undesirable. Hydrogen sulfide smells like rotting eggs and is more toxic than hydrogen cyanide. 1-butanethiol is partially responsible for the odor of a skunk. Since the odors of thiols are so readily detectable in trace amounts, they are often deliberately added to heating gas as a safety measure against leaks.

Thiols are prepared by reaction of HS^- with alkyl iodides. Sulfides are prepared as are ethers by reaction of the thiol salt with alkyl bromides.

$$R{-}CH_2{-}I + H{-}S^-Na^+ \longrightarrow R{-}CH_2{-}SH$$
$$R{-}S^- + R{-}Br \longrightarrow R{-}S{-}R'$$

The reaction of thiols with weak oxidizing agents produces disulfides. Conversely, disulfides can be reduced back to thiols by treatment with reducing agents.

$$2R{-}S{-}H \underset{\text{Reduction}}{\overset{\text{Oxidation}}{\rightleftharpoons}} R{-}S{-}S{-}R$$

These processes are extremely important biologically as they are involved in the formation of the overall structure of many proteins and enzymes. The amino acid cysteine is a thiol; when subjected to weak oxidizing agents, two cysteine units combine through a disulfide bridge to make cystine, a disulfide.

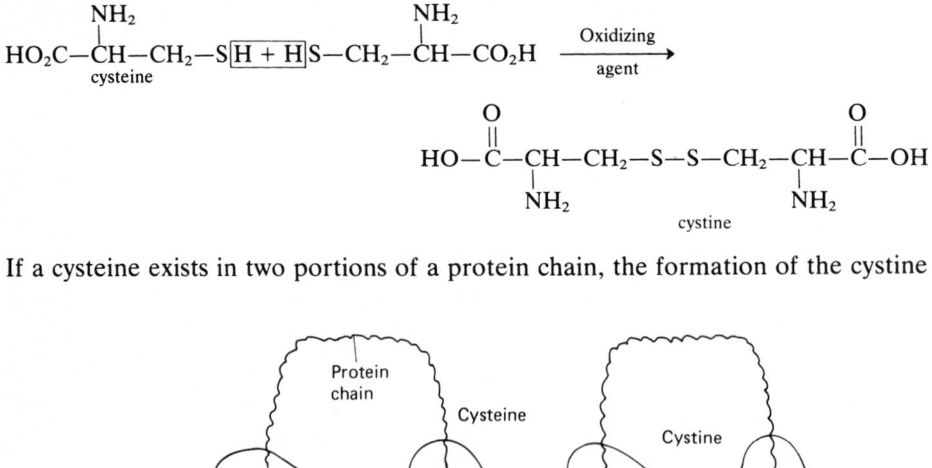
cysteine
Oxidizing
agent
cystine

If a cysteine exists in two portions of a protein chain, the formation of the cystine

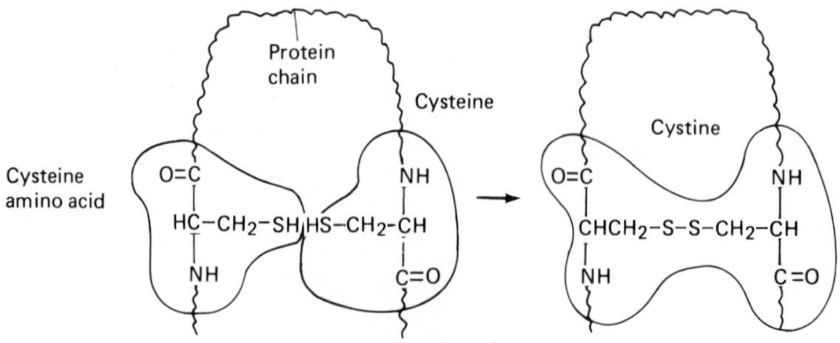
Protein
chain
Cysteine
Cysteine
amino acid
Cystine

Figure 30–1

—S—S— link will bond these portions of the chain together (see Figure 30–1). These types of disulfide bonds occur quite often in proteins. The cleavage and formation of multiple disulfide bonds is also the basis for curling of hair by permanent wave lotion. First, the disulfide bonds are cleaved and the hair is set in curlers. Next neutralizer is added to allow the disulfide bonds to reform and hold the protein molecules in the configuration set by the rollers (see Figure 30–2).

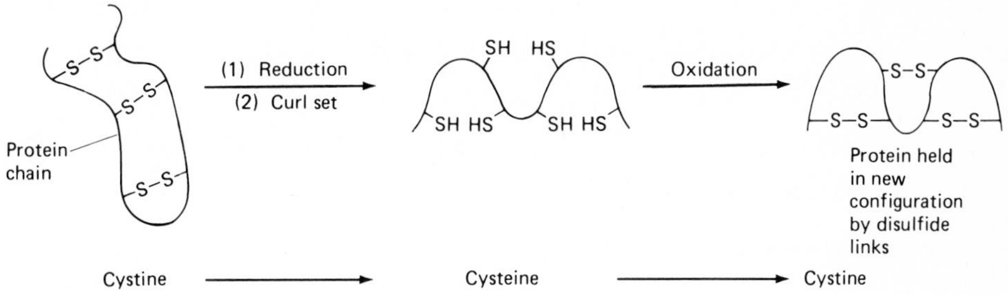

Protein chain
(1) Reduction
(2) Curl set
Oxidation
Protein held in new configuration by disulfide links
Cystine
Cysteine
Cystine

Figure 30–2

UNIT 30 Problems and Exercises

30–1. Name the following compounds.

(a) $CH_3CH_2-O-CH_3$

(b) $CH_3-CH(CH_3)-O-CH_2CH_3$

(c) 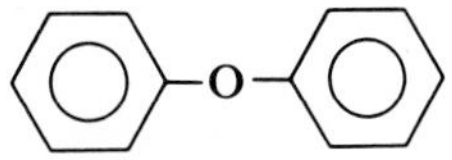

(d) $CH_3-CH=CH-CH_2-O-CH_3$

(e) 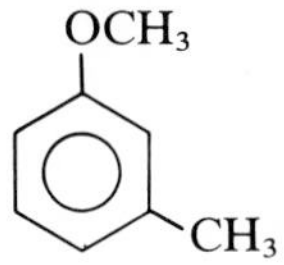

(f) $CH_3CH_2CH_2-S-H$

(g) 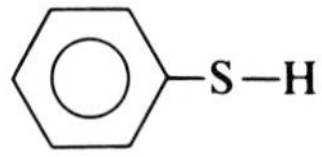

(h) $CH_3-S-CH_2CH_2CH_3$

(i) $CH_3CH_2CH_2-S-S-CH_3$

(j) 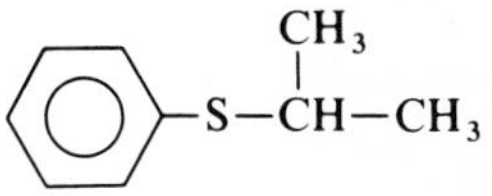

30–2. Draw structures for the following compounds.

(a) Phenyl ethyl ether
(b) 3-methoxypentane
(c) *sec*-butyl *n*-pentyl ether
(d) *p*-ethoxyaniline
(e) 1-butanethiol
(f) Diisopropyl sulfide
(g) Diethyl sulfide
(h) 4-*t*-butoxy-1-hexene

30–3. Rank the following compounds in terms of increasing acidity.

CH_3-O-CH_3	CH_3-CH_2-OH	CH_3-CH_2-SH
(a)	(b)	(c)

30–4. Rank the following compounds in terms of increasing boiling points and explain your reasoning.

$$\underset{(a)}{CH_3CH_2CH_2OH} \qquad \underset{(b)}{CH_3CH_2CH_2{-}SH} \qquad \underset{(c)}{CH_3CH_2{-}O{-}CH_3}$$

30–5. Complete the following reactions.

(a) $2CH_3OH \xrightarrow{H^+}$

(b) $CH_3{-}\overset{\displaystyle CH_3}{\underset{\displaystyle CH_3}{\overset{|}{\underset{|}{C}}}}{-}OH + Na \longrightarrow$

(c) 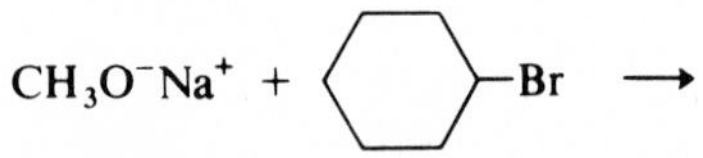

(d) $CH_3\overset{\displaystyle OH}{\overset{|}{C}}HCH_3 \xrightarrow{Na} A \xrightarrow{CH_3CH_2Br} B$

(e) $CH_3CH_2{-}Br + HS^-Na^+ \longrightarrow$

(f) 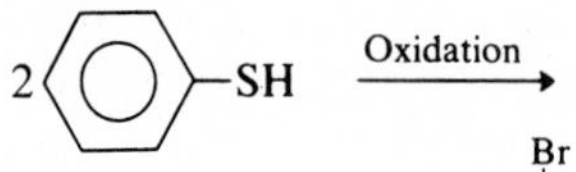

(g) 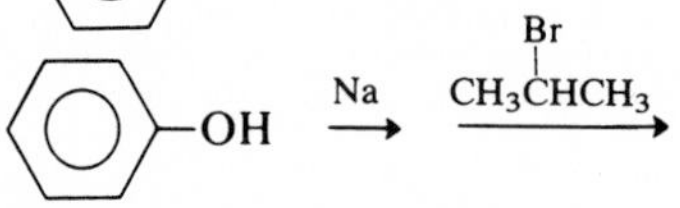

30–6. Write equations for the Williamson ether synthesis of methyl ethyl ether from the following.
(a) Ethyl alcohol and methyl bromide
(b) Methyl alcohol and ethyl bromide

30–7. Devise a simple chemical test to help you distinguish between an ether and an alcohol. Tell how the results of the test would help you to decide.

30–8. What simple method could you use to distinguish between an ether and a thiol?

30–9. Give the structures of the reactant alcohol and alkyl bromide necessary to synthesize the following ethers by the Williamson ether synthesis. In each case there are two possible sets of reactants.

(a) $CH_3CH_2{-}O{-}\overset{\displaystyle CH_3}{\overset{|}{C}}H{-}CH_3$

(b) 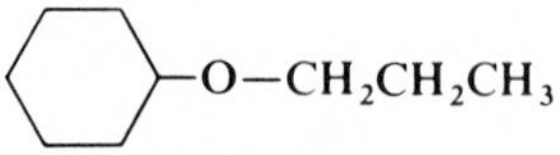

(c) *n*-butyl-ethyl ether

(d) 3-methoxypentane

(e) $CH_3{-}\overset{\displaystyle CH_3}{\underset{\displaystyle CH_3}{\overset{|}{\underset{|}{C}}}}{-}O{-}\overset{\displaystyle CH_3}{\overset{|}{C}}H{-}CH_2CH_3$

30–10. Use the Williamson ether synthesis to make the following compounds from the reactants given. Several steps may be required before using the Williamson ether synthesis step.

(a) Make 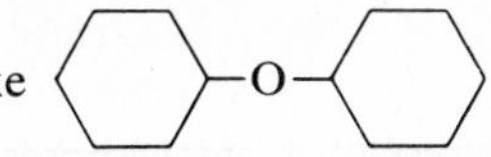 from 2 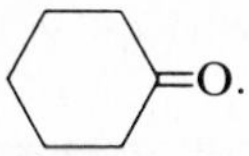

(b) Make $CH_3{-}\overset{\displaystyle CH_3}{\overset{|}{C}}H{-}O{-}CH_2CH_2CH_3$ from $2CH_3{-}CH_2{-}CH_2{-}Br$.

(c) Make $CH_3{-}\overset{\displaystyle CH_3}{\overset{|}{C}}H{-}O{-}\overset{\displaystyle CH_3}{\overset{|}{C}}H{-}CH_3$ from $2(CH_3CH_2\overset{\displaystyle H}{\overset{|}{C}}{=}O)$.

(d) Make 2-methoxybutane from alcohols.

30–11. A compound with the formula of $C_4H_{10}O$ does not react with sodium to give off hydrogen gas. It can be synthesized from methyl alcohol and *n*-propyl alcohol. Give its structure and show how it was synthesized.

30–12. A compound was formed by treatment of *n*-propyl alcohol with acid. It was analyzed and found to contain 70.6% carbon, 13.7% hydrogen, and 15.7% oxygen. Give a possible structure for the compound.

30–13. A vile-smelling compound was formed in a reaction mixture by addition of Na^+HS^-. This compound was analyzed and found to contain 53.3% carbon, 11.1% hydrogen, and 35.5% sulfur. Upon treatment with a mild oxidizing agent, it formed a second compound, which was analyzed and found to contain 53.9% carbon, 10.1% hydrogen, and 35.9% sulfur. Its molecular weight was 178 g. Give the structures of both compounds.

UNIT 31 AMINES

So far we have considered organic compounds containing mostly carbon, oxygen, and hydrogen. Another important member of this select group is nitrogen. Alcohols and ethers are derivatives of water; thiols and sulfides are derivatives of hydrogen sulfide; amines are derivatives of ammonia. Notice that the term **amines** encompasses all derivatives of ammonia, regardless of how many hydrogens have been replaced.

H—O—H water	R—O—H alcohol		R—O—R ether
H—S—H hydrogen sulfide	H—S—R thiols		R—S—R sulfides
NH_3 ammonia	$R{-}NH_2$	$R{-}\overset{\displaystyle R}{\overset{|}{N}}H$	$R{-}\underset{\displaystyle R}{\underset{|}{N}}{-}R$

amines (brace under $R{-}NH_2$, R_2NH, R_3N)

Amines are classified into several separate groups depending upon the type of attachments present. **Primary**, **secondary**, and **tertiary** amines have one, two, and three carbon atoms directly attached to nitrogen. **Quaternary ammonium salts** have four groups on nitrogen and thus are derivatives of ammonium ion. Aromatic and complex ring systems that contain nitrogen are called **heterocyclic** aromatic amines. Figure 31–1 gives some examples. Notice that the terms primary, secondary, and tertiary have slightly different meanings in amine chemistry than they do in alcohol chemistry.

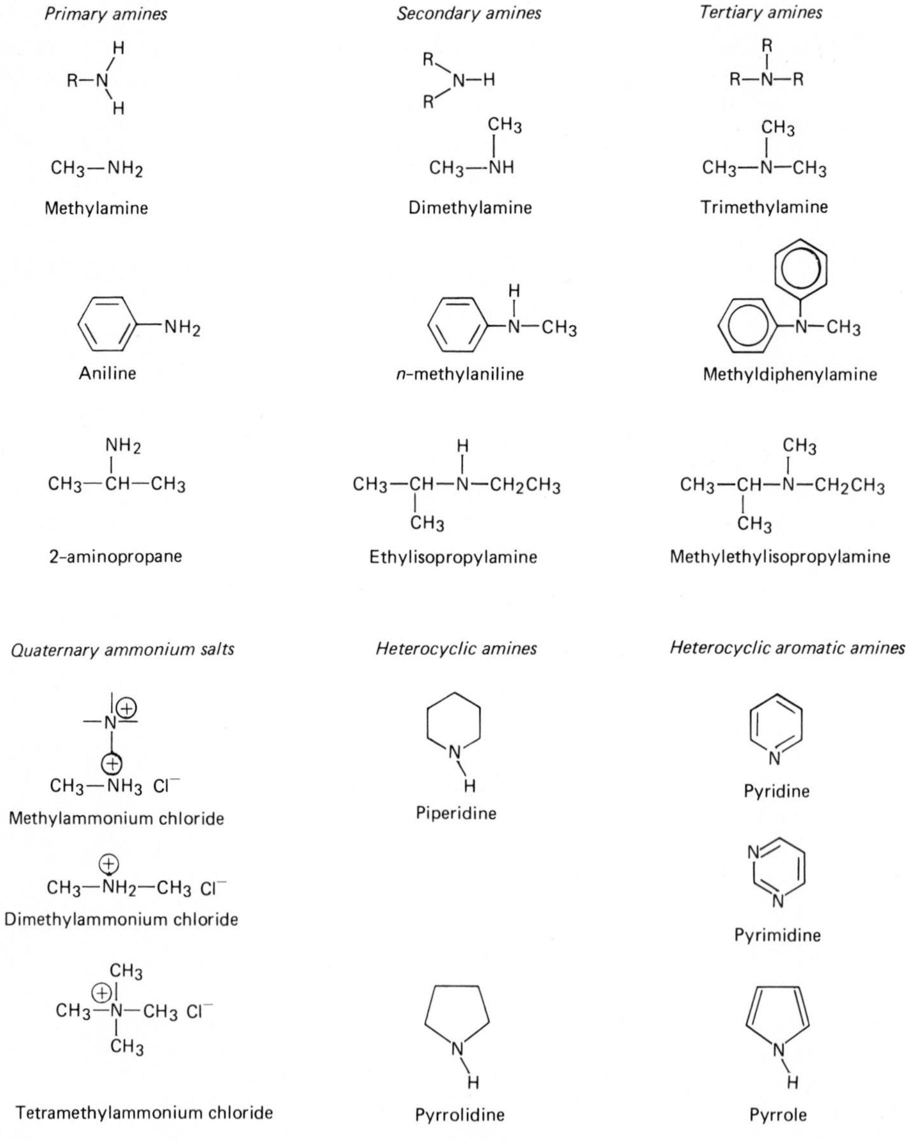

Primary amines
Secondary amines
Tertiary amines
Methylamine
Dimethylamine
Trimethylamine
Aniline
n-methylaniline
Methyldiphenylamine
2-aminopropane
Ethylisopropylamine
Methylethylisopropylamine
Quaternary ammonium salts
Heterocyclic amines
Heterocyclic aromatic amines
Methylammonium chloride
Piperidine
Pyridine
Dimethylammonium chloride
Pyrimidine
Tetramethylammonium chloride
Pyrrolidine
Pyrrole

Figure 31–1 Types of amines.

Nomenclature of Amines

Simple amines, for which all groups attached to the nitrogen have simple names, are often named like ethers. The names of the substituents precede the word *amine.*

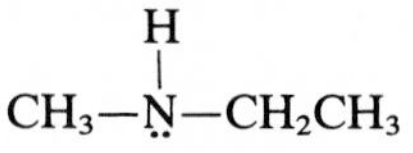

methylethylamine

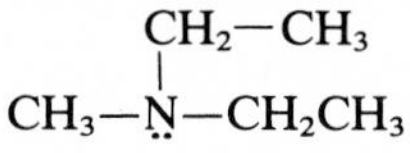

methyldiethylamine

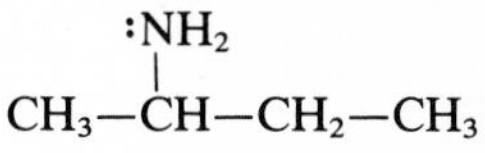

sec-butylamine

More complex amines are named like alcohols. The prefix *amino-* or suffix *-amine* may be placed before or after the parent chain. The location of nitrogen is designated by the carbon number to which it is attached.

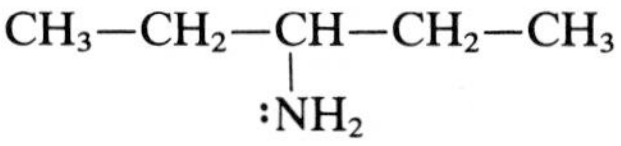

3-aminopentane or
3-pentanamine

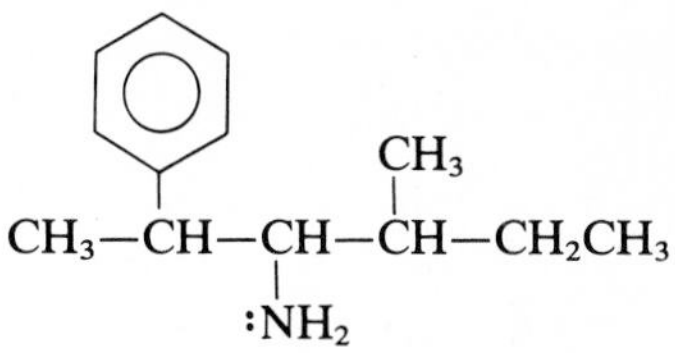

4-methyl-2-phenyl-3-aminohexane

When substituents occur on nitrogen rather than carbon, an **N-** is placed in front of that substituent to note its location.

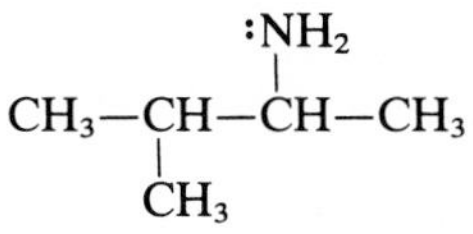

3-methyl-2-aminobutane

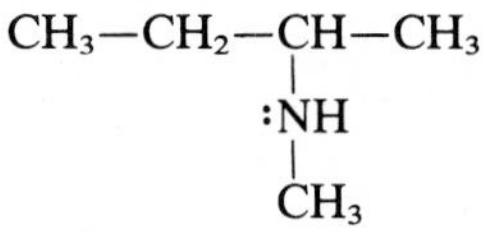

N-methyl-2-aminobutane

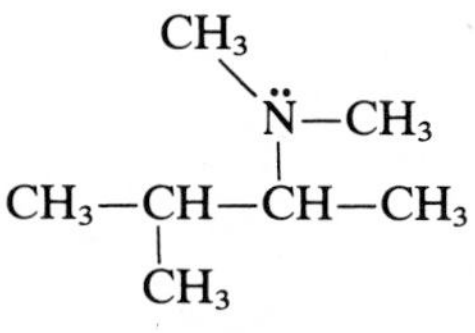

3-methyl-*N,N*-dimethyl-2-aminobutane

Benzene rings to which a nitrogen is attached are often named as derivatives of aniline.

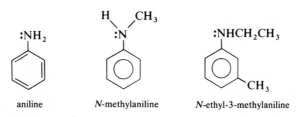

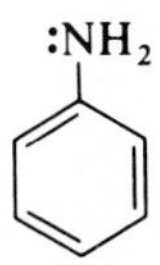

aniline

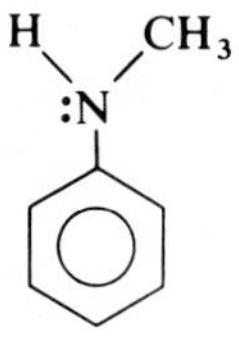

N-methylaniline

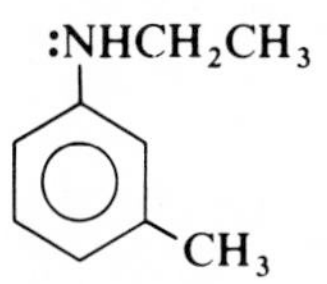

N-ethyl-3-methylaniline

Sources of Amines: Nitrogen Fixation

By far the largest fraction of nitrogen near the earth's surface is present in the form of nitrogen gas, N≡N. As N_2, nitrogen is relatively inert, but certain organisms have the ability to convert N_2 into ammonia.

$$N_2 + 6H \xrightarrow[\text{in plants}]{\text{Microorganisms}} 2NH_3$$

This process, called **nitrogen fixation**, provides the source of nitrogen needed to sustain life. The ammonia formed is converted into amines and other nitrogen compounds. Industrially, ammonia is synthesized by high-temperature hydrogenation of N_2. The NH_3 is then converted into amines by other processes:

$$N_2 + 3H_2 \xrightarrow[\text{High pressure, 500°C}]{\text{Metal catalysts}} 2NH_3$$

Physical Structure and Properties of Amines

The nitrogen atom in most amines is sp^3 hybridized. Three of the orbitals form sigma bonds to carbon or hydrogen, while the fourth contains the unshared pair of electrons. Since nitrogen is not nearly as electronegative as oxygen or fluorine, this unshared pair of electrons is much freer to participate in chemical reactions than are unshared electron pairs in alcohols (see Figure 31–2). Indeed much of the chemistry of amines results from the involvement of this unshared pair of electrons.

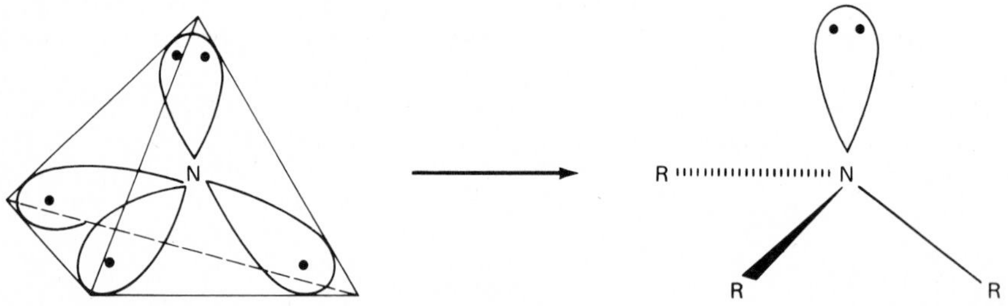
N
R
N
R
R

Figure 31–2

Amines are one of the few types of organic compounds that are *basic*. In the presence of water and other acids, they are converted to ammonium ions. The unshared pair of electrons provides both electrons for the new covalent bond and thus nitrogen must have a positive charge on it.

$$\text{H—}\ddot{\text{N}}\text{(H)—H} + \text{HCl} \longrightarrow \text{H—N}^+\text{(H)(H)—H}\ \ \text{Cl}^-$$

In water, the reaction is reversible so we have to be aware of equilibrium constants.

$$R{-}NH_2 + H_2O \rightleftharpoons R{-}\overset{+}{N}H_3 + OH^-$$

$$K_b = [H_2O]K_{eq} = \frac{[R{-}\overset{+}{N}H_3][OH^-]}{[R{=}\ddot{N}H_2]} \sim 10^{-5}$$

The equilibrium constants for amines show that they are stronger bases than water. When working with amines, we have to be careful to note the pH of the solution, since either the free amine or the ammonium salt may be present. At pH values less than 9, [[OH^-] < 10^{-5} M], over 50% of the amine is present as the ammonium salt rather than the free amine. This basicity of amines provides one qualitative test for their presence. Any organic compound insoluble in water but soluble in dilute acid is probably an amine. The acid converts the amine into the more soluble ammonium salt.

$$\underset{\text{insoluble in } H_2O}{R{-}\ddot{N}H_2} + HCl \longrightarrow \underset{\text{soluble in } H_2O}{R{-}\overset{+}{N}H_3\ \ Cl^-}$$

Like alcohols, amines can hydrogen bond and thus are more soluble in water and have higher boiling points than non-hydrogen-bonding compounds. The polarity of the >N—H bond is not as great as the O—H bond, so the effect of hydrogen bonding in amines is not as pronounced as in alcohols. Note in Table 31–1 that amines do not boil at nearly as high temperatures as alcohols of similar molecular weight. Note also that boiling points of amines decrease as the ability to hydrogen bond decreases. Primary amines boil at higher temperatures than secondary and tertiary amines of the same molecular weight.

Amines differ dramatically from alcohols in their odor, and the smaller amines have quite disagreeable odors. Trimethylamine smells like dead salmon. Two

Table 31–1 Boiling points of some amines

	Name	BP (°C)	Comparison with alcohols and ethers	BP (°C)
Primary amines				
$CH_3{-}\ddot{N}H_2$	Methylamine	−6.3	$CH_3{-}OH$	67
$CH_3CH_2{-}\ddot{N}H_2$	Ethylamine	16.6	$CH_3CH_2{-}OH$	78
$CH_3CH_2CH_2CH_2CH_2CH_2{=}\ddot{N}H_2$	1-hexylamine	130	$CH_3(CH_2)_4CH_2{-}OH$	158
Secondary amines				
$CH_3{-}\ddot{N}H{-}CH_3$	Dimethylamine	7.4	$CH_3{-}O{-}CH_3$	−24
$CH_3{-}CH_2{-}\ddot{N}H{-}CH_2CH_3$	Diethylamine	56.3	$CH_3CH_2{-}O{-}CH_2CH_3$	36
$CH_3CH_2CH_2{-}\ddot{N}H{-}CH_2CH_2CH_3$	Di-n-propylamine	110		
Tertiary amines				
$CH_3{-}\underset{..}{N}(CH_3){-}CH_3$	Trimethylamine	2.9		
$(CH_3CH_2)_3\ddot{N}$	Triethylamine	89.3		

diamines, putrasine and cadaverine, were so named because of the odors that they give to decaying flesh. Both of these amines are formed by degradation of amino acids.

$$NH_2-CH_2-CH_2-CH_2-CH_2-CH_2-NH_2 \qquad NH_2-CH_2-CH_2-CH_2-CH_2-NH_2$$

1,5-diaminopentane (cadaverine) — 1,4-diaminobutane (putrasine)

Biological Properties of Amines

It is impossible to describe all the biological functions of compounds that contain nitrogen, but we shall briefly mention some of the more obvious ones.

Proteins, the building blocks of the body, are a vital component of all life. They are giant molecules formed by the coupling of groups of molecules called **amino acids**. In this coupling the $-NH_2$ of one amino acid interacts with the $-\overset{O}{\overset{\|}{C}}-OH$ portion of another amino acid to form an **amide** linkage. When this process is repeated over and over, long chains called **proteins** are formed.

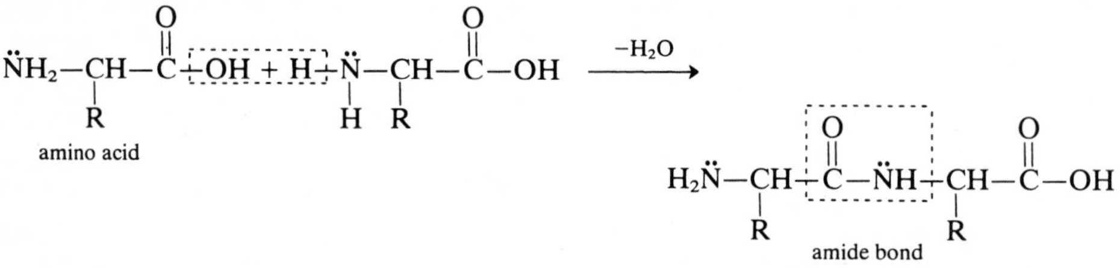

amino acid
−H₂O
amide bond

Another essential group of compounds in the body are the **nucleic acids**. DNA is the master molecule of life: it contains all the genetic information and in essence directs all growth. Both DNA and RNA contain four essential heterocyclic amines. RNA differs from DNA in that it contains uracil rather than thymine. All these compounds contain nitrogen as amines or amides.

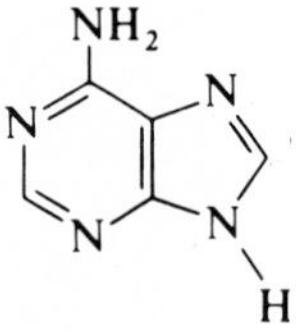
adenine

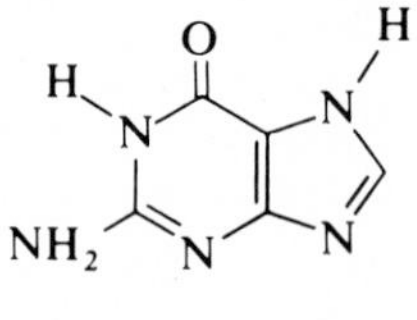
guanine

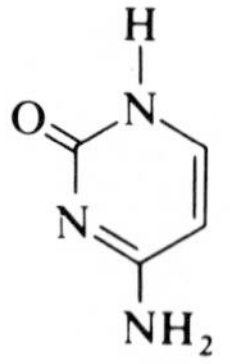
cytosine

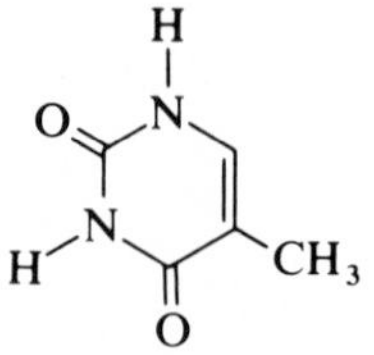
thymine

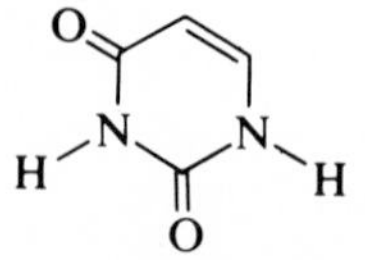
uracil

Many drugs also contain the amine group. Table 31–2 lists some examples.

Table 31–2 Drug compounds that contain amines

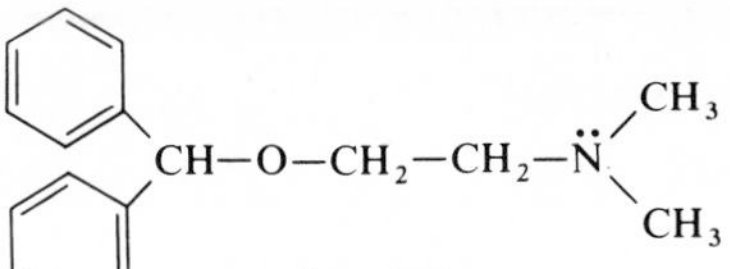

benadryl
(an antihistamine)

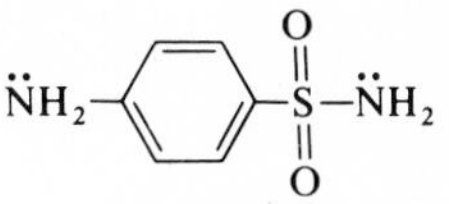

sulfanilamide
(antibacterial agent)

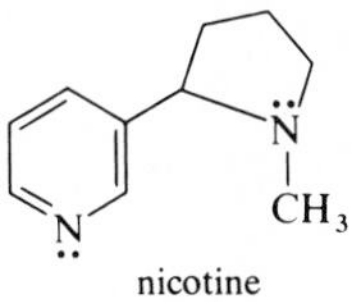

nicotine
(from tobacco)

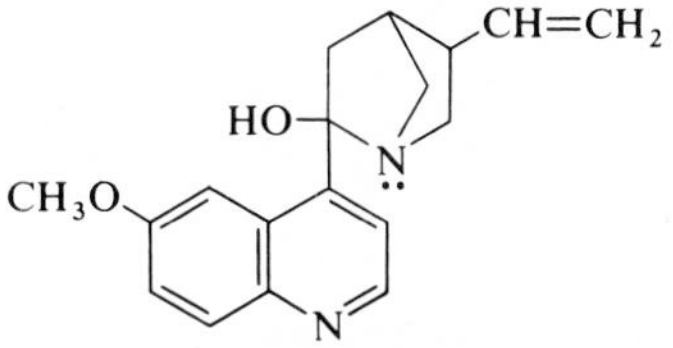

quinine
(used in treatment
of malaria)

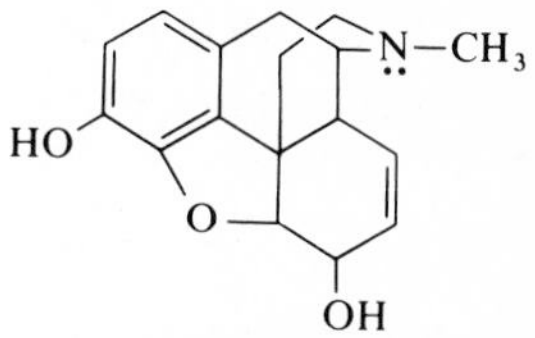

morphine
(active component
of opium)

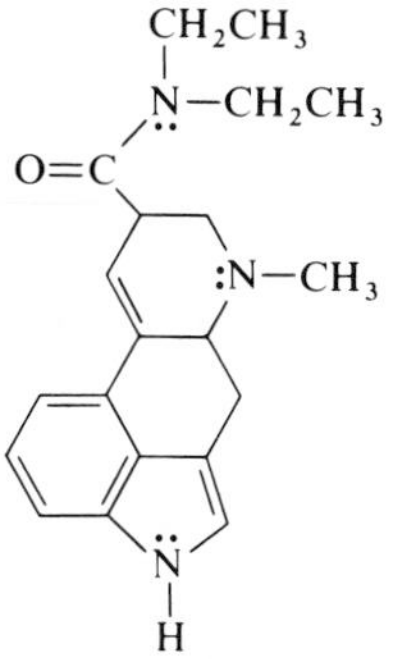

LSD

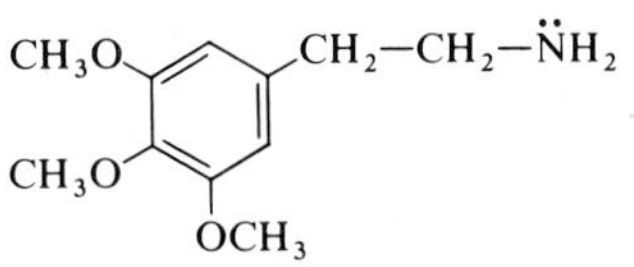

mescaline
(from peyote cactus)

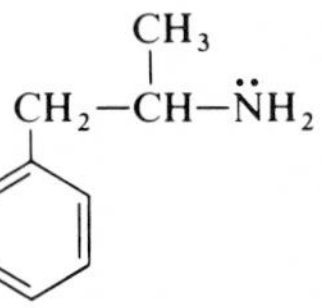

amphetamine
("speed")

Synthesis and Reactions of Amines

Basicity: Amines are basic and react with acids. This process is reversible, so the treatment of ammonium salts with base regenerates the amine. We shall see the importance of this reaction in later units when amino acids are studied.

$$\underset{\text{a free amine}}{R{-}\ddot{N}H_2} \underset{OH^-}{\overset{HCl}{\rightleftharpoons}} \underset{\text{an ammonium salt}}{R{-}\overset{+}{N}H_3\ Cl^-}$$

Substitution:

$$R'X + R_2\ddot{N}H \longrightarrow R'{-}\overset{+}{N}(H)R_2 \xrightarrow{OH^-} R'{-}\ddot{N}R_2$$

The unshared electron pair on nitrogen acts to displace halide from alkyl halides to form a new C—N bond. This reaction was introduced in unit 29 as a reaction of alkyl halides, but it can also be classified as a method of synthesis of amines. These substitution reactions can be used to synthesize primary, secondary, tertiary, or quarternary amines.

Synthesis of a primary amine

$$\underset{\text{ammonia}}{R{-}X + {:}NH_3} \longrightarrow [R{-}\overset{+}{N}H_3\ \ X^-] \xrightarrow{OH^-} R{-}\ddot{N}H_2$$

Synthesis of a secondary amine

$$R'{-}X + \underset{\text{use a 1° amine}}{\ddot{N}H_2{-}R} \longrightarrow [R'{-}\overset{+}{N}H_2R] \xrightarrow{OH^-} R'{-}\ddot{N}H{-}R$$

Synthesis of a tertiary amine

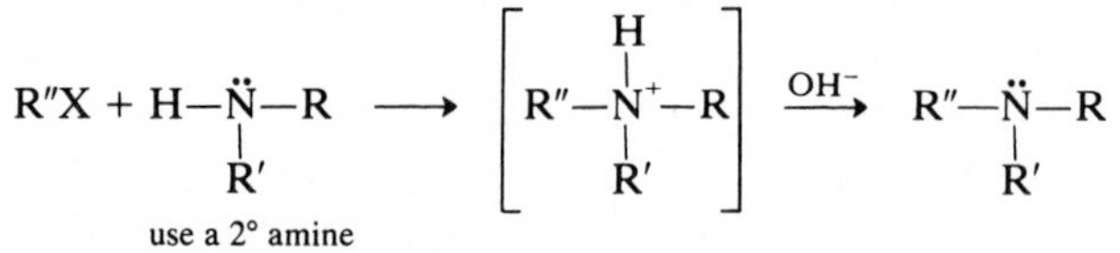

use a 2° amine

Synthesis of a quaternary ammonium salt

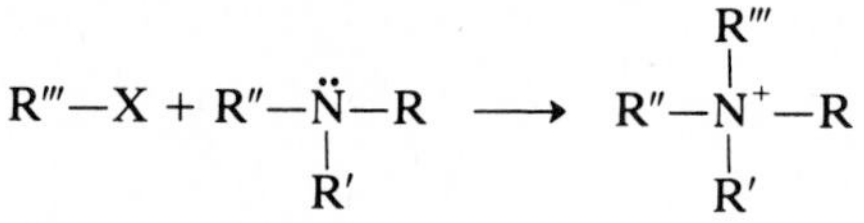

use a 3° amine

The synthesis of amines is quite similar to the synthesis of ethers, as the following examples show. Remember that the nitrogen becomes fastened to the carbon that originally held the halogen.

Example 1: Synthesize $CH_3{-}\ddot{N}H{-}CH(CH_3){-}CH_3$.

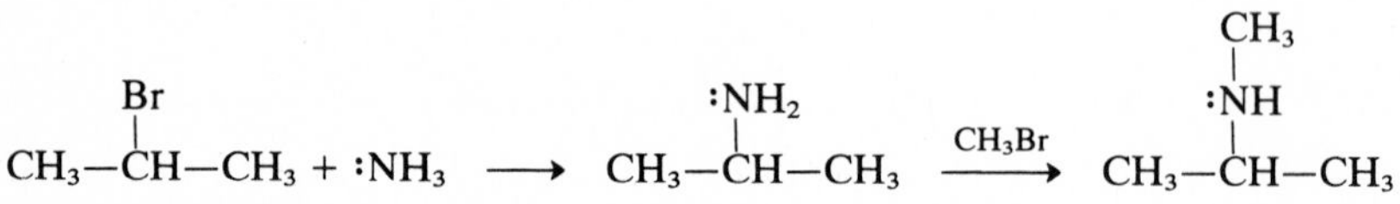

Example 2: Make *N,N*-dimethylamine from aniline.

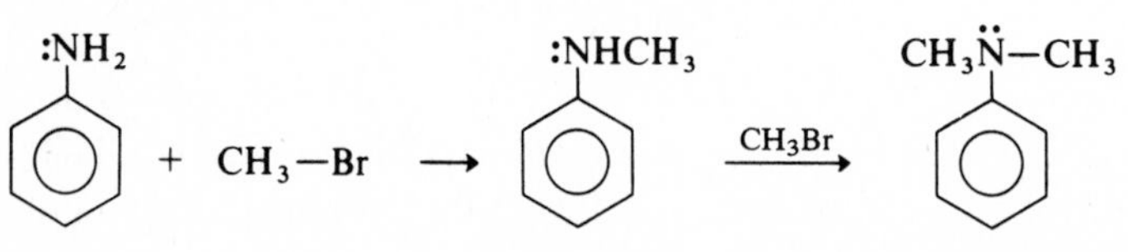

Example 3: Synthesize ethylmethylisopropylamine.

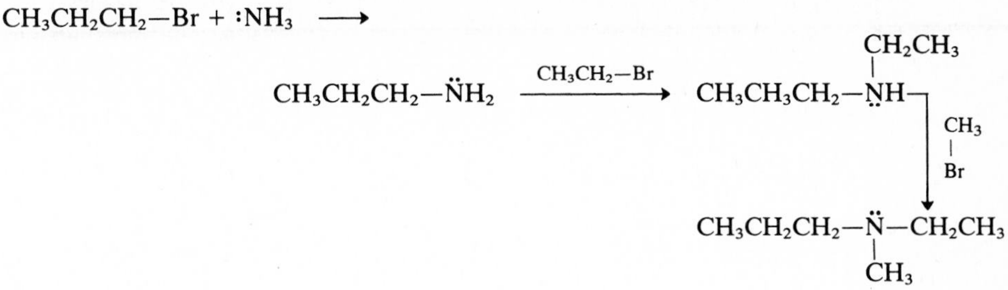

Reduction of Cyanides

Addition of hydrogen to cyanide groups results in formation of primary amines of the form R—CH_2—NH_2. The reduction is similar to reduction of π-bonds in alkenes and aldehydes or ketones. This reaction is often used in complex synthesis for increasing the length of the carbon chain.

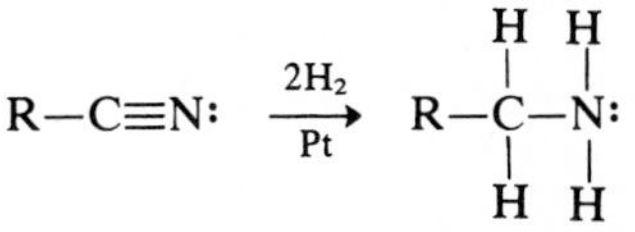

$$CH_3CH_2CH_2{-}Br \xrightarrow{CN^-} CH_3CH_2CH_2{-}C{\equiv}N \xrightarrow[Pt]{2H_2} CH_3CH_2CH_2CH_2{-}NH_2$$

1-bromopropane — 1-cyanopropene — 1-aminobutane

UNIT 31 Problems and Exercises

31–1. Name the following compounds.

(a) $CH_3CH_2\ddot{N}H{-}CH_2CH_3$

(b) $CH_3CH_2{-}\ddot{N}H{-}CH_3$

(c) 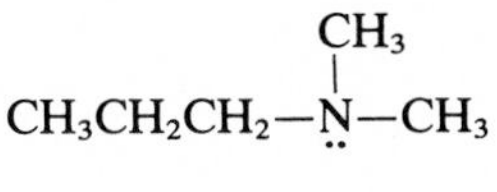

(d) 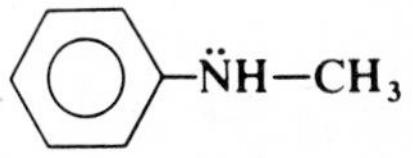

(e) 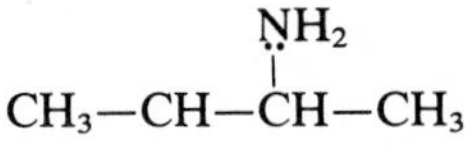

(f) 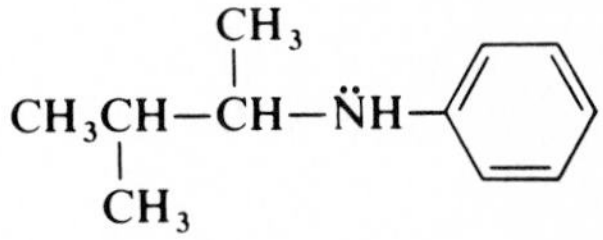

(g) 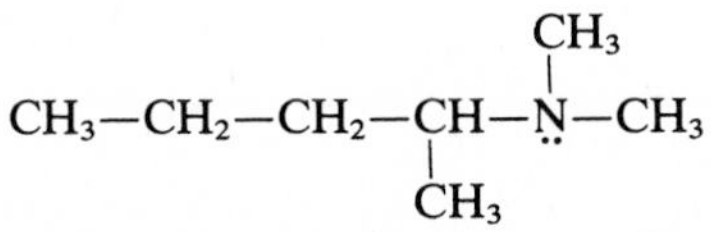

(h) 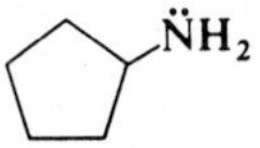

(i) 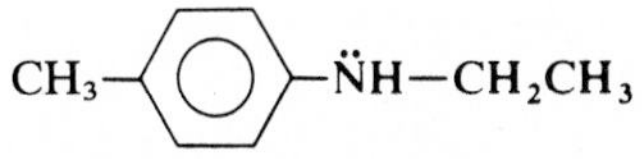

(j) 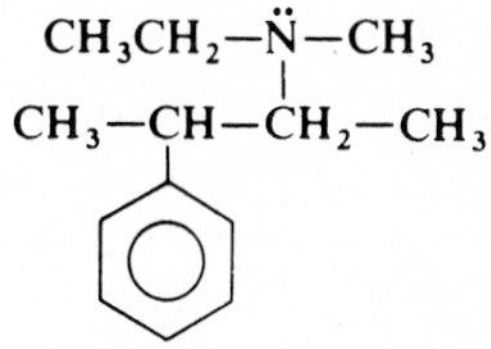

31–2. Draw the structures of the following amines.

(a) Isopropylamine
(b) *sec*-butylmethylamine
(c) 2-aminobutane
(d) *N*-methyl-2-butanamine
(e) *N*-methyl-*p*-ethylaniline
(f) *N,N*-dimethyl-2-hexanamine
(g) Triethylamine
(h) *N*-ethyl-4-methyl-2-ethylaniline

31–3. Label each of the amines in problems (31–1) and (31–2) as primary, secondary, or tertiary amines.

31–4. Rank the following compounds in terms of increasing solubility in water and explain your ranking.

(a)	(b)	(c)	(d)
$CH_3CH_2CH_2OH$	$CH_3CH_2{-}O{-}CH_3$	$CH_3CH_2CH_2\ddot{N}H_2$	$CH_3CH_2\ddot{N}HCH_3$

31–5. Complete the following reactions.

(a) $CH_3{-}\underset{\large Br}{\underset{|}{C}H}{-}CH_3 + \ddot{N}H_3 \longrightarrow$

(b) $CH_3{-}\underset{\large Br}{\underset{|}{C}H}{-}CH_3 + CH_3\ddot{N}H_2 \longrightarrow$

(c) $CH_3{-}\underset{\large Br}{\underset{|}{C}H}{-}CH_3 + CH_3{-}\ddot{N}H{-}CH_3 \longrightarrow$

(d) $CH_3{-}\underset{\large Br}{\underset{|}{C}H}{-}CH_3 + CH_3{-}\underset{\large CH_3}{\underset{|}{\ddot{N}}}{-}CH_3 \longrightarrow$

31–6. Complete the following reactions.

(a) 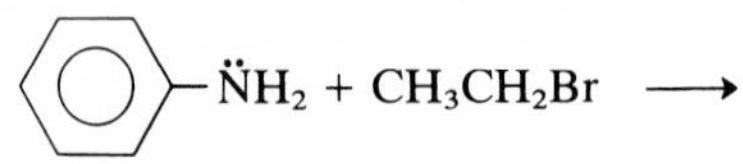

(b) $CH_3CH_2Br + CN^- \longrightarrow$

(c) $CH_3{-}CH_2{-}CH_2{-}CN \xrightarrow[Pt]{2H_2}$

(d) $CH_3\overset{:NH}{\overset{|}{C}}HCH_2CH_3$ + excess $CH_3{-}I \longrightarrow$

31–7. What primary amine and alkyl halide would you use to synthesize the following secondary amines? There are two separate pairs of reactants possible for each product.

(a) $CH_3CH_2\ddot{N}HCH_3$

(b) 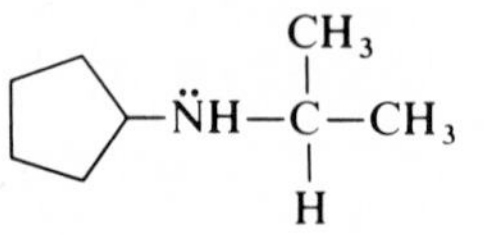

(c) 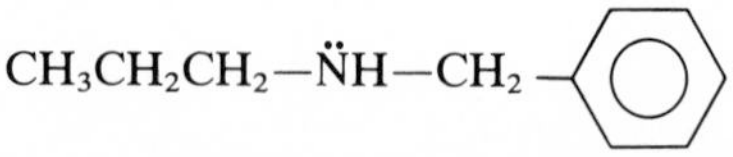

(d) 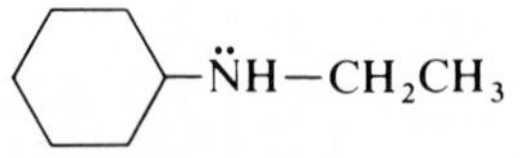

31–8. Outline a method of synthesis of the following amines, starting with the reactants given and any other reagents that you need. All require more than one step.

(a) Make $CH_3CH_2{-}\ddot{N}H{-}CH_2CH_3$ from $2CH_2{=}CH_2$.

(b) Make $CH_3CH_2CH_2{-}\ddot{N}H{-}CH(CH_3){-}CH_3$ from $2CH_3CH_2{-}C({=}O){-}H$.

(c) Make $CH_3CH_2CH_2{-}\ddot{N}H_2$ from $CH_3CH_2{-}Br$.

(d) Make *N*-methylpropanamine from methyl bromide and ethene.

31–9. Suggest a short chemical experiment that could be used to identify the presence of an amine functional group.

31–10. How could you use short chemical tests to quickly distinguish between the following pairs of compounds?

(a) $CH_3{-}OH$ and $CH_3{-}NH_2$

(b) $CH_2{=}CH_2$ and $CH_3CH_2NH_2$

(c) $CH_3{-}O{-}CH_3$ and $CH_3{-}NH{-}CH_3$

31–11. An organic compound with the formula of $C_3H_9\ddot{N}$ reacts with HCl to become soluble in water. It also reacts with 2 mol of methyl iodide to give an ammonium salt, $C_5H_{14}N^+I^-$. Give a structure for each compound and name them.

UNIT 32 MECHANISMS OF SUBSTITUTION REACTIONS

Many of the reactions discussed in units 28 through 31 can be grouped in the general class of **substitution** reactions. Most substitution reactions occur by one of two mechanisms. In this unit we shall discuss these mechanisms and the various factors that affect them.

Nucleophilic Substitution

The word *nucleophile* comes from Greek and means *nucleus loving*. Thus a nucleophile is any group that is attracted to the positive charge of a nucleus. Nucleophiles are either negatively charged or have unshared pairs of electrons that can be used in bond formation. In nucleophilic substitution, the nucleophile attacks an sp^3 carbon atom and displaces some group from it.

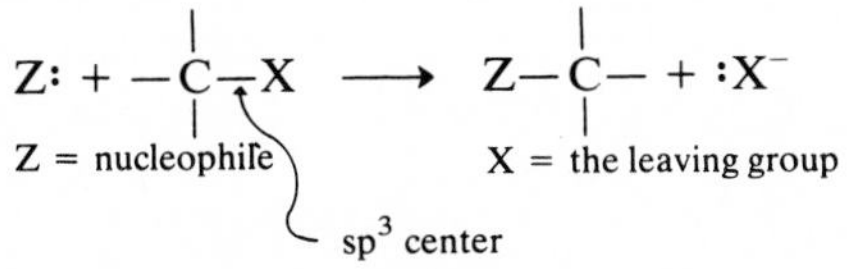
Z: + —C—X ⟶ Z—C— + :X⁻
Z = nucleophile
X = the leaving group
sp3 center

SN-l Mechanism

In one mechanism of nucleophilic substitution, the leaving group, X, is removed by the solvent before the nucleophile enters. The resulting carbonium ion then becomes bonded to the nucleophile in the second step of the reaction:

$$-\overset{|}{\underset{|}{\mathrm{C}}}-\mathrm{X} + \text{solvent} \xrightarrow[\text{Slow step}]{} -\overset{|}{\underset{|}{\mathrm{C}}}{}^{\oplus} \xrightarrow[\text{Fast step}]{\mathrm{Z}:} -\overset{|}{\underset{|}{\mathrm{C}}}-\mathrm{Z}$$

This type of mechanism is called **nucleophilic substitution of the first order** (SN-1), since its rate depends only upon the concentration of the starting $-\overset{|}{\underset{|}{\mathrm{C}}}-\mathrm{X}$ group and not on the concentration of the nucleophile.

$$\text{Rate} \simeq k \cdot [-\overset{|}{\underset{|}{\mathrm{C}}}-\mathrm{X}]$$

The slow step in this reaction is the loss of X to form the carbonium ion. Once formed, the carbonium ion reacts rapidly with the nucleophile.

SN-2 Mechanism

A second mechanism of substitution reactions does not involve carbonium ions. In the SN-2 mechanism the nucleophile bonds to the carbonium atom *at the same time that the group X leaves.* There is no intermediate formed and the process is a gradual one of bond forming and bond breaking:

$$\mathrm{Z}: + -\overset{|}{\underset{|}{\mathrm{C}}}-\mathrm{X} \longrightarrow [\mathrm{Z}\cdots\overset{|}{\underset{/\ \backslash}{\mathrm{C}}}\cdots\mathrm{X}] \longrightarrow \mathrm{Z}-\overset{|}{\underset{/\ \backslash}{\mathrm{C}}} + :\mathrm{X}^-$$

This process is called **nucleophilic substitution of the second order** (SN-2), since the rate of the reaction is dependent on the concentrations of the nucleophile and the $-\overset{|}{\underset{|}{\mathrm{C}}}-\mathrm{X}$ group.

$$\text{Rate} = k[\mathrm{Z}:]\cdot[-\overset{|}{\underset{|}{\mathrm{C}}}-\mathrm{X}]$$

Factors That Influence the SN-1 and SN-2 Mechanisms

Both mechanisms can occur in substitution reactions, but generally one mechanism predominates in a particular reaction. The factors that determine which mechanism predominates include the structure of the sp^3 carbon center, the type of nucleophile Z, and the type of leaving group X.

Structure of the Carbon Center

In the SN-1 mechanism, a carbonium ion is formed at the carbon center during the reaction. Any groups that will stabilize a carbonium ion will increase the rate of SN-1 reactions. *Tertiary centers are more likely to react by the SN-1 mechanism than primary centers,* which have no groups to stabilize the carbonium ion.

Rate of SN-1 reactions:

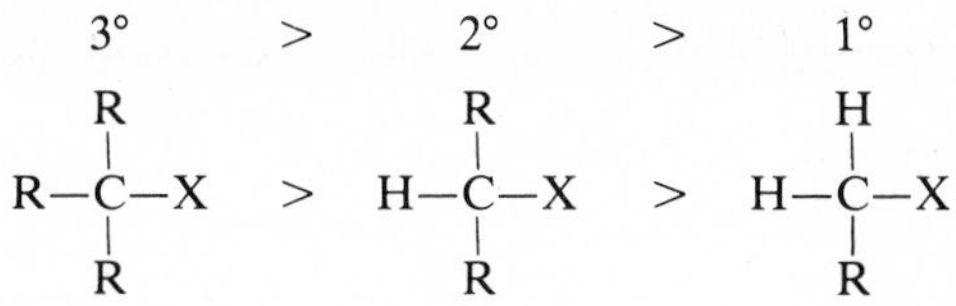

In the SN-2 mechanism, the nucleophile must get close enough to the carbon center to bond with it for reaction to occur. If this center has many large groups attached to it, the nucleophile will have difficulty bonding. *SN-2 reactions occur most rapidly at primary centers* where the small hydrogens do not hinder the approach of the nucleophile.

Rate of SN-2 reactions:

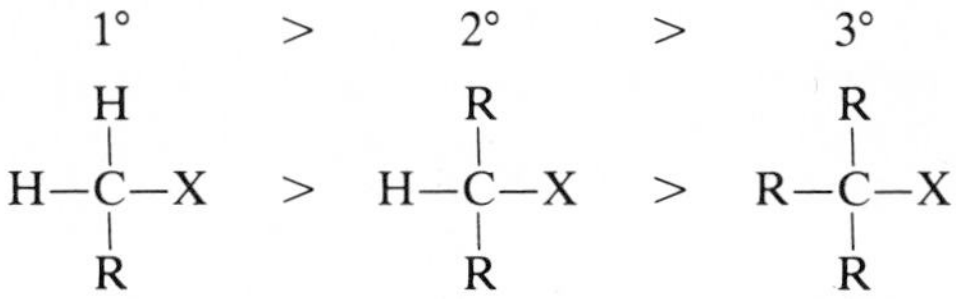

Effects of the Nucleophile

In an SN-1 reaction, the nucleophile enters only after the carbonium ion is formed. It is not involved in the initial rate-determining step, so changes in the nucleophile will not dramatically affect the rate of an SN-1 reaction.

In an SN-2 mechanism the nucleophile must be able to form a strong bond with the carbon center for reaction to occur. The ability of the nucleophile to attack the carbon center is roughly proportional to its basicity. A strong base is quite reactive and bonds readily, whereas a weak base reacts more slowly. Table 32–1 lists common nucleophiles in order of their strengths.

Table 32–1 Strengths of nucleophiles

Strongest bases:			
best nucleophiles	R—O$^-$	(R—OH)	Weakest acids
	OH$^-$	(H_2O)	
	CN$^-$	(HCN)	
	HS$^-$	(H_2S)	
	NH_3	(NH_4^+)	
	H_2O	(H_3O^+)	
Weakest bases:	Cl$^-$	(HCl)	Strongest acids
poorest nucleophiles			

Effects of the Leaving Group

For SN-1 and SN-2 reactions to occur, the group X must leave the carbon center. Usually, it leaves as an anion; the more stable it is as a negative ion the better able it will be to leave the carbon center. Conjugate bases of strong acids are the most stable anions and are the best leaving groups. Both SN-1 and SN-2 reactions occur most rapidly if the leaving group is Br^-, Cl^-, or I^-. It is difficult to dislodge $R—O^-$, OH^-, or NR_2^- groups as they are not very stable anions.

Leaving group abilities:

$$I^- > Br^- > Cl^- > H_2O > HS^- > HO^- > R—O^-$$

Sometimes the ability of a group to leave is enhanced by reaction with acid. Treatment of alcohols with HBr goes smoothly even though HO^- is a poor leaving group. In this reaction the H^+ first protonates the alcohol to convert —OH into $—\overset{\oplus}{O}H_2$, which is a good leaving group.

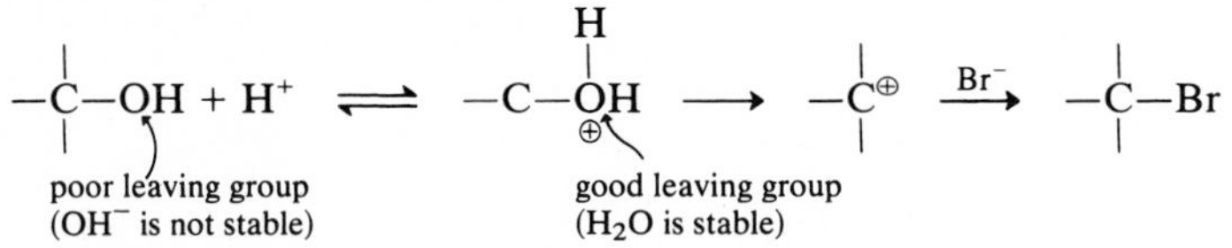

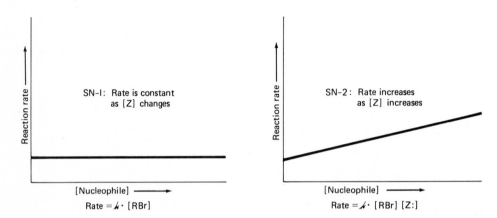

poor leaving group
(OH⁻ is not stable)
good leaving group
(H₂O is stable)

Distinguishing Between SN-1 and SN-2 Reactions

Reaction Rate Dependence

The rate of reaction of SN-2 type reactions is proportional to the concentrations of both the alkyl halide and the nucleophile. Changes in the concentration of the nucleophile will cause the rate of SN-2 reactions to change. In contrast, the rate of reaction for SN-2 type reactions is proportional only to the concentration of the alkyl halide; changes in the concentration of the nucleophile do not affect the rate of the reaction. One method of determining the mechanism for a substitution, then, is to measure the rate of the reaction using several different concentrations of the nucleophile. If the rate remains constant, the mechanism must be SN-1. If the rate increases with increasing concentration of the nucleophile, the mechanism must be SN-2 (see Figure 32–1).

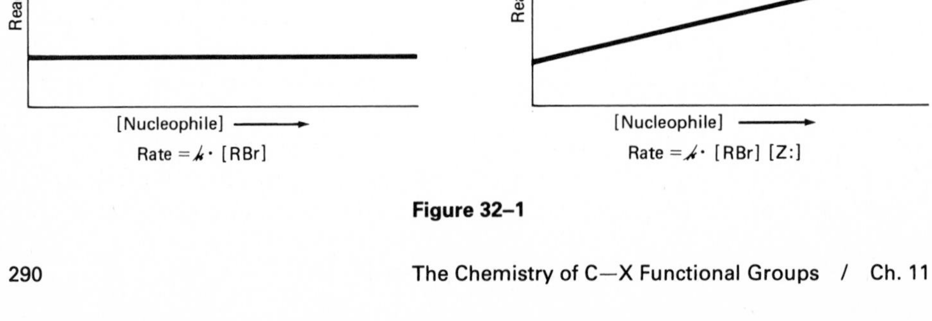

Reaction rate
SN-I: Rate is constant
as [Z] changes
[Nucleophile]
Rate = k· [RBr]
Reaction rate
SN-2: Rate increases
as [Z] increases
[Nucleophile]
Rate = k· [RBr] [Z:]

Figure 32–1

Inversion at Chiral Centers

The four groups attached to a sp^3 hybridized carbon center are arranged in a tetrahedral geometry around that carbon center. If these four groups are all different from each other, two separate tetrahedral structures are possible (Figure 32–2).

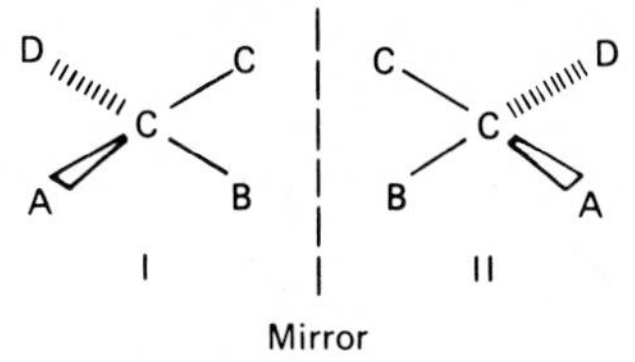

D
C
C
A
B
I
C
D
C
B
A
II
Mirror

Figure 32–2

Structures I and II are not identical; they are **mirror images** of each other. As long as the four groups (A, B, C, and D) are different from each other, structures I and II cannot be superimposed on each other. These types of carbon centers are called **chiral centers**, and structures I and II together are called a pair of **enantiomers** or **mirror images**. They have many similar physical and chemical properties, but there is at least one method of distinguishing between them. Each structure will affect polarized light in a different manner. We shall discuss this subject in detail later in the text, but for now it is sufficient to know that the two structures can be distinguished from each other.

In an SN-2 reaction, the nucleophile enters and bonds to the carbon center from the side *opposite* that of the leaving group. In an SN-2 reaction, the geometry of a chiral carbon center is *inverted*; it changes from structure III to structure IV (Figure 32–3).

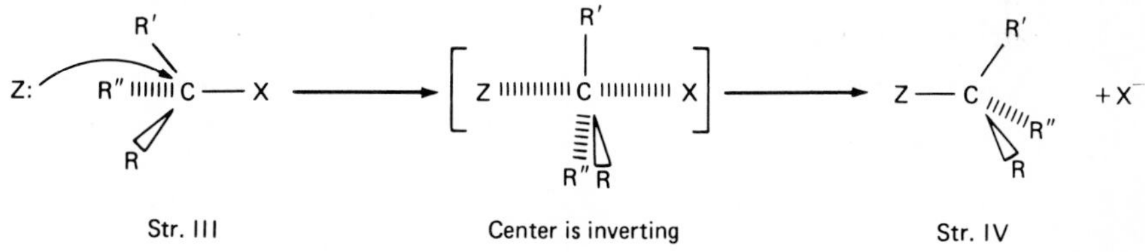

Z:
R′
R″
C
X
R
Str. III
R′
Z
C
X
R″
R
Center is inverting
Z
C
R′
R″
R
+X⁻
Str. IV

Figure 32–3

In the SN-1 mechanism, the carbonium ion formed is a planar ion, so the Z: group can attack at either side. The products can have either geometry, and structures III and IV will be formed in equal amounts (Figure 32–4).

If we start with a chiral carbon center and add a nucleophile, we can tell which mechanism occurred by examining the structures of the products. If the product has a structure that is inverted with respect to the reactant, the reaction mechanism must have been SN-2. If the product molecules have both structures (inverted and noninverted) in equal amounts, the mechanism must have been SN-1. Finally, if both structures are present, but the inverted structure is present in the larger amount,

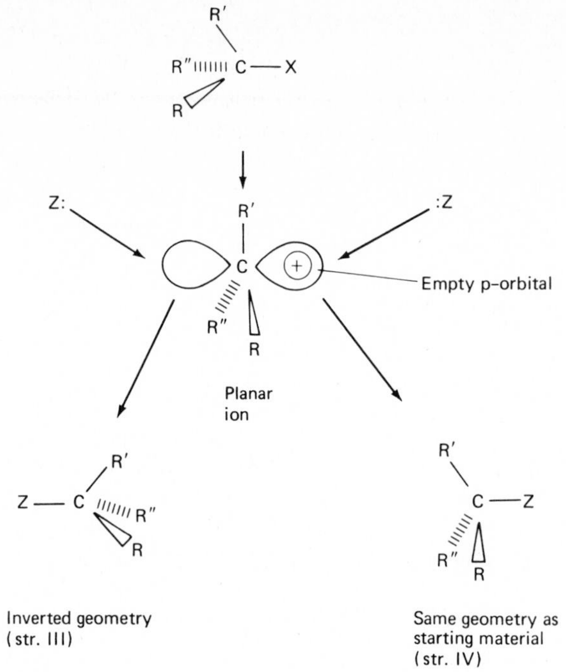
R'
R''
C
X
R
Z:
:Z
Empty p-orbital
Planar ion
Inverted geometry (str. III)
Same geometry as starting material (str. IV)

Figure 32–4

both mechanisms could have occurred. This can be expected to occur at secondary centers.

In conclusion, there are two main mechanisms for substitution reactions. The products are the same for both mechanisms if the geometries of the products are neglected, but those products were formed by different mechanisms. In an SN-1 reaction, the nucleophile waits for the carbonium ion to form before bonding to it. In an SN-2 reaction, the nucleophile is the instigator; it causes the reaction to occur and helps to force the departure of the leaving group.

UNIT 32 Problems and Exercises

32–1. Rank the following compounds in terms of increasing rate of reaction by the SN-1 mechanism. Explain your ranking.

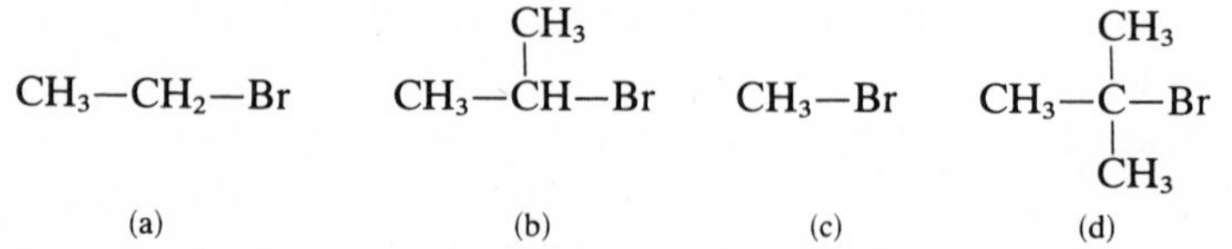

(a) (b) (c) (d)

32–2. Rank the compounds listed in problem (32–1) in terms of increasing rate of reaction by the SN-2 mechanism. Explain your ranking.

32–3. Show all the steps involved in treatment of t-butyl chloride with OH^-. What type of mechanism is it?

32–4. Give the mechanism for the reaction of ethyl bromide with CH_3O^-. What type of mechanism is it?

32–5. Which of the following would react most rapidly? Explain your answers.

    (a) $CH_3-Br + OH^-$ or $CH_3-I + OH^-$

    (b) $\overset{\displaystyle CH_3}{\underset{\displaystyle CH_3}{CH_3-\overset{|}{\underset{|}{C}}-SH}} + CH_3O^-$ or $\overset{\displaystyle CH_3}{\underset{\displaystyle CH_3}{CH_3-\overset{|}{\underset{|}{C}}-Cl}} + CH_3O^-$

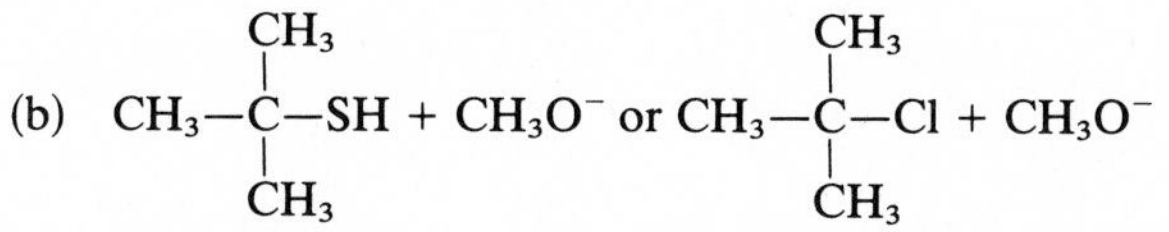

32–6. The rate of reaction (1) does not change when the concentration of CN^- is increased, but the rate of reaction (2) dramatically increases as $[CN]$ increases. Explain these facts.

(1)   C₆H₅—CH(Br)—CH₃ + CN⁻ → C₆H₅—CH(CN)—CH₃

(2) $CH_3-\overset{\displaystyle Cl}{\overset{|}{CH}}-CH_3 + CN^- \longrightarrow CH_3-\overset{\displaystyle CN}{\overset{|}{CH}}-CH_3$

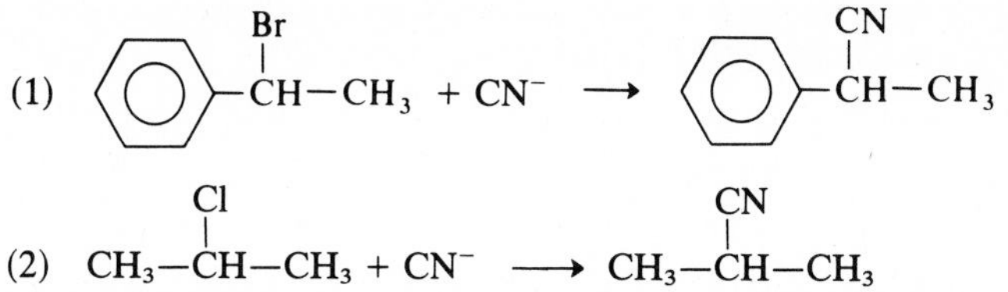

32–7. Which of the following SN-2 reactions would you expect to occur most rapidly? Most slowly? Explain your reasoning.

    (a) $CH_3-\overset{\displaystyle H}{\underset{\displaystyle CH_3}{\overset{|}{\underset{|}{C}}}}-Br + OH^- \longrightarrow CH_3-\overset{\displaystyle H}{\underset{\displaystyle CH_3}{\overset{|}{\underset{|}{C}}}}-OH + Br^-$

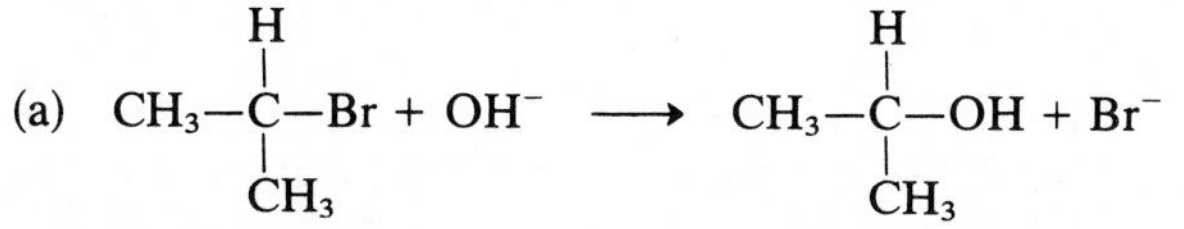

    (b) $CH_3-Br + NH_3 \longrightarrow CH_3-NH_2 + HBr$

    (c) $CH_3-Br + CH_3O^- \longrightarrow CH_3-O-CH_3 + Br^-$

32–8. Can you explain why the nature of the nucleophile dramatically affects the rate of SN-2 reactions but is unimportant in SN-1 reactions?

32–9. Outline steps that you would take to determine whether the following reaction was SN-1 or SN-2.

$$CH_3-\overset{\displaystyle Br}{\underset{\displaystyle H}{\overset{|}{\underset{|}{C}}}}-CH_2CH_3 + OH^- \longrightarrow CH_3-\overset{\displaystyle OH}{\underset{\displaystyle H}{\overset{|}{\underset{|}{C}}}}-CH_2CH_3$$

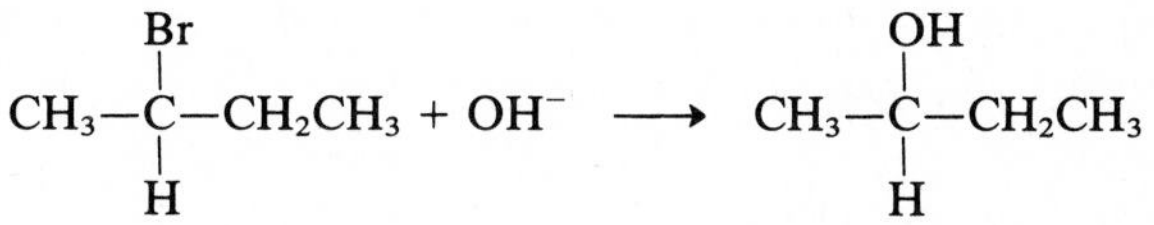

CHAPTER 12

the chemistry of >C=O functional groups

The **carbonyl group** (a >C=O center) is one of the most important reactive sites in organic molecules. Several types of organic molecules contain this center, and many different reactions are possible. In unit 33 we present the chemistry of carbonyl groups as they are found in **aldehydes** and **ketones**. In units 34 and 35, we present the chemistry of carbonyl groups found in **carboxylic acids** and their derivatives.

In unit 36, we shall try to summarize the reactions of carbonyl groups, to explain the differences in reactions of aldehydes and ketones as opposed to carboxylic acids.

Units 33 through 36 are the final units in our study of organic chemistry. After that we shall begin to study the chemistry of larger, more complex molecules. The reactive sites and reactions will be similar to those previously covered, but the molecules will contain more reactive centers and the processes will be somewhat more involved. We shall see how the general and organic chemistry introduced so far is involved in the functioning of living species.

UNIT 33 ALDEHYDES AND KETONES

Carbonyl Group

A **carbonyl group** consists of a carbon and oxygen linked by a double bond. It is present in aldehydes, ketones, carboxylic acids, esters, amides, acid chlorides, and anhydrides. In this unit we shall learn some of the reactions of carbonyl groups present in aldehydes and ketones.

A carbonyl group has nearly the same type of bonding as an alkene group. Both the carbon and oxygen are sp^2 hybridized, but two of the sp^2-orbitals on the oxygen atom contain unshared pairs of electrons rather than covalent bonds (see Figure 33–1).

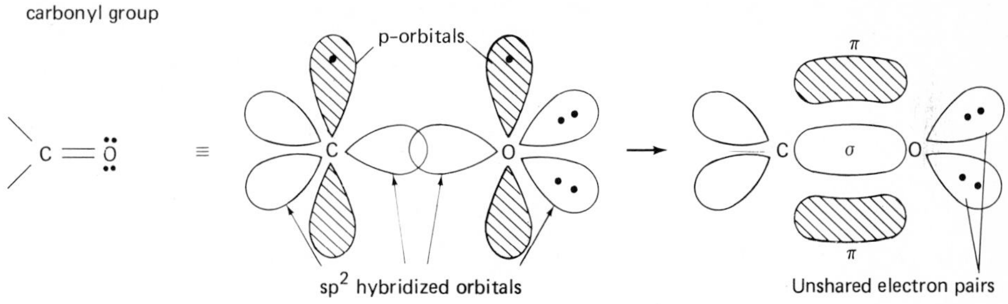
carbonyl group
p-orbitals
C
O
σ
π
π
sp² hybridized orbitals
Unshared electron pairs

Figure 33–1

One important difference between a carbonyl group and an alkene bond is polarity. Oxygen is much more electronegative than carbon, so the electrons in the C=O bonds are more attracted to oxygen than carbon. This sets up a strong dipole, the carbon being δ^+ and oxygen being δ^-. Reagents that are negatively charged will be attracted to the carbon; positively charged reagents will attack the oxygen.

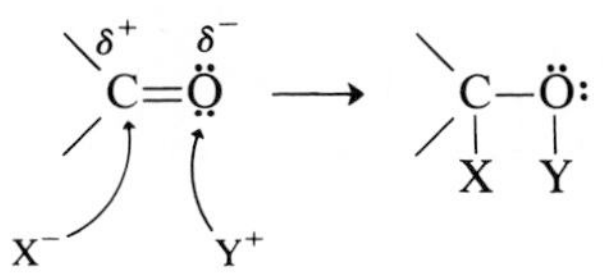
δ+
δ−
C=Ö
C—Ö:
X
Y
X−
Y+

Nomenclature

1. Aldehydes have the general formula R—C(=O)—H, where one of the groups bonded to the carbonyl group is hydrogen. In ketones there are two carbon atoms and no hydrogens bonded to the carbonyl group.

R—C(=O)—H (aldehydes)   R—C(=O)—R (ketones)

2. In the formal IUPAC method of naming, the ending on the parent for aldehydes is **-al**; that for ketones is **-one**. Numbering starts with the carbonyl carbon in aldehydes, but in ketones numbering begins at the end closest to the carbonyl group.
3. Substituents on the parent chain are named as for the other classes of organic compounds.
4. Common names: One common method of naming aldehydes is to use the ending *aldehyde* instead of the ending *-al*. Thus propanal becomes propanaldehyde. Ketones can be named like ethers; the two groups bonded to the carbonyl carbon are named and the word *ketone* is added. So 2-propanone becomes dimethyl ketone.

In addition to IUPAC and common names, some historical names have also remained. Table 33–1 gives examples of all three types of names.

Table 33–1 Names for aldehydes and ketones

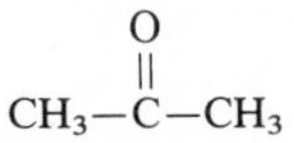

2-propanone
dimethyl ketone
acetone

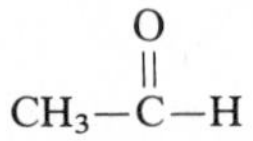

ethanal
acetaldehyde

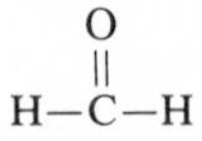

methanal
formaldehyde

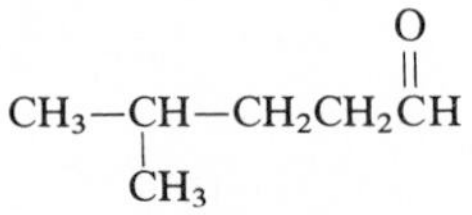

4-methylpentanal
isohexylaldehyde

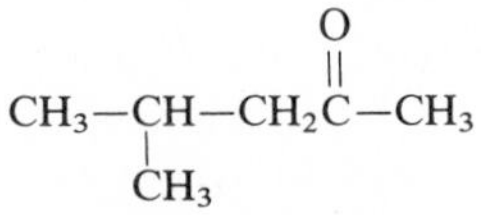

4-methyl-2-pentanone
methylisobutyl ketone

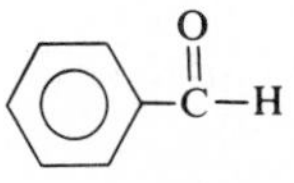

phenylmethanal
benzaldehyde

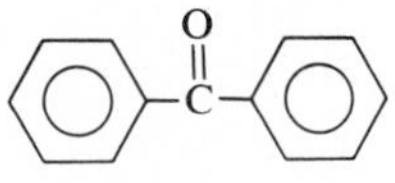

diphenyl ketone
benzophenone

Sources and Uses of Aldehydes and Ketones

Most aldehydes and ketones are synthesized industrially by oxidation of alcohols but a number are also isolated from plants. Table 33–2 gives some common names of aldehydes and ketones that are isolated from natural sources.

A great number of ketones and aldehydes are used industrially in the synthesis of other compounds. Formaldehyde is polymerized with phenol to yield Bakelite, a hard plastic used as an insulator and as handles for pots and pans. Formaldehyde is also used in the synthesis of other plastics like Formica. Acetone and other low-molecular-weight ketones are widely used as solvents.

Table 33–2 Some commonly occurring aldehydes and ketones

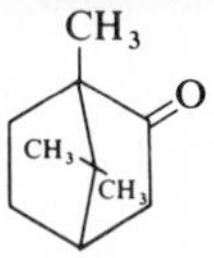 camphor (camphor tree)	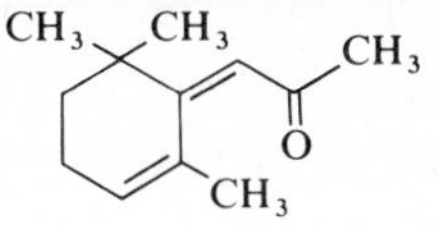 irone (violet)	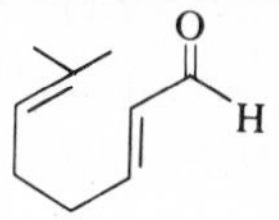 citral (lemon grass oil)
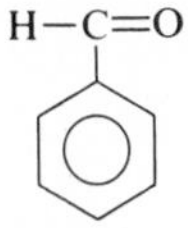 benzaldehyde (bitter almond)	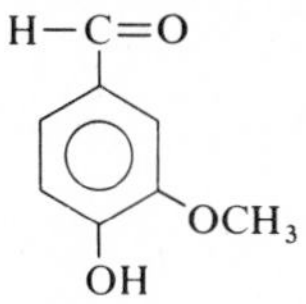 vanillin (vanilla bean)	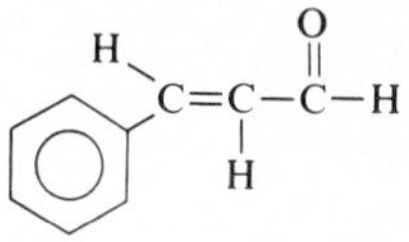 cinnamaldehyde (oil of cinnamon)

Physical Properties

Aldehydes and ketones are moderately polar compounds. They are more polar than ethers but not as polar as alcohols. The polar carbonyl group does participate in hydrogen bonding, but not nearly as strongly as do alcohols. Aldehydes and ketones thus boil at higher temperatures than ethers but at much lower temperatures than alcohols of similar molecular weight.

Low-molecular-weight aldehydes and ketones are soluble both in water and nonpolar solvents; solubility in water decreases as the molecular weight increases. Acetone is often used as a solvent; it dissolves many nonpolar compounds and is also soluble in water.

Aldehydes and ketones have pleasant odors, and are often used in perfumes and other substances. Part of the aroma of many plants comes from ketones.

Acidity of Alpha-Hydrogen Atoms

Carbon atoms adjacent to carbonyl groups are often called **alpha-carbon atoms**, and hydrogens attached to these alpha-carbons are called **alpha-hydrogen atoms**:

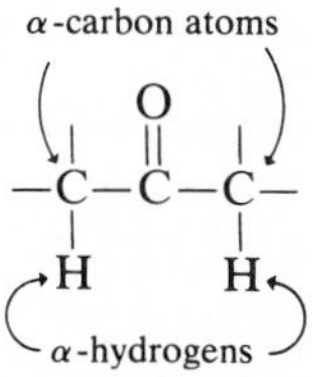
α-carbon atoms
α-hydrogens

Alpha-hydrogen atoms are different from most hydrogens covalently bound to carbon in that they are much more acidic than other hydrogen atoms. The K_a values for alpha hydrogens are 10^{-20}, whereas K_a values for hydrogens bound to other carbons are about 10^{-50}. One reason for this increased acidity is that the anion formed when alpha-hydrogen ions leave is much more stable than other types of carbanions. It is stabilized because some of the negative charge can be moved to the oxygen of the carbonyl group through *resonance*. In resonance, we move electrons between two or more atoms and thus spread out the negative charge. This provides for a more stable species.

$$R-CH_2-H \rightleftharpoons R-CH_2^- + H^+ \qquad K_a = 10^{-50}$$

In this reaction, the carbon must bear the entire negative charge, an unstable situation. The equilibrium is shifted to the left.

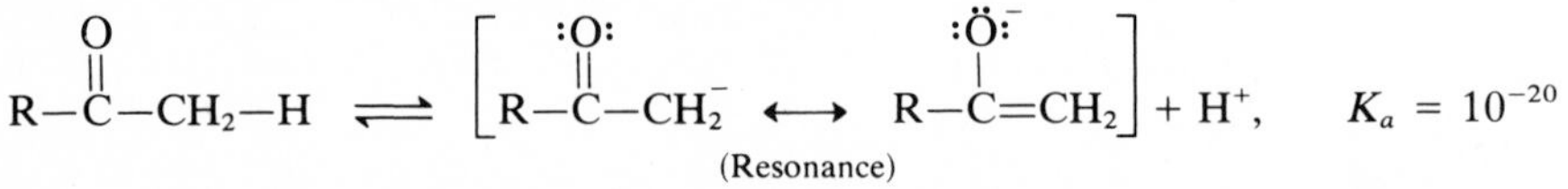

$$R-\overset{\overset{\displaystyle O}{\|}}{C}-CH_2-H \rightleftharpoons \left[R-\overset{\overset{\displaystyle :O:}{\|}}{C}-CH_2^- \longleftrightarrow R-\overset{\overset{\displaystyle :\ddot{O}:^-}{|}}{C}=CH_2 \right] + H^+, \qquad K_a = 10^{-20}$$
(Resonance)

Here the negative charge is spread over two atoms, a more stable situation. The equilibrium is shifted farther to the right than in simple alkanes.

The acidity of alpha-hydrogen atoms allows us to create carbanions at alpha-carbon atoms fairly easily. These carbanions can act as nucleophiles in substitution and addition reactions just like the other types of nucleophiles presented in units 29 through 32.

$$R-\overset{\overset{\displaystyle O}{\|}}{C}-\underset{\overset{\displaystyle |}{H}}{CH_2} \xrightarrow{\text{Base}} R-\overset{\overset{\displaystyle O}{\|}}{C}-\overset{\ominus}{CH_2}$$

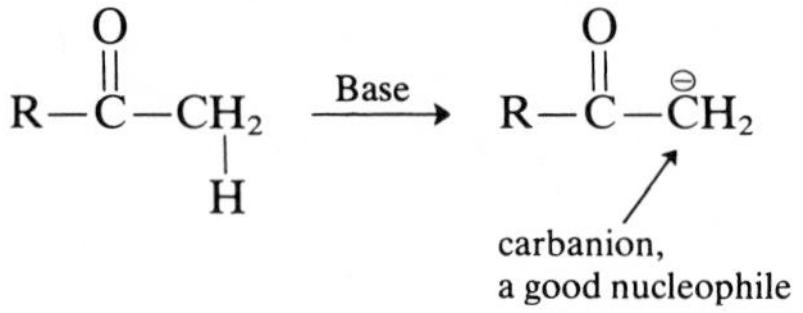

carbanion,
a good nucleophile

$$R-\overset{\overset{\displaystyle O}{\|}}{C}-CH_2^{\ominus} + C-X \longrightarrow R-\overset{\overset{\displaystyle O}{\|}}{C}-CH_2-C + X:^-$$

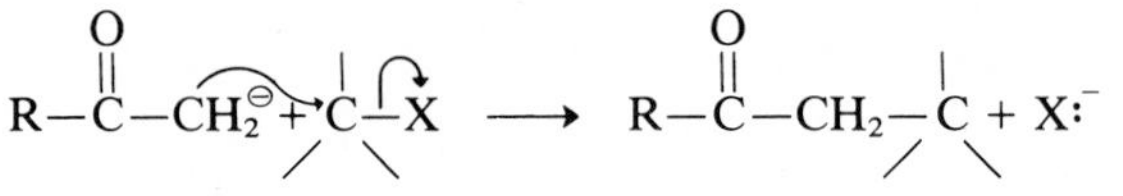

Synthesis and Reactions of Aldehydes and Ketones

Oxidation-Reduction

Carbonyl groups are often synthesized by oxidation of the appropriate alcohol with oxidizing agents such as potassium permanganate, sodium dichromate, or oxygen gas. Oxidation of a primary alcohol yields an aldehyde. But since aldehydes themselves are susceptible to further oxidation; they are converted into carboxylic acids. It is difficult to isolate aldehydes in good yield when they are formed by oxidation.

$$R-\underset{\overset{\displaystyle |}{H}}{\overset{\overset{\displaystyle H}{|}}{C}}-O-H + \text{oxidizing agent} \longrightarrow \left[R-\overset{\overset{\displaystyle H}{|}}{C}=O \right] \xrightarrow{(Ox)} R-\overset{\overset{\displaystyle OH}{|}}{C}=O$$

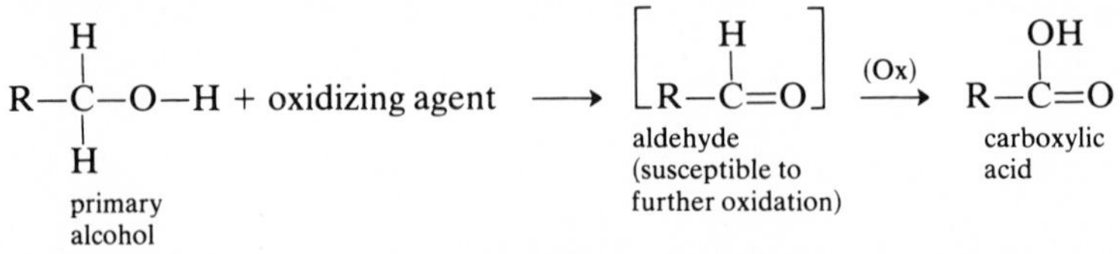

primary
alcohol

aldehyde
(susceptible to
further oxidation)

carboxylic
acid

Oxidation of secondary alcohols yields ketones. Ketones are generally resistant to further oxidation, so this is a good method of synthesizing ketones.

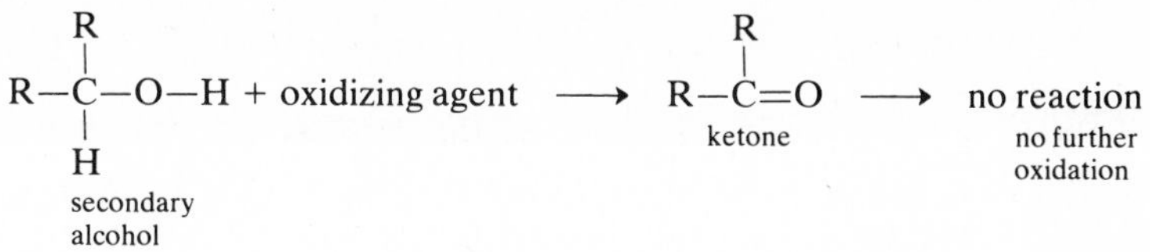
oxidizing agent
ketone
no reaction
no further oxidation
secondary alcohol

The reverse of oxidation is reduction, so reduction of aldehydes yields 1° alcohols and reduction of ketones yields 2° alcohols:

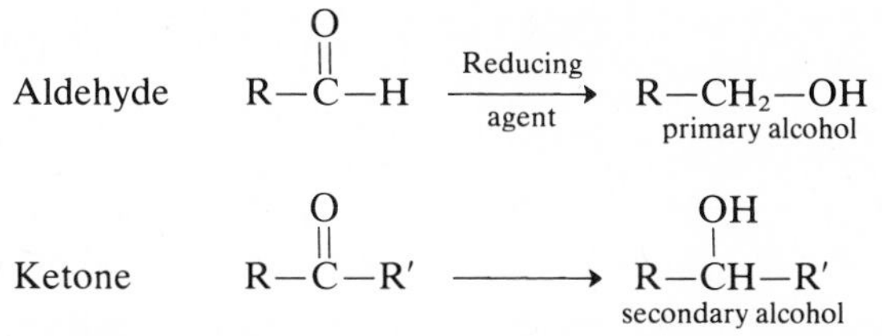
Aldehyde
Reducing agent
primary alcohol
Ketone
secondary alcohol

Reducing agents: H_2 + Pt, $LiAlH_4$

Many of the metabolic cycles present in living organisms use oxidation and reduction as a means of processing food and obtaining energy. Two examples are:

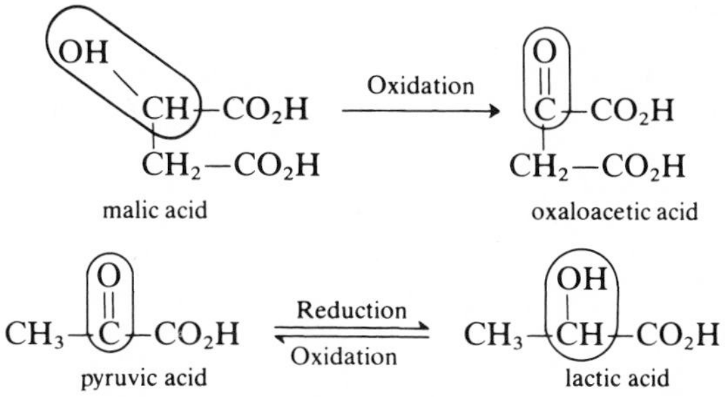
Oxidation
malic acid
oxaloacetic acid
Reduction
Oxidation
pyruvic acid
lactic acid

Detection of Aldehydes and Ketones

2,4-dinitrophenylhydrazones: Both aldehydes and ketones react with 2,4-dinitrophenylhydrazine to yield highly colored solids, 2,4-dinitrophenylhydrazones (2,4-DNP's). These 2,4-DNP's form quite rapidly and range from bright yellow to deep red in color. The hydrazine reagent does not react in similar fashion with other functional groups, so this is a good test for the presence of aldehyde or ketone groups within a molecule.

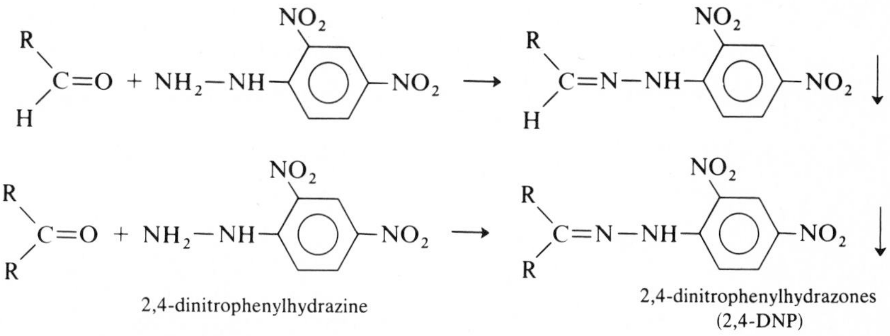
2,4-dinitrophenylhydrazine
2,4-dinitrophenylhydrazones
(2,4-DNP)

Tollens' test: Since aldehydes are susceptible to further oxidation and ketones are not, we can use oxidation reactions to distinguish between the two groups. In the Tollens' test, silver ion is used as the oxidizing agent. During the oxidation–reduction reaction, it is converted to metallic silver, which plates out in the test tube as a silver mirror. The formation of this silver mirror is indication of the presence of an aldehyde. No other groups are oxidized by the reagent.

$$R-\overset{\overset{\displaystyle O}{\|}}{C}-H + Ag(NH_3)_2^+ \, OH^- \longrightarrow R\overset{\overset{\displaystyle O}{\|}}{C}-O^- + Ag°\downarrow + 2NH_3 + H_2O$$

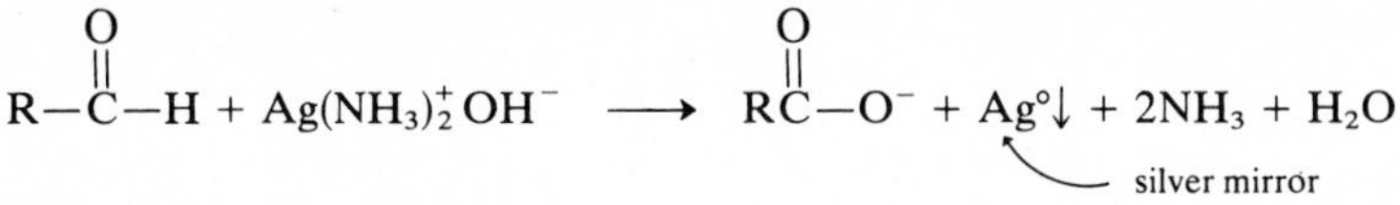

silver mirror

Additions to the $>C=O$ Group

The carbonyl group in aldehydes and ketones is polar, so polar reagents will add to it readily. The negative portion of the reagent (the portion with nonbonding electrons) always ends up at the carbon atom, and the positive portion attacks oxygen. There is a variety of different reagents for these additions but the mechanisms for the additions are quite similar.

$$\overset{\delta^+}{\underset{/}{\diagdown}}C\overset{\delta^-}{=}O + H^{\delta^+} \cdots A^{\delta^-} \longrightarrow \overset{\diagdown}{\underset{/}{}}\underset{A}{\overset{|}{C}}-\underset{H}{\overset{|}{O}}$$

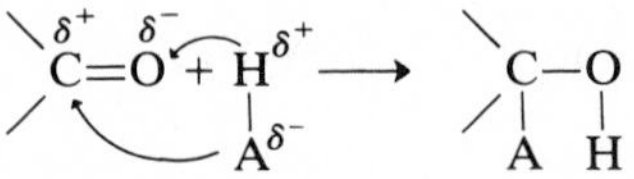

Hydration

$$R-\overset{\overset{\displaystyle O}{\|}}{\underset{H}{C}} + OH_H \overset{H^+}{\rightleftharpoons} R-\underset{H}{\overset{O-H}{\overset{|}{C}}}-OH$$

$$R-\overset{\overset{\displaystyle O}{\|}}{\underset{R}{C}} + OH_H \overset{H^+}{\rightleftharpoons} R-\underset{R}{\overset{OH}{\overset{|}{C}}}-OH$$

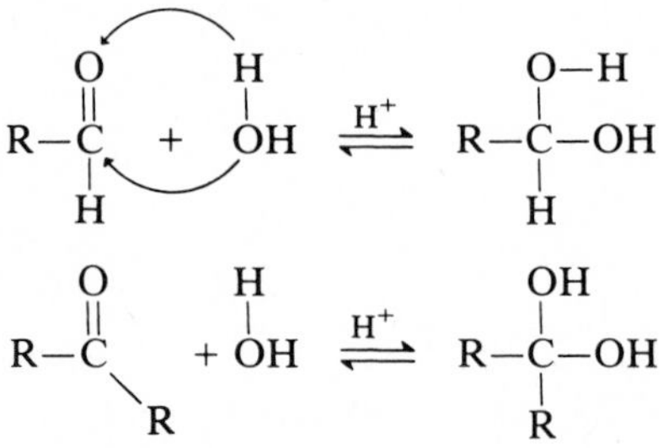

Hydration is a reversible, acid-catalyzed reaction. Normally, the equilibrium lies on the side of the aldehyde or ketone.

Hemiacetals and Acetals

Alcohols can add reversibly to aldehydes in the presence of acid catalysts. A **hemiacetal** group is defined as a carbon atom bearing both an alcohol (—OH) and ether (—O—R') group. Since the formation of hemiacetals is reversible, a hemiacetal is often considered a potential aldehyde group. Exchange of the free OH group with a second —OR group converts a hemiacetal to an **acetal** (a diether). Note again that the reaction is reversible; in fact, to isolate acetals, we must carefully remove all water from the system.

$$CH_3-CH_2-\overset{\overset{\displaystyle O}{\|}}{C}-H + CH_3OH \overset{H^+}{\rightleftharpoons} CH_3CH_2-\underset{OCH_3}{\overset{OH}{\overset{|}{C}}}-H \xrightarrow[\text{+ CH}_3\text{OH}]{} CH_3CH_2-\underset{OCH_3}{\overset{O-CH_3}{\overset{|}{C}}}-H + H_2O$$

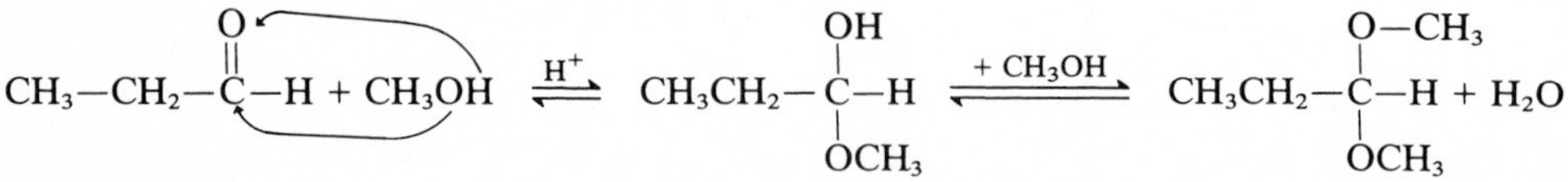

a hemiacetal   an acetal

Hemiacetals can form whenever an aldehyde and alcohol are present together in solution. When the >C=O and R—OH groups are present in the same molecule, internal hemiacetal formation can occur, resulting in the formation of rings. This sort of ring formation occurs readily in carbohydrates.

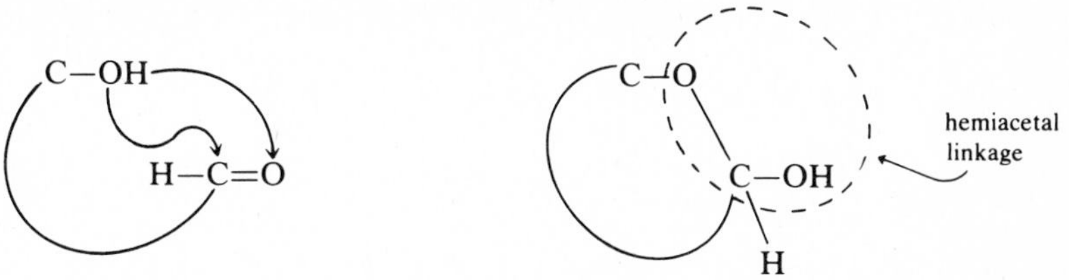
C—OH
H—C=O
C—O
C—OH
H
hemiacetal
linkage

Acetals are readily converted back to aldehydes if treated with acid and water, but they are stable in neutral or basic media. Often the acetal group is used to prevent a carbonyl group from reacting while other groups within the molecule react. For example, if you wanted to make pentanal from 3-pentenal you could not do it by treatment with H_2 since both the >C=C< and >C=O groups would be reduced. First, you should make the acetal and then reduce the >C=C< group with hydrogen. Later the aldehyde center can be regenerated.

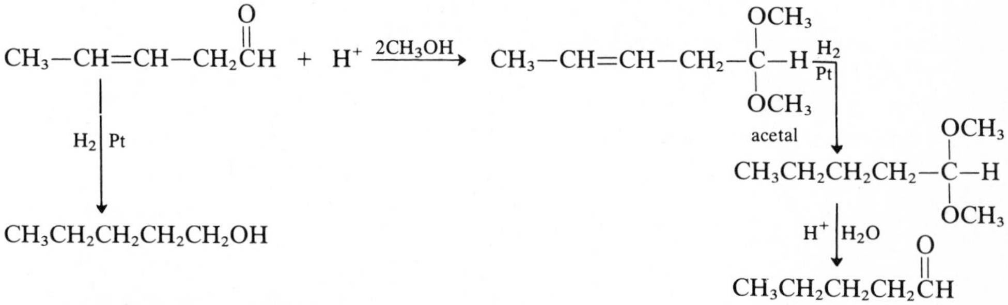
2CH₃OH
H₂ Pt
acetal
H⁺ H₂O
CH₃CH₂CH₂CH₂CH₂OH

Hemiketals and Ketals

Aldehydes react with alcohols to form hemiacetals and acetals; ketones react to form **hemiketals** and **ketals**. The main difference is that hemiketals and ketals do not form as readily as hemiacetals and acetals.

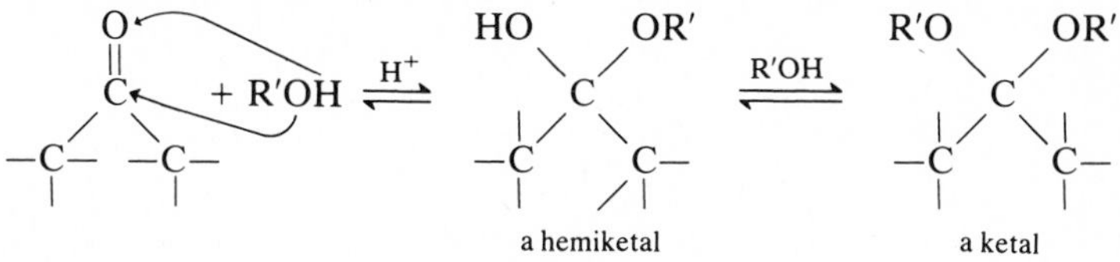
+ R'OH
H⁺
R'OH
a hemiketal
a ketal

Addition of HCN to Aldehydes and Ketones

Hydrogen cyanide adds to aldehydes and ketones in much the same manner that alcohols add to them. The products are hydroxycyanides:

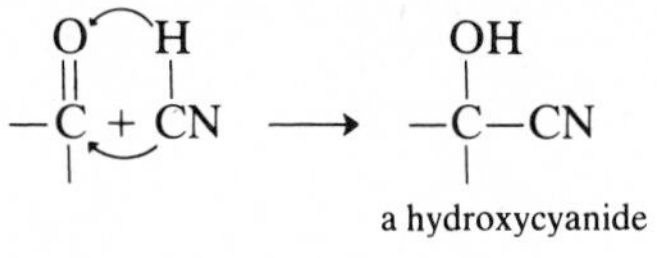
a hydroxycyanide

This reaction is often useful in synthesis because the hydroxycyanides can be converted into several other types of compounds:

$$\underset{\underset{H}{|}}{\overset{\overset{OH}{|}}{R-C-CN}} \;\rightarrow \rightarrow \rightarrow\; \underset{\underset{H}{|}}{\overset{\overset{NH_2}{|}}{R-C-CO_2H}}$$

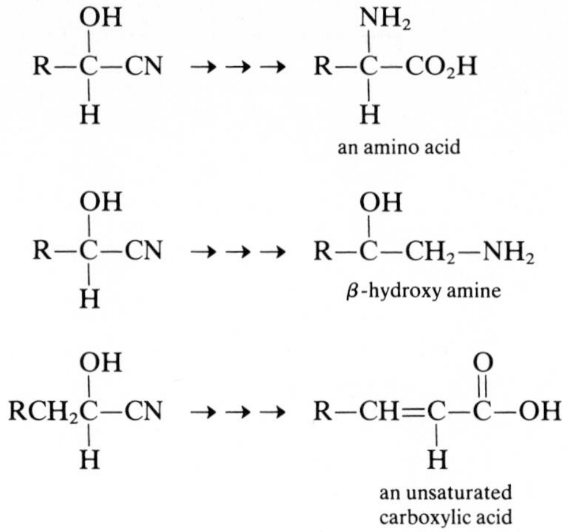

an amino acid

$$\underset{\underset{H}{|}}{\overset{\overset{OH}{|}}{R-C-CN}} \;\rightarrow \rightarrow \rightarrow\; \underset{}{\overset{\overset{OH}{|}}{R-C-CH_2-NH_2}}$$

β-hydroxy amine

$$\underset{\underset{H}{|}}{\overset{\overset{OH}{|}}{RCH_2C-CN}} \;\rightarrow \rightarrow \rightarrow\; \underset{\underset{H}{|}}{R-CH=C-\overset{\overset{O}{||}}{C}-OH}$$

an unsaturated
carboxylic acid

Grignard Reactions

When organo halides are treated with elemental magnesium, the halide is removed and the carbon bearing the halide becomes a carbanion:

$$R-CH_2-X + Mg \longrightarrow R-\ddot{C}H_2^- MgX^+ \quad \text{(a Grignard reagent)}$$

These types of carbanions are called **Grignard reagents** and are often used as nucleophiles to attack the carbonyl groups of aldehydes and ketones:

$$R-\ddot{C}H_2^- + \overset{|}{\underset{|}{C}}=\ddot{O}: \quad R-CH_2-\overset{|}{\underset{|}{C}}-\ddot{O}:^- \xrightarrow{H_2O} R-CH_2-\overset{|}{\underset{|}{C}}-O-H$$

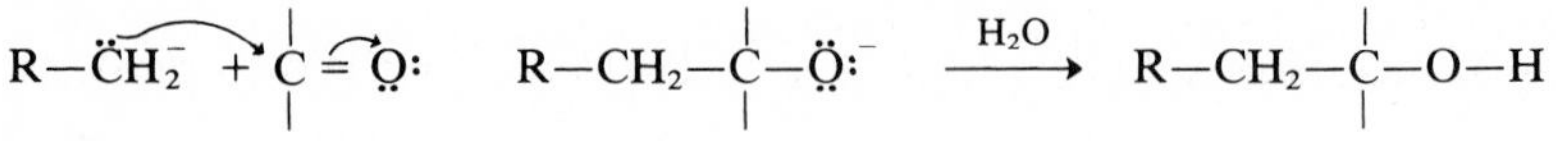

These types of additions are good synthetic methods for making alcohols and enlarging the molecule. Some examples are shown below.

$$CH_3CH_2CH_2-Br \xrightarrow{Mg} CH_3CH_2\bar{C}H_2\overset{+}{Mg}Br \xrightarrow[CH_3]{\overset{\overset{H}{|}}{\underset{}{C=O}}} CH_3CH_2CH_2-\underset{\underset{CH_3}{|}}{\overset{\overset{H}{|}}{C}}-O^-$$

$$\Big\downarrow H_2O$$

$$CH_3CH_2CH_2-\underset{\underset{CH_3}{|}}{\overset{\overset{H}{|}}{C}}-OH$$

$$\underset{\underset{CH_3}{|}}{\overset{}{\bigcirc}\text{—}CH-Br} \xrightarrow{Mg} \underset{\underset{CH_3}{|}}{\ddot{C}H^-\overset{+}{Mg}Br} \xrightarrow[CH_3]{\overset{\overset{CH_3}{|}}{\underset{\underset{CH_3}{|}}{\overset{}{C=\ddot{O}:}}}{CH_2}} \underset{\underset{CH_2}{\underset{\underset{CH_3}{|}}{|}}{\overset{}{CH-\overset{\overset{CH_3}{|}}{\underset{\underset{CH_3}{|}}{C}}-O^-}} \xrightarrow{H_2O} \underset{\underset{CH_2}{\underset{\underset{CH_3}{|}}{|}}{\overset{}{CH-\overset{\overset{CH_3}{|}}{\underset{\underset{CH_3}{|}}{C}}-OH}}$$

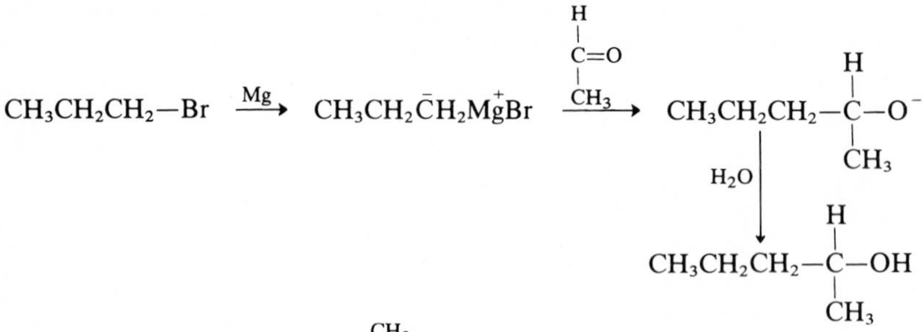

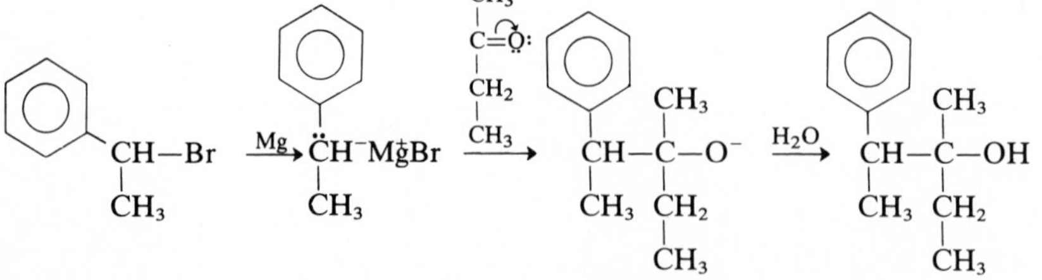

Aldol Condensation

The **aldol condensation** is a reaction similar in many respects to other addition reactions. The main difference is that the molecule that attacks the aldehyde is another aldehyde. The products, hydroxyaldehydes, are much more stable than hemiacetals.

$$R-\overset{\overset{O}{\|}}{C}-H + R'CH_2\overset{\overset{O}{\|}}{C}-H \xrightarrow{\text{Base}} R-\overset{\overset{OH}{|}}{\underset{\underset{H}{|}}{C}}-\overset{\overset{}{}}{\underset{\underset{R}{|}}{CH}}-\overset{\overset{O}{\|}}{C}-H$$

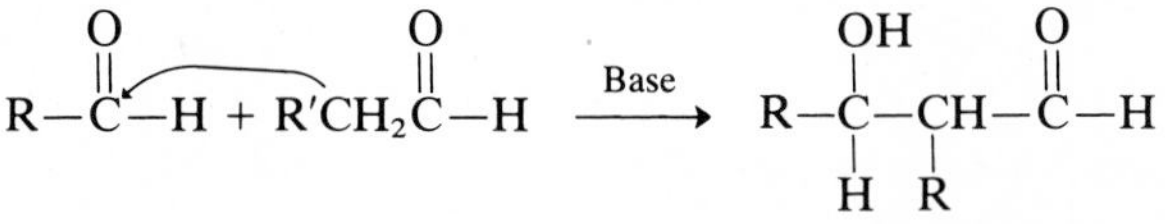

a β-hydroxyaldehyde

This reaction is illustrative of a number of reactions in which *a carbon adjacent to the carbonyl group (the α-carbon) of one molecule becomes attached to the carbonyl carbon of a second molecule.* It involves attack by a carbanion at an alpha-carbon atom on the carbon of the carbonyl group in a second aldehyde molecule:

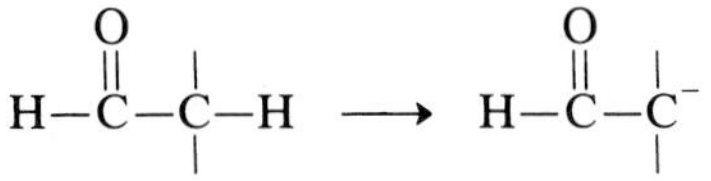

$$H-\overset{\overset{O}{\|}}{C}-\overset{\overset{}{|}}{\underset{\underset{}{|}}{C}}-H \longrightarrow H-\overset{\overset{O}{\|}}{C}-\overset{\overset{}{|}}{C^-} \quad (\alpha\text{-carbanion})$$

$$H-\overset{\overset{O}{\|}}{C}-\overset{\overset{H}{|}}{\underset{\underset{-C-}{|}}{C^-}}+C{=}\ddot{O}: \longrightarrow H-\overset{\overset{O}{\|}}{C}-\overset{\overset{H}{|}}{\underset{\underset{-C-}{|}}{C}}-\overset{}{\underset{\underset{}{|}}{C}}-O^- \xrightarrow{H_2O} H-\overset{\overset{O}{\|}}{C}-\overset{\overset{H}{|}}{\underset{\underset{-C-}{|}}{C}}-\overset{}{\underset{\underset{}{|}}{C}}-OH$$

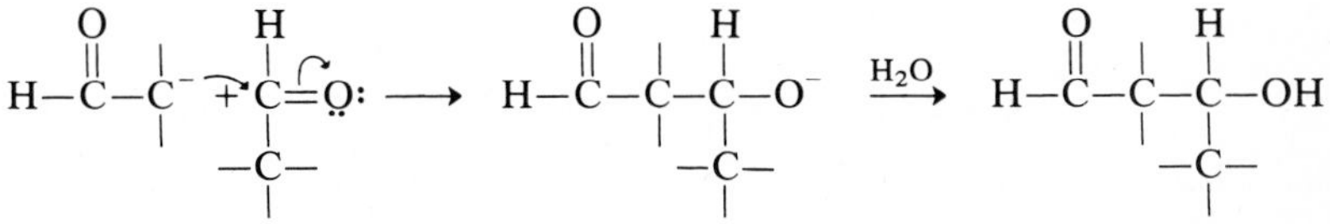

In all cases, the alpha-carbon atom of one aldehyde molecule becomes covalently bound to the carbonyl carbon atom of a second aldehyde. The carbonyl group of that *second* aldehyde molecule is converted into an alcohol group. Notice that two different products can form if each aldehyde has an α-hydrogen.

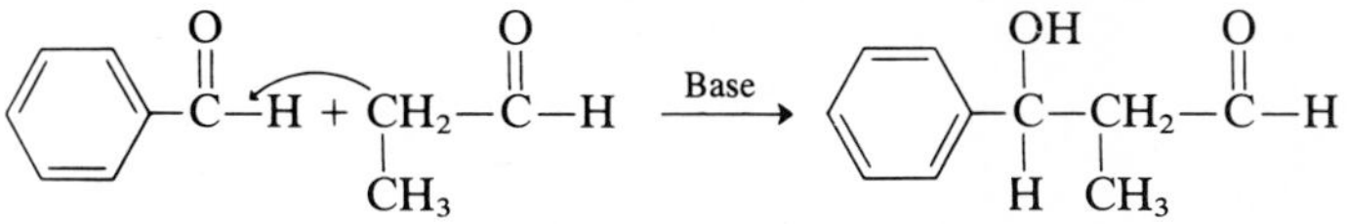

$$\bigcirc-\overset{\overset{O}{\|}}{C}-H + \overset{\overset{O}{\|}}{\underset{\underset{CH_3}{|}}{CH_2}}-C-H \xrightarrow{\text{Base}} \bigcirc-\overset{\overset{OH}{|}}{\underset{\underset{H}{|}}{C}}-\overset{}{\underset{\underset{CH_3}{|}}{CH_2}}-\overset{\overset{O}{\|}}{C}-H$$

(no α-hydrogens)   (α-hydrogens)

$$CH_3\overset{\overset{O}{\|}}{C}-H + CH_3-\overset{\overset{O}{\|}}{C}-H \xrightarrow{\text{Base}} CH_3-\overset{\overset{OH}{|}}{\underset{\underset{H}{|}}{C}}-\overset{}{\underset{\underset{CH_3}{|}}{CH_2}}-\overset{\overset{O}{\|}}{C}-H$$

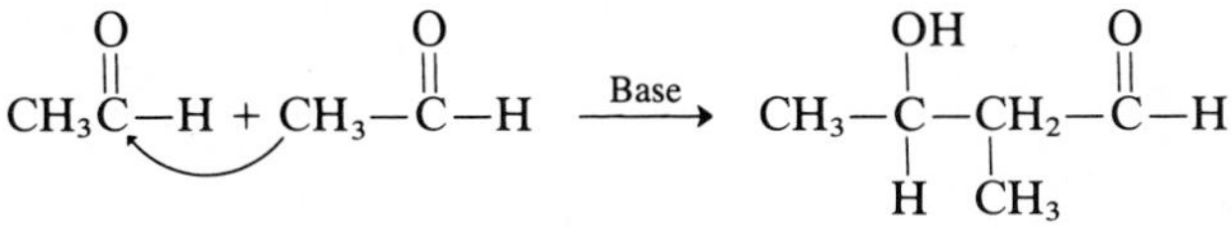

or (both have α-hydrogens):

$$CH_3-\overset{\overset{O}{\|}}{C}-H + CH_3CH_2\overset{\overset{O}{\|}}{CH} \xrightarrow{\text{Base}} CH_3CH_2\overset{\overset{OH}{|}}{\underset{\underset{H}{|}}{C}}CH_2\overset{\overset{O}{\|}}{C}-H$$

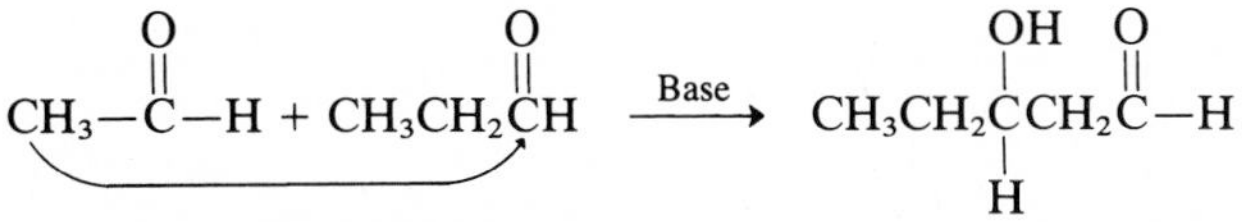

Aldol condensations can also occur between two molecules of the same aldehyde or ketone. This is called **self-condensation**.

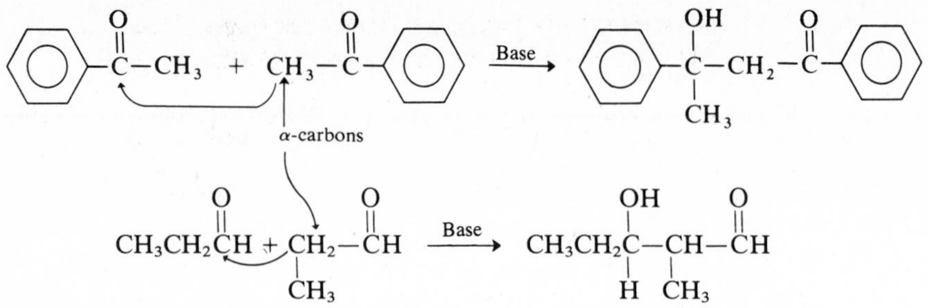
Base
α-carbons
Base

Reverse aldol condensations can also occur. In these reactions, a hydroxyaldehyde can cleave to form two aldehydes.

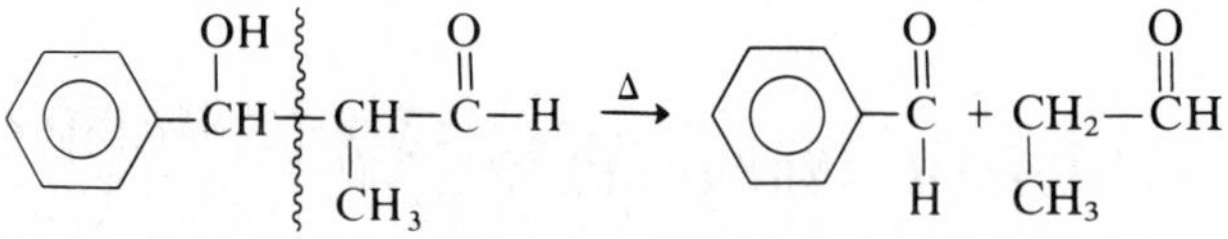

Metabolically, one method of carbohydrate synthesis occurs by aldol condensation between dihydroxyacetone phosphate and glyceraldehyde.

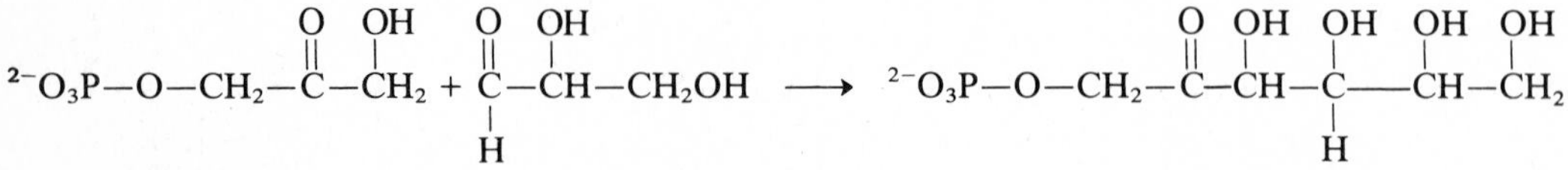

dihydroxyacetone phosphate anion — glyceraldehyde — a precursor to fructose

UNIT 33 Problems and Exercises

33–1. Name the following compounds.

(a) 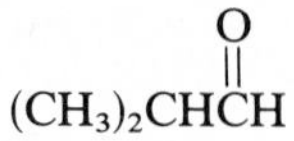

(b) 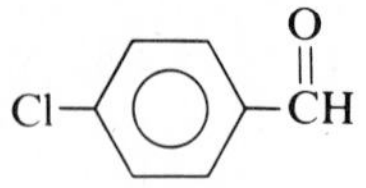

(c) 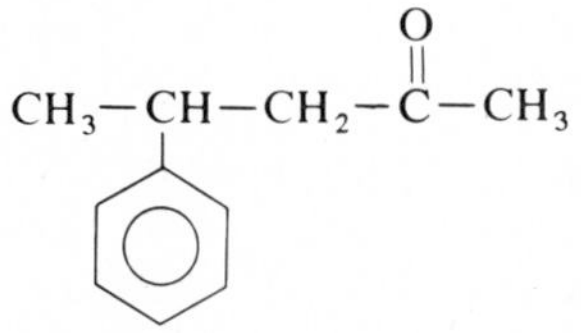

(d) 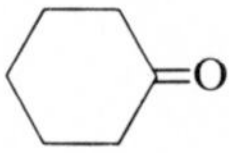

(e) 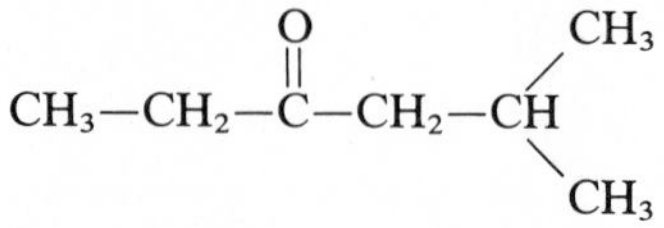

(f) 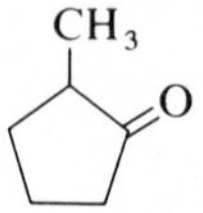

(g) 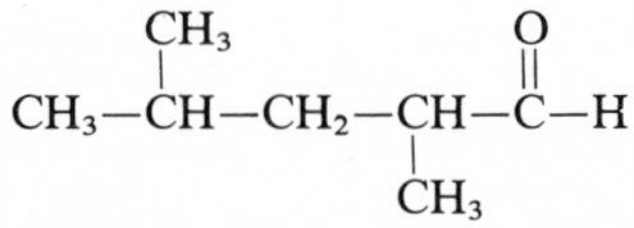

33–2. Draw structures for the following compounds.

(a) Acetone
(b) Benzaldehyde
(c) Formaldehyde
(d) Methyl-*sec*-butyl ketone
(e) 3-methylpentanal
(f) 3-methyl-2-hexanone
(g) 4,4-diphenyl-1-butanal
(h) 1,2-diphenyl-2-propen-1-one
(i) 2-pentenal
(j) 3-methoxy-2-heptanone
(k) 3-iodo-2-ethylheptanal

33–3. Draw the structural formulas, give the common names, if any exist, and give the IUPAC names for the following.

(a) Seven carbonyl compounds with the formula $C_5H_{10}O$
(b) Five carbonyl compounds with the formula C_8H_8O that contain a benzene ring

33–4. Explain why aldehydes and ketones are less polar than alcohols but more polar than alkenes.

33–5. Complete the following reactions.

(a) $CH_3{-}\overset{\text{OH}}{\overset{|}{C}H}{-}CH_3 + KMnO_4 \longrightarrow$

(b) Butanal + $KMnO_4 \longrightarrow$

(c) Acetone + $KMnO_4 \longrightarrow$

(d) 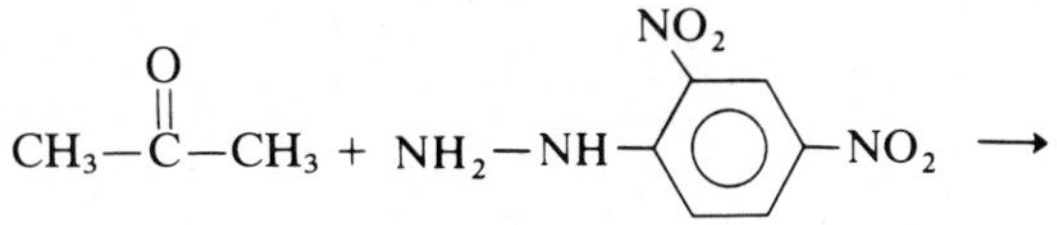

(e) 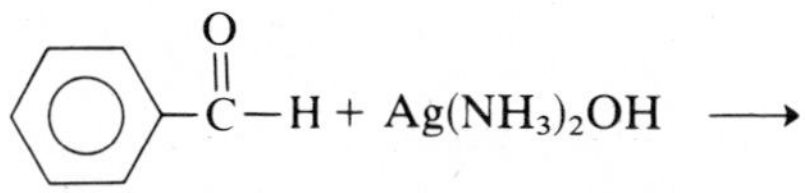

(f) 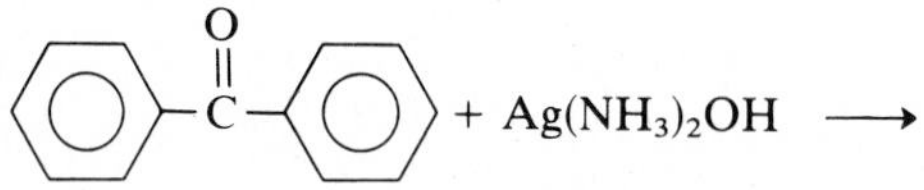

(g) Benzaldehyde + $2CH_3CH_2OH \xrightarrow[\Delta]{H^+}$

(h) $CH_3CH_2{-}\overset{\text{O}}{\overset{||}{C}}{-}CH_3 + H_2 \xrightarrow{\text{Pt}}$

(i) $CH_3{-}CH_2{-}\overset{\text{O}}{\overset{||}{C}}{-}CH_3 + HCN \longrightarrow$

33–6. Give the structures of the alcohols that will result from the following Grignard addition reactions.

(a) 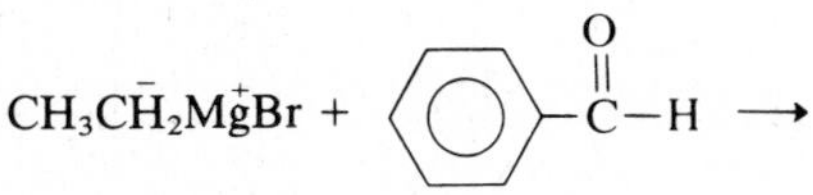

(b) 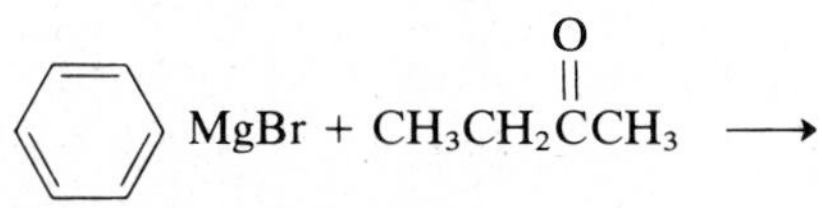

(c) $CH_3\overset{-}{C}H(CH_3)M\overset{+}{g}Br + CH_3{-}\overset{O}{\overset{\|}{C}}{-}CH_2CH_3 \longrightarrow$

33–7. Choose Grignard reagents and aldehydes or ketones that will yield the following alcohols. More than one choice is possible.

(a) $CH_3{-}CH_2{-}\underset{}{\overset{CH_3}{\overset{|}{C}H}}{-}OH$ (b) 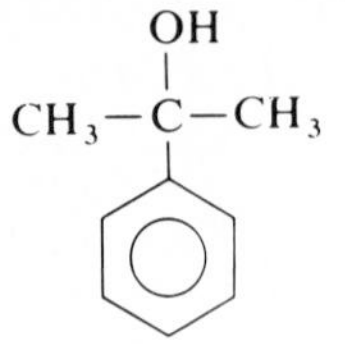 (c) $CH_3CH_2CH_2CH_2\overset{OH}{\overset{|}{C}H_2}$

33–8. Use acetone as starting material and write equations for the synthesis of the following compounds. Use any other reagents that you need.

(a) 2-propanol (b) 2-bromopropane (c) $(CH_3)_2CH{-}CN$

(d) 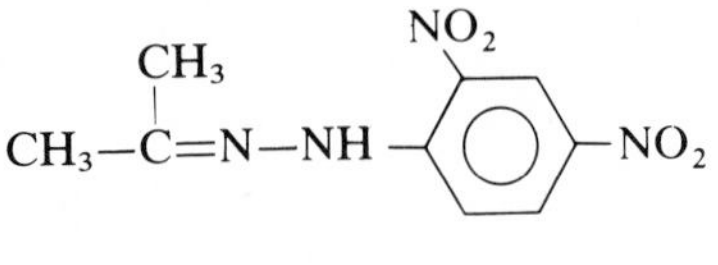 (e) $CH_3{-}\overset{OH}{\overset{|}{C}}(CH_3){-}CH_2{-}\overset{O}{\overset{\|}{C}}{-}CH_3$

33–9. Use propanol as starting material and write equations for the synthesis of the following. Use any other reagents that you need.

(a) *n*-propanol (b) Isopropanol (c) Propene

(d) $CH_3CH_2\overset{O}{\overset{\|}{C}}{-}OH$ (e) 1-cyanopropane (f) *N*-propyl-1-butanamine

(g) $CH_3CH_2\overset{OH}{\overset{|}{C}H}{-}\underset{H-C=O}{\underset{|}{C}HCH_3}$

33–10. How would you synthesize the following compounds from the reactants listed?
(a) Make acetone from 1-bromopropane. (b) Make butanal from 1-bromobutane.

33–11. Write the equations and the hemiacetal and acetal forms for the following.

(a) $CH_3CH_2\overset{O}{\overset{\|}{C}}H + 2CH_3OH$ (b) 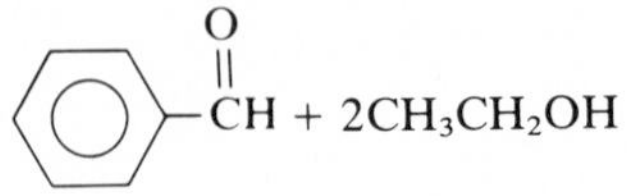

33–12. Write the hemiketal and ketal forms as well as the equations for the following.

(a) Acetone + $2CH_3OH$

(b) 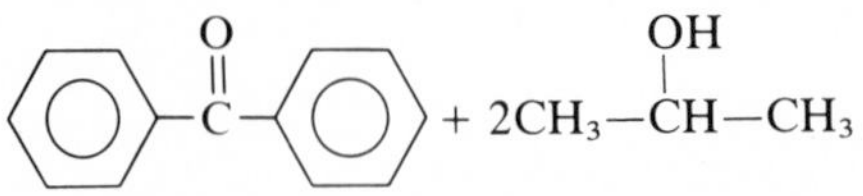

33–13. Complete the following aldol condensation reactions.

(a) 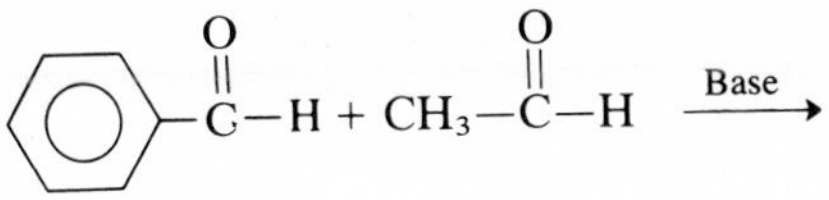

(b) $CH_3CH_2\overset{O}{\overset{\|}{C}}-H + CH_3CH_2\overset{O}{\overset{\|}{C}}-H \xrightarrow{\text{Base}}$

(c) 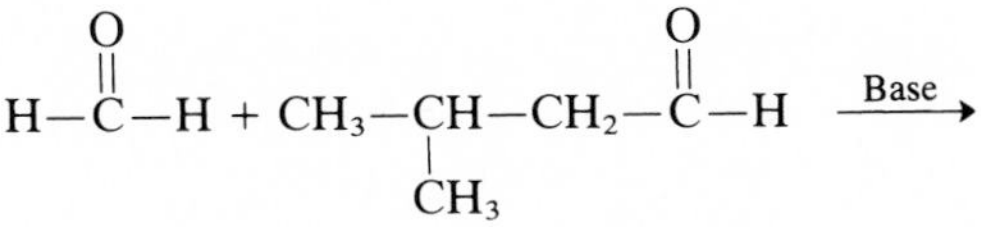

(d) 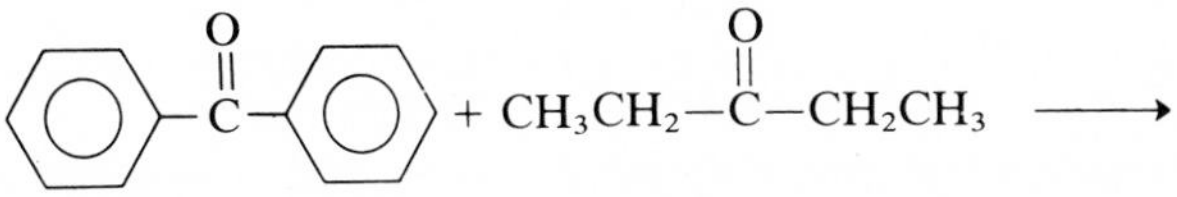

33–14. Draw all the possible aldol condensation products resulting from self-condensation of

$$CH_3-CH_2-\overset{O}{\overset{\|}{C}}-CH_2-CH_2-CH_3$$

33–15. Draw all the possible aldol condensation products resulting from reaction of 2-butanone and 2-pentanone with each other.

33–16. What aldehydes must you begin with to synthesize the following by the aldol condensation?

(a) 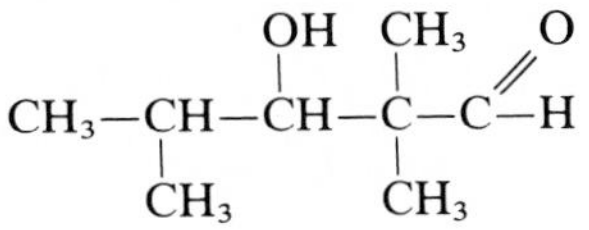

(b) 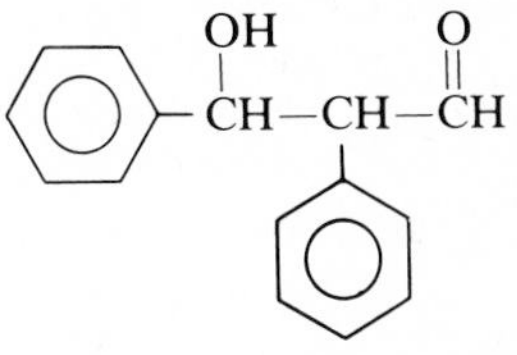

(c) 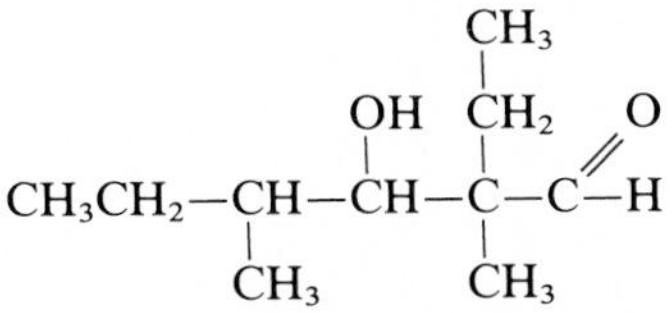

33–17. What ketones must you begin with to synthesize the following by the aldol condensation?

(a) 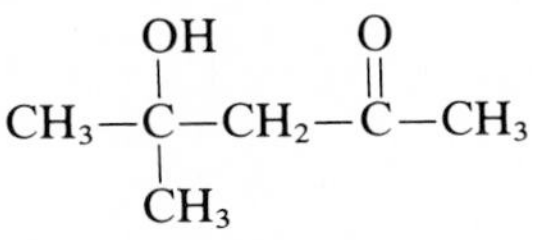

(b) 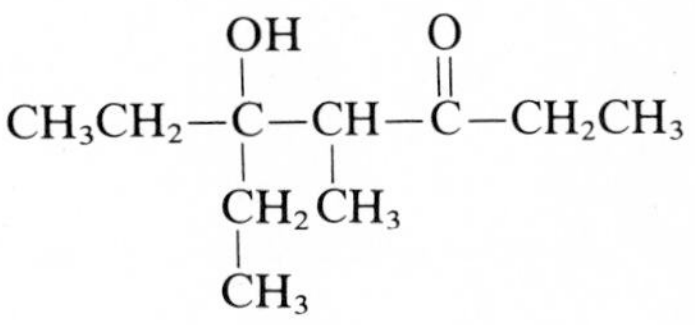

(c) 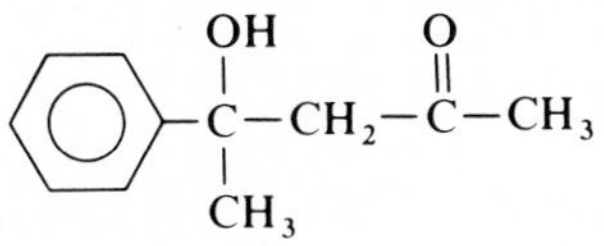

(d) 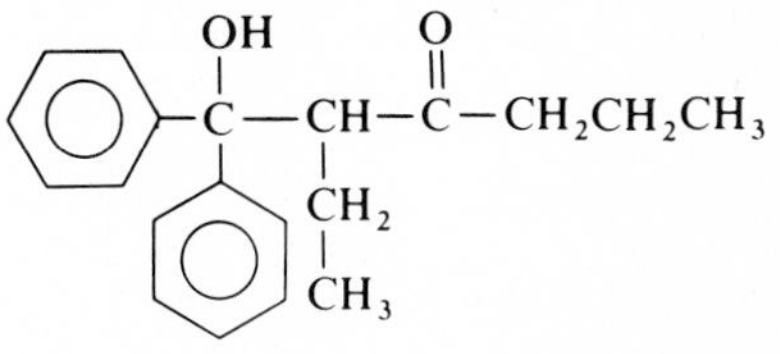

33–18. Devise short chemical tests to help you distinguish between the following pairs of compounds. Tell how your tests would help you to decide.
(a) Propane and acetone
(b) Propanal and l-propanol
(c) Acetone and propanal
(d) 2-butanone and 2-butanol

33–19. An organic compound with the formula $C_5H_{10}O$ reacts readily with 2,4-DNP reagent But not with Tollens' reagent. It will give but one aldol self-condensation product when treated with base. Give a possible structure for the compound and its aldol product.

33–20. An organic compound was analyzed and found to contain 71.4% carbon, 9.5% hydrogen, and 19.0% oxygen. It has a molecular weight of 84, reacts readily with Tollens' reagent, 2,4-DNP, and bromine. It does not possess cis or trans isomers. Give a structure for the compound.

UNIT 34 CARBOXYLIC ACIDS

In addition to aldehydes and ketones, other molecules that contain carbonyl groups are the **carboxylic acids** and their derivatives. All differ from aldehydes and ketones in that they have electronegative groups attached to the carbonyl carbon atom. A carboxylic acid has an —OH group directly attached to the carbonyl carbon, and the derivatives have OR′, Cl, N̈<, or —O—C(=O)—R attached.

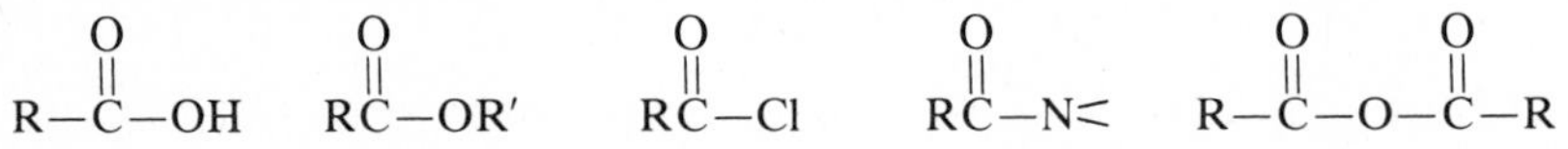

carboxylic acid ester acid chloride amide anhydride

Nomenclature of Carboxylic Acids

Carboxylic acids are named much like aldehydes. *The carbonyl group carbon atom is always carbon number 1.* The parent suffix is *-oic* acid. Substituents are named as usual.

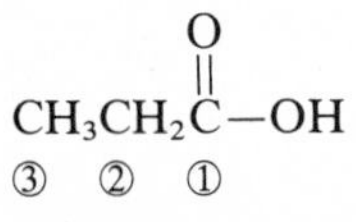

propanoic acid

$CH_3CH_2CH_2CH_2CO_2H$
⑤ ④ ③ ② ①

pentanoic acid

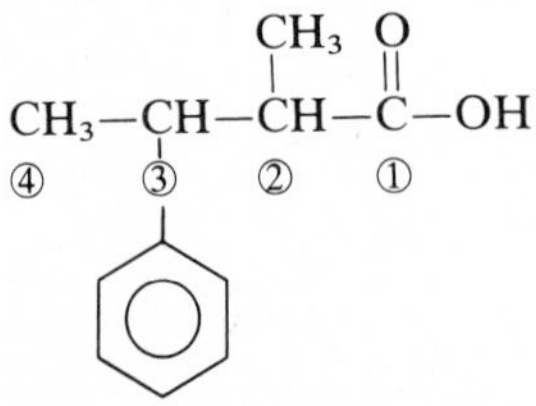

2-methyl-3-phenylbutanoic acid

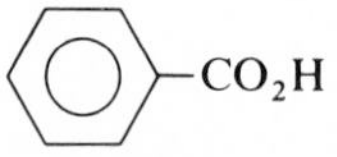

benzoic acid

One problem many students encounter is that they forget to count the carboxyl carbon in the parent and thus give the acid the name for a one-carbon-shorter chain.

A second method of locating positions in the parent chain is by use of Greek letters rather than by numbers. This method is still used so we must be familiar with it.

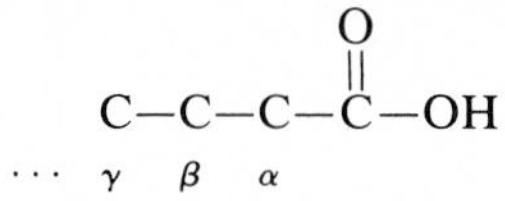

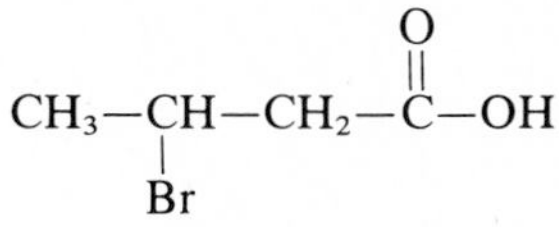

β-bromobutanoic acid

As with aldehydes, a large number of common names exist for carboxylic acids. Most of them are derived from sources from which the acids were isolated. Table 34–1 gives some of the common derivations.

Table 34–1 Common names of carboxylic acids

IUPAC name	Common name	Formula	Derivation
Methanoic	Formic	HCO_2H	Latin, *formica* (ant)
Ethanoic	Acetic	CH_3CO_2H	Latin, *acetum* (vinegar)
Propanoic	Propionic	$CH_3CH_2CO_2H$	Greek, *protos pion* (first fat)
Butanoic	Butyric	$CH_3(CH_2)_2CO_2H$	Latin, *butyrium* (butter)
Pentanoic	Valeric	$CH_3(CH_2)_3CO_2H$	Latin, *valere* (powerful)
Hexanoic	Caproic	$CH_3(CH_2)_4CO_2H$	Latin, *caper* (goat)
Octanoic	Caprylic	$CH_3(CH_2)_6CO_2H$	Latin, *caper* (goat)
Decanoic	Capric	$CH_3(CH_2)_8CO_2H$	Latin, *caper* (goat)
Dodecanoic	Lauric	$CH_3(CH_2)_{10}CO_2H$	Laurel
Hexadecanoic	Palmitic	$CH_3(CH_2)_{14}CO_2H$	Palm
Octadecanoic	Stearic	$CH_3(CH_2)_{16}CO_2H$	Greek, *stear* (tallow)

Physical Properties of Carboxylic Acids

A carboxylic acid is not simply a combination of a carbonyl group and a hydroxyl group. It has properties quite different from aldehydes and alcohols. The OH group is attached to a C=O group, and this interaction gives rise to several new properties. This new group is called a **carboxyl** group and may be represented as

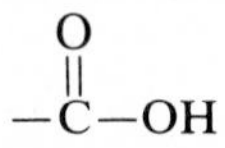 or —COOH or $—CO_2H$

In alcohols the polar O—H bond results in increased boiling points and water solubility. The O—H bond in carboxylic acid is even more polar than in alcohols

because of the presence of the C=O group. Thus carboxylic acids are the most polar of the organic compounds and as a result have the highest boiling points and water solubility. Table 34–2 compares carboxylic acids with alcohols.

Table 34–2 Comparison of boiling points and solubilities of carboxylic acids and alcohols of similar molecular weight

Carboxylic acid	BP	Solubility in H_2O	Alcohol	BP	Solubility in H_2O
Formic	101	Infinite	Ethanol	78	Infinite
Acetic	119	Infinite	l-propanal	97	Infinite
Propionic	141	Infinite	l-butanol	117	8 g/100 mℓ
Butyric	163	5.6 g/100 mℓ	l-pentanol	138	2.2 g/100 mℓ
Valeric	186	3.7 g/100 mℓ	l-hexanol	158	0.7 g/100 mℓ

A number of carboxylic acids have rather strong disagreeable odors, which are associated with goats, rancid butter, certain cheeses, vinegar, and sweaty locker rooms. It is fortunate that these acids have high boiling points, or we might all be inundated by the odors.

Acid Behavior of Carboxylic Acids

Carboxylic acids are more acidic than any of the other types of organic molecules. They have K_a values ranging from 10^{-2} to 10^{-6}. In the presence of bases, carboxylic acids release H^+ ions and become **carboxylate** ions.

$$\underset{\text{carboxylic acids}}{\mathrm{RC(=O)-OH}} \rightleftharpoons \underset{\text{carboxylate ions}}{\mathrm{RC(=O)-O^-}} + \mathrm{H^+}, \qquad K_a \sim 10^{-5}$$

$$\underset{\text{alcohols}}{\mathrm{R-OH}} \rightleftharpoons \underset{\text{alkoxide ions}}{\mathrm{R-O^-}} + \mathrm{H^+}, \qquad K_a \sim 10^{-20}$$

There are two reasons for the increased acidity of carboxylic acids. The electron-withdrawing ability of the adjacent >C=O group greatly increases the polarity of the O—H bond. Second, the resulting carboxylate ion is **resonance stabilized**. In this process the negative charge does not exist simply on one atom but is spread over both oxygen atoms by resonance. The lessening of the charge on single atoms greatly increases their stability.

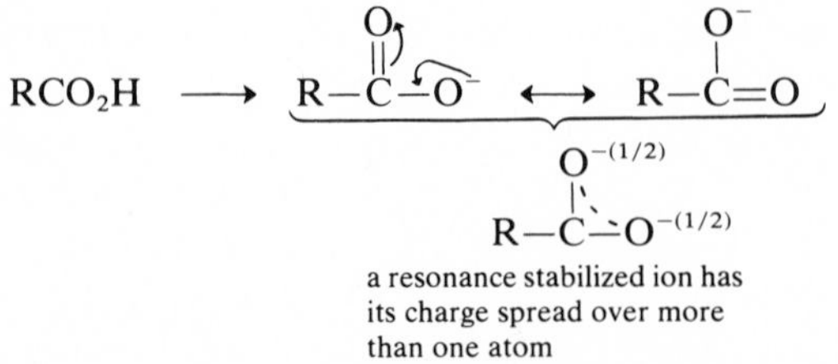

a resonance stabilized ion has its charge spread over more than one atom

Sources and Biological Uses of Carboxylic Acids

Carboxylic acids can be formed by oxidation of aldehydes and primary alcohols. When a wine goes bad, some of the ethyl alcohol present has been converted to acetic acid. A wine connoisseur always smells the cork of an opened bottle of wine to try to detect if the odor of acetic acid is present.

Dilute solutions of acetic acid act as inhibitors of microbial growth, so vinegar (~ 5% acetic acid in water) has often been used as a preservative for meat, fish, and pickled vegetables.

A diverse number of compounds contain the carboxyl group. The amino acids are most prevalent, but compounds like aspirin also contain carboxylic acid groups. Lysergic acid, the precursor of LSD, contains a carboxylic group, as does 2,4-D, a herbicide.

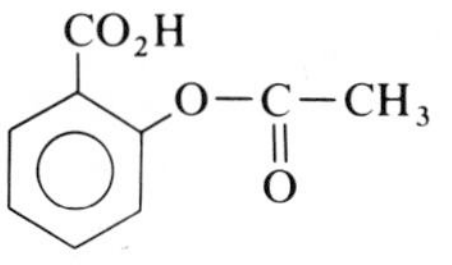

acetylsalicyclic acid
(aspirin)

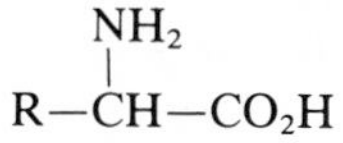

amino acids

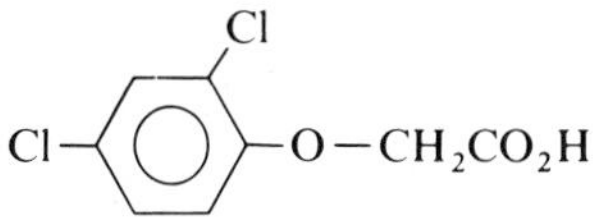

2,4-dichlorophenoxyacetic acid
(2,4-D)

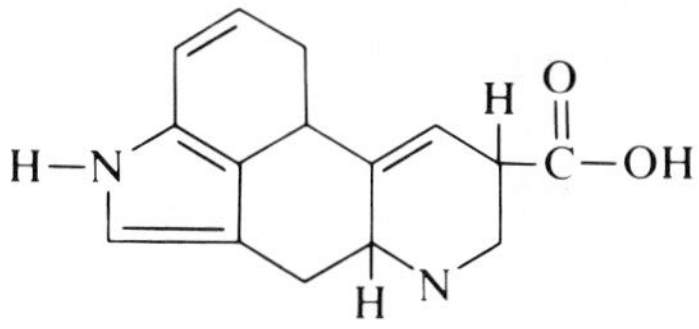

lysergic acid

Synthesis and Reactions of Carboxylic Acids

Two important methods of synthesizing carboxyl groups are oxidation of primary alcohols and hydrolysis of cyano groups. In oxidation, a primary alcohol is first converted to an aldehyde, which is further oxidized to a carboxylic acid. Biologically, this oxidation is catalyzed by enzymes and often involves further oxidation of the carboxyl group to carbon dioxide. This is one of the main sequences by which the body obtains energy from foods.

$$RCH_2OH \xrightarrow[\text{Enzymes}]{[Ox]} R\overset{O}{\overset{\|}{C}}-H \xrightarrow[\text{Enzymes}]{[Ox]} R\overset{O}{\overset{\|}{C}}-OH \xrightarrow[\text{Enzymes}]{[Ox]} CO_2$$

Cyano groups are often added to an organic molecule by substitution reactions on alkyl halides. Treatment of a cyano group with water in the presence of acid or base catalysts results in formation of a carboxyl group:

$$R-Br + CN^- \xrightarrow{\text{Substitution}} R-CN \xrightarrow[H^+ \text{ or } OH^-]{H_2O} R-CO_2H$$

Hydrolysis

$$CH_3CH_2-Br + CN^- \longrightarrow CH_3CH_2CN \xrightarrow[H^+]{H_2O} CH_3CH_2CO_2H + NH_4^+$$

This reaction is important to organic chemists as it is a good method of increasing the length of the carbon chain by one carbon atom.

Reactions With Bases

Carboxylic acids react rapidly with basic compounds. A chemical test often used to detect the presence of a carboxyl group is treatment with sodium bicarbonate. The vigorous evolution of carbon dioxide is taken as proof of the presence of a carboxyl group:

$$RCO_2H + NaHCO_3 \longrightarrow RCO_2^-Na^+ + H_2O + CO_2\uparrow$$

Molecular weights of carboxylic acids can often be determined by titrating the carboxylic acid with base.

Reduction: Reaction With Reducing Agents

Carboxylic acids can be converted to primary alcohols if they are treated with reducing agents such as hydrogen gas and a platinum catalyst, or lithium aluminium hydride, $LiAlH_4$. In these reductions, the carboxylic acid is first converted to the aldehyde, which is then reduced to the primary alcohol:

$$\underset{\text{acid}}{R{-}CO_2H} \xrightarrow[(LiAlH_4;\ H_2,\ Pt)]{\text{Reducing agent}} \underset{\text{aldehyde}}{(R{-}\overset{\overset{\displaystyle O}{\|}}{C}{-}H)} \xrightarrow[(LiAlH_4;\ H_2,\ Pt)]{\text{Reducing agent}} \underset{1^\circ\ \text{alcohol}}{R{-}CH_2{-}OH}$$

Formation of the Derivatives of Carboxylic Acids

The reactions of carboxylic acids that are of principal concern to us can be classed as substitution reactions in which some group, G, takes the place of the OH in the carbonyl group. The overall reaction is one of substitution, but it may also be classed as addition-elimination.

$$R{-}\overset{\overset{\displaystyle O}{\|}}{C}{-}OH \longrightarrow R\overset{\overset{\displaystyle O}{\|}}{C}{-}G \quad \text{(overall reaction is substitution)}$$

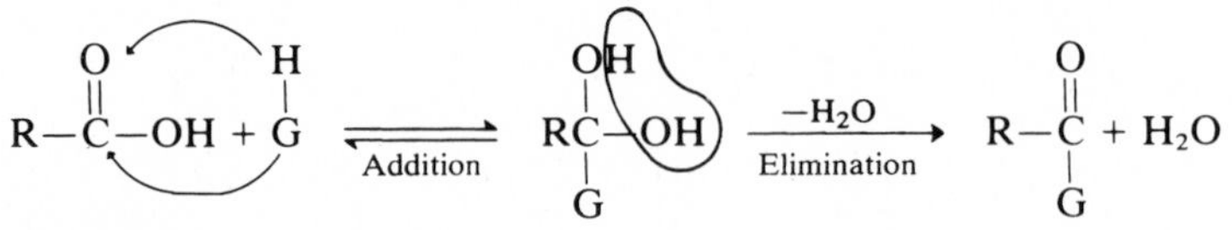

O
H
R—C—OH + G
Addition
OH
RC—OH
G
−H₂O
Elimination
O
R—C + H₂O
G

Acid chlorides:

$$R\overset{\overset{\displaystyle O}{\|}}{C}{-}OH + \xrightarrow[\text{or } PCl_3 \text{ or } PCl_5]{SOCl_2} R\overset{\overset{\displaystyle O}{\|}}{C}{-}Cl$$

Acid chlorides are generally synthesized by treatment of the corresponding carboxylic acids with thionyl chloride ($SOCl_2$), phosphorous trichloride (PCl_3), or phosphorous pentachloride (PCl_5). The following are examples:

$$CH_3CO_2H + SOCl_2 \longrightarrow \underset{\text{acetyl chloride}}{CH_3\overset{\overset{\displaystyle O}{\|}}{C}{-}Cl} + SO_2\uparrow + HCl\uparrow$$

$$CH_3CH_2CH_2CO_2H + PCl_3 \longrightarrow CH_3CH_2CH_2\overset{O}{\overset{\|}{C}}{-}Cl + H_3PO_3$$

butanoic acid — butanoyl chloride

$$CH_3\underset{}{\overset{CH_3}{\overset{|}{C}}}H{-}CO_2H + PCl_5 \longrightarrow CH_3\underset{\underset{CH_3}{|}}{C}H{-}\overset{O}{\overset{\|}{C}}{-}Cl + POCl_3 + HCl$$

2-methylpropanoic acid — 2-methylpropanoyl chloride

Esters: In the presence of acid catalysts, carboxylic acids react with alcohols to yield **esters**. This is in essence a substitution reaction where the OH group from the carboxylic acid is replaced by the R—O— group from the alcohol:

$$R'{-}\overset{O}{\overset{\|}{C}}{-}\boxed{OH + H}{-}OR \overset{H^+}{\rightleftharpoons} R'\overset{O}{\overset{\|}{C}}{-}O{-}R + H_2O$$

water is eliminated

For example,

$$CH_3{-}\overset{O}{\overset{\|}{C}}{-}\boxed{OH + H}{-}O{-}CH_3 \overset{H^+}{\rightleftharpoons} CH_3{-}\overset{O}{\overset{\|}{C}}{-}O{-}CH_3 + H_2O$$

acetic acid — methanol — methyl acetate

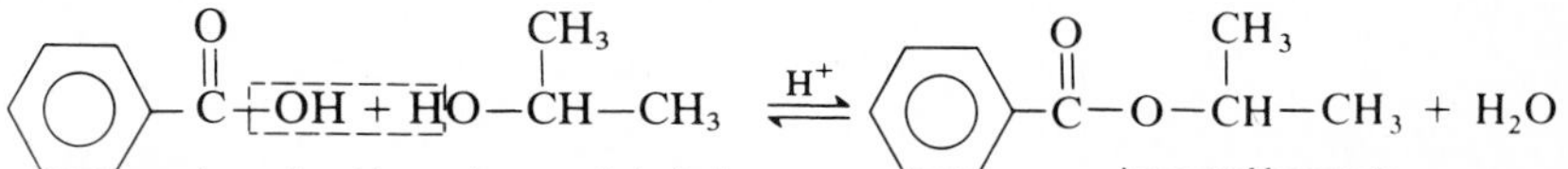

benzoic acid — isopropyl alcohol — isopropyl benzoate

Formation of esters in this manner is a *reversible* reaction. Thus, in order to convert all of a carboxylic acid into an ester, you must remove the water from the reaction mixture as it forms. As long as water is present in the reaction mixture, some of the ester can be converted back into a carboxylic acid.

Amides: Under appropriate conditions, primary and secondary amines react with carboxylic acids to form **amides**. Water is eliminated in this reaction just as it was in ester formation.

$$R{-}\overset{O}{\overset{\|}{C}}{-}\boxed{OH + H}{-}N\!< \rightleftharpoons R\overset{O}{\overset{\|}{C}}{-}\ddot{N}\!< + H_2O$$

water is eliminated — an amide

The following are examples:

$$CH_3\overset{O}{\overset{\|}{C}}{-}\boxed{OH + H}{-}\overset{H}{\overset{|}{N}}{-}CH_2CH_3 \longrightarrow CH_3\overset{O}{\overset{\|}{C}}{-}\overset{H}{\overset{|}{N}}{-}CH_2CH_3 + H_2O$$

acetic acid — ethylamine — *N*-ethylacetamide

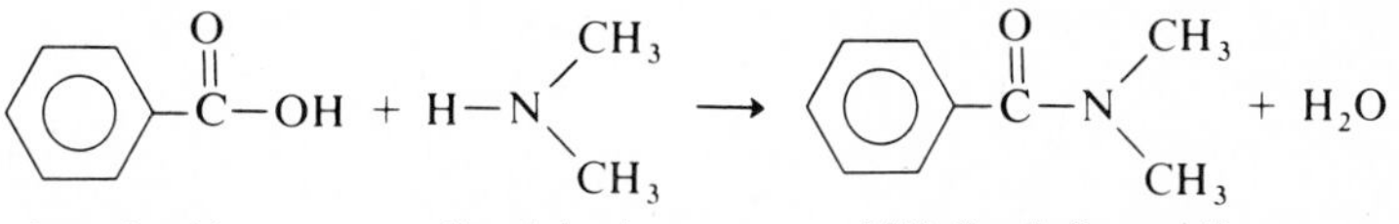

benzoic acid — dimethylamine — *N,N*-dimethylbenzamide

These reactions look similar to ester formation, but they occur much less easily. The reaction temperatures must be high, and usually an excess of acid must be present. Formation of amides occurs readily in body cells, however, but here enzymes provide the catalysis.

Acid anhydrides: Acid anhydrides are formed by the elimination of water from two molecules of carboxylic acid; but the process is slow and yields are low.

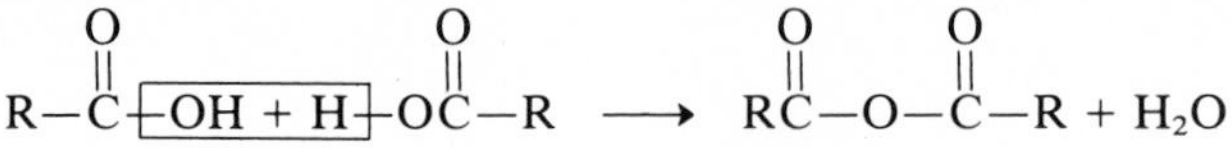

an acid anhydride

A better method of preparing them involves treatment of carboxylate ion with an acid chloride. This process will be discussed in the next unit.

UNIT 34 Problems and Exercises

34–1. Name the following carboxylic acids.

(a) 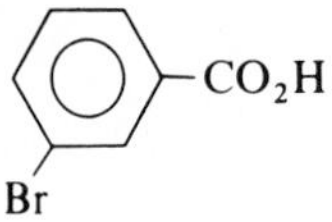

(b) HCO_2H

(c) CH_3CO_2H

(d) 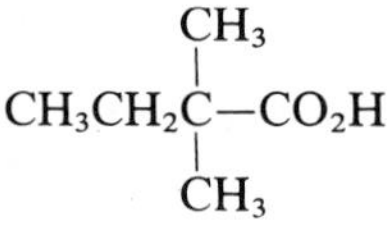

(e) 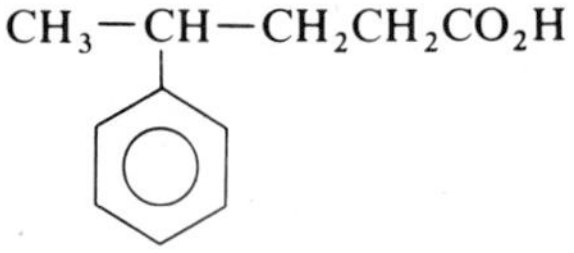

(f) 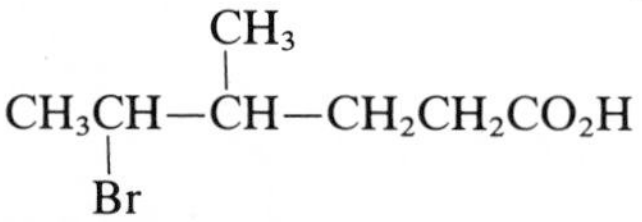

(g) $CH_3{-}CH{=}CH{-}CO_2H$

34–2. Draw the structures of all the carboxylic acids with the formula $C_6H_{12}O_2$ and name them.

34–3. Draw the structures for the following carboxylic acids.
(a) Trimethylacetic acid
(b) Isovaleric acid
(c) β-bromopropanoic acid
(d) 2-phenylbutanoic acid
(e) 3-methylvaleric acid
(f) *p*-isopropylbenzoic acid

34–4. Rank the following compounds according to predicted solubility in water. Explain your reasoning.

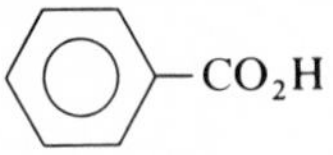

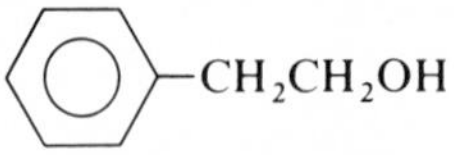

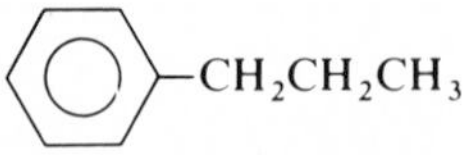

34–5. Explain why carboxylic acids are more acidic than alcohols.

34–6. Complete the following reactions.

(a) $CH_3CH_2OH \xrightarrow[KMnO_4]{Excess}$

(b) $CH_3CH_2CN \xrightarrow[H^+]{H_2O}$

(c) 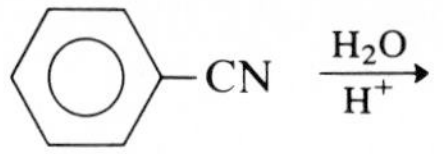

(d) 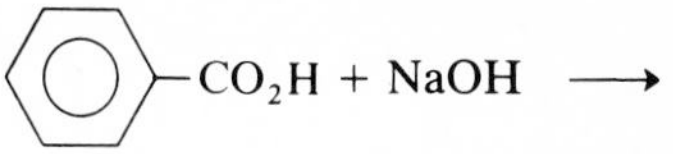

(e) $CH_3CO_2H + NaHCO_3 \longrightarrow$

(f) 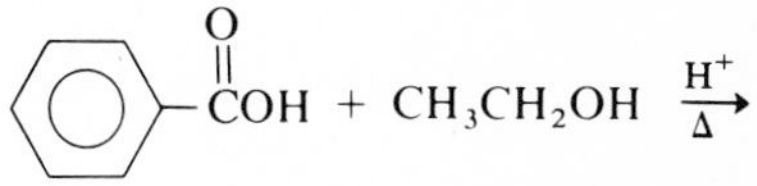

(g) $CH_3CO_2H \xrightarrow{SOCl_2}$

34–7. Give the alcohol and carboxylic acid necessary to synthesize the following esters.

(a) $CH_3CH_2\overset{\overset{O}{||}}{C}-OCH_3$

(b) 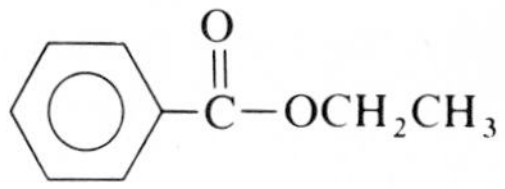

(c) 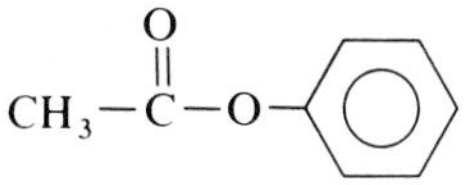

(d) $CH_3\underset{}{\overset{\overset{CH_3}{|}}{C}H}-O-\overset{\overset{O}{||}}{C}-CH_3$

(e) $CH_3\underset{\underset{CH_3}{|}}{C}H-\overset{\overset{O}{||}}{C}-O-CH_2CH_3$

34–8. Perform the following synthesis, starting with the material given and any other reagents necessary.

(a) Make propanoic acid from bromoethane.

(b) Make $CH_3CH_2\overset{\overset{O}{||}}{C}-O-CH_2CH_2CH_3$ from 2 mol of 1-propanol.

(c) Make $CH_3\overset{\overset{O}{||}}{C}-O-CH_2CH_3$ from 2 mol of acetic acid.

(d) Make $CH_3CH_2\overset{\overset{O}{||}}{C}-O-CH_2CH_3$ from 2 mol of ethanol.

34–9. Devise a quick chemical test to help you identify the presence or absence of a carboxylic acid group. Tell how the test would work.

34–10. How could you use simple chemical tests to distinguish between the following pairs of compounds?

(a) Ethanol and acetic acid

(b) Acetaldehyde and acetic acid

(c) Acetone and acetic acid

(d) Benzoic acid and 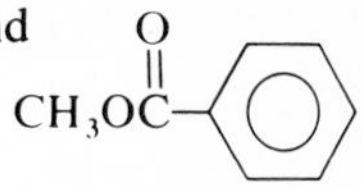

34–11. Calculate the molecular weight of a carboxylic acid if 0.44 g was neutralized by the addition of 50 mℓ of 0.1 *M* NaOH.

UNIT 35 DERIVATIVES OF CARBOXYLIC ACID

In this unit we shall investigate the nomenclature, properties, and reactions of the derivatives of carboxylic acids: **acid chlorides, anhydrides, esters,** and **amides**.

Nomenclature

All the derivatives are named starting from the name of the parent carboxylic acid. Generally, the ending of the parent carboxylic acid is changed to indicate the presence of a derivative.

Acid chlorides: Acid chlorides are formed by replacing the OH of the carboxyl group with Cl. They are named by replacing the *-oic acid* with *-oyl chloride*. See the examples in Table 35–1.

$$R-\overset{\displaystyle O}{\overset{\|}{C}}-OH \longrightarrow R-\overset{\displaystyle O}{\overset{\|}{C}}-Cl$$

a carboxylic acid   an acid chloride

$$CH_3CH_2CO_2H \qquad CH_3CH_2\overset{\displaystyle O}{\overset{\|}{C}}-Cl$$

propan*oic acid*   propan*oyl chloride*

Anhydrides: The combination of two molecules of carboxylic acid accompanied by the loss of water yields an anhydride.

$$R\overset{\displaystyle O}{\overset{\|}{C}}-(OH \ + \ H)O-\overset{\displaystyle O}{\overset{\|}{C}}-R \longrightarrow R-\overset{\displaystyle O}{\overset{\|}{C}}-O-\overset{\displaystyle O}{\overset{\|}{C}}-R$$

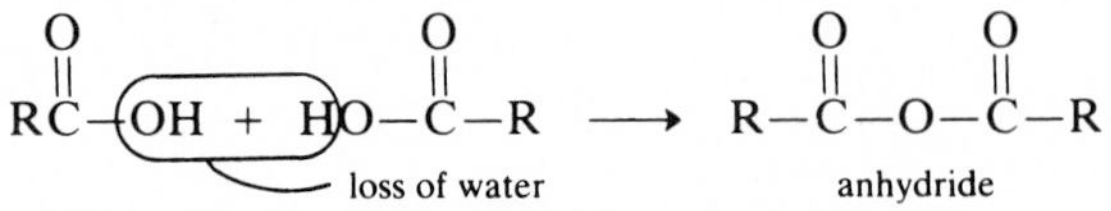

loss of water   anhydride

The word *anhydride* means without water. Anhydrides are named in a fashion similar to ethers. The name of each carboxylic acid that makes up the anhydride is followed by the word *anhydride*. If the two acids are identical, the name of the acid is used only one time.

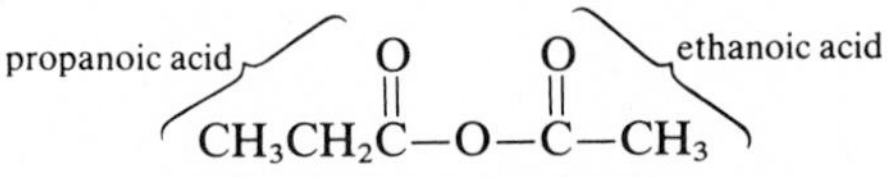

propanoic acid   ethanoic acid

$$CH_3CH_2\overset{\displaystyle O}{\overset{\|}{C}}-O-\overset{\displaystyle O}{\overset{\|}{C}}-CH_3$$

ethanoic propanoic anhydride

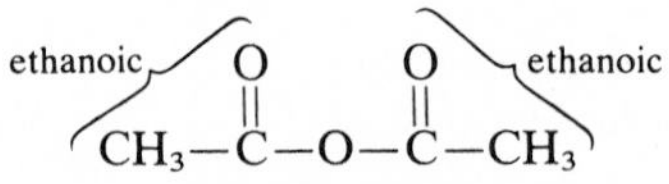

ethanoic   ethanoic

$$CH_3-\overset{\displaystyle O}{\overset{\|}{C}}-O-\overset{\displaystyle O}{\overset{\|}{C}}-CH_3$$

ethanoic anhydride
(acetic anhydride)

Esters: The combination of a carboxylic acid and an alcohol yields an ester. The ester portion, —OR, which comes from the alcohol reactant, is presented first in

Table 35–1 Naming of derivatives of carboxylic acids

Carboxylic acid (-oic acid)	Acid chlorides and anhydrides (-oyl chloride; -oic anhydrides)	Esters (alcohol . . . ate)	Amides (*N* . . . ______ amide)
CH_3CO_2H ethanoic acid (acetic acid)	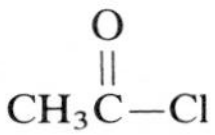 ethanoyl chloride (acetyl chloride)	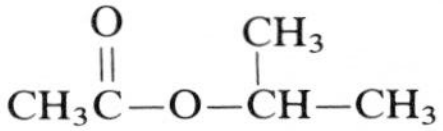 isopropyl ethanate (isopropyl acetate)	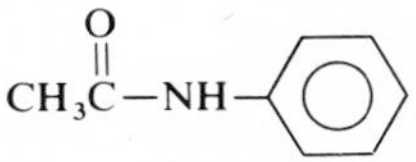 *N*-phenylethanamide (*N*-phenylacetamide)
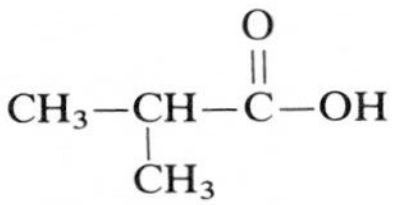 2-methylpropanoic acid	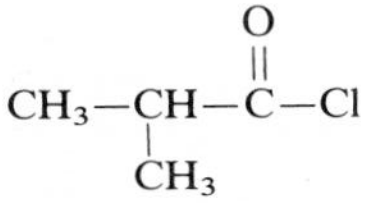 2-methylpropanoyl chloride	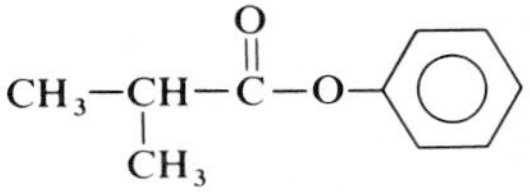 phenyl 2-methylpropanate	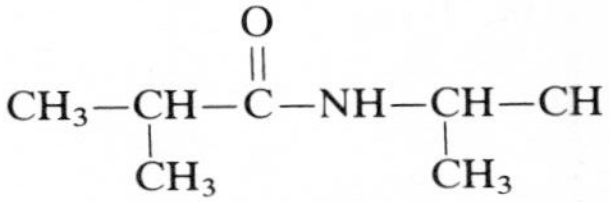 *N*-isopropyl-2-methyl propanamide
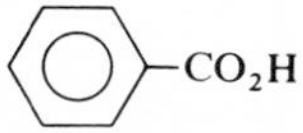 benzoic acid	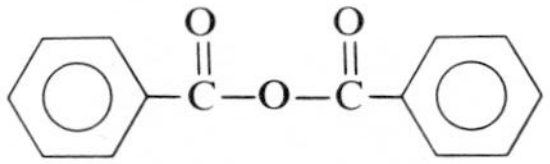 benzoic anhydride	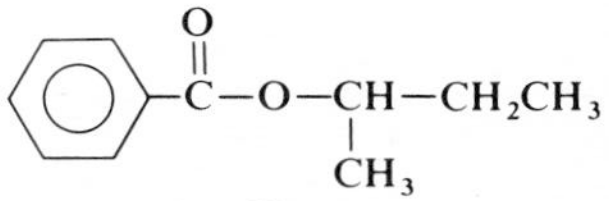 *sec*-butyl benzoate	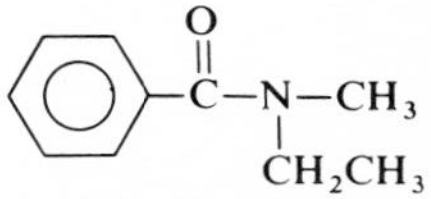 *N*-methyl-*N*-ethyl benzamide
	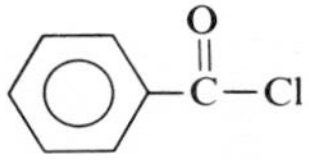 benzoyl chloride		
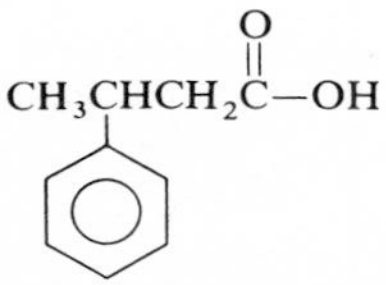 3-phenylbutanoic acid	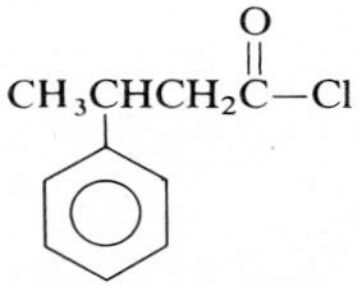 3-phenylbutanoyl chloride	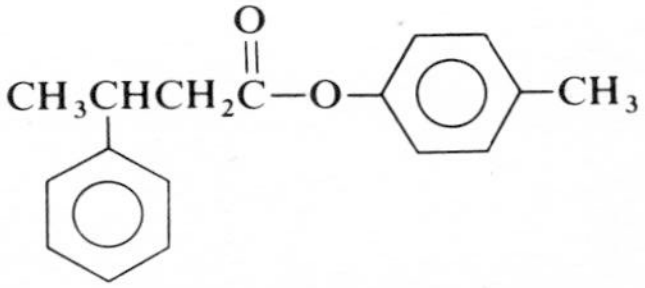 *p*-methylphenyl 3-phenylbutanate	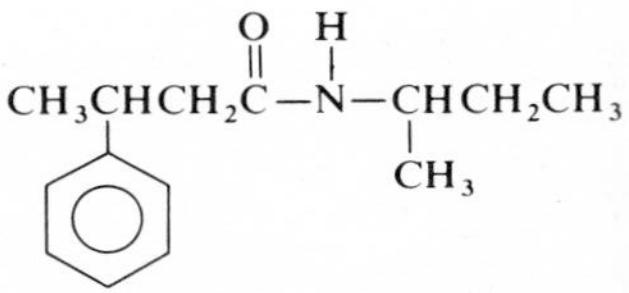 *N*-*sec*-butyl-3-phenyl butanamide

the name. It is named like other substituents (-yl ending), but it is a separate word. The name of the carboxylic acid parent follows, except the -oic acid is replaced by -ate.

$$CH_3CH_2\overset{\displaystyle O}{\overset{\|}{C}}\boxed{-O-H \quad H}-O-CH_2CH_3 \longrightarrow CH_3CH_2\overset{\displaystyle O}{\overset{\|}{C}}-O-CH_2CH_3$$

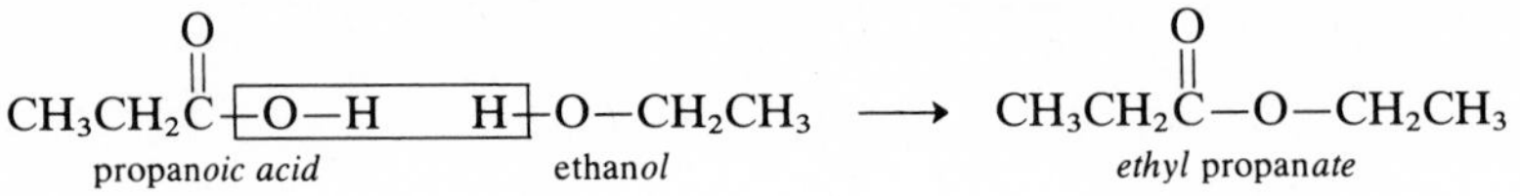

propanoic acid | ethanol | ethyl propanate

$$CH_3-\overset{\displaystyle CH_3}{\overset{|}{CH}}-CH_2\overset{\displaystyle O}{\overset{\|}{C}}-OH + CH_3OH \rightleftharpoons CH_3-\overset{\displaystyle CH_3}{\underset{|}{CH}}-CH_2\overset{\displaystyle O}{\overset{\|}{C}}-O-CH_3$$

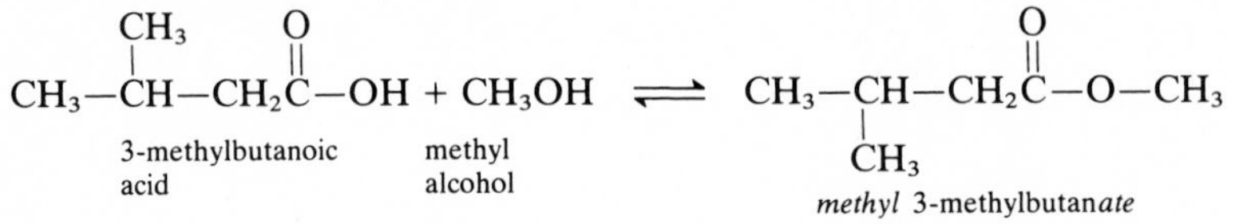

3-methylbutanoic acid | methyl alcohol | methyl 3-methylbutanate

Amides: Carboxylic acids and amines react to form amides. In naming them, replace the -oic acid with *amide*. Any organic groups attached to nitrogen are placed at the beginning of the name and are preceded by the prefix *N*-.

$$CH_3CH_2\overset{\displaystyle O}{\overset{\|}{C}}-OH + \overset{\displaystyle H}{\overset{|}{HN}}-CH_3 \longrightarrow CH_3CH_2\overset{\displaystyle O}{\overset{\|}{C}}-\overset{\displaystyle H}{\overset{|}{N}}-CH_3$$

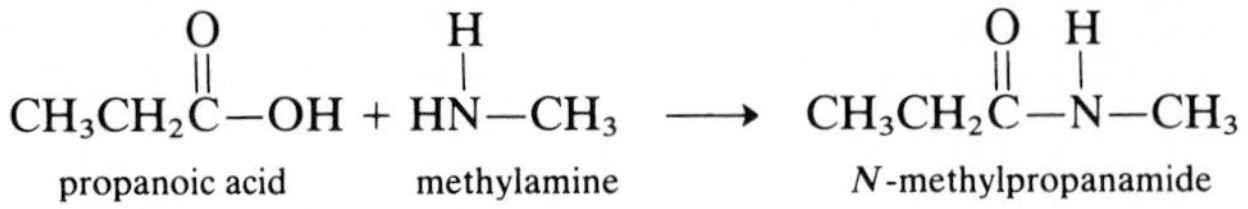

propanoic acid | methylamine | *N*-methylpropanamide

Physical Properties of the Derivatives

The main derivatives, acid chlorides, esters, and amides, differ substantially from carboxylic acids in polarity, solubility, and boiling points. Acid chlorides and esters no longer have highly polar functional groups that can hydrogen bond strongly; they have lower boiling points and water solubilities than carboxylic acids. Amides are quite polar, can hydrogen bond, and generally have boiling ranges higher or comparable to acids. Table 35–2 lists some examples.

Biological Uses

Neither acid chlorides nor anhydrides are present in nature because they are so reactive with water, but their reactions are used as model reactions for many enzymatic reactions. The —Cl and $-O\overset{\displaystyle O}{\overset{\|}{C}}R$ groups are readily displaced by a variety of reagents, so they are called good leaving groups. This great ability to be replaced makes acid chlorides quite useful in synthesis:

$$R\overset{\displaystyle O}{\overset{\|}{C}}-OH \longrightarrow R\overset{\displaystyle O}{\overset{\|}{C}}-Cl \xrightarrow{Z:} R\overset{\displaystyle O}{\overset{\|}{C}}-Z + Cl^-$$

$$R-\overset{\displaystyle O}{\overset{\|}{C}}-OH \longrightarrow R\overset{\displaystyle O}{\overset{\|}{C}}-O\overset{\displaystyle O}{\overset{\|}{C}}R \xrightarrow{Z:} R\overset{\displaystyle O}{\overset{\|}{C}}-Z + {}^-O\overset{\displaystyle O}{\overset{\|}{C}}R$$

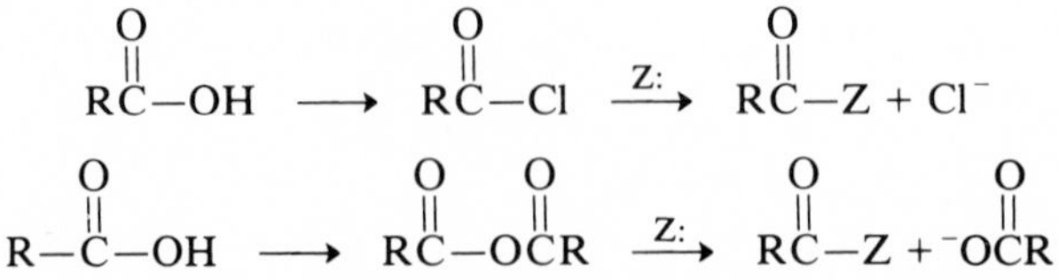

Table 35–2 Physical properties of carboxylic acids and derivatives

Structure	Name	MP (°C)	BP (°C)	Solubility in H_2O
CH_3CO_2H	Acetic acid	17	118	Completely
$CH_3\overset{\overset{O}{\|}}{C}-Cl$	Acetyl chloride	−112	51	Reacts
$CH_3\overset{\overset{O}{\|}}{C}-OCH_3$	Methyl acetate	−99	57	24 g/100 mℓ
$CH_3\overset{\overset{O}{\|}}{C}-\underset{\underset{H}{\mid}}{N}-CH_3$	*N*-methylacetamide	28	206	Soluble
$CH_3\overset{\overset{O}{\|}}{C}-NH_2$	Acetamide	82	222	Soluble
$CH_3(CH_2)_2\overset{\overset{O}{\|}}{C}OH$	Butanoic acid	−6	164	Soluble
$CH_3(CH_2)_2\overset{\overset{O}{\|}}{C}Cl$	Butanoyl chloride	−89	102	Reacts
$CH_3(CH_2)_2\overset{\overset{O}{\|}}{C}-OCH_2CH_3$	Ethyl butanate	−93	120	0.5 g/100 mℓ
$CH_3(CH_2)_2\overset{\overset{O}{\|}}{C}-NH_2$	Butanamide	115	216	Soluble
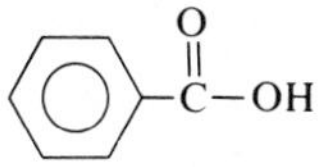	Benzoic acid	122	249	0.3 g/100 mℓ
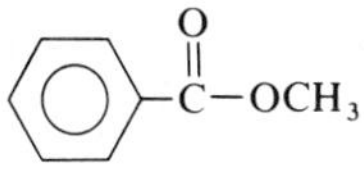	Methyl benzoate	−12	199	Insoluble
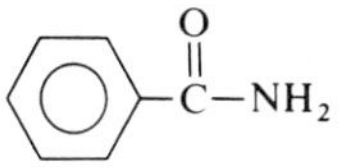	Benzamide	133	290	Slightly soluble

The biochemical equivalent of acid chloride and anhydride reactions is

$$R\overset{\overset{O}{\|}}{C}-OH + \text{enzyme (activated)} \longrightarrow R\overset{\overset{O}{\|}}{C}-\text{enzyme} \xrightarrow{Z:} R\overset{\overset{O}{\|}}{C}-Z + \text{enzyme}$$

Esters are widely found in nature. Unlike acids, esters generally have quite pleasant odors. They are responsible in part for the aromas of many perfumes, flowers, and fruits. Many artificial fruit flavors and odors are comprised of esters. Table 35–3 lists some examples.

Ester linkages are also prevalent in many polymers. **Dacron** is made by copolymerizing ethylene glycol and terephthalic acid. The key here is that there are two functional groups on each molecule, so bonds can be made at both ends.

Table 35–3 Artificial fragrances and flavors associated with esters

Structure	Name	Aroma or flavor
$HC(=O){-}OCH_2CH_3$	Ethyl formate	Rum
$HC(=O){-}O{-}CH_2CH(CH_3)CH_3$	Isobutyl formate	Raspberries
$CH_3C(=O){-}O{-}CH_2CH_2CH_2CH_2CH_3$	*n*-pentyl acetate	Bananas
$CH_3C(=O){-}O{-}CH_2CH_2CH(CH_3)CH_3$	Isopentyl acetate	Pears
$CH_3C(=O){-}O{-}CH_2(CH_2)_6{-}CH_3$	*n*-octyl acetate	Oranges
$CH_3CH_2CH_2C(=O){-}OCH_2CH_3$	Ethyl butanate	Pineapples, peach flavoring
$CH_3CH_2CH_2C(=O){-}O{-}CH_2CH_2CH_2CH_2CH_3$	*n*-pentyl butanate	Apricots
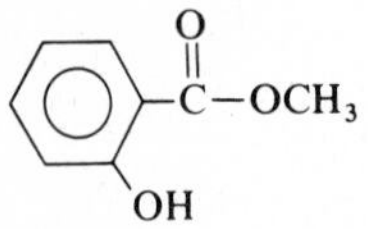	Methyl salicate	Oil of wintergreen
$CH_3C(=O){-}OCH_2{-}CH{=}C(CH_3)_2$	3-methyl-2-butenyl acetate	Chewing gum
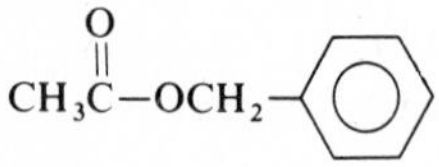	Benzyl acetate	Jasmine
$CH_3(CH_2)_4{-}C(=O){-}OCH_2CH_3$	Ethyl hexanoate	Wine
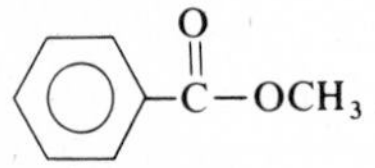	Methyl benzoate	Cologne

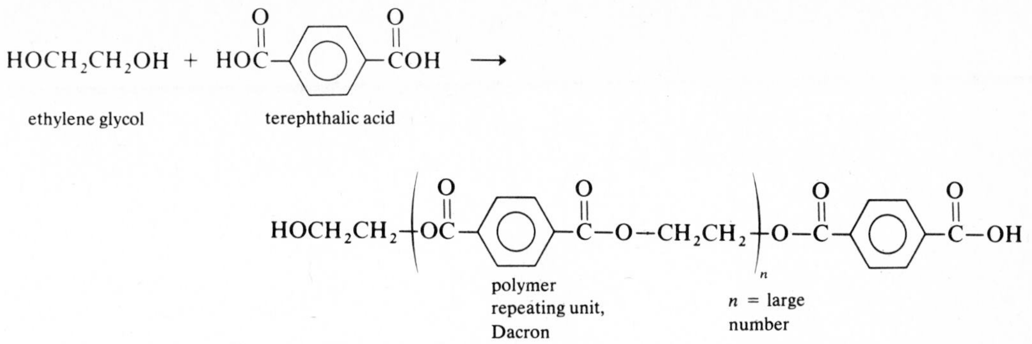

$$HOCH_2CH_2OH \;+\; HOC\!-\!\!\!\bigcirc\!\!\!-\!COH \;\longrightarrow$$

ethylene glycol   terephthalic acid

$$HOCH_2CH_2\!\!\left(\!OC\!-\!\!\!\bigcirc\!\!\!-\!C\!-\!O\!-\!CH_2CH_2\!\right)_n\!O\!-\!C\!-\!\!\!\bigcirc\!\!\!-\!C\!-\!OH$$

polymer
repeating unit,
Dacron

n = large
number

Amide functional groups are some of the most prevalent groups found in nature. A host of drugs and biologically active compounds contain amide groups. All proteins are polyamides. Many polymers are synthesized by copolymerizing dicarboxylic acids and diamines into polyamides. Obviously, the range of compounds that contains the amide linkage is too broad to list, but we have given some examples in Figure 35–1.

$NH_2\!-\!C\!-\!NH_2$

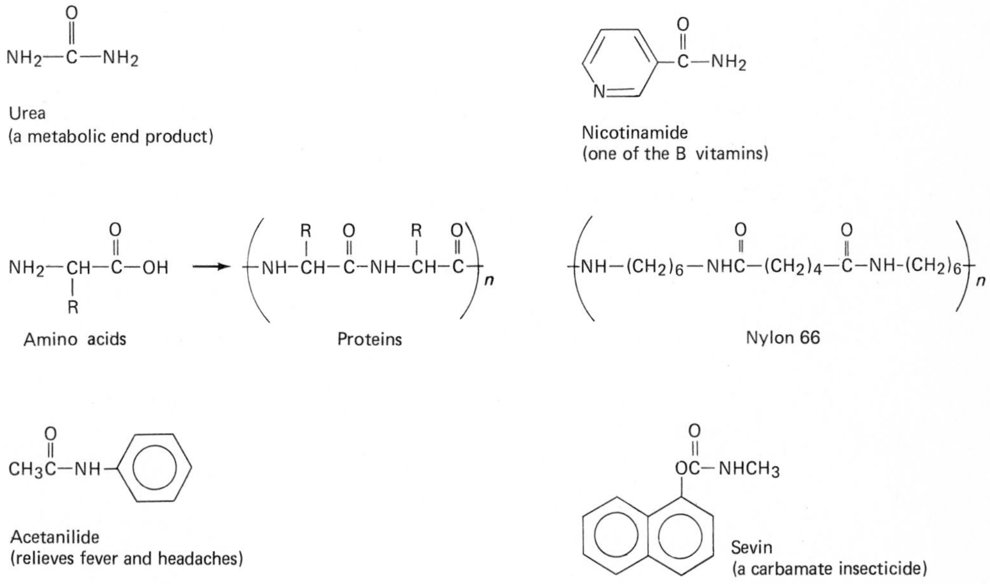

Urea
(a metabolic end product)

Nicotinamide
(one of the B vitamins)

$NH_2\!-\!CH\!-\!C\!-\!OH \longrightarrow$

Amino acids

Proteins

Nylon 66

$CH_3C\!-\!NH\!-\!\bigcirc$

Acetanilide
(relieves fever and headaches)

Sevin
(a carbamate insecticide)

Figure 35–1   Examples of compounds containing the amide group.

The esters and anhydrides of **phosphoric acid** also deserve mention because they are exceedingly important in the whole of biological chemistry. The combination of organic molecules with phosphoric acid (H_3PO_4) is an important metabolic reaction because many organic substances can be metabolized only in the phosphorylated state.

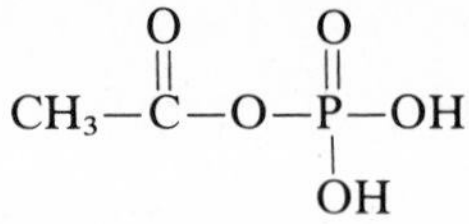

$$\underset{\text{a phosphate anhydride,}}{\underset{\text{acetyl phosphate}}{CH_3-\overset{\overset{\displaystyle O}{\|}}{C}-O-\underset{\underset{\displaystyle OH}{|}}{\overset{\overset{\displaystyle O}{\|}}{P}}-OH}} \qquad \underset{\text{a phosphate ester,}}{\underset{\text{trimethyl phosphate}}{CH_3-O-\underset{\underset{\displaystyle OCH_3}{|}}{\overset{\overset{\displaystyle O}{\|}}{P}}-OCH_3}}$$

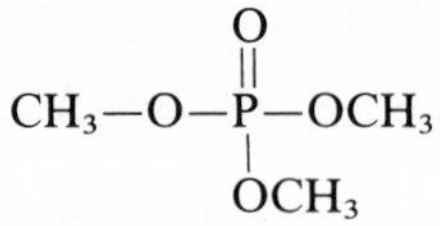

Synthesis and Reactions of Derivatives of Carboxylic Acids

Acid chlorides and anhydrides play important roles in synthesis because the other derivatives are usually synthesized from them. Acid chlorides are synthesized by treatment of carboxylic acids with $SOCl_2$, PCl_3, or PCl_5. Anhydrides can be made by vigorous heating of carboxylic acids or by treatment of acid chlorides with carboxylate salts.

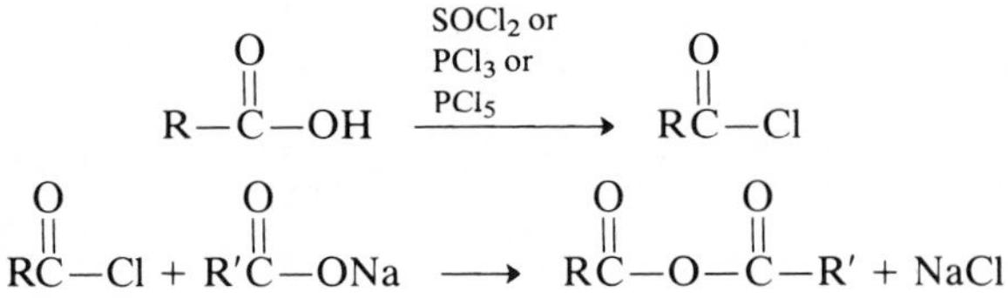

$$R-\overset{\overset{\displaystyle O}{\|}}{C}-OH \xrightarrow{\substack{SOCl_2 \text{ or}\\ PCl_3 \text{ or}\\ PCl_5}} R\overset{\overset{\displaystyle O}{\|}}{C}-Cl$$

$$R\overset{\overset{\displaystyle O}{\|}}{C}-Cl + R'\overset{\overset{\displaystyle O}{\|}}{C}-ONa \longrightarrow R\overset{\overset{\displaystyle O}{\|}}{C}-O-\overset{\overset{\displaystyle O}{\|}}{C}-R' + NaCl$$

One of the main synthetic routes to esters and amides is by reaction of alcohols and amines with acid chlorides or anhydrides.

Esters: An acid chloride or anhydride and an alcohol are used to make an ester. For example,

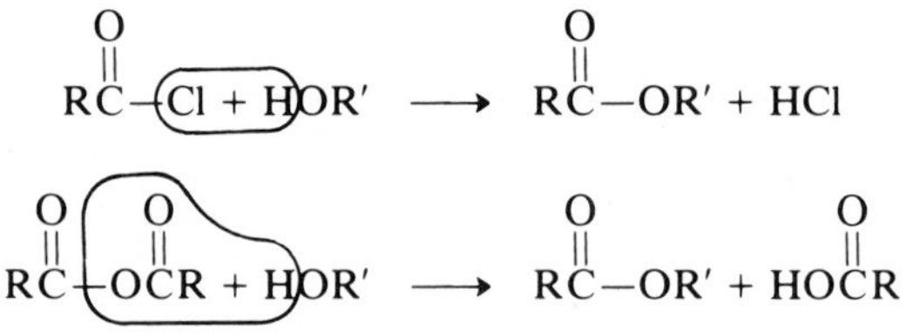

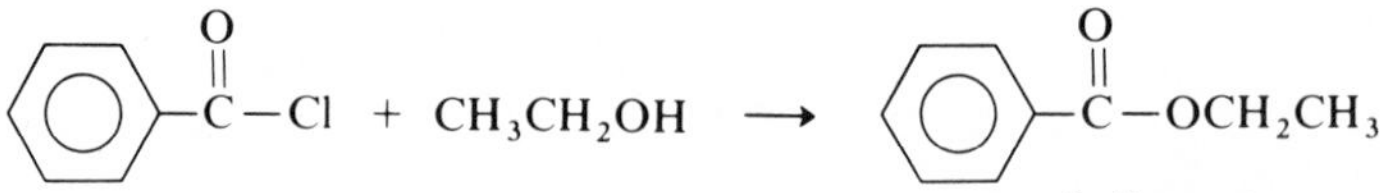

$$R\overset{\overset{\displaystyle O}{\|}}{C}-(Cl + H)OR' \longrightarrow R\overset{\overset{\displaystyle O}{\|}}{C}-OR' + HCl$$

$$R\overset{\overset{\displaystyle O}{\|}}{C}-(O\overset{\overset{\displaystyle O}{\|}}{C}R + H)OR' \longrightarrow R\overset{\overset{\displaystyle O}{\|}}{C}-OR' + HO\overset{\overset{\displaystyle O}{\|}}{C}R$$

$$C_6H_5-\overset{\overset{\displaystyle O}{\|}}{C}-Cl + CH_3CH_2OH \longrightarrow \underset{\text{ethyl benzoate}}{C_6H_5-\overset{\overset{\displaystyle O}{\|}}{C}-OCH_2CH_3}$$

$$CH_3\overset{\overset{\displaystyle O}{\|}}{C}-O-\overset{\overset{\displaystyle O}{\|}}{C}CH_3 + C_6H_5-OH \longrightarrow \underset{\text{phenyl acetate}}{CH_3\overset{\overset{\displaystyle O}{\|}}{C}-O-C_6H_5}$$

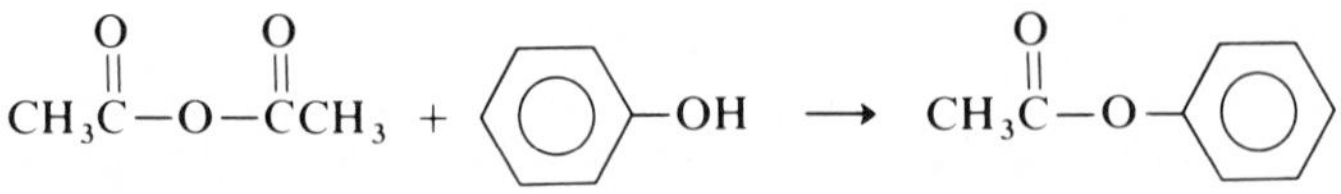

Amides: An acid chloride or anhydride and an amine are used to make an amide. For example,

$$R\overset{\overset{\displaystyle O}{\|}}{C}-Cl + H-NR'_2 \longrightarrow R\overset{\overset{\displaystyle O}{\|}}{C}-NR'_2 + HCl$$

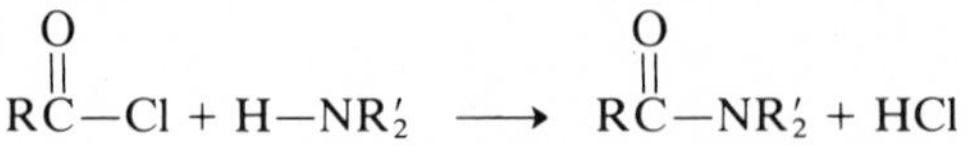

$$\mathrm{R\overset{O}{\overset{\|}{C}}{-}O{-}\overset{O}{\overset{\|}{C}}R + H{-}NR'_2 \longrightarrow R\overset{O}{\overset{\|}{C}}{-}NR'_2 + HO\overset{O}{\overset{\|}{C}}R}$$

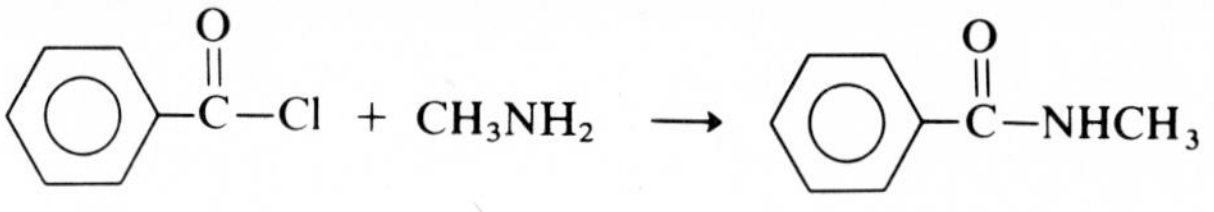

benzoyl chloride N-methylbenzamide

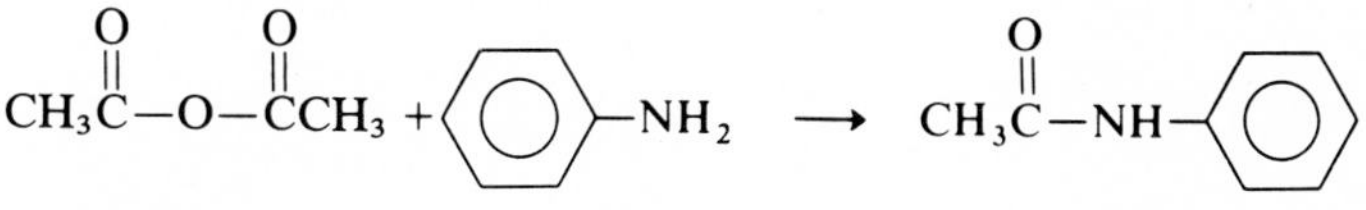

acetic anhydride aniline acetanilide

In making amides, base catalysts are often added to neutralize the HCl as it is formed and to keep the amine in a neutral, nonprotonated form.

$$\mathrm{R\overset{O}{\overset{\|}{C}}Cl + NH_2R' \longrightarrow R\overset{O}{\overset{\|}{C}}NR'_2 + HCl}$$

$$\mathrm{HCl + NH_2R' \longrightarrow \overset{+}{N}H_3R'} \quad \text{(protonated form)}$$

$$\mathrm{\overset{+}{N}H_3R + base \longrightarrow NH_2R'} \quad \text{(nonprotonated form; reactive form)}$$

Esters can also be made by heating alcohols and carboxylic acids in the presence of acid catalysts. As was mentioned in unit 34, this reaction is reversible; to synthesize the ester in good yield we must either remove the water or use a large excess of the acid or alcohol.

$$\mathrm{R\overset{O}{\overset{\|}{C}}OH + R'OH \overset{H^+}{\rightleftharpoons} R\overset{O}{\overset{\|}{C}}OR' + H_2O}$$

In summary, all derivatives can be synthesized from the carboxylic acid, and usually the preferred route is through the acid chloride.

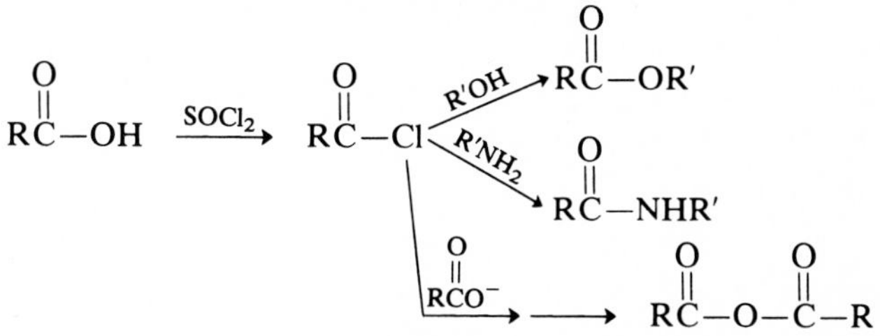

The reverse reactions can also occur. Treatment of any of the derivatives with water in the presence of acid or base catalysts results in the formation of a carboxylic acid or carboxylate ion.

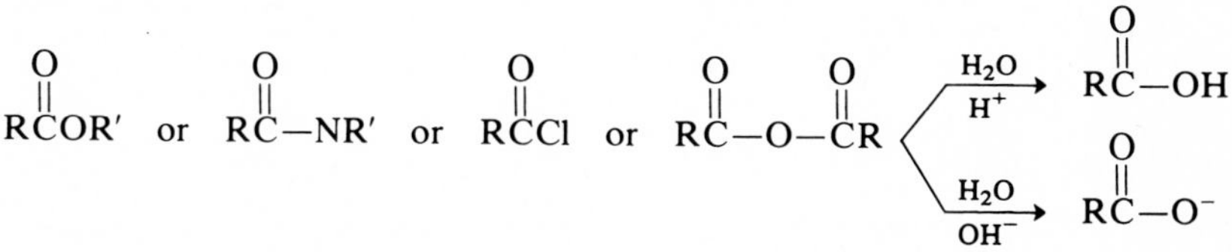

Claisen Condensation

In unit 33 we discussed the aldol condensation, a reaction in which two aldehyde molecules are joined to form a β-hydroxy aldehyde. In the **Claisen condensation** the same sort of joining of two molecules occurs, only this time the products are **β-keto esters.** *The carbon atom adjacent to the carbonyl group of one ester becomes attached to the carbonyl carbon of the second ester. The alcohol portion of the second ester is lost.*

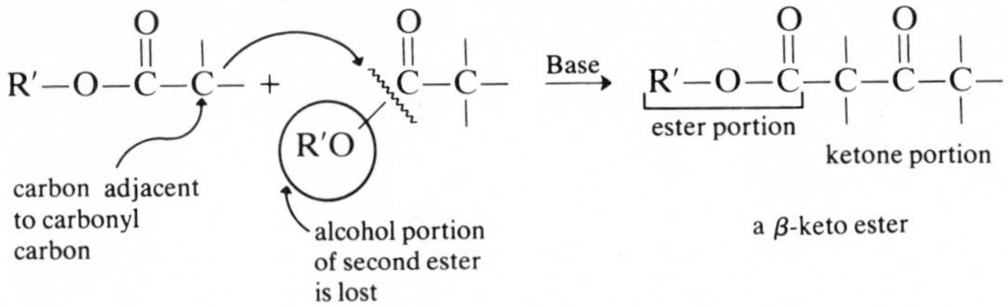

Base
ester portion
ketone portion
carbon adjacent to carbonyl carbon
alcohol portion of second ester is lost
a β-keto ester

For example,

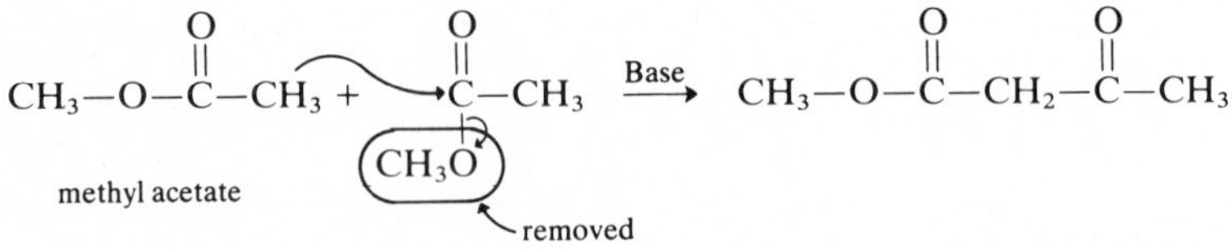

Base
methyl acetate
removed

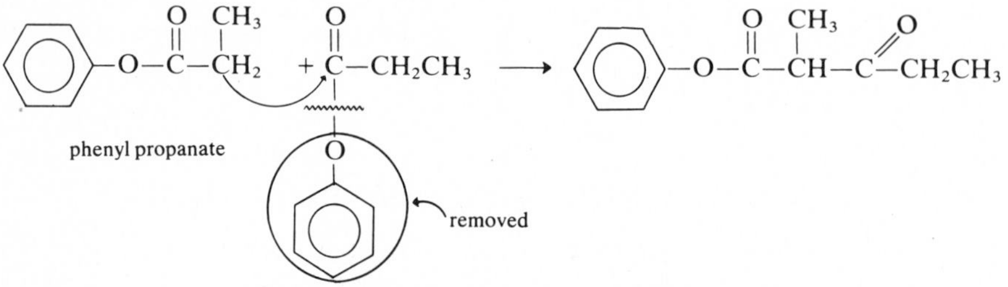

phenyl propanate
removed

One major use of the Claisen condensation is in the synthesis of ketones. Once the β-keto ester is formed in the Claisen condensation, it can be converted to a β-keto carboxylic acid. Upon heating, the β-keto acid loses carbon dioxide to yield a ketone. In the following example, acetone is synthesized from two molecules of acetic acid:

$$2CH_3-\overset{\overset{O}{\|}}{C}-OH + \underset{\text{an alcohol}}{ROH} \xrightleftharpoons{H^+} 2CH_3\overset{\overset{O}{\|}}{C}-OR' \quad \text{(formation of esters)}$$

$$R'-O-\overset{\overset{O}{\|}}{C}-CH_3 + \underset{OR'}{\overset{\overset{O}{\|}}{C}}-CH_3 \xrightarrow{\text{Base}} R'-O-\overset{\overset{O}{\|}}{C}-CH_2\overset{\overset{O}{\|}}{C}CH_3 \quad \text{(Claisen condensation)}$$

$$R'-O-\overset{\overset{O}{\|}}{C}-CH_2\overset{\overset{O}{\|}}{C}CH_3 \xrightarrow[H_2O]{H^+} HO-\overset{\overset{O}{\|}}{C}-CH_2\overset{\overset{O}{\|}}{C}CH_3 \quad \text{(hydrolysis of the ester to form the } \beta\text{-keto acid)}$$

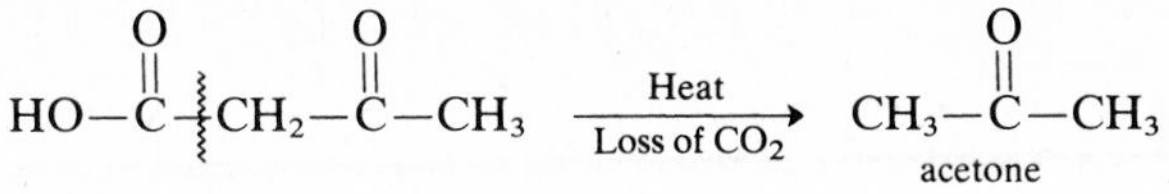
Heat
Loss of CO2
acetone

In our bodies, these four steps are done enzymatically in the synthesis of lipids.

UNIT 35 Problems and Exercises

35–1. Name the following compounds.

(a) 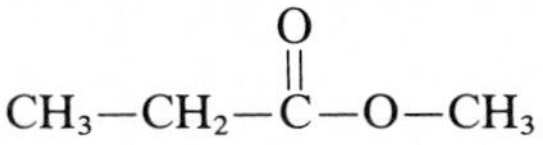

(b) 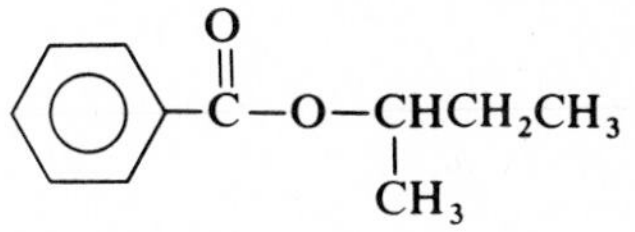

(c) 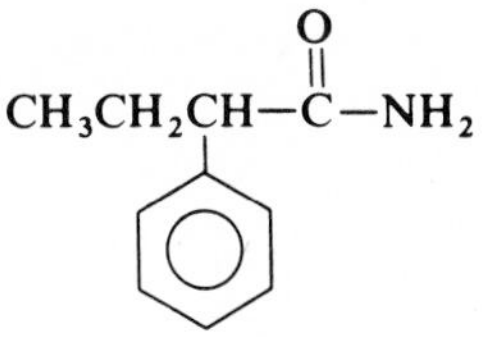

(d) 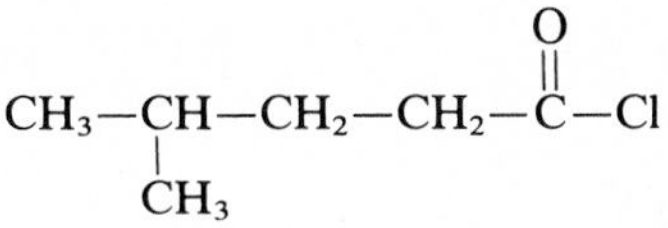

(e) 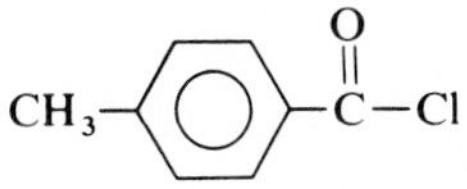

(f) 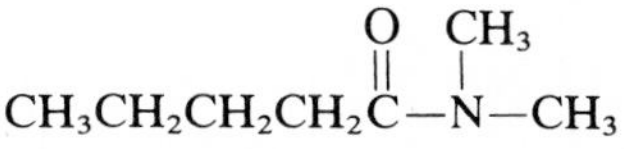

(g) 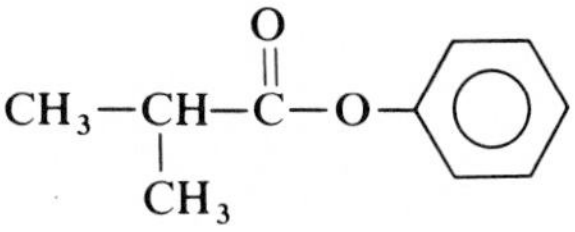

(h) 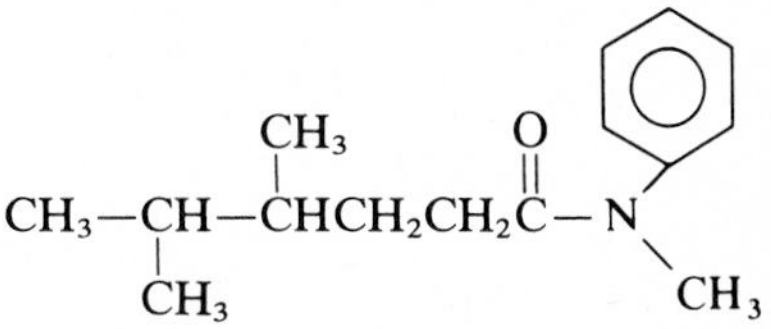

(i) 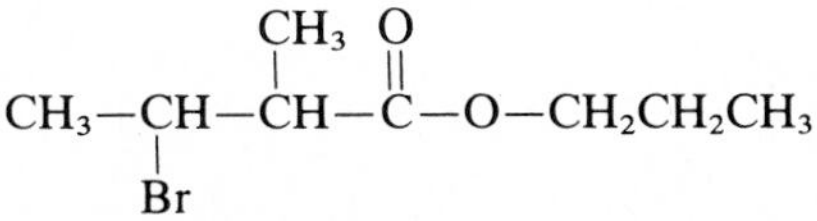

35–2. Draw the structures for the following compounds.

(a) Methyl benzoate
(b) *N*-methylbenzamide
(c) *n*-propyl acetate
(d) Hexanoyl chloride
(e) Phenyl acetate
(f) *N*-methyl-*N*-phenylpropanamide
(g) 3-methyl-pentanoyl chloride
(h) Isopropyl formate
(i) *N*-propylformamide

35–3. Draw and name all the esters that have a molecular formula of $C_5H_{10}O_2$.

35–4. Draw and name all the amides that have a formula of $C_5H_{11}NO$.

35–5. Complete the following reactions.

(a) 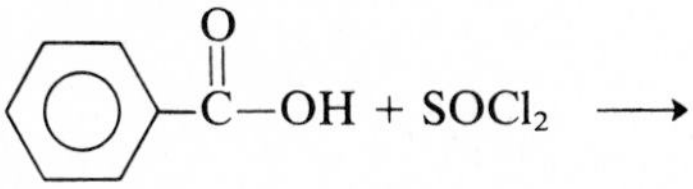

(b) 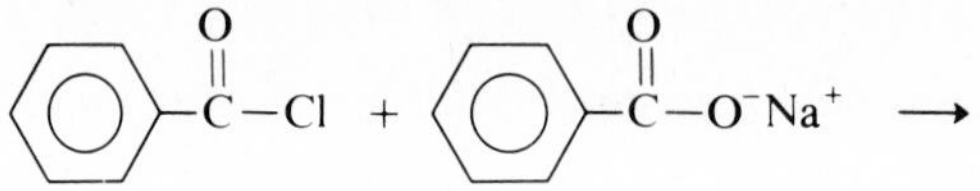

(c) $CH_3C(=O)-Cl + NH_2-CH_2CH_3 \longrightarrow$

(d) $CH_3C(=O)-Cl$ + isopropyl alcohol $\longrightarrow$

(e) Acetic anhydride + phenol $\longrightarrow$

(f) Propanoyl chloride + *n*-propyl alcohol $\longrightarrow$

(g) Benzoyl chloride + ethylmethylamine $\longrightarrow$

(h) Ethyl benzoate + $H_2O \xrightarrow{H^+}$

(i) 2-methylpropanamide + $H_2O \xrightarrow{H^+}$

35–6. Perform the following synthesis reactions, starting with the reactant given and any other necessary materials.

(a) Make ethyl acetate from 2 mol of ethyl alcohol.

(b) Make *N*-ethylacetamide from 2 mol of ethyl bromide.

(c) Make benzoic anhydride from 2 mol of toluene.

(d) Make ethyl propanate from 2 mol of ethylbromide.

(e) Make *t*-butyl 2-methylpropanate from 2 mol of isobutyl alcohol.

35–7. How could you distinguish an ester from an acid or amide?

35–8. Complete the following Claisen condensation reactions. Stop at the β-keto ester.

(a) $CH_3-O-C(=O)-CH_3 + CH_3-O-C(=O)-CH_3 \xrightarrow{\text{Base}}$

(b) $CH_3-O-C(=O)-CH_3 + CH_3-O-C(=O)-CH_2CH_3 \xrightarrow{\text{Base}}$

(c) 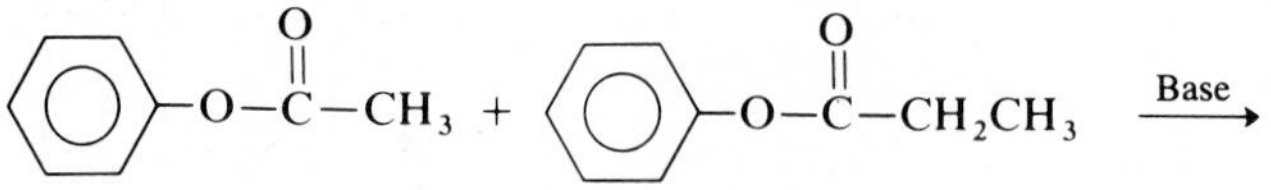

(d) $CH_3-O-C(=O)-CH(CH_3)_2 + CH_3-O-C(=O)-CH(CH_3)_2 \xrightarrow{\text{Base}}$

35–9. What ester would you use to make the following Claisen condensation products?

(a) $CH_3-O-C(=O)-CH_2-C(=O)-CH_3$

(b) 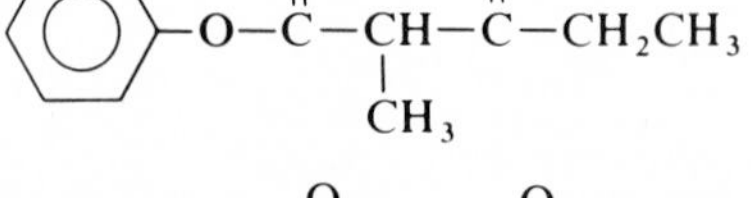

(c) $CH_3CH_2-O-C(=O)-CH(CH_3)-C(=O)-CH_2CH_3$

(d) 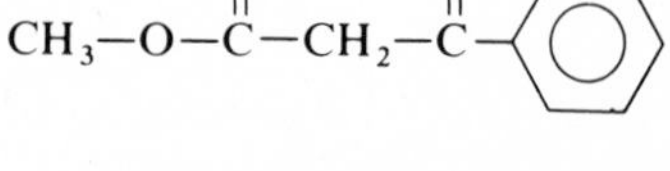

(e) 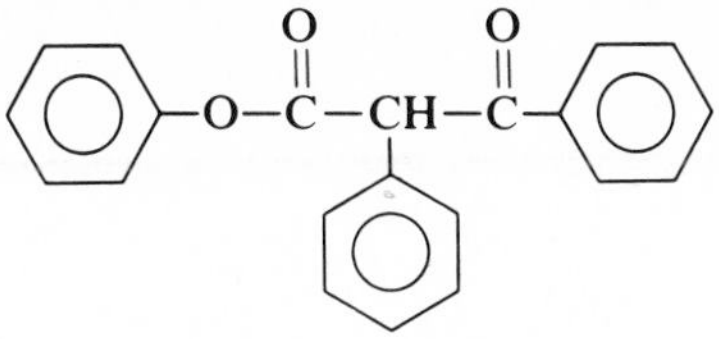

(f) 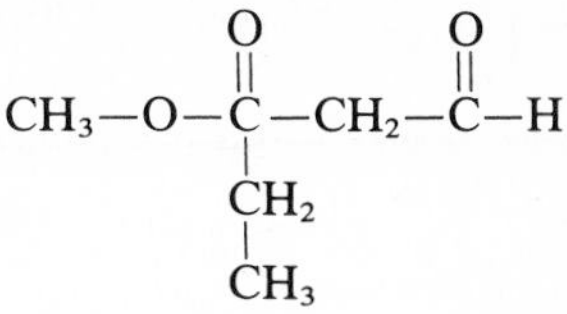

35–10. Outline all the steps required to make 2-pentanone from 2 mol of methyl propanate.

35–11. Complete the Claisen condensations in problem (35–8) by hydrolyzing the ester and heating to lose CO_2.

35–12. Compound A with the formula $C_7H_{14}O$ has a very pleasant odor. Treatment of this compound with H^+ and H_2O resulted in the formation of compounds B and C. Compound B reacted readily with $NaHCO_3$ and was found to contain 54% C, 9% H, and 36.4% O. Compound C reacted with Na and cannot be oxidized to carboxylic acids. Give structures for A, B, and C.

UNIT 36 SUMMARY OF THE REACTIONS AT THE >C=O CENTER

Many of the reactions that occur at the carbonyl group center have now been introduced. It is interesting that the reactions of carbonyl groups in aldehydes and ketones are generally quite different from the reactions of the carbonyl groups in carboxylic acids and their derivatives. In this unit we shall try to rationalize these differences and present some of the reasons for the differences in reactivity.

Two types of attack can occur at a carbonyl group. Since there is a great difference in the electronegativity of the oxygen and carbon atoms in the carbonyl group, that center is quite polar: the oxygen atom is partially negative and the carbon atom is partially positive. Therefore, any reagents that interact with the carbonyl group will generally attack either the oxygen end or the carbon end, depending on their charge. A reagent bearing a positive charge (an **electrophile**) will be attracted to the oxygen end of the carbonyl group; reagents bearing negative charges or unshared electron pairs (**nucleophiles**) will primarily attack the carbon end of the carbonyl group. These two modes of attack are virtually the same for all compounds containing the carbonyl center. The difference in reactivity between aldehydes and ketones versus carboxylic acids comes in what happens after the initial electrophilic or nucleophilic attack.

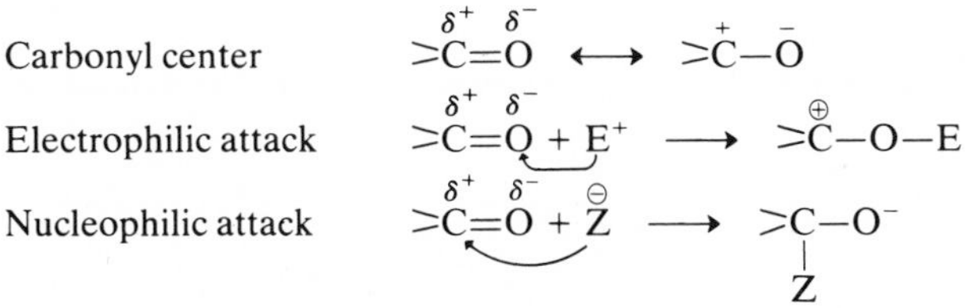
Carbonyl center
Electrophilic attack
Nucleophilic attack

Substitution Versus Addition

After nucleophilic addition occurs, the carbonyl center can react in two ways: it can remain as an sp^3 center or eliminate some other group to re-form the carbonyl group

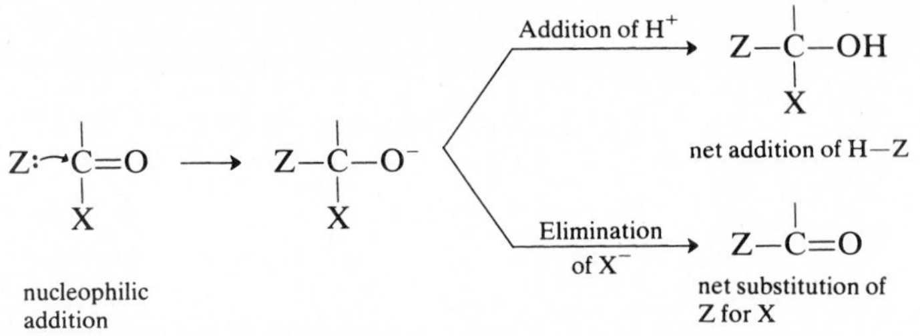

Addition of H+
net addition of H—Z
Elimination of X−
net substitution of Z for X
nucleophilic addition

The process that dominates is determined largely by the ability of the X group to leave as X^- or HX. This is partly determined by the basicity of the X anion. The more basic it is, the poorer it will function as a leaving group. The basicities of some leaving groups are listed in Table 36–1.

Table 36–1 Leaving group basicity

Strongest base						Weakest base
H^-	$-\overset{\vert}{\underset{\vert}{C}}^-$	$R-N^-$	$R-O^-$	OH^-	$R\overset{O}{\overset{\Vert}{C}}-O^-$	Cl^-
Poorest leaving group						Best leaving group

Reactions of Aldehydes With Ketones: Addition

In aldehydes and ketones the atoms attached to the carbonyl carbon are hydrogen and carbon atoms. From Table 36–1 we see that these atoms are poor leaving groups, so we predict that *substitution reactions* at the carbonyl center should not occur readily. Therefore, *the major mode of reaction of aldehydes and ketones should be addition.* Generally, this involves nucleophilic attack followed by addition of a hydrogen ion:

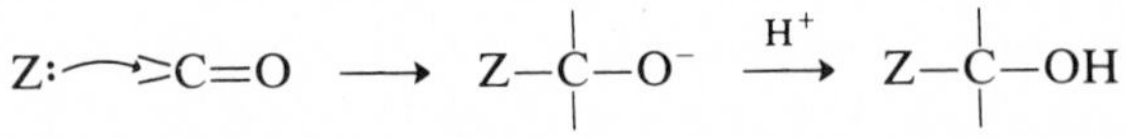

Let's review some of the reactions of aldehydes and ketones and look at the addition reactions.

Addition of cyanide: The carbon atom of the cyanide is negatively charged, so it adds to the carbonyl center.

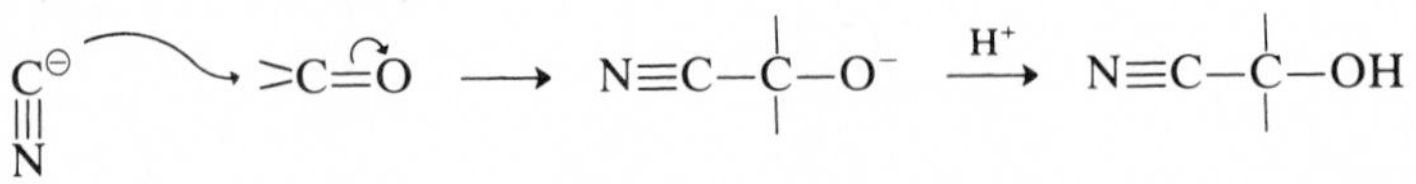

Hemiacetal formation: Here the acid catalyst first attacks the oxygen of the carbonyl center. Then the alcohol oxygen adds to the carbon atom of the carbonyl group.

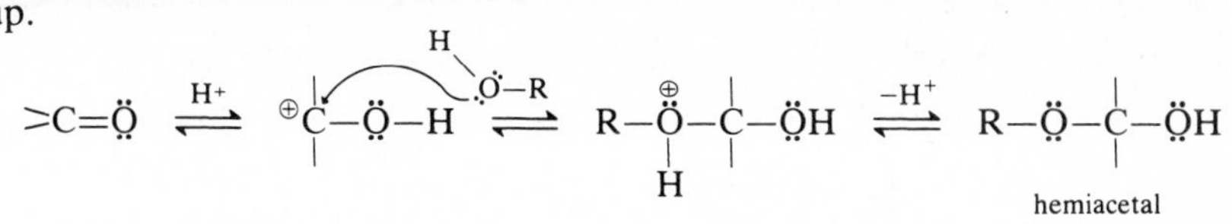

$$\text{>C=\ddot{O}} \underset{}{\overset{H^+}{\rightleftharpoons}} \overset{\oplus}{C}-\ddot{O}-H \rightleftharpoons R-\overset{\oplus}{\underset{H}{O}}-\overset{|}{\underset{|}{C}}-\ddot{O}H \underset{}{\overset{-H^+}{\rightleftharpoons}} R-\ddot{O}-\overset{|}{\underset{|}{C}}-\ddot{O}H$$

hemiacetal

Aldol condensation: In aldehydes and ketones, hydrogen atoms attached to alpha-carbon atoms are more acidic than the other hydrogen atoms. They can be removed by treatment with strong base:

$$R-\overset{O}{\overset{||}{C}}-CH_3 + OH^- \longrightarrow \underbrace{R-\overset{O}{\overset{||}{C}}-\overset{\ominus}{C}H_2 \longleftrightarrow R-\overset{O^-}{\overset{|}{C}}=CH_2}$$

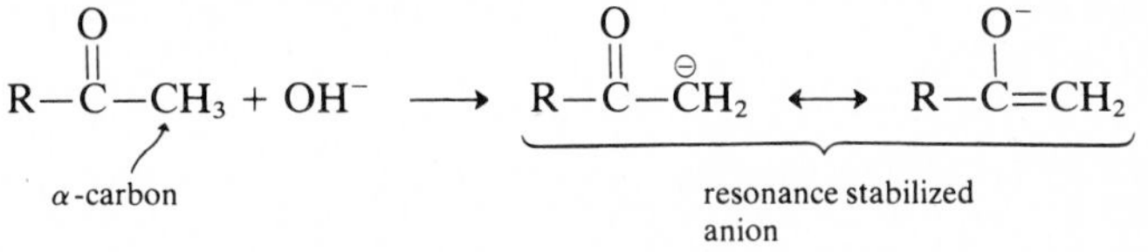

α-carbon   resonance stabilized anion

The anion is a good nucleophile and can attack the carbonyl carbon just like cyanide.

$$R-\overset{O}{\overset{||}{C}}-\overset{\ominus}{C}H_2 \overset{|}{\underset{|}{C}}=\overset{\frown}{O} \longrightarrow R-\overset{O}{\overset{||}{C}}-CH_2-\overset{|}{\underset{|}{C}}-O^- \overset{H^+}{\longrightarrow} R-\overset{O}{\overset{||}{C}}-CH_2-\overset{|}{\underset{|}{C}}-OH$$

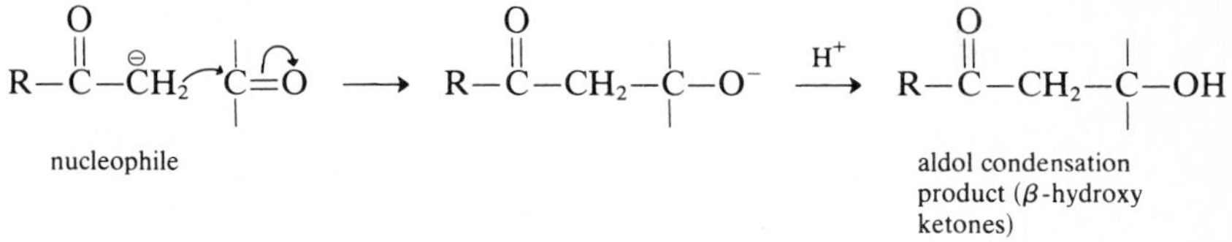

nucleophile   aldol condensation product (β-hydroxy ketones)

Carboxylic Acids and Derivatives: Substitution

Carboxylic acids and their derivatives differ from aldehydes or ketones in that they have electronegative groups attached to one end of the carbonyl center instead of a C or H. Many of these groups are good leaving groups, so *the major mode of reaction is substitution rather than addition.* This is really an addition–elimination reaction.

$$R-\overset{O}{\underset{L}{\overset{||}{C}}}\overset{H}{\overset{|}{\cdots}}\ddot{Z} \underset{}{\overset{\text{Addition}}{\rightleftharpoons}} R-\overset{O^-}{\underset{L}{\overset{|}{\underset{|}{C}}}}-\overset{|}{\underset{H}{Z}} \rightleftharpoons R-\overset{O}{\underset{\overset{\oplus}{L}-H}{\overset{|}{C}}}-Z \overset{\text{Elimination}}{\longrightarrow} R-\overset{O}{\overset{||}{C}}-Z + HL$$

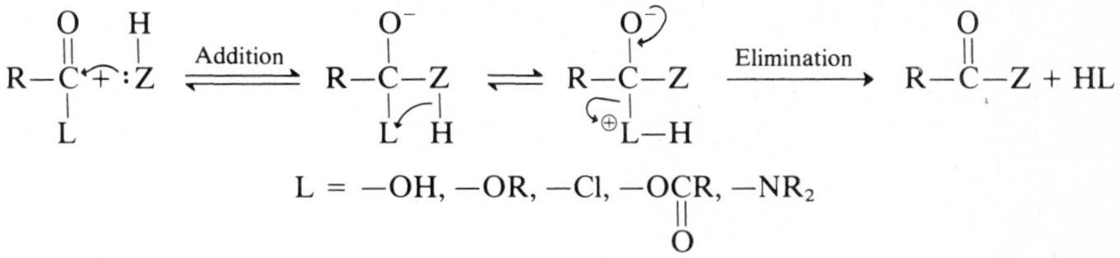

$$L = -OH, -OR, -Cl, -O\overset{O}{\underset{||}{C}}R, -NR_2$$

The nature of the leaving group L (a base) has a large effect on the rate at which the elimination occurs. The more stable it is as a base, the better it is as a leaving group. By using the leaving group rankings presented in Table 36–1, it is possible to rank the derivatives in order of reactivity toward substitution.

$$R\overset{O}{\overset{||}{C}}-Cl \; > \; R\overset{O}{\overset{||}{C}}-O\overset{O}{\overset{||}{C}}R \; > \; R\overset{O}{\overset{||}{C}}-OH \; > \; R\overset{O}{\overset{||}{C}}-OR' \; > \; R\overset{O}{\overset{||}{C}}-NHR$$

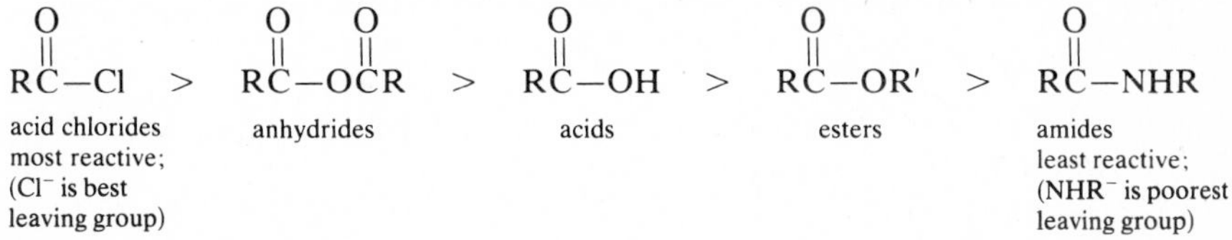

acid chlorides most reactive; (Cl⁻ is best leaving group)   anhydrides   acids   esters   amides least reactive; (NHR⁻ is poorest leaving group)

Acid chloride reactions: Since chloride ion is such a good leaving group, substitution reactions on acid chlorides proceed quite rapidly, sometimes explosively. Acid chlorides are rapidly converted into acids, esters, and amides, and the reactions are not reversible, since chloride is a poor nucleophile and does not attack the carbonyl center.

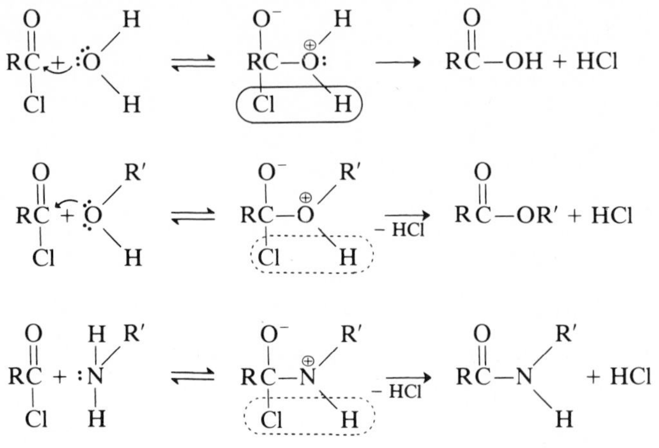

Anhydrides: The $\mathrm{R\overset{O}{\overset{\|}{C}}{-}O^{-}}$ group is a reasonably good leaving group but not as good as chloride. The same substitutions occur on anhydrides as on acid chlorides, but at slower rates.

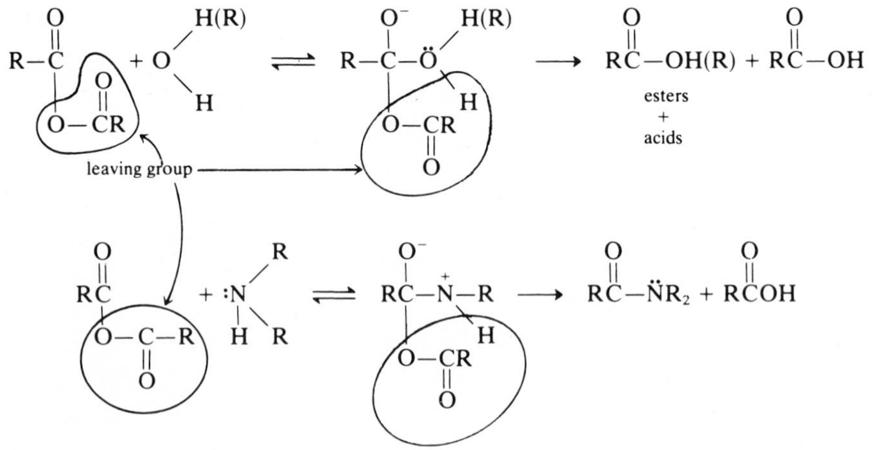
leaving group
esters
+
acids

Esters: Alcohols are not particularly good leaving groups, so substitution reactions of esters are usually acid or base catalyzed. The acid catalyst makes the carbonyl carbon more susceptible to attack. Esters can be converted into carboxylic acids and amides.

Acid-catalyzed substitutions:

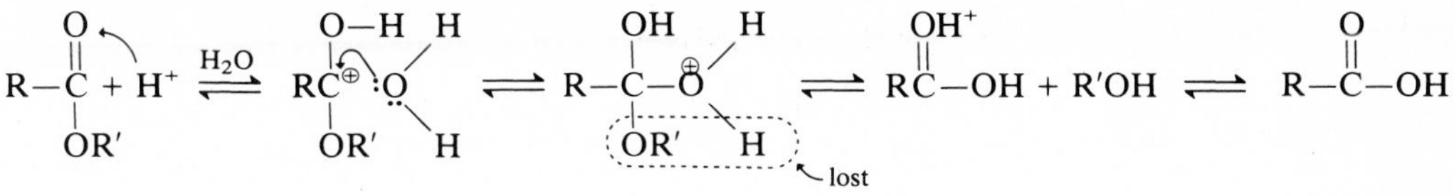
lost

Notice that the reaction is completely reversible. The mechanism for acid-catalyzed conversion of carboxylic acids into esters is simply the reverse of the above mechanism.

Base-catalyzed substitutions:

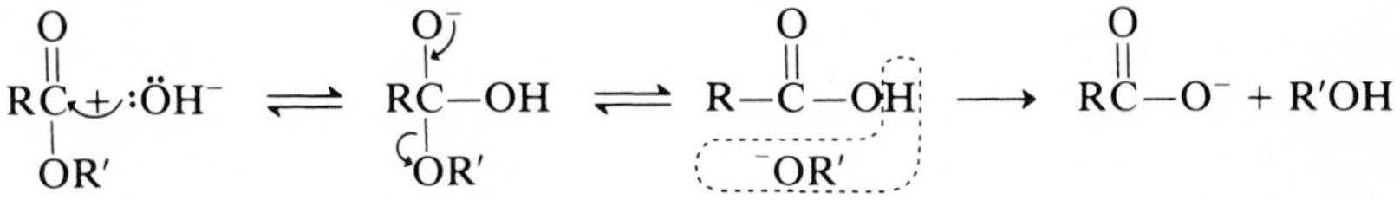

Base-catalyzed reactions are not reversible because of the final step. The hydrogen exchange between $R'O^-$ and $RC(=O)OH$ is not reversible because the carboxylate group is a much weaker base than the alkoxide ion, $R'O^-$.

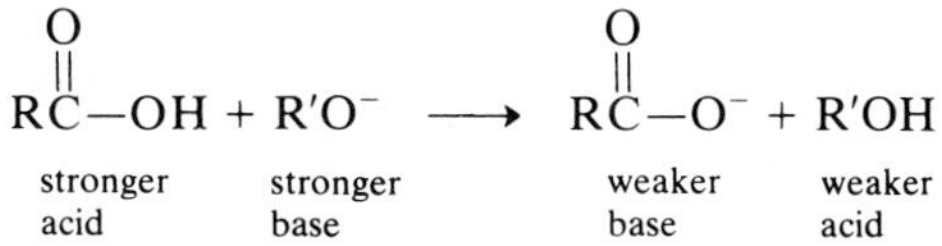
stronger acid
stronger base
weaker base
weaker acid

Claisen Condensation of Esters

The hydrogens on α-carbon atoms of esters are acidic, so treatment of an ester with base can create a nucleophile at the α-carbon. This nucleophile can participate in substitution reactions just like the other nucleophiles. The Claisen condensation is simply a more exotic example of substitution reactions on esters:

nucleophile

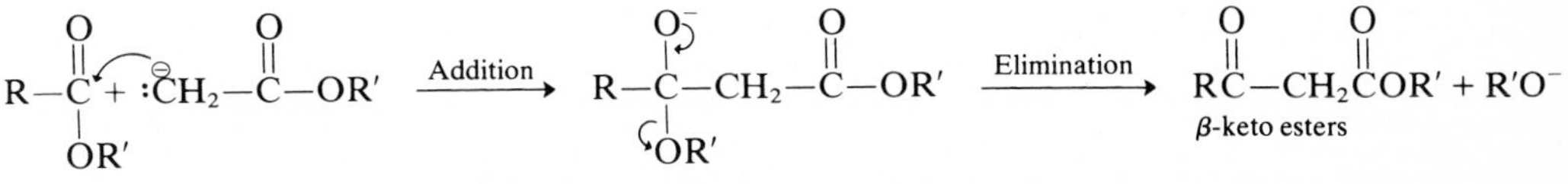
Addition
Elimination
β-keto esters

Amides: Amides react the slowest of the derivatives toward substitution since the R_2N^- groups are not good leaving groups. In addition, the unshared pair of electrons on the nitrogen is capable of reducing the δ^+ charge on the carbonyl carbon, so nucleophilic attack is slowed. The only substitution reaction that generally occurs is conversion to carboxylic acids.

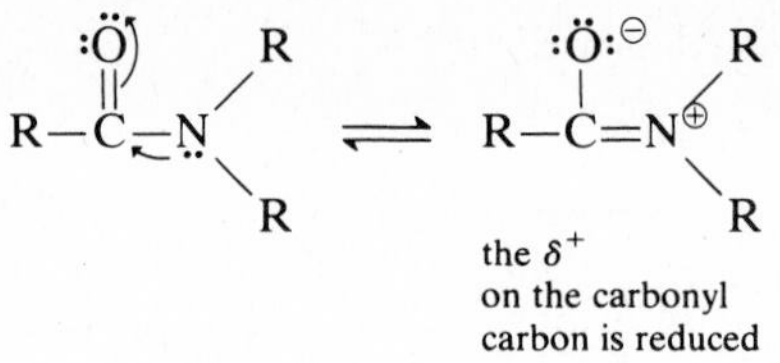
the δ+
on the carbonyl
carbon is reduced

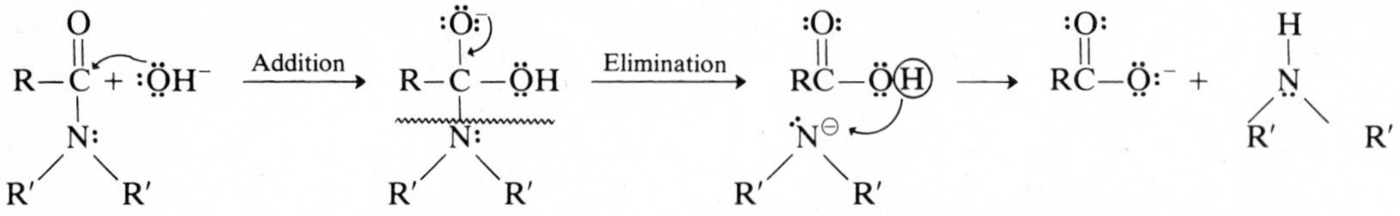
Addition
Elimination

In conclusion, aldehydes and ketones generally undergo addition reactions. A nucleophile will attack the carbonyl carbon, whereas an electrophile will add to the carbonyl oxygen. Substitution reactions are the predominant mode of reaction of carboxylic acids and their derivatives. In these reactions nucleophilic or electrophilic addition is followed by elimination to yield a net substitution.

UNIT 36 Problems and Exercises

36–1. Explain why aldehydes and ketones undergo predominantly addition reactions and not substitution reactions.

36–2. Explain why carboxylic acids and their derivatives undergo predominantly substitution reactions.

36–3. Rank the following compounds in terms of increasing rate of substitution reaction. Explain your reasoning.

(a) $CH_3{-}\overset{\overset{\displaystyle O}{\|}}{C}{-}CH_3$ (b) $CH_3{-}\overset{\overset{\displaystyle O}{\|}}{C}{-}OH$ (c) $CH_3{-}\overset{\overset{\displaystyle O}{\|}}{C}{-}Cl$ (d) $CH_3{-}\overset{\overset{\displaystyle O}{\|}}{C}{-}OCH_3$

(e) $CH_3{-}\overset{\overset{\displaystyle O}{\|}}{C}{-}NH_2$

36–4. Give the mechanism for hemiacetal formation between benzaldehyde and isopropyl alcohol.

36–5. Give the stepwise mechanism for the aldol condensation between two molecules of 2-pentanone.

36–6. Give the mechanisms for the following conversions:
(a) Benzoyl chloride → methyl benzoate
(b) Benzoyl chloride → *N*-methyl benzamide
(c) Methyl acetate → acetic acid

36–7. Using reaction mechanisms and theory, explain why acid chlorides are the most reactive of the derivatives of carboxylic acids.

36–8. Why is acid-catalyzed hydrolysis of esters reversible, but base-catalyzed hydrolysis is not?

36–9. Give the stepwise mechanism for the Claisen condensation between methyl benzoate and ethyl acetate.

PART THREE

Biochemistry

CHAPTER 13

carbohydrates

One of the important classes of compounds in nature is known as the carbohydrates, saccharides, or sugars. These names can be used interchangeably. They are usually defined as polyhydroxyaldehydes, polyhydroxyketones, or compounds that can be hydrolyzed to yield them. We already know about alcohols, aldehydes, and ketones, but now we must consider a combination of these functional groups within a single molecule. We shall note that in some cases this leads to new chemical and biological relationships.

The central role of the carbohydrates in nature can be seen in both their formation and their function. In the grand scheme of the carbon cycle on this planet, CO_2 has a rather unique position. It is the stable end product of the animal metabolism. We shall consider the step-by-step mechanism by which our foodstuffs, including our carbohydrates, undergo chemical reactions that yield CO_2 and energy. Throughout the world, carbohydrates (starches) from rice, corn, and wheat, supply a large portion of human energy requirements.

Carbohydrates are very important because they comprise almost three-quarters of the solid material in plants. They are the first organic compounds that are formed as CO_2 becomes fixed into organic molecules in the process of photosynthesis. In this reaction, plants use solar energy to reverse the metabolic reactions of animals by building carbohydrates from CO_2.

From a chemical point of view, carbohydrates offer good examples of biological specificity. This specificity is seen in the form of isomerization possibilities, which in turn depend on the structural arrangement of the molecules. We shall review a number of types of isomerisms, and consider the subtle differences of the stereochemistry that deals with molecular structure in three dimensions.

UNIT 37 ISOMERISM, OPTICAL ACTIVITY, AND BIOCHEMICAL SPECIFICITY

Isomers are different kinds or forms of compounds with the same molecular or empirical formulas. Recall that the tetrahedral carbon provided the basis for specific and unique structural possibilities. The four bonds from the carbon atom are identical in the methane molecule, and are directed toward the hydrogen atoms at their positions of optimum overlap of the atomic orbitals. Any way you turn the methane molecule in your mind or any way you tumble a space-filling model of methane, each hydrogen is equivalent to every other hydrogen. The structural uniqueness appears when we substitute a variety of other atoms or groups for each of the hydrogen atoms, or build longer carbon chains with carefully defined specific linkages.

Structural Isomers

Structural isomers can be formed with rather simple alkanes containing only carbon and hydrogen atoms in different arrangements. **Structural isomers** are two or more compounds with the same molecular formula but a different structural formula. Butane was the first example with either of the following structural representations:

$CH_3CH_2CH_2CH_3$
n-butane (C_4H_{10})

$$\begin{array}{c} CH_3\underset{|}{C}HCH_3 \\ \;\;CH_3 \end{array}$$

isobutane
(2-methylpropane) C_4H_{10}

These isomers are different in both their chemical and physical properties.

Another case was encountered when we considered compounds with other atoms. There are two different alcohols with the molecular formulas C_3H_8O, 1-propanol and 2-propanol. The first two structures of propanol are exactly equivalent, but the third representation is a structural isomer:

$HOCH_2CH_2CH_3$ $\qquad$ $CH_3CH_2CH_2OH$ $\qquad$ $\begin{array}{c} CH_3\underset{|}{C}HCH_3 \\ \;\;OH \end{array}$

The first two are the same, only written in reverse order on the page. This illustrates one problem in dealing with real three-dimensional structures using only the two dimensions of the printed page. The first two representations of 1-propanol are the same because we can superimpose one upon the other simply by rotating one of the models in the plane of the paper. The second representation would look just like the first one if you could see it through the back of this page. This will be an important way of looking at molecules later in this unit.

Different structural isomers are formed when two or more substituents are added to a benzene ring. This can be observed in the case of three isomers of dibromobenzene, which are very different compounds:

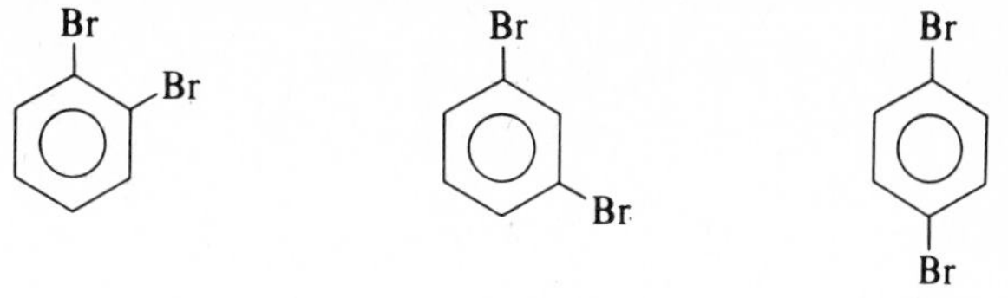

o-dibromobenzene m-dibromobenzene p-dibromobenzene

Geometrical Isomers

Another type of isomerism is **geometrical isomerism**. Here there is restricted rotation about two carbon atoms which are double bonded (alkene linkage) to each other. This π-bond does not allow free rotation of one carbon relative to the other carbon atom to which it is bonded. This restriction yields a rather fixed position in space for the other two remaining atoms attached to each of the carbons in the π-bond. The four atoms attached to the two carbons lie in a single plane. If the two groups attached to each carbon are different from each other, two geometric isomers are possible. If similar groups were on the same side of the double bond, a *cis* relationship exists. Similar groups on opposite sides of the double bond are in a *trans* configuration. Cis $\rightleftharpoons$ trans interconversions are involved in the visual process as previously discussed.

Another good example of isomerization is observed if succinic acid is dehydrogenated, forming two different molecules:

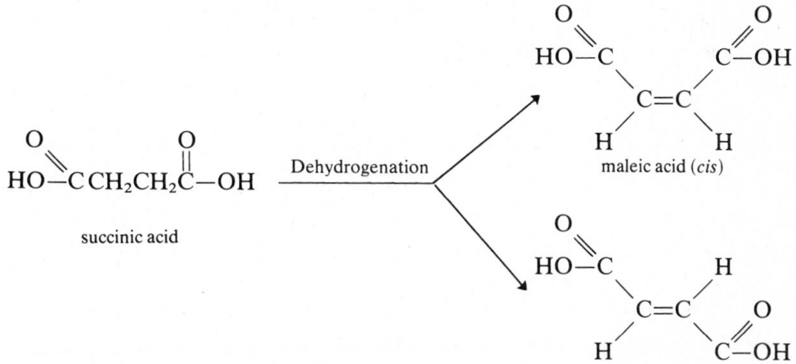
Dehydrogenation

succinic acid

maleic acid (*cis*)

fumaric acid (*trans*)

Fumaric and maleic acids are separate and distinct geometric isomers. The positioning of their respective atoms in space accounts for their uniqueness.

Other cases of restricted rotation can lead to geometric isomers. This is true in the case of substitutions on cyclic hydrocarbons. The usual rotation about the single C—C bonds can be restricted in a cyclic molecule. These can be designated as cis-trans isomerization in the case of a disubstituted cyclohexane:

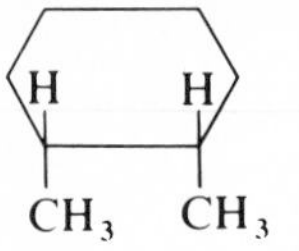

cis-1,2-dimethylcyclohexane

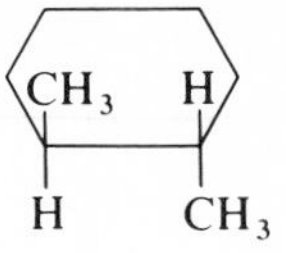

trans-1,2-dimethylcyclohexane

Stereoisomers

The two cases of geometric isomers represent stereoisomers, or molecules identical except for their orientation in space. Another important type of stereoisomerism is **optical isomerism**. This type of isomerism appears to be very subtle, but it has a profound influence upon the biochemical activity of molecules. This type of isomerism is most commonly studied with a *polarimeter*, an instrument that uses what is known as plane polarized light. An optically active molecule will change the way this kind of light acts (is rotated) within this instrument.

Just as in the cases above, optical isomers are made up of pairs of compounds. One of the pair of optical isomers can rotate the plane of light to the right or clockwise (dextrorotatory = +); the other member of the pair will cause rotation to the left or counterclockwise (levorotatory = −).

How does the structure and spatial arrangement of molecules become involved in this optical activity? The entire answer to this question is highly complex, but it is dependent upon the tetrahedral structure of the carbon atom. Molecules with a certain composition are said to exhibit a property of *chirality* or *handedness*. Just as our right hand is a mirror image of our left (try it), the arrangement of some substituents on a given carbon atom can lead to one of two possible distinct arrangements. For our purposes, this arrangement will involve one common atom that has four different atoms or groups attached to it. Such a carbon atom is said to be an **asymmetric** carbon atom.

An illustration of this asymmetric carbon as a single carbon atom substituted with four different groups is shown in Figure 37–1. This figure also shows what could be viewed as its *mirror image* as well. It is a bit difficult to make this visualization using only the two dimensions of the printed page; try to do it in your mind's eye. These are actually two distinct organic molecules. One molecule cannot be placed upon the other, or superimposed, to end up with only one form. This handedness can be illustrated if you try to superimpose your right hand on your left hand. This cannot be done either. You have two different hands and two different isomeric forms of the molecule. Most molecules of interest to us will exhibit optical activity if they possess an asymmetric carbon. There are exceptions, but in general they will not be in our scope of interest.

The two different mirror images making up a pair of compounds are called **enantiomers**. Enantiomers of a compound have identical physical properties (BP, MP) but different chemical properties, as well as a difference in the direction in which they rotate a plane of polarized light. Enantiomers rotate the plane of polarized light the same amount, but in different directions, + or −.

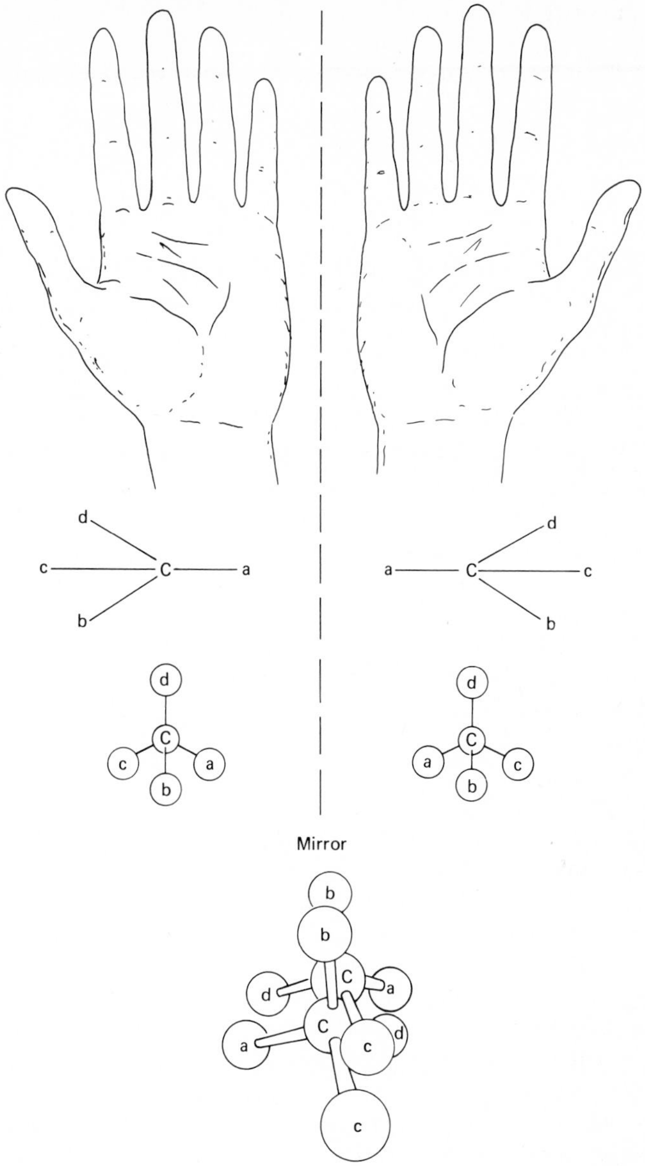
d
c
C
a
b
d
a
C
c
b
d
C
c
a
b
d
C
a
c
b
Mirror
b
b
C
d
a
C
a
d
c
c

Figure 37–1 The pair of hands and an asymmetric carbon with its mirror image are shown at top. The inability to superimpose the mirror images is shown below.

If equal amounts of the enantiomers are mixed together, an optically inactive or **racemic mixture** results. This equimolar mixture would not yield a rotation of the plane of light. Optical isomers may be designated by + or − preceding their names; the racemic mixture is preceded by the symbol ±. There are other cases when apparently optically active compounds do not experimentally exhibit optical activity.

If a molecule had two optically active centers, they could compensate for each other and cancel out the optical effect.

Although a number of different nomenclature systems are used to denote the specific orientation about an asymmetric carbon atom, one is widely used in biochemical studies. The two forms of the glyceraldehyde molecule illustrate this:

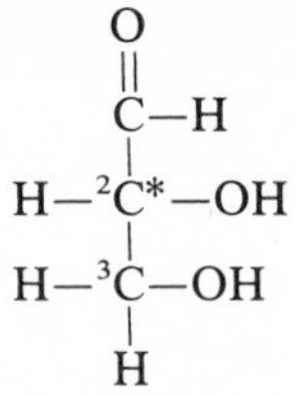

D-glyceraldehyde

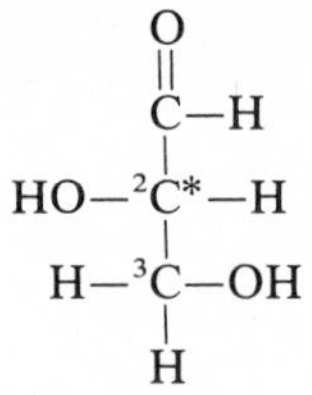

L-glyceraldehyde

Only the number 2 carbon has four different substituents and is asymmetric. This is the only carbon atom that bears four different groups. Although the carbonyl carbon can be written in several different ways, there is actually only one form. The number 3 carbon does not have four different groups attached to it. This carbon atom could be written with two hydrogen and one hydroxyl group in any position you choose. For glyceraldehydes, there is only one asymmetric carbon atom with a spatial arrangement that is unique.

Since two enantiomers of glyceraldehyde must be specified, a nomenclature system was arbitrarily established. A German chemist, Emil Fischer (Nobel laureate, 1902), proposed that the glyceraldehyde is said to have a **D configuration** if the hydroxyl group on the asymmetric carbon is written to the right. If the hydroxyl is written to the left in this presentation (the mirror image), it is said to be of the **L configuration**. This accounts for the way in which the compounds are labeled above. One should not confuse these arbitrary D and L designations with the way in which light is rotated, as previously discussed.

As indicated, the D- and L-glyceraldehyde molecules have the same physical properties. They would rotate a plane of polarized light in equal but opposite directions if placed under the same experimental conditions. They would both undergo many of the same chemical reactions. However, these molecules would each undergo unique biochemical reactions as they become involved in living systems. For example, it is D-glyceraldehyde that is involved in the reactions of the carbohydrates within our cells. Biochemical reactions have great selectivity in choosing between the enantiomers. This is part of the uniqueness and beauty of nature that will be explored as we examine the living processes at the molecular level.

UNIT 37 Problems and Exercises

37–1. Draw all the possible structural isomers of butanol.

37–2. Draw the aldehyde, 2-butenal, in both its cis and trans forms.

37–3. If both maleic acid and fumaric acid were hydrogenated, would one or two different compounds form? Explain.

37–4. Circle the asymmetric carbons in the following compounds.

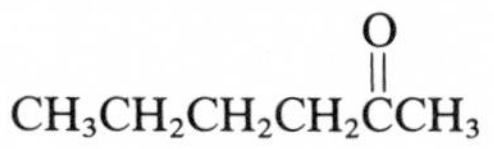

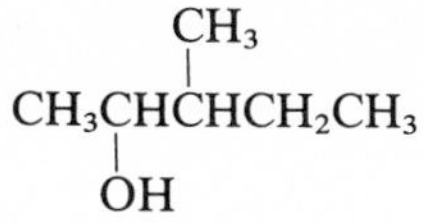

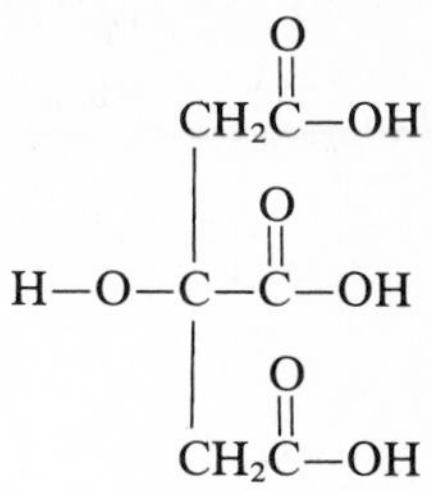

37–5. Can any of the possible forms of bromobutane have an asymmetric carbon? Explain.

37–6. Draw all the possible structural isomers of aminobenzoic acid.

37–7. If an equimolar mixture of D-glyceraldehyde and L-glyceraldehyde was prepared, would this mixture rotate a plane of polarized light?

37–8. Draw the structure of the enantiomers of lactic acid, $CH_3CHOHC(=O){-}OH$.

UNIT 38 MONOSACCHARIDES AND THEIR REACTIONS

The simplest carbohydrates are called **monosaccharides**. These compounds cannot be hydrolyzed to yield simpler compounds. The monosaccharides are important themselves and are the basic repeating units of the more complex types of carbohydrates. Two monosaccharides may combine to yield a **disaccharide** with the loss of a water molecule. A **polysaccharide** is a carbohydrate that results from the combination of many monosaccharide molecules. We shall study each of these types and examine their roles in biochemical systems.

Carbohydrates are involved as nutritional sources, structural materials, fabrics, and fibers, including the very paper of this book. Typically, about 70% of the mass of the average diet is composed of carbohydrates.

Naming

Carbohydrates are polyhydroxy compounds that contain other functional groups, usually a carbonyl group. We will be interested only in compounds with one alcohol group per carbon atom. If a carbohydrate contains an aldehyde group, it is known as an **aldose**. If a ketone is present, the molecule is a **ketose**. Often the letters *ul* will be inserted in the name of the ketoses. Notice the ending *-ose*, which is common to all the carbohydrates. Some general names are also useful in describing the carbohydrates; they are based upon the number of carbon atoms in the molecule. Triose, tetrose, pentose, hexose, and heptose would be given to carbohydrates with three to seven carbon atoms. These names may be used in combination, such as aldopentose for a five-carbon sugar with an aldehyde functional group.

Let us start with the least complicated carbohydrate molecules and build up to the compounds of great interest in living systems. Two trioses that have useful common names, glyceraldehyde and dihydroxyacetone, are shown along with their mirror images:

Aldose | Ketose

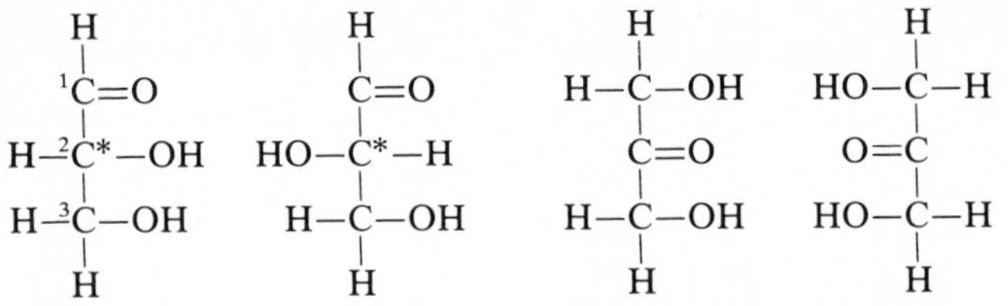

D-glyceraldehyde | L-glyceraldehyde | dihydroxyacetone | dihydroxyacetone

different compounds | same compounds

These examples can be used to illustrate some important concepts of optical activity. Of these two trioses, only glyceraldehyde has an asymmetric carbon atom with four different groups (shown with asterisk). The two different enantiomers are shown as mirror images. Neither carbon atom number 1 nor number 3 of the glyceraldehyde is asymmetric. Dihydroxyacetone does not possess any asymmetric carbons. Although the imaginary mirror is depicted, both structures of the ketose are the same and represent only one molecule. If there are not four different groups on a single carbon atom, the molecule can be drawn in a number of equivalent ways to represent only one actual structure.

Note the numbering system used to designate each carbon. Number 1 is written to the top in this representation, and increasing numbers will be used going down the chain. The carbonyl carbon is given the smallest number possible. Unless specified otherwise, the highest-numbered carbon will be a primary alcohol. The orientation of the highest-numbered asymmetric carbon, sometimes called the **penultimate** (next-to-last) carbon, will specify the D or L family. The D prefix is used when the hydroxyl group is written to the right side of this asymmetric carbon. The L prefix is used for the mirror image, which has the hydroxyl group to the left of this carbon. In the case of glyceraldehyde, with only one asymmetric carbon, the number 2 atom specifies this configuration.

Some examples of tetroses and pentoses of interest are shown using an abbreviation where a bar (–) indicates a hydroxyl group. The hydrogen atoms are not shown but are assumed to be present and to complete the necessary tetrahedral bonding positions.

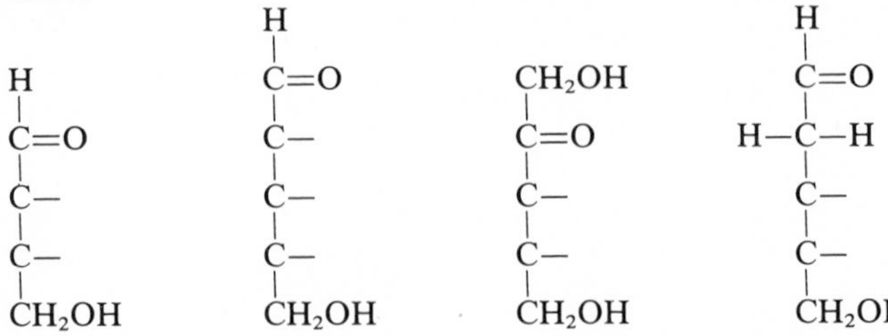

D-erythrose (aldotetrose) | D-ribose (aldopentose) | D-ribulose (2-ketopentose) | D-deoxyribose

There are a number of other isomers of each of these compounds. All the above are of the D-family, which is of greatest general importance in nature. Examples of two important isomeric aldo- and keto-pentoses are shown. The D-deoxyribose is not an isomer of the other pentoses because it has an empirical formula of $C_5H_{10}O_5$. The deoxy (without oxygen) ribose is an important constituent of one of the large families of nucleic acids (DNA). Any carbon atoms that lack the oxygen functional group will always be specified.

Hexoses

The hexoses are very important biochemically. Three of the most common members of this group are

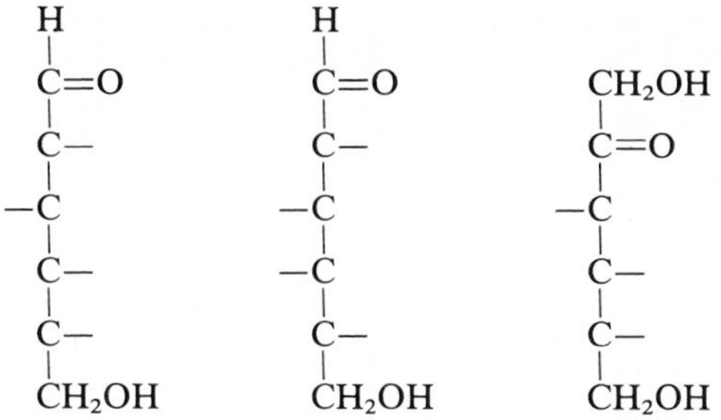

D-glucose D-galactose D-fructose

There are eight different aldohexoses in this D family. Since there is an L form (mirror image) equivalent to each D member of the pair, there is a total of 16 aldohexoses. An examination of all the possibilities indicates that there should be a total of 2^n optically active isomers, where n is the number of asymmetric carbons.

Two of these aldohexoses are of special interest to us. **D-glucose** is the most abundant hexose as it is known to occur as part of other more complicated molecules, including table sugar and the starches. **D-galactose** is a part of milk sugar and thus is of special importance in infant nutrition. Recall that these are all the D family, since the highest-numbered asymmetric carbon is shown to the right in this drawing. Since D-glucose and D-galactose differ from each other in only the orientation of groups about one carbon atom, they are said to be **epimers** of each other. The carbon atom involved in this different spacial orientation is known as the epimeric carbon atom.

D-fructose is an important constituent of table sugar and is present in polysaccharides from plant sources. Note that D-fructose has the identical orientation as D-glucose in the four highest-numbered carbon atoms.

Hemiacetal

It has been observed experimentally that there are actually two distinct optical isomers of D-glucose which can be isolated in different crystalline forms. A solution of one form rotates a plane of polarized light to a greater extent than a comparable solution of the other form when studied in a polarimeter. An interesting event occurs, however. If separate solutions of both of these forms are allowed to stand for a time, they will each finally yield the same rotation value. What does this experimental observation mean? This first indicates that there must be an additional asymmetric site in the D-glucose that is not apparent from our structural representations. How else would the two optical isomers be explained? In addition, these two forms must be interrelated in some chemical way when the molecules are in solution. An equilibrium situation must exist between the two different forms.

To explain the existence of these two nonidentical forms of D-glucose, a new structural representation must be used. We have studied the reaction between a carbonyl group and an alcohol whereby either an acetal or hemiacetal is formed (a ketal or hemiketal is formed when a ketone is used). As you examine a molecule like D-glucose, you will note that both groups for this reaction are available within the same molecule. An **intramolecular hemiacetal** occurs as the more favored and stable form of the molecule in solution. The number 1 carbonyl carbon reacts with the alcohol group on the number 5 carbon to yield a six-membered ring. This then makes a new asymmetric site at the number 1 position, since there are four different substituents attached to this carbon atom. There are two forms of the cyclic D-glucose, known as α and β, which can be shown by a great distortion of the new bonds formed:

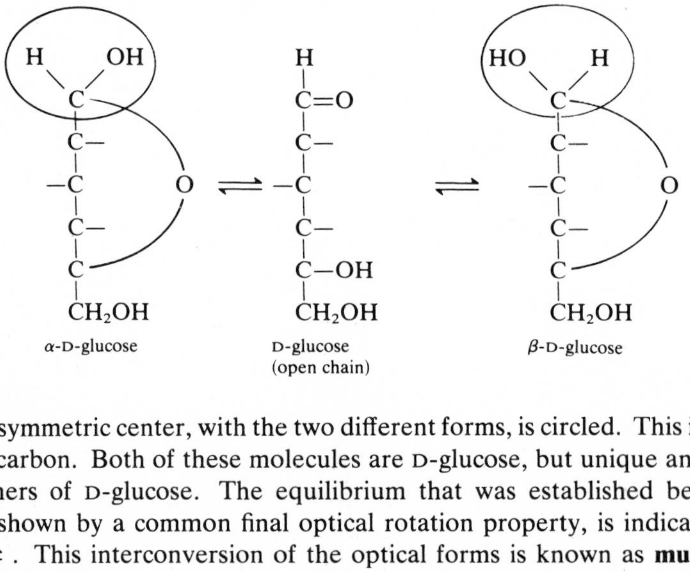

α-D-glucose D-glucose (open chain) β-D-glucose

The new asymmetric center, with the two different forms, is circled. This is called an **anomeric** carbon. Both of these molecules are D-glucose, but unique and separate stereoisomers of D-glucose. The equilibrium that was established between the forms, as shown by a common final optical rotation property, is indicated by the arrows $\rightleftharpoons$. This interconversion of the optical forms is known as **mutarotation**.

The **equilibrium** aspects of mutarotation deserve consideration. Although one form may be present in greatest concentration, at least a few molecules of each form are present in a solution. If a reaction of interest would involve the open chain form, that form would react. Once a given molecule reacted, the equilibrium could be reestablished with the formation of another molecule in the open-chain molecular form. It would be possible to convert all molecules to this open-chain form if a reaction were involved that used this molecule in an irreversible fashion.

The molecular formulas we have used up to this point are called **Fischer projections**. These representations were quite lacking because of distorted bond lengths in the cyclic forms. The cyclic forms are especially well suited to another way of writing the structures, called **Haworth projections**. In this system, the cyclic forms are drawn as pentagons (five-membered rings) or hexagons (six-membered rings). The five-membered rings are known as **furanoses**; the six-membered rings are **pyranoses**. The five- or six-membered rings will form depending upon the relative stability of the systems involved. The heterocyclic rings should be imagined as rather planar molecules, with the bottom part of the ring coming toward you from the plane of the paper. They will often be shown as a heavy line to indicate this effect. If you imagine this planar situation, all substituents must be positioned either above or below the plane of this ring. The highest-numbered carbon atom is always written above the ring for D sugars in this system. The rule is that everything written to the *right* in the Fischer projection is *below* the plane of the ring or down. Conversely, substituents written to the *left* in the Fischer system are shown *above* the plane of the ring or up. Some examples are shown with their complete chemical names:

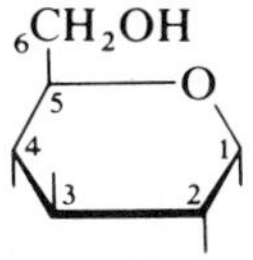

α-D-glucopyranose

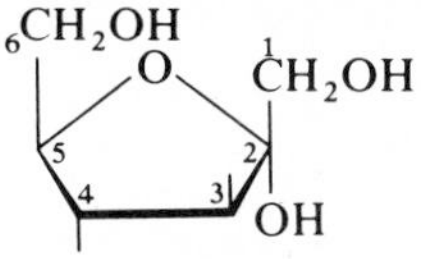

α-D-fructofuranose

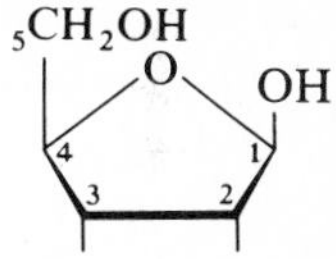

β-D-ribofuranose

Although the Fischer and Haworth representations are very useful, they are not completely accurate. A more accurate representation depicts the various bonds in less "strained" conformations, α-D-glucose is shown in this representation in its so-called **chair** form:

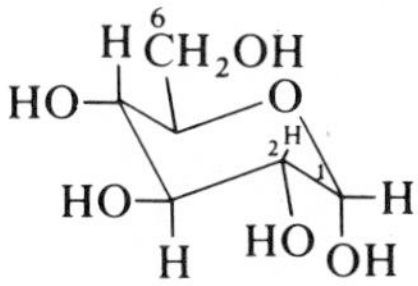

Derivatives and Reactions of Monosaccharides

The various functional groups on the monosaccharide can undergo their usual chemical reactions. Among the reactions of general interest to us are oxidation and reduction at the end carbons of the molecule. In Figure 38–1, some of these possible reactions of glucose are shown. The names given are general and would be applied to derivatives of other monosaccharides. The prefixes would be changed to specify the carbohydrate undergoing the reaction.

The easy oxidation reactions of the carbohydrates have been used as the basis for some quantitative carbohydrate assays. Since the carbonyl carbon of many carbohydrates undergoes ready oxidation, carbohydrates with a free carbonyl group are often called **reducing sugars**. The carbonyl group must be free and unattached to any other group in order to act as a reducing agent. This will be useful in the next

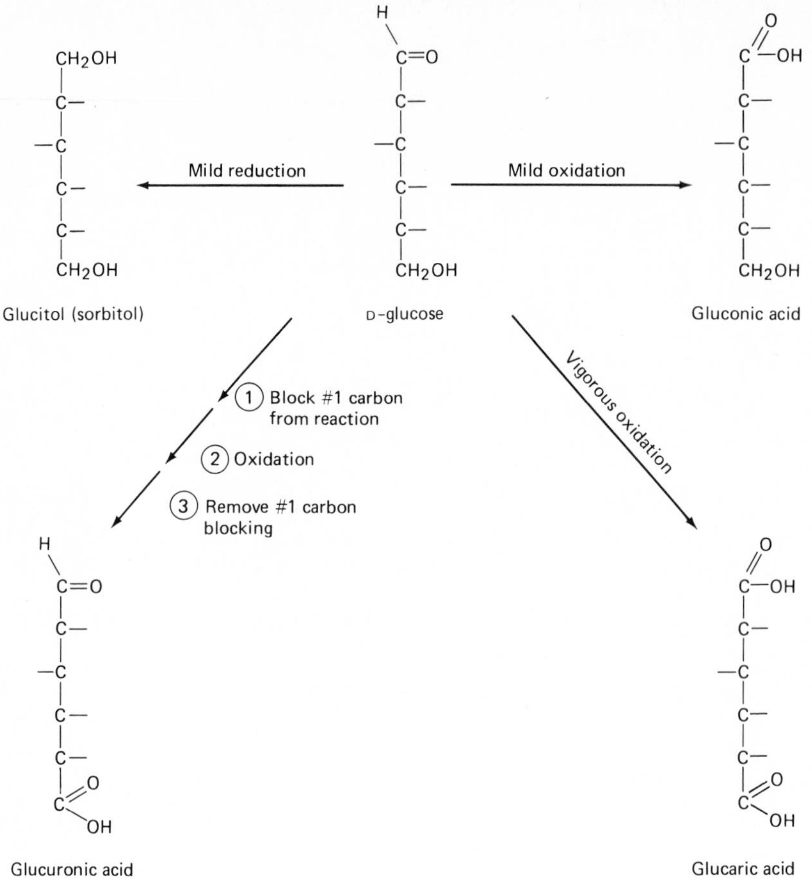

Mild reduction
Mild oxidation
Glucitol (sorbitol)
D-glucose
Gluconic acid
1 Block #1 carbon from reaction
2 Oxidation
3 Remove #1 carbon blocking
Vigorous oxidation
Glucuronic acid
Glucaric acid

Figure 38–1 Common oxidation and reduction reactions for glucose and other monosaccharides.

chapter as we consider the linkages involved in carbohydrates containing two or more repeating units. All the common monosaccharides that we have studied are considered to be reducing sugars. The reducing sugars can readily reduce Ag^+ to Ag^0 (Tollens' test) or Cu^{2+} to Cu^+ (Fehlings' test).

If the carbonyl group is not free to undergo ready oxidation, the carbohydrate cannot be considered a reducing sugar. An example of this might be in the case of the methyl **glucosides**. In this case 1 mol of methyl alcohol is allowed to react with 1 mol of glucose (consider the open-chain form of glucose for the reaction). This is an intermolecular reaction in which an acetal is formed and a molecule of water is lost. There would be two different methyl glucosides formed, depending upon the orientation of the substituents around the acetal carbon. Acetals are not in equilibrium with open-chain forms. Since the methyl glucoside would not possess the free carbonyl group, it would not be a reducing sugar. These two different methyl glucosides are shown:

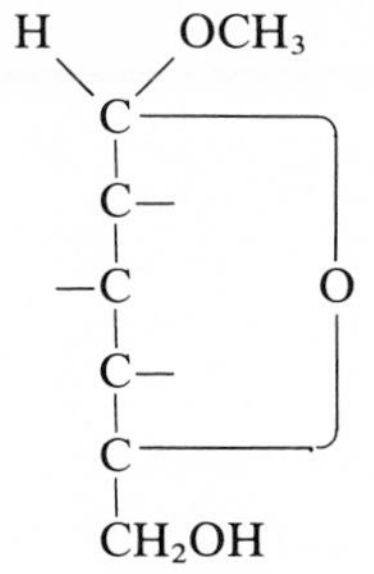

H
OCH_3
C
C
C
O
C
C
CH_2OH

α-methyl glucoside
(Fischer representation)

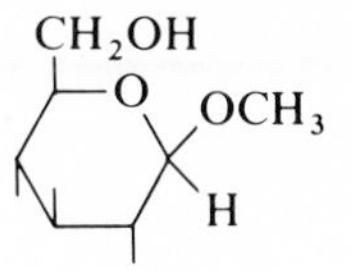

CH_2OH
O
OCH_3
H

β-methyl glucoside
(Haworth representation)

These compounds are derivatives of a carbohydrate and a noncarbohydrate molecule. The disaccharides and polysaccharides are products of two or more carbohydrates combining with each other. Other derivatives of the carbohydrates include the phosphate esters that form from the reaction with phosphoric acid. A variety of different phosphate esters could be formed from glucose, depending upon which hydroxyl group was involved. These phosphate esters will be important as we study the fate of carbohydrates in biochemical reactions.

UNIT 38 Problems and Exercises

38–1. Draw the structure of L-erythrose, using the open-chain form.

38–2. Draw the structure of a D-aldoheptose.

38–3. Would you classify glycerol as a carbohydrate? Why?

38–4. Using Haworth representations, show the various forms of D-glucose as it undergoes mutarotation.

38–5. Would you expect methyl glucoside to undergo mutarotation? Explain.

38–6. Draw all the possible D-aldopentoses in the open-chain form. How many aldopentose isomers would you expect could exist?

38–7. If D-galactose was oxidized at both the number 1 and 6 carbons, a compound known as mucic acid would be formed. Draw the structure of this compound.

38–8. Would you expect galactose to be a reducing sugar? Explain.

38–9. If D-fructose were allowed to undergo a transformation to form its corresponding aldohexoses, two compounds could be formed. Draw the structures of these two resulting aldohexoses, using both the Fischer and Haworth representations.

38–10. Dihydroxyacetone phosphate is formed when phosphoric acid forms an ester with one of the alcohol groups of dihydroxyacetone. Draw the structure of this phosphorylated derivative. Does this derivative possess an asymmetric carbon? Explain.

UNIT 39 DISACCHARIDES AND POLYSACCHARIDES

The monosaccharides are the basic building blocks of both disaccharides (two units) and large polysaccharides (many units). The *condensation* of monosaccharides to disaccharides is brought about by two monomeric units losing a molecule of water as a new chemical bond is formed between them. The reverse of this process, *hydrolysis*, is the addition of a water molecule to provide two new functional groups following the disaccharide scission. Whereas the disaccharides can be cleaved to form only two monosaccharides, the polysaccharides can undergo hydrolysis at a number of different points to yield a number of different products:

$$\text{Disaccharides} + H_2O \xrightleftharpoons{\text{Hydrolysis}} \text{2 monosaccharides}$$

$$\text{Polysaccharide} + H_2O \rightleftharpoons \text{monosaccharide} + \text{disaccharides} + \text{smaller polysaccharides}$$

The chemical linkage involved in joining most of the di- and polysaccharides is the *acetal* linkage. The hemiacetal carbon is bonded to one of the oxygens of an alcohol in an adjacent monosaccharide to form an acetal. Hydrolysis of the acetal linkage is readily catalyzed by acids or by specific enzymes. In all cases, the chemical reaction involved is the addition of a molecule of water.

Disaccharides

Maltose is a disaccharide in which the anomeric carbon on one D-glucopyranose is joined to make an acetal with the number 4 carbon of an adjacent D-glucopyranose. The bond from the anomeric carbon can be either in the α or β orientation. This linkage is of critical importance in the di- and polysaccharides. In the case of maltose, the anomeric carbon is in the α configuration and the linkage is symbolized as α $(1 \rightarrow 4)$. Note that it is only necessary to specify the orientation of the anomeric carbon in the linkage. The orientation of the number 4 carbon of the second molecule is already specified when we indicated that it is a glucose molecule. The complete name for maltose would be 4-O-α-D-glucopyranosyl-β-D-glucose, with a structure as shown:

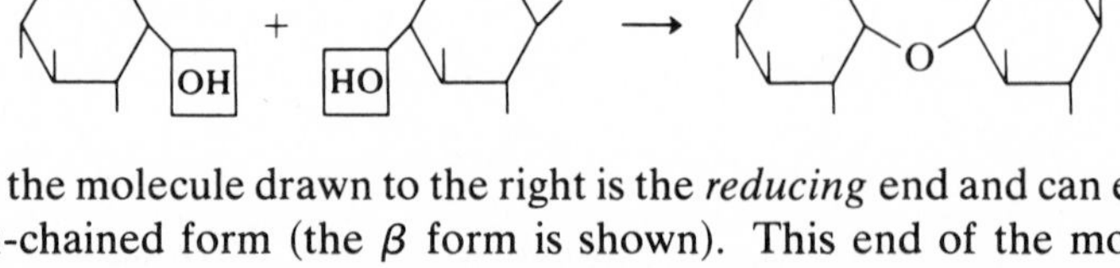

CH2OH
O
OH
+
HO
CH2OH
O
→
CH2OH
O
O
CH2OH
O

The end of the molecule drawn to the right is the *reducing* end and can exist in the α, β, or open-chained form (the β form is shown). This end of the molecule could undergo oxidation just as do the monosaccharides considered earlier. The *nonreducing* end, which does not undergo ready oxidation, is pictured on the left side. Although the reducing end of a disaccharide can have the hemiacetal carbon equilibrium with an open-chain structure, the acetal linkage joining the two monosaccharides absolutely cannot. If this were to occur, the disaccharide would no longer be a disaccharide but two monosaccharides.

Isomaltose is another disaccharide that has an α $(1 \rightarrow 6)$ linkage, using the earlier symbolism. This disaccharide is not abundant in nature as a free molecule, but is a repeating unit in branched polysaccharides. Its structure can be shown:

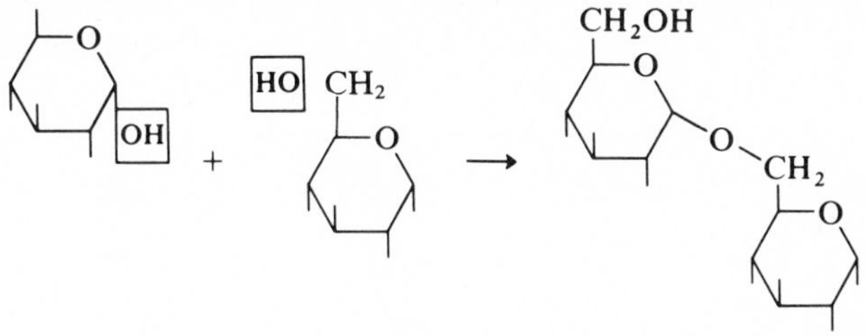

O
OH
+
HO
CH2
O
CH2OH
O
O
CH2
O

Cellobiose is a disaccharide that differs from maltose only in linkage configuration between the two glucose monomers. This is a highly critical difference, however. In the case of cellobiose it remains a $(1 \rightarrow 4)$ intermolecular linkage, but the glycosidic linkage is in the β form. This is seen in the structure for cellobiose:

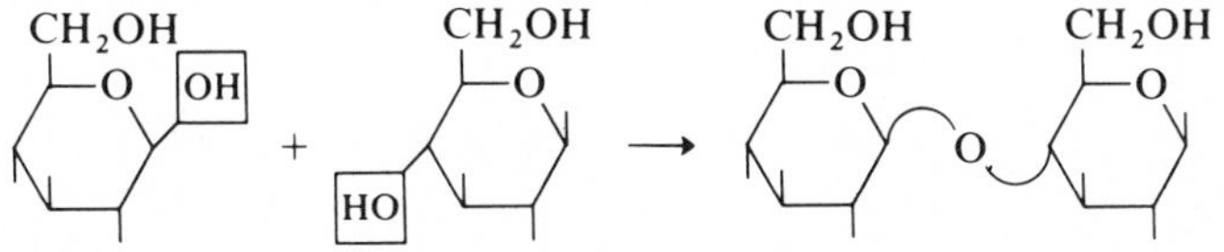

CH2OH
O
OH
+
HO
CH2OH
O
CH2OH
O
O
CH2OH
O

This molecule represents the major repeating unit in the polysaccharide cellulose. The subtle difference in the way in which the glucose molecules are joined together imparts tremendous differences in the resulting polysaccharides.

Two disaccharides of nutritional importance are sucrose and lactose. Both of these exist as free disaccharides, and are responsible for a significant amount of our dietary intake of carbohydrates at various stages of our lives. Lactose is sometimes known as milk sugar because it is found to be present up to about 5% in milk. Sucrose is common table sugar and is isolated from beets and cane.

Lactose is a galactoside composed of D-glucose and D-galactose. The linkage between the monomeric units is β $(1 \rightarrow 4)$, which would give a name of 4-O-β-D-galactopyranosyl-D-glucopyranose:

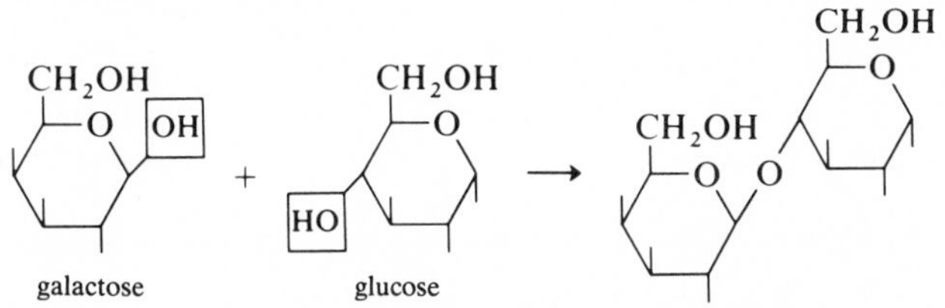

CH2OH
O
OH
galactose
+
HO
CH2OH
O
glucose
CH2OH
O
O
CH2OH
O

α-lactose

Both α-lactose and β-lactose are known to exist. These differ only in the configuration of the anomeric carbon at the reducing end of the molecule.

The fact that this molecule contains galactose is important in certain disease conditions. In certain situations, infants lack the necessary systems to metabolize galactose. This is a hereditary disorder known as galactosemia and will be discussed later.

Sucrose is the most abundant disaccharide in nature; it is composed of glucose and fructose. The intramolecular linkage is somewhat unique compared to the other disaccharides. The anomeric carbons of both monosaccharides are joined together. This makes a labile bond that can undergo hydrolysis very easily. This also means that sucrose is *not* a reducing sugar, since there are no free hemiacetal groups in the molecule. The glucopyranose is an α configuration; the fructofuranose is β, as shown:

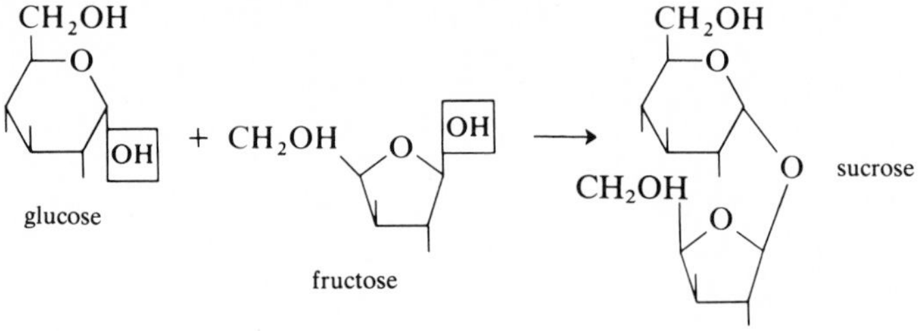
CH2OH
O
OH
glucose
+ CH2OH
O
OH
fructose
CH2OH
O
CH2OH
O
O
sucrose

Polysaccharides

The polysaccharides are macromolecules or biopolymers composed of a great many repeating monosaccharides as the monomeric units. In some cases, **homopolysaccharides**, only one kind of monosaccharide is involved as the repeating monomer. **Heteropolysaccharides** yield mixtures of different monomeric units. In some cases, polysaccharides are found in conjunction with proteins (glycoproteins) and lipids (glycolipids). In other cases, atoms other than C, H, and O are involved in the structure of complex polysaccharides. The functions of the polysaccharides vary as widely as do their chemical structures. Polysaccharides have important nutritional (starch), structural (cellulose), and regulatory (heparin) functions.

Starch is the homopolysaccharide found in plants that is of great nutritional significance. One form of starch, **amylose**, is a linear polymer of glucose in α $(1 \rightarrow 4)$ linkages. Actually, one could consider maltose to be the repeating unit. As is true of polysaccharides in general, the lengths of the chains that can be isolated for study are not uniform. Amylose may be 50,000 or more in molecular weight. This would correspond to nearly 300 glucose molecules in a linear arrangement. The second form of starch is known as **amylopectin**. This is also a high-molecular-weight compound with a main chain of glucose in the α $(1 \rightarrow 4)$ linkage plus *branching* points of α $(1 \rightarrow 6)$ linkages. This second or branching linkage corresponds to isomaltose. This branching in amylopectin tends to occur about every 25 to 30 glucose residues along the α $(1 \rightarrow 4)$ backbone. Pictorial representations of both of these starch molecules are seen in Figure 39–1.

Glycogen is the storage form of glucose in animals that corresponds to starch in plants. It is found in the liver (up to 8% of wet weight) and muscles (up to 1% of wet weight) of normal individuals in a healthy nutritional state. Glycogen is similar to amylopectin but is even more highly branched, with the α $(1 \rightarrow 6)$ branches occurring every 8 to 10 glucose units along the α $(1 \rightarrow 4)$ main chain. In our study of

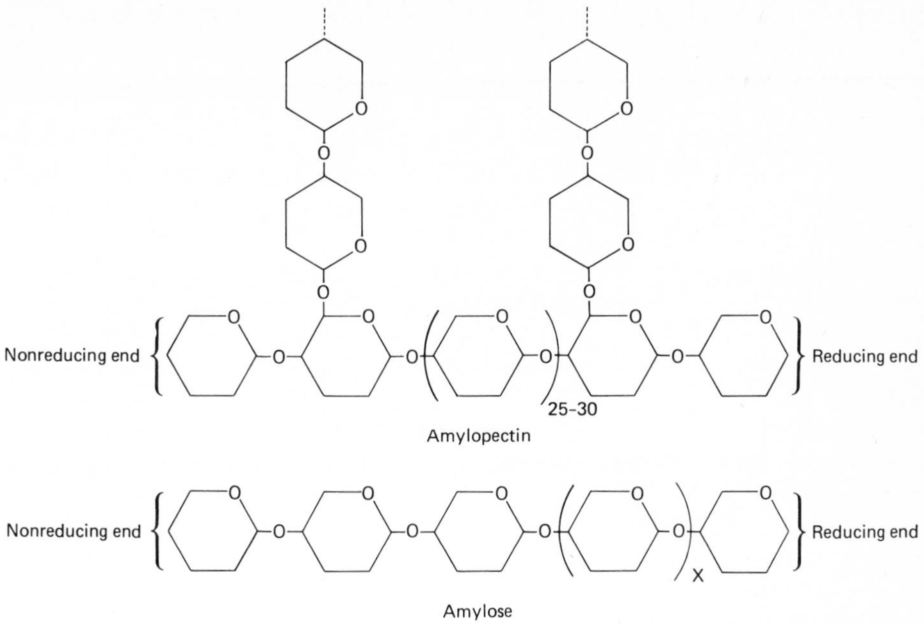
Nonreducing end
Reducing end
25-30
Amylopectin
Nonreducing end
Reducing end
X
Amylose

Figure 39–1 Representatives of amylose and amylopectin. 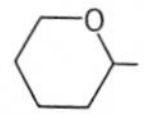 represent (1 → 4) linkages, 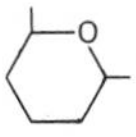 represent (1 → 4) and (1 → 6) branches.

metabolism, the mechanism will be considered by which glucose is stored as glycogen, or glycogen is hydrolyzed to glucose. In this way the glucose levels of the blood are carefully maintained and controlled.

It is interesting to consider why nature chooses to synthesize these macromolecules as a storage form. Why not store the glucose as glucose and eliminate all this additional trouble of forming and breaking down these polysaccharides? One answer must lie in the regulation of glucose as an energy source in animals. Another reason may have its basis in the physiochemical action of molecules. The osmotic pressure within a cell is dependent upon the number of molecules present, regardless of their molecular weights. Many glucose molecules may thus be stored without upsetting this important balance that must be maintained within cells.

No discussion of starch would be complete without mentioning one of its chemical reactions. Linear starch will react with iodine to produce a dark purple color; branched starches and glycogen will produce a reddish-purple color with iodine. This simple reaction provides the basis for the detection of the presence of starch in many laboratory experiments.

The most widely distributed polysaccharide in nature is **cellulose**, since nearly half of all the carbon in the plant kingdom is in the form of cellulose. Cotton is an excellent source of this polysaccharide, as it is about 90% cellulose. This is a linear polymer of glucose molecules with β (1 → 4) linkages. The repeating unit is really cellobiose. The molecular weights of cellulose molecules may be up to almost half a million, corresponding to just under 3000 glucose monomers. The difference between amylose and cellulose is the α versus the β linkage at the anomeric carbon. Although this does not seem to be great, it actually makes a tremendous difference in the resulting macromolecule. Consider the influence of the α and β configuration on the water solubility of the resulting polysaccharide. Starch has a reasonable solubility in water, whereas cellulose is insoluble. From a nutritional point of view, humans can metabolize starch, but we do not possess the necessary catalysts (enzymes) to degrade cellulose. Certain microorganisms found in the rumin of cattle and in termites do possess this ability, however.

Man has been able to use cellulose directly in cotton clothing and paper products and also as a starting material for a number of compounds. As is usually true of man's activities, some of the products of his ingenuity are better than others. If cellulose is reacted with a mixture of nitric and sulfuric acids, it is nitrated and converted to cellulose nitrate or guncotton. This is a basic ingredient in smokeless gunpowder. On the other hand, cellulose can be treated with carbon disulfide and sodium hydroxide to form a viscous dispersion called **viscose**. If this viscose is forced through a spinnerette into an acid bath, rayon thread is formed. If the viscose is forced through a narrow slit, cellophane is formed.

There are a number of heteropolysaccharides that have interesting structures and important uses. These polymers may contain oxidized derivatives of monosaccharides, as well as amine and sulfate groups. A few of these polysaccharides and

Table 39–1 Some heteropolysaccharides and their uses

Name	Constituents	Function
Heparin	Sulfate derivatives of glucuronic acid and glucosamine	Blood anticoagulant
Hyaluronic acid	Glucuronic acid *N*-acetyl glucosamine	Joint fluid and found in vitreous and synovial fluids
Chitin	*N*-acetyl-glucosamine	Exoskeleton of insects and crustaceans
Pectin	Galacturonic acid Galacturonic acid methyl esters	Structural components of fruits and berries; used in making jelly
Agar	Galactose Galactose sulfates	Found in seaweed; useful for microbiological cultures
Mureins	*N*-acetyl-glucosamine *N*-acetyl muramic acid + peptides	Bacterial cell walls

their uses are shown in Table 39–1. The linkages are not specified in this table, but they are unique for the compounds involved. An examination of this table clearly shows the widespread importance of polysaccharides in nature.

UNIT 39 Problems and Exercises

39–1. Maltose can undergo mutarotation. Draw this disaccharide in the various forms in which it would exist undergoing this reaction.

39–2. Explain why sucrose is not a reducing sugar.

39–3. Explain why amylose could or could not be considered a reducing carbohydrate.

39–4. Gentiobiose is a disaccharide with the systematic name 6-O-β-D-glucopyranosyl-D-glucose. Draw the structure of this compound in its α form.

39–5. Using a schematic representation for glucose, depict the structure of glycogen.

39–6. A carbohydrate named panose was found to yield only glucose upon complete hydrolysis. Panose has a molecular weight of approximately 500. Using several controlled hydrolysis procedures, either maltose or isomaltose was detected, in addition to glucose. Using skeleton structures, give the structure of this carbohydrate.

39–7. If pectin was hydrolyzed to its component monosaccharide, draw the structures of the resulting compounds.

CHAPTER 14

Lipids

When cells are extracted with any one of several *organic* solvents such as ether, benzene, chloroform, ethanol, or acetone, the material that is *dissolved* is referred to as a **lipid** or a **fat**. We shall use the terms lipid and fat interchangeably. Since the materials that dissolve under these extraction conditions have great variety, the lipids are composed of a group of what may appear to be unrelated chemical structures. The solubility characteristic is one of the common factors that joins these materials together.

Lipids are among the most common of the types of biological materials that we encounter in daily life. Anyone interested in personal weight control is keenly conscious of the caloric content of fats. The average diet contains between 80 to 150 g of fat per day, which may account for 30 to 40% of the caloric intake. Upon metabolism, lipids yield almost twice the energy on a per gram basis as do either carbohydrates or proteins.

Another major factor of great importance relates to the solubilities of the lipids. Since they have low water solubility, the lipids tend to isolate themselves from water. In this way animals can carry less weight to keep a maximum energy reserve, and seeds can be as light as possible for the amount of energy stored. Also, lipids function as important constituents of membranes that keep cells separated from the aqueous environment surrounding them. In addition, lipids are important constituents of waxes, paints, soaps, and synthetic detergents.

Although considerable diversity and overlap are found in nature, the lipids can be considered under three general topics: simple lipids, complex lipids and steroids, and the fat-soluble vitamins (A, D, E, and K).

UNIT 40 SIMPLE LIPIDS

Wax

Waxes are simple esters of long-chain carboxylic acids (up to 36 carbon atoms) and monohydroxy alcohols. A common wax is beeswax, which has as one of its constituents ceryl myristate:

$$CH_3(CH_2)_{12}\overset{\displaystyle O}{\overset{\displaystyle \|}{C}}-O-CH_2(CH_2)_{24}CH_3$$

Carnauba wax, a component of many polishes, is also a mixture of esters of high molecular weights.

Waxes melt in the range from 60 to 90°C, and are generally of a harder consistency than lower-molecular-weight lipids. Since waxes have extremely low water solubility and undergo almost no hydrolysis or digestion in our intestinal tract, they are not of great nutritional significance.

Triglycerides

The **triglycerides** are nearly colorless, odorless, and tasteless if pure and have specific gravities of less than 1.0. We know from experience that oils generally float on the surface of water. When the triglycerides react with bases such as KOH or NaOH in a process called **saponification**, a mixture of salts of carboxylic acids and glycerol result. Thus the triglycerides are built in the form of triesters from three separate molecules of carboxylic acids and one molecule of glycerol. The common feature in this linkage, as in the case of disaccharides, involves loss of a water molecule in joining functional groups to yield a complex molecule. Conversely, the breakdown of the complex molecule to its chemically distinct units is accomplished by the addition of the elements of water. This can be expressed as

$$
\begin{array}{lll}
\alpha & R-\overset{O}{\overset{\|}{C}}-O-CH_2 + HOH & R-\overset{O}{\overset{\|}{C}}-OH \quad HO-CH_2 \\[2mm]
\beta & R-\overset{O}{\overset{\|}{C}}-O-CH + HOH \longrightarrow & R-\overset{O}{\overset{\|}{C}}-OH + HO-CH \\[2mm]
\alpha' & R-\overset{O}{\overset{\|}{C}}-O-CH_2 + HOH & R-\overset{O}{\overset{\|}{C}}-OH \quad HO-CH_2
\end{array}
$$

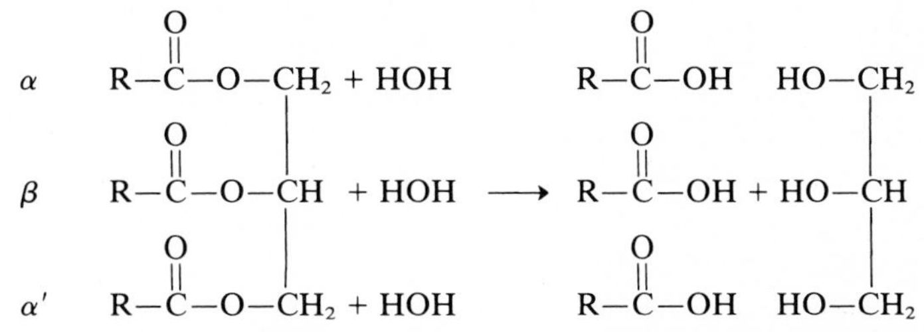

a triglyceride   carboxylic   glycerol
acids

Glycerol has been discussed as a representative of the polyhydroxy alcohols. The three carbons are designated by the use of lowercase Greek letters, α, α', and β. The acids that result from the hydrolysis of the triglyceride vary considerably, accounting for the differences in both the chemical and biological properties of the triglycerides. In general, the fatty acids are linear and aliphatic, but may be unsaturated to a greater or lesser degree. A few branched and cyclic fatty acids are known to exist, but we will not include them here. It is possible to have only one or two of the hydroxyl groups of glycerol esterified, and these are known as mono- or diglycerides, respectively.

The saturated acids that are of greatest importance are listed in Table 40–1.

Table 40–1 Saturated acids

Common name	Systematic name	Formula
Acetic acid	Ethanoic acid	CH_3COOH
Butyric acid	Butanoic acid	$CH_3(CH_2)_2COOH$
Caproic acid	Hexanoic acid	$CH_3(CH_2)_4COOH$
Caprylic acid	Octanoic acid	$CH_3(CH_2)_6COOH$
Capric acid	Decanoic acid	$CH_3(CH_2)_8COOH$
Lauric acid	Dodecanoic acid	$CH_3(CH_2)_{10}COOH$
Myristic acid	Tetradecanoic acid	$CH_3(CH_2)_{12}COOH$
Palmitic acid	Hexadecanoic acid	$CH_3(CH_2)_{14}COOH$
Stearic acid	Octadecanoic acid	$CH_3(CH_2)_{16}COOH$

The unsaturated acids of interest are listed in Table 40–2.

Table 40–2 Unsaturated acids

Common name	Systematic name	Formula
Palmitoleic acid	9-hexadecanoic acid	$CH_3(CH_2)_5CH{=}CH(CH_2)_7COOH$
Oleic acid	9-octadecanoic acid	$CH_3(CH_2)_7CH{=}CH(CH_2)_7COOH$
Linoleic acid	9,12-octadecadienoic acid	$CH_3(CH_2)_4CH{=}CHCH_2CH{=}CH(CH_2)_7COOH$
Linolenic acid	9,12,15-octadecatrienoic acid	$CH_3CH_2CH{=}CHCH_2CH{=}CHCH_2$- $CH{=}CH(CH_2)_7COOH$
Arachidonic acid	5,8,11,14-eicostetrenoic acid	$CH_3(CH_2)_4CH{=}CHCH_2CH{=}CHCH_2$- $CH{=}CHCH_2CH{=}CH(CH_2)_3COOH$

The most abundant saturated fatty acids found in triglycerides from animals are palmitic (C_{16}) and stearic (C_{18}). The most abundant unsaturated fatty acid is oleic acid. Fatty acids with fewer than 10 carbon atoms are not very common in animal tissues, but they do occur in milk. Typical American diets have switched from materials rich in the saturated fatty acids to those containing larger quantities of the unsaturated acids, especially the polyunsaturated fatty acids. The fatty acid content of some common foods is shown in Table 40–3. The triglycerides contain a wide variety of fatty acids, and our normal diet is composed of many different lipids.

The range of fatty acids contained in human body fat varies quite considerably from individual to individual. The range of values is shown in Table 40–4.

In addition to providing energy, certain specific fatty acids are **essential** and must be supplied in the diet. Two or three grams of linoleic acid per day should supply this specific need for a healthy young adult. One chicken leg or 3 oz of tuna would just

Table 40–3 Fat content of common foods*

Food	Approximate fat content (g/100 g of food)	Saturated fatty acids	Polyunsaturated fatty acids	Monounsaturated fatty acids
Eggs	12	34	14	52
Beef	25	43	3	54
Chicken	13	30	22	48
Pork	24	39	10	51
Halibut	8	25	75	0
Peanuts	44	23	31	46
Coconut oil	100	92	2	6
Corn oil	100	12	55	33
Olive oil	100	9	5	86
Safflower oil	100	7	75	18

* Expressed as gram fatty acid per 100 g of fatty acid.

Table 40–4 Percentage of fatty acids in triglycerides of human body fat

Fatty acid	Percent
Lauric	0.1–1.7
Myristic	1.5–5.9
Palmitic	20.8–25.0
Stearic	2.2–8.4
Oleic	38.7–46.9
Linoleic	0.0–24.8
Unsaturated	1.5–8.3
Other unsaturated	5.0–17.0

about meet this need. A majority of the other fatty acids can be synthesized by our metabolic system.

It is rather unusual when a triglyceride contains three molecules of only a single fatty acid. Most of the common oils yield a variety of fatty acids. Milk will have up to 15 different fatty acids in a variety of triglycerides. The triglycerides of peanut oil contain mainly oleic (56%), linoleic (26%), and palmitic (8%) acids, as well as six or seven others in smaller amounts.

Reactions of Triglycerides

The physical and chemical properties and reactions of the triglycerides are dependent upon the fatty acids of which they are composed. The melting point of the fat and its physical state at room temperature vary with its composition. Unsaturated fatty acids have lower melting points than saturated fatty acids of the same molecular weight. In the production of solid shortenings, referred to as **hardening**, mixtures of unsaturated liquid vegetable oils react with hydrogen to become partially saturated. Triolein can be hydrogenated to form tristearin:

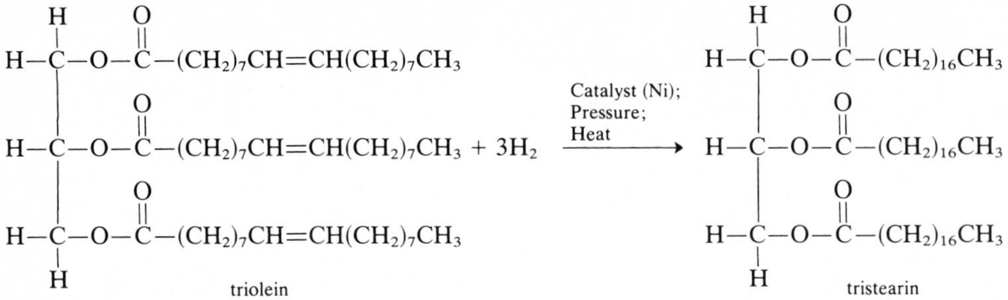
Catalyst (Ni);
Pressure;
Heat

triolein — tristearin

Another reaction of the double bonds involves the addition of a halogen to the alkene linkage. This reaction is the basis of the determination of the **iodine number**. Although the specific definition of the iodine number may only be of interest to lipid chemists, it is important to understand its general meaning. Since halogens like iodine or bromine are known to add quantitatively to the alkene group, it stands to reason that the amount of halogen that can react with a fat should be directly related to the number of double bonds that it contains. This experiment is easy to conduct since bromine solutions are reddish brown in color, whereas the halogenated derivatives of the fatty acids are generally colorless. To determine the degree of unsaturation of a fat, Br_2 is added to the oil until the color of bromine persists. If a saturated fat was used, the brown color would be observed with the first addition of Br_2. The saturated fat would have an iodine number of 0 as there would be no bromine uptake. If a polyunsaturated system was used, a considerable quantity of the bromine solution would have to be added before all the available alkene linkages reacted and an excess of the bromine could be observed. The greater the number of double bonds present, the greater the iodine number. Safflower oil has an iodine number of about 150; coconut oil has a value of less than 10 on the same scale.

The oxidation reactions that can occur at double bonds are important and have both desirable and undesirable implications. Fats that contain the double bonds can become spontaneously oxidized in the presence of air. This can lead to the **rancidification** of butter. However, these same chemical reactions can be used to an advantage in the preparation of **drying oils**. Many of the polyunsaturated oils react with oxygen to form a dry film and are thus called drying oils. This is not the simple drying or evaporation that we usually encounter. In this case the lipid undergoes oxidation followed by polymerization to make a chemical network of molecules that forms the thin film. Oil-base paints and varnishes contain drying oils that react to form the desired protective coating. Linseed oil, derived from flaxseed, is one of the most common examples of a drying oil. As expected, linseed oil is rich in linolenic and linoleic acids and has a high iodine number (about 170 on our relative scale).

In the **saponification** reaction, a known amount of triglyceride reacts with a measured amount of KOH or NaOH according to the following:

$$
\begin{array}{ccc}
\underset{H}{\overset{H\quad O}{H-C-O-C-R}} \ \ KOH & & H \\
& & H-C-OH \quad O \\
\underset{H}{H-C-O-C-R} + KOH & \longrightarrow & H-C-OH + 3RC-O^-\ K^+ \\
& & H-C-OH \\
H-C-O-C-R \ \ KOH & & H \\
H & &
\end{array}
$$

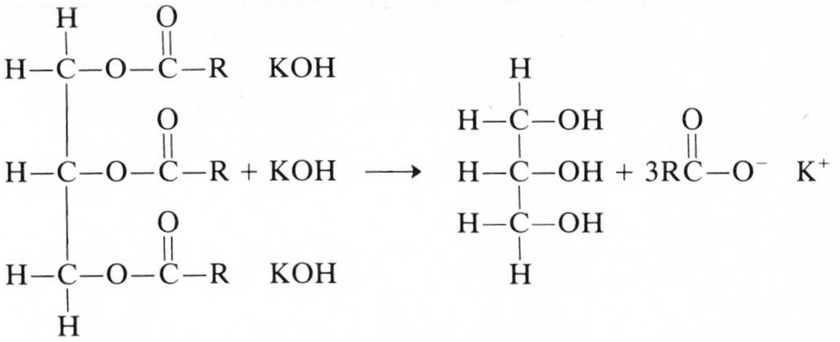

Frequently, the reaction is carried out in an alcoholic solution to increase the solubility of the triglyceride. The amount of KOH used in the reaction depends on the number of moles of triglyceride in the original reaction mixture. Since the glycerol portion is common to all triglycerides, the **saponification number** is actually related to the molecular weight of only the fatty acids of which the triglyceride is composed. For example, 1 g of a triglyceride containing low-molecular-weight fatty acids would be made up of a larger number of molecules of a triglyceride than the same weight of a triglyceride composed of fatty acids of a higher molecular weight. The lower the molecular weight of the fatty acid in the triglyceride, the larger the amount of KOH necessary to saponify a given weight of it, and the greater the saponification number. The number is actually defined as the milligrams of KOH necessary to saponify 1 g of fat or oil. Calculated values for a number of triglycerides are shown in Table 40–5.

Table 40–5 Saponification numbers for some triglycerides

Triglyceride	Saponification number	MW
Tributyrin	557	302
Tripalmitin	208	806
Triolein	190	884
Tristearin	189	890

Soaps and Detergents

When triglycerides undergo either acid-catalyzed hydrolysis or saponification by a base, fatty acids are produced. Under these two different sets of conditions, the ionic state of the resulting carboxylic acid is different. Acid hydrolysis yields an un-ionized carboxyl group, and basic saponification yields a carboxylate anion. These **ionic forms** are reversible depending upon the pH of the media. This can be illustrated as

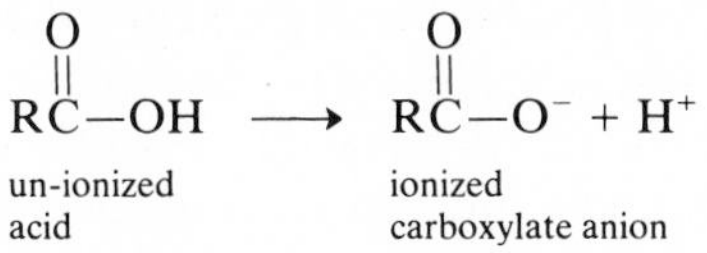

$$
\underset{\substack{\text{un-ionized} \\ \text{acid}}}{RC-OH} \ \longrightarrow \ \underset{\substack{\text{ionized} \\ \text{carboxylate anion}}}{RC-O^-} + H^+
$$

The ionic form of the carboxyl group has a very important influence on the solubility of the fatty acid. At any pH above the pK_a of the acid group, the ionized portion of the molecule tends to interact very readily with polar or ionic solvents like water. The R group of the fatty acid tends to interact very well with nonpolar solvents and other lipid material. This is the chemical basis for the action of both soaps and detergents. A single molecule possesses both **hydrophilic** (water-loving) and **hydrophobic** (water-hating) characteristics. For the long-chain fatty acids, the alkane or alkene tail is hydrophobic and the carboxylate head is hydrophilic, as shown for stearic acid:

$$\underbrace{CH_3CH_2CH_2CH_2CH_2CH_2CH_2CH_2CH_2CH_2CH_2CH_2CH_2CH_2CH_2CH_2CH_2}_{\text{hydrophobic}}-\underbrace{\overset{\overset{\displaystyle O}{\|}}{C}-O^-}_{\text{hydrophilic}}$$

Soaps are usually potassium or sodium salts of the fatty acids. The K^+ salts form softer and more soluble soaps and frequently are incorporated in liquid soaps. The cleansing action involves the attraction of oils to the hydrophobic end of the molecule while the charged end of the molecule remains strongly attracted by water. The soap reacts at the oil-water interface and allows the oils, with the dissolved dirt, to be swept away. The major disadvantage to the use of these soaps is that they will also react with divalent cations like Ca^{2+} and Mg^{2+}. The fatty acid salts that result from the reaction of these divalent cations are water insoluble. Since such cations are present in hard water, these simple fatty acid soaps lead to the formation of scum and bath tub rings and are no longer of great general use by themselves.

Because of our desire to keep ourselves and our wearing apparel clean (over \$1 billion worth of detergents are sold annually), other synthetic detergents have been developed. The general principle has always been the same: develop a molecular structure that consists of both hydrophobic and hydrophilic groups. Two specific examples include:

1. Sodium alkyl aryl sulfate: 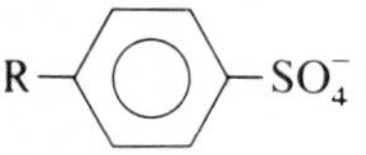 Na^+, where R is frequently a hydrocarbon of 12 carbon atoms.
2. Sodium lauryl sulfate: $CH_3(CH_2)_{11}-SO_4^-\ \ Na^+$

Although these synthetic detergents are especially desirable since they do not form insoluble salts with Ca^{2+} or Mg^{2+}, the first does have an undesirable characteristic. A problem results from the disposal and degradation of the used detergents. As such detergents are used, they must either build up in our disposal systems or be broken down by natural means. Microorganisms are known to metabolize straight-chain aliphatic hydrocarbons (**biodegradation**), but they cannot metabolize branched-chain or aromatic compounds to any great extent. The result of using tons of highly branched or aromatic detergents leads to polluted streams and lakes.

The phosphate in many detergents caused additional problems. Polyphosphate molecules are used to tie up Ca^{2+} and Mg^{2+} and thus remove the undesirable qualities of hard water. Unusually high levels of phosphates in streams and lakes cause water plants and algae to grow in an uncontrolled fashion. A series of reactions then occurs in which sediments build up to the detriment of the normally plentiful species of fish and plants. It is because of this that phosphate-containing detergents have been banned in many parts of the world.

Man remained ingenious, however, and switched to another detergent additive, nitrilotriacetic acid (NTA). Unfortunately, NTA has recently been linked to cancer and genetic abnormalities in laboratory animals. Thus the ideal detergent has not yet been developed. Chemists continue to apply their skills to develop better systems without sacrificing the long-range needs of man. Unfortunately, our demands are always high, and our commercial chemical experiments are on such a scale that they can go beyond our ability to curb them before damage is inflicted on our environment.

Prostaglandins

Another special group of fatty acids has gained considerable attention recently in the study of biochemistry and medicine. This is the **prostaglandin** family of compounds, each a fatty acid containing 20 carbon atoms. The individual members of the family differ from one another in the number and position of double bonds and in oxygen functional groups. Two examples are

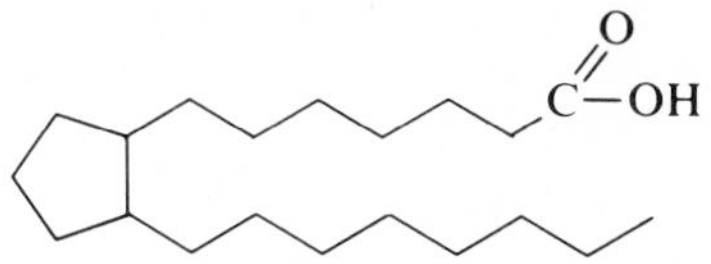
O
C—OH

prostanoic acid

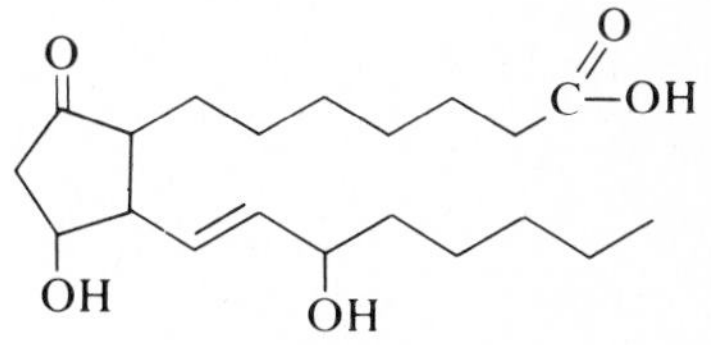
O
O
C—OH
OH
OH

prostaglandin E_1 (PGE)

The prostaglandins were discovered about 40 years ago as components present in semen but have subsequently been found in minute quantities in a wide range of tissues in both male and female animals. The prostaglandins seem to act as regulators of a variety of biochemical functions, including muscle extension and contraction, lipid metabolism, cardiovascular effects, nervous control, and reproductive physiology. They are very potent materials; for example, a 1-ng/mℓ solution of prostaglandin is known to bring about the contraction of smooth muscles. It is likely that as the action of these compounds becomes better understood they will develop into important pharmacological agents.

UNIT 40 Problems and Exercises

40–1. Spermaceti, a wax from the sperm whale, has as its chief component cetyl palmitate. Draw the structure of this wax. (Cetyl alcohol is $C_{16}H_{34}O$.)

40–2. Draw the structure of a monoglyceride of stearic acid.

40–3. Draw the structure of trilaurin and trilinolein.

40–4. Which of the two triglycerides, trilinolein or triolein, would have the greater iodine number?

40–5. Which would more likely be a solid at room temperature, peanut oil or coconut oil? Why?

40–6. Calculate the saponification number of trimyristin (MW 723).

40–7. Would you expect trimyristin to have a higher or lower saponification number than tripalmitin? Why?

40–8. Explain why acetate salts might not be effective soaps, whereas stearate salts would be effective.

40–9. Briefly describe some of the desirable properties of a synthetic detergent.

40–10. What is meant by the term *essential* fatty acids? If the body contains a variety of fatty acids, explain why they are not all essential.

UNIT 41 COMPLEX LIPIDS AND STEROIDS

Phospholipids

A class of lipids that contains an element other than C, H, and O is the **phospholipids**, which contains the element phosphorus in addition to the others. The phospholipids are natural components of all cells.

Most phospholipids are composed of phosphoric acid esters of glycerol. The parent molecule is phosphatidic acid, which contains the phosphate ester of glycerol and two long-chain fatty acids, as follows:

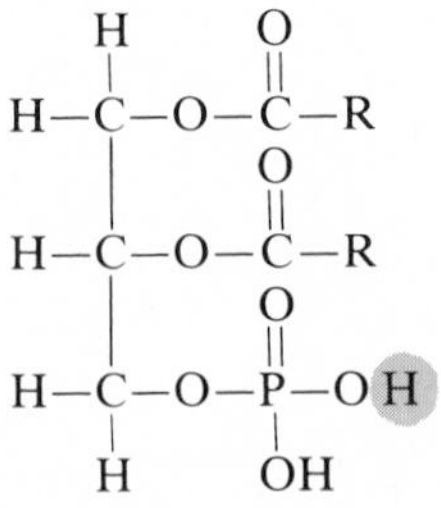

Usually this phosphatidic acid molecule has an ionic or hydrophilic group substituted for the shaded H in the above structure. The differences in the substituted molecule account for the variety of the phospholipids. The long-chain fatty acids impart a hydrophobic character to the molecule, whereas this ionic substituent gives the molecule a hydrophilic group.

One member of the family of phospholipids is known as the **cephalins**. Either ethanolamine or the amino acid serine can become part of the molecule, as follows:

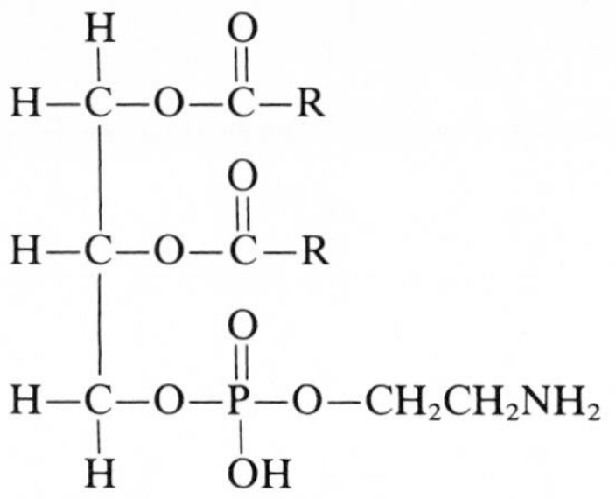
phosphatidyl ethanolamine

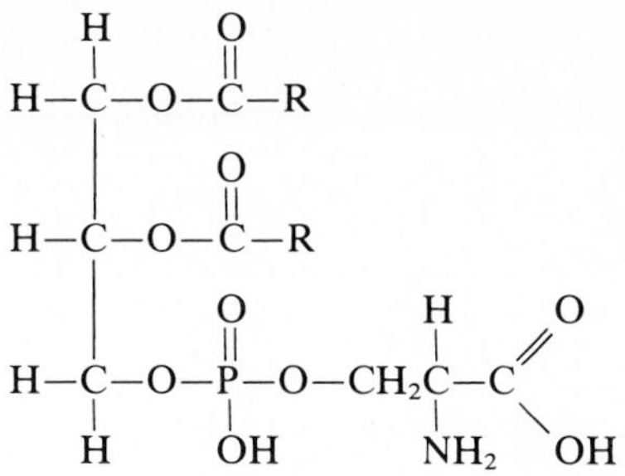
phosphatidyl serine

The brain and spinal cord are excellent sources of either of these cephalins.

The un-ionized structures shown are not very likely forms at normal physiological pH. At physiological pH, the phosphoric acid would be dissociated, and the molecule would exist as a negatively charged anion. The nitrogen can accept an additional proton and become positively charged. This forms a hydrophilic portion of a molecule with detergent characteristics:

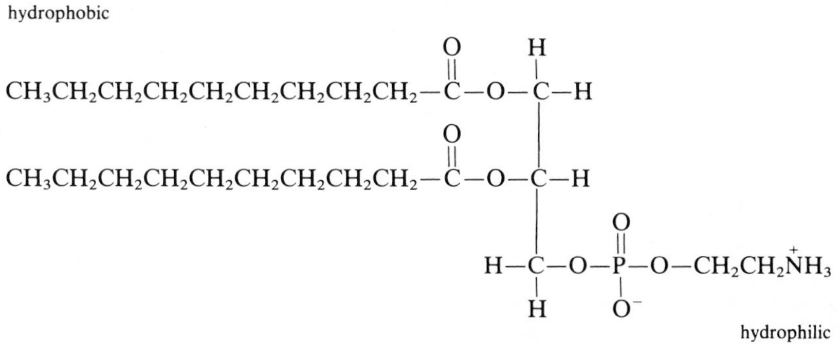
hydrophobic
hydrophilic

Another group of phospholipids are the **lecithins** in which the phosphatidic acid is combined with choline (β-hydroxyethyl trimethylammonium hydroxide). The phospholipid can be shown in the ionized form, which would predominate at physiological pH:

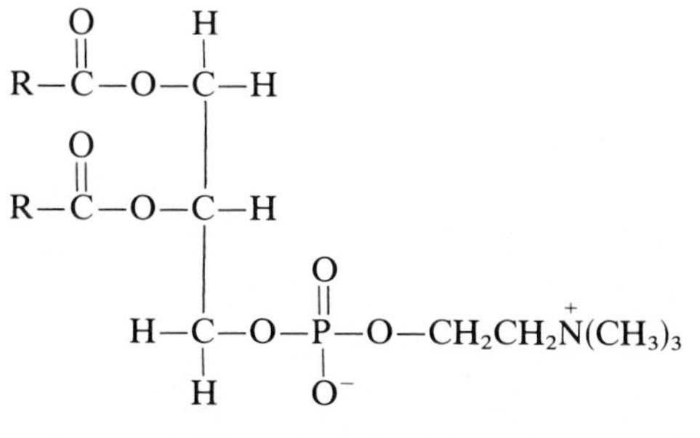

The lecithins are frequently added to foods as emulsifiers, especially chocolates, to keep the lipids in a more palatable condition. Check the ingredients of your next candy bar for the presence of lecithin.

Cell Membranes

The cell membrane is more than merely a barrier; it is a functioning component of the cell. For example, the membrane is involved in an **active-transport** process whereby sugars, amino acids and other organic molecules, and ions of interest are moved into the cell, many times against a concentration gradient. This requires energy and will be considered later. Some enzymes increase their activity by attaching themselves to specific positions on some membranes.

Membranes may be composed of phospholipids forming a **double layer**. The ionic portions are located on the outside so as to be in contact with the water. The nonionic portions form a hydrophobic layer in the interior of the membrane, as shown in the cross-sectional model of Figure 41–1. This figure also depicts globular proteins embedded within the double layer. Proteins also have both charged and uncharged portions with the same molecule.

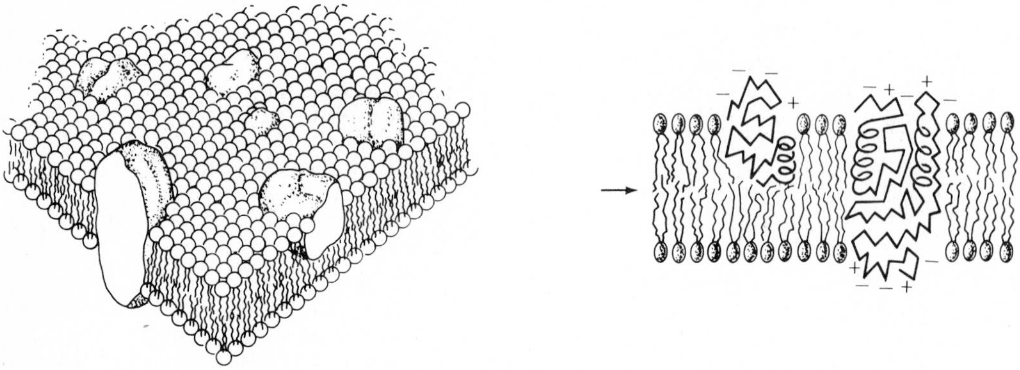

Figure 41–1 Lipid-protein mosaic model of membrane double layer. The drawing on the left shows the globular protein molecules embedded in the lipid bilayer. Here the hydrophobic portions are turned in while the hydrophilic portions are on the exterior of the membrane in contact with water. The drawing on the right is an enlargement also showing that the proteins have charged portions. (From S. J. Singer and G. L. Nicholson, *Science*, 175: 720, 1972. Copyright © 1972 by The American Association for the Advancement of Science.)

Cholesterol

Another class of lipids are the **steroids**. Although the steroids have a variety of biochemical functions, they do have a common chemical structural feature. They are all members of the family containing four fused (bearing letters A, B, C, and D) ring systems. Each of the carbon atoms is numbered according to its position in the molecule, as follows:

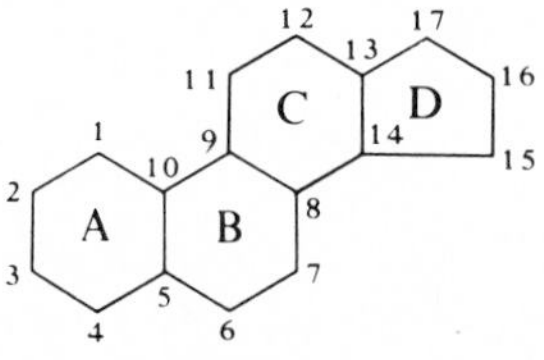
A
B
C
D
1
2
3
4
5
6
7
8
9
10
11
12
13
14
15
16
17

Most steroids have an oxygen group at C-3 and another substituent at C-17, in addition to other specific structural features.

One of the most abundant animal steroids known is **cholesterol**:

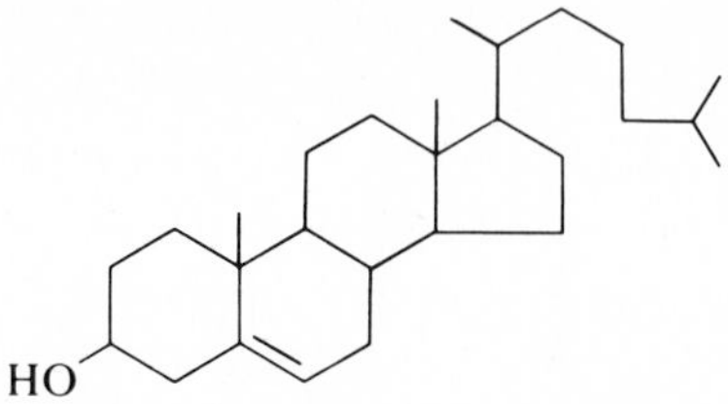
HO

A hydroxyl group is on C-3 and an eight-carbon aliphatic side chain is at position C-17. There is a double bond between C-5 and C-6, and two methyl groups attached to C-10 (called number 19) and C-13 (called number 18), respectively.

Cholesterol is an important constituent of blood plasma, where almost two thirds of it is present as an esterified derivative at C-3. In arteriosclerosis, cholesterol-containing plaques are deposited within the blood vessels. Cholesterol is also an important constituent of normal cell membranes, and makes up about 80% of gallstones.

Bile Acids

Cholesterol is metabolized to form an important class of compounds known as the **bile acids**. The bile acids and their salts are synthesized in the liver and travel through the gall bladder into the intestine where they have a significant role in fat digestion and absorption. A major bile acid is **cholic acid**, which is often found conjugated with other molecules like the amino acid glycine to form glycocholic acid.

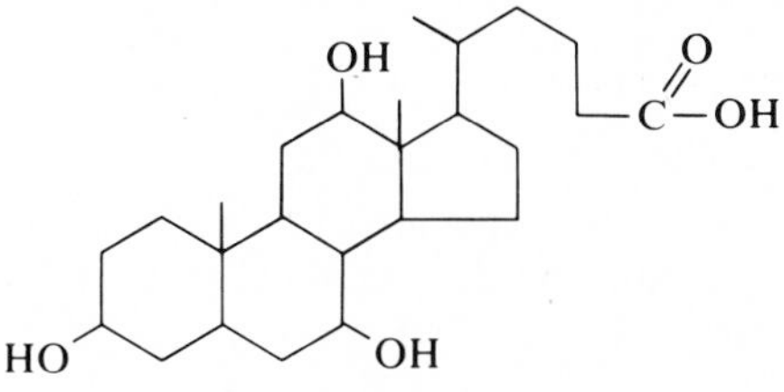
OH
O
C—OH
HO
OH

cholic acid

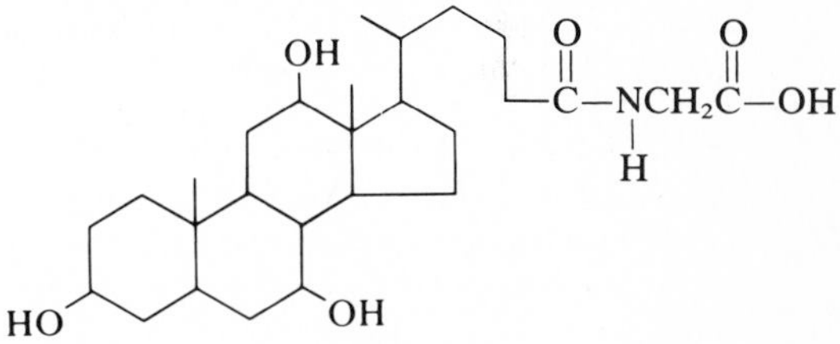
OH
O
O
C—NCH₂C—OH
H
HO
OH

glycocholic acid

The detergent properties of these kinds of molecules aid the emulsification of fats in the intestinal tract. The bile acids are partially reabsorbed into the portal blood and pass through the liver to be resecreted into the bile. Some of the bile acids are lost into the feces, and this amount must be synthesized from cholesterol.

Androgens

Cholesterol can also be converted into a number of steroid hormones. Among these are the **androgens**, or male sex hormones, with **testosterone** being a good example:

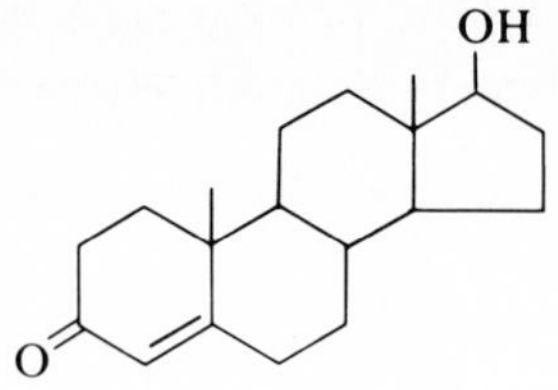
OH
O

Testosterone is one of the major hormones that functions to bring about the development of the male reproductive organs, as well as to maintain the secondary male characteristics. These functions range from sustaining spermatogenesis to the development of a deep voice and facial whiskers.

The androgens are synthesized by the testes and adrenals and, interestingly, also in ovarian tissue in the female. The male and female steroid hormones are present in individuals of both sexes. The carefully controlled domination of one group of compounds over the other accounts for the final observed differences between the two sexes. A common embryological development of the above-mentioned organs undoubtedly accounts for their ability to secrete compounds with greatly differing biochemical properties.

Estrogens

One type of female sex hormone, or **estrogen**, is illustrated by **estradiol**:

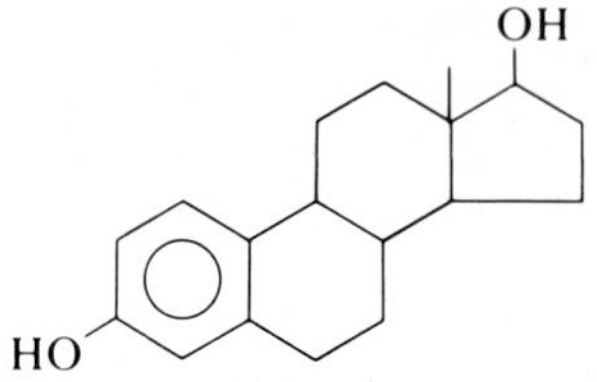
OH
HO

This compound has the same basic steroid nucleus, but only 18 carbon atoms and an A ring that is aromatic. The number 10 position cannot accommodate the extra methyl group because of the aromatic characteristic of the A ring. The aromaticity makes this C-3 hydroxyl group phenolic rather than simply alcoholic. The phenolic character accounts for the fact that estrogens in general are rather soluble in strong aqueous bases.

Estrogens are present in very small amounts. The first isolation of estradiol (12 mg) came from an extraction procedure starting with four tons of pig ovaries. This is an excellent example of the problems facing early biochemists in their quest to discover the chemical basis of living cells.

Estrogens and androgens arise from the common precursor cholesterol. The close relationship of these diverse molecules is observed by noting that, although the ovaries may be the major site of estrogen synthesis, the estrogenic content of horse testes is reported to be higher than that of any other endocrine system.

Estradiol is one of the estrogens that brings about the development of the female reproductive organs and supportive glands, as well as maintaining the secondary female characteristics. Estrogens exhibit a primary influence on both the estrus and reproductive cycles. Estrogens act in concert with progesterone if pregnancy develops.

Progesterone

Progesterone is a steroid hormone secreted by several different endocrine tissues, including the ovary (corpus luteum), the adrenals, and the placenta during pregnancy.

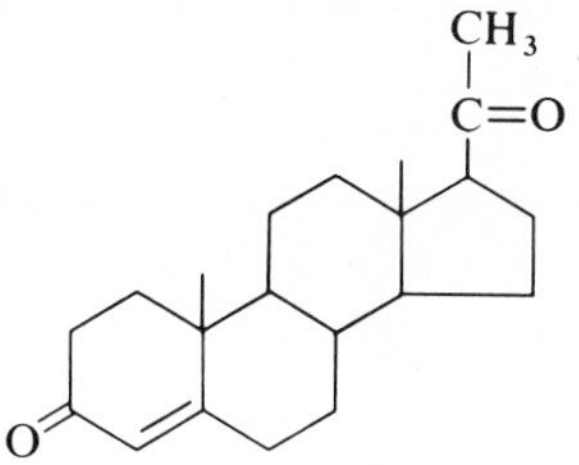
CH3
C=O
O

This hormone acts on the uterus, in concert with the estrogens, to prepare it for the maintenance of the embryo and fetus. It also acts in the preparation of the mammary gland for the secretion of milk following the birth of the young. The secretion of these hormones in the normal female system is under further control of protein hormones secreted by the hypophysis. The system must act in precise sequence if pregnancy and reproduction of the species is to be completed.

Corticosteroids

A portion of the adrenal gland, known as the adrenal cortex, secretes many steroids that are similar to those we have already studied. One is **cortisol** or hydrocortisone:

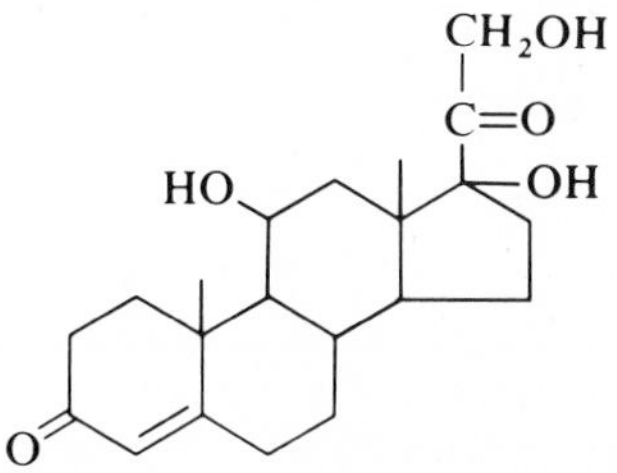
CH2OH
C=O
HO
OH
O

About 20 mg of this compound is secreted each day by a healthy adult. The adrenal corticosteroids have a great influence upon both mineral, water, and carbohydrate metabolism. Since approximately 30 steroids have been isolated from the adrenals, a detailed discussion of their specific activities will not be covered here.

Synthetic Drugs Related to Steroids

Antifertility drugs have generated considerable interest in biomedical research. The expansion of the world's population has become a problem in which medical, social, ethical, and economic factors are interlaced. Steroids related to the female hormones have been studied in great detail in an effort to limit the population explosion. One constituent of Enovid, a birth control pill, is **norethynodrol**:

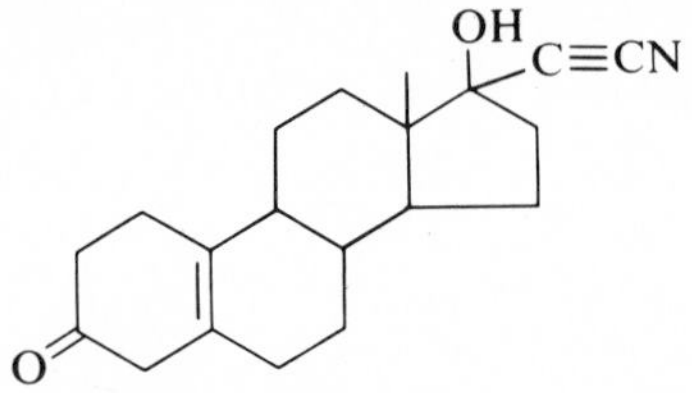

OH
C≡CN
O

Synthetic compounds such as this have solubility characteristics that allow them to be taken orally. They initiate physiological activities that cause the body to mimic pregnancy, thus preventing ovulation.

Another synthetic drug, **stilbestrol** (sometimes called diethylstilbestrol, DES), has estrogen-like activity when administered orally. It is not a steroid but has the following structure.

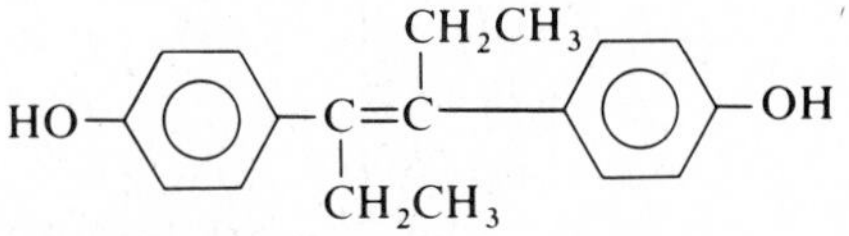

CH2CH3
HO
C=C
OH
CH2CH3

DES has been used to stimulate more rapid production of beef when added to cattle feeds. Although its estrogenic action cannot be completely explained, the aromatic hydroxyl groups do have molecular resemblance to those of the estrogens. Calculations have been made that show a similarity of the overall chemical dimensions of DES with the estrogens. DES has come under governmental control because of its chemical stability and incorporation in the food chains of various species, including man. Because of their biochemical potency and important physiological response, the hormones deserve careful study and judicious application.

UNIT 41 Problems and Exercises

41–1. Which would you expect to be most soluble in water, a triglyceride or a phospholipid? Explain.

41–2. Draw phosphatidyl serine in the structural form that you would expect at the pH of the intestinal tract, which is slightly above 7.

41–3. Briefly explain why you might expect membranes to be composed of a double layer.

41–4. Draw the structure of cholesterol and number each carbon atom.

41–5. Explain why estradiol does not have a methyl group at carbon 10, whereas testosterone does have a methyl at this position.

41–6. Draw the structure of estradiol as you would expect to find it in the presence of a strong base. Would testosterone be in the same form under the basic conditions?

41–7. What significance can you attach to the observation that individuals on a synthetic cholesterol-free diet may still have quite high blood plasma cholesterol levels?

41–8. Explain from a chemical point of view why bile salts and bile acids would be better emulsifiers than cholesterol itself.

41–9. Explain from a biochemical point of view why older women sometimes develop whiskers and older men sometimes develop enlarged breasts.

UNIT 42 FAT-SOLUBLE VITAMINS

Vitamins

In the first decade of the twentieth century a British biochemist, F. G. Hopkins, initiated studies on what are known today as vitamins when he wrote, "The animal body is adjusted to live either upon plant tissues or the tissues of other animals, and these contain countless substances other than proteins, carbohydrates, and fat." Prior to the twentieth century, the concept of a nutrient deficiency was not readily understood. Many vitamin deficiencies (avitaminosis) were previously attributed to unknown "toxic" elements or undesirable microorganisms.

A Dutch biochemist, Christiaan Eijkman, was the first to purposely induce a vitamin deficiency in animals. He fed chickens white polished rice and induced what is known as beriberi. He found this syndrome could be cured by changing the diet of the chickens to brown rice, which contained the rice hulls. The hulls were thus shown to contain the factor that prevented the deficiency. The demonstration of a cure for an induced deficiency was the key that led to much of the later work in this area. In this case the *processing* or *refining* of rice led to a much less wholesome material than rice in its natural form.

It was study on this same problem that led to the coining of the word *vitamin.* In 1912, a Polish biochemist, Casimǐr Funk, worked on purifying the antiberiberi substance present in rice polishings. He called the curative material he isolated *vitamine*, a substance *vital* to life, which had characteristics of the class of compounds known as *amines.* In 1926 the specific antiberiberi factor was first isolated by Dutch and German workers, and in 1937 the actual chemical structure of thiamine was determined by R. R. Williams, an American biochemist. In 1929, the importance of this work was acknowledged when Hopkins and Eijkman shared the Nobel prize for their persistence in searching for these previously unknown substances, vitamins.

Vitamins can be divided into two major classes: **water soluble** and **fat soluble**. The fat-soluble vitamins include those designated by the letters A, D, E, and K. These vitamins are not readily soluble in aqueous media in most of their structural forms, but are readily soluble in nonpolar organic solvents. Fat-soluble vitamins also share another common factor of being derived from a common precursor, **isoprene**:

$$\begin{array}{c} C{=}C{-}C{=}C \\ \hspace{-2.2em}| \\ \hspace{-2.2em}C \end{array}$$

Vitamin A

Vitamin A itself occurs only in animal tissues, although related compounds with potent vitamin A activity are present in plants as **carotenes**. Carotenes can act as precursors of vitamin A in animals. There are several forms of vitamin A, the most common in higher animals being vitamin A_1. Vitamins occurring in structurally related multiple forms are called **vitamers**. Vitamin A_1 is a highly unsaturated alcohol. The five-carbon skeleton of the isoprene units are outlined in this C-20 compound. In Figure 42–1, all the double bonds are in the trans configuration. It is the isomerization of one of these double bonds to its cis form that is involved in the photochemical process of vision. The molecule actually thought to be involved in the visual process in the rods of the eye is the aldehyde derivative of vitamin A_1, sometimes called **retinal** or **retinene**. This aldehyde is known to exist with one double bond, the one at the C-11 position, in the cis configuration. A protein known as **opsin** can combine with the cis retinal to yield a complex molecule known as **rhodopsin**, as shown in Figure 42–2. When light strikes this molecule, energy is absorbed and the cis configuration is converted to the trans configuration. Somehow this conformational change is involved with the excitement of the nerve cells of the rods and accounts for the origin of visual sensations. The all-trans aldehyde detaches itself from the protein opsin. The cycle is complete when the all-trans retinal is converted to the C-11 cis molecule.

It is obvious that a deficiency of vitamin A should lead to impaired vision. Since some of the vitamin A is destroyed in this process, a constant exogenous supply is needed. Other manifestations of vitamin A deficiencies include a general lack of growth of the young, undesirable skin changes, and generalized lesions of the eye (unrelated to the visual process) known as *xerophthalmia*.

Experiments on human subjects indicate that from 600 to 700 μg of vitamin A is a sufficient supply on a daily basis. It is estimated however that up to 500 μg of vitamin A may be stored in each gram of liver tissue. These small weights are frequently replaced by the term **international unit** (IU). One IU is defined as the vitamin A equivalent to the activity of 0.6 μg of β-carotene, which has the structure

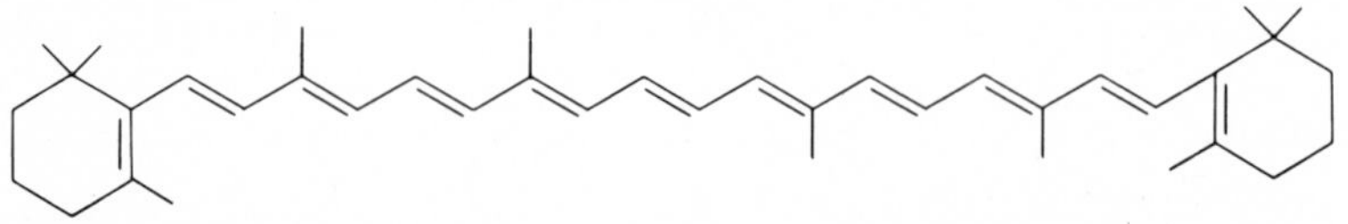

It is apparent that this molecule consists of the equivalent of two vitamin A molecules fused together. Although this is the case from a structural point of view, β-carotene is related to vitamin A on a one-to-one ratio when its activity is expressed on a molar basis. **Carotene** is a yellow pigment found in plants that supplies much of our dietary intake of vitamin A.

A note of warning must also be given when dealing with the intake of vitamin A because of its toxicity in excessive amounts. The human body does not have an

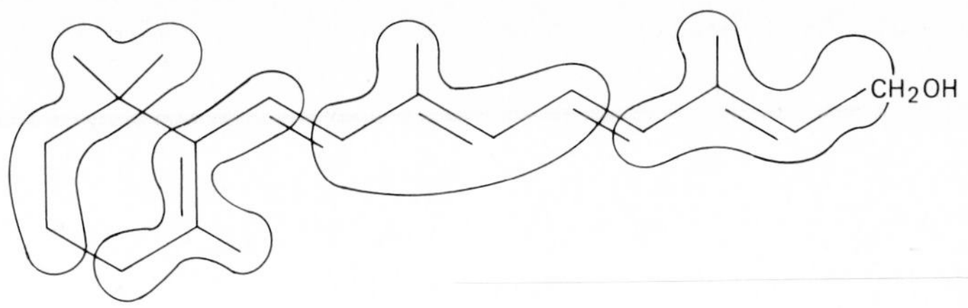
CH2OH

Figure 42–1 Vitamin A_1.

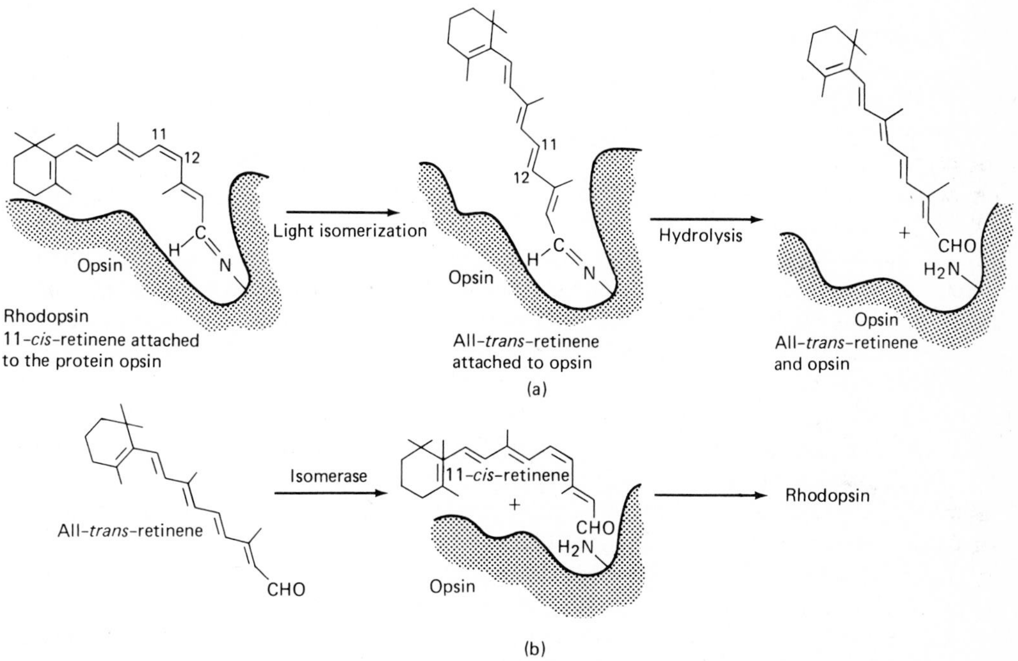
11
12
Light isomerization
Opsin
Rhodopsin
11-*cis*-retinene attached
to the protein opsin
Opsin
All-*trans*-retinene
attached to opsin
(a)
Hydrolysis
CHO
H2N
Opsin
All-*trans*-retinene
and opsin
Isomerase
All-*trans*-retinene
CHO
11-*cis*-retinene
CHO
H2N
Opsin
Rhodopsin
(b)

Figure 42–2 Reactions of the visual cycle. (a) The photochemical reaction and (b) regeneration of rhodopsin.

adequate mechanism to metabolize or eliminate huge quantities of vitamin A or other fat-soluble vitamins, and excesses can lead to serious health problems. A balanced diet of green and yellow vegetables should supply adequate amounts of vitamin A for healthy individuals. The recommended daily allowance is 5000 IU.

Vitamin D

Vitamin D is derived from a steroid, but is not a true steroid because it does not have a closed B ring. A mixture of the D vitamins was used in early studies until separation techniques allowed the individual specific members of the family of vitamins to be isolated. We shall consider only vitamin D_2, calciferol:

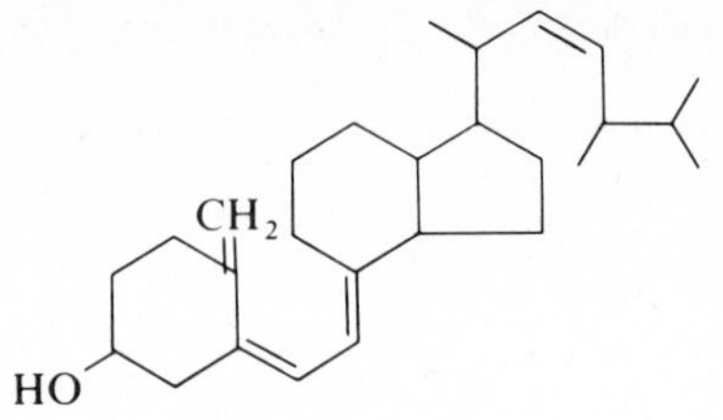

In this form, the steroid relationship is obvious.

Vitamin D is most often involved as the curative agent for rickets. Rickets has been a common disease for centuries and was prevalent in certain areas of the United States as late as 1917. During the early part of this century, rickets was shown to be a vitamin-deficiency disease curable by an adequate diet. It was first assumed that vitamin A would cure rickets. However, in 1922, Hopkins discovered that vitamin A could be rather easily destroyed by oxidization. Since a mixture of oxidized lipids could cure rickets, another nutritional factor was actually involved in the curative process. The fact that the amount of sunlight falling on the skin influenced the control of this disease also complicated early studies on this vitamin, which became known as the "sunshine vitamin." The irradition of a cholesterol derivative in the skin actually accounts for the formation of antirachitic compounds.

The biochemical basis of rickets implicates vitamin D and the metabolism of Ca^{2+} in the bones in the final stages of their growth. Normal bone development requires that Ca^{2+} be incorporated in an organic matrix. The mechanism of this action involves a metabolite of vitamin D with an induction of the synthesis of a specific calcium-carrying protein. The complete calcium balance within the body is not only influenced by vitamin D and dietary Ca^{2+}, but by other protein hormones known as *parathyroid* and *calcitonin*. Both of these hormones are elaborated by the parathyroid glands. The involvement of vitamin D in curing rickets can be visualized in the *line test* shown in Figures 42–3 and 42–4.

Since vitamin D is so intricately related to Ca^{2+} deposition and the formation of normal bone structures, excessive amounts of the vitamin have undesirable effects. Abnormally high amounts of vitamin D in the diet over prolonged periods of time can lead to mobilization of the Ca^{2+} and demineralization of the bones. This may result in the weakening of bones and even fractures in severe cases.

Measurements of vitamin D are sometimes given in unit equivalents. The international unit (IU) is related to the biological activity of a given amount of cod-liver oil and is equivalent to 0.025 μg of pure vitamin D_2. The recommended daily allowance is 400 IU for humans when exposure to sunlight is restricted. Four milliliters of cod-liver oil has this activity.

Vitamin E

This family of vitamins was given the general name of **tocopherols** (Greek for "childbearing alcohols") in 1936. The fertility of both male and female rats is impaired when this factor is omitted from the diet. The deficiency of tocopherols in humans usually includes such diverse manifestations as impaired lipid absorption,

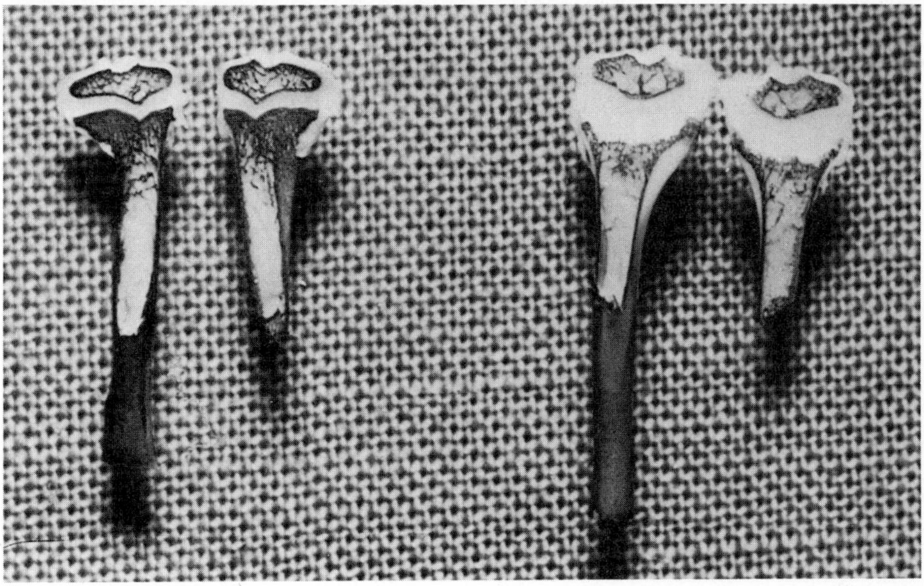

(a) (b)

Figure 42–3 Line test for rickets: (a) normal, (b) rachitic. (Courtesy of the Upjohn Company.)

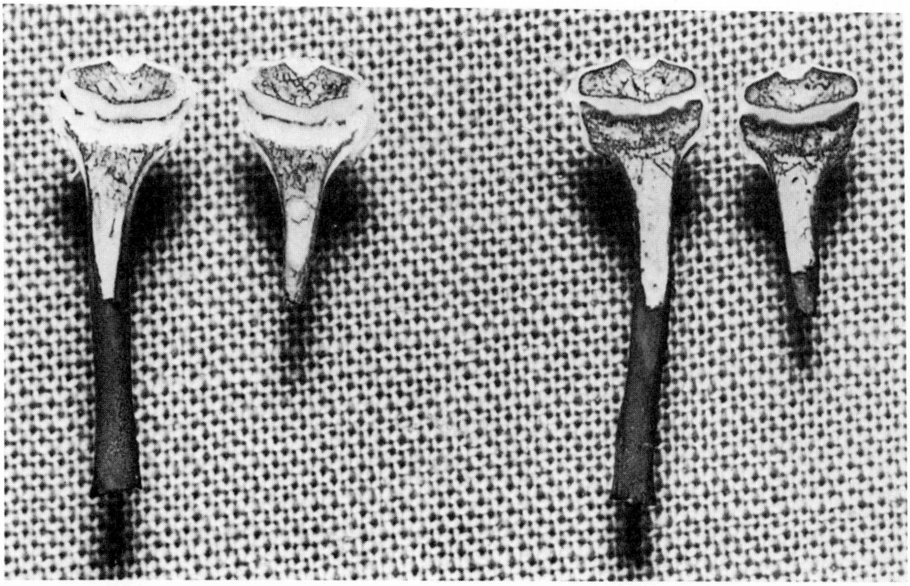

(a) (b)

Figure 42–4 Line test for rickets (cont'd): (a) healing, (b) healed. (Courtesy of the Upjohn Company.)

red cell fragility, and muscular weakness. There is no clear evidence that vitamin E acts as an antisterility factor in humans as it does in rats. Vitamin E deficiencies are species dependent. For example, chickens show capillary damage, monkeys may exhibit a form of anemia, and lambs appear to have muscular problems when fed diets deficient in this vitamin.

The structure of the most active form of this group, α-tocopherol, is:

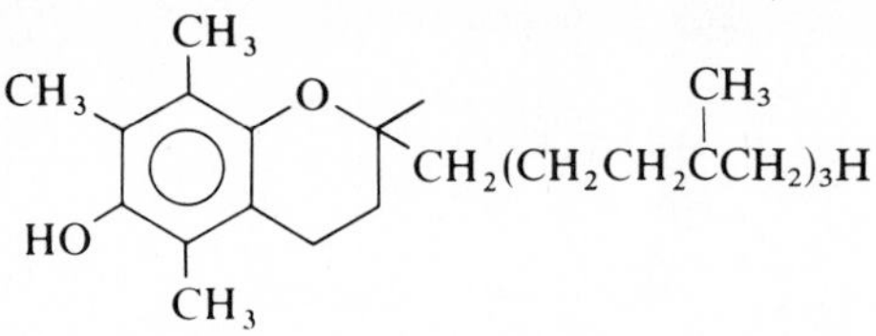
CH3
CH3
O
CH3
CH2(CH2CH2CCH2)3H
HO
CH3

The repeating isoprene structure can be identified in this compound. The other members of this group differ in chemical structure by alterations in the methyl substituents on the aromatic ring.

The function of vitamin E may be related to its ability to undergo oxidation–reduction reactions after a bond cleavage adjacent to the aromatic ring. This can be shown in skeleton form as

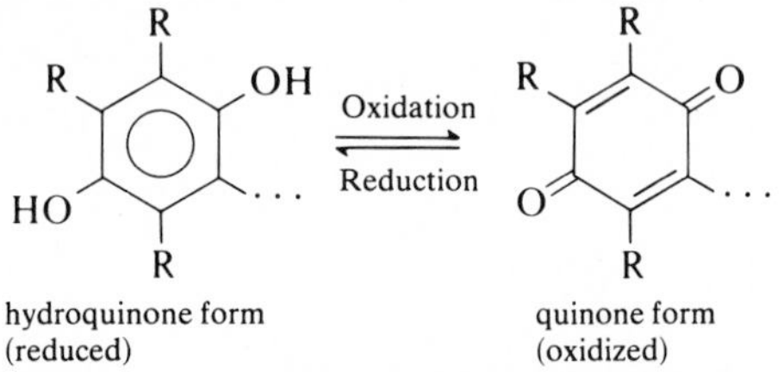
R
R
OH
HO
R
Oxidation
Reduction
R
R
O
O
R
hydroquinone form
(reduced)
quinone form
(oxidized)

The oxidation of vitamin E can take place very easily under biological conditions within the cell, and particularly within cell membranes. Thus this molecule might undergo ready oxidation and might spare other molecules from oxidative damage. For example, vitamin E can prevent the oxidation of vitamin A in natural fats. This may be the basis of vitamin E protection of lungs in a smog-filled atmosphere.

The recommended daily dietary intake of vitamin E for humans ranges from 5 IU for infants to 25 IU for mature adults. One milligram of a mixture of *dl*-α-tocopherol acetate is equal to 1 IU. Since the free forms of the tocopherols undergo such ready oxidation, this vitamin is usually handled as the acetate derivative, which is unaffected by oxidizing agents. Plant oils are rich dietary sources of vitamin E. Fish oils, so important in vitamin A and D studies, are deficient in vitamin E. Excessive amounts of vitamin E may lead to undesirable conditions, as in the case of other fat-soluble vitamins.

Vitamin K

Vitamin K was the last of the fat-soluble vitamins to be discovered. In the 1930's, Danish workers discovered that, when chickens were fed a certain synthetic diet, they would develop a condition that would cause easy hemorrhage and a prolonged blood-clotting time. A material isolated from alfalfa was shown to prevent this condition and was called vitamin K, from the Danish word for coagulation. The member of this family of vitamins known to have the greatest physiological activity in this connection is K_2:

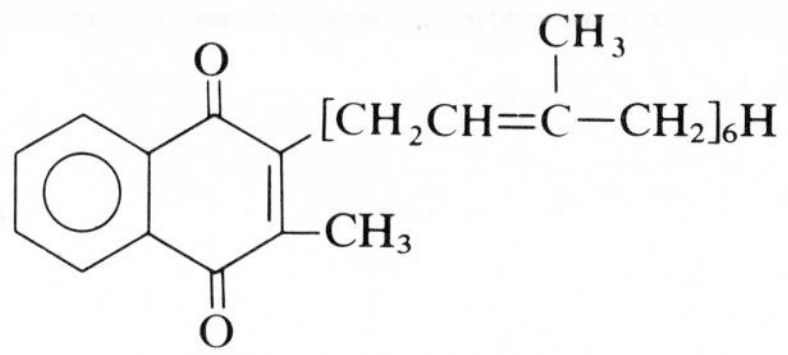

The repeating isoprene unit is obvious. The ring portion of the molecule is known as a *naphthoquinone.* There are chemically related vitamers that have similar structures, the differences usually being in the aliphatic side chains, such as menadione or vitamin K_3:

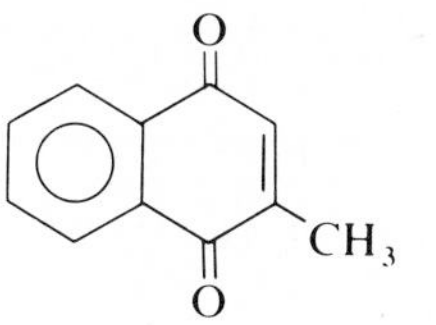

Vitamin K is involved in the biosynthesis of prothrombin, one of the components in the **blood clotting process**. Although a number of specific factors are known to function in what has been called a *cascade* sequence of reactions, the basic blood coagulation scheme can be summarized:

$$\text{Prothrombin} \xrightarrow[\text{Ca}^{2+}]{\text{Thromboplastin}} \text{thrombin}$$

$$\text{Fibrinogen} \xrightarrow{\text{Thrombin}} \text{fibrin} \xrightarrow{\text{Polymerized}} \text{clot}$$

The fibrinogen is a soluble protein produced in the liver, which leads to the formation of the insoluble protein fibrin at the site of the clot. Thrombin is a protein that acts in an enzymatic role to catalyze this event. The fibrin undergoes a polymerization, which eventually forms a polymeric fibrous material known as a hard clot.

Thrombin, which is a catalyst in the clotting event, is not a constituent of normal blood, but rather is formed from another protein, prothrombin. Besides requiring Ca^{2+}, this conversion requires another protein catalyst known as thromboplastin. The *cascade effect* is used to describe the total clotting process because other catalysts are required to form thromboplastin.

The Ca^{2+} is essential in this process. Clotting can be inhibited by removing the naturally occurring Ca^{2+} upon drawing the blood specimen. Some decalcifying agents, which act to complex or tie up the Ca^{2+} in an inactive form, include oxalate, citrate, and fluoride anions. In clinical measurements of blood clotting time, Ca^{2+} is removed initially and then added back under controlled conditions. The time necessary for clotting under these controlled conditions can be measured and is frequently reported as prothrombin time.

In certain medical situations, such as thrombophlebitis, treatments sometimes involve the administration of chemically related analogues of vitamin K. These act to inhibit the formation of prothrombin to thus reduce the undesirable formation of clots within the veins. One of these antagonists to vitamin K activity is coumadin or warfarin, which has a structural resemblance to the vitamin:

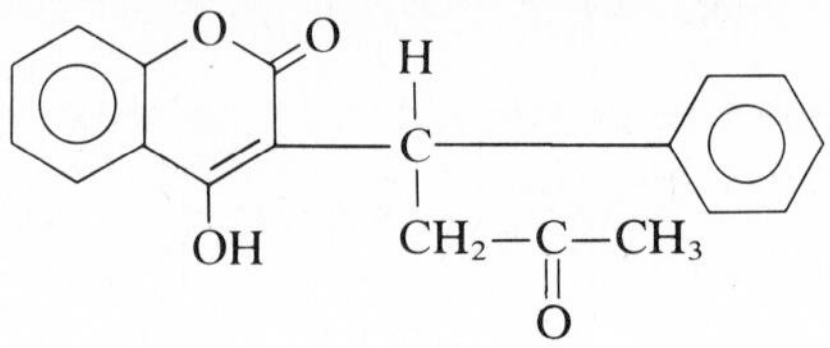

Vitamin K deficiencies are not common in laboratory animals such as rats because the intestinal flora supply the minimal amounts necessary. However, newborn human infants may show some signs of vitamin K deficiency until their bacterial flora become established. The total amount of vitamin K required is not known, but problems are not usually encountered in humans if the intestinal flora are healthy and intestinal lipid absorption is not impaired.

UNIT 42 Problems and Exercises

42–1. Look up the chemical structure of thiamine to verify why it was initially named *vitamine.*

42–2. Would you expect a mixture of fat-soluble vitamins to react with bromine in a reaction similar to the one we considered previously in the iodine number determination? Explain.

42–3. List two vegetables that you would expect to be rich sources of vitamin A (or carotenes) and two vegetables that would be poor sources of this vitamin.

42–4. If a cup of green beans supplies approximately 20% of the minimum daily requirement of vitamin A, calculate the weight of carotene (in milligrams) in this sample of vegetables. Assume that all the vitamin A activity is in the form of carotene.

42–5. Prove to yourself that vitamin D_2 could have been formed from isoprene precursor units.

42–6. Why might Eskimos require diets containing fish-liver oils, whereas individuals living near the equator may not?

42–7. How might the vitamin D requirement of individuals have been involved in the development of pigmentation of individuals living near the equator?

42–8. Explain how vitamin E may act to protect other vitamins from being destroyed by oxidative reactions.

42–9. Would you expect vitamins K_2 or K_3 to have the greater water solubility?

42–10. Briefly explain why coumadin administration may not have an immediate effect to inhibit the coagulation of the blood.

CHAPTER 15

Amino Acids and Proteins

Proteins are polymers of amino acids linked together by an amide-type linkage known as the **peptide bond**. The term *peptide*, or *polypeptide*, is sometimes used to describe cases where two or more amino acids are joined together in these covalent linkages. Once again, large molecules are formed by joining together simpler constituents, while removing the elements of water.

The very term *protein* describes the important role this group of compounds plays in nature. The Greek word *protos*, meaning first, was aptly given. Proteins occupy a central position in both the structure and functions of all living systems. Proteins are the most abundant organic materials in animal cells, making up 50% or more of the dry weight of the cell. Their importance in structural aspects can readily be seen in the cases of collagen of connective tissues, structural proteins in membranes, and the fibrous keratins of hair. The contractional process of muscles involves the biochemistry of the changes that take place in proteins like myosin and actin. Both enzymes and hormones are often composed of a variety of highly specific peptides and proteins. Besides these cases, numerous other important examples of proteins could be given, such as the oxygen-carrier hemoglobin and the various antibodies involved with the immunological defense mechanisms of the body.

Before we can hope to really understand the variety of biochemically important functions that proteins carry out, we must start by understanding what these molecules are like from a chemical point of view.

UNIT 43 AMINO ACIDS

Once proteins are purified, chemical analysis indicates that they are composed of carbon, hydrogen, oxygen, and nitrogen. Some proteins are also known to contain sulfur, phosphorus, iron, zinc, and other trace metals.

Proteins are polymers of **amino acids**. The hydrolysis of a protein under catalytic control yields amino acids:

$$\text{Proteins} \xrightleftharpoons{+H_2O,\ \text{catalyst}} \alpha \text{ amino acids}$$

The amino acids from proteins are compounds containing carboxyl groups with an amino group in the alpha (α) position. There are 20 different amino acids commonly found in proteins.

All the amino acids have a common structural feature, as shown in the shaded area. The difference in amino acids is in the side chain or *R-group* that is attached:

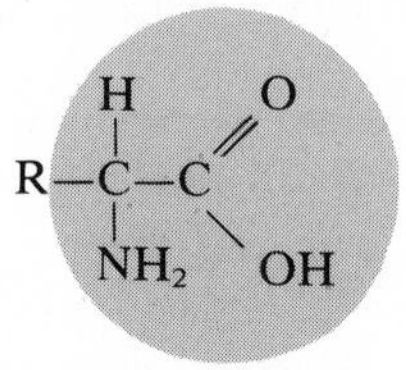

The first (1820) amino acid isolated from a protein hydrolysate turned out to be the simplest, glycine (amino acetic acid). In this case the R group is simply hydrogen. As more complex substituents are added, the variety of amino acids is formed. There are several ways in which amino acids have been classified based upon the chemical nature of the R group. Table 43–1 shows the structure, symbol, and name of the amino acids classified by the following scheme:

1. Aliphatic amino acid.
2. Aromatic amino acid.
3. Acidic amino acid.
4. Basic amino acid.
5. Sulfur-containing amino acid.
6. Imino acid.

For the amino acids more complex than glycine, the alpha carbon bears four different functional groups. This asymmetric center then can yield two different optically active **stereoisomeric forms** called D and L forms. These two forms have unique biological properties. Living systems are capable of differentiating between these two forms. Some amino acids have more than one center of asymmetry, but this is beyond the scope of our discussion. All the amino acids from proteins are in the L form; L-serine is shown:

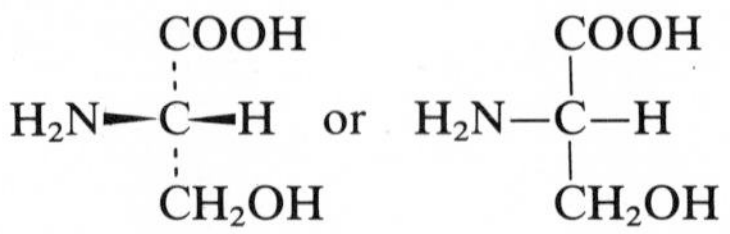

L-serine

The D amino acids do exist in nature, in cell walls, antibiotics, and other drugs. For all that follows, however, we shall assume that the amino acid is in the L form unless otherwise indicated.

Table 43–1 Common amino acids

Name	Symbol	Structure
1. Aliphatic		
Glycine	Gly	$H_2C(^+NH_3)-C(=O)-O^-$
Alanine	Ala	$CH_3CH(^+NH_3)-C(=O)-O^-$
Valine	Val	$(CH_3)_2CHCH(^+NH_3)C(=O)-O^-$
Leucine	Leu	$(CH_3)_2CHCH_2CH(^+NH_3)C(=O)-O^-$
Isoleucine	Ile	$CH_3CH_2CH(CH_3)CH(^+NH_3)C(=O)-O^-$
Serine	Ser	$HO-CH_2-CH(^+NH_3)-C(=O)-O^-$
Threonine	Thr	$CH_3CH(OH)CH(^+NH_3)C(=O)-O^-$
2. Aromatic		
Phenylalanine	Phe	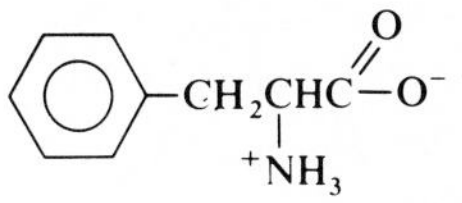
Tyrosine	Tyr	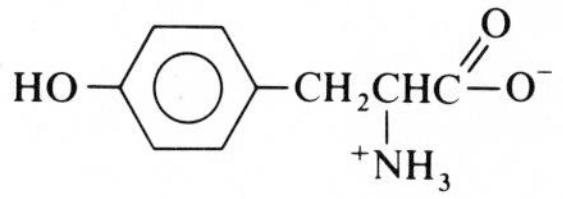
Tryptophan	Trp	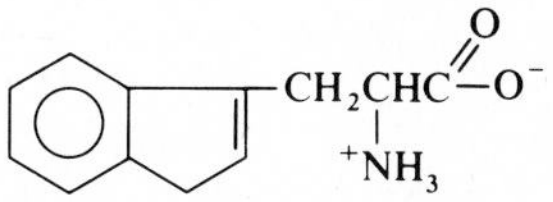

Table 43–1 Common amino acids (cont.)

Name	Symbol	Structure
3. Acidic		
Aspartic acid	Asp	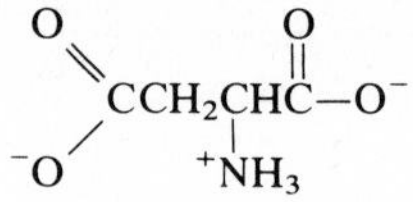
Glutamic acid	Glu	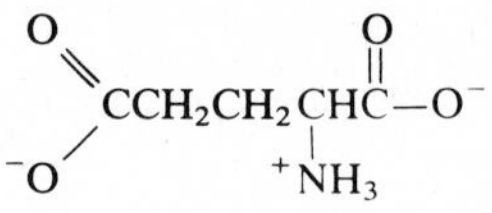
4. Basic		
Lysine	Lys	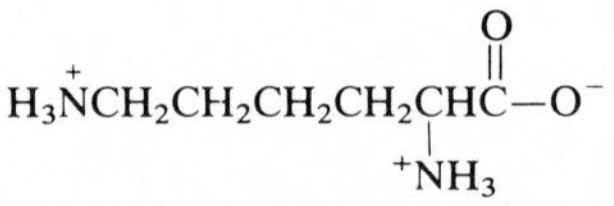
Arginine	Arg	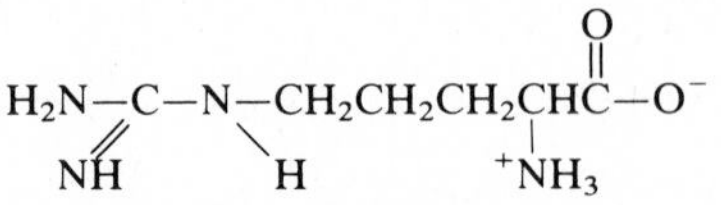
Histidine	His	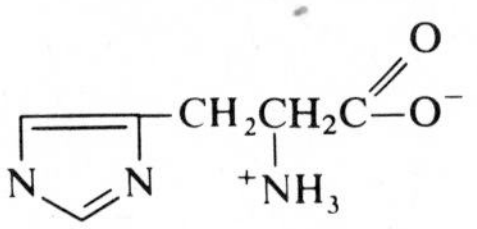
5. Sulfur-containing		
Cysteine	Cys	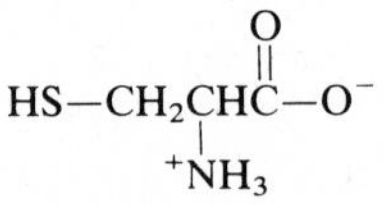
Cystine	$(Cys)_2$	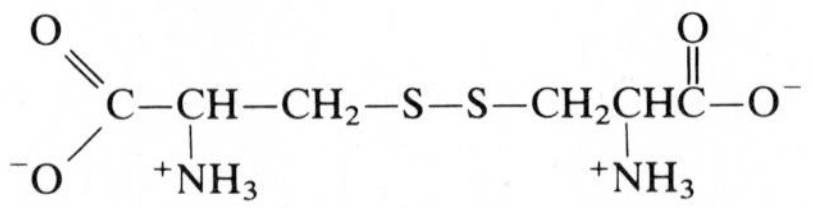
Methionine	Met	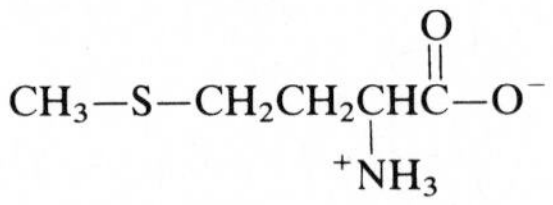
6. Imino		
Proline	Pro	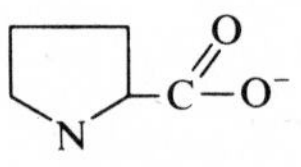
Hydroxyproline	Hpro	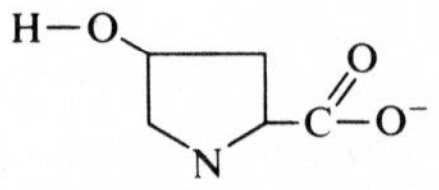

Ionized Forms

The **ionic form** of an amino acid is very important. At biological pH conditions slightly above 7, the carboxyl group would be found in the anion form 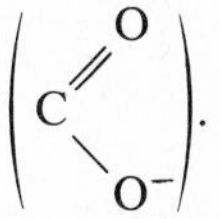.

Likewise, the amino group would have a tendency to attract the proton under these pH conditions and be found as a positively charged amino form ($-NH_3^+$). Therefore, in living cells an amino acid like glycine exists as a doubly charged molecule, sometimes called a **dipolar ion** or **Zwitter ion**, rather than an uncharged molecule.

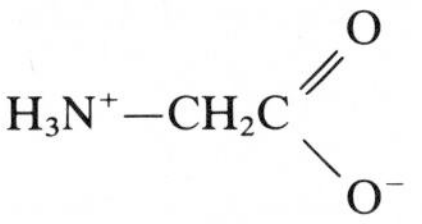

dipolar or dissociated form

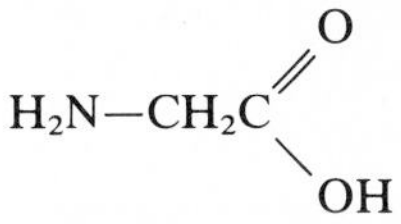

uncharged form

Since the molecule has two unlike charges, it is still electrically neutral or is said to have a net charge of zero (Z = net charge = 0). If such a molecule was placed in a direct-current electrical field, it would not be attracted toward either the + or − electrode. Under these conditions it can be called **isoelectric**. Sometimes the term **isoionic** is used. Although these terms are not exactly the same under all conditions, we will not differentiate between them. An amino acid would be in the isoelectric form if it was placed in a solution of pure water. We shall use the designation pI to indicate the isoelectric point, that is, the pH at which Z = 0.

Since amino acids are charged, many of their properties are more salt-like than might be expected from organic molecules. For example, most of the amino acids are generally quite soluble in water and insoluble in nonpolar solvents.

Amino acids, at their isoelectric point, react with either acids or bases. In strong acid solution, the carboxyl group accepts a proton to become uncharged. The amino group would yield its proton under strongly basic conditions, resulting in an uncharged primary amine group. These reactions are reversible and yield the following molecular forms under various conditions of pH:

Acidic conditions (low pH) — Basic conditions (high pH)

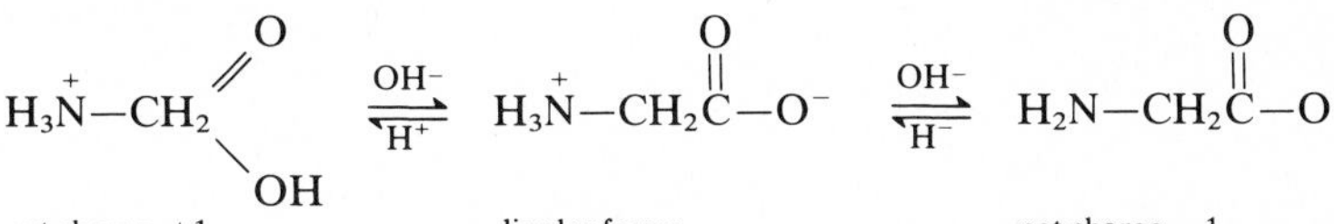

net charge, +1 — dipolar form; net charge, 0 — net charge, −1

The complete titration curve of glycine using a pH meter is shown in Figure 43–1. Assume that 1 mol of strong acid is added to 1 mol of glycine at its isoelectric point. This gives the ionic form at the extreme left side of the graph. Upon the addition of base to this positively charged form, the curve can be followed to the dipolar ion form, and finally to the negatively charged form at the far right side of the graph. The pI point on the curve is shown when the dipolar form of the molecule is reached. Two

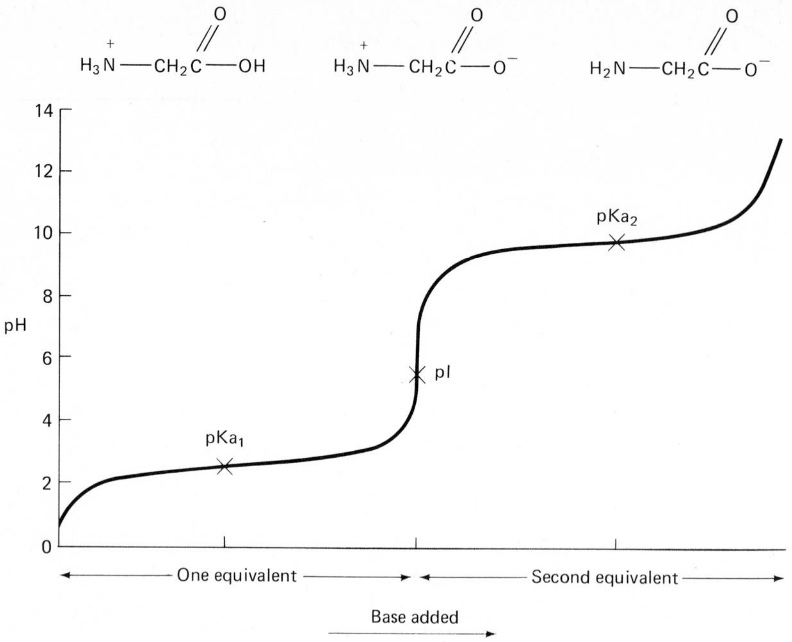

$\overset{+}{H_3N}$—CH_2C(=O)—OH
$\overset{+}{H_3N}$—CH_2C(=O)—O^-
H_2N—CH_2C(=O)—O^-
pH
14
12
10
8
6
4
2
0
pKa$_2$
pI
pKa$_1$
One equivalent
Second equivalent
Base added

Figure 43–1 Titration curve for glycine, showing the molecular forms at several points.

different **p*K*$_a$**'s are shown, which correspond to the halfway point in the titration of each of the titratable groups. The pK_{a1} is for the carboxyl group titration from $-\overset{\overset{O}{\|}}{C}-OH$ to $-\overset{\overset{O}{\|}}{C}-O^-$, and the pK_{a2} is for the amino group titration from $-NH_3^+$ to $-NH_2$. The Henderson-Hasselbach equation relates the pH, p*K*, and ratio of ionized and un-ionized forms of a system. For the titration of the carboxyl group of glycine, this would be:

$$pH = pK_{a1}(\text{glycine}) + \log \frac{[\overset{+}{H_3N}CH_2\overset{\overset{O}{\|}}{C}-O^-]}{[H_3N^+CH_2\overset{\overset{O}{\|}}{C}-OH]}$$

When $[H_3N^+CH_2\overset{\overset{O}{\|}}{C}-OH] = [H_3^+NCH_2\overset{\overset{O}{\|}}{C}-O^-]$, the log term becomes zero (log of $1 = 0$). Here the pH would be equal in value to the first or pK_{a1} of glycine. Note that this is very similar to the simple titration of acetic acid. The pK_{a1} of glycine and the pK_a of acetic acid would have slightly different values because of chemical differences between these two compounds. A similar equation can be written for the second ionization:

$$pH = pK_{a2}\,(\text{glycine}) + \log \frac{[H_2NCH_2\overset{\displaystyle O}{\overset{\|}{C}}-O^-]}{[H_3N^+CH_2\overset{\displaystyle O}{\overset{\|}{C}}-O^-]}$$

This shows why molecules like acetic acid have only one pK_a value, whereas amino acids have two separate pK_a values. Each functional group has a unique pK_a value. The titration curve of glycine shows that it could act as a buffer to resist rapid changes in pH upon addition of acids and bases at these two points. The pK_a values for some representative amino acids are shown in Table 43–2.

Table 43–2 Approximate pK_a and pI values for some common amino acids

Amino acid	pK_{a1} (1st carboxyl)	pK_{a2}	pK_{a3}	pI
Alanine	2.3	9.7 (NH_3^+)	—	6.0
Glycine	2.3	9.6 (NH_3^+)	—	5.9
Valine	2.3	9.6 (NH_3^+)	—	5.9
Aspartic acid	1.8	3.6 (COOH)	9.6 (NH_3^+)	2.7
Lysine	2.2	8.9 (NH_3^+)	10.5 (NH_3^+)	9.7

The pH at which the dipolar ion (pI) form exists is dependent upon the ionization constants and the pK_a values on either side of this point. In the case of glycine, the pI value (Z = 0) is midway between the only two pK_a values and is calculated as follows:

$$pI = \frac{pK_{a1} + pK_{a2}}{2} = \frac{2.3 + 9.7}{2} = 6.0$$

Note that the pI is not necessarily at 7 but is dependent upon the specific pK_a values. This in turn is due to the way in which the protons are attracted to the groups involved. For the more complicated amino acids, the form of the molecule must be determined that would yield a net charge of zero. The isoelectric point in these cases would be midway between the pK_a values on either side of the pI. For aspartic acid the dipolar ion form of the molecule would have the following structure:

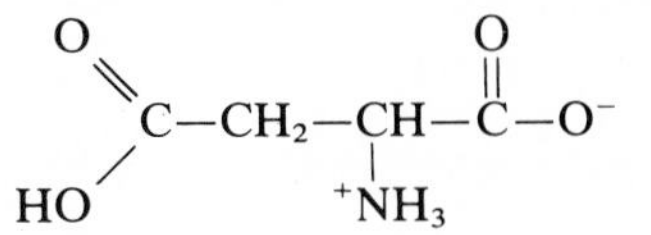

This would yield a pI of

$$2.7 = \frac{1.8 + 3.6}{2}$$

This molecule is classified as an acidic amino acid since a solution of pure aspartic acid in water would yield an acidic pH value.

Lysine, on the other hand, has a basic pI value, because the molecular form having a net charge of zero is the isoelectric form:

$$H_3N^+{-}CH_2CH_2CH_2CH_2\underset{\substack{|\\ {}^+NH_2}}{CH}\overset{\substack{O\\ \|}}{C}{-}O^-$$

Although the determination of the pK_a and pI values of amino acids is important, these values reach their full significance when they are considered as constituents of proteins. The functional groups of the side chains of the amino acids making up the protein molecules determine how the protein behaves as a charged molecule in solution. This is used in separating proteins in a technique known as **electrophoresis**. Disease conditions like agammaglobulinemia, multiple myeloma, Hodgkin's disease, and cirrhosis of the liver may be detected or confirmed using this technique based upon the differences in ionic forms of the various proteins in the blood.

Separation and Analysis of Amino Acids

The detection and quantitative estimation of amino acids in solutions is a very important procedure. For example, nutritionists may be interested in the amount of a specific amino acid that is present in a specialized diet or nutrient medium. Clinicians are often interested in the presence of amino acids in blood or urine, as in cystinuria, where cystine appears in the urine in abnormal concentrations. The detection of the amino acids, which are colorless in solution, can be carried out by a reaction with ninhydrin. When **ninhydrin** and amino acids are reacted together under proper conditions, a vivid color develops. The colored product of this reaction provides a convenient method of assay, because the intensity of the color developed is related to the concentration of the amino acid available for the reaction with ninhydrin.

One of the most convenient techniques for the separation of amino acids, and other compounds as well, is thin-layer or paper **chromatography**. The separation of the various amino acids depends upon the interaction of their side chains with various solvents. At least two different systems or phases are presented to the amino acid, which then distributes itself between the phases according to its chemical nature. Each amino acid would have a unique way of interacting with the different systems. In the case of thin-layer chromatography, a thin film of an absorbent material is placed on a glass plate. An amino acid mixture is placed near one end of this absorbent material, and a solvent is then allowed to flow across the thin layer. The amino acids distribute themselves along the line of flow of the solvent. The plate is allowed to dry, and the amino acids can be visualized after reacting the plate with a solution of ninhydrin. This process can be seen in Figure 43–2. In a controlled system of absorbent material and solvent, any given amino acid will move in a consistent manner. Each amino acid has a characteristic **migration rate**, or what is known as an R_f value. The R_f value is a ratio of the distance moved by the material under consideration to the distance the solvent is allowed to move.

$$R_f = \frac{\text{distance solute moved beyond point of application}}{\text{distance solvent moved beyond point of solute application}}$$

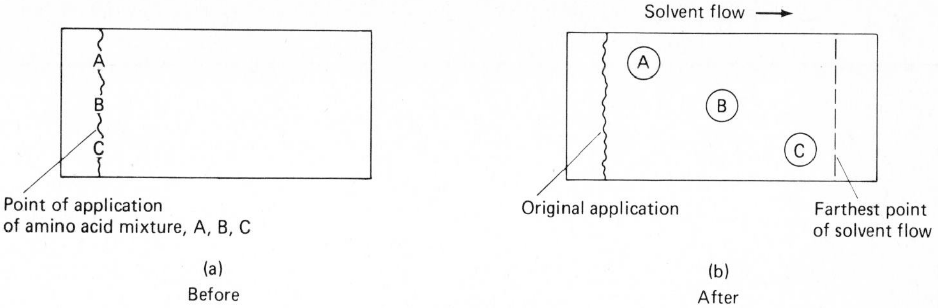
A
B
C
Point of application
of amino acid mixture, A, B, C
(a)
Before
Solvent flow
A
B
C
Original application
Farthest point
of solvent flow
(b)
After

Figure 43–2

Obviously, if the solute moved with the solvent front, it would have $R_f = 1$. In Figure 43–2, material B would have $R_f \simeq 0.5$. Paper can also be used rather than the thin film on a glass plate.

Another chromatographic technique that is commonly applied to the separation of mixtures of amino acids relies on ion-exchange resins that are packed in columns. In this case, resins are placed in glass columns, and a solution of amino acid mixtures is allowed to percolate through. This technique relies heavily on the different charged groups of the amino acids reacting with countercharged groups on the supporting resin in the column. The acidity of the various charged groups of amino acids as well as the additional affinity of the side chains for the resin allows very good separation of all 20 common amino acids. The results of such a separation can be automatically recorded on a continuous graph, as shown in Figure 43–3.

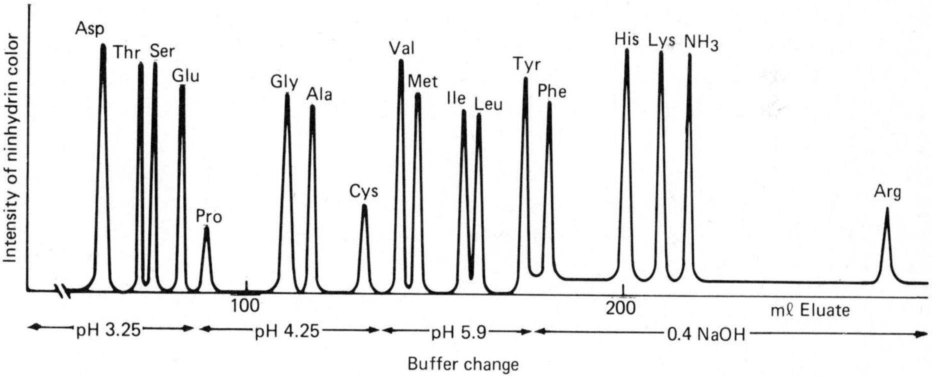
Intensity of ninhydrin color
Asp
Thr Ser
Glu
Pro
Gly
Ala
Cys
Val
Met
Ile
Leu
Tyr
Phe
His Lys NH3
Arg
100
200
mℓ Eluate
pH 3.25
pH 4.25
pH 5.9
0.4 NaOH
Buffer change

Figure 43–3 An amino acid separation resulting from the use of column chromatography. Ninhydrin is used to develop a color with the amino acids.

Derivatives of Amino Acids

We have concentrated our attention on the amino acids found in proteins; however, others have significant biochemical importance. Two different amide derivatives of previously described amino acids are **asparagine** and **glutamine**:

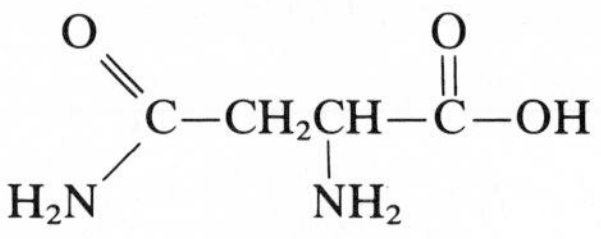

asparagine (asn)

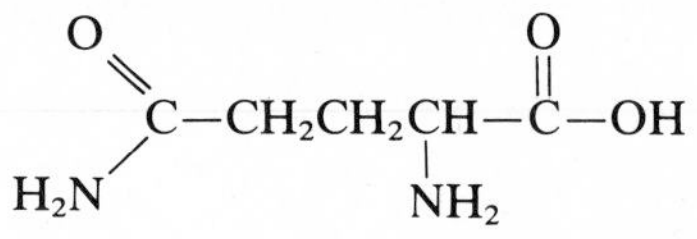

glutamine (gln)

These amides are found as components of numerous peptides and can be isolated as such if the proteins are hydrolyzed by enzymatic means. If the proteins containing these amides were hydrolyzed by either acid or basic catalysis, ammonia and the corresponding amino acid would be released. Because of ammonia toxicity, these amides might function as innocuous carriers for nitrogen. It is interesting to note that the normal level of glutamine in human blood plasma is 5 to 10 mg%. Since the sum of all the amino acids is normally only 35 to 65 mg%, glutamine must be important.

Two other closely related compounds that do not appear in proteins but have a significant role in muscle metabolism are **creatine** and **creatinine**.

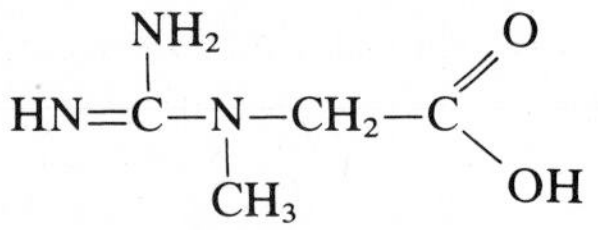

creatine

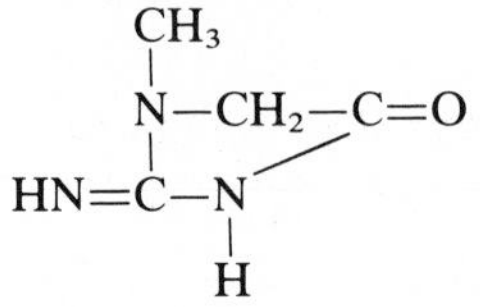

creatinine

The sum of the creatine + creatinine levels in mammalian skeletal muscle is about 0.25–0.60 g/100 g of wet weight. The creatine is involved as an energy-storage system and is converted to creatinine by a phosphorylated derivative. Creatinine is eliminated in the urine. The creatinine output is usually quite constant and is directly related to the active muscle mass of an individual. The daily output of creatinine ranges from 20 to 30 mg/kg for normal males and 10 to 25 mg/kg for normal females. Since the creatinine output of an individual is normally constant, this is often a useful index used to check for the completeness of the collection of a 24-hr urine sample.

All these compounds and the free amino acids themselves have unique and important biochemical roles, but it is the fact that amino acids are the building blocks of proteins that make them of even greater significance in nature.

UNIT 43 Problems and Exercises

43–1. Insulin is a protein hormone that must be administered by injection rather than through an oral route. Explain why.

43–2. When hair is burned, a characteristic odor of sulfur compounds is given off. From this observation, what amino acids would you expect to find in abundance in hair?

43–3. Explain why we should use the designation L-alanine but not L-glycine.

43–4. Would you expect alanine to be more water soluble than leucine? Why?

43–5. Draw the structure of valine as you would expect to find it under isoelectric conditions.

43–6. Draw the charged forms of aspartic acid as you would expect to find it under the following pH conditions: (a) pH = 1.0, (b) pH = 2.7, (c) pH = 6.6, (d) pH = 11.6.

43–7. Calculate the pI for lysine.

43–8. Would you expect phenylalanine to have a larger or smaller R_f than glutamic acid if a thin-layer solvent rich in nonpolar materials was employed?

43–9. Suggest a reason why normal-level ranges of creatinine in females may be below normal-level ranges in males.

UNIT 44 PEPTIDES AND PROTEINS

When the carboxyl group of one amino acid is joined with the amino group of a second amino acid, a **peptide bond** is formed upon the loss of HOH. The new compound is a **peptide**. Small peptides are named depending upon the number of amino acids involved, such as dipeptide, tripeptide, and tetrapeptide. When a larger number of amino acids are linked together, the resulting polymer is known as a **polypeptide**. Generally, the term protein is reserved for the polymer resulting from the linkage of a great many amino acids to form a compound of high molecular weight. These macromolecules are conveniently described as the mass of a single molecule using the term **dalton**, which is defined as equal to 1 on the atomic mass scale.

The Peptide Bond

The peptide bond is the primary linkage in protein chemistry and could be formed from the amino acids glycine and alanine. Two different dipeptides could result.

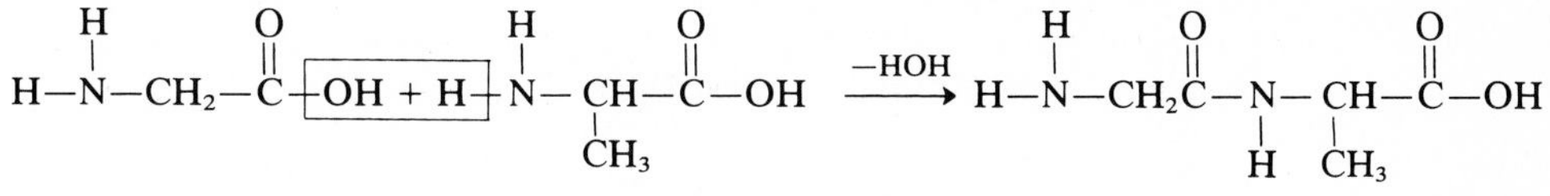
−HOH

glycylalanine

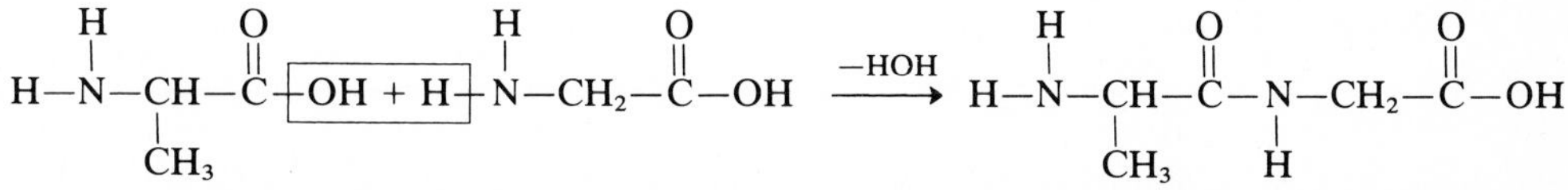
−HOH

alanylglycine

The peptide or protein formed depends on the order or sequence in which the amino acids are linked. This is referred to as the **primary structure** of the protein, and is sometimes given the symbol **I°**. The names of the peptides depend upon the amino acids involved and upon their sequence. By convention, the amino end (N-terminal) is written to the left side, and the molecule is constructed until the free carboxyl group (C-terminal) is reached. Since such molecules could be described as derivatives of the carboxyl group, the peptides use the names of each amino acid with a suffix *−yl*. Some letters are often dropped for convenience or ease in pronunciation. Obviously, when the number of the amino acids linked together becomes very large, such a nomenclature system would become quite complicated. In these cases, specific names have been assigned to specific proteins. These names often do not have chemical significance, but frequently are dependent upon their biological role, location, or function. Table 44–1 lists a few representative proteins with some of their characteristics.

Table 44–1 Some common proteins

Name	MW (daltons)	Source	Biological role
Ribonuclease	12,600	Pancreas	Hydrolysis of RNA
Cytochrome C	13,400	Aerobic cells	Electron-transport system
Trypsin	24,000	Small intestine	Digestion of proteins
Pepsin	36,000	Stomach	Digestion of proteins
Zein	40,000	Corn	Nutritional
Thyroglobin	55,000	Thyroid gland	I carrier, regulator
Hemoglobin	64,500	Red blood cell	Oxygen transport
Serum albumin	68,000	Blood serum	Maintain osmotic pressure
Gammaglobin	160,000	Blood serum	Immune reactions
Collagen	300,000	Skin	Structural
Fibrinogen	330,000	Blood	Clotting of blood
Myosin	480,000	Muscle	Muscular activity

It is easy to see that as the number of amino acids being joined together becomes larger, a great many choices might exist for both composition and sequence. This great diversity accounts for the many different proteins. A tripeptide, for example, could be glycylalanylphenylalanine, or it could be any one of six possible arrangement patterns for these three amino acids.

gly-ala-phe	ala-gly-phe	phe-gly-ala
gly-phe-ala	ala-phe-gly	phe-ala-gly

A peptide composed of 10 different amino acids could have a staggering number of different arrangements. Proteins containing up to 20 different amino acids have even greater diversity.

Terminal Amino Acids

The complete amino acid sequence, or primary sequence, of a great many proteins has been determined. Although a complete discussion of the I° structure of proteins is beyond our scope, the **N-** and **C-terminal analysis** illustrates some of the reactions involved.

A determination of the specific amino acid that occupies the N-terminal position of a single protein chain was discovered by Frederick Sanger (Nobel laureate, 1958). He found that dinitrofluorobenzene (DNFB) would react quite specifically with free amino groups. If DNFB was allowed to react under controlled conditions with one of the tripeptides given above, a yellow-colored derivative of the following structure would result:

$$O_2N-\bigcirc-F + H\overset{H}{N}-gly-ala-phe \xrightarrow{-HF} O_2N-\bigcirc-\overset{H}{N}-gly-ala-phe$$
$$\qquad\qquad NO_2 \qquad\qquad\qquad\qquad\qquad\qquad NO_2$$

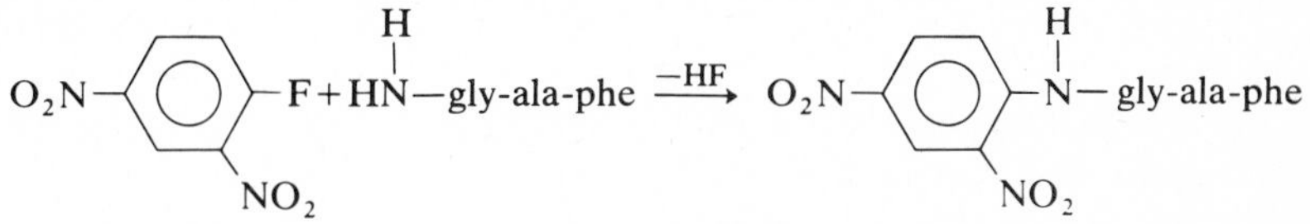

This yellowed-colored derivative could be isolated and then subjected to hydrolysis of its peptide bonds. Since dinitrobenzene forms a bond that is stable toward some hydrolytic conditions that cleave peptide bonds, the following compounds result from hydrolysis:

$$O_2N-\bigcirc-\overset{H}{N}-CH_2-\overset{O}{\underset{\|}{C}}-OH + ala + phe$$
$$\qquad NO_2$$

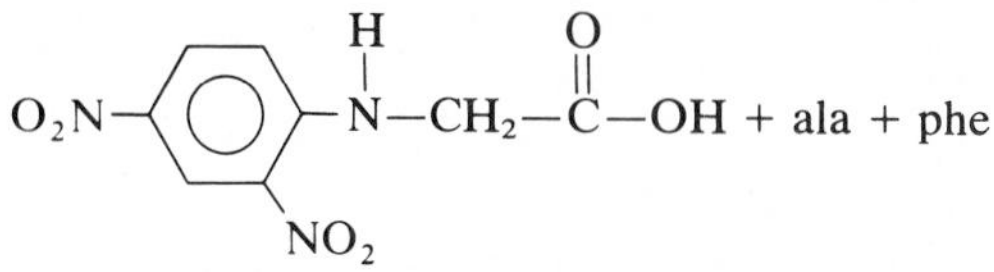

All three of these compounds can be separated by either thin-layer or paper chromatography. The unique R_f values of each compound can be compared with compounds of known structure to determine the exact results of the hydrolytic procedure. Only the amino acid bearing the free amino group, the N-terminal amino acid, keeps the dinitrobenzene attached to it. This allows the N-terminal end of proteins to be determined.

Controlled enzymatic hydrolysis can be used for end-group analysis. For example, an enzyme known as **aminopeptidase** has specificity toward the hydrolysis of the peptide bond adjacent to the free amino group. When this enzyme is allowed to catalyze the hydrolysis of peptides or proteins, the first amino acid released is the one occupying the N-terminal position. The rates of release of the amino acids reacting with this enzyme system can be measured, and this gives information as to the I° structure of the original peptide. Another different enzyme, **carboxypeptidase**, has selectivity for cleaving the peptide bond adjacent to the free carboxyl group. The enzymes that attack only the ends of proteins are known as **exopeptidases**.

Chemical methods can also be applied to C-terminal analysis. One such method that could be used would be to subject the peptide to a controlled reduction. If a peptide was reduced, the free carboxyl group would be reduced to a primary alcohol. The hydrolysis of the reduced peptide would release amino acids except for the amino acid at the C-terminal position of the original peptide. The C-terminal amino acid would have been converted to an amino alcohol. This amino alcohol could be separated and identified by chromatographic methods.

The entire sequence of very long protein molecules can be determined by using a combination of techniques. The large molecule can be broken down to short fragments by special methods. This may involve the use of enzymes that hydrolyze selective peptide bonds in the interior of the peptide chain (**endopeptidase**). The sequence of the fragments can be determined and then fit together like a jigsaw puzzle to yield the complete primary structure of the protein.

The Disulfide Bond

The peptide bond is the fundamental linkage in proteins, forming the backbone of the molecule and accounting for its I° structure. Another covalent linkage in proteins contributes to the specific structural stability of the molecule. This is the **disulfide bond** or bridge, which joins the two sulfhydryl groups of cysteine to form a cystine molecule. This disulfide bond may form between two cysteines in the same peptide chain (intramolecular) or between different chains on separate peptides (intermolecular). The disulfide bond contributes to the way in which separate protein chains may be joined together or to the specific orientation of a single chain. Both cases can be seen in the hormone insulin, which consists of two different chains, each with its own I° structure. Three disulfide bonds can form between the six cysteine residues present. In the A chain there is an intramolecular disulfide bond that gives this chain a unique orientation. There are also two different disulfide bridges between the A and B chains, which result in the formation of the complete biologically active molecule. Both the I° structure and disulfide linkages of bovine insulin are shown in Figure 44–1. Insulin isolated from different species contains

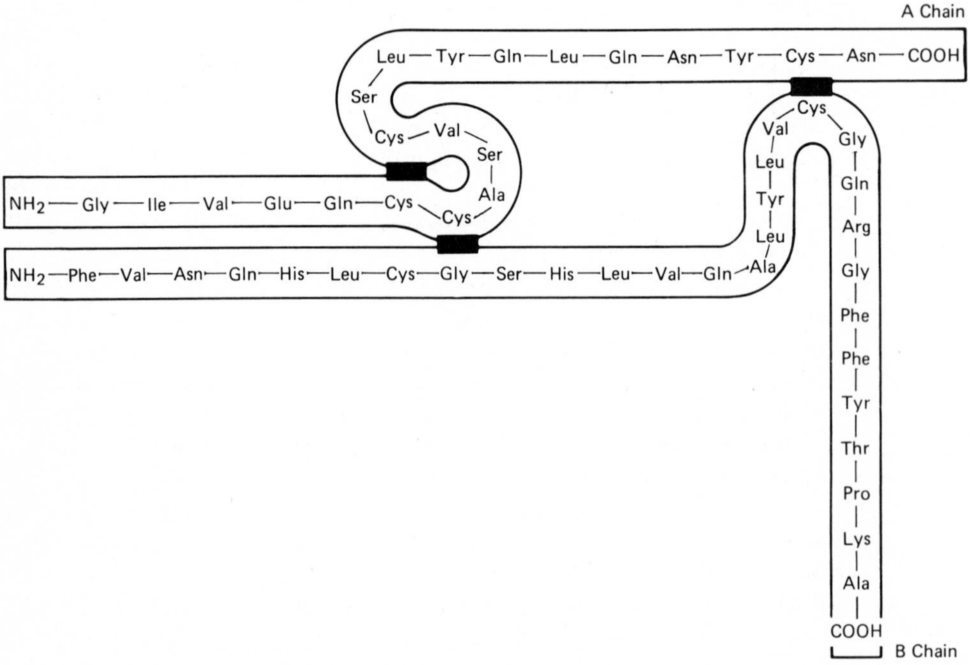
A Chain
Leu—Tyr—Gln—Leu—Gln—Asn—Tyr—Cys—Asn—COOH
Ser
Cys—Val
Ser
Ala
Cys
NH2—Gly—Ile—Val—Glu—Gln—Cys
NH2—Phe—Val—Asn—Gln—His—Leu—Cys—Gly—Ser—His—Leu—Val—Gln—Ala
Leu
Tyr
Leu
Val
Cys
Gly
Gln
Arg
Gly
Phe
Phe
Tyr
Thr
Pro
Lys
Ala
COOH
B Chain

Figure 44–1 The structure of the hormone insulin from bovine sources. Note both the intramolecular and intermolecular disulfide bonds.

slightly different amino acids. For example, human insulin has thr—ser—ile in the positions 8, 9, 10 of the A chain.

The disulfide bond is formed by the oxidation of two properly placed sulfhydryl groups of cysteine. The disulfide bond can be broken by reductive reactions. This can be seen by the *reversible* reaction

$$\underset{\text{two sulfhydryl groups}}{\text{R—S—H + H—S—R}} \underset{\text{Reduction}}{\overset{\text{Oxidation}}{\rightleftharpoons}} \underset{\text{disulfide}}{\text{R—S—S—R}}$$

Protein Structure

If the amino acid sequence was the only factor to consider, the proteins would be long random chains as free to move about as a flag in a breeze. Physical studies, especially those involving the use of X-rays, have shown that proteins actually have a specific orientation or conformation. The peptide bond has a unique almost double bond characteristic that limits the way in which the protein can exist. Under a given set of conditions, a protein will assume what is known as its **native** conformation. When the protein is not in this native conformation, it is said to be **denatured**.

The Helical Structure

Although proteins may be twisted or folded in a variety of positions, experiments have shown that one frequently found arrangement leading to structural stability is that of a **helix**. This is even more precisely described as an α-helix when the peptide chain assumes a coil containing about 3.6 amino acid monomers per turn of the helix. This rather compact conformation is favored in some cases because it allows for intrachain **hydrogen bonds** to be formed between oxygen and nitrogen atoms at proper distances in the adjacent loops of the helix. These hydrogen bonds are parallel to the major axis of the helix, as shown in Figure 44–2. The helix itself is seen in the shaded line running through the molecule. The large number of these hydrogen bonds in the helix makes this conformation quite stable. Since the helix is such a compact structure, certain amino acids are not compatible with this conformation. The side chains of the amino acids are in a position to extend outward from the main backbone of the helix. Amino acids with certain bulky side chains or those with high charges, as well as both proline and hydroxyproline, will not be found in this helical arrangement, and are called **helix breakers**. Since this conformation is dependent upon the composition and sequence of the amino acids involved, any large protein molecule may have only certain portions within it in the form of a helix. Nevertheless, this type of secondary structure contributes to the overall structure of many proteins.

A different helical form of fibrous protein is found to be present in the collagen molecule, which is a constituent of skin and tendons. The amino acids that make up a large part of this protein include glycine, proline, and hydroxyproline. The peptide chains of collagen contain about 1000 amino acids and have molecular weights in the neighborhood of 100,000. This protein exists in a helical form in which the chains

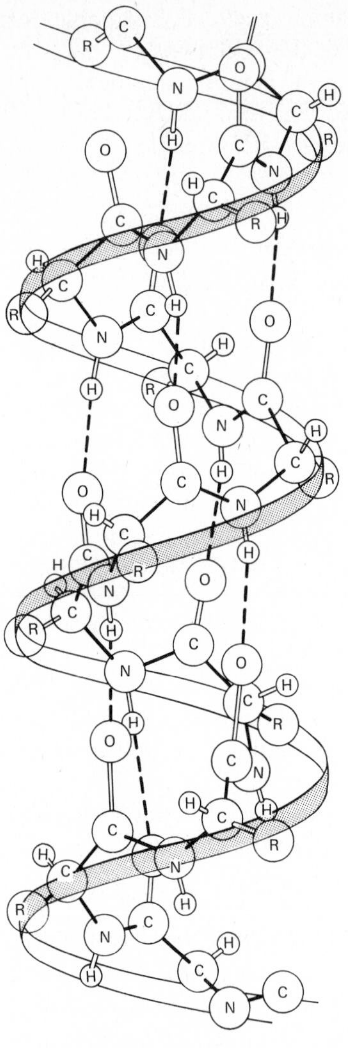

Figure 44–2 A portion of a protein chain in a helix conformation. The heavily shaded line follows the backbone of the helix. The hydrogen bonds are shown by dashed lines.

are intertwined in groups of three to form a triple helix. This triple helical arrangement is called **tropocollagen**, which forms overlapping strands of the structural fibrils of the skin. Figure 44–3 shows various magnifications of calfskin.

The Sheet Structure

Another conformation of proteins is different from either of these two types of helixes. In this conformation the polypeptide chains lie adjacent to each other and are stabilized by hydrogen bonds between the neighboring chains. The chains are

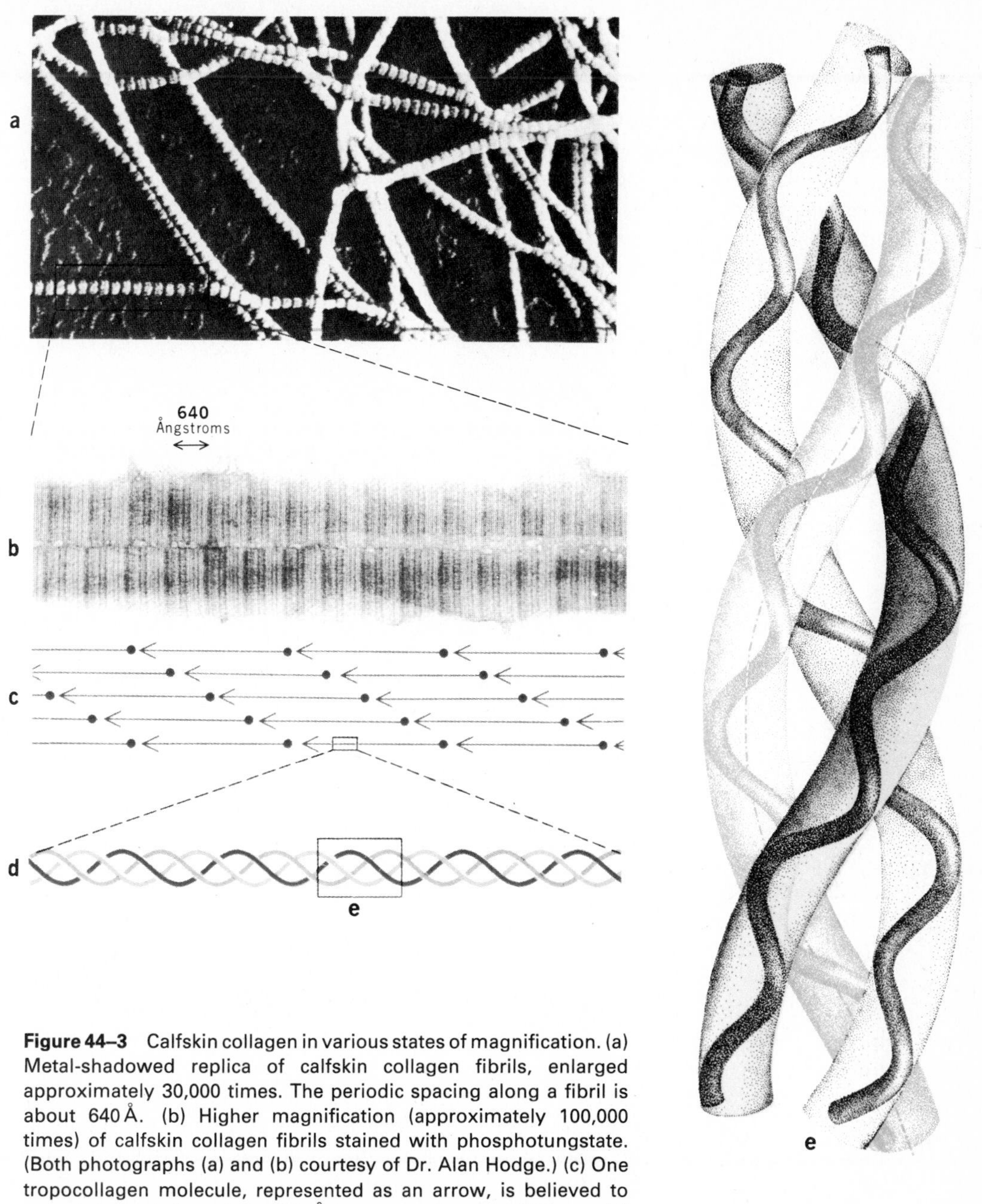
a
640
Ångstroms
b
c
d
e
e

Figure 44–3 Calfskin collagen in various states of magnification. (a) Metal-shadowed replica of calfskin collagen fibrils, enlarged approximately 30,000 times. The periodic spacing along a fibril is about 640 Å. (b) Higher magnification (approximately 100,000 times) of calfskin collagen fibrils stained with phosphotungstate. (Both photographs (a) and (b) courtesy of Dr. Alan Hodge.) (c) One tropocollagen molecule, represented as an arrow, is believed to extend through four of the 640-Å long sets of cross-striations. (d) The triple helix of the tropocollagen molecule. (e) The section above, enlarged at the right, shows how three left-handed helices are given a right-handed twist to form a three-fold superhelix. (Parts (c), (d), and (e) from R. E. Dickerson and I. Geis, *The Structure and Action of Proteins*, W. A. Benjamin, Inc., Menlo Park, N.J. Copyright 1969 by Dickerson and Geis.)

said to be antiparallel, which means they are running in opposite directions, if alternate strands are considered, from the N-terminal or C-terminal ends. If many of these chains are to be taken together, this arrangement leads to the formation of **sheets**. These sheets are stabilized by numerous hydrogen bonds that are perpendicular to the long axis of the peptide backbone of the molecule. The amino acids glycine, alanine, and serine predominate in proteins with this type of conformation, a short segment of which is shown in Figure 44–4. This is called the β sheet form and is one of the principal conformations of proteins found in silk. These stabilized sheets account for the tensile strength of silk fibers.

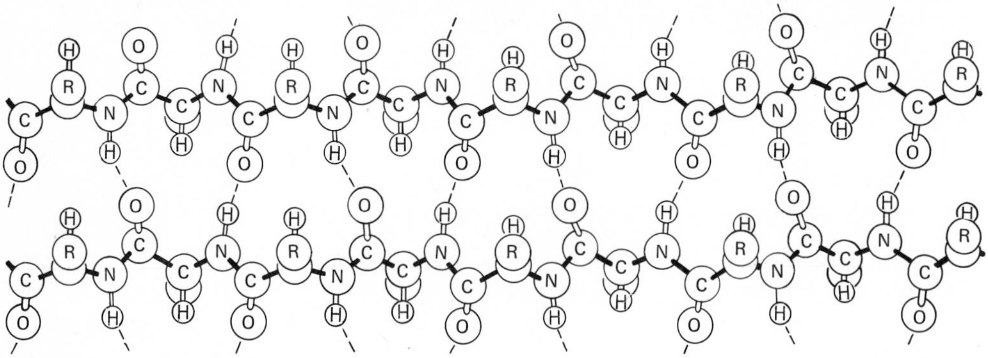

Figure 44–4 Pleated sheet segment of a protein. (From Paul Doty, *Proteins*. Copyright © 1957 by Scientific American, Inc.)

Tertiary Structure

The helical conformations and sheet arrangements are known as the **secondary structures** (II°) of proteins. These conformations are of special importance in fibrous proteins. However, in globular proteins, a variety of different conformational forms may be involved in a single protein molecule, including helical and β sheet segments. A three-dimensional or **tertiary structure** (III°) results from the specific arrangement of all the parts of the globular molecule. In addition to the stabilization due to hydrogen bonding, many proteins contain intrachain cross-linkages of disulfide linkages. Since this is a covalent linkage, it gives added stability to the tertiary conformation of the molecule. The compact molecular form of greatest stability usually also involves a folding process in which a majority of the hydrophobic (nonpolar) side chains are buried in the interior of the molecule, while the polar and charged side chains (hydrophilic) are free to extend into the aqueous environment surrounding the molecule.

A two-dimensional picture of one of the protein chains of hemoglobin (β chain) is shown in Figure 44–5. The helical segments, which account for about 10% of the molecule, are labeled, and the random sections of the polypeptide backbone that connects these segments can also be seen. The heme is the iron-containing compound that is attached to the protein and accounts for the oxygen-carrying ability

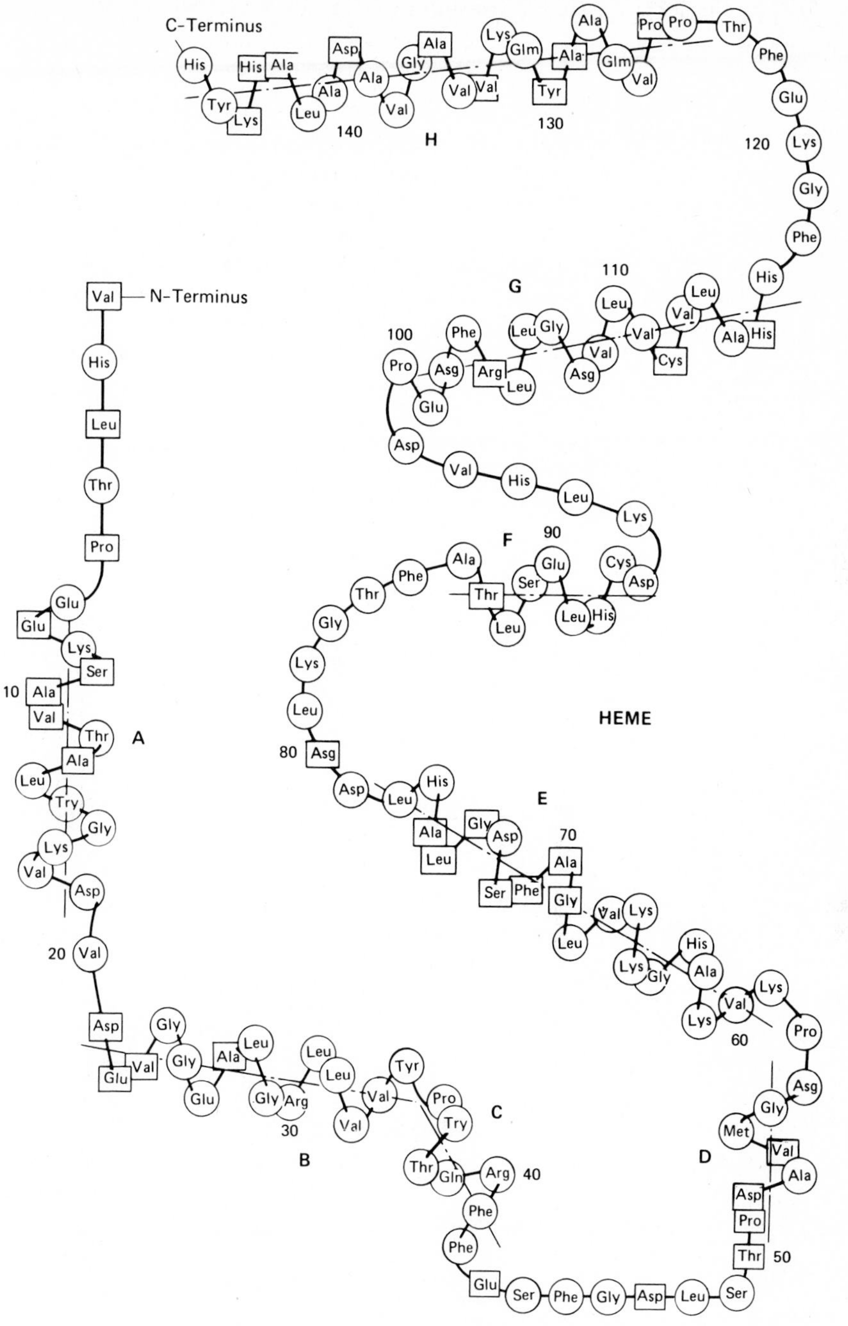
C-Terminus
N-Terminus
HEME
A
B
C
D
E
F
G
H
10
20
30
40
50
60
70
80
90
100
110
120
130
140

Figure 44-5 One of the chains of the hemoglobin molecule from human sources. The regions of secondary structure are shown. (From "The Hemoglobins," W. A. Schroeder, *Ann. Rev. Biochem. 32*, 301, 1963.)

of hemoglobin. This flat presentation cannot fully give the illusion of a globular molecule, which is actually nearly spherical in solution.

Quaternary Structure

Some proteins assume yet another higher level of ordered arrangement, known as the **quaternary structure** (IV°). Individual protein molecules, each with its own characteristic tertiary structure, interact in specific ways to form a complex of ordered multiple subunits. A number of protein systems are known to be built of **multiple subunits**, including hemoglobin. Hemoglobin in the red blood cell is composed of four separate peptide chains, each bearing an iron-containing heme molecule. There are two α chains and two β chains in the tetrameric structure. These

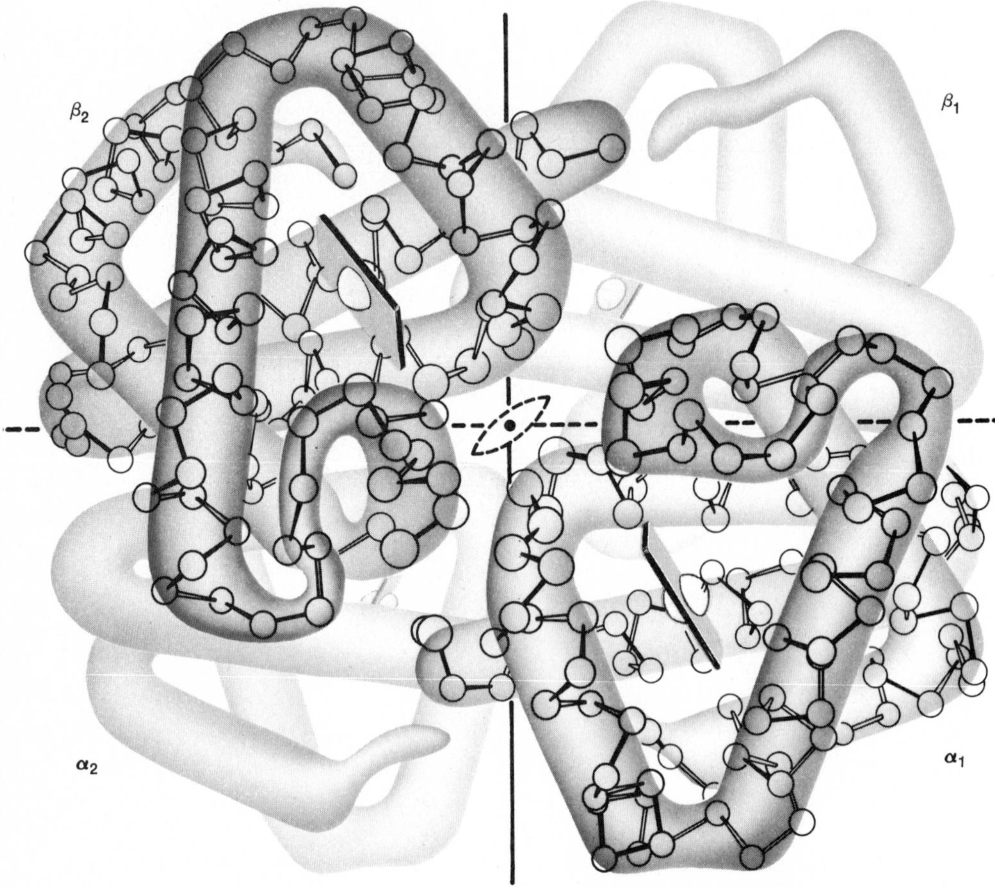

β_2
β_1
α_2
α_1

Figure 44–6 A model of the four separate monomeric chains, α_1, α_2, β_1, β_2, in the quaternary structure of hemoglobin. (From C. David Gutsche and Daniel J. Pasto, *Fundamentals of Organic Chemistry*. Copyright © 1975 by Prentice-Hall, Inc. By permission.)

two chains are quite similar in size (α has 141 amino acids and β has 146) and tertiary structure. When the interaction between the four chains is correct, the quaternary structure results, which is shown for oxyhemoglobin in Figure 44–6. This has an overall spherical structure with molecular dimension of $64 \times 55 \times 50$ Å. The molecule pictured is oxyhemoglobin with the O_2 bound to the heme groups. When the O_2 is not attached to the hemoglobin, the quaternary structure of the molecule undergoes some small change. The tertiary structure of the individual subunits remains about the same in either case. Protein molecules are neither inert nor rigid, but rather have the ability to undergo slight but significant conformation changes as needed.

Denaturation

The uniqueness of the molecular architecture of the proteins cannot be overemphasized. The amino acid sequence in the protein actually dictates the three-dimensional structure as it assumes its most energetically favorable orientation. It is possible to disrupt this native state, yielding a denatured molecule with reduced biological activity. The **denaturation** leads to disruption of the various levels of organization beyond the primary level. Denaturation does not involve hydrolysis of peptide bonds, but leads to disruption of weaker bonds or forces like hydrogen bonds or hydrophobic interactions. Sometimes the denaturation process can be reversible, and the native conformation can be partly or completely reassumed.

Both chemical and physical factors are known to cause denaturation. The physical process of beating egg whites (egg albumin) leads to an irreversible denaturation. Heat also leads to denaturation of the protein in the egg. Many chemical agents, including acids, bases and organic solvents, lead to denaturation. Sour milk can be prepared when it is needed in a recipe by adding vinegar (acetic acid) to fresh milk. The acid acts to denature or curdle the milk protein. Other chemicals, like urea ($H_2N\overset{\overset{O}{\|}}{C}NH_2$), are useful to bring about a reversible denaturation with proper handling. Upon the removal of a reversible denaturing agent like urea, there is frequently a restoration to the native state of the disrupted protein structure. Denaturation is a conformational change of a protein without hydrolysis of peptide bonds.

UNIT 44 Problems and Exercises

44–1. Draw the peptide alanylcysteinylphenylalanine as you would expect to find it (a) under oxidized conditions, and (b) under reduced conditions.

44–2. Suppose that you allowed the peptide in problem (44–1) to react with Sangers reagent. If the reacted peptide was isolated and carefully hydrolyzed, two amino acids and one other compound could be isolated. Name the amino acids that could be isolated. Draw the structure of the other compound that would result.

44–3. Suppose that you reduced the peptide in problem (44–1) and then carefully hydrolyzed it. Name the two amino acids that would result, and draw the structure of the amino alcohol you would expect to find.

44–4. When a purified sample of the β chains of hemoglobin was subjected to hydrolysis catalyzed by aminopeptidase, one amino acid was promptly released and another amino acid was freed after some time. What are these two amino acids?

44–5. If insulin is reduced and then reoxidized, it sometimes yields a molecule with considerably less hormonal activity. Briefly explain how you might account for this.

44–6. Draw the structure of a tetrapeptide composed of four alanines as you would expect to find it in an α helix.

44–7. Using a sketch, account for the fact that the proteins in silk yield a material that can be used to form threads.

44–8. Would you expect a denatured β chain of hemoglobin to have a different molecular weight from the native form of this protein? Explain.

44–9. List three amino acids you would expect to find in the interior of a globular protein molecule in its native tertiary structure. List three amino acids you would expect to find in exterior positions in such a protein.

44–10. A hexapeptide yielded the following dipeptides using one hydrolysis procedure:

ala-leu, gly-ser, val-asp, leu-gly

This same hexapeptide yielded the following two tripeptides using another hydrolytic procedure:

asp-ala-leu and val-asp-ala

Draw the structure of the entire hexapeptide.

UNIT 45 METHODS OF STUDYING PROTEINS

Before studying proteins, they must be **separated** from other components of a natural mixture and purified. The cells or tissues from which proteins are obtained also contain a myriad of other substances. The proteins of interest must have at least minimal water solubility. Although this may sound trivial at first, the **solubility** of the proteins in water or dilute salt solution is of great importance. It is also important that every effort be made to maintain the protein in its native state during any of the processes of separation, purification, and characterization. Although it must be assumed that a protein in solution in the test tube bears great resemblance to a protein in its natural environment in a living system, this certainly would not be true if denaturation was allowed to take place.

Since most proteins are colorless in solution, a procedure must be available that is specific and quantitative for the presence of proteins. The **biuret** test meets both of these requirements. In this test, compounds with multiple peptide bonds yield a rather distinct violet color when treated with copper sulfate under basic conditions. Another assay method is based upon the determination of the amount of nitrogen in a sample containing a protein. It has been found that many proteins contain about 16% nitrogen.

Some protein purification can be made based upon differences in solubility. For example, the name albumin has been given to a number of proteins that are quite soluble in pure water. Serum albumin and egg albumin are good examples. The globulins are another group of proteins that are not very soluble in pure water, but show increased solubility (up to a point) if a salt is added to the water. These categories are actually rather broad and are no longer widely used.

Once a protein is partially separated from other cellular materials, it can be subjected to more refined methods of analysis. A great deal of information can be obtained about the size, shape, charge, and molecular weight of proteins. A few representative methods are given that might be selected for either purification or characterization (or both) of a protein.

Chromatography

One of these separation techniques involves column chromatography or what is sometimes known as either **gel filtration** or **molecular sieve chromatography**. In one use of this technique, glass columns are filled with especially prepared porous beads made from a complex polysaccharide. The beads become filled with solvent and then act as momentary traps for molecules of specific sizes or diameters. It is assumed that proteins are nearly spherical and that the size of the protein is a good estimate of its molecular weight. The protein is allowed to percolate through this column of porous beads as shown in Figure 45–1. Very large molecules are excluded from the beads entirely and quickly eluted from the column. The very small molecules readily penetrate into the beads through the small pores. The rate of percolation down the column, as additional fresh solvent is added on the top, is slow because of numerous encounters with the beads in the length of the column. A larger protein does not encounter quite as many possibilities for penetration into the porous beads and moves through the column at a greater rate. A protein passes through the column and is eluted according to its size or molecular weight. If a series of containers were placed below the column at various time intervals, the larger molecules would be found in the early tubes, the medium-sized molecules in the middle tubes, and the smaller molecules in the last tubes to be considered. This technique can be used for both the separation of mixtures of proteins as they might be found in natural mixtures, as well as the estimation of molecular weights of proteins.

Molecular Weight of Proteins

The molecular weight of a protein is one of its more important characteristics. A **minimal molecular weight** of a protein can be estimated if the amino acid composition and content of the protein is known. For example, insulin contains about 2.0%

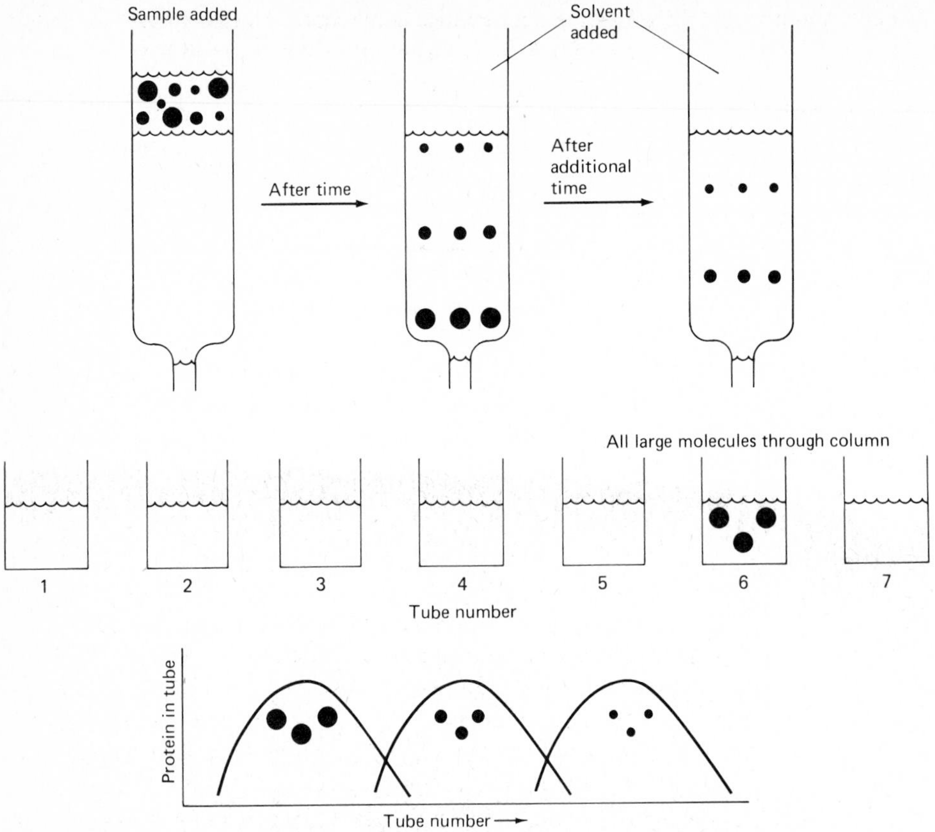
Sample added
Solvent added
After time
After additional time
All large molecules through column
1
2
3
4
5
6
7
Tube number
Protein in tube
Tube number

Figure 45–1 The separation of a mixture of three proteins of different sizes by means of gel filtration.

(by weight) of the amino acid threonine. It also has been determined that only one threonine residue is present in each insulin molecule. Since the molecular weight of threonine is about 120, a minimal value for the molecular weight of insulin can be estimated:

$$\text{Minimal MW} = \frac{\text{MW of constituent}}{\text{\% of constituent}} \times 100$$

$$= \frac{120 \times 100}{2\%} = 6000$$

A precise determination of the molecular weight of insulin shows it to be slightly less than 6000. However, we did not take into account the water that was lost as the peptide bond formed. This is a minimal molecular weight based on the fact that only one constituent is present in the proteins. Since many proteins contain several of any one amino acid, the true molecular weight will be some multiple of this minimal value. This same reasoning could be applied to the estimation of a minimal molecular weight of a protein if the concentration of some other constituent or element was known.

Ultracentrifugation

The ultracentrifuge can also be used to determine the molecular weights of proteins. An ultracentrifuge may generate a force 400,000 times normal gravity. The rate at which a protein moves under the gravitational force is dependent upon a number of experimental factors, as well as its shape, density, and molecular weight. In general, the greater the molecular weight, the more rapid the **sedimentation**.

Several different applications of this technique may be used. In the **sedimentation velocity** method, the rate of the movement of a protein in the centrifuge tube is measured directly and the molecular weight can be calculated.

Another method involves a **density gradient** or density difference in the solvent system within the centrifuge tube during the centrifugation study. In this case, the protein is placed in a tube containing a solution of increasing solvent concentrations from the top to the bottom of the tube. Sucrose and cesium chloride solutions can be used. The sedimentation of the protein will continue down the tube, upon the application of a centrifugal force, until it reaches an equilibrium when the gravitational force (downward) is balanced by the buoyancy of the solvent (upward). At equilibrium, the protein will move neither up nor down the tube. These methods are compared in Figure 45–2.

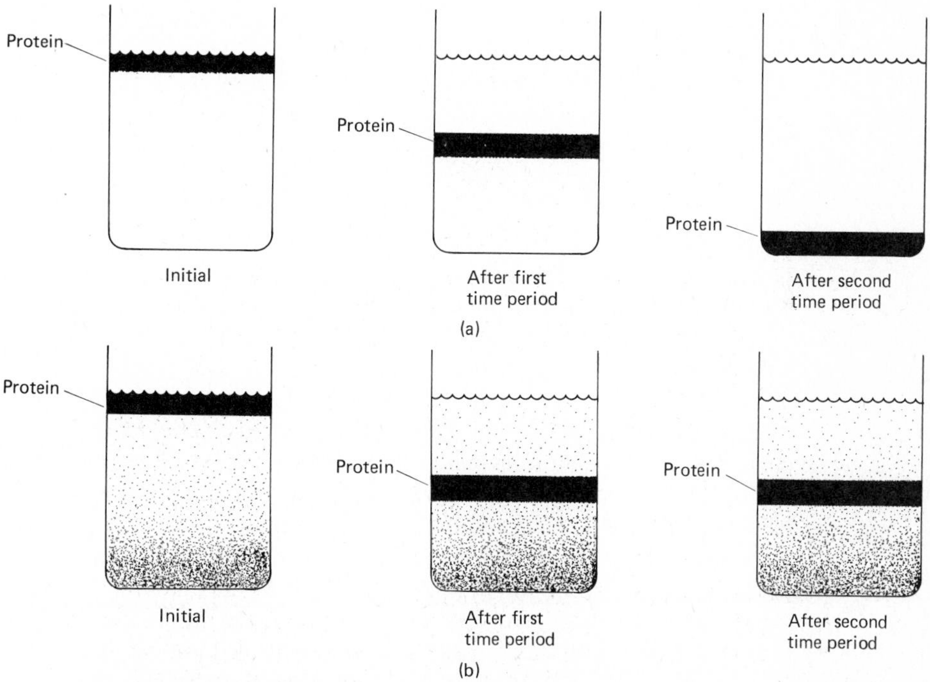

Protein
Protein
Protein
Initial
After first
time period
After second
time period
(a)
Protein
Protein
Protein
Initial
After first
time period
After second
time period
(b)

Figure 45–2 (a) Examples of initial and final distributions of protein molecules using sedimentation velocity. The solvent is homogeneous throughout the tube. (b) Examples of the initial and final distribution of protein molecules in density gradient centrifugation. Note that, in this case, the density of the solvent changes from the top to the bottom of the centrifuge tube.

Electrophoresis

Electrophoresis is another technique that has been widely used in biochemistry and medicine for the separation and characterization of proteins. Since proteins have a net surface charge at all pH values except at the isoelectric point, they would migrate or move if placed in a polarized (+ or −) electric field. Since different proteins would have different numbers of charged groups, they would have different directions and rates of movement in the electric field. The shape and size of the protein may have some influence on the electrophoresis, but the major factor is the net charge on the protein molecule at a given pH value. If the protein has an excess of groups with a negative charge at a certain pH value, like ionized carboxyl groups, it would migrate toward the positive pole (anode). On the other hand, if an excess of positively charged amino groups were to predominate, the protein would migrate toward the negative pole (cathode). The greater the number of charged groups, the greater the rate of the movement in general. If the protein were studied at its isoelectric condition (zero net charge), it would not move in either direction. In modern electrophoresis, the proteins are placed on some sort of supporting material like cellulose acetate. An illustration of a separation of three proteins is shown in Figure 45–3.

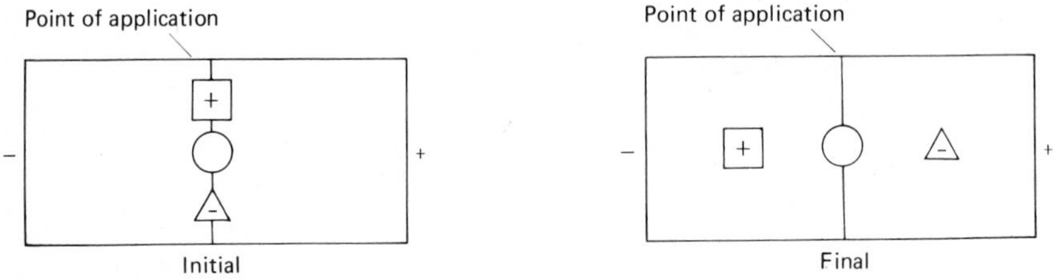
Point of application
−
+
Initial
Point of application
−
+
Final

Figure 45–3 Schematic separation of three proteins by electrophoresis. Let ⊞ represent a protein with a net positive charge; △, a protein with a net negative charge; and ○, a protein at its isoelectric point. Note that the ⊞ and △ proteins might not migrate the same distance.

Electrophoresis studies are commonly applied to the separation of blood proteins. When normal human plasma proteins are subjected to electrophoresis, at least six different proteins are visualized following a staining technique, as shown in Figure 45–4. When stains are employed, it is assumed that the color imparted by the stain is directly related to the concentration of the protein. The densitometer used in this figure is an instrument that measures and records the amount of dye attached to the proteins. The greater the area under the individual peaks in the scan, the greater the amount of that individual protein constituent. A more sensitive and specific detection system known as immunoelectrophoresis can also be used.

As biochemists develop new methods of studying proteins, the techniques will surely be applied directly to problems influencing human health and well-being. It is difficult to know just which experiments will prove fruitful. It is doubtful that when electrophoresis experiments were first carried out in 1937 anyone could foresee the

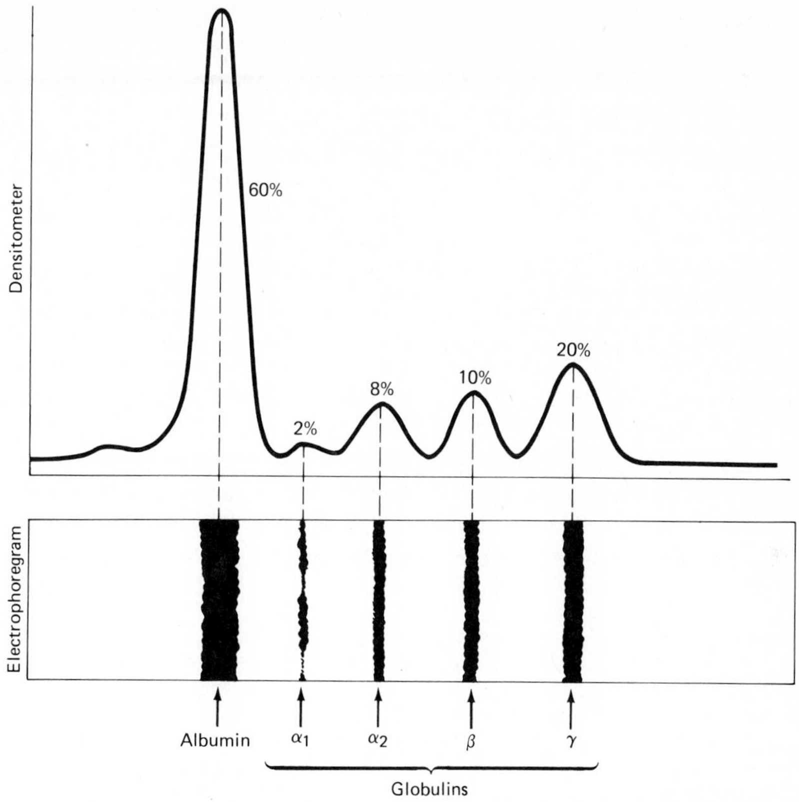

Densitometer
60%
2%
8%
10%
20%
Electrophoregram
Albumin
α1
α2
β
γ
Globulins

Figure 45–4 Electrophoresis of human blood serum protein: above, the color density of each component peak as determined by the densitometer; below, the pattern on the strip. Note that the area under each peak is proportional to component concentration.

importance of this procedure in modern clinical medicine. The basic science must develop before the refined application can be available to solve our many problems and meet our future needs.

UNIT 45 Problems and Exercises

45–1. Suppose you determined that 1000 g of milk contained about 5.1 g of nitrogen. Assume that most of the nitrogen in milk is in the form of protein. Estimate the amount of protein that would be present in the amount of milk.

45–2. Suppose that you had separate solutions of glucose, glycine, and ribonuclease. Indicate what chemical tests or methods you might use to distinguish each of these solutions.

45–3. Human serum albumin has been studied and is shown to contain one tryptophan (MW 204) in its structure. An assay system on a sample of albumin indicates that it is 0.3% tryptophan. Calculate the minimal molecular weight of serum albumin.

45–4. Hemoglobin is known to contain 0.33% iron. Iron has an atomic weight of 55.9. Calculate the minimum molecular weight of hemoglobin. Since we know that there are actually four iron atoms for each complete hemoglobin, estimate the actual molecular weight of hemoglobin.

45–5. Suppose that you were given a mixture of three proteins, ribonuclease, serum albumin, and thyroglobin, which were present in concentration ratios of 3 : 1 : 1, respectively. If these were subject to gel filtration chromatography, draw the elution profile that you might expect to find (see Table 44–1 for molecular weights).

45–6. Suppose that these same three proteins were studied using sedimentation velocity centrifugation. Draw a centrifuge tube and mark the positions where you might expect the three proteins to be at the conclusion of the study.

45–7. Would you expect that hemoglobin and serum albumin would be easy to separate using an ultracentrifuge? Briefly explain.

45–8. Various hemoglobin molecules are known to exist that contain substitutions of one amino acid for another. In one case, known as hemoglobin A, a glutamic acid is in a certain position. In hemoglobin S (sickle cell), there is a valine substituted for this glutamic acid. If these two hemoglobins were subjected to electrophoresis at pH 8.6, indicate where you would expect to find each sample at the conclusion of the electrophoresis.

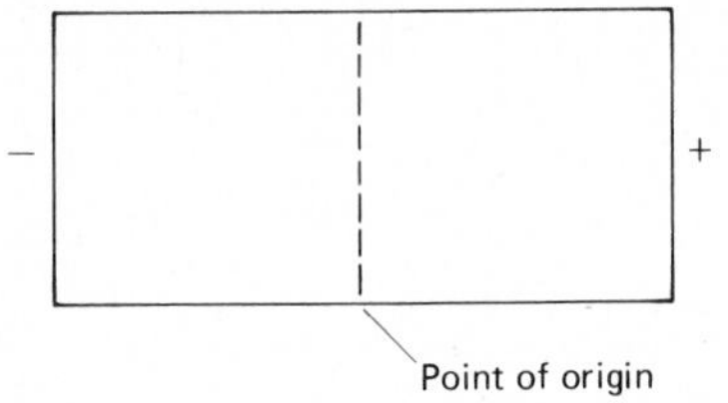
–
+
Point of origin

CHAPTER 16

cells and the study of metabolism

The biological process of metabolism is actually a composite of anabolism and catabolism. **Anabolism** involves energy-requiring reactions that lead to the synthesis of large molecules from simpler precursor molecules. **Catabolism** involves degradative reactions whereby large molecules are broken down to simpler and more stable ones. Catabolic reactions yield energy and frequently involve oxidation schemes with end products of CO_2 and H_2O. Although anabolic and catabolic pathways are not always the same, portions of these schemes may be shared or have dual functions. Shared pathways are known as **amphilbolic.**

Whereas chemical reactions often involve a reactant going to a major product, metabolic reactions in living cells usually occur as a pathway or series of linked reactions. The product of the first reaction becomes the reactant in a series that can be coupled and combine a great many steps. The overall reaction of A $\longrightarrow$ F could be broken down to a metabolic series:

$$A \longrightarrow B \longrightarrow C \longrightarrow D \longrightarrow E \longrightarrow F$$

Each step must be considered as a distinct process with separate energy considerations and separate enzyme control.

Two major aspects of catabolic reactions involve the generation of biochemically useful energy and the synthesis of a variety of new molecules. The variety of different molecules that make up the pathways of the metabolic schemes can be studied by using appropriate isotopic *tags*. The metabolic processes must finally be integrated into their functions in subcellular components, intact cells, and whole organisms.

UNIT 46 CELL CONSTITUENTS

Water

All reactions of living systems involve the solvent water. The importance of water is obvious in the human body in which the total fluids make up about 70% of the body weight. About 5% of the adult body weight is present as blood plasma and another 15% or so is made up of other extracellular body juices or interstitial fluids such as lymph. The major portion of the body fluids is said to be intracellular or within cells of the body. The rest of the weight of a typical 160-lb man is made up of about 30 lb of protein, 25 lb of fat, 5 lb of minerals, 1 lb of carbohydrates, and a few ounces of other constituents.

The water balance in humans is very critical, and its regulation is under careful biochemical control, especially by the kidney. For a normal 70-kg adult, daily water losses may amount to approximately 2.5 ℓ. The water in the urine may account for half of this, with sweat, respiration, and stool losses making up the remainder. Water losses are dependent upon both the physiological state of the individual (body temperature) and the environment in which he is placed. Naturally, there must be water replacement to maintain the balance. About 90% of this replacement water comes directly from the beverages that we drink and the water contained in solid foods. Another portion comes from the metabolism of foodstuffs within the body as part of the equation

$$O_2 + \text{food} \longrightarrow CO_2 + \text{energy} + \text{water}$$

The actual amount of metabolic water may vary depending upon the composition and amount of food ingested, but approximately 300 mℓ of metabolic water is generated each day by an adult.

Water is important as both the solvent used to bathe the tissues and to carry away the waste products of metabolism. If intake does not match water loss, **dehydration** results. This is especially important in children and infants for whom daily fluid losses represent a much greater percentage of total body fluids. As much as 50% of the extracellular fluid of an infant must be replaced on a daily basis, as compared to a much smaller percentage for an adult.

The biochemical reactions controlling water and electrolyte balance (especially Na^+) involve a variety of enzymes and hormones. Diuretic agents are available to physicians to control water retention and losses. General diuretic drugs known as natriuretic agents stimulate the urinary secretion of Na^+ and water. The name of these agents arises from the Latin word for sodium, *natrium*, from which the chemical symbol for sodium, Na^+, is also derived.

Cells

Although water may be highly mobile within the various components of the cell, many other biochemically important molecules are not. Just as higher organisms have assigned body functions to various organs, the more complex cells also are compartmentalized. The small simple cells are known as **procaryotic** cells. They have only a single membrane surrounding the cell components. Although some molecules or systems may be somewhat localized within these cells, they are not clearly separated into discrete subcellular entities or organelles. Probably the most widely studied cell of this type is the bacterium *Escherichia coli*, found in the intestinal tract of man. This cell is only about 2 μ long and 1 μ in diameter, yet it is capable of nearly every biochemical reaction known. These cell types are considered by some to be the earliest to arise in biological evolution.

Eucaryotic cells, found in higher organisms, are generally larger and more complex than procaryotic cells. Probably the most widely studied eucaryotic cell is that found in the mammalian liver. This cell is about 20 μ in diameter. Eucaryotic cells have subcellular levels of specialized structures, which are sometimes enclosed in a defined membrane.

Subcellular Components

Subcellular components can be separated from each other by using a centrifuge to sediment the heaviest components, leaving the lighter ones suspended. The separation is carried out by increasing the gravitational force is a stepwise fashion, each time removing the sedimented materials. The various components of a liver cell are shown with their important specialized enzymes and metabolic sequences in Figure 46–1.

The heaviest component, which is most readily sedimented by the centrifuge, is the **nucleus**. The DNA is located mainly in the nucleus, and part of the RNA may be located there as well. These nucleic acids are usually found in association with proteins. The proteins known as **histones** are associated with DNA to make up the genetic material known as **chromosomes**. A localized concentration of the RNA, along with the DNA, is found in the **nucleolus**. The enzymes influencing the metabolism of nucleic acids are found in the nucleus.

The **mitochondria** sediment from a nucleus-free material when a greater centrifugal force is applied. Each mitochondion is somewhat smaller than the nucleus; it is the cellular powerhouse. These organelles are rich in oxidative enzyme systems that capture energy by forming ATP molecules.

The microbodies, golgi apparatus, and lysozomes are sites of a number of proteins and enzymes, especially the digestive and degradative enzymes.

The endoplasmic reticulum and ribosomes are extremely important as sites of synthetic reactions. The former bodies also participate in some phases of lipid synthesis. **Ribosomes** are the major sites of protein synthesis within the cell. Forces of as much as 100,000 × gravity are often used to sediment these particles.

The material that remains suspended after the centrifugation of these various organelles is called the **cytoplasm**, or cystosol. The catabolism of carbohydrates, **glycolysis**, takes place in the cytoplasm. The enzymes directing the synthesis of fatty acids are also found here. The formation of urea, the important end product of human nitrogen metabolism, takes place in the cytoplasm of liver cells.

Although the separation of subcellular components is important for their study, the reactions of the whole cell are greater than the sum of its parts. As the individual

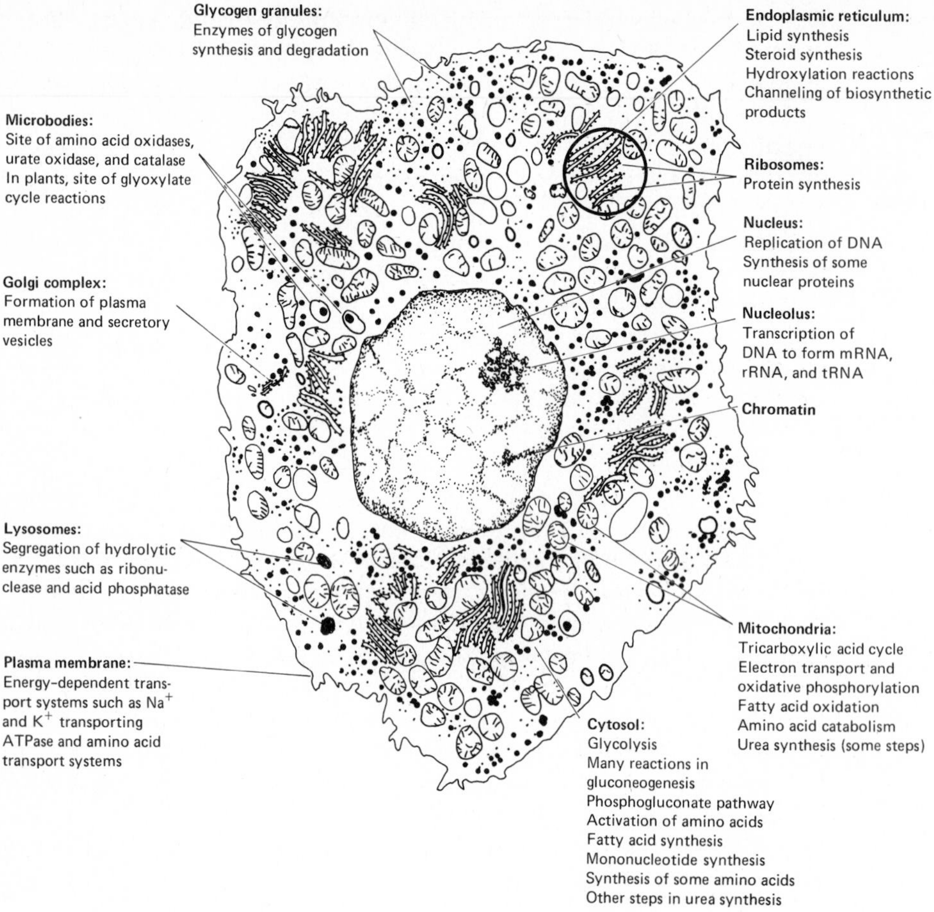
Glycogen granules:
Enzymes of glycogen synthesis and degradation
Endoplasmic reticulum:
Lipid synthesis
Steroid synthesis
Hydroxylation reactions
Channeling of biosynthetic products
Microbodies:
Site of amino acid oxidases, urate oxidase, and catalase
In plants, site of glyoxylate cycle reactions
Ribosomes:
Protein synthesis
Nucleus:
Replication of DNA
Synthesis of some nuclear proteins
Golgi complex:
Formation of plasma membrane and secretory vesicles
Nucleolus:
Transcription of DNA to form mRNA, rRNA, and tRNA
Chromatin
Lysosomes:
Segregation of hydrolytic enzymes such as ribonuclease and acid phosphatase
Mitochondria:
Tricarboxylic acid cycle
Electron transport and oxidative phosphorylation
Fatty acid oxidation
Amino acid catabolism
Urea synthesis (some steps)
Plasma membrane:
Energy-dependent transport systems such as Na+ and K+ transporting ATPase and amino acid transport systems
Cytosol:
Glycolysis
Many reactions in gluconeogenesis
Phosphogluconate pathway
Activation of amino acids
Fatty acid synthesis
Mononucleotide synthesis
Synthesis of some amino acids
Other steps in urea synthesis

Figure 46–1 A typical liver cell. (From Albert L. Lehninger, *Biochemistry*, Worth Publishers, New York, 1970, page 284.)

parts and reactions of the cell are studied, we must employ a sense of perspective. We cannot always be certain that the individual pieces of the puzzle fit together in nature as we would hope. The scientific investigator must use a variety of techniques for study, but frequently must apply some restraint to avoid overinterpretation of the data.

Organs

In higher species, the individual organ represents an important functioning unit. A good example of this is found in the studies proving that the liver is the principal site of the formation of urea. Since urea is found in urine at high levels (12 to 40 g/day in adults), the kidney must be involved in its secretion and could possibly be involved in its formation. However, in 1924 it was shown that hepatectomized (removal of liver) dogs could not form urea. It was further noted that even in hepatectomized dogs the blood urea levels continued to decrease. However, if the kidneys in these animals were ligated, the urea level of the blood would not fall. This clearly points up the role of the kidney in the elimination of this metabolite and the role of the liver in its biosynthesis. Investigators were then able to concentrate on the liver to find and study the enzymes responsible for urea formation. From these studies, the entire biochemical reaction sequence of urea synthesis was completed.

Isotopes in Biochemistry

It would be desirable to follow the digestion and metabolism of any given food from the point of ingestion to its final elimination by the body. Furthermore, it would be interesting to follow these reactions on a molecule-by-molecule or atom-by-atom basis. With modern biochemical techniques and the use of isotopes, this can nearly be accomplished. In earlier sections, **isotopes** were defined as forms of the elements with the same atomic numbers but different atomic weights.

A few stable isotopes (do not emit radiation) have been employed extensively in biochemical studies. These are ^{2}H, ^{15}N, and ^{18}O, which can replace ^{1}H, ^{14}N, and ^{16}O, the isotopes of greatest natural abundance or concentration. The disadvantage of using these particular isotopes lies in the experimental techniques that must be used to detect them. Since they do not emit a detectable radiation, their presence must be measured by mass spectrometry, a rather difficult and expensive procedure.

The more frequently used isotopes are those that emit detectable radiation. Table 46–1 gives some common radioactive isotopes and their biological application.

Table 46–1 Some radioactive isotopes of importance in biochemical and medical studies

Natural abundant isotope	Radioactive isotope	Half-life	Utilization
$^{1}_{1}H$	$^{3}_{1}H$	12.3 yr	Autoradiography
$^{12}_{6}C$	$^{14}_{6}C$	5700 yr	Dating studies
$^{23}_{11}Na$	$^{24}_{11}Na$	15 hr	Extracellular space studies
$^{31}_{15}P$	$^{32}_{15}P$	14.3 days	Leukemia therapy
$^{32}_{16}S$	$^{35}_{16}S$	88 days	Amino acid labels
$^{40}_{20}Ca$	$^{45}_{20}Ca$	152 days	Inorganic metabolism
$^{56}_{26}Fe$	$^{59}_{26}Fe$	45 days	Blood iron studies
$^{59}_{27}Co$	$^{60}_{27}Co$	5.24 yr	Radiation therapy source
$^{127}_{53}I$	$^{131}_{53}I$	8 days	Thyroid function tests
$^{197}_{79}Au$	$^{198}_{79}Au$	2.7 days	Cancer treatments

The usefulness of a radioactive isotope in a biological system is dependent upon both the energy of its radiation and its half-life. The type and energy of the radiation must be compatible with the measuring techniques available. The half-life must be long enough to allow the desired biochemical reaction to take place and laboratory measurements to be completed. Since the **half-life** is the time for half of a given sample to decay, a sufficient amount of radioactive material must remain at the end of the time of the experiment to allow it to be quantified. The choice of the isotope may be made on this basis. For example, ^{131}I has a half-life of 8.1 days, whereas ^{130}I has a half-life of 12.5 hours. If a given experiment would require several days to be completed, ^{131}I would probably be the isotope of choice. Sometimes isotopes with shorter half-lives are chosen in order to minimize radiation hazards to the study system.

Isotopes are useful because it is assumed that every atom of a given element is biochemically indistinguishable from every other atom of that element. Therefore, the reactions of $^{14}_{6}C$ in a molecule are assumed to be chemically the same as the $^{12}_{6}C$. In effect, this technique gives the biochemist a *tag* or an "atomic tweezers" to pick and hold a specific atom in a compound under study. Actually, only a very few of the atoms of interest within a molecule need to bear the label to make the study useful.

In most cases it is important to know the exact position of the label or tracer in the molecule. For example, two molecules of acetic acid could be prepared, each with a different carbon atom labeled

$$^{14}CH_3COOH \qquad CH_3^{14}COOH$$

This means that some of the carbon atoms of the number 2 carbon of the molecule on the left side are $^{14}_{6}C$ instead of $^{12}_{6}C$. The unmarked atoms are assumed to be the most abundant isotope. The molecule on the right side has an enrichment of the radioactive isotope on the number 1 or carboxyl carbon. If you considered a large number of each of these molecules, only a fraction of the molecules would actually have to bear the label shown to be useful in experiments. If acetic acid labeled with these two different forms was presented in separate experiments to a biological system, different results could be expected. For example, a different distribution of isotopes would result in a cholesterol molecule synthesized from these two differently labeled acetic acids.

Melvin Calvin and coworkers were able to use $^{14}CO_2$ to trace its incorporation into organic molecules in photosynthetic reactions. They identified the first organic molecule bearing a ^{14}C level following the use of $^{14}CO_2$. They then knew what sort of metabolic schemes to expect, both preceding and following this incorporation step.

Isotopes are useful to determine the rates of metabolic activities of specific atoms in a molecule. For example, if glucose was prepared with two different labeling patterns, differing rates of release of $^{14}CO_2$ could be observed if these glucose samples were placed in contact with different metabolic systems. Figure 46–2 indicates that more than one metabolic system must be functioning in this tissue. From this we notice that each atom in a molecule may have a unique fate. The isotopic tracer is useful in following a metabolic sequence and in isolating the various participants along the metabolic pathway.

Stable ^{15}N tracers have been used to establish the dynamic turnover or exchange of nitrogen compounds of some cellular components. Instead of a static condition, the total amount of nitrogenous compounds remains rather constant within the body, but many of the individual molecules or components of the molecules undergo a

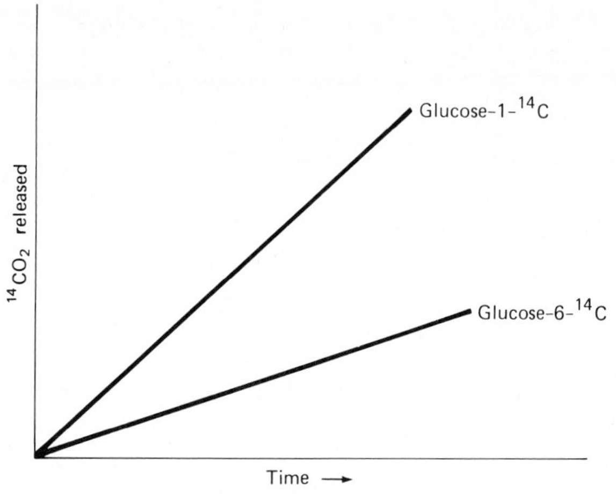
Glucose-1-^{14}C
Glucose-6-^{14}C
$^{14}CO_2$ released
Time →

Figure 46–2 The rates of loss of $^{14}CO_2$ from liver systems using glucose-1-^{14}C and glucose-6-^{14}C.

change. Some ^{15}N-labeled amino acids were found to be incorporated into the proteins of the liver of mature animals. The liver did not change in size or overall composition during this uptake of a new material, however. It must be that the molecules taken in were balanced by molecules removed. When the loss of nitrogen exactly balances the uptake, the organism is said to be in a dynamic steady state of **nitrogen equilibrium**. If an individual suffers from a wasting disease or starvation, the net loss of nitrogen results in a condition known as negative nitrogen balance. On the other hand, during the growth of the young or in some cases of convalescence, an individual may be in what is known as a positive nitrogen balance. Using this technique, it has been estimated that a 70-kg adult man must synthesize and catabolize about 400 g of protein each day to remain in nitrogen equilibrium or at a steady state. Without tracers it would have been nearly impossible to show the very dramatic turnover values of some specific tissue components.

Tracers can be used to follow the chemical details of many reactions. For example, ^{18}O may be used as a label to indicate which oxygen atoms are involved in the esterification reaction. If the alcohol is labeled with the isotope, the label will appear in the ester that is formed:

$$R{-}\overset{\overset{\displaystyle O}{\|}}{C}{-}OH + R'{-}^{18}OH \rightleftharpoons R{-}\overset{\overset{\displaystyle O}{\|}}{C}{-}^{18}OR'$$

In this way the atom-by-atom details of a chemical reaction may be uniquely considered.

UNIT 46 Problems and Exercises

46–1. List each of the subcellular organelles that you would expect to find in a mammalian cell, and give one biochemical system that is known to occur in each.

46–2. Starting with a broken eucaryotic cell, sketch a flow diagram whereby the subcellular organelles would be separated.

46–3. Estimate the weight of water present in your own blood plasma.

46–4. If you eat a large portion of ham, you frequently become quite thirsty. Account for this in a biochemical way.

46–5. Indicate some of the advantages of using the isotope ^3H over the isotope ^2H in some biochemical studies.

46–6. Indicate some of the advantages of using the isotope ^2H over the isotope ^3H in some biological studies.

46–7. What might be a good reason for choosing an isotope of a shorter half-life or one having a long half-life in a biochemical study?

46–8. Draw the structure of glucose-3-^{14}C and indicate what is meant by the designation.

46–9. Describe the physiological situation when an individual is said to be in negative nitrogen balance.

46–10. Given a sample of glycerol phosphate labeled as shown,

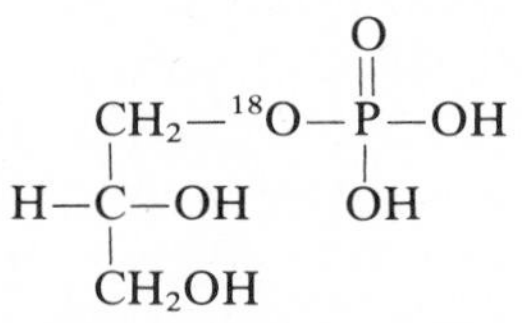

O
CH₂—¹⁸O—P—OH
H—C—OH
OH
CH₂OH

draw the structures and mark all the oxygen atoms that could possibly be labeled with ^{18}O if glycerol phosphate was hydrolyzed to glycerol and phosphoric acid using H_2O.

UNIT 47 BIOCHEMICAL MINERAL METABOLISM

In studying organic chemistry, the central position of carbon is emphasized. The carbon compounds often contain five other major elements to build biochemically important molecules: hydrogen, oxygen, nitrogen, phosphorus, and sulfur. These are the important elements, but animal life would not be possible without at least 18 other elements in varying amounts. Unfortunately, biochemical research has rather neglected the study of mineral metabolism and its relation to nutrition.

If an investigation sets out to determine if a given element is required by a living system, a number of problems arise. The diet and water supply must be highly purified and accurately known. The air supply must be controlled, since trace amounts of material may be carried on dust or other particulate matter. Containers must be especially designed, for common cage metals usually contain a wide variety of different elements. Assuming all these experimental difficulties can be worked out, what definition does one use to describe essential? Growth of the young may be a possibility; maintenance of the adult may be another. Variations may be encountered from species to species. Since the elements are often closely related chemically, their metabolic interrelationships add another difficult dimension.

Two techniques that are widely used for quantitative studies of inorganic elements are **atomic absorption spectroscopy** and **flame photometry**. In flame photometry, a solution containing the element of interest is atomized into a flame under controlled conditions. The atoms become excited by absorbing energy from the flame. The excited atoms then emit unique radiation as they return to their more stable state. A solution of Na^+ ions in a flame yields a bright yellow color, copper will yield a green color, and calcium a reddish-orange color. Although more than 35 metals can be studied using this technique, it is most frequently used for the quantitative measurements of sodium, potassium, and calcium that occur in body fluids.

Another technique, atomic absorption, uses a somewhat similar system to study up to 60 elements. If radiation of a specific wavelength is passed through the flame containing an atomized element, some of the atoms will absorb the radiant energy. The amount of energy absorbed by the element can be measured and displayed on a meter or chart. This technique is so sensitive that it can detect the presence of some elements to as low as 1 part per billion (ppb). The detection of the presence of mercury in tuna came about through the use of this technique by which as little as 0.01 part per million (ppm) of mercury was found.

Magnesium, Calcium, and Phosphorus

One very interesting aspect of mineral biochemistry involves the specific distribution of the different ions. For some reason the interior of cells is rich in Mg^{2+} and K^+. The fluids that surround cells, such as blood plasma, are rich in Ca^{2+} and Na^+. The monovalent cations, along with Cl^- ions, will be considered later under transport processes.

Magnesium, calcium, and phosphorus are interrelated. Mg^{2+} is required in a number of enzyme systems and often complexes with the phosphates of ATP. Magnesium is related to reactions of the central nervous system; high concentrations may reduce the heart beat rate and ultimately lead to cardiac arrest. In the plant kingdom, Mg^{2+} is the central metal atom of the chlorophyll molecule. Serum Mg^{2+} levels in normal adult blood is about 2 mEq/ℓ, including that bound to blood proteins. The Mg^{2+} concentration in cells is usually considered to be 16 mEq/ℓ.

Calcium and phosphorus make up about one fourth of the hard structures of bones and teeth. Calcium plays an important role in blood clotting, muscle action, and nerve transmission. Plasma Ca^{2+} levels are normally about 5 mEq/ℓ in adults. The Ca^{2+} levels are under control by both vitamin and hormone systems. Derivatives of vitamin D are involved in the absorption and transport of the ion. The parathyroid secretes two hormones, parathormone and calcitonin, that influence Ca metabolism. Calcitonin decreases the loss of Ca^{2+} from the bone by inhibiting its transfer into the bloodstream; parathormone has the opposite effect.

Serum calcium and phosphate normally show a reciprocal relationship. When the blood level of one increases, the other decreases. The phosphorus is in the ionic form HPO_4^{2-} and $H_2PO_4^{1-}$. Under the normal pH conditions of the plasma, a 4:1 ratio of these forms is found. The organic phosphate esters are very common in nucleic acids, coenzymes, and carbohydrate metabolism.

Iron

Iron is unique in many ways, although there is probably less than 5 g of it in the entire body (a nickel weighs about 5 g). Although a daily intake of 12 mg/day for adults and up to 15 mg/day for children may be recommended as sufficient, individual needs show great variation. The female has special needs during the time of her life when the menstrual cycle alone may require an average of 0.6 mg/day. This is compared with a total need of an adult man of approximately 1 mg/day to maintain an adequate balance. Except for special needs, disease, or hemorrhage, the body has a special way of recycling iron. Studies using radioactive ^{59}Fe indicate that red blood cells have a definite life span before they are replaced by new cells, but even then much of the iron is recycled.

A great deal of the iron in the blood is held in a tight complex in the hemoglobin molecule within the red blood cell. Another special protein, **transferrin**, carries most of the iron in the plasma. Since the iron is usually bound to a protein, this may account for its very small loss through the kidneys. Some iron, about 1 mg/day, may be lost in the fecal material. Since there is not a very active pathway to remove iron from the body, excess of iron can prove to be a real health problem. Prolonged periods of high levels of iron within the body can actually lead to severe liver damage. Iron metabolism is not like that of many other nutrients, because here the absorption process seems to control the iron levels within the body.

Iron absorption is highly dependent upon its form in the food we ingest. For some as yet unexplained reason, some foods seem to have the iron in a much more available form. Meats represent a good source of iron, and frequently yield their iron much more readily to absorption processes than do many plant forms. Milk is a notoriously poor source of iron, and special attention needs to be given to infants to assure an adequate supply of iron. The ferrous (Fe^{2+}) form of iron seems to be more readily absorbed than does the ferric (Fe^{3+}) form. A protein, **ferritin**, which has been found to contain over 20% iron by weight, is involved with the absorption of iron through the intestine. The presence of another element, copper, may be involved in the normal absorption and transport of iron. Large amounts of inorganic phosphate in the diet may also impair iron absorption because of the formation of poorly absorbed iron phosphates.

Iron is involved in important enzyme and protein systems. For example, **cytochromes** are iron-containing proteins that are involved in the process of energy production. The mitochondria of aerobic cells are particularly rich in these iron-bearing proteins. The reaction involves an electron exchange in the oxidation–reduction of iron:

$$Fe^{3+} + e^- \rightleftharpoons Fe^{2+}$$

Iodine

Iodine is another element known to be needed for the maintenance of health. Almost 80% of the iodine in the adult is found in the thyroid gland where it is bound to several organic compounds and found as part of the protein thyroglobin. The two specific compounds that have been of greatest interest are thyroxine and 3,5,3′-triiodothyronine, the latter of which has the greatest biological potency. They are closely related to the amino acid tyrosine and have the following structures:

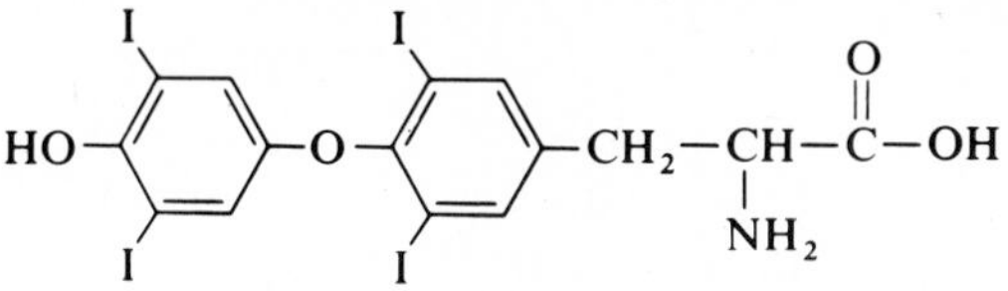

thyroxine

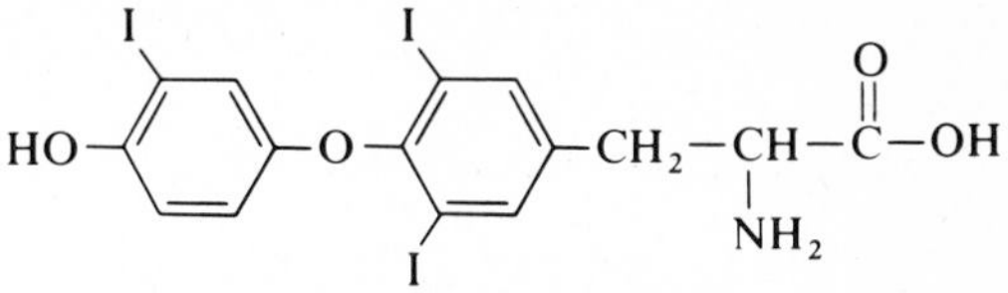

3,5,3′-triiodonthyronine

Iodine is found in the plasma in concentrations normally ranging from 8 to 15 $\mu g/100\ m\ell$. About half of this is in the form of protein-bound iodine (PBI), which is of clinical use as a measure of circulating thyroid hormone. The recommended dietary allowance of iodine is about 100 μg/day for adults. Seafoods are a rich source of iodine, as well as fruits and vegetables grown in soil that contains

Table 47–1 Other trace elements with biochemical functions thought to be essential to the growth of young animals

Element	Atomic number	Role or function in living systems
Fluorine	9	May be important in dental health; suggested as growth factor for cats
Silicon	14	Shown to be essential in chickens; may be involved in a structural role; recently implicated in cholesterol metabolism
Vanadium	23	Important in marine animals and cats; some implication in dental health
Chromium	24	Reported to be essential for carbohydrate metabolism in cats; may work in conjunction with insulin
Manganese	25	Associated with many enzymes, including muscle ATPase and system associated with urea formation; deficiency studies of widespread importance in cats, chickens, pigs
Cobalt	27	Part of vitamin B_{12} molecule; participates in some enzyme systems; may be involved in certain anemia situations
Copper	29	Involved in iron uptake and metabolism; implicated in some species with ricket-like syndromes; excess copper deposits in man in Wilson's disease
Zinc	30	Associated with Ca and Fe in human deficiency cases; constituent of numerous enzyme systems
Selenium	34	Required in liver function of rats; implicated in some studies in reactions with vitamin E and sulfur-containing amino acids
Molybdenum	42	Implicated with several metabolic enzymes, especially oxidative ones
Tin	50	Required for growth of young rats; found in skin and tongue especially

iodine. In some cases, iodide is added to drinking water supplies or as a supplement in table salt. Simple goiters have been attributed to lack of sufficient dietary iodine and have been especially widespread in iodine-deficient areas.

The thyroid hormone increases metabolic reactions in practically every tissue of the body that has been studied. Specifically, increased basal metabolic rates, increased O_2 consumption, and accelerated growth are noticed upon administration of thyroid hormones. The clinical syndrome known as Graves' disease is involved with the hyperfunction of the thyroid. A hypoactive function of the thyroid is known clinically as myxedema in humans and has the opposite manifestations.

The other trace elements now thought to be essential for the growth of young animals are shown in Table 47–1. In some cases the exact function of the element is not yet known. In other cases, the proof of the essential nature of the element may be confined to only a limited number of species. Future studies and new techniques may add to this list.

UNIT 47 Problems and Exercises

47–1. Account for the presence of the five most abundant elements in the human body.

47–2. Which two cations are considered to be predominantly extracellular? Which two are predominantly intracellular?

47–3. There may be some antagonism between Ca^{2+} and Mg^{2+}. Why might this be expected?

47–4. Why might you expect iron deficiency to be more prevalent in the very young?

47–5. Why is the presence of iron of great importance within the mitochondria of many of our cells?

47–6. Briefly explain what is meant by the term *protein-bound iodine.*

47–7. Cobalt is unique as it is actually a part of a vitamin. Name this vitamin.

47–8. If you wished to determine the presence or concentration of one of the trace metals present in living sytems, what laboratory instrumentation might you employ?

UNIT 48 ENZYMES

All enzymes are proteins. All that is known about the I°, II°, III°, and IV° structures of proteins applies to enzymes. The enzymes vary in molecular weight from relatively small proteins like lysozyme (MW 14,400) to very large molecules like urease (MW 480,000). Enzymes may exist as multicomplexes, like pyruvate dehydrogenase (MW about 10^7), to bring about a series of integrated reactions.

Enzymes may not be composed exclusively of proteins, however. In some cases metal ions are attached to the protein and are absolutely essential for the catalytic activity. In some cases other ions or molecules may function as helping groups called

cofactors or **prosthetic groups**. Sometimes organic molecules are required in addition to the protein to cause it to act as the biocatalyst. These organic molecules are relatively small as compared to proteins. They are usually modified water-soluble vitamins. Since they are absolutely necessary to help the protein portion of the enzyme carry out the catalytic function, they are called **coenzymes**. The protein portion of the enzyme system is the **apoenzyme**. The complete functioning unit is the **holoenzyme**.

Apoenzyme +	coenzyme	⟶	holoenzyme
protein portion	derivative of vitamin		active catalytic unit

The word *enzyme* will sometimes be used to denote the catalytic unit even though in many cases it may actually refer to a holoenzyme.

Enzymes are more than mere globs of proteins. They are very carefully designed molecules that act in specific ways to influence the rates of reactions. The enzyme lysozyme, shown in Figure 48–1, can be used as a three-dimensional model.

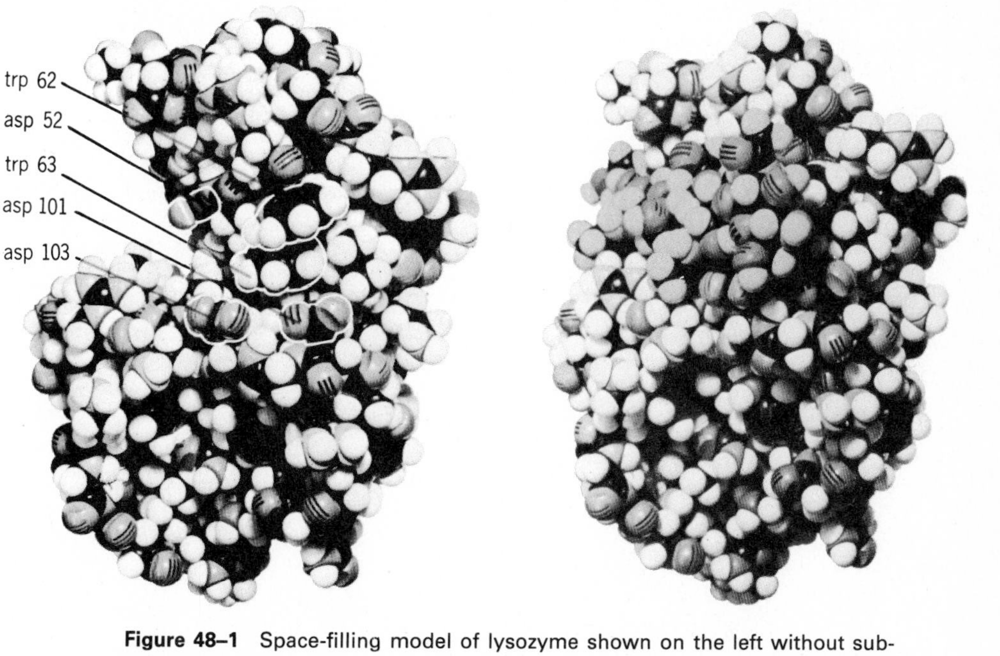
trp 62
asp 52
trp 63
asp 101
asp 103

Figure 48–1 Space-filling model of lysozyme shown on the left without substrate, and on the right, with substrate in position at the active site. (From R. E. Dickerson and I. Geis, *The Structure and Action of Proteins*, W. A. Benjamin, Inc., Menlo Park, N.J. Copyright 1969 by Dickerson and Geis.)

Lysozyme

Lysozyme is an enzyme that was discovered in 1922 as an antibacterial agent. It is found widely distributed in nature. For example, tears and egg whites are rich sources of lysozome. The enzyme acts as an antibacterial agent to inhibit the possible invasion of bacteria into either the eye or the egg.

First we shall examine lysozyme from a chemical point of view. Lysozyme has a molecular weight of about 14,600 and contains 129 amino acids. There is only one terminal amino acid, lysine. Thus, lysozyme is one single continuous chain of amino acids as shown in Figure 48–2. The lysozyme molecule is stabilized in its three-dimensional structure by four intrachain disulfide linkages. X-ray analysis shows that about 30% (of the amino acids) is in a helical arrangement, and one section is

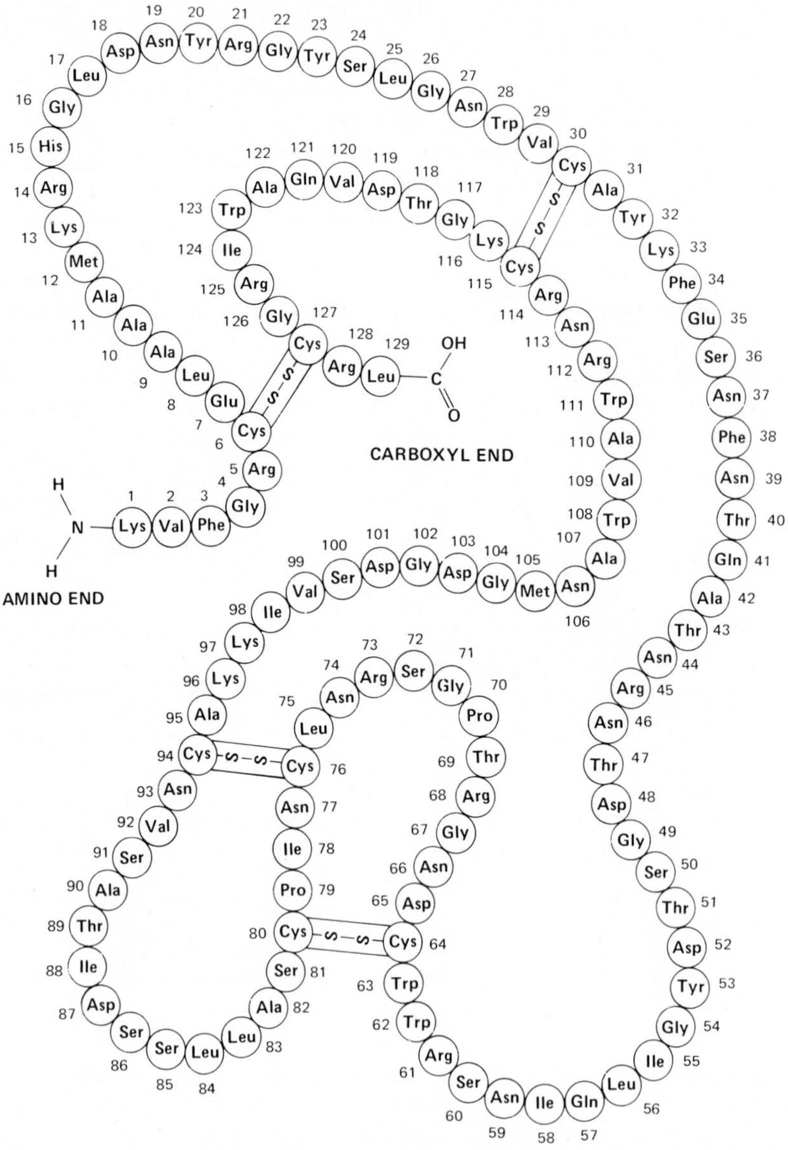

H
N
H
AMINO END
1 Lys
2 Val
3 Phe
4 Gly
5 Arg
6 Cys
7 Glu
8 Leu
9 Ala
10 Ala
11 Ala
12 Met
13 Lys
14 Arg
15 His
16 Gly
17 Leu
18 Asp
19 Asn
20 Tyr
21 Arg
22 Gly
23 Tyr
24 Ser
25 Leu
26 Gly
27 Asn
28 Trp
29 Val
30 Cys
31 Ala
32 Tyr
33 Lys
34 Phe
35 Glu
36 Ser
37 Asn
38 Phe
39 Asn
40 Thr
41 Gln
42 Ala
43 Thr
44 Asn
45 Arg
46 Asn
47 Thr
48 Asp
49 Gly
50 Ser
51 Thr
52 Asp
53 Tyr
54 Gly
55 Ile
56 Leu
57 Gln
58 Ile
59 Asn
60 Ser
61 Arg
62 Trp
63 Trp
64 Cys
65 Asp
66 Asn
67 Gly
68 Arg
69 Thr
70 Pro
71 Gly
72 Ser
73 Arg
74 Asn
75 Leu
76 Cys
77 Asn
78 Ile
79 Pro
80 Cys
81 Ser
82 Ala
83 Leu
84 Leu
85 Ser
86 Ser
87 Asp
88 Ile
89 Thr
90 Ala
91 Ser
92 Val
93 Asn
94 Cys
95 Ala
96 Lys
97 Lys
98 Ile
99 Val
100 Ser
101 Asp
102 Gly
103 Asp
104 Gly
105 Met
106 Asn
107 Ala
108 Trp
109 Val
110 Ala
111 Trp
112 Arg
113 Asn
114 Arg
115 Cys
116 Lys
117 Gly
118 Thr
119 Asp
120 Val
121 Gln
122 Ala
123 Trp
124 Ile
125 Arg
126 Gly
127 Cys
128 Arg
129 Leu
S—S
OH
C
O
CARBOXYL END

Figure 48–2 Chain model of lysozyme.

arranged in a *pleated sheet* form. Naturally, these structural features are stabilized by hydrogen bonding.

We can consider lysozyme in three dimensions by returning to Figure 48–1. A hydrophobic environment is buried within the molecule. The amino acids that have side chains that can ionize are generally found in the outer parts of the molecule, where they would be in contact with water (hydrophilic). One very important structural feature on the surface of the molecule is a distinct cleft or crevice. This cleft is the part of the molecule that is involved in its enzymatic role and is termed the **active site**. To understand how this active site becomes involved in antibacterial action, return to the chemistry of the system.

The bacterial cell wall has polysaccharides as a major constituent. The polysaccharide is made up of several different repeating monomeric units. One of the units is called *N*-acetylglucosamine (NAG), and another is *N*-acetylmuramic acid (NAM). A hydrolysis reaction is required to break down this polysaccharide. The lysozyme acts as an enzyme by accelerating the rate of hydrolysis of the polysaccharide. The polysaccharide upon which the enzyme acts is given the general designation **substrate**. Once hydrolysis takes place, the structural integrity of the cell wall is destroyed and the interior of the bacteria is subject to further destruction or **lysis**.

In this reaction, a portion of the polysaccharide fits into the cleft of the lysozyme molecule where it is momentarily held for hydrolysis. To complete the reaction, the hydrolyzed portions of the substrate molecule move away from the surface of the enzyme to make room for the next intact polysaccharide destined to be hydrolyzed.

Enzyme–Substrate Complex

The fact that the enzyme provides the site or position of action is the general scheme of enzymatically controlled reactions. There must be a momentary coming together of the enzyme and substrate to form what is known as an **enzyme–substrate complex**.

$$\text{Enzyme} + \text{substrate} \longrightarrow \text{enzyme substrate complex} \longrightarrow \text{enzyme} + \text{product}$$

$$E + S \longrightarrow ES \longrightarrow E + P$$

The attraction of the substrate to the enzyme brings about some important structural changes. First, the formation of the ES complex may be brought about by what is known as **induced fit**. This may be compared to a lock and key relationship. The complex forms only after some configurational changes of the protein portion of the enzyme takes place. Next, part of the substrate molecule may become distorted and in doing so becomes more vulnerable to the hydrolysis. Thus the enzyme has made the already feasible reaction more rapid. The enzyme is regenerated as the ES complex breaks up and the desired product is formed. The molecular specificity becomes apparent, since the enzyme must have the proper sequence of amino acids and final three-dimensional conformation necessary to play its role.

The enzymatic activity of lysozyme illustrates how the hydrolysis is accomplished on a molecular level. It should be noted that this hydrolysis is a favorable chemical reaction in the first place. The degradation of the polysaccharide to its smaller and simpler units would take place without the enzyme, if enough time were allowed. The catalyst is only involved in increasing the rate of the reaction to make it feasible within the time limits necessary for biological systems. For example, each molecule of catalase, the enzyme responsible for the breakdown of hydrogen peroxide, has been established to react with 40,000 molecules of H_2O_2 per second. In general, the more critical the enzyme is in essential reactions, the faster its rate of catalysis.

In hydrolysis reactions the substrate molecule must come in contact with the water molecule. In addition, the reactants must form an **activated complex** and overcome an activation energy barrier for the reaction to take place. The rate of the reaction is actually dependent upon the concentration of reactants reaching a given activation energy. This is shown in graphical form in Figure 48–3. In this figure the

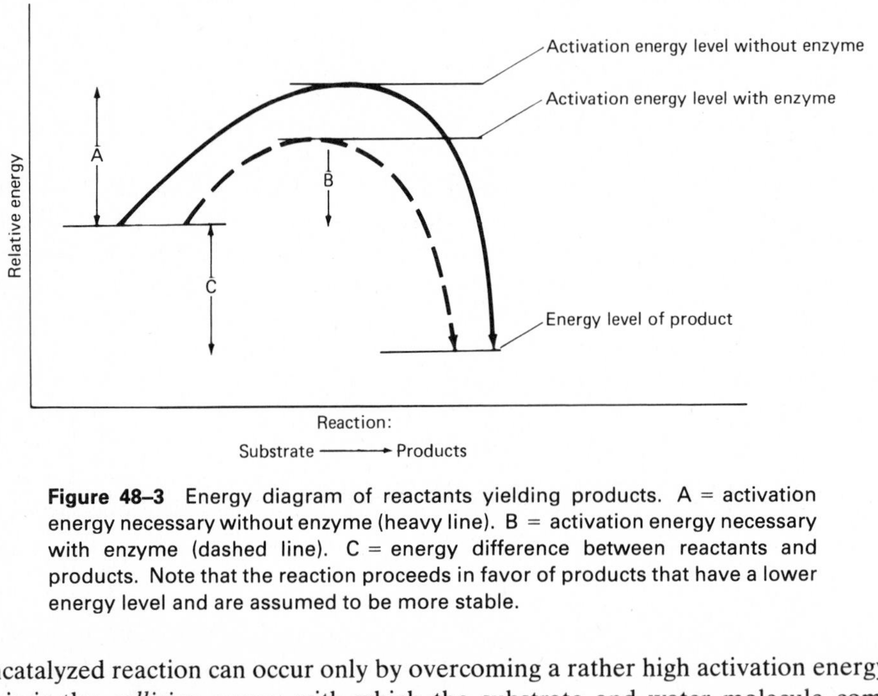

Activation energy level without enzyme
Activation energy level with enzyme
Energy level of product
Relative energy
A
B
C
Reaction:
Substrate → Products

Figure 48–3 Energy diagram of reactants yielding products. A = activation energy necessary without enzyme (heavy line). B = activation energy necessary with enzyme (dashed line). C = energy difference between reactants and products. Note that the reaction proceeds in favor of products that have a lower energy level and are assumed to be more stable.

uncatalyzed reaction can occur only by overcoming a rather high activation energy. This is the *collision energy* with which the substrate and water molecule come together to bring about the hydrolysis of the polysaccharide. If the two reactants do not interact with sufficient energy to overcome the barrier of the activation energy, no products will form. There would be no evidence of a reaction in this case. This would be as if the reaction started up the energy barrier along the heavy line but did not reach the summit (activation energy). The water and polysaccharide would be disengaged from each other and no reaction would be observed. If any collisions of water and polysaccharide are of sufficient energy to meet or exceed the activation energy, the reaction occurs. This accounts for the observations that heating a reaction usually increases its rate. Increased temperature is directly related to the increased velocity of molecules in the system. The faster a population of molecules moves, the more energy there is in their collisions. It follows that, if the activation energy is lowered, a greater proportion of any population of molecules (at a given temperature) could meet or exceed that barrier. Enzymes act to lower the activation energy. This is done by providing the specific site favorable to the reaction, and stabilizing the reaction intermediates for very short periods of time. This lowering of the activation energy allows a greater portion of reactants to form products.

It must be pointed out that the net energy difference between the substrate and products is unaffected by the activation energy. Also, this energy difference between products and reactants is the same in both directions of the reaction. The products shown here are more stable and are shown at a lower energy level. The energy-level difference between the products and reactants is independent of the rate of the reaction. The two aspects of rate and energy must be considered separately. The energetics of the reactions are based upon careful thermodynamic studies of the initial and final states of the molecules. The interest in the catalytic event is centered upon the rate or the kinetics of the reaction.

Nomenclature

Many of the enzymes known to be involved in living systems have been isolated and can be obtained commercially. One of the simplest nomenclature systems for enzymes is to use the name of the substrate and add the suffix *ase*. For example, maltase increases the rate of hydrolysis of maltose, and a peptidase increases the rate of hydrolysis of peptide bonds. Another naming system involves common or trivial names that were assigned to some enzymes many years ago and often remain the names of choice today. For example, pepsin is a digestive enzyme found in peptic juices, and papain is another proteolytic enzyme found in the papaya (the latter enzyme is often found as a component of commercial meat tenderizer).

As the number of enzymes under study increased, a more systematic nomenclature scheme was developed. Each enzyme was given four numbers which are keyed to classes and subclasses in the scheme. These classes are shown with representative examples.

1. Oxidoreductases or dehydrogenases (oxidation–reduction catalysis):

$$CH_3CH_2OH \longrightarrow CH_3\overset{\overset{H}{|}}{C}{=}O + 2H$$

2. Transferases (transfer of functional groups):

$$R-\underset{\underset{O}{\|}}{C}-\overset{\overset{O}{\|}}{C}-OH + R'-\underset{\underset{NH_2}{|}}{CH}-\overset{\overset{O}{\|}}{C}-OH \longrightarrow R-\underset{\underset{NH_2}{|}}{CH}\overset{\overset{O}{\|}}{C}-OH + R'-\underset{\underset{O}{\|}}{C}-\overset{\overset{O}{\|}}{C}-OH$$

3. Hydrolases (hydrolysis):

$$R-\overset{\overset{O}{\|}}{C}-O-R' + H_2O \longrightarrow R-\overset{\overset{O}{\|}}{C}-OH + HOR'$$

4. Lyases (removal of parts of molecules by nonhydrolysis):

$$CH_3\overset{\overset{\displaystyle O}{\|}}{C}-\overset{\overset{\displaystyle O}{\|}}{C}-OH \longrightarrow CH_3\overset{\overset{\displaystyle O}{\|}}{C}-OH + CO_2$$

5. Isomerases (intramolecular rearrangement):

$$\text{Glucose phosphate} \rightleftharpoons \text{fructose phosphate}$$

6. Ligases (bond formation coupled with ATP utilization):

$$\text{Amino acid} + \text{ATP} \longrightarrow \text{activated amino acid}$$

Isozymes

Some enzymes have been found to exist in multiple molecular forms within the same organism or even within the same cell. These multiple forms are called **isozymes** or **isoenzymes**. Although a given enzyme may catalyze a single reaction, the various different forms of the isozymes may be unique and located in different tissues. A good example of this is lactic acid dehydrogenase (LDH). This enzyme is involved in catalyzing the interchange between lactic acid and pyruvic acid. Some forms of LDH are unique to the heart muscles; others are localized in the smooth white muscles. The characteristic differences of isoenzymes have been used to advantage in a number of important clinical diagnostic studies.

Enzymes of Clinical Interest

Enzyme tests play a key role in a number of clinical diagnoses, including myocardial infarction (heart attack). The use of enzyme tests is especially important here because both angiographic findings and electrocardiographic findings may not be as complete as desired. Damage to a tissue such as the heart can release its unique enzymes into the serum. The enzymes frequently studied in the case of myocardial infarction are the following:

Aminotransferase (GOT, glutamic-oxaloacetic transaminase): This enzyme is elevated in most cases shortly after the onset of the attack. The level of the enzyme usually reaches a peak within 6 to 8 hr and returns to normal within several days. The levels of this enzyme are included in the total enzyme study of Figure 48–4.

Phosphotransferase (CPK, creative phosphokinase): Myocardial infarction leads to increased serum levels of this enzyme in about 90% of these cases. The levels of this enzyme increase quite rapidly, then slowly decline, as shown in Figure 48–4. CPK levels can be found to be elevated in several other disease situations as well. However, raised CPK levels are not found in liver diseases. This then aids in distinguishing cardiac disorders from liver diseases, which will not interfere with this diagnostic study. Also, normal CPK levels are found in the case of pulmonary embolism, which might mimic a heart attack.

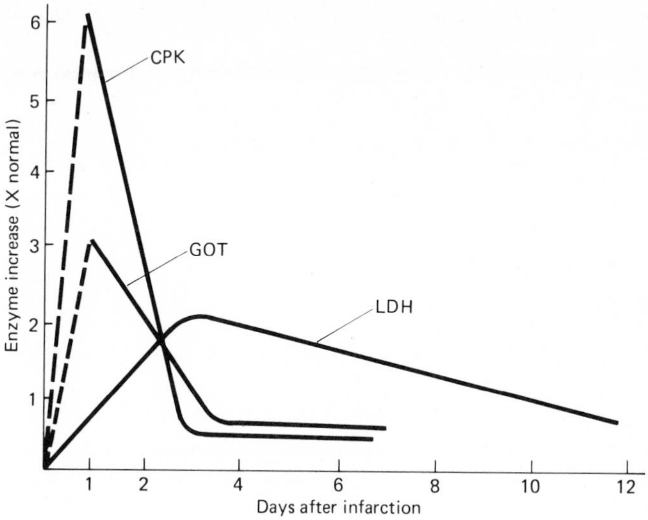

CPK
GOT
LDH
Enzyme increase (X normal)
Days after infarction

Figure 48–4 Typical changes in some serum enzyme levels following myocardial infarction.

Dehydrogenase (LDH, lactic acid dehydrogenase): The level of the activity of the specific heart isozymes of this enzyme is almost always elevated from two to ten times after a myocardial infarction. The increase in activity is at a slow rate, but usually remains abnormally high for longer periods of time. Therefore, the determination of the levels of this enzyme is important if blood samples are not available for some time after the initial heart attack.

As valuable as these enzyme studies are in clinical diagnosis, they have their limitations. These studies are indicative of the state of health of a patient and of the direct application of basic biochemistry to clinical investigations.

UNIT 48 Problems and Exercises

48–1. If an apoenzyme and a coenzyme were placed inside of a dialysis bag, which of the two would pass through to the outside?

48–2. Since lysozyme can destroy some bacteria, comment on its use as a potential oral antibiotic.

48–3. If lysozyme was subjected to an oxidizing environment, would you expect any possible changes in the molecular form of the enzyme? Explain.

48–4. Sometimes metal surfaces can act as catalytic sites. Contrast this catalysis with that brought about by an enzyme.

48–5. If a patient entered a hospital several days after the onset of a heart attack, which enzyme assay would be used in the diagnosis?

48–6. Holoenzymes and isoenzymes both consist of two or more parts. How could you distinguish if a two-part system was either a holoenzyme or an isoenzyme?

48–7. Give a specific name that would describe the enzyme which would catalyze the reaction shown in the first nomenclature class.

48–8. Which nomenclature class would be appropriate for a peptidase?

48–9. Draw a graph similar to the one shown in Figure 48–3 and label it with the specific participants of the lysozyme-catalyzed reaction.

CHAPTER 17

Biocatalysts and reaction control

When the great variety of reactions is considered, it is obvious that cells are controlled and organized in a highly specialized way. Reactions can be accelerated or speeded up by living systems. For example, a sugar solution may stand for long periods of time (assuming no contamination). However, once the sugar solution is ingested, it undergoes a reaction due to biocatalytic activity. Heat can also speed up reactions. Cooking may accelerate many chemical reactions, but living systems cannot tolerate very wide variations in temperature. Strong acids or bases commonly increase the rates of some chemicals, but cells cannot tolerate very great changes in pH. Living systems must have unique control mechanisms. **Enzymes** are biocatalysts produced by living cells that alter the rates of the reactions that are chemically possible. Since they are catalysts, they are not consumed in the reactions with which they are involved.

We all have experienced a great period of growth within our lifetime. It was necessary to accelerate many of our biochemical reactions to provide this growth. What made this growth stop at a given point? Many of us have had small growths or tumors. In most cases the growths come under some sort of control. The key is reaction control at a molecular level.

Nature has selected the protein molecule to participate in this catalytic activity. The great diversity of biochemical reactions requires a great diversity of controlling factors. The 20 different amino acids make up the several million different proteins. Biochemists have identified over 1000 different enzymes and have purified about 200. If we can gain some understanding of how and why these remarkable molecules work, we will have opened the door to understanding the unique chemical and physical processes of life itself.

UNIT 49 FACTORS INFLUENCING THE RATES OF ENZYME-CATALYZED REACTIONS

Enzymes function only to alter the rates of reactions. They participate as catalysts in the reaction, but are not involved as either the substrate or ultimate product of the reaction. Because of this, only very small amounts of enzymes are necessary to bring about the desired response. This causes difficulties in determining the presence of an enzyme in a complex system of study. Since enzymes are proteins, it might seem that you should be able to determine the protein concentration present and thus know the enzyme concentration. Often this cannot be done. What must be done in most cases is to determine the changes in concentration of the substrates or products. The decrease in substrate concentration in a given period of time is given the symbol $-dS/dT$. The S refers to the concentration of the substrate, usually measured in molar terms. The T is a measure of the time involved. The d symbol is used to denote a small change. Since the quantity of substrate decreases as the reaction proceeds, the negative sign is used. The amount of product formed in a given period of time is given the symbol dP/dT. Note the new dimension: **time**.

Both dS/dT and dP/dT are changes in concentration within a given time interval. These specify the **rate** or **velocity** of a reaction, for example, the change in substrate concentration per minute. Notice that this is a rate in the same sense that in driving a car, we use miles per hour. This means change of position, from one point to another, in a given period of time. The enzymatically catalyzed reactions are measured in the same way. Velocity denotes the activity of the enzyme, as shown in Figure 49–1.

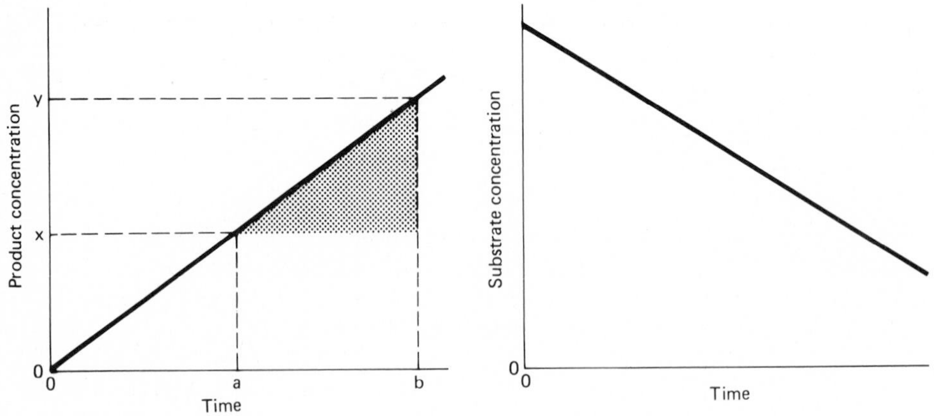

Product concentration
y
x
0
0
a
b
Time
Substrate concentration
0
0
Time

Figure 49–1 The formation of products or the loss of substrate as a reaction proceeds in time.

The product is absent (zero concentration) at the beginning of the reaction (zero time). As the reaction proceeds, the product forms. In the time interval from time a to time b, the concentration changes from value x to value y. The shaded triangle shows these differences. The velocity or rate of the reaction is given by

$$v = \text{velocity} = \frac{dP}{dT} = \frac{(y - x)\ \text{moles}}{(b - a)\ \text{minutes}}$$

The rate of loss of the substrate ($-dS/dT$) is assumed to be equal to the amount of product formed during this time. In measuring the activities of enzyme-catalyzed reactions, the initial velocities are most important. This involves only what happens during the first part of the reaction.

Activity Units

The standard term used to characterize the activity of an enzyme is the unit of activity. One **unit** (U) of activity of any enzyme is the amount that will catalyze the transformation of 1 μmol of substrate /min. The term **specific activity** is used to take into consideration the protein nature of enzymes. The specific activity is expressed as units of enzyme per milligram of protein. This term can be useful in studies involving the purification of a protein from a natural source. A given sample of tissue would have a total number of units of activity within it. The sample of tissue would have a great many proteins in it besides the enzyme of interest, however. As the purification of the enzyme is carried out, the specific activity increases. The increase is due to getting rid of other proteins until only a purified enzyme remained. A sample of pure enzyme would not show an increase in specific activity upon an attempt to purify it.

Another very useful term for the comparison of activities of different enzymes is the **turnover number**, which is defined as the moles of substrate undergoing reaction per mole of enzyme per minute under carefully defined experimental conditions. These numbers vary from several thousand to several million for the very active enzymes. From this it is obvious how a few enzyme molecules influence the reaction of an enormous number of substrate molecules.

Temperature Effects

One important factor that influences the rate of an enzyme reaction is the temperature. An increase in temperature generally increases the rate of a chemical reaction. This is seen as the rising part of the curve in Figure 49–2. An optimum temperature is reached, as shown by the maximum in this figure. At some temperature above this point, the rate of the reaction is seen to decrease sharply. This is due to a denaturation of the enzyme brought about by the elevated temperatures. No actual temperature values are given on this graph because the activity of an enzyme varies with the enzyme being considered. Most enzymes of importance to us have reasonable activity at body temperatures. It is possible to find a rather broad range of optimal temperatures for various enzymes found in nature. For example, enzymes are active in microorganisms found in pools near geysers and boiling mud pots. On the other hand, living systems are known to remain viable at quite low temperatures.

Most enzymatic reactions are carried out at constant temperatures. This is of great importance in the study of enzymes of clinical significance when a comparison is

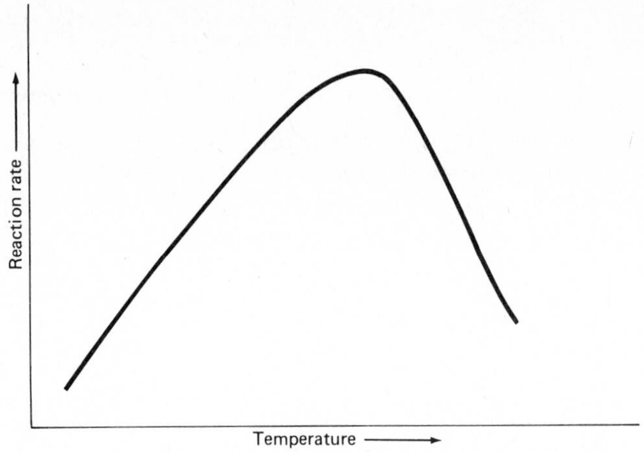
Reaction rate
Temperature

Figure 49–2 The influence of temperature on the rate of an enzymatic reaction.

made between normal and abnormal samples. As might be expected, enzymes that are studied at somewhat higher temperatures may have a more limited time of action. Even moderate temperatures may lead to the gradual denaturation of some enzymes.

Heat has often been used as a simple test for the presence of enzymatic activity. If a reaction is suspected of being catalyzed by an enzyme, a simple test can prove it. Place the reaction under study in a boiling water bath for a short period of time. If an enzyme is present, usually it will be denatured and the reaction rate will drop sharply. Generally, reactions catalyzed by nonenzymatic catalysts can withstand such treatment.

pH Effects

The pH and ionic composition of the reaction mixture also influence the rate of an enzymatic reaction. Most enzymes have an optimum pH value shown as the maximum of the curve in Figure 49–3. This curve is usually found to be rather symmetric about the pH optimum. The control of pH is necessary to maintain the proper charge of the protein portion of the molecule and the substrate itself. Once again, no specific pH values are given on this figure as each enzyme has its own specific pH-dependent profile. Many enzymes of interest to us do have an optimal value near pH 7, but there are some notable exceptions. For example, the stomach contains an enzyme, pepsin, that acts to hydrolyze proteins. The pH of the stomach fluid is acidic, and pepsin has a pH optimum ranging from 1.5 to 4.0. Nature seems to have designed enzymes for their specific locale of action.

Enzyme-Concentration Effect

The rate of an enzymatic reaction is dependent upon the concentration of the enzyme as shown in Figure 49–4. The rate of the reaction is directly related to the amount of enzyme present, if there are no other limiting factors. The substrate must be present

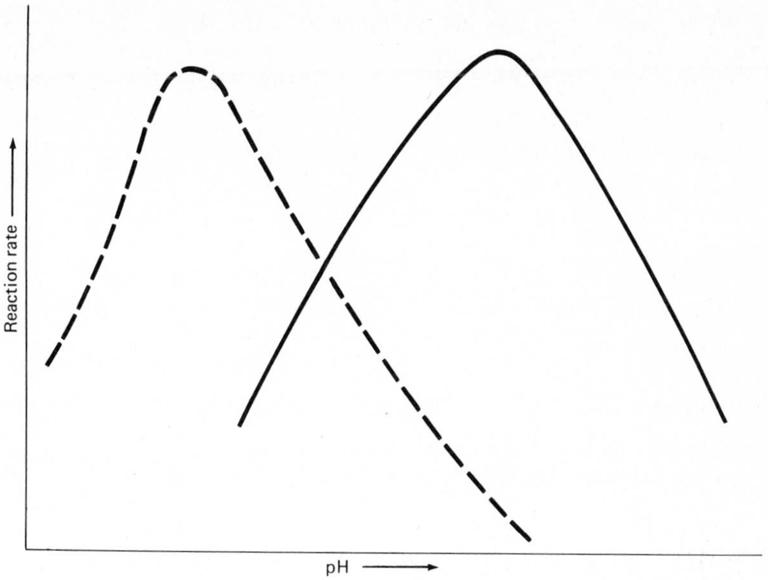
Reaction rate
pH

Figure 49–3 The influence of pH on the rate of an enzymatic reaction. The solid line represents pH rate profile for trypsin and the dashed line represents a pH rate profile for pepsin.

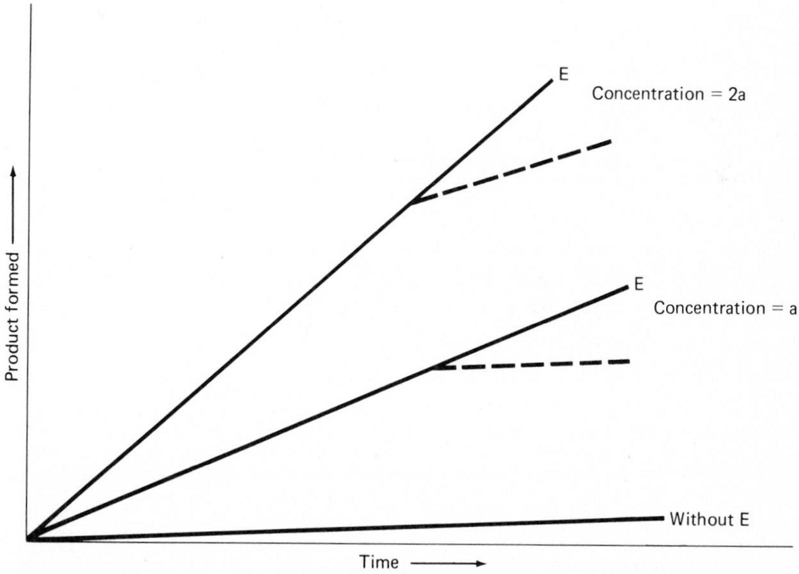
E
Concentration = 2a
E
Concentration = a
Without E
Product formed
Time

Figure 49–4 The influence of two enzyme concentrations on the observed reaction rates. A comparison is made with the reaction rates without an enzyme. The dashed lines indicate the expected results if the reaction rate becomes limited as the substrate becomes used up.

in sufficient amounts so that its concentration will not limit the rate of the reaction. The temperature and pH must be carefully controlled for reproducibility.

The influence of two enzyme concentrations is given in Figure 49–4. The solid lines show that an increase in enzyme concentration increases the reaction rate. The dotted lines illustrate what might be expected if the substrate supply became exhausted as the reaction progressed. Generally, only the early part of the reaction where a straight line is found is important (initial velocity).

Substrate-Concentration Effect

As observed, the substrate concentration can influence the rate of an enzymatic reaction. If a series of reactions is carried out under carefully controlled conditions where all other factors are held constant, the increase in substrate concentration would increase the reaction rates as shown in Figure 49–5. Note that, as the substrate

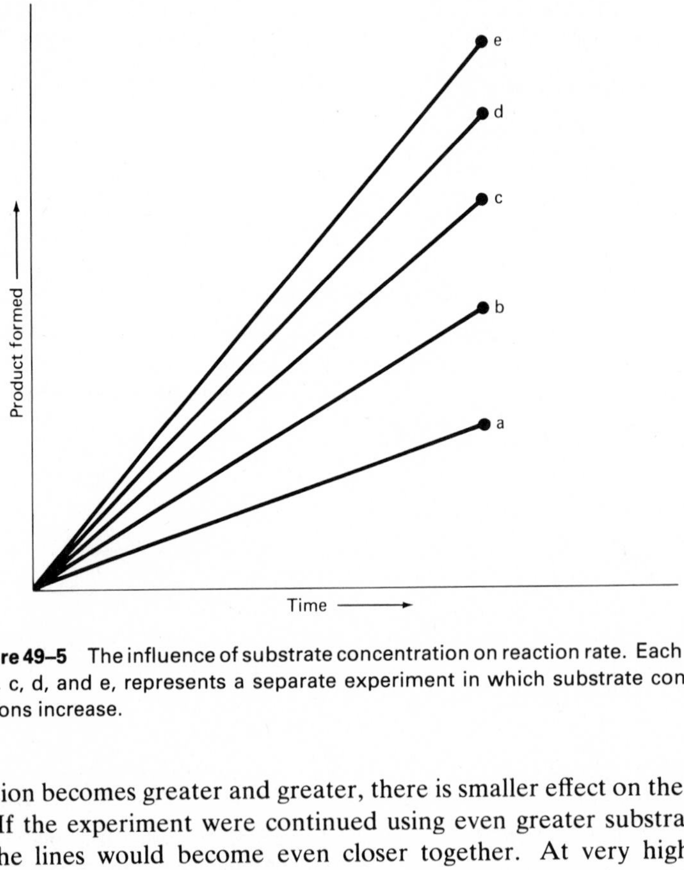
e
d
c
b
a
Product formed
Time

Figure 49–5 The influence of substrate concentration on reaction rate. Each line, a, b, c, d, and e, represents a separate experiment in which substrate concentrations increase.

concentration becomes greater and greater, there is smaller effect on the rate of the reaction. If the experiment were continued using even greater substrate concentrations, the lines would become even closer together. At very high substrate concentrations, additional amounts of substrate would not lead to increased reaction rates (all other factors must be held constant).

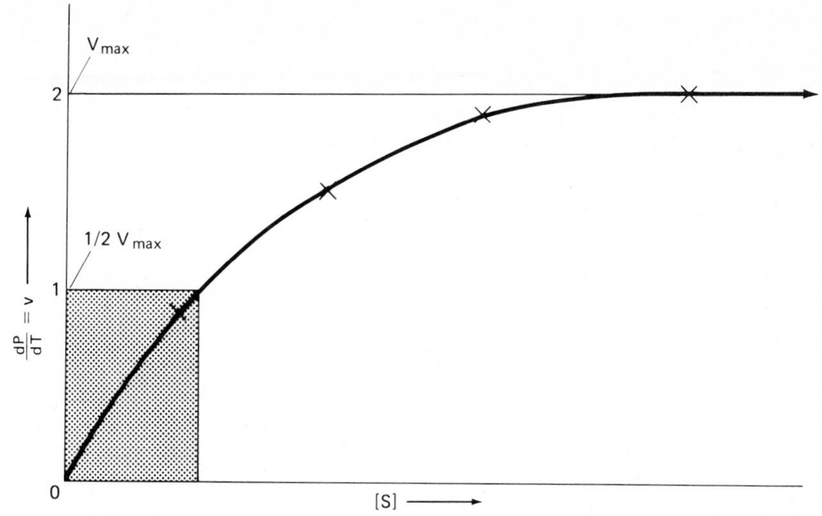
V max
2
1/2 V max
1
0
dP/dT = v
[S]

Figure 49–6 Effect of substrate concentration. S = molecular concentration versus initial reaction velocity (v). The origin represents a zero velocity when no substrate was used ($S = 0$). The other points marked on the curve represent the increasing substrate concentrations used in Figure 49–5.

This can be visualized if the initial velocity (dP/dT) is plotted against increasing substrate concentrations, as shown in Figure 49–6. All factors such as temperature, pH, and enzyme concentration remain constant during these experiments; only the substrate concentration is changed. Note again that at very high substrate concentrations there does not appear to be a very great increase of the rate of reaction. At this high substrate concentration, the reaction is said to reach a **maximum velocity** (V_{max}). At this point the enzyme is said to be *saturated* by the substrate. The enzyme molecules are all occupied with the catalytic action, and there are essentially no enzyme molecules free of substrate.

To better understand this situation, return to the simplified situation:

$$E + S \rightleftharpoons ES \longrightarrow E + P$$

At V_{max} the maximum amount of enzyme is converted to ES. There can be no change in rate upon the addition of more substrate. This is known as a **zero-order** condition for the substrate. It is at this point that clinical assays are conducted. In the case of clinical investigations the enzyme must be operating under its optimal conditions, unlimited by the amount of substrate present. This is now a point where activity is no longer dependent upon small changes in substrate concentration.

During the first part of the study in Figure 49–6, the increased substrate concentrations lead to increased velocity. In the simplest case, when there is a direct relationship between substrate concentration and velocity, the reaction is said to be **first order** with respect to substrate concentration, and can be given as

$$v = k[S]$$

(the k is a rate constant for this first-order reaction).

Michaelis Constant

The reaction can be pursued one more step to determine an important factor used to characterize an enzymatic reaction. At the point of $v = 1/2\,V_{\max}$, shown in the shaded part of Figure 49–6, the substrate concentration is equal to the **Michaelis constant** (K_m). This can also be shown by mathematical expressions to be equal to the substrate concentration that yields half of the maximum velocity. The K_m provides information about an enzymatic reaction in the same general way that an equilibrium constant describes a chemical reaction. In the case of enzyme reactions,

1. Small K_m values indicate that the enzyme requires only small amounts of substrate to become saturated.
2. Large K_m values indicate that a great amount of substrate is necessary to reach the V_m.
3. When related substrates are used in an enzymatic system, the substrate with the smallest K_m is usually assumed to be the "best" or most natural one for the system.

Tables of K_m values are available to describe enzyme systems.

Inhibitors

Enzymes are known to be influenced by the presence of inhibitors. These are factors that decrease the rate of the reaction or even stop the reaction completely. Although there are several types, inhibitors may be placed into two broad categories: reversible and irreversible.

One type of reversible inhibitor is known to display **competitive** inhibition. This can be illustrated by adding a competitive inhibitor (I_c) in the reaction scheme:

$$E + S \rightleftharpoons ES \longrightarrow E + P$$

$$E + I_c \rightleftharpoons EI_c \quad \text{(no product formation possible)}$$

The I_c forms a complex EI_c with the enzyme in a reversible manner. The amount of enzyme tied up in this complex is not available to form the ES necessary as an intermediate in the desired reaction. If a lower concentration of enzyme is available for the reaction, the rate of the reaction decreases. Both the substrate (S) and this inhibitor (I_c) compete for a position at the active side of the enzyme. In this competition the rate of the desired reaction is dependent upon the relative concentrations of substrate and inhibitor. At very low I_c concentrations, the competition of the inhibitor has very little influence on the reaction rates. As the I_c concentration increases, most of the enzyme can be combined in an inactive EI_c complex, which slows or even stops the reaction. In most of these cases, the competitor is quite closely related to the substrate on a molecular level. The classic case of a competitive inhibitor is malonic acid as it competes with the natural substrate succinic acid in the succinic dehydrogenase system of the Krebs cycle. An examination of the structures of these compounds shows how an enzyme might recognize either molecule at its active site.

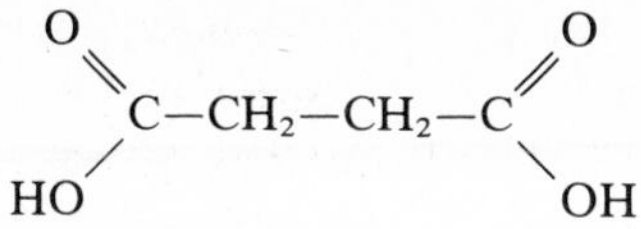

succinic acid
(natural substrate)

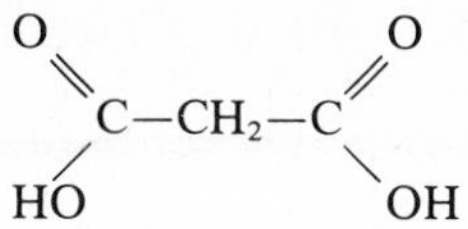

malonic acid
(competitive inhibitor)

Noncompetitive inhibition is not reversible by increasing substrate concentration, even though it may involve a binding of the inhibitor to the enzyme in a reversible fashion. In this case, the noncompetitive inhibitor cannot be readily displaced by increasing the substrate concentration. One example of this type of inhibition is shown by reagents when they attach themselves to the sulfhydryl groups (−SH) at the active sites of enzymes. The sulfhydryl groups would be the side chains of the amino acid cysteine on the surface of the protein portion of the enzyme. Heavy-metal ions (Ag^+, Hg^{2+}) also act as inhibitors of this type according to the following:

$$\text{E–S–H} + Ag^+ \rightleftharpoons \text{E–S–Ag} + H^+$$

The E–S–Ag would not be able to become involved in the normal ES complex necessary to bring about the desired reaction. To reverse this inhibition, it is necessary to remove the Ag^+ by some chemical method.

An irreversible inhibitor is seen in the case of a toxic organophosphorous compound that is classed as a nerve poison. It reacts to form a complex that is irreversibly inactive

$$E + I_{irr} \longrightarrow EI_{irr} \quad \text{(inactive)}$$

These compounds may react irreversibly with serine side chains that are known to be involved in enzymatic reactions of nerve transmission. This inhibitor is extremely potent; less than 1 mg would be the lethal dose for an average adult.

Proenzymes or Zymogens

A dilemma arises when some degradative enzyme systems are considered. For example, trypsin is an enzyme that catalyzes the hydrolysis of proteins in the intestinal tract. This enzyme is synthesized in the pancreas and transferred to the intestine, where it is involved in proteolytic activity. The problem is: What keeps trypsin from bringing about the digestion of the cells responsible for its formation? Nature has designed a specific way to circumvent this problem by the formation of proenzymes or zymogens. The **proenzymes** are inactive precursors of the active enzyme systems. They usually are given the name of the active enzyme with the added suffix *-ogen*. Trypsinogen is the proenzyme of trypsin. Trypsinogen is a single peptide chain with the same composition as trypsin, plus as additional hexapeptide at the amino-terminal end of the molecule. The proenzyme is the form in which the enzyme system is actually formed within the pancreas and transported to the intestine. As long as the hexapeptide remains attached, no proteolytic activity is possible. As soon as it is removed, by a simple hydrolysis reaction, the active trypsin molecule is available for normal reactions. Once again, specific molecular architecture allows for the necessary reactions to take place in nature in the right way at the right time.

UNIT 49 Problems and Exercises

49–1. What would the term dES/dt represent? How would this be related to dP/dt?

49–2. Explain why we measure the presence of an enzyme in units rather than in terms of molar concentrations.

49–3. Why are we interested in initial velocities rather than the velocities near the completion of a reaction controlled by an enzyme?

49–4. Explain how the units of enzymatic activity of a given sample could remain constant upon the purification of the enzyme, whereas the specific activity could increase.

49–5. Could you think of a situation in which the units of activity could seem to increase during a purification process without changing the specific activity?

49–6. Would you expect that digestive enzymes would have a higher or lower turnover number than those enzymes involved in respiration processes?

49–7. Briefly explain why the pH influence graph is symmetrical, and why the temperature influence graph is not symmetrical.

49–8. It is known that the product of some enzymatic reactions can act as inhibitors. Draw a graph of product formed versus time to show this possibility.

49–9. An enzyme is known that catalyzes a certain reaction on any one of several monosaccharides. The K_m value for this reaction using glucose is 1×10^{-5}, and using fructose it is 1×10^{-3}.

(a) Draw a graph of initial velocity versus substrate concentration for each of these monosaccharides.

(b) Which of these monosaccharides would be considered the "best" for the reaction?

49–10. How would you determine the difference between a competitive and noncompetitive inhibitor in the laboratory?

UNIT 50 COENZYMES AND WATER-SOLUBLE VITAMINS

Vitamins are small organic molecules that must be supplied in the diet. Those vitamins soluble in the nonpolar solvents are designated fat-soluble vitamins (A, D, E, and K). The other major class of vitamins is water-soluble in most of their chemical forms. This distinction may not seem important at first, but there is some nutritional basis for it. For example, low-fat diets may be sufficient in water-soluble vitamins but deficient in fat-soluble vitamins. Because of the common solubility properties, some studies involving the isolation of the water-soluble vitamins yield complex mixtures of many molecules. An example of this is the B vitamin group, which is now known to be composed of many different vitamins. The following are members of the water-soluble group: niacin (B_5), riboflavin (B_2), thiamine (B_1), pyridoxine (B_6), ascorbic acid (C), pantothenic acid (B_3), biotin (B_7), folic acid (B_9), *p*-aminobenzoic acid, and cobalamin (B_{12}). The chemical names are now most frequently used for classification.

Niacin

Around the turn of the last century, biochemists became able to characterize **dietary deficiency** diseases. At about this same time, coenzyme functions were identified as participants in enzymatic reactions. The first coenzyme was isolated by the British biochemists, Harden and Young, early in this century. This coenzyme was named cozymase, and was identified chemically as **nicotinamide adenine dinucleotide** (NAD) in the 1930's by Otto Warburg and coworkers in Germany. The link between vitamins and coenzymes became established chemically when it was recognized that one portion of the NAD, nicotinic acid, was essential in animal nutrition. It is now known that coenzymes are the chemically altered forms or derivatives of the vitamins that participate in enzymatic reactions.

The vitamin that is part of the NAD complex is **niacin**. Another related form of this vitamin is **nicotinic acid**.

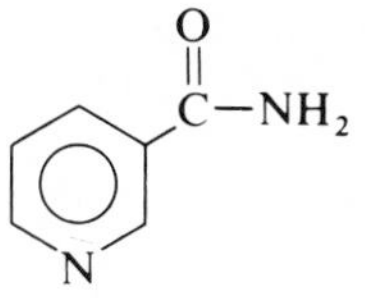

O
C—NH2
N

nicotinamide

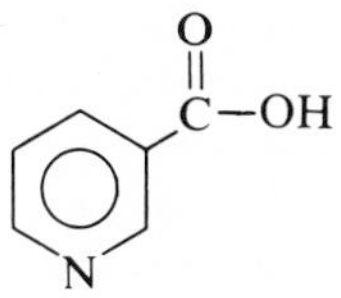

O
C—OH
N

nicotinic acid

The heterocyclic ring is known as pyridine, and sometimes this name has been applied to enzymes bearing this vitamin.

Pellagra, a specific disease related to the deficiency of this vitamin, has been recognized for two centuries. It may be hard to believe that pellagra was rather widespread in the United States as late as the first quarter of this century. The recommended daily allowance of 20 mg of niacin should be ample for healthy adults. Rich sources of this vitamin include both plant and animal tissues, with meat products being especially important. There is a complicating factor in evaluating nutritional requirements of this vitamin. It has been shown that the amino acid tryptophan can be converted to niacin in limited amounts. This then provides a **sparing effect** of this amino acid for this vitamin. If a diet is especially deficient in proteins that contain tryptophan, special care must be taken to ensure adequate niacin intake.

Niacin has been found to be the active component of two coenzymes, NAD and NADP (**nicotinamide adenine dinucleotide phosphate**). Both coenzymes function in redox reactions involving the transfer of electrons. The structure of NAD in its oxidized and reduced forms is

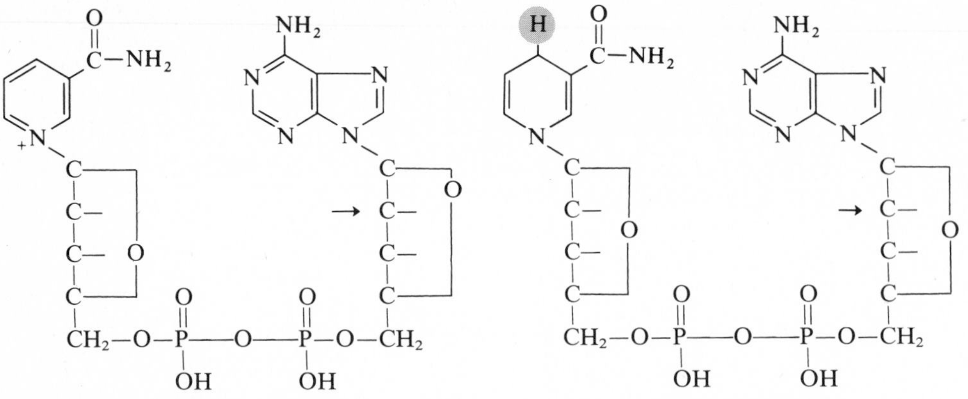

H

NAD^+ (oxidized form) NADH (reduced form)

The coenzyme is bound to a specific site on the protein portion of an enzyme (apoenzyme) to make the complete functioning holoenzyme. In some cases, divalent metal ions have been found to bind the coenzyme to the apoenzyme portion of the system. This is a good illustration of the way in which the inorganic ions and organic molecules act in concert to provide the highly complex systems necessary in living processes.

It should be pointed out that in the reduction of NAD to NADH there is a transfer of one hydrogen (shown as shaded portion of NADH) and a pair of electrons. Many different substrate molecules react with the specific dehydrogenase enzymes bearing the NAD. The substrates undergo oxidation as this enzyme accepts the electrons and becomes reduced. The oxidized form of the coenzyme is regenerated by a subsequent reaction passing the electrons on to another acceptor molecule. The coenzyme acts in a catalyic role as a transfer agent for the electrons and hydrogen.

The second member of this family is nicotinamide adenine dinucleotide phosphate (NADP). NADP differs from NAD in that a third phosphate is involved. The 2′ hydroxyl of the ribose linked to the adenine is replaced by a phosphate group in NADP. This is marked with an arrow on the structure of NAD and NADH. The NADP can also undergo reversible oxidation–reduction reactions:

$$NADP^+ \text{ (oxidized)} + 2e^- + H^+ \rightleftharpoons NADPH \text{ (reduced)}$$

NADPH plays an important role as a biological reducing agent. It is especially involved in biosynthetic pathways in which oxidized molecules must be reduced. Some pathways using this reducing ability of NADPH include the formation of saturated fatty acids and steroids. It is interesting to notice how the reduced NADPH is formed in the first place. Some catabolic pathways involving the oxidation of substrates seem to prefer NADP rather than NAD as the primary electron acceptor. This is true of the metabolic pathway of glucose metabolism known as the pentose phosphate pathway. As expected, this pathway of glucose metabolism is found in tissues where NADPH is needed. The liver (fatty acid biosynthesis), adrenal glands (steroid biosynthesis), and mammary glands (milk production) have the ability to form and utilize NADPH.

Riboflavin

Another pair of coenzymes involved in oxidation reduction reactions are the **flavin** dehydrogenases. In this case, the vitamin is **riboflavin** (B_2). The structure of this yellow-colored vitamin is

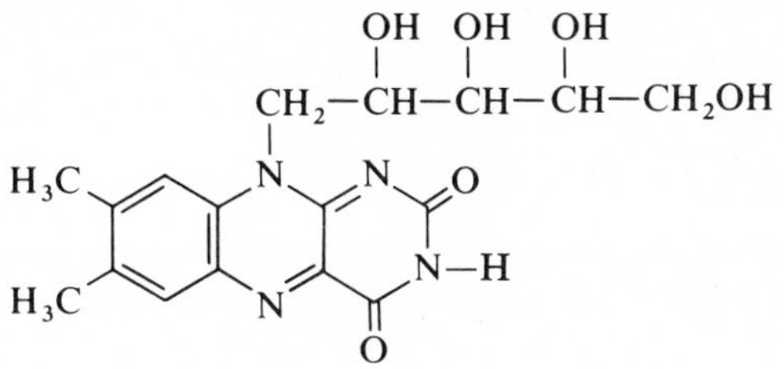

The specific disease or clinical manifestation of the avitaminosis associated with this vitamin is not easy to describe because there are multiple signs associated with its deficiency. This is often true for the other water-soluble vitamins. Some of the general symptoms usually involve poor growth, impaired reproductive functions, rough skin, and problems involving the eyes. Because there is a problem of precisely defining the avitaminosis, it is also difficult to specify exactly the amount to be included as an adequate dietary intake. The recommended daily allowance for B_2 is usually given as between 1 and 2 mg for adults. Good dietary sources include meat (especially liver), yeast, eggs, and leafy vegetables.

In the mid 1930's, Warburg and his coworkers were responsible for the chemical identification of the coenzymes derived from this vitamin. Two distinct coenzymes were ultimately characterized. One is known as **flavin mononucleotide** (FMN) and is shown in both its oxidized and reduced forms:

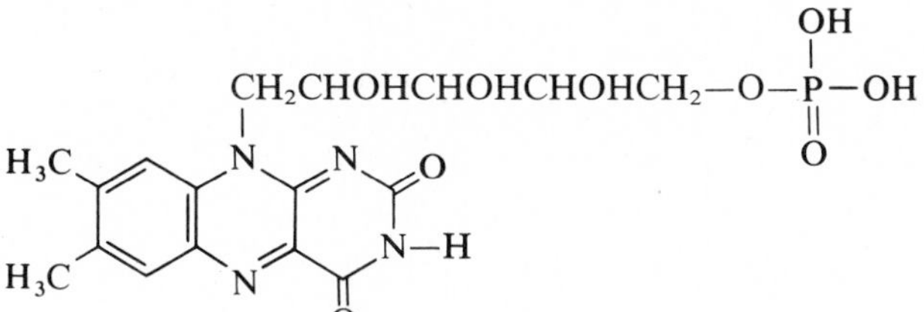

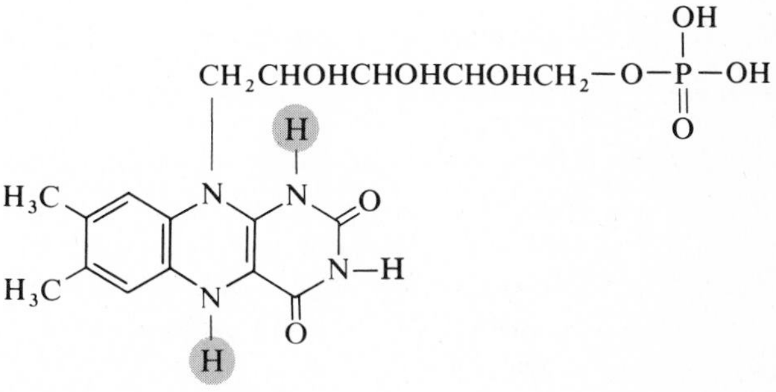

In this case the reduction of the molecule involves accepting two electrons and two hydrogens. The added hydrogens are shaded in the reduced form. One of the most important sources of these electrons in this reduction process is NADH. This is a portion of an electron-transport scheme that will be considered later. Other common organic molecules may directly contribute their electrons to flavin coenzymes. Examples include succinic acid in the Krebs cycle and fatty acids undergoing dehydrogenation in the β-oxidation spiral.

A second form of this coenzyme is known as **flavin adenine dinucleotide** (FAD). The structure of FAD can be visualized by adding an adenine nucleotide portion to the phosphate of FMN. This is related to the structure of the dinucleotide form of NAD. The same oxidation–reduction reactions occur in FAD as in FMN. Metal ions, especially Fe and Mo, are frequently found in the complete and active holoenzyme forms of these coenzymes.

Thiamine

Thiamine (B_1) was described previously as an essential nutrient. Beriberi in human subjects and polyneuritis of birds characterize this particular avitaminosis. The recommended daily allowance of this vitamin is about 0.5 mg/1000 Cal of dietary intake for adults. The direct relationship between the requirement for this vitamin and its coenzyme role in metabolism is seen by specifying the daily allowance in this manner. Whole grains and meat products, especially pork, are rich sources of this vitamin.

Both the vitamin and coenzyme forms are given as

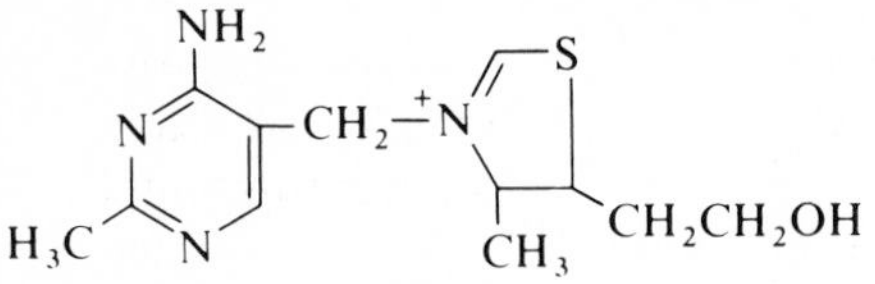

NH2
N
CH2
N
S
H3C
N
CH3
CH2CH2OH

thiamine

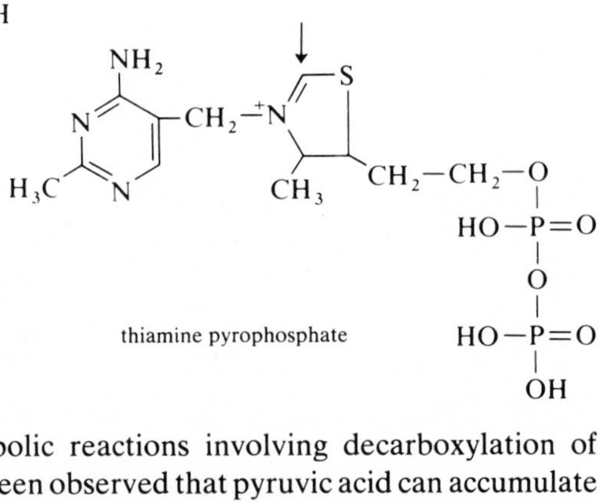

NH2
N
CH2
N
S
H3C
N
CH3
CH2—CH2—O
HO—P=O
O
HO—P=O
OH

thiamine pyrophosphate

This coenzyme is involved in metabolic reactions involving decarboxylation of α-keto acids like pyruvic acid. It has been observed that pyruvic acid can accumulate in the blood and tissues of thiamine-deficient animals. The involvement of the coenzyme is at the reactive site shown by the arrow on the structure. This will be considered in carbohydrate metabolism in which pyruvic acid is one of the specific intermediates.

Pantothenic Acid

The specific nutritional role of **pantothenic acid** has not been well documented for man, even though its need seems well established. Its role as part of a coenzyme is certainly well established. Like other water-soluble vitamins, meat and eggs are rich sources that supply the average 5 to 10 mg usually taken in by adults. This level seems to be adequate.

Pantothenic acid is part of the **coenzyme A** molecule. The vitamin is shown as the encircled portion of coenzyme A:

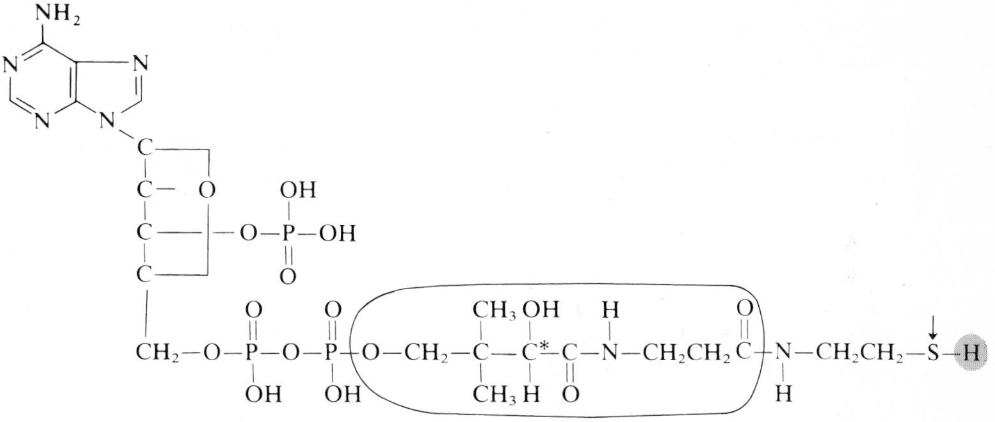

Part of the pantothenic acid molecule is ***β*-alanine**, a different form of an amino acid. The other portion of the pantothenic acid molecule is called **pantoic acid**. It should be pointed out that only one of the possible isomeric forms (at the position marked by the asterisk) yields a coenzyme A molecule with biochemical activity. The adenine nucleotide appears again in this coenzyme. One uniquely placed phosphate group on the ribose is also needed for biochemical potency. The chemically active portion of the molecule involves the sulfhydryl end marked by an arrow. Coenzyme A is involved in the transport of the acetyl group. In this case a $CH_3\overset{O}{\overset{\|}{C}}—$ group is substituted for the shaded hydrogen of the thioethanol amine. The coenzyme A is then converted to **acetyl coenzyme A**, or **active acetate** as it is sometimes known. The $CoA—S—\overset{O}{\overset{\|}{C}}—CH_3$ is involved as a key intermediate in a great variety of both catabolic and anabolic reaction schemes. Among the important metabolic reactions involving coenzyme A are the reactions involving degradation and biosynthesis of fatty acids, the reactions involved with the formation of citric acid in the Krebs cycle, and the reactions leading to the biosynthesis of the steroids. As a rule, coenzyme A is a participant in reactions involving the metabolism of two carbon units.

Pantothenic acid also appears to be a functional part of a system involved with the biosynthesis of fatty acids. In this case, pantothenic acid is attached directly to the enzyme at the active site of the reaction.

Coenzyme A can be used to illustrate an economic aspect of biochemistry. Since it functions in a catalytic capacity, only very small amounts are needed for biochemical studies. This is fortunate, since the current price for coenzyme A is \$50/100 mg (about \$225,000/lb). If we want to continue the research that is essential to our understanding of living processes and ultimately to the improved state of the health of mankind, there is a price tag to consider.

Pyridoxine

Another important member of the B vitamin family is **pyridoxine**. There are several forms of this B_6 family:

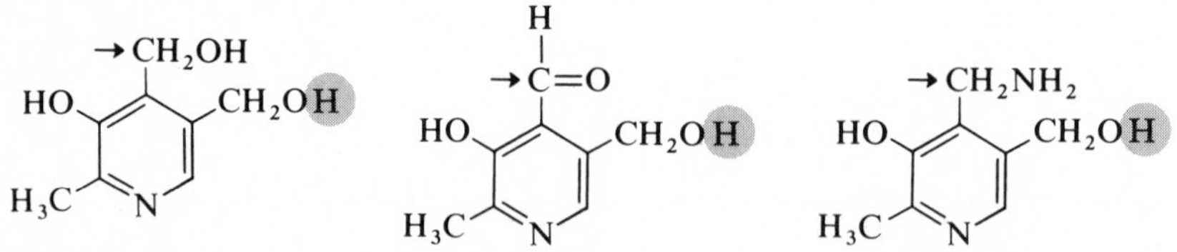

pyridoxine (pyridoxol) — pyridoxal — pyridoxamine

All three forms seem to be effective in animal nutrition, although there may be some species preferences for one form over another. Because of this interchangeable nature, none of these vitamins is *the* vitamin B_6. The first member of this family was isolated and synthesized in 1938. Since that time it has been established that a deficiency of this vitamin leads to a wide variety of undesirable results, including impaired growth and dermatitis in rats, anemic conditions in dogs, and atherosclerotic lesions in rats. No specific disease syndrome has been exclusively assigned to the avitaminosis of the vitamin B_6 family in man. Nevertheless, it is a necessary component of our diets at about 2 mg/day. Since the coenzymes derived from this vitamin are involved with amino acid metabolism, high-protein diets require a somewhat greater B_6 intake. As usual for the B complex vitamins, meat, eggs, and the germs of various grains are good sources.

The coenzyme forms derived from this vitamin have a phosphate group replacing the hydrogen atom, which is shaded on the structures shown. This would lead to pyridoxal phosphate, for example, as a coenzyme. The carbon marked by an arrow is the part of the molecule actually involved in the reactions of amino acid metabolism.

Biotin

Biotin is a vitamin with a rather remarkable characteristic. A biotin deficiency cannot be produced in most animals, including man, merely by a dietary deficiency. The bacterial flora of the intestinal tract seem to be able to supply our minimal needs under normal conditions. Besides this, biotin is found rather widely distributed in many common foods. To induce a biotin deficiency experimentally, special action must be taken. The intestinal tract can be sterilized to inhibit the microbial source. In other cases, a special chemical can be used that reacts uniquely with biotin to inactivate it. One material that is known to react specifically to inactivate biotin is the protein avidin. Avidin is found in egg white; it tightly complexes with biotin. Since cooking the egg denatures this protein and destroys its ability to complex with biotin, this usually poses no nutritional problem.

The structure of biotin is

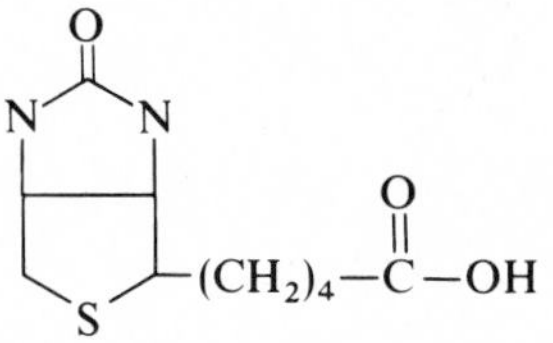
O
N
N
S
(CH$_2$)$_4$—C—OH
O

Biotin is required for biosynthesis of fatty acids.

Folic Acid

Folic acid, a **pteroylglutamic acid** (PGA), is a vitamin with the following structure:

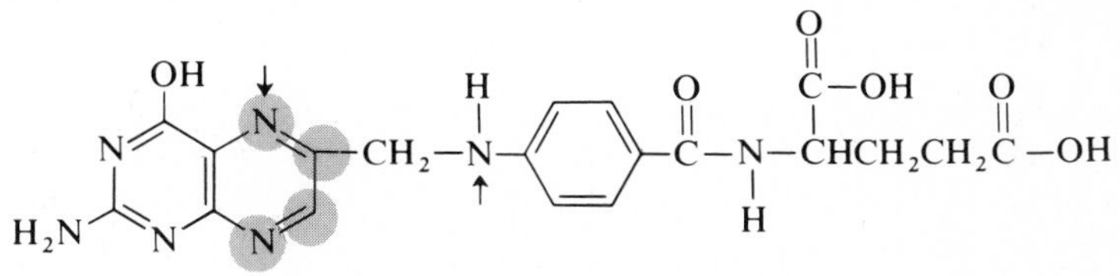
OH
N
N
N
N
H$_2$N
CH$_2$—N
H
O
C—N—CHCH$_2$CH$_2$C—OH
H
C—OH
O
O

Some vitamin forms are known with more than one glutamic acid residue in the molecule. The deficiency of this vitamin in mammals leads to impaired growth and various forms of anemia. Some organisms, especially microorganisms, do not require folic acid for normal growth, but only require the *p*-aminobenzoic acid portion. In these cases, one of the forms of a sulfa drug, sulfanilamide, can act as a competitive inhibitor to the action of this vitamin.

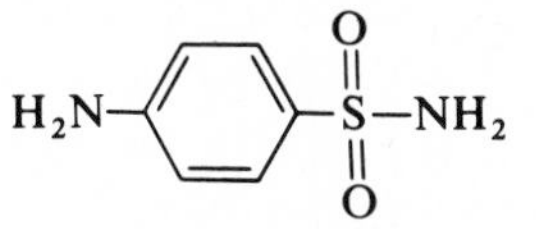
H$_2$N
S—NH$_2$
O
O

sulfanilamide

This molecule acts in a competitive fashion with *p*-aminobenzoic acid to inhibit the formation of a factor necessary for the microorganisms. Since man requires the intact folic acid as a vitamin, the essential reaction for the microorganism is blocked while the metabolism of the host is not greatly affected. This is the basis of the **antibiotic** effect of these kinds of compounds (antimetabolic).

The recommended allowance of folic acid is about 0.4 mg/day. It is rather widely distributed in meats and green leafy vegetables.

The coenzyme form of the vitamin is known as tetrahydrofolic acid (THFA) with four hydrogens attached to the atoms shaded in the structure shown for folic acid. Some of the important reactions catalyzed by this coenzyme involve the transport of single carbon atoms. Just as coenzyme A is involved with two-carbon metabolism, THFA is involved with one-carbon metabolism.

Vitamin B_{12}

One of the most recently discovered vitamins is vitamin B_{12}. In 1948 vitamin B_{12} was crystallized, and in 1957 its structure was elucidated. Vitamin B_{12} has several components. One is a nitrogen ring system that is similar to a nucleotide. A second part of the molecule is a four-membered ring system somewhat similar to that found in the heme portion of hemoglobin. This is called a **corrin** ring system. The name cobalamin is sometimes given to vitamin B_{12}.

Vitamin B_{12} deficiency is usually associated with pernicious anemia, but in a rather complicated manner. It is not a deficiency of vitamin B_{12} in the diet but a lack of absorption of the vitamin through the intestinal tract that is important. A special protein called the **intrinsic factor** is necessary in the intestine to cause the proper absorption of vitamin B_{12}. An individual could be vitamin B_{12} deficient even though there was an abundant supply of this vitamin in the diet. It is necessary to inject vitamin B_{12} to treat this form of anemia.

The efficiency of vitamins as coenzymes is most easily seen in the case of vitamin B_{12}. It has been estimated that the human body content of this vitamin is about 2 μmol. This could be less than 3 mg, or in other terms, less weight than one tenth of one drop of water!

Vitamin C

Vitamin C (**L-ascorbic acid**) is one of the most widely discussed vitamins in recent years. The deficiency of this vitamin produces the disease state known as scurvy. This syndrome encompasses undesirable changes of the gums, a tendency toward hemorrhage, and generally poor wound healing. All these are attributed to the role of vitamin C in maintaining the proper formation of collagen, the protein of connective tissue.

There has been a certain amount of controversy surrounding desirable or adequate amounts of intake of this vitamin. Although much higher values have been suggested by some scientists, 30 to 50 mg/day is probably sufficient to maintain adequate levels in adults. Larger amounts are undoubtedly excreted in the urine. It may be that there is a physiological level for the maintenance of normal levels, and a pharmacological level involved with the treatment or prevention of some diseases.

Fresh fruits and vegetables are excellent sources of this vitamin. One orange may contain 50 to 75 mg of a vitamin C. It is interesting that although vitamin C is a dietary requirement for man, other primates, and the guinea pig, it does not need to be supplied in the diet of other animals because they can synthesize vitamin C with their own enzyme systems.

In 1932 vitamin C was isolated, and its structure has been found to exist in both an oxidized and reduced form:

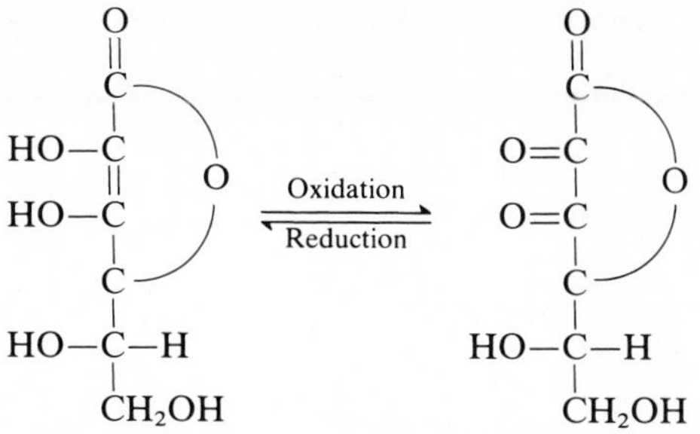
Oxidation
Reduction

Since ascorbic acid is known to undergo a redox reaction very easily, biochemists have investigated this as a possible role for the vitamin. Unfortunately few specific reactions have yet been shown to be specifically dependent upon ascorbic acid.

UNIT 50 Problems and Exercises

50–1. Draw the complete structure of NADH.

50–2. Indicate some cases where the specific composition of the diet can influence the requirements of a vitamin intake.

50–3. Write the reaction in which the appropriate form of the niacin-containing coenzyme reduces the flavin-containing coenzyme.

50–4. Suppose that the following compound was found in a growth medium for a microbiological system:

$$H_2NCH_2CH_2CH_2\overset{\overset{\displaystyle O}{\|}}{C}{-}OH$$

In which vitamin might this compound become incorporated?

50–5. Explain why it is proper to abbreviate acetyl coenzyme A as $CoA{-}S{-}\overset{\overset{\displaystyle O}{\|}}{C}{-}CH_3$ rather than just $CoA{-}\overset{\overset{\displaystyle O}{\|}}{C}{-}CH_3$.

50–6. Draw the coenzyme pyridoxal phosphate.

50–7. If a certain biochemical system could be inhibited by adding avidin, what vitamin would you expect to be operating in this system?

50–8. Explain why, although biotin may be a required vitamin by humans, it might not necessarily be required in the diet.

50–9. Which coenzyme is tied with one-carbon metabolism? Which with two-carbon metabolism?

50–10. Suggest a ready biochemical precursor of vitamin C in those species for which it is not required in the diet.

CHAPTER 18

Bioenergetics

Water runs downhill. Heat moves from a hot body to a colder one. An ice cube will melt if we hold it in our hand. Natural materials decay upon exposure to the atmosphere over a sufficient period of time. All these observations are examples of energy changes. These changes are governed by the same rules of nature that must be followed by the reactions that take place within our bodies. The name **energetics** might be used to describe the overall bookkeeping system that relates these changes. Chemists usually use the name **thermodynamics** to describe some types of energy transformations; we will choose the term **bioenergetics** to describe the energy exchanges that take place in living systems. Since living systems are so complex, simplifying assumptions are necessary.

We have just finished considering enzymes that influence the rate of reactions. In bioenergetics we shall not be interested in the rate but rather in the energy exchanges of reactions. This means that we shall be interested in differences in energy levels of reactants and products of reactions. Within living systems, we must have reactions that require energy coupled with reactions that supply energy. Special molecules, like ATP, are important in this way as an energy-exchange currency.

The problem that we encounter in bioenergetics is quite obvious. How do we couple the energy gained from the food we eat with all the energy-demanding processes that living systems require? Somehow last night's dinner must drive today's muscular contractions, nerve transmission, biosynthetic processes, and so on. We must develop a systematic way of dealing with these exchanges before we can understand or appreciate biochemical systems.

UNIT 51 GENERAL BIOENERGETICS

It was shown earlier that the free energy change of a reaction (ΔG) was important from several points of view. The relationship of free energy with other energy factors was shown by the Gibb's equation:

$$\Delta G = \Delta H - T\,\Delta S$$

The ΔH, or **enthalpy** change, represented the heat exchange during the reaction. Under the restricted conditions of most biological situations, ΔH could also be considered the total energy exchange in the reaction. This can be visualized by rearranging the above equation to

$$\Delta H = \Delta G + T\,\Delta S$$

Here ΔH is the sum of two components, the **free** or **available energy** change (ΔG) and the **entropy** change (ΔS) factor at a constant temperature (°K). The entropy change is related to the changes in the order (randomness) during the reaction and represents energy unavailable for doing work. The natural changes that are spontaneous have positive entropy changes ($\Delta S = +$).

These factors relate to a reaction and the change that takes place between the products and reactants. All factors have the same relationship:

$$\Delta G = \text{free energy of products} - \text{free energy of reactants}$$

$$\Delta H = \text{heat energy of products} - \text{heat energy of reactants}$$

$$\Delta S = \text{entropy of products} - \text{entropy of reactants}$$

From this it follows that to reverse the direction of a reaction we exchange the positions of the products and the reactants. This would change the sign of the factor considered, but the numerical value would remain the same regardless of the direction of the reaction. Recall that spontaneous reactions have negative ΔG values.

The Meaning of $\Delta G^{0\prime}$

Let us consider the term $\Delta G^{0\prime}$ before we look at the actual values for these reactions. The term $\Delta G^{0\prime}$ is defined as the **standard free energy change** at pH 7.0. This denotes several things:

1. The values are only true when the pH is 7.0 and all products and reactants are in the proper state of ionization for this pH value.
2. The $\Delta G^{0\prime}$ is related to the equilibrium constant for a reaction under these conditions. This can be given for a model reaction at equilibrium:

$$\text{A—B} + H_2O \longrightarrow \text{AH} + \text{BOH}$$

$$\Delta G^{0\prime} = -(\text{constant})(\text{temperature}) \log \frac{[\text{AH}][\text{BOH}]}{[\text{A—B}]}$$

Note that the value of water can be eliminated from these calculations. This is because the concentration of water does not change in a significant way as the molecules of interest undergo their reactions. To make this simplification we must always consider molecules reacting in dilute aqueous solutions.

The $\Delta G^{0\prime}$ simply defines the idealized or standard conditions at the more appropriate value to use for biochemical reactions. This then allows us to determine the ΔG value, which has significance in determining the direction of a spontaneous reaction. The observed ΔG is still dependent upon the two components, the standard free energy change under biochemical conditions and the term relating to the concentrations of the products and reactants under any given conditions of study.

$$\Delta G = \Delta G^{0\prime} + \left[\text{constant} \times \text{temperature} \times \log \frac{[\text{AH}][\text{BOH}]}{[\text{AB}]}\right]$$

In biochemical systems, especially, this concentration term may be as important as the $\Delta G^{0\prime}$ value in determining the ΔG of any reaction. This is because biochemical systems frequently involve a long series of sequential reactions in metabolism.

Bond Energy and Group Transfer Potentials

From an energetic point of view, a reaction is the sum of chemical bonds broken and bonds formed as the reactants go to products. For example, a model compound A—B may be hydrolyzed to yield a given free energy change of −4 kcal/mol.

$$\text{A—B} + H_2O \longrightarrow \text{AH} + \text{BOH}, \qquad \Delta G^{0\prime} = -4 \text{ kcal/mol}$$

In this reaction A—B and H—O bonds are broken and new A—H and B—O bonds are formed. The value of $\Delta G^{0\prime}$ will depend upon the kind of molecule undergoing hydrolysis. The $\Delta G^{0\prime}$ values for most hydrolysis reactions range from nearly zero to as much as −13 kcal/mol. The term **high-energy bond** was established many years ago to describe a reaction such as this hydrolysis,when the energy yield ($\Delta G^{0\prime}$) was −7 kcal or a smaller number. (Remember that we are dealing with negative numbers, and −8 kcal/mol shows a greater difference between products and reactants than −7 kcal/mol.) This would mean that a reaction with $\Delta G^{0\prime}$ of −13 kcal/mol would result from the hydrolysis of an extremely high-energy bond. These high-energy bonds of molecules were often designated by a squiggle (~). For example, A ~ B would denote a high-energy bond, while the normal A—B would indicate a low-energy bond. Although this is a rather convenient designation that is still often used, it is actually an error from an energetics point of view. From the term *high-energy bond*, you would be led to believe that this particular bond was just bursting with energy and about to pop apart all by itself. This is not so because from a chemical point of view it *requires* energy to break a bond. It is the difference between the bond energy of the products and reactants that is important. Since spontaneous reactions occur as energy is released, this occurs when a reaction goes from less stable reactants to more stable products. This is like saying that water will run downhill spontaneously, but not uphill.

Upon reflection this means that reactant molecules with high-energy bonds (less stable) would actually require less energy for bond breakage than product molecules with low-energy bonds (more stable). To get around this problem, we should more properly consider the reaction as a **group transfer potential**, which can be defined as the change in free energy ($\Delta G^{0\prime}$) when a substituent group of 1 mol of a reactant molecule is transferred to a standard acceptor, such as water, under what is known as standard conditions. Now we are back on sound chemical principles. The concept of high-energy bonds can now be related to high group transfer potential.

The Phosphate Ester Bond

A phosphate group may be attached to a glucose molecule at any of several different points. Each of these compounds, or derivatives of glucose, is a separate and distinct molecule, as shown:

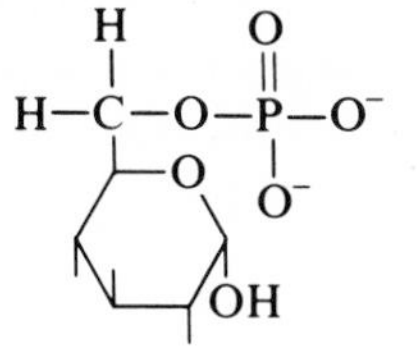

α-D-glucose-6-phosphate

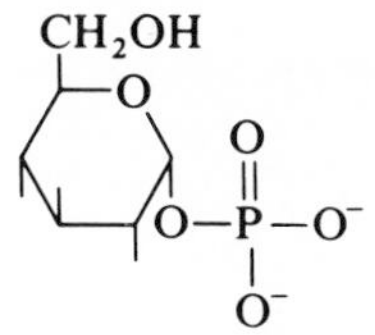

α-D-glucose-1-phosphate

The group transfer potential for each hydrolysis could be determined and would yield values as shown.

$$\text{glu-1-phos} + H_2O \longrightarrow \text{glu} + \text{phosphate}, \quad \Delta G^{0\prime} = -4.7 \text{ kcal/mol}$$

$$\text{glu-6-phos} + H_2O \longrightarrow \text{glu} + \text{phosphate}, \quad \Delta G^{0\prime} = -3.3 \text{ kcal/mol}$$

The only factor that differs in the hydrolysis of glu-1-phos and glu-6-phos is the placement of the phosphate on the glucose. All other products and reactants are exactly the same. A closer look at the linkages involved shows one to be a simple phosphate ester, whereas the other involves the hemiacetal linkage.

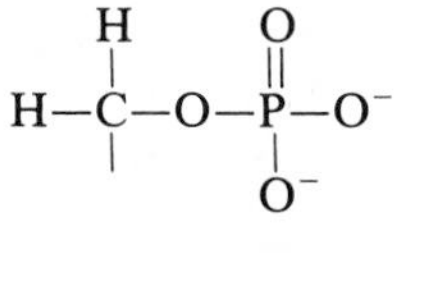

phosphate ester (glu-6-phos)

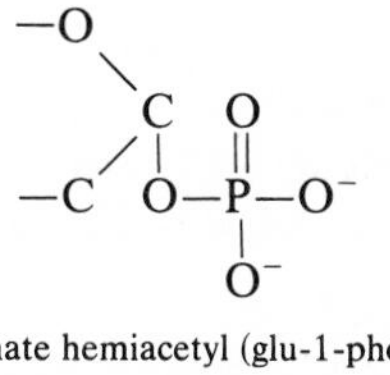

phosphate hemiacetyl (glu-1-phos)

Since the $\Delta G^{0\prime}$ values for these two hydrolyses are different, this can only be due to the differences in bond energies in the two reactants.

Let us look at the hydrolysis of another phosphate, acetyl phosphate:

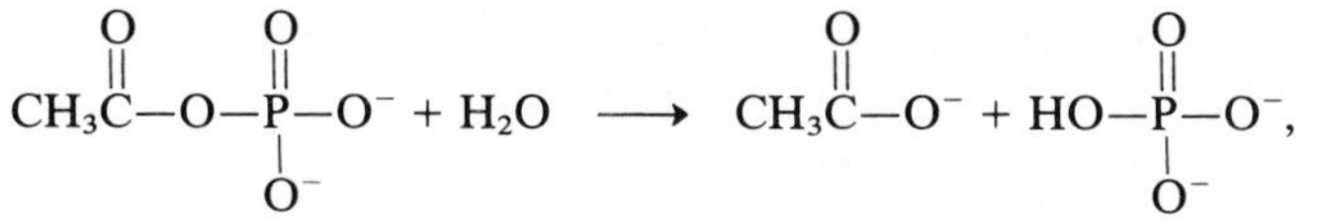

$G^{0\prime} = -10.3$ kcal/mol

This is hydrolysis of what is known as a mixed anhydride of a carboxylic acid and a phosphoric acid. Tremendous energy is released as this exergonic reaction proceeds.

The numerical values of the $\Delta G^{0\prime}$ is related to the way the phosphate is attached to the organic molecule. As a general rule, simple phosphate esters have low group transfer potentials ($\Delta G^{0\prime}$ are small negative numbers), and anhydrides have large negative values. Compounds with other chemical structures will have their own unique values for their free energy of hydrolysis.

The Pyrophosphate Bond

The anhydride of phosphoric acid is of great importance in bioenergetics. The pyrophosphate linkages have a high group transfer potential:

$$^{-}O-\underset{O^{-}}{\underset{|}{\overset{O}{\overset{\|}{P}}}}-O-\underset{O^{-}}{\underset{|}{\overset{O}{\overset{\|}{P}}}}-O^{-} + H_2O \longrightarrow 2HO-\underset{O^{-}}{\underset{|}{\overset{O}{\overset{\|}{P}}}}-O^{-}, \qquad \Delta G^{0\prime} = -8.0 \text{ kcal/mol}$$

This pyrophosphate linkage is very important because it is the action portion of adenosine triphosphate (ATP). ATP is one of a family of molecules known as nucleotides. ATP and similar molecules will be considered in much greater detail later since they are related to constituents of nucleic acids.

ATP is composed of a specific organic molecule (sometimes called an organic base), a five-carbon sugar ribose, and three phosphate groups. The phosphate groups are linked with two pyrophosphate linkages. Several different organic bases can be involved to yield a number of different compounds. Figure 51–1 shows the structure of ATP and two other organic bases. If these bases were attached to the

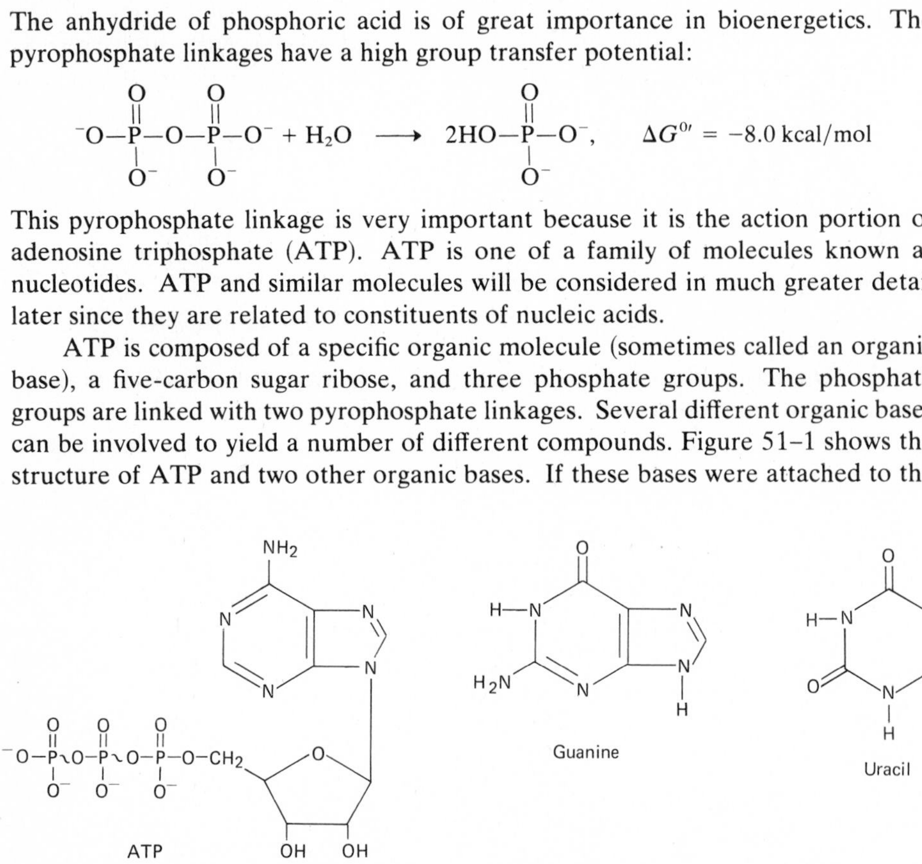
ATP
Guanine
Uracil

Figure 51–1 The structure of ATP and organic bases that form related triphosphates.

ribose in place of adenine, this would result in the formation of guanosine triphosphate (GTP) and uridine triphosphate (UTP), respectively. The nomenclature system uses the term adenosine to denote the organic base adenine combined with the ribose.

In Figure 51–1 there are two squiggles to represent two separate phosphate linkages with high group transfer potential. This means that hydrolysis of the first phosphate yields a reaction with a relatively large free energy, $\Delta G^{0\prime} = -7.3$ kcal/mol as follows:

$$\text{Adenosine}-\text{O}-\overset{\overset{\text{O}}{\|}}{\underset{\underset{\text{O}^-}{|}}{\text{P}}}-\text{O}\sim\overset{\overset{\text{O}}{\|}}{\underset{\underset{\text{O}^-}{|}}{\text{P}}}-\text{O}\sim\overset{\overset{\text{O}}{\|}}{\underset{\underset{\text{O}^-}{|}}{\text{P}}}-\text{O}^- + \text{H}_2\text{O} \longrightarrow \text{adenosine}-\text{O}-\overset{\overset{\text{O}}{\|}}{\underset{\underset{\text{O}^-}{|}}{\text{P}}}-\text{O}\sim\overset{\overset{\text{O}}{\|}}{\underset{\underset{\text{O}^-}{|}}{\text{P}}}-\text{O}^- + \text{P}_i$$

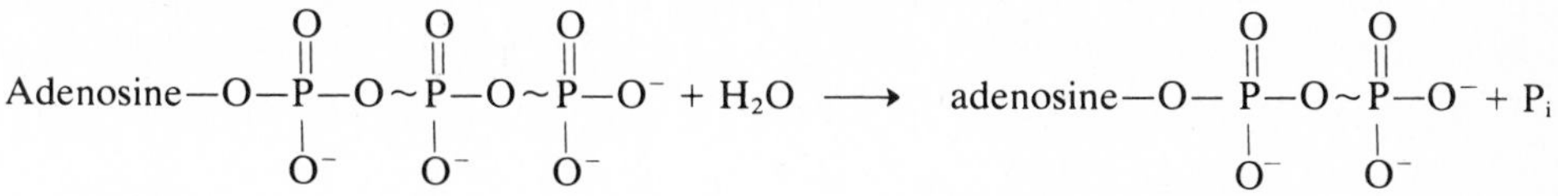

ATP   adenosine diphosphate   phosphate
                                                 (ADP)

We shall use the symbol P_i to denote the orthophosphate group.

Likewise, the hydrolysis of the terminal phosphate on ADP yields a relatively large free energy. Since the pyrophosphate linkage is the same as in the case of ATP, we shall use the same $\Delta G^{0\prime}$ of -7.3 kcal/mol for the same reaction:

$$\text{Adenosine}-\text{O}-\overset{\overset{\text{O}}{\|}}{\underset{\underset{\text{O}^-}{|}}{\text{P}}}-\text{O}\sim\overset{\overset{\text{O}}{\|}}{\underset{\underset{\text{O}^-}{|}}{\text{P}}}-\text{O}^- + \text{H}_2\text{O} \longrightarrow \text{adenosine}-\text{O}-\overset{\overset{\text{O}}{\|}}{\underset{\underset{\text{O}^-}{|}}{\text{P}}}-\text{O}^- + \text{P}_i$$

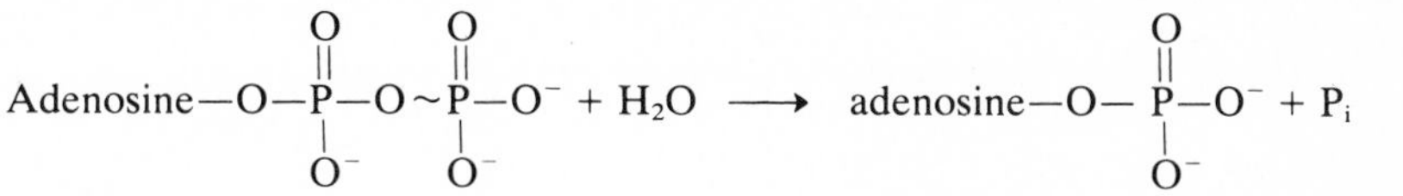

ADP   adenosine monophosphate   phosphate
                                                 (AMP)

The hydrolysis of the phosphate of AMP is different from the previous two reactions because the phosphate of AMP is a simple ester linkage. No squiggle is shown in AMP, which has a free energy of hydrolysis of -3.4 kcal/mol.

$$\text{Adenosine}-\text{O}-\overset{\overset{\text{O}}{\|}}{\underset{\underset{\text{O}^-}{|}}{\text{P}}}-\text{O}^- + \text{H}_2\text{O} \longrightarrow \text{adenosine} + \text{P}_i$$

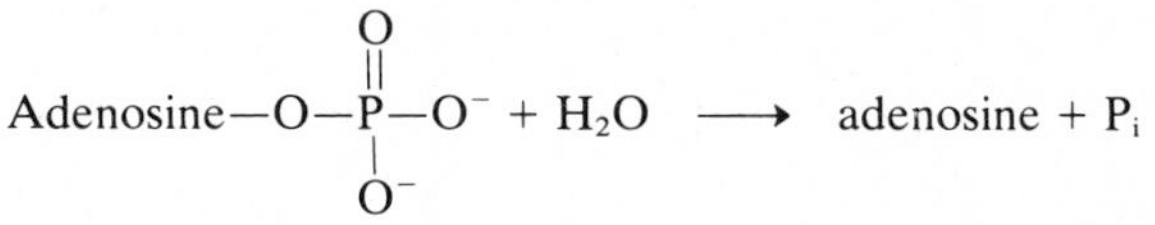

The $\Delta G^{0\prime}$ values that we have shown here have been experimentally determined under controlled laboratory conditions. In all our further studies in this text we shall use these standardized values. It may be that within specific living cells, with their own unique conditions, different numerical values for the actual ΔG would be found.

Coupled Reactions

If you have a series of biochemical reactions such as

$$X \rightleftharpoons Y \rightleftharpoons Z$$

the $\Delta G^{0\prime}$ values are additive. This means that, although $X \rightleftharpoons Y$ may have a $\Delta G^{0\prime}$ of $+3$ kcal/mol (which means it is a nonspontaneous reaction), the overall reaction of $X \rightleftharpoons Z$ could proceed. $X \rightleftharpoons Y$ would have to be coupled with a spontaneous reaction, $Y \rightleftharpoons Z$, to allow the overall reaction to proceed. Reactions like the hydrolysis of ATP are frequently coupled with other reactions to provide the desired outcome. A good example is the phosphorylation of glucose, which is an initial step in the metabolism of this important compound:

$$\text{Glucose} + \text{phosphate} \longrightarrow \text{glu-6-phos} + \text{H}_2\text{O}, \quad \Delta G^{0\prime} = +3.3 \text{ kcal/mol}$$

We can use ATP hydrolysis as a driving force and add the reactions and $\Delta G^{0\prime}$:

$$ATP + H_2O \longrightarrow ADP + P_i, \qquad \Delta G^{0\prime} = -7.3 \text{ kcal/mol}$$

which will yield the following net spontaneous reaction after simplification:

$$\text{Glucose} + ATP \longrightarrow \text{glu-6-phos} + ADP, \qquad \Delta G^{0\prime} = -4.0 \text{ kcal/mol}$$

The reaction can proceed at the expense of 1 mol of ATP.

Note that the hydrolysis of glu-6-phos to glucose and phosphate would be a spontaneous reaction. This is in accord with our general concepts that catabolic or breakdown reactions occur in a spontaneous manner with some release of energy. The entropy change factor may be a contributor to the negative ΔG.

Concentration Effect

We should not leave this topic before recalling the importance of the concentration factor in determining the ΔG for a reaction. High concentrations of either glucose or ATP will cause the ΔG of the phosphorylation reaction shown above to become more negative:

$$\text{Glucose} + ATP \longrightarrow \text{glu-6-phos} + ADP, \qquad \Delta G^{0\prime} = -4 \text{ kcal/mol}$$

$$\Delta G = -4 \text{ kcal/mol} + \left[\text{constant} \times \text{temperature} \times \log \frac{[\text{glu-6-phos}][ADP]}{[\text{glu}][ATP]}\right]$$

Anything that enlarges the denominator (or decreases the numerator) of the concentration term will cause the ΔG of a reaction to become more negative. It is the ΔG value that will be crucial in determining if a reaction is spontaneous or not.

UNIT 51 Problems and Exercises

51–1. What sign would you expect to find for the overall free energy change of the following systems at 37°C?

(a) Denaturation of a protein
(b) Biosynthesis of triglyceride
(c) Contraction of a muscle
(d) Catabolism of glucose to CO_2
(e) Melting of ice

51–2. Show the specific bonds broken and bonds formed in the following processes.

(a) Hydrolysis of a peptide
(b) Denaturation of a protein
(c) Hydrolysis of a triglyceride

51–3. Would you expect the $\Delta G^{0\prime}$ for the hydrolysis of sucrose to be the same as maltose? How about lactose versus maltose?

51–4. In some reactions ATP is hydrolyzed as

$$ATP + H_2O \longrightarrow AMP + P_iP_i$$

Estimate a $\Delta G^{0\prime}$ value for this reaction, and explain your reasoning.

51–5. There is often an exchange of phosphates between nucleosides.

$$ADP + GTP \longrightarrow ATP + GDP$$

Estimate a $\Delta G^{0\prime}$ value for this reaction and explain your reasoning.

51–6. Would the reaction moving the phosphate from position 6 to position 1 be a spontaneous reaction?

$$\text{Glu-6-phos} \longrightarrow \text{glu-1-phos}, \qquad \Delta G^{0\prime} = ?$$

Calculate the value for $\Delta G^{0\prime}$. Can you imagine any biochemical condition that could make this reaction possible?

51–7. Creatine phosphate is an important molecule for the storage of some reserve energy in mammalian muscle. It can contribute to the synthesis of ATP by the following reactions:

$$\text{Creatine phos} + \text{ADP} \longrightarrow \text{ATP} + \text{creatine}, \qquad \Delta G^{0\prime} = -3\ \text{kcal/mol}$$

Using coupled reactions, calculate the $\Delta G^{0\prime}$ value for the hydrolysis of creatine phosphate (group transfer potential).

51–8. Write the expression of ΔG for the formation of ATP from creatine phosphate. How would high concentrations of ATP influence the ΔG of the reaction? Since the muscle uses ATP and forms ADP, what is the influence of the concentration of ADP in this synthesis of ATP?

UNIT 52 ENERGY-GENERATING SYSTEMS IN BIOCHEMISTRY

We have considered the hydrolysis of ATP and the liberation of available energy as this reaction proceeds. The reverse reaction forming ATP requires energy. The problem in a biological system is to utilize the energy of chemical reactions (chemosynthesis) or the energy from some other source like the sun (photosynthesis) to drive the formation of this high-energy compound. We must couple energy-yielding systems, with their $-\Delta G$ values, with the endergonic formation of ATP. In this way, ATP can act as an energy intermediate or as an energy exchange currency in biochemistry. Nature seems to have chosen the addition or deletion of the phosphate group to provide this very important function. The phosphate exchange, or the group transfer potential of the phosphate, lies at the very center of the process of bioenergetics.

There are three general mechanisms by which phosphorylation of ADP to ATP can be accomplished: (1) substrate phosphorylation, (2) oxidative phosphorylation, and (3) photosynthetic phosphorylation.

Substrate Phosphorylation

In these reactions ATP is formed by the direct transfer of a phosphate from an organic molecule to ADP. Since the overall energetics of the coupled reaction must be favorable ($-\Delta G$), the phosphate on the organic substrate molecule must have a high group transfer potential to synthesize ATP:

$$X \sim P_i + ADP \longrightarrow XOH + ATP, \qquad \Delta G = -\text{value}$$

This is a molecule-to-molecule transfer. In other words, it takes one substrate molecule with a high group transfer potential to form one ATP in this system. No systems are known whereby several substrates acting together can form only one ATP. However, there may be several different steps within a single metabolic sequence where this substrate phosphorylation can take place. This is why the metabolism of a molecule like glucose is a multistep process that can yield several ATP molecules.

Oxidative Phosphorylation

Oxidative phosphorylation involves the synthesis of ATP with the energy supplied by an oxidation process (a loss of electrons) involving a carrier system. As organic molecules are oxidized, they usually release their electrons two at a time:

$$\text{RCH}_2\text{OH} \longrightarrow \text{RC(H)=O} \quad + \text{ 2 electrons}$$

$$-\text{CH}_2\text{CH}_2- \longrightarrow -\text{CH=CH}- \quad + \text{ 2 electrons}$$

reduced molecules $\quad$ oxidized molecules $\quad$ + electrons

In living systems these electrons are ultimately involved with oxygen and protons to form water:

$$\tfrac{1}{2}O_2 + 2H^+ + 2 \text{ electrons} \longrightarrow H_2O$$

Oxidation reactions yield energy as well as electrons. The oxidation of fuels provides the energy to heat our homes or power our cars just as readily as the oxidation of food provides for our biochemical energy. The energy of oxidation provides the driving force for the ATP synthesis in **coupled** oxidative phosphorylation:

$$\text{Reduced molecules} + O_2 \longrightarrow \text{oxidized molecules } (\uparrow CO_2) + \text{electrons}$$

$$\text{electrons} \xrightarrow{\text{Carrier system}\ (ADP + P_i \rightarrow ATP)} H_2O$$

Net reaction:

$$\text{Reduced molecules} + O_2 + ADP + P_i \longrightarrow CO_2 + H_2O + ATP$$

Many molecules of ATP can be formed in this coupled phosphorylation process from the oxidation of a single molecule like glucose. The higher living systems would undoubtedly not survive without the large amounts of energy made available through oxidative phosphorylation.

The first problem here is to oxidize an organic molecule to provide the electrons. The second problem is to provide a molecule that can accept these electrons. This is an important function of some of the coenzymes considered previously. Actually, there are a number of different coenzymes involved in the process that accept the electrons from reduced foods or substrates. These coenzymes are arranged in a series whereby the electrons can be passed from one coenzyme to another. This results in a series of oxidation–reduction reactions before the electrons are finally utilized with oxygen in the production of water. In this scheme, each coenzyme is involved in only a temporary way with the electrons. A coenzyme would first be reduced as it accepted electrons, and then oxidized as it passed the electrons to the next coenzyme member of the series. This series of electron transfers is called the **electron-transport system** (ets) or the **respiratory chain**. The electron flow can be coupled to the phosphorylation process, which leads to the production of ATP as shown in Figure 52–1.

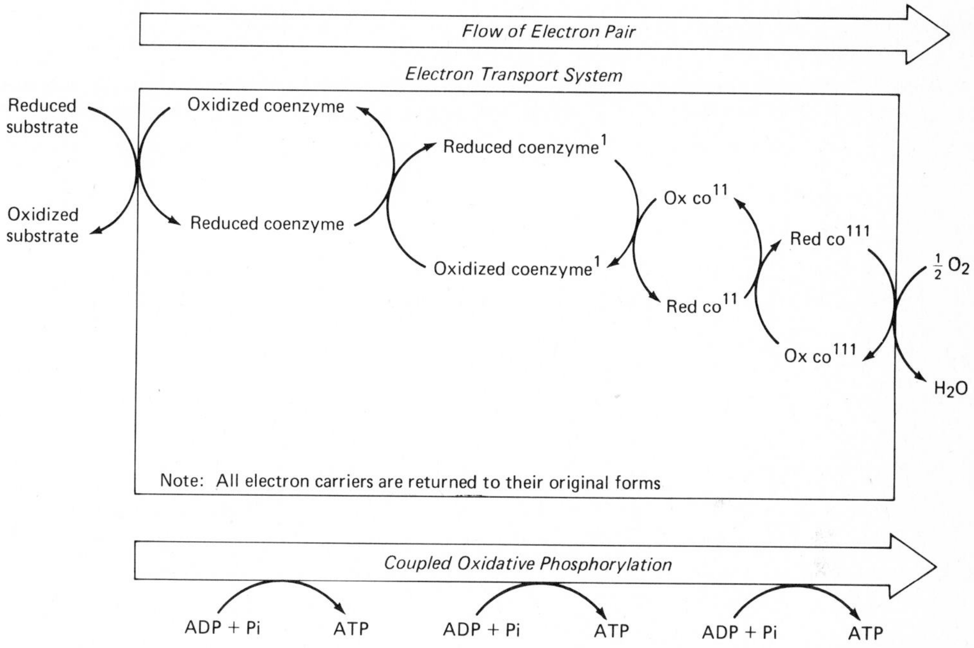

Flow of Electron Pair
Electron Transport System
Reduced substrate
Oxidized substrate
Oxidized coenzyme
Reduced coenzyme
Reduced coenzyme1
Oxidized coenzyme1
Ox co11
Red co11
Red co111
Ox co111
½ O2
H2O
Note: All electron carriers are returned to their original forms
Coupled Oxidative Phosphorylation
ADP + Pi
ATP
ADP + Pi
ATP
ADP + Pi
ATP

General net Rx:

Reduced substrate + O_2 + ADP + Pi $\longrightarrow$ Oxidized substrate + H_2O + ATP

Specific net Rx:

$$RCH_2OH + \frac{1}{2}O_2 + 3\,ADP + 3\,Pi \longrightarrow R{-}\overset{\displaystyle H}{\overset{|}{C}}{=}O + H_2O + 3\,ATP$$

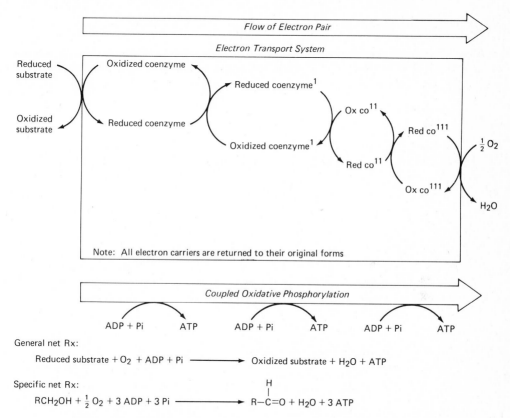

Figure 52-1 Schematic flow of a pair of electrons from a reduced substrate to water through an electron transport system. ATP molecules are synthesized by oxidative phosphorylation through a coupled reaction system.

Let us look at the electron-transport system on a molecular basis. Within the subcellular organelles, known as the mitochondria, we find the electron-transport coenzymes and the oxidative phosphorylation system. The participants in order of their involvement with the electrons in a tissue such as the heart are usually given as

$$NAD \longrightarrow FAD \longrightarrow \text{coenzyme Q} \longrightarrow \text{cytochromes} \longrightarrow H_2O$$

We have already considered the specific oxidation–reduction reactions of both NAD and FAD in molecular detail in unit 50. Because of the important role of the ets, we can show the necessity of understanding how these coenzyme molecules function on a molecular basis.

In the first reaction a reduced substrate may react with a coenzyme like NAD, as in the oxidation of an alcohol molecule by an enzyme alcohol dehydrogenase:

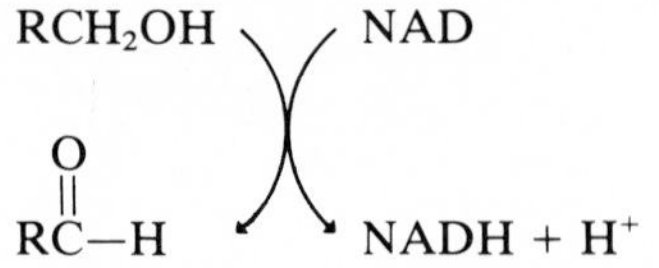
RCH₂OH
NAD
O
RC—H
NADH + H⁺

The NADH must then be reoxidized by the next coenzyme in the series if we are to follow the model shown in Figure 52–1.

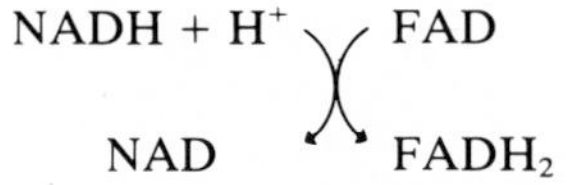
NADH + H⁺
FAD
NAD
FADH₂

The next participating coenzyme is generally thought to be coenzyme Q. There are actually several slightly different molecules in a family known as coenzyme Q. The Q stands for quinone, which is the action center of the molecule for all the members of this family of compounds. This coenzyme can be shown in a reversible redox reaction:

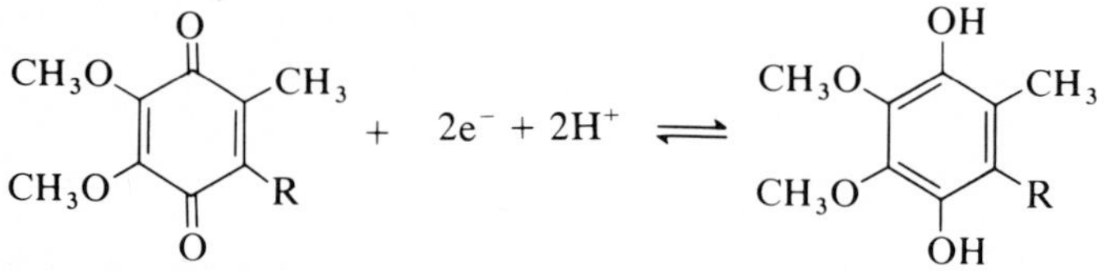
O
CH₃O
CH₃
CH₃O
R
O
+ 2e⁻ + 2H⁺
OH
CH₃O
CH₃
CH₃O
R
OH

quinone
(co Q, oxidized)

hydroquinone
(co Q, reduced)

There are several different members of the cytochrome family that act as the next participants in the ets. The cytochromes are similar to hemoglobin, since they contain an iron atom that is held in a ring system known as **porphyrin** or **heme**. Although more than 20 different cytochromes are known to exist, we will be interested in only a few: cytochrome a and a_3, cytochrome b, and cytochrome c.

Cytochrome c is shown in Figure 52–2. The iron atom is shown surrounded by a heme group. This group is attached to a protein that contains 104 or more amino acids. The exact primary structure of the protein part of this molecule is slightly different in different species. One common feature of the cytochromes is that they all have an iron that can undergo the reversible oxidation–reduction reaction:

$$Fe^{3+} + e^- \longrightarrow Fe^{2+}$$

This is the basic reaction of the cytochromes as they are involved in the electron-

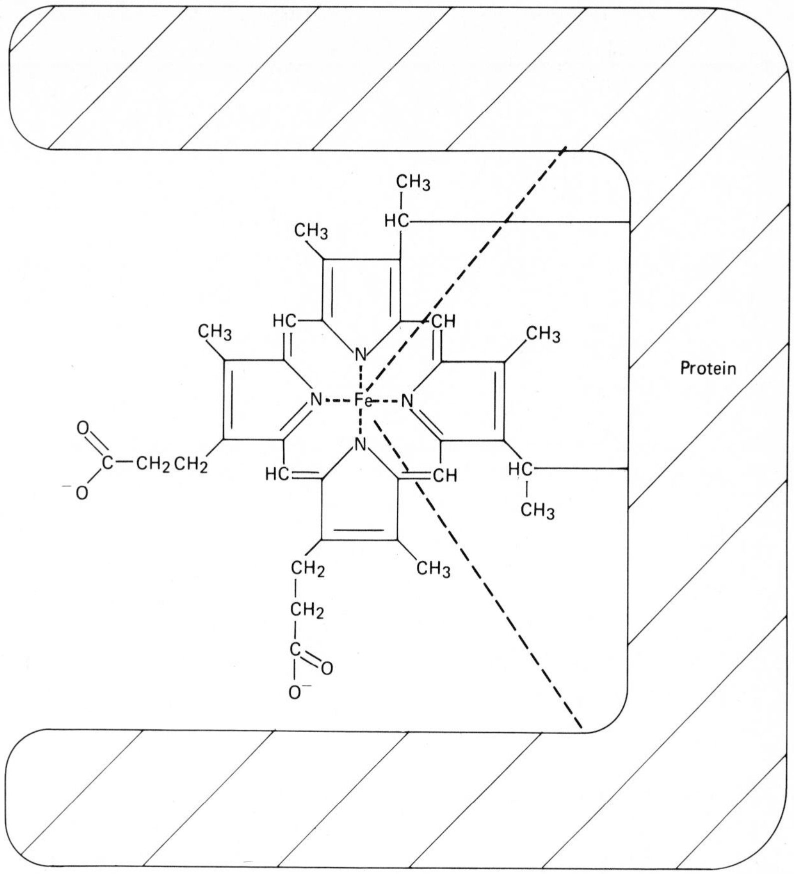
CH3
HC
CH3
HC
CH
CH3
CH3
N
N
Fe
N
N
Protein
O
C—CH2CH2
O
HC
CH
HC
CH3
CH2
CH3
CH2
C
O
O

Figure 52–2 Cytochrome c. The iron is attached to four parts of the heme molecule and two parts of the protein. The iron can exist as Fe^{2+} or Fe^{3+}.

transport system. The complex heme molecule provides the proper environment and position for the iron atom.

Note that the reaction of iron involves only one electron exchange. Although the organic molecules lose a pair of electrons in oxidation, the inorganic iron atom is involved with a transfer of only one electron at a time. Just how nature handles this problem of electron flow has yet to be discovered through research.

The sequence or order in which the coenzymes undergo their oxidation–reduction reactions is related to their **standard reduction potential**, $E^{0\prime}$, at biological conditions and at pH 7. This $E^{0\prime}$ is defined as a relative ability to gain electrons and thereby become reduced. By our convention, the greater the positive value, the greater is the tendency to accept electrons and undergo reduction. A series of half-reactions for the oxidation and reduction of coenzymes is shown in Table 52–1. For example, FAD tends to become reduced more readily than NAD. Finally, O_2 tends to form water very readily according to this principle. These values, which

Table 52–1 Approximate standard reduction potentials under biochemical conditions (pH 7, 30°C)

	$E^{o\prime}$(V)
NAD ⟶ NADH	−0.32
FAD ⟶ $FADH_2$	−0.06
Coenzyme Q (ox) ⟶ coenzyme Q (red)	0.00*
Cytochrome b (Fe^{3+}) ⟶ cytochrome b (Fe^{2+})	+0.04
Cytochrome c (Fe^{3+}) ⟶ cytochrome c (Fe^{2+})	+0.25
Cytochrome a (Fe^{3+}) ⟶ cytochrome a (Fe^{2+})	+0.25
Cytochrome a_3 (Fe^{3+}) ⟶ cytochrome a_3 (Fe^{2+})	+0.5*
$\frac{1}{2}O_2$ ⟶ H_2O	+0.82

* Considerable uncertainty.

were determined through experiments, lead us to construct a flow of electrons in the ets as follows:

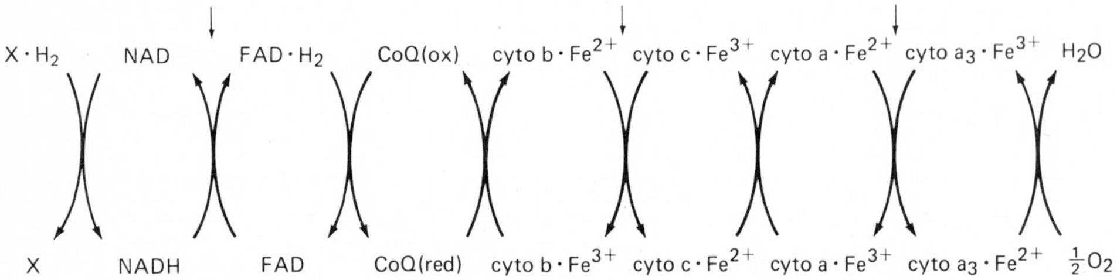
X·H2
NAD
FAD·H2
CoQ(ox)
cyto b·Fe2+
cyto c·Fe3+
cyto a·Fe2+
cyto a3·Fe3+
H2O
X
NADH
FAD
CoQ(red)
cyto b·Fe3+
cyto c·Fe2+
cyto a·Fe3+
cyto a3·Fe2+
½O2

Each of the coenzymes can be returned to its original state with the final formation of water. This allows the coenzymes to function in a catalytic manner.

The energy released during oxidation can be calculated. For two electron exchanges, the standard free energy under biochemical conditions can be determined from the differences ($\Delta E^{0\prime}$) in the reduction potentials of the half-reductions involved:

$$\Delta G^{0\prime} = -46{,}000\,\frac{\text{cal}}{\text{V}} \times \Delta E^{0\prime} \quad \text{(in volts)}$$

For the overall reaction of NADH forming water,

$$\Delta G^{0\prime} = -46{,}000\,\frac{\text{cal}}{\text{V}} \times [+0.82 - (-0.32)]$$

$$= -46{,}000\,\frac{\text{cal}}{\text{V}} \times 1.14\text{ V} = -52{,}440\text{ cal} = -52.4\text{ kcal}$$

This energy yield could provide the energy to drive the synthesis of many ATP molecules from ADP and P_i. Careful biochemical studies have shown that only three ATP's are generated from the oxidation of NADH by the ets.

P/O Ratio

The energy generation can be designated as the **P/O ratio**. This can be defined as the number of moles of P_i that are converted to ATP for every gram atom of $O(\frac{1}{2}O_2)$ that is consumed. Since the oxidation of NADH provides the energy for three ATP's to be synthesized, this would have a P/O ratio of 3. In some cases other coenzymes participate as the initial electron acceptors from reduced molecules. One case of this is the oxidation of succinate, which can directly transfer its electrons to FAD. The resulting $FAD{\cdot}H_2$ has been found by experiments to provide for the synthesis of only two ATP's. Thus the $FAD \cdot H_2$ has a P/O ratio of 2.

To utilize the energy of the respiratory chain, there must be some special mechanism within the cell to form the ATP molecules. This ets coupled with ATP generation is the process of oxidative phosphorylation. There is still controversy as to how this is accomplished, and two different mechanisms have been proposed.

A **chemical coupling mechanism** postulates that a high-energy intermediate is formed as the result of the electron transport. The high-energy intermediate transfers a phosphate to ADP to form ATP. There are three specific phosphorylating sites, as shown by arrows, on the complete ets that operate between the following pairs of half-reactions:

1. NADH $\longrightarrow$ FAD.
2. Cytochrome b (reduced) $\longrightarrow$ cytochrome c (oxidized).
3. Cytochrome a (reduced) $\longrightarrow$ cytochrome a_3 (oxidized).

The second proposed mechanism is called the **chemiosmotic hypothesis**. In this mechanism, protons are moved across the inner membrane of the mitochondria as the result of electron transfer. This causes a pH difference on the two sides of a membrane, which in turn provides the energy to generate the ATP. Various ionic pumps are known where ATP is utilized to move ions across a membrane. This energy-generating process can be likened to a reverse of the pump processes.

Both of these proposals have been suggested by reputable biochemists who have different interpretations of the available experimental results. This is a good example of how science grows. The speculation as to how this process operates stimulates new ideas for research. As future experiments become more refined and definitive, the operations of living systems will become better known.

It is known that the respiratory chain and the phosphorylation system are highly interdependent. When these two processes are operating together, as they normally do, they are said to be **tightly coupled**. ADP and P_i must be available to allow the electron exchange to proceed. This means that the ratio of ADP to ATP could regulate the transport of electrons. This can be demonstrated experimentally, since it is known that relatively high levels of ADP increase O_2 consumption and ATP generation. This regulation scheme is known as **respiratory control** of tightly coupled systems.

There are some situations where these two systems can be uncoupled. There is a complex hormonal mechanism that operates within living systems to balance energy production with the utilization of reduced molecules. Certain chemicals, like

2,4-dinitrophenol are also known as effective uncouplers and may be useful in experimental systems.

Photosynthetic Phosphorylation

The process of photosynthesis actually consists of two major functions. In one case, CO_2 must be assimilated and converted to reduced organic molecules. This reduction requires energy, which must be supplied as the second function of this process. The ultimate energy source is light from the sun, which must be trapped by chlorophyll and other compounds before it can be converted to ATP.

The crucial process involves the photolytic cleavage of water according to the following reaction:

$$\text{Light energy} + 2H_2O + 2\text{NADP} \xrightarrow[\text{apparatus}]{\text{Photosynthetic}} 2\text{NADPH} + 2H^+ + O^{2-}$$

It is beyond our present interest to consider the biochemisry of individual electron-transport reactions in the photosynthetic apparatus. The same general chemical principles are involved here as we found in electron-transport systems. In the case of photosynthesis, the light energy that is absorbed activates acceptor molecules and provides the energy to transfer electrons toward more negative reduction potentials. As the reduced coenzymes pass their electrons to other coenzymes with more positive reduction potentials, energy is made available. This energy can be captured to drive the synthesis of ATP from ADP and P_i. Although the molecular architecture may have some specific differences in the plant and animal kingdom, the common threads of biochemistry appear most vividly in both of them. The unity of nature remains intact.

UNIT 52 Problems and Exercises

52–1. The compound 1,3-diphosphoglyceric acid (1,3-dga) is known to transfer one of its two phosphates to ADP to form ATP. These two reactions could be written as:

(1) 1,3-dga + ADP → 1-phosphoglyceric acid + ATP, $\Delta G^0 = +3.5$ kcal/mol

(2) 1,3-dga + ADP → 3-phosphoglyceric acid + ATP, $\Delta G^0 = -4.5$ kcal/mol

(a) Would you expect reaction 1 or reaction 2 to be involved in the generation of ATP by substrate phosphorylation?

(b) Estimate the group transfer potential for each of the phosphate groups of 1,3-dga.

52–2. Briefly explain why substrate phosphorylation can yield only one ATP per substrate, whereas oxidative phosphorylation can yield a number of ATP's per substrate molecule oxidized.

52–3. Show the structures (active portion only) of NAD as it is reduced to form NADH.

52–4. Suppose that you had to fit a newly found coenzyme, cytochrome c_1, into the respiratory chain. If cytochrome c_1 has a $E^{0\prime}$ value of +0.22, sketch the respiratory chain with cytochrome c_1 in its proper position.

52–5. Would you expect the new respiratory chain with the cytochrome c_1 to yield more than three ATP molecules if it was coupled with an adequate phosphorylating system? Why?

52–6. If P_i was eliminated from a tightly coupled phosphorylating system, what result would you expect?

52–7. If an enzyme ATPase, which causes the reaction

$$ATP + H_2O \longrightarrow ADP + P_i$$

was added to a tightly coupled oxidative phosphorylating system, would you expect this to stimulate or inhibit the flow of electrons in the ets?

52–8. If a system was discovered with a P/O ratio of 1, explain what this would mean.

52–9. If someone suggested that one of the specific phosphorylating sites in oxidative phosphorylation was between cytochrome c (red) → cytochrome a (ox), how might you prove this was in error?

52–10. Briefly, what are the two major functions of photosynthesis?

CHAPTER 19

carbohydrate metabolism

A number of different carbohydrates are found in food; once eaten, they become available for metabolic reactions within our bodies. Large molecules, like disaccharides and polysaccharides, must first be converted into monosaccharides by digestion. The monosaccharides are then absorbed into the bloodstream and carried to the various organs and tissues, where they are metabolized.

Once a compound like glucose is available within a cell, it is subject to several different possible metabolic reaction systems or pathways. Several different pathways may be required to meet the special needs of the tissue in which the reactions are taking place. Regardless of the pathways involved, the initial reaction of glucose always involves a phosphorylation leading to the formation of glucose-6-phosphate. This ATP-requiring reaction seems to "lock" the glucose into the metabolic system. We shall consider some of these metabolic systems in detail.

The two major functions of metabolic reactions are to provide both **energy** and the different **carbon skeletons** for the variety of molecules needed in living cells. We have considered some of the energy needs previously and already recognize the importance of ATP as an exchange molecule in bioenergetics. The *interrelationships* that exist between classes of compounds, like carbohydrates and fats, can only be understood as we unravel the intricacies of metabolism. For example, we know that a high-carbohydrate diet can lead to the formation of fatty deposits in our bodies. An examination of the chemical structures of these two classes of compounds does not show their close chemical relationship. These relationships can only be understood by studying the metabolism of these compounds on a biochemical basis.

UNIT 53 DIGESTION AND GLYCOLYSIS

Polysaccharides and disaccharides make up the major fraction of carbohydrates in our diet. The most common **digestible** carbohydrates are starch, glycogen, sucrose, maltose, and lactose. The **nondigestible** polysaccharides, such as cellulose, are not subject to attack by the hydrolytic enzymes that humans possess. We shall consider the digestion and subsequent metabolism of the monosaccharides resulting from the hydrolysis of these compounds in some detail.

The digestible polysaccharides are composed of glucose molecules joined principally by $\alpha(1 \rightarrow 4)$ linkages with some $\alpha(1 \rightarrow 6)$ branching. The hydrolysis of these molecules is initiated by an α-amylase found in the saliva of humans. This enzyme attacks the $\alpha(1 \rightarrow 4)$ linkages in a rather random fashion to initiate the breakdown of the complex polysaccharides. This enzyme may have a limited influence because of the short time the food is in the mouth. Once the food passes into the stomach, this particular digestion process is inactivated because of the low pH. The acid in the stomach may catalyze some hydrolysis of both polysaccharides and disaccharides, but this is undoubtedly a minor event in the digestive process.

As the carbohydrates move into the small intestine, they are subjected to major hydrolysis. The pancreatic juice containing another amylase (pH optimum about 7.1) is added at this point. Enzymatic action results in the formation of maltose, isomaltose, and glucose. These amylases do not have the capacity to catalyze the hydrolysis of $\beta(1 \rightarrow 4)$ glucosidic linkages and thus cannot bring about the hydrolysis of cellulose. This small but specific chemical difference in the polysaccharides illustrates once again the importance of considering molecules on a structural basis. Obviously, it is the uniqueness of the biochemically important enzyme systems that is so critical.

Maltose and isomaltose, along with sucrose and lactose, make up the disaccharides of interest in human nutrition. The hydrolysis of disaccharides is catalyzed by enzymes of the mucosal cells rather than within the intestinal lumen itself. A number of different enzymes have been isolated that bring about the hydrolysis of specific disaccharides to form D-glucose, D-fructose, and D-galactose. Other monosaccharides may be present in small amounts but will not be of interest to us at this time. The actual mechanism by which the monosaccharides are absorbed into the bloodstream is a quite complex energy-requiring process.

Normally, carbohydrates are digested and absorbed rather rapidly. The blood sugar levels usually reach a maximum within 1 hr after eating the sugar. The blood sugar level then normally falls to a more constant level within 2 hr. The rate at which the glucose we eat can enter the blood can be determined by a **glucose tolerance test**. This test measures the blood glucose levels at various time intervals following the ingestion of a controlled amount of glucose and is often of use in confirming diabetes.

Glycolysis

The metabolism of glucose, fructose, and galactose is carried out by a common pathway. Under normal circumstances, glucose can enter the cells of the various organs in a controlled manner. In this case, the hormone insulin facilitates the movement of glucose into muscle and adipose cells. It is of interest that, although insulin influences the entry of glucose into muscle cells, it may have little influence on this process in liver cells. This points up the need of specifying the conditions of metabolism.

The first metabolic reaction that glucose is known to undergo is an addition of a phosphate (**phosphorylation**) by an enzyme of the family known as the **kinases.** Specific kinases act upon separate monosaccharides: glucokinase, fructokinase, and galactokinase. More general hexokinases can bring about the phosphorylation of any of the common monosaccharides.

The kinases transfer the terminal phosphate of ATP to either the number 1 or number 6 position of the monosaccharide. For glucose, this results in the formation of glucose-6-phosphate:

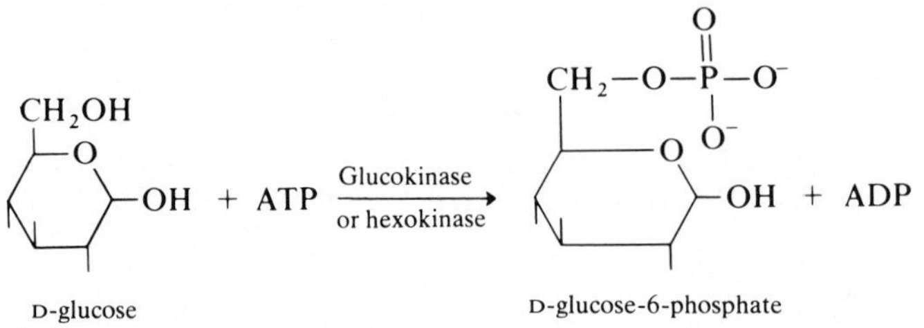

CH₂OH
O
OH + ATP
Glucokinase
or hexokinase
CH₂—O—P—O⁻
O
O⁻
OH + ADP
D-glucose
D-glucose-6-phosphate

This phosphate ester is the *metabolic form* through which glucose enters a variety of possible pathways. This key reaction also traps the glucose within the cell as the phosphate ester is not readily transported back out through cell membranes. It may be significant that this reaction adds a charged group to the previously neutral glucose. We shall not make an issue of this charge hereafter, but will use an abbreviation, Ⓟ, to denote the phosphate group.

The hexokinases may be involved in metabolic control since they are subject to product inhibition. High levels of glucose-6-phosphate produced by the enzyme may inhibit its further action. This is an example of an **allosteric effect** whereby the cell can control the entry of glucose into its metabolic system.

Glucose-6-phosphate reacts in a number of different pathways, as outlined in Figure 53–1. These reactions may not all occur at the same time or be present in all tissues or cells. Each system will be treated separately.

The hydrolysis of glucose-6-phosphate to glucose is really a nonproductive reaction controlled by glucose phosphatase:

$$\text{Glu-6-phos} + H_2O \xrightarrow[\text{phatase}]{\text{Glucose phos-}} \text{glucose} + P_i$$

Note that this reaction does *not* regenerate ATP but is simply a hydrolysis. By using a coupled reaction, glucose may be regenerated:

$$\begin{array}{ll} \text{Hexokinase:} & \text{glucose} + \text{ATP} \longrightarrow \text{glu-6-phos} + \text{ADP} \\ \text{Phosphatase:} & \text{glu-6-phos} + H_2O \longrightarrow \text{glucose} + P_i \\ \hline \text{Net reaction:} & \text{ATP} + H_2O \longrightarrow \text{ADP} + P_i \end{array}$$

The futility of this series of reactions is seen in that it only nets the loss of an ATP.

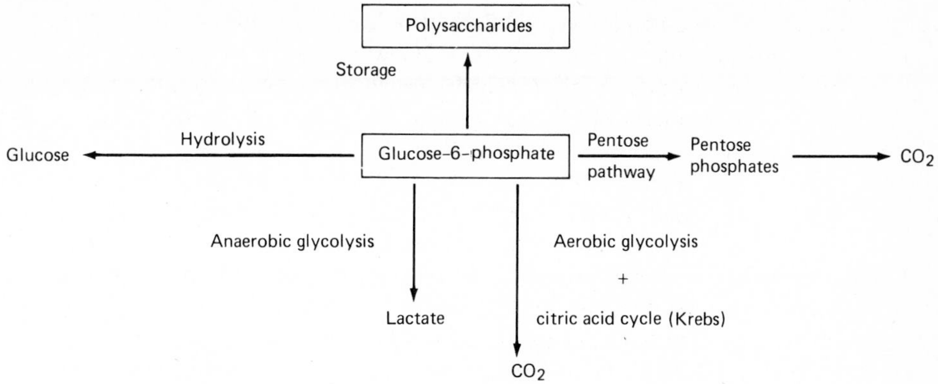
Polysaccharides
Storage
Hydrolysis
Glucose
Glucose-6-phosphate
Pentose
pathway
Pentose
phosphates
CO_2
Anaerobic glycolysis
Aerobic glycolysis
+
Lactate
citric acid cycle (Krebs)
CO_2

Figure 53–1 The possible metabolic pathways of glucose-6-phosphate.

This is just the opposite of what we have set out to do, that is, to generate ATP from the catabolism of glucose. These coupled reactions could allow glucose to return to the bloodstream from organs like the liver if the physiological need arose. This may be part of the overall mechanism maintaining blood sugar levels.

Glucose can undergo reactions of **anaerobic** (without O_2) and **aerobic** (with O_2) glycolysis. Although many of the reactions in these two systems are similar, different end products are formed. In either type of glycolysis, glucose-6-phosphate is readily converted to fructose-6-phosphate by an enzyme **phosphoglucose isomerase**. This isomerase reaction involves a reversible interconversion of an aldehyde and ketone, which may be illustrated in simplified open-chain form as follows:

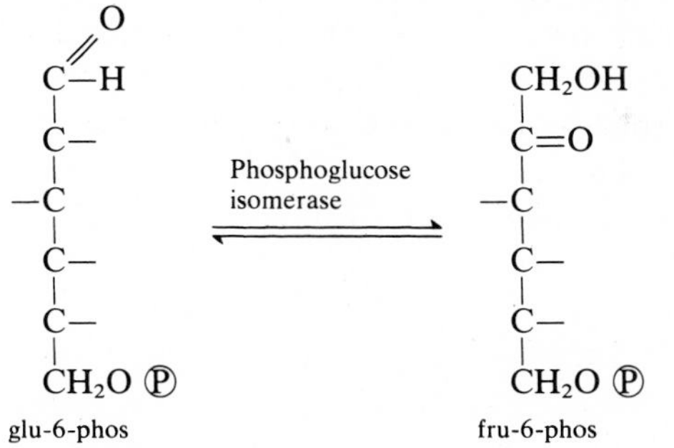
Phosphoglucose
isomerase

glu-6-phos fru-6-phos

The next reaction in this series involves the utilization of yet another molecule of ATP. Fructose-1,6-diphosphate is formed using the enzyme phosphofructokinase and Mg^{2+}:

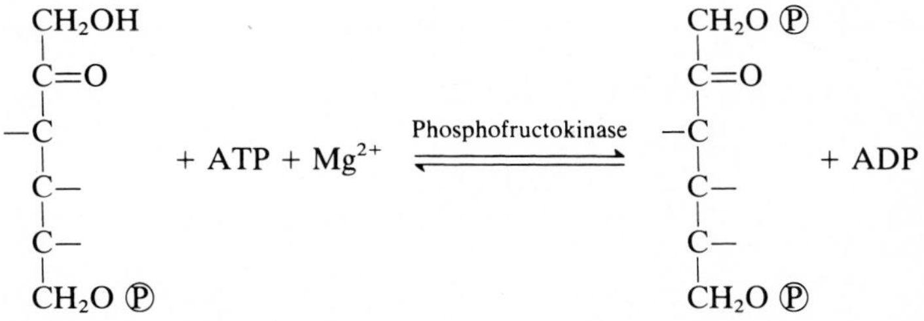
+ ATP + Mg^{2+}
Phosphofructokinase
+ ADP

Another molecule of ATP has been consumed in this step. This enzyme has an important control function through **allosteric inhibition**. Obviously, the cell needs a reasonable level of ATP present to allow this reaction to proceed. In a sense this ATP acts as one of the substrates. We know that reaction rates are usually dependent upon substrate concentrations. The allosteric influence is involved at very high ATP concentrations when this enzyme can slow down this reaction. At high levels, ATP binds to a second or allosteric site on the enzyme to alter it in a subtle fashion. The altered enzyme no longer is able to catalyze this reaction, and subsequent reactions are not able to operate. This inhibition can stop the reaction series when ATP levels are high and ATP production is not needed.

The next reaction involves a cleavage of the carbon–carbon bond of the original hexose to yield two triose phosphates. The enzyme aldolase yields:

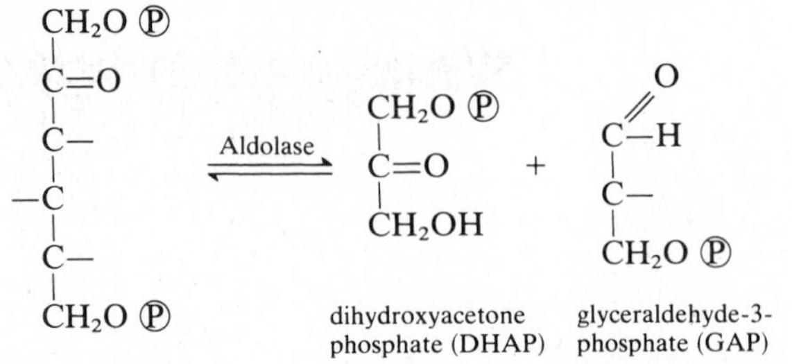
Aldolase
dihydroxyacetone phosphate (DHAP)
glyceraldehyde-3-phosphate (GAP)

The reaction is displayed in this manner to emphasize the point of cleavage between carbon 3 and carbon 4 of the hexose diphosphate.

Both trioses are available for subsequent reactions. Studies have shown that only the aldose form is actually involved in further reactions in this series, however. The problem then is to bring about an interconversion of the trioses so that both molecules can continue in this metabolic sequence. This is accomplished by triose phosphate isomerase:

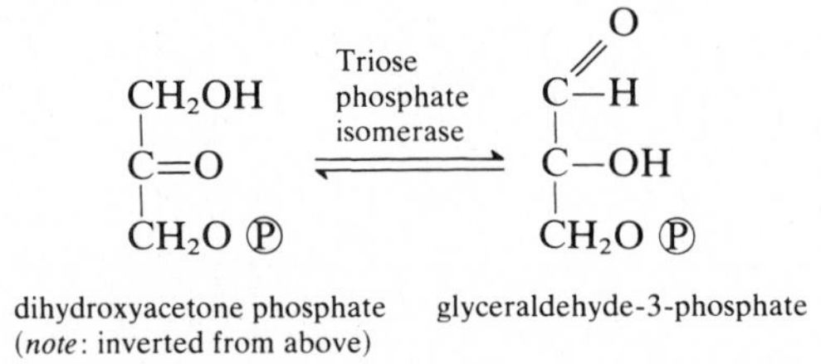
Triose phosphate isomerase
dihydroxyacetone phosphate
(*note*: inverted from above)
glyceraldehyde-3-phosphate

We shall see later that the ketose form has a key position in relating lipid metabolism with carbohydrate metabolism.

The next reaction in glycolysis continues with the aldose, GAP. We must remember that two triose molecules are proceeding through the reaction sequence for each hexose originally subjected to metabolism. This is especially important when we consider the energetics of this sequence. The aldose phosphate reacts with the coenzyme NAD, and an inorganic phosphate becomes incorporated into the molecule. The enzyme glyceraldehyde-3-phosphate dehydrogenase catalyzes the overall reaction:

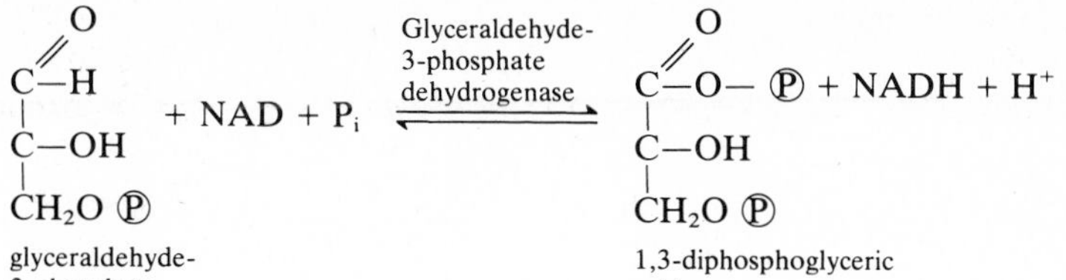

glyceraldehyde-3-phosphate — 1,3-diphosphoglyceric acid

The active side of this hydrogenase enzyme has been studied in considerable detail. The S—H group of a cysteine of the enzyme is involved. At one stage in this reaction, the carbonyl group of the aldose becomes bound to the enzymes as shown:

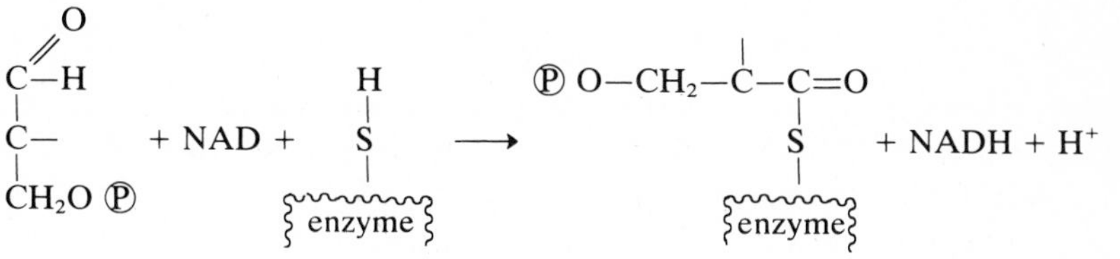

The next stage of the reaction uses the inorganic phosphate to break the C—S bond and form 1,3-diphosphoglyceric acid.

The NADH is now available for a number of different fates. Each NADH could move through the electron-transport scheme to yield three ATP molecules by the process of **oxidative phosphorylation**. This process requires O_2, and this would be the situation in aerobic glycolysis. If O_2 were not available, the NADH would have a different fate. It is important to recall that NADH is a cofactor, derived from a vitamin, and thus is present only in small amounts. The NAD $\rightleftharpoons$ NADH cycle must take place to allow the cofactor to be used repeatedly.

The following reaction in this sequence involves the process of **substrate phosphorylation**. The phosphate attached to the carboxyl position of the glyceraldehyde has a high group transfer potential. In this case a different kinase enzyme transfers the phosphate group to ADP to yield ATP:

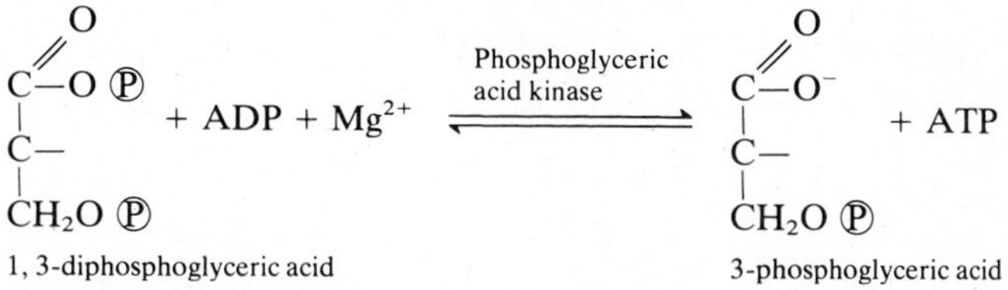

1, 3-diphosphoglyceric acid — 3-phosphoglyceric acid

Consider the energetic situation at this point where ATP is first generated. We have utilized 2 mol of ATP per hexose previously. Under aerobic conditions we now could produce 6 mol of ATP (2 mol of NADH) from oxidative phosphorylation and 2 mol from substrate phosphorylation to have a net yield of six ATP molecules for every glucose used. Anaerobic glycolysis would not have produced any net ATP to this point, but would have just broken even.

The next reaction involves the transfer or migration of the remaining phosphate group from one position to another. In this case, phosphoglyceromutase catalyzes the reaction:

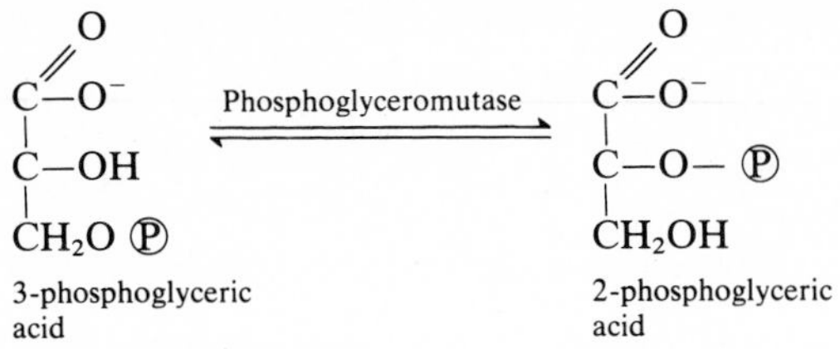

The pathway continues as this reactant undergoes a loss of a water (dehydration) in the presence of the enzyme enolase:

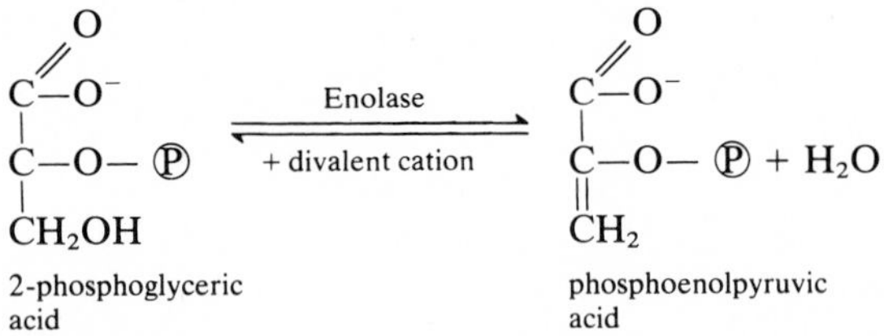

This is the enzyme that is directly influenced by a powerful inhibitor of glycolysis, the fluoride ion.

Phosphoenolpyruvate also has the phosphate group in a configuration yielding a high group transfer potential. This phosphate group can be transferred, by a coupled reaction, to an ADP molecule. The substrate phosphorylation process leads to the formation of another ATP through the action of the enzyme pyruvate kinase:

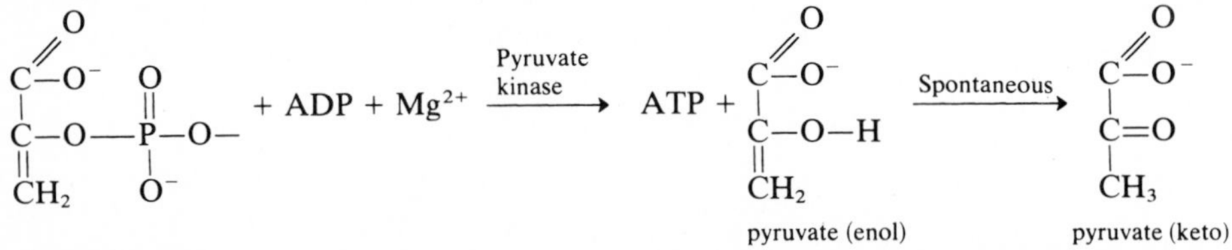

This reaction shows the phosphate transfer and the formation of pyruvate. The intermediate enol form of pyruvate is shown to more easily visualize the reaction. This form of pyruvate is quickly and nonenzymatically shifted to the more stable keto form as shown.

Consider the energetic situation to this point. Since the two triose phosphates per original hexose would now have gone through two substrate phosphorylations, a net yield of two ATP's per hexose results from anaerobic glycolysis. Under anaerobic conditions the NADH can be used as a reducing agent. Under aerobic conditions, the substrate phosphorylation is augmented by the NADH, which can drive oxidative phosphorylation. This latter case would show a net of eight ATP's generated as a hexose was transformed to yield two pyruvate molecules in the total scheme shown in Figure 53–2.

The pyruvate represents a **junction** point in this metabolic scheme. The pyruvate can react to yield several different products depending upon which tissue or system is involved and in what environment the reaction takes place.

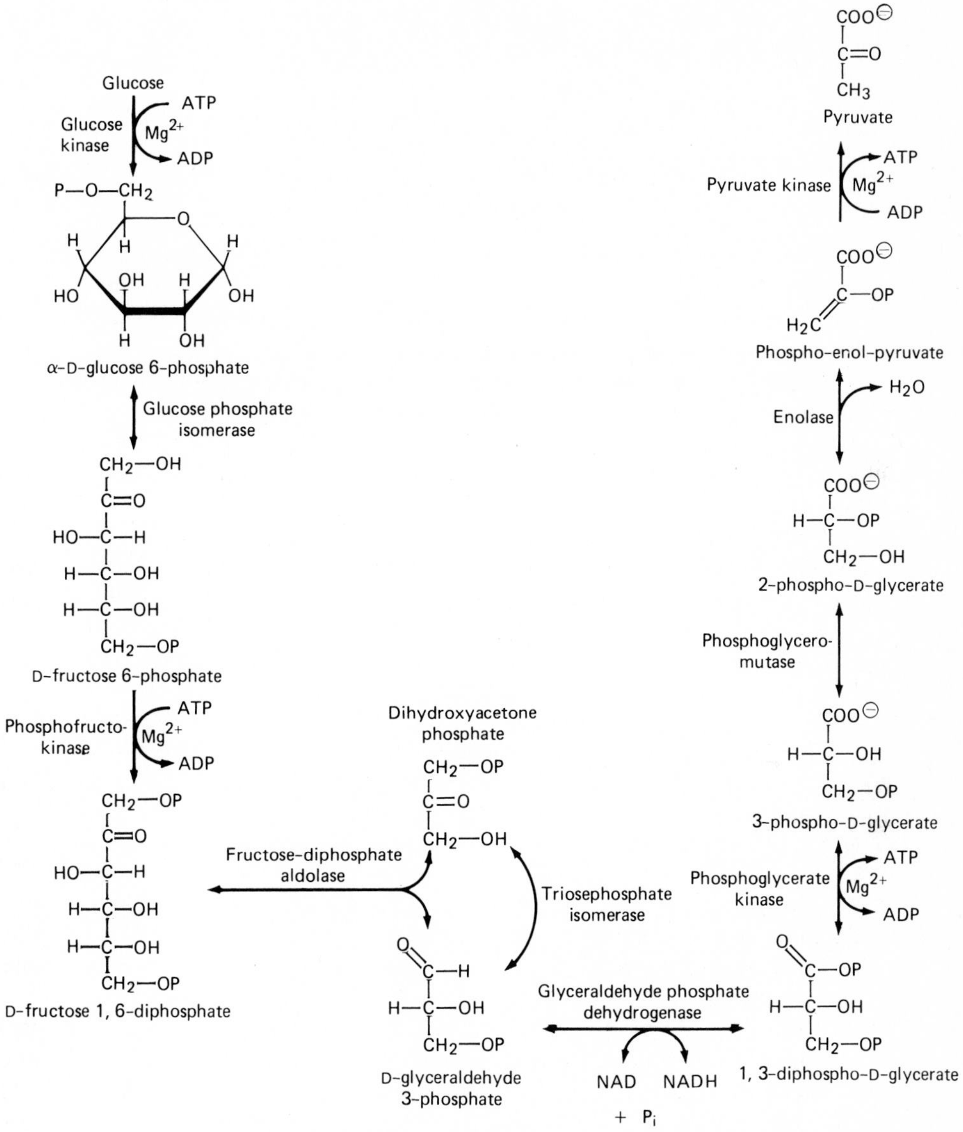
Glucose
ATP
Glucose kinase
Mg2+
ADP
α-D-glucose 6-phosphate
Glucose phosphate isomerase
D-fructose 6-phosphate
Phosphofructo-kinase
D-fructose 1, 6-diphosphate
Fructose-diphosphate aldolase
Dihydroxyacetone phosphate
Triosephosphate isomerase
D-glyceraldehyde 3-phosphate
Glyceraldehyde phosphate dehydrogenase
NAD + Pi
NADH
1, 3-diphospho-D-glycerate
Phosphoglycerate kinase
3-phospho-D-glycerate
Phosphoglycero-mutase
2-phospho-D-glycerate
Enolase
H2O
Phospho-enol-pyruvate
Pyruvate kinase
Pyruvate

Figure 53–2 The glycolytic scheme of glucose to pyruvate.

Lactate Metabolism

Under anaerobic conditions, such as a muscle operating with an insufficient oxygen supply (oxygen debt), lactate can be formed by the enzyme lactate dehydrogenase:

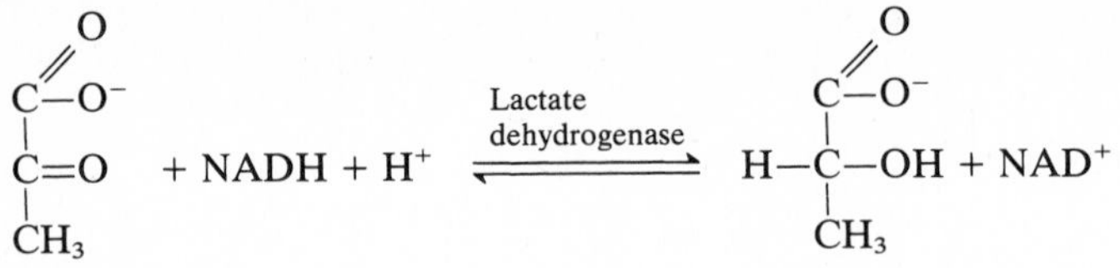

Lactate
dehydrogenase
+ NADH + H⁺
+ NAD⁺

pyruvate lactate

This reaction is accompanied by the regeneration of NAD and provides a mechanism to re-form the oxidized coenzyme in the absence of oxygen. The metabolic sequence can continue and yield a limited amount of ATP. This process will sustain some muscle activity when the oxygen is limited.

Lactate dehydrogenase exists as a tetramer in its active form. This enzyme is one of the best-studied cases of **isoenzymes** (isozymes). The subunits that make up the active tetrameric form may be of two distinct types. The M type predominates in the skeletal muscle; the H type predominates in the heart muscle. All possible combinations of the tetramer that are known to exist are:

H_4 M_1H_3 M_2H_2 M_3H_1 M_4

The H_4 form is readily inhibited by pyruvate; the M_4 form is not. The muscle seems to have this special form of the isozyme in order to provide for continual activity under oxygen debt. The heart, which normally has a rich O_2 supply, prefers to oxidize the pyruvate through reactions considered later. It is also important to note that the only metabolic reaction of lactate is its reconversion back to pyruvate by oxidation. Fortunately, both pyruvate and lactate are quite readily diffusible through cell membranes. Increased levels of lactate are noted in the blood after significant muscular activity. This provides for the completion of a loop of the metabolic sequence at this point in animals:

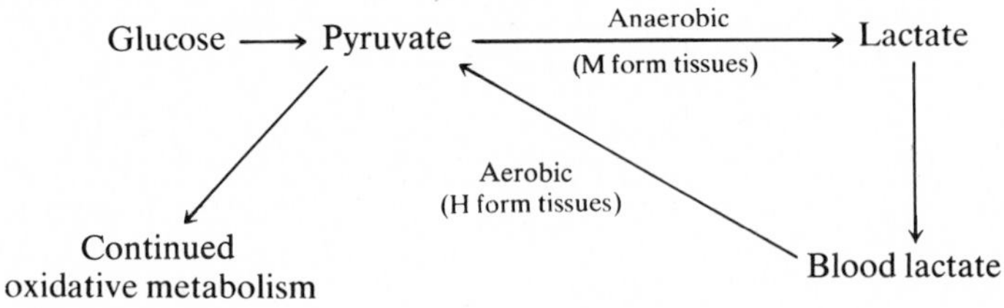

Glucose
Pyruvate
Anaerobic
(M form tissues)
Lactate
Aerobic
(H form tissues)
Continued
oxidative metabolism
Blood lactate

Ethanol Synthesis

Lactate itself is also the end product of some microbial systems under anaerobic conditions. Probably the most important fate of pyruvate in microbial metabolism is the anaerobic reaction series that yields ethanol and CO_2. The reaction to form ethanol actually is a two-step process in yeast:

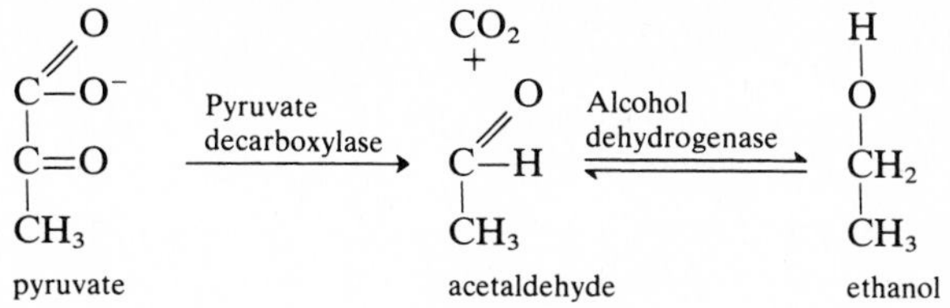

Pyruvate
decarboxylase
Alcohol
dehydrogenase
pyruvate
acetaldehyde
ethanol

The alcohol dehydrogenase reaction requires a reducing agent. NADH, formed from the glyceraldehyde-3-phosphate dehydrogenase system, acts as the reducing agent under anaerobic conditions and leads to alcohol production:

$$\text{Acetaldehyde} + \text{NADH} + \text{H}^+ \longrightarrow \text{ethanol} + \text{NAD}^+$$

This would complete the overall reaction of this fermentation process. The NAD $\rightleftharpoons$ NADH cycle would not even appear in the net reaction, which is

$$\text{Glucose} + 2\text{ADP} + 2\text{P}_i \longrightarrow 2\ \text{ethanol} + 2\text{CO}_2 + 2\text{ATP}$$

UNIT 53 Problems and Exercises

53–1. Draw pH profiles (activity versus pH) of salivary and pancreatic amylase. Assume the stomach to be about pH 2 and the intestine about pH 8. Label the graph accordingly.

53–2. Suggest names for the enzymes of the intestinal mucosal cells that are important in carbohydrate metabolism.

53–3. Citrate is a molecule formed in subsequent metabolic systems using pyruvate. Citrate is also known to be an allosteric effector for the phosphofructokinase reaction. Would you expect high levels of citrate to be a positive or negative effector?

53–4. Assume that you were given a sample of glucose labeled with the ^{14}C isotope at the number 6 position on the molecule.
(a) Where would the ^{14}C be in the molecule glyceraldehyde-3-phosphate?
(b) Where would the ^{14}C be in the molecule dehydroxyacetone phosphate?
(c) Where would the ^{14}C be in the molecule pyruvate?

53–5. What would be the net ATP yield from the catabolism of one molecule of glucose to form two molecules of 3-phosphoglyceric acid:
(a) Under aerobic conditions?
(b) Under anaerobic conditions?

53–6. Lactate is often administered in intravenous injections. Would lactate be a better or poorer energy source than pyruvate for aerobic tissues?

UNIT 54 OXIDATIVE METABOLISM AND THE TCA CYCLE

Pyruvate may be metabolized by any of several different pathways. The reactions of glycolysis yield pyruvate within the cytoplasm of our cells. Pyruvate then passes into the mitochondria, where a multienzyme complex, pyruvate dehydrogenase, is initially involved in a process of oxidative decarboxylation. Finally, the tricarboxylic acid cycle completes the process of converting the carbohydrate to CO_2.

Oxidative Decarboxylation

The first reaction in this complex involves the enzyme pyruvate decarboxylase and a coenzyme, thiamine pyrophosphate (TPP). As the name of the enzyme suggests, this reaction involves a loss of CO_2 and the formation of a two-carbon compound. This actually proceeds by the pyruvate first attaching itself to the TPP coenzyme, which is part of the complex:

$$CH_3-\overset{\overset{O}{\|}}{C}-\overset{\overset{O}{\|}}{C}-O^- + TPP \longrightarrow CH-\overset{\overset{OH}{|}}{\underset{\{\ TPP\ \}}{C}}-\overset{\overset{O}{\|}}{C}-O^-$$

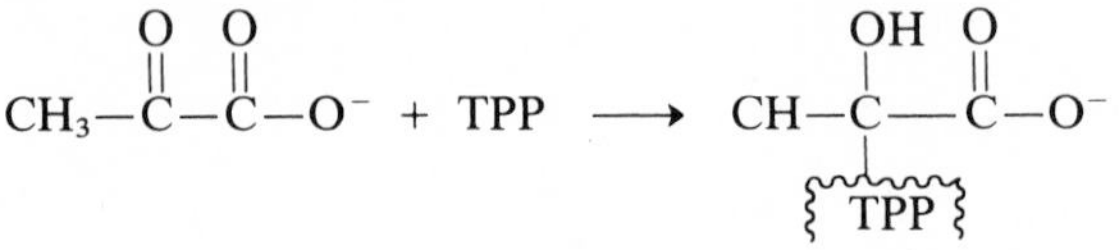

CO_2 is then lost to yield *active acetaldehyde*:

$$CH-\overset{\overset{OH}{|}}{\underset{TPP}{C}}-\overset{\overset{O}{\|}}{C}-O^- \longrightarrow CO_2 + CH-\overset{\overset{OH}{|}}{\underset{TPP}{C}}-H$$

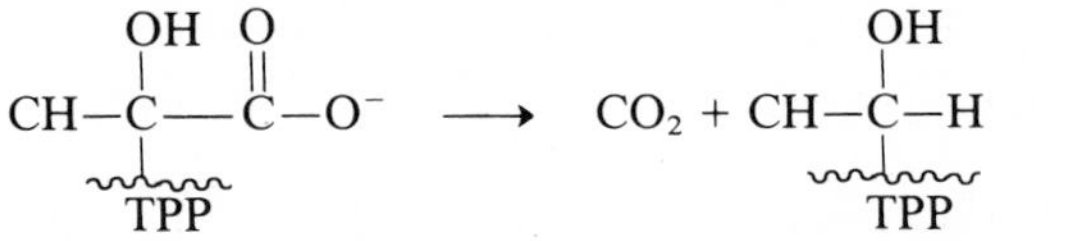

active acetaldehyde

The next step involves the transfer of the acetaldehyde to a second acceptor molecule. The active site of this acceptor is the oxidized form of a lipoic acid molecule that is bound to the surface of the enzyme complex. The transfer is accomplished by the enzyme lipoate acetyltransferase:

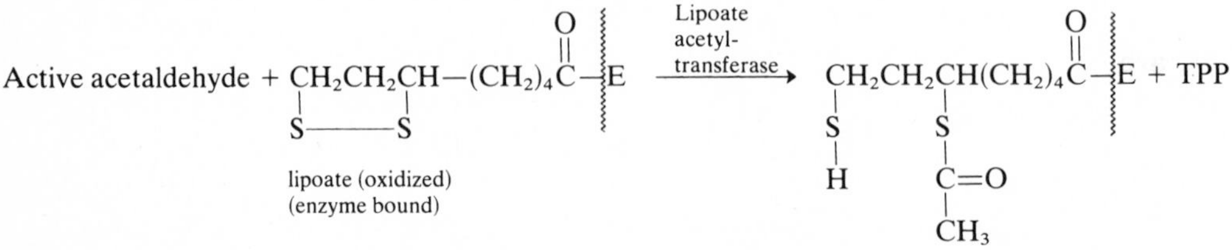

$$\text{Active acetaldehyde} + \underset{\underset{lipoate\ (oxidized)}{\underset{(enzyme\ bound)}{}}}{\overset{\overset{O}{\|}}{CH_2CH_2CH-(CH_2)_4C-E}} \underset{S\text{---}S}{} \xrightarrow[\text{transferase}]{\substack{Lipoate\\ acetyl-}} \underset{\underset{H}{|}\ \underset{\underset{CH_3}{|}}{\overset{|}{C=O}}}{\overset{\overset{O}{\|}}{CH_2CH_2CH(CH_2)_4C-E}} + TPP$$

The released TPP is free to react with another pyruvate.

The next step involves yet another acceptor for the two-carbon compound, coenzyme A. This transfer moves the acetyl group from one sulfhydryl to another:

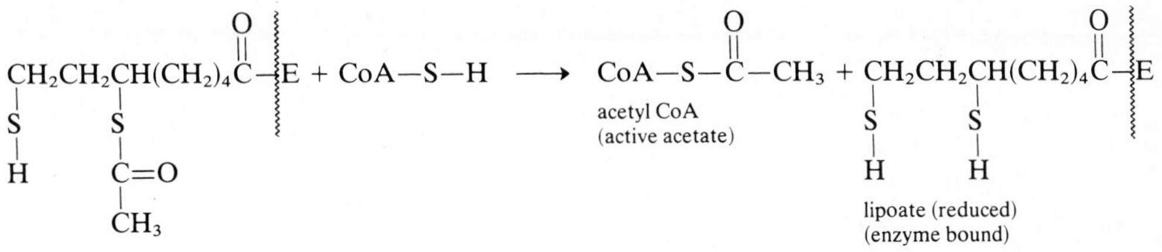
acetyl CoA
(active acetate)
lipoate (reduced)
(enzyme bound)

Acetyl CoA or active acetate is formed. The active prefix is used because of the high group transfer potential that exists if the C—S bond of this molecule is broken. The acetyl CoA again represents a **junction** point in metabolic sequences. There are a number of ways in which this molecule can be formed, as well as a number of ways in which it can be metabolized. These possibilities will be considered separately.

The lipoate must not be left in this reduced form. Coenzymes must always be regenerated to their original form to function as catalysts. Through their regeneration, nature allows a few coenzyme molecules to become involved in the metabolic sequences of tremendous numbers of substrate molecules. Within the mitochondria, the electron-transport system is immediately available for the oxidation of the reduced lipoate:

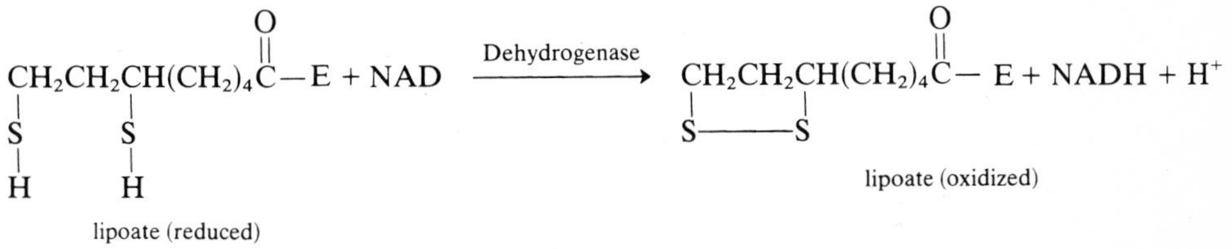
Dehydrogenase
lipoate (reduced)
lipoate (oxidized)

This NADH can drive the formation of three ATP molecules through oxidative phosphorylation. The net multistep oxidative decarboxylation yields:

$$CH_3\overset{\overset{O}{\|}}{C}-\overset{\overset{O}{\|}}{C}-O^- + 3ADP + 3P_i + CoA-S-H \longrightarrow 3ATP + CO_2 + CoA-S-\overset{\overset{O}{\|}}{C}-CH_3$$

Tricarboxylic Acid Cycle

A principal route of metabolism for the acetyl CoA is catabolism through the **citric acid cycle**. This route is also known as the **tricarboxylic acid cycle** (TCA) or **Krebs cycle**, named after Sir Hans Krebs (Nobel laureate, 1953). This cycle is a catalytic unit that acts to oxidize the two carbons of acetyl CoA to CO_2 and H_2O while providing tremendous energy.

The first reaction uses the acetate portion of acetyl CoA to form citric acid. The enzyme citrate synthetase catalyzes the reactions of acetyl CoA with an oxaloacetate to yield citrate:

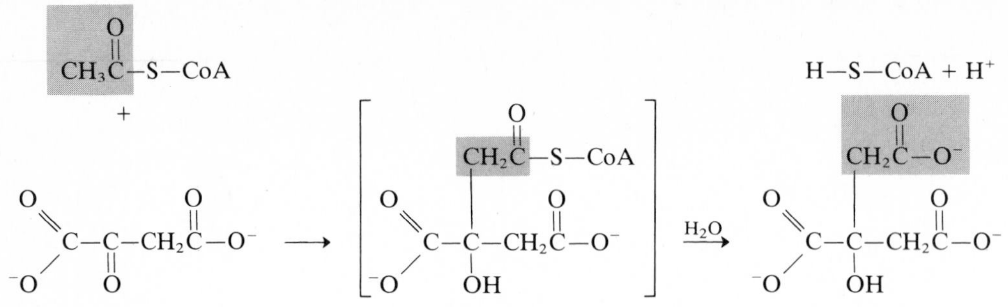

oxaloacetate citrate

This reaction involves the formation of a new C—C bond and the release of coenzyme A in an essentially irreversible fashion. The acetate is committed to utilization in the TCA cycle. In the release of coenzyme A, the coenzyme is also restored.

The citrate synthetase enzyme is subject to a number of control processes. This might be expected because of its operation at the point of entry into this whole scheme of metabolic reactions. These control aspects will be considered after the entire reaction process has been covered.

Citrate is next converted to isocitric acid. This involves the migration of the hydroxyl group from one carbon atom to another. This series of reactions takes place on the surface of the enzyme aconitase:

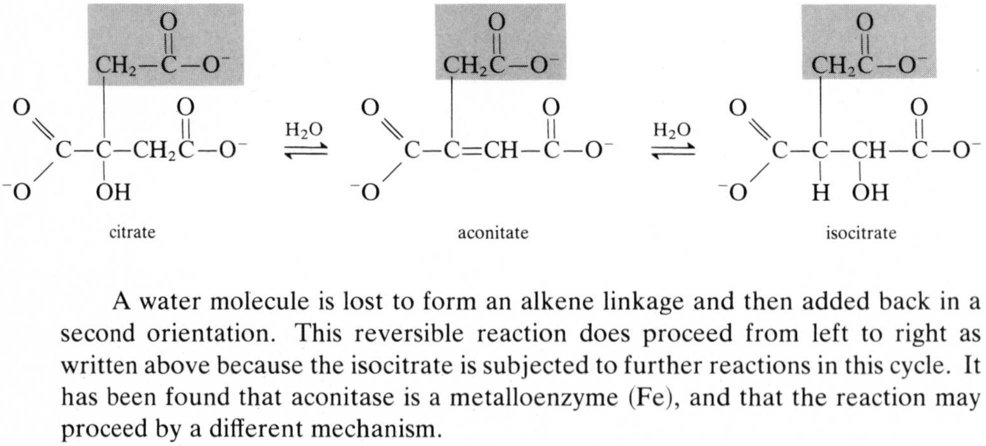

citrate aconitate isocitrate

A water molecule is lost to form an alkene linkage and then added back in a second orientation. This reversible reaction does proceed from left to right as written above because the isocitrate is subjected to further reactions in this cycle. It has been found that aconitase is a metalloenzyme (Fe), and that the reaction may proceed by a different mechanism.

The two $-CH_2COO^-$ groups of citrate are not structurally equivalent owing to the specific interactions and attachments of the substrate and enzyme. Since this is not easily visualized, the two carbon groups from acetyl CoA will be shown within a shaded box.

The stereospecific isocitrate molecule is then subjected to both a loss of CO_2 and a dehydrogenase reaction involving the conversion of a secondary alcohol to a ketone. The loss of CO_2 makes this an irreversible reaction. Isocitrate dehydrogenase involves NAD:

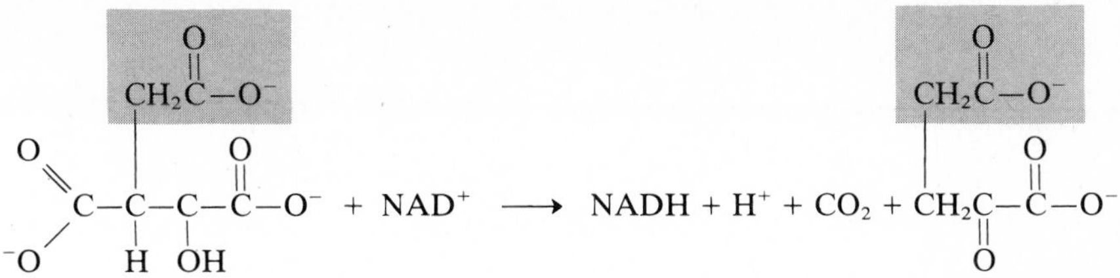

isocitrate α-ketoglutarate

The NADH produced here can be reoxidized through the electron-transport system to drive the formation of three ATP's.

The isocitrate dehydrogenase is another important control point in the TCA. This allosteric enzyme appears to be quite specifically activated by ADP. It is interesting that this positive modification occurs when isocitrate itself is present in low concentrations. In some manner, ADP alters the enzyme to increase the attractiveness of the substrate (isocitrate) to it. This lowers the K_m for isocitrate, as shown in Figure 54–1.

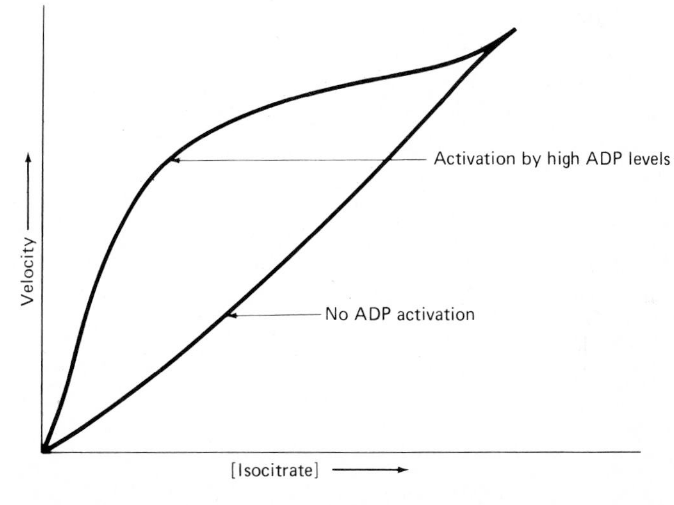
Activation by high ADP levels
Velocity
No ADP activation
[Isocitrate]

Figure 54–1 ADP activation of isocitrate dehydrogenase. Note that the V_{max} does not appear to change but that the enzyme is more active at low substrate concentrations.

NADH and ATP may act as negative effectors for this enzyme. Thus at low ATP levels the TCA cycle is stimulated and energy is generated. As NADH and ATP accumulate within the cell, the TCA system slows down to conserve its substrate molecules.

The oxidative decarboxylation of α-ketoglutarate proceeds in a mechanism similar to the one discussed for pyruvate. Cofactors such as thiamine pyrophosphate, lipoate, NAD, and the coenzyme A react under the control of a multienzyme complex of α-ketoglutarate dehydrogenase:

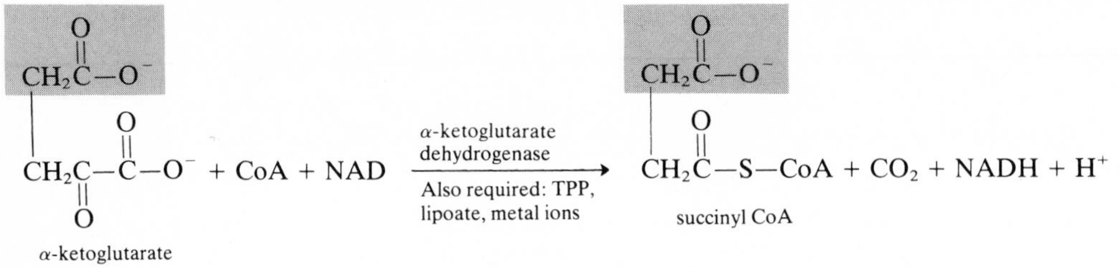

$$\underset{\alpha\text{-ketoglutarate}}{\underset{\displaystyle \overset{\text{O}}{\underset{\displaystyle \|}{\text{CH}_2\text{C}-\overset{\text{O}}{\underset{\|}{\text{C}}}-\text{O}^-}}}{\overset{\displaystyle \overset{\text{O}}{\underset{\|}{\text{CH}_2\text{C}-\text{O}^-}}}{}} + \text{CoA} + \text{NAD} \quad \xrightarrow[\substack{\text{Also required: TPP,}\\ \text{lipoate, metal ions}}]{\substack{\alpha\text{-ketoglutarate}\\ \text{dehydrogenase}}} \quad \underset{\text{succinyl CoA}}{\overset{\displaystyle \overset{\text{O}}{\underset{\|}{\text{CH}_2\text{C}-\text{O}^-}}}{\underset{\displaystyle \overset{\text{O}}{\underset{\|}{\text{CH}_2\text{C}-\text{S}-\text{CoA}}}}{}} + \text{CO}_2 + \text{NADH} + \text{H}^+$$

Succinyl CoA could be designated *active succinate* as we did for active acetate. In the reaction using acetate, the energy released during this bond cleavage was used to bring about the C—C bond formation leading to citrate. In the case of succinyl CoA, the energy is trapped by a type of substrate phosphorylation.

An enzyme, succinyl thiokinase, is able to couple the loss of coenzyme A with the phosphorylation of a nucleotide. ATP is not formed directly, but GTP acts as an intermediate. One mole of ATP can be formed here for each mole of succinate formed. These two coupled reactions may be summarized:

$$\overset{\displaystyle \overset{\text{O}}{\underset{\|}{\text{CH}_2\text{C}-\text{O}^-}}}{\underset{\displaystyle \overset{\text{O}}{\underset{\|}{\text{CH}_2\text{C}-\text{S}-\text{CoA}}}}{}} + \text{GDP} + \text{P}_i \quad \xrightarrow{\substack{\text{Succinyl}\\ \text{thiokinase}}} \quad \overset{\displaystyle \overset{\text{O}}{\underset{\|}{\text{CH}_2\text{C}-\text{O}^-}}}{\underset{\displaystyle \overset{\text{O}}{\underset{\|}{\text{CH}_2\text{C}-\text{O}^-}}}{}} + \text{CoA} + \text{GTP}$$

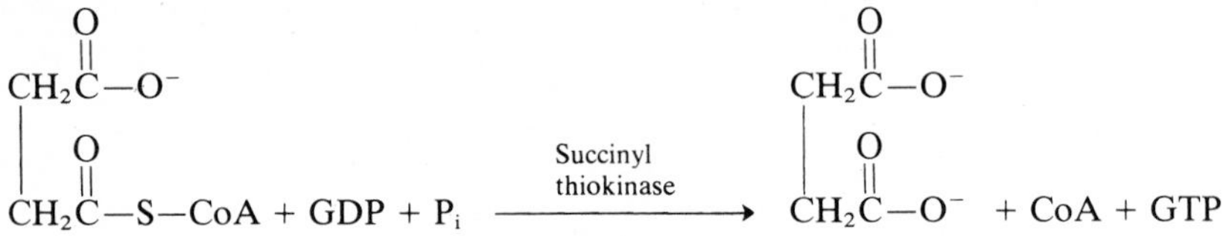

$$\text{GTP} + \text{ATP} \quad \xrightarrow{\text{Nucleotide diphosphokinase}} \quad \text{GDP} + \text{ATP}$$

or

$$\text{Succinyl CoA} + \text{P}_i + \text{ADP} \quad \longrightarrow \quad \text{succinate} + \text{CoA} + \text{ATP}$$

At this point, 2 mol of CO_2 have been formed from the acetate portion of acetyl CoA originally entering the TCA cycle. The stereospecific way in which the enzymes carried the original acetate carbons has been shown in the shaded box. From this it appears that neither of the two carbons of acetate was actually converted to CO_2 molecules. The cycle must continue for another complete turn before these specific carbon atoms become oxidized. This emphasizes the **cyclic** nature of this process as well as the great degree of specificity in which substrates are acted upon by enzymes. Isotopic tracer studies have clearly proved that this is indeed the way in which acetate molecules undergo these reactions. Once the succinate molecule is formed, it does not exhibit this specificity. Unlike the previous molecules considered in the TCA cycle, the succinate is truly a symmetrical molecule. The carboxyl groups of succinate are both chemically and enzymatically indistinguishable. We cannot use the shaded box to define a portion of the molecule in subsequent reactions.

The TCA cycle continues with a somewhat different dehydrogenase reaction. An alkene linkage is formed from an alkane by the action of the succinate dehydrogenase. The electron acceptor in this case is FAD, rather than NAD. Fumarate forms

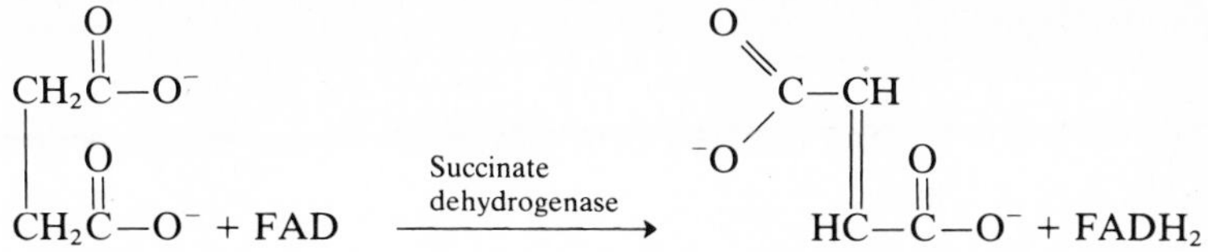

Succinate
dehydrogenase

succinate fumarate

Note that the carboxyl groups are in the trans orientation in reference to the alkene linkage. The cis product does not form by the action of this enzyme.

The $FADH_2$ formed here can become oxidized and provide the energy necessary to drive the formation of 2 mol of ATP by oxidative phosphorylation. This is a case where the P/O ratio is 2.

Succinate dehydrogenase has been shown to be inhibited by competitive inhibitors such as malonate. Malonate is not normally present within cells in significant amounts, however. A more important competitive inhibitor from a physiological point of view is oxaloacetate. Oxaloacetate is the first member of this cycle. When oxaloacetate is present in relatively high levels, it acts to inhibit the succinate dehydrogenase and thus influences its own formation. This allosteric effector exercises an important level of control.

The TCA cycle continues with the addition of water (hydration) to the alkene linkage of fumarate:

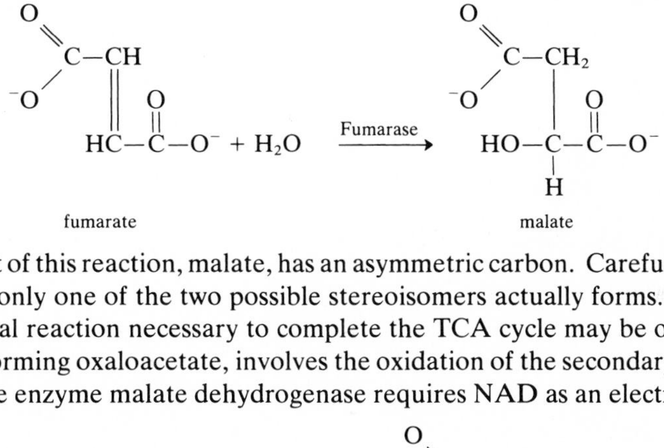

Fumarase

fumarate malate

The product of this reaction, malate, has an asymmetric carbon. Careful studies have shown that only one of the two possible stereoisomers actually forms.

The final reaction necessary to complete the TCA cycle may be obvious. This next step, forming oxaloacetate, involves the oxidation of the secondary alcohol to a ketone. The enzyme malate dehydrogenase requires NAD as an electron acceptor.

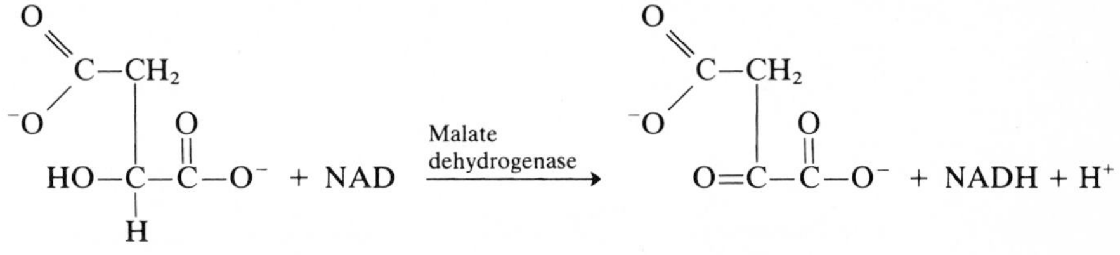

Malate
dehydrogenase

malate oxaloacetate

The oxaloacetate formed is available to condense with another active acetate and start the cycle over again, as summarized in Figure 54–2.

One turn of the TCA cycle yields 2 mol of CO_2 and a tremendous energy yield. There are three dehydrogenase reactions involving NAD as the electron acceptor and one reaction involving FAD. These reduced coenzymes can drive the formation

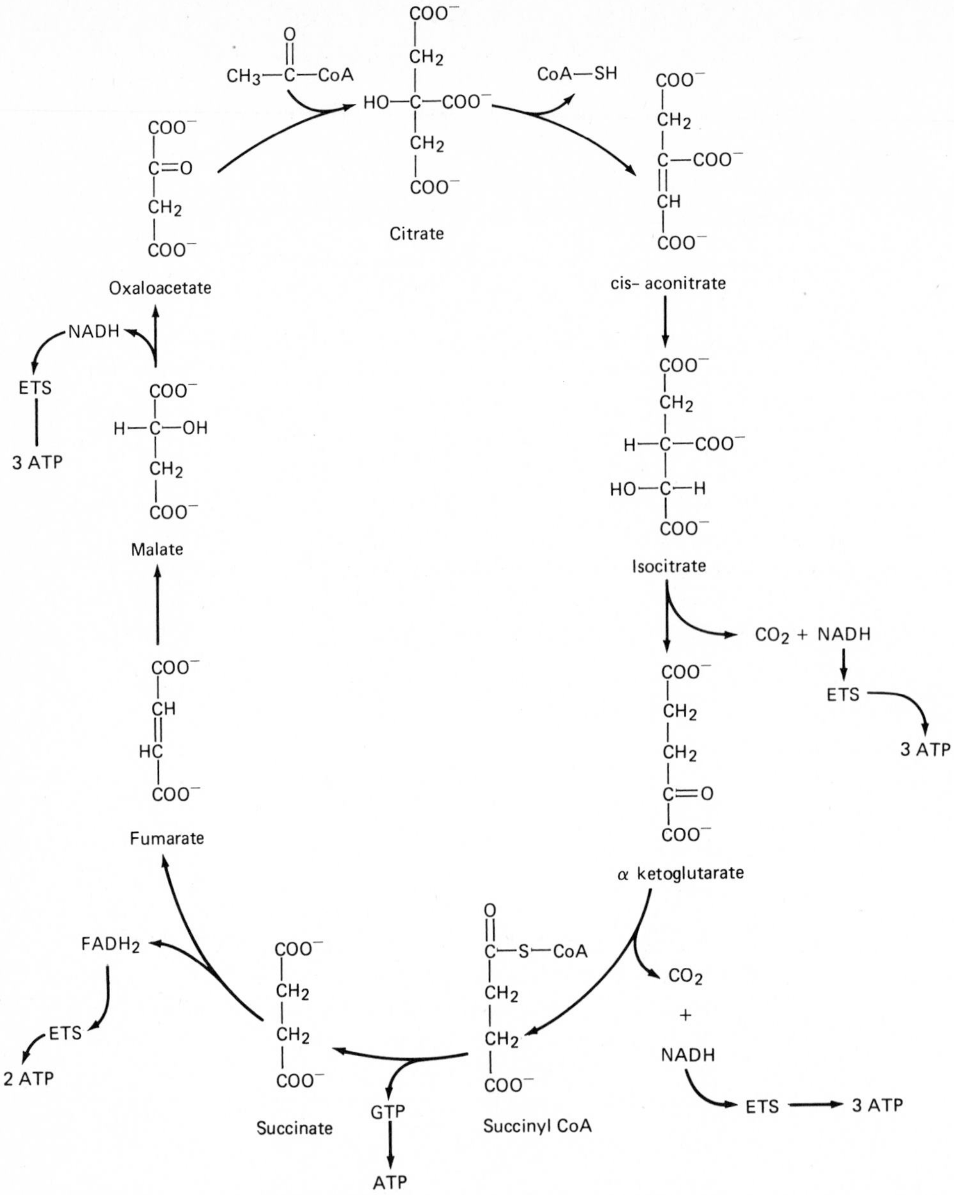
Citrate
CoA—SH
cis- aconitrate
Isocitrate
CO_2 + NADH
ETS
3 ATP
α ketoglutarate
CO_2
+
NADH
ETS
3 ATP
Succinyl CoA
GTP
ATP
Succinate
$FADH_2$
ETS
2 ATP
Fumarate
Malate
NADH
ETS
3 ATP
Oxaloacetate

Figure 54–2 The tricarboxylic acid cycle (TCA).

of 11 ATP through the process of oxidative phosphorylation. The one substrate phosphorylation adds another ATP for a *total of* 12 *ATP molecules* formed in one turn of the cycle. It now seems most appropriate to label the mitochondrion as the powerhouse of the cell.

The TCA cycle has other important functions besides providing energy. In some cases the various intermediates of the cycle are precursors of molecules needed by the cell, especially amino acids. The TCA cycle can either provide the carbon skeleton for some amino acid biosynthesis or provide a method for their catabolism. The TCA cycle not only has widespread use in many species throughout nature, it also has a central role in many metabolic systems.

Pyruvate Carboxylase

One question remains for this catalytic system. How does nature provide the molecules to start the reaction? In the liver and kidney an enzyme, pyruvate carboxylase, can form oxaloacetate from pyruvate in a rather complex reaction:

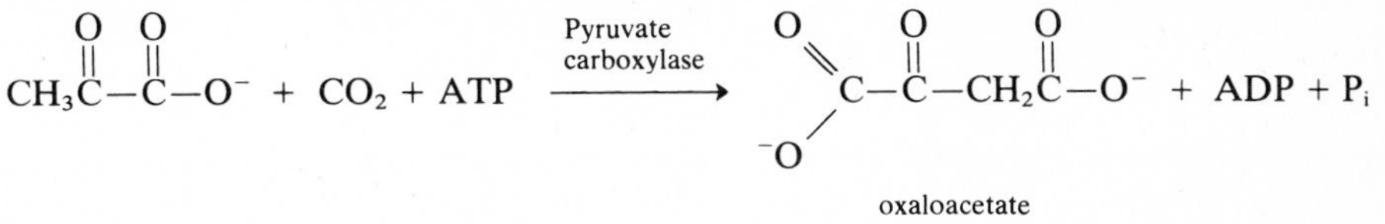

Pyruvate carboxylase
oxaloacetate

This reaction actually occurs on a multiunit enzyme and requires acetyl CoA as a positive allosteric effector. The vitamin biotin is involved in this CO_2 fixing reaction. Biotin becomes attached to the enzyme, and CO_2 becomes attached to one of the nitrogen atoms of the biotin. This is one of the few cases where CO_2 is fixed into an organic molecule in animal cells.

UNIT 54 Problems and Exercises

54–1. Show an example of a reaction in which each of the following vitamins is involved: (a) niacin, (b) riboflavin, (c) biotin.

54–2. Show the reactions in the TCA cycle where CO_2 is expelled.

54–3. How many ATP molecules could be formed from one glucose if it were catabolized by a combination of aerobic glycolysis and the TCA cycle?

54–4. In animal systems, the TCA cycle is unidirectional. Which reactions are involved in accounting for the unidirectional nature of this cycle?

54–5. When the CoA portion is removed from succinyl CoA, substrate phosphorylation results. No phosphorylation accompanies the loss of the CoA portion of acetyl CoA as this molecule enters the TCA cycle. Account for this difference.

54–6. Both malonate and oxaloacetate are known to inhibit succinate dehydrogenase. Indicate how these molecules might act as competitors of succinate, and why they cannot undergo the dehydrogenase reaction.

54–7. Given a sample of labeled pyruvate, $^{14}CH_3\overset{\overset{O}{\|}}{C}-\overset{\overset{O}{\|}}{C}-O^-$,

(a) Show which carbon(s) of succinate would bear the ^{14}C label if this pyruvate was introduced into a TCA system that had been inhibited by malonate.

(b) Show which carbon atoms of citrate would bear the ^{14}C label if this pyruvate was introduced into a liver system that had a specific inhibitor for the aconitase system.

UNIT 55 ALTERNATIVE SCHEMES OF CARBOHYDRATE METABOLISM

A great variety of different molecules was encountered along the pathway just considered, and different end products resulted in some cases. There are several unanswered questions remaining that are of biochemical interest in carbohydrate metabolism.

Fructose Metabolism

Only the metabolism of glucose has been considered, but there are other monosaccharides of importance in our diets. How do these other monosaccharides enter into the metabolic pathways? Fructose is metabolized by the rather nonspecific hexokinase:

$$\text{D-fructose} + \text{ATP} \xrightleftharpoons{\text{Hexokinase}} \text{D-fructose-6-phosphate} + \text{ADP}$$

Since this is an intermediate in the pathway already considered, the rest of the metabolism of fructose proceeds by this scheme. The other intermediates, end products, and energy yields would be the same as from glucose.

There is another pathway for fructose in the liver that is somewhat unique. A specific enzyme, fructokinase, is able to phosphorylate D-fructose at a different point on the molecule:

$$\text{D-fructose} + \text{ATP} \xrightleftharpoons{\text{Fructokinase}} \text{D-fructose-1-phosphate} + \text{ADP}$$

Since this molecule was not on the pathway previously covered, a new entry system must be involved. It has been shown that the fructose-1-phosphate can be cleaved by a special aldolase found in the liver:

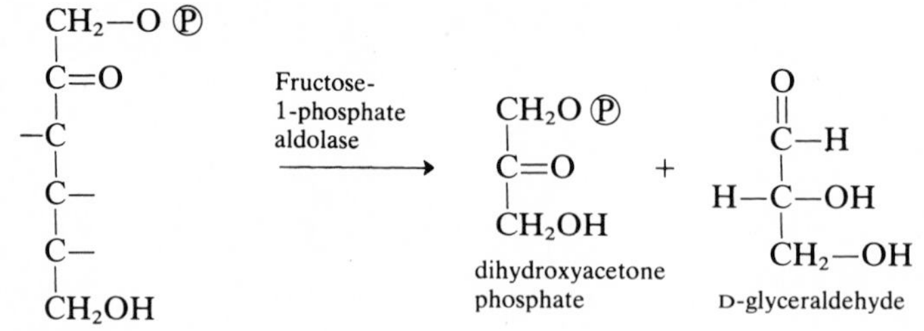
Fructose-
1-phosphate
aldolase
dihydroxyacetone
phosphate
D-glyceraldehyde

fructose-1-phosphate

D-glyceraldehyde must be phosphorylated by ATP before it can enter the metabolic system previously considered.

Galactose Metabolism

The metabolism of fructose closely parallels that of glucose, but galactose has a rather different metabolic sequence. Galactose, present as a constituent of the milk sugar lactose, appears to have a special system for entering the glycolytic scheme. Galactose enters the liver and is phosphorylated by ATP using the enzyme galactokinase:

$$\text{D-galactose} + \text{ATP} \xrightarrow{\text{Galactokinase}} \text{D-gal-1-phos} + \text{ADP}$$

In the livers of adults, the enzyme uridine diphosphate galactose pyrophosphorylase uses galactose-1-phosphate with uridine triphosphate (UTP):

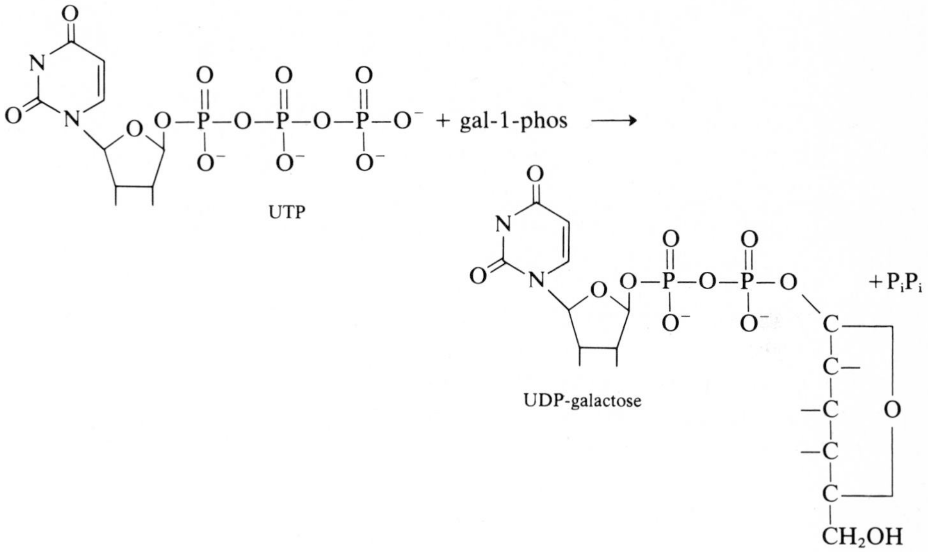
+ gal-1-phos
UTP
UDP-galactose
$+P_iP_i$

The UDP-galactose undergoes an epimerization at the number 4 carbon atom of the galactose, converting it to glucose:

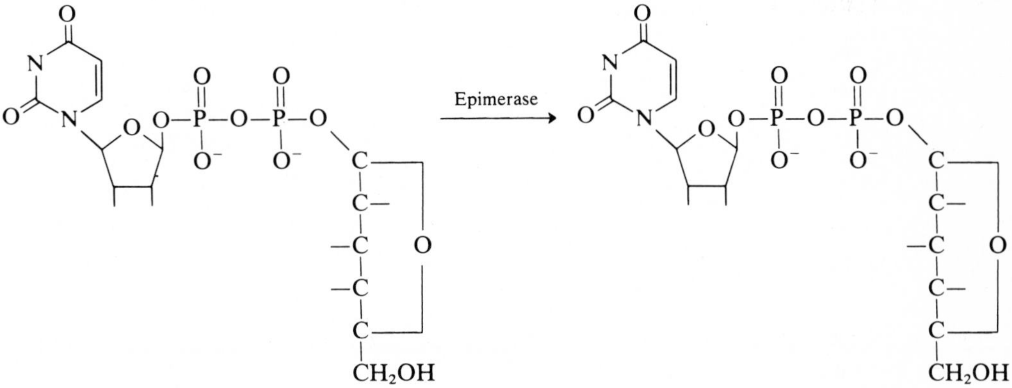
Epimerase

UDP-galactose UDP-glucose

The UDP glucose reacts with another molecule of galactose-1-phosphate and the enzyme galactose phosphate uridyl transferase:

$$\text{UDP-glu} + \text{gal-1-phos} \xrightarrow[\text{transferase}]{\text{Gal-phos uridyl}} \text{UDP-gal} + \text{glu-1-phos}$$

This series of reactions essentially converts the galactose to a glucose phosphate:

Recycle (UDP-gal → UDP-gal)

$$\text{Galactose} \rightarrow \text{gal-1-phos} \rightarrow \text{UDP-gal} \rightarrow \left.\begin{array}{r}\text{UDP-glu} \rightarrow \\ \\ \text{gal-1-phos} \rightarrow\end{array}\right\} \begin{array}{c}\text{UDP-gal} \\ + \\ \text{glu-1-phos}\end{array}$$

or an overall net reaction

$$\text{D-galactose} + \text{ATP} \longrightarrow \text{glucose-1-phos} + \text{ADP}$$

A deficiency of the uridyl transferase enzyme causes a disease known as **galactosemia**. This disease results from a specific genetic defect and is found in about 1 in every 70,000 people. If infants lacking this transferase are fed milk (lactose), the galactose levels of the blood and urine will increase dramatically. High levels of galactose in the blood have severe consequences, such as nerve tissue damage, growth failure, and cataract formation. Generally, this enzyme deficiency disease can be successfully overcome by restricting the milk intake of the individual involved. It is usually the case that as the child matures alternative routes for the metabolism of galactose develop.

The glucose-1-phosphate formed from galactose then enters the normal glycolytic pathways. An enzyme, phosphoglucomutase, catalyzes the reversible reaction involving the migration of the phosphate group on the glucose molecule:

$$\text{Glu-1-phos} \xrightleftharpoons{\text{Phosphoglucomutase}} \text{glu-6-phos}$$

Starch and Glycogen Metabolism

The control and regulation of cellular and blood levels of carbohydrates involves two distinct processes: (1) the synthesis of glycogen (**glycogenesis**), and (2) its breakdown (**glycogenolysis**). The balance between these two processes is normally under hormonal control.

The synthesis of glycogen in animal cells can be traced from glucose-6-phosphate to glucose-1-phosphate. The glucose-1-phosphate reacts to form a nucleoside phosphate derivative, UTP:

$$\text{Glu-1-phos} + \text{UTP} \xrightleftharpoons[]{\substack{\text{UDP-glucose} \\ \text{pyrophosphorylase}}} \text{UDP-glu} + P_iP_i$$

The glucose portion of the UDP-glucose supplies the repeating monomeric unit of the polysaccharide glycogen. Glucose is transferred from the UDP to the nonreducing end of a growing chain of glucose molecules. The glucose is added by an $\alpha(1 \rightarrow 4)$ linkage by the enzyme glycogen synthetase. This can be visualized as follows:

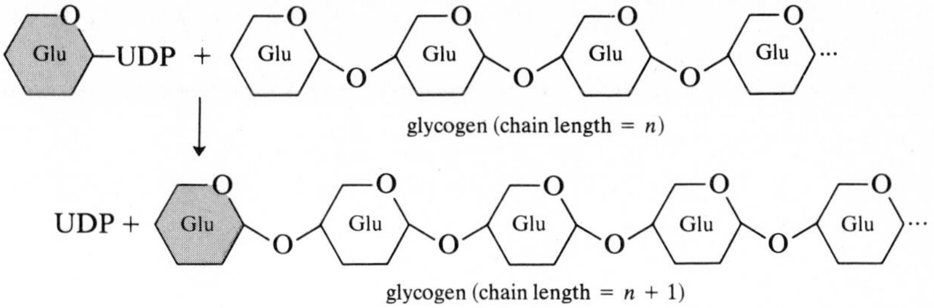
Glu
UDP +
Glu
Glu
Glu
Glu
glycogen (chain length = n)
UDP +
Glu
Glu
Glu
Glu
Glu
glycogen (chain length = n + 1)

Summary:

$$\text{UDP-glu} + (\text{glu})_n \longrightarrow \text{UDP} + (\text{glu})_{n+1}$$

The branching structures are brought about by an enzyme known as a transglycosidase or **branching enzyme**. This enzyme removes a fragment from a growing end of an $\alpha(1 \rightarrow 4)$ chain and reattaches it to a number 6 position at spaces of about 8 to 12 units apart. This is an $\alpha(1 \rightarrow 6)$ linkage, which makes a branch point for further elongation of an $\alpha(1 \rightarrow 4)$ linkage:

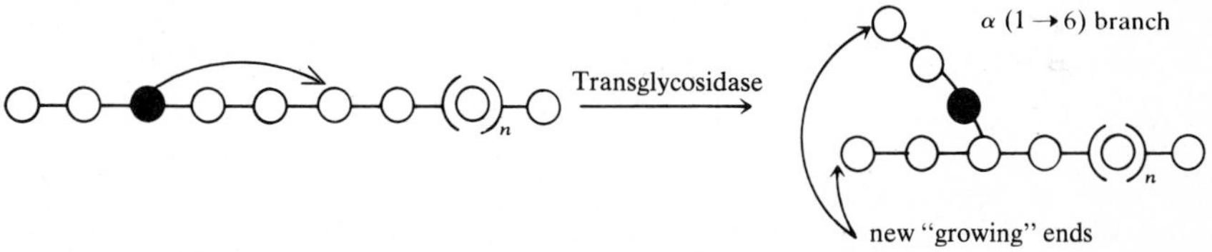
Transglycosidase
α (1 → 6) branch
new "growing" ends

In plants, where starch biosynthesis is of extreme importance, an analogous biosynthetic system is involved. Here, the enzyme catalyzing the synthetic process is known as **amylose synthetase**. Other nucleoside phosphate derivatives may be involved in plants, like ADP-glucose.

Glycogenolysis involves the breakdown of the polymeric glycogen. Each glucose moiety of glycogen is released from glycogen by a phosphorolysis rather than by a simple hydrolysis. The enzyme glycogen phosphorylase catalyzes this phosphorolysis. Inorganic phosphate is added to the terminal glucose at the nonreducing end of the glycogen to yield a glucose phosphate:

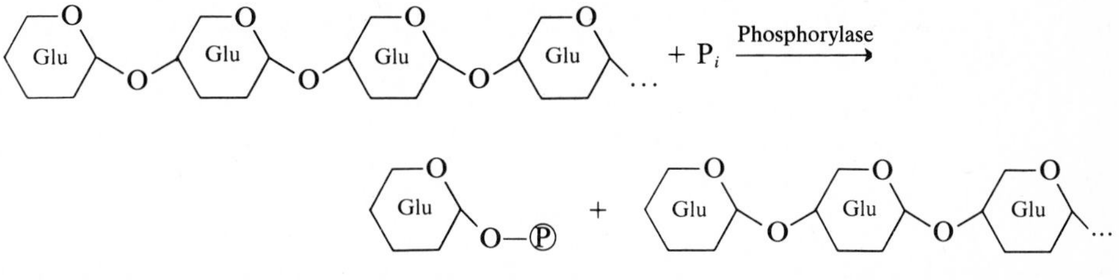
Glu
Glu
Glu
Glu
+ P_i
Phosphorylase
Glu
O—Ⓟ
+
Glu
Glu
Glu

Summary:

$$(\text{glucose})_n + P_i \longrightarrow (\text{glucose})_{n-1} + \text{glu-1-phos}$$

The phosphorylase enzyme can continue to attack the $\alpha(1 \rightarrow 4)$ linkages of the newly exposed glucose until it reaches a point at which it can no longer react. This remaining macromolecule is sometimes known as **limit dextrin**. At or near the $1 \rightarrow 6$ branch of the glycogen another process becomes involved, since phosphorylase cannot act upon these linkages. A special debranching enzyme, $\alpha(1 \rightarrow 6)$ glucosidase, is necessary.

At one time it was thought that the phosphorylase enzyme could control both the synthetic and degradative processes. Although it is possible to force phosphorylase to act in either direction under proper experimental circumstances, the living cell uses this enzyme for degradation. Phosphorylase has two forms, the active form (a) and a much less active form (b). The rather inactive (b) form consists of two separate but identical protein subunits of molecular weight of about 100,000 each. The most common way of converting form (b) to the highly active form (a) involves four phosphorylation processes, each using an ATP. This is mediated by an enzyme, phosphorylase kinase. The result of this is to form an active tetramer, which has four phosphate groups attached. This can be shown as

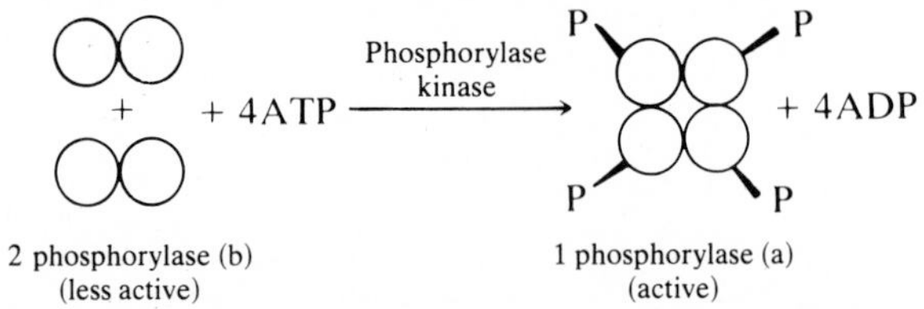
+
+ 4ATP
Phosphorylase kinase
P
P
P
P
+ 4ADP
2 phosphorylase (b)
(less active)
1 phosphorylase (a)
(active)

The activation of this kinase is dependent upon the ATP levels present within the cell and is under additional control. Phosphorylase kinase itself has an active and an inactive form. A regulator molecule, **cyclic AMP**, stimulates the activation of the phosphorylase by activating yet another enzyme kinase. The levels of cyclic AMP are subject to hormonal control. Cyclic AMP may be formed from ATP by an enzyme adenylate cyclase. The synthesis of the cyclic AMP is triggered in the muscle by epinephrine and glucagon. These hormones, produced in the adrenal and pancreas, respectively, have an indirect but profound influence on the availability of glucose to the cell. Although this appears to be a rather complex series of reactions within the body, it can be seen as a rapidly occurring cascade effect in Figure 55–1.

This series of activation steps also acts as an **amplifier** system. Each reaction has a catalytic influence upon the following one. Only a few molecules of epinephrine or glucagon are necessary to catalyze the initial reaction. The influence of the hormone is multiplied many times before it actually reaches the process of releasing glucose from its storage form.

Phosphorylase (b) can exhibit a significant degree of activity in the presence of relatively high levels of AMP. This represents a separate control process from the one described above and is desirable when the cell is in need of energy.

Glycogen synthetase also is known to exist in both active and inactive forms. In the liver and the resting muscle, one form is active in the presence of relatively high concentrations of glucose-6-phosphate. The storage of glycogen is promoted when the glucose-6-phosphate levels are high. In the contracting muscle and in the liver under energy demand conditions, the synthetic system is effectively shut down, and the degradation of glucose is accelerated. The scheme is summarized in Figure 55–2.

It is necessary to provide molecules for the biosynthetic reactions leading to the formation of glycogen. The glucose could come from an exogenous (outside) source, or it could arise by a reversal of some of the catabolic processes previously

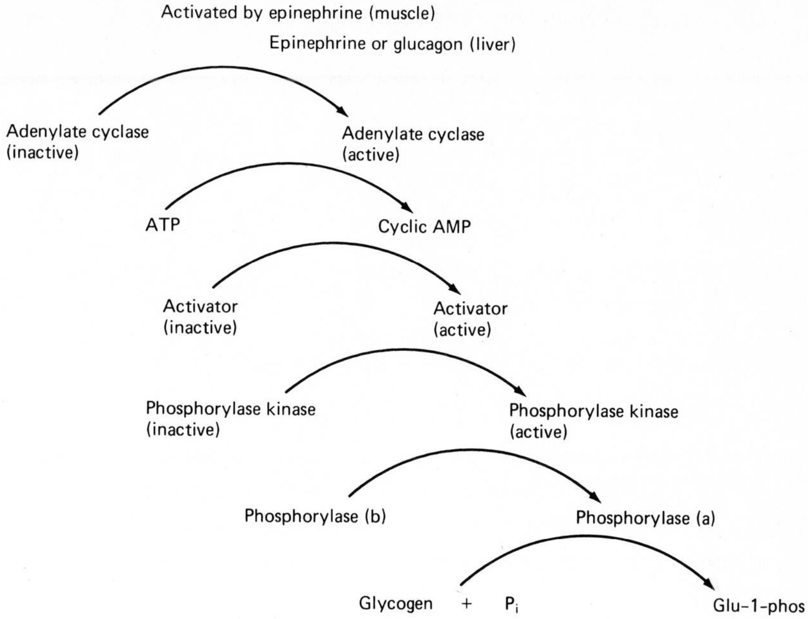

Activated by epinephrine (muscle)
Epinephrine or glucagon (liver)
Adenylate cyclase (inactive)
Adenylate cyclase (active)
ATP
Cyclic AMP
Activator (inactive)
Activator (active)
Phosphorylase kinase (inactive)
Phosphorylase kinase (active)
Phosphorylase (b)
Phosphorylase (a)
Glycogen + P_i
Glu-1-phos

Figure 55–1 The cascade effect of processes controlling the activity of glycogen phosphorylase, leading to the formation of glu-1-phos.

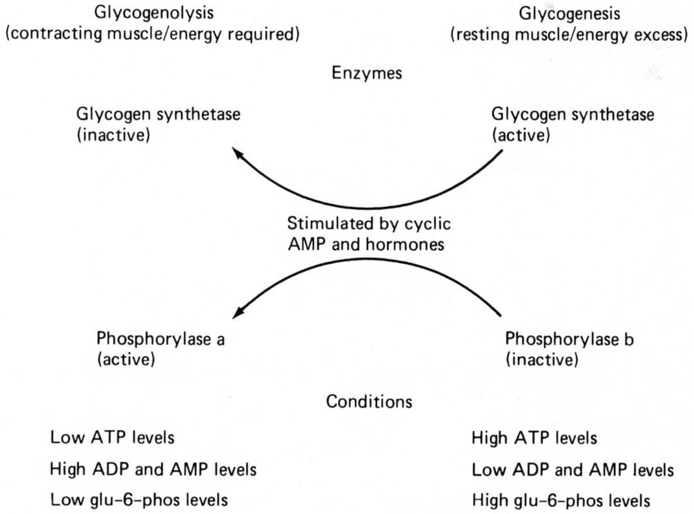

Glycogenolysis (contracting muscle/energy required)
Glycogenesis (resting muscle/energy excess)
Enzymes
Glycogen synthetase (inactive)
Glycogen synthetase (active)
Stimulated by cyclic AMP and hormones
Phosphorylase a (active)
Phosphorylase b (inactive)
Conditions
Low ATP levels
High ADP and AMP levels
Low glu-6-phos levels
High ATP levels
Low ADP and AMP levels
High glu-6-phos levels

Figure 55–2 Regulation of glycogenesis and glycogenolysis in the muscle and liver.

considered. The formation of a new sugar molecule from other molecules is known as **gluconeogenesis**.

Reverse Glycolysis

The glycolytic pathway and the TCA cycle are nearly unidirectional. Although many reactions of these two systems can operate in either direction, the following do not:

1. Phosphoenol pyruvate $\longrightarrow$ pyruvate.
2. Fructose-6-phosphate + ATP $\longrightarrow$ fructose-1,6-diphosphate + ADP.
3. Glucose + ATP $\longrightarrow$ glucose-6-phosphate + ADP.
4. Glycogen + P_i $\longrightarrow$ glucose-1-phosphate.

A problem arises as to how these can be circumvented to allow the reverse process of glycolysis to occur. These can be considered in reverse order:

4. The special systems of glycogen synthetase operate in the synthetic direction and thus circumvent the irreversibility of phosphorylase.

3. The special phosphatase has already been considered, which catalyzes

$$\text{glucose-6-phosphate} + H_2O \longrightarrow \text{glucose} + P_i$$

2. A similar reaction system has been found to occur in the case of the fructose diphosphate. A separate enzyme, diphosphofructose phosphatase, catalyzes

$$\text{fructose-1,6-diphosphate} + H_2O \longrightarrow \text{fructose-6-phosphate} + P_i$$

1. The remaining irreversible step is the more complex one. This involves a reversal or bypass whereby phosphoenolpyruvate can be formed from pyruvate. The first step involves a reaction that takes place within the mitochondria. An enzyme, pyruvate carboxylase, can fix a CO_2 into an organic molecule:

$$CH_3\overset{\overset{\displaystyle O}{\|}}{C}-\overset{\overset{\displaystyle O}{\|}}{C}-O^- + CO_2 + ATP \longrightarrow \quad {}^-O\underset{\diagup}{\overset{\diagdown}{C}}-CH_2\overset{\overset{\displaystyle O}{\|}}{C}-\overset{\overset{\displaystyle O}{\|}}{C}-O^- + ADP + P_i$$

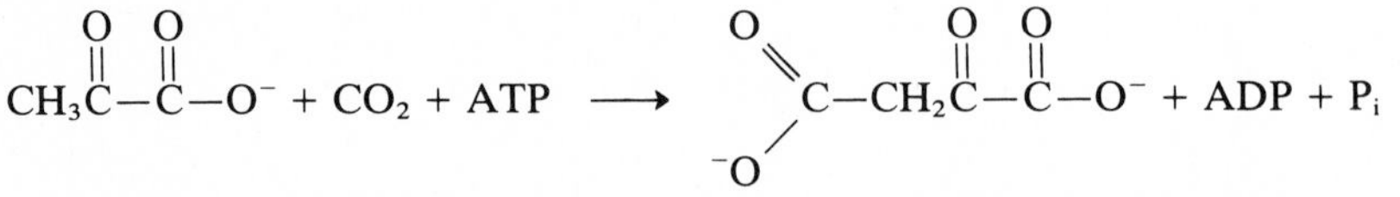

pyruvate   oxaloacetate

The oxaloacetate is reduced to malate, which is able to diffuse out of the mitochondria into the cytoplasm. The malate is then reoxidized into oxaloacetate. This is just a reversible set of reactions to allow the molecule to move from one compartment to another within a cell.

The bypass reaction system continues by using an enzyme phosphoenolpyruvate carboxykinase:

$$ {}^-O\underset{\diagup}{\overset{\diagdown}{C}}-CH_2\overset{\overset{\displaystyle O}{\|}}{C}-\overset{\overset{\displaystyle O}{\|}}{C}-O^- + GTP \longrightarrow CH_2\overset{\overset{\displaystyle \textcircled{P}}{\underset{\displaystyle |}{O}}}{C}-\overset{\overset{\displaystyle O}{\|}}{C}-O^- + CO_2 + GDP$$

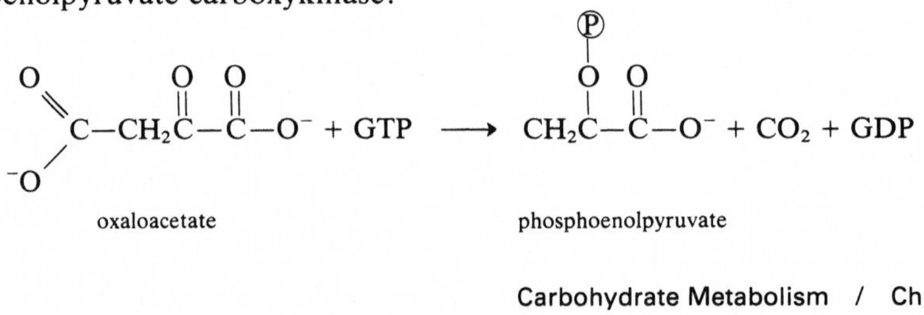

oxaloacetate   phosphoenolpyruvate

The phosphoenolpyruvate then follows a reverse set of reactions "back up" the glycolytic pathway.

Pentose Phosphate Pathway

Although glycolysis and the TCA cycle are important, they are not the only reaction pathways known. If that were so, how would we account for the formation of five carbon sugars, like ribose and deoxyribose? Many cells have another pathway, the **pentose phosphate pathway** (PPP), which is found within the cytoplasm. This pathway is especially important in the liver, adrenal gland, and mammary gland, but is lacking in the skeletal muscle.

This pathway starts with glucose-6-phosphate and an enzyme, glucose-6-phosphate dehydrogenase, which catalyzes a redox reaction specifically involving NADP:

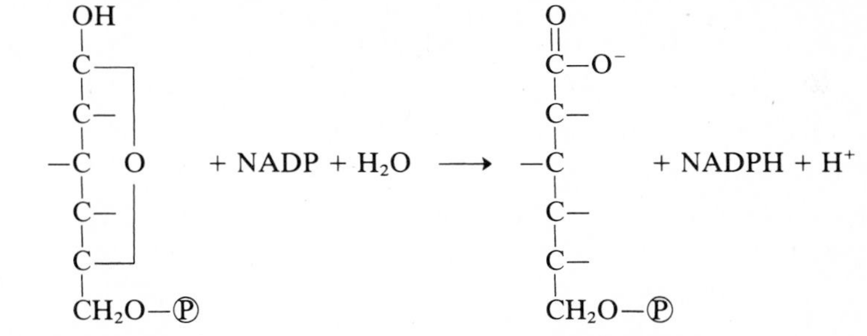
+ NADP + H₂O
+ NADPH + H⁺

glucose-6-phosphate   6-phosphogluconate

A second dehydrogenase follows this reaction, again using NADP. An enzyme, 6-phosphogluconate dehydrogenase, catalyzes

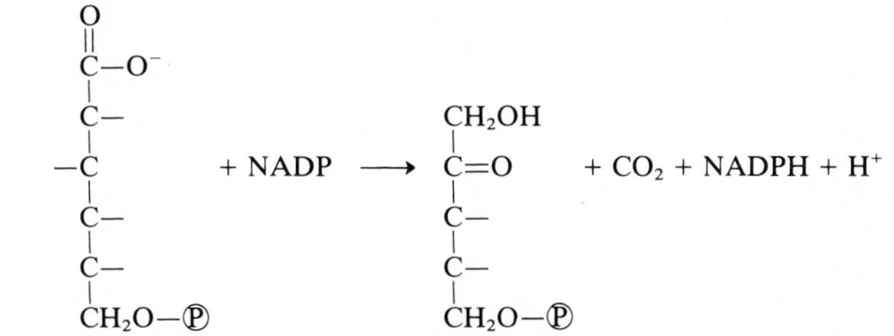
+ NADP
+ CO₂ + NADPH + H⁺

6-phosphogluconate   ribulose-5-phosphate

The reactions to this point in the PPP are very important for several reasons.

1. This is the only point in this system where CO_2 is released.
2. These are the only two reactions where redox reactions occur. Two NADPH molecules are generated for each glucose-6-phosphate taken to this point. These two NADPH molecules could be considered energetically equivalent to six ATP if combined with an oxidative phosphorylation system. This is not the way in which these NADPH molecules are usually

used, however. NADPH is the biochemical reducing agent of choice. The NADPH is reoxidized by coupled reduction of other substrates. Tissues that have the PPP are those which are involved in many synthetic reactions involving reductive steps, liver (fatty acids), adrenal (steroids), and mammary (fatty acids).

3. This is the first synthesis of a sugar containing five carbon atoms.
4. All reactions beyond this point in the PPP only involve reversible interconversions of the skeleton or the substituents of carbohydrate molecules.

Ribulose-5-phosphate may become involved in one of two possible pathways. One of these is an isomerization and the other an epimerization. These interconversions and those taking place beyond this step are shown in Figure 55–3. The specific three-, four-, and seven-carbon sugars are not shown. The main point of this phase of the pathway is to convert the various intermediates, not needed in other

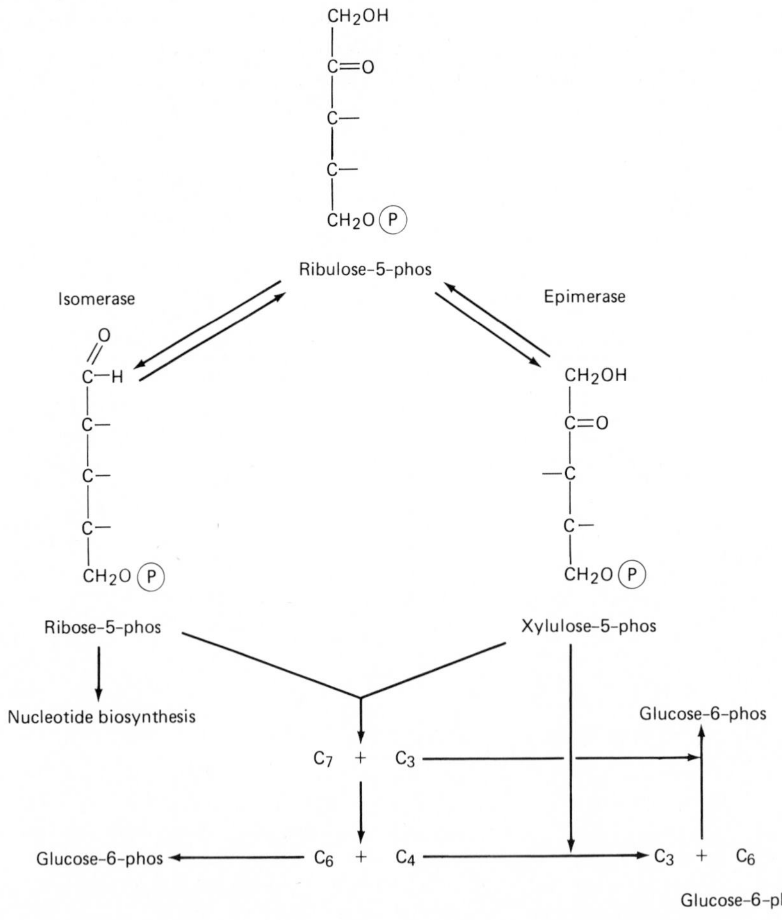

Ribulose-5-phos
Isomerase
Epimerase
Ribose-5-phos
Xylulose-5-phos
Nucleotide biosynthesis
Glucose-6-phos
C7 + C3
Glucose-6-phos
C6 + C4
C3 + C6
Glucose-6-phos

Figure 55–3 Interconversions of pentose phosphate pathway (PPP).

ways, back to glucose-6-phosphate. A reconversion of all these intermediates back to a hexose phosphate is possible. If all the reactions were summed, they would yield

$$\text{glu-6-phos} + 2\text{NADP} \longrightarrow CO_2 + \text{ribulose-5-phos} + 2\text{NADPH} + 2H^+$$

$$\text{X6} = (6\,\text{glu-6-phos} + 12\text{NADP} \longrightarrow 6CO_2 + 6\,\text{ribulose-5-phos} + 12\text{NADPH} + 12H^+)$$

To continue the reaction:

$$\underset{(6\times5=30)}{6\ \text{ribulose-5-phos}} \longrightarrow \underset{(5\times6=30)}{5\ \text{glu-6-phos}}$$

Although this pathway could be used for the generation of ATP, its more likely role within the cells is to generate NADPH and five carbon sugars.

Photosynthesis

We would be remiss if we did not consider the reaction whereby CO_2 becomes fixed into organic molecules in green plants. The entire plant and animal kingdom is dependent upon the photosynthetic process. It has been found that ribulose-5-phosphate, formed in the PPP in plants, leads to ribulose-1,5-diphosphate, which is the actual acceptor of the fixed CO_2. The enzyme ribulose diphosphate carboxylase catalyzes the reaction:

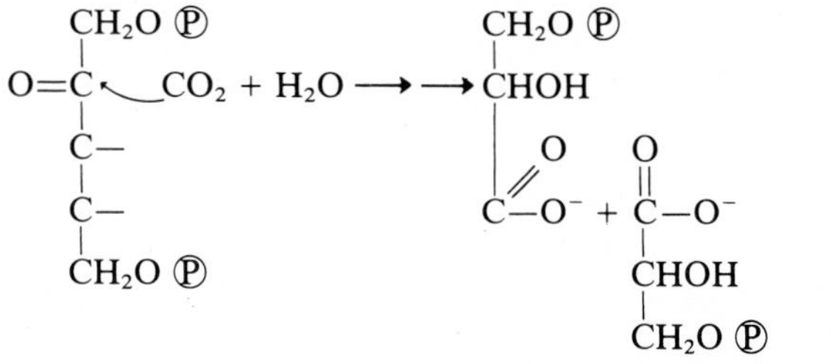

ribulose-1,5-diphosphate 2 molecules of 3-phosphoglycerate

Careful studies have shown that in the photosynthetic process (Melvin Calvin, Nobel laureate, 1961) 3-phosphoglycerate is the first organic molecule to form from the CO_2 in the atmosphere. The glycerate phosphate can be reduced to triose phosphate and then enter into reactions similar to those already considered. For example, two trioses could condense to form a hexose. Through reactions such as these, nature completes the entire carbon cycle.

UNIT 55 Problems and Exercises

55–1. Compare the number of ATP molecules formed from the glycolytic reactions of both glucose and fructose to yield 3-phosphoglyceric acid.

55–2. If the glucose–galactose interconversion was known to require NAD as a cofactor, suggest an intermediate in this epimerization.

55–3. Explain why infants suffering from galactosemia might successfully be fed sucrose.

55–4. Explain why glycogenolysis would be expected in a contracting muscle, whereas glycogenesis would be expected in a resting muscle.

55–5. Draw the structure of the three glucose molecules that would be involved at a junction point of glycogen.

55–6. Which could ultimately provide a cell with the greatest energy yield: (a) a molecule of free glucose or (b) a molecule of glucose that is a monomeric unit in starch?

55–7. Explain what functions a regulator molecule like cyclic AMP has in the process of glucose level regulation.

55–8. Would you expect to find PPP in a cell that did not undergo division or become involved in any significant number of biosynthetic reactions? Why?

55–9. Suppose that $^{14}CO_2$ was introduced in a photosynthesis system. If the trioses became involved in an aldolase reaction, which carbon(s) of the hexose would you expect to find bearing the ^{14}C?

55–10. Using the information you now possess, explain the significance of Figure 46–2 in a biochemical manner.

CHAPTER 20

Lipid and steroid metabolism

The lipids or fats constitute a large category of chemical materials found in living systems. Their unique chemistry in an environment where water is the universal solvent gives the lipids a special place in biochemistry. This can present both chemical problems and opportunities. It is obvious that there may be some problems in the metabolic system whereby the lipids are broken down and enter our organs and tissues. The advantages of lipids in providing parts of structural barriers between aqueous environments that need to be separated is equally obvious. Structural characteristics also allow lipids to become highly specific biochemical messengers (hormones).

We shall need to examine how the lipids, especially the triglycerides, enter our bodies and provide us with an efficient fuel source. Living systems have selected lipids as the most efficient energy-storage material. It is only through a careful chemical investigation of their metabolism that an understanding of the importance of lipids in bioenergetics can be obtained.

As we have noted, all the classes of compounds that we separated for convenience of study must become integrated with each other in the living cell. The close metabolic interrelationships between the metabolism of the lipids and the carbohydrates are excellent cases in point. In both normal and disease situations these separate metabolic systems must be understood in their biochemical context in order to appreciate these interrelations in living processes.

UNIT 56 LIPID METABOLISM

Lipids commonly make up one fifth to one third of the calories that we ingest in our diets. This dietary intake is composed mostly of triglycerides, but also includes varying amounts of phospholipids and cholesterol (and its esters). Upon ingestion, the lipids must be emulsified before they can undergo normal digestion and absorption. Tiny aggregates or **micelles** are formed, which are composed of hydrophilic lipid emulsifiers, like the bile salts, and some water. The lipids are largely carried in this manner in the digestive process. The micelles provide the site for hydrolytic reactions on the lipids. The most common biochemical reaction of digestion is **hydrolysis**. The major problem remaining is to move the lipid molecules across the intestinal barrier of the digestive tract and into the blood and other organs of the body.

Very little digestive activity of the lipids is found in the stomach. However, several hydrolytic enzymes are produced in the pancreas and are secreted into the duodenum, where the digestive process begins. Pancreatic lipase can bring about a partial hydrolysis of triglycerides. This enzyme prefers to catalyze the hydrolysis of long-chain fatty acids at the α and α' position of the triglycerides. The products of this reaction are two long-chain fatty acids and a monoglyceride, which is esterified at the β position. Other esterase enzymes are also found in the pancreatic juice. These enzymes can catalyze the hydrolysis of triglycerides composed of short-chain fatty acids, other monoesters, cholesterol esters, and some phospholipids.

There are two different routes of lipid absorption. The triglycerides containing the short-chain fatty acids can be completely hydrolyzed to free fatty acids and glycerol. Since glycerol is water soluble, it can be directly absorbed. The shorter-length fatty acids, as such, pass into the portal vein and are transported directly to the liver for utilization. The fatty acids may be bound to the serum albumin as they are transported in the blood.

The longer-chain fatty acids have a more complex mechanism for absorption. The β-monoglycerides and the long-chain fatty acids resulting from the lipase action pass into the mucosal cells of the intestine. Within these cells, triglycerides are re-formed. These resynthesized triglycerides next combine to form a complex lipoprotein. The lipoproteins are released into the lymph system, and from this system finally enter into the venous blood.

The lipid level of the blood rises shortly after a fatty meal. The increase in blood levels is called **lipemia**. In normal individuals the blood lipid levels return to base-line values within 6 or 8 hr after a meal.

The lipids of the blood are mainly taken up by the adipose tissue for storage or by the liver for metabolism. In either case, there are some special mechanisms that must be involved to allow the lipid to pass through a membrane and into a cell. These involve hydrolytic processes and are known to be under hormonal control. Cyclic AMP is known to be involved in the control of the mobilization of lipids.

Metabolism of Triglycerides

The catabolism of triglycerides is of interest because of their high caloric yield. We shall now examine how the conversion of fats to CO_2 can yield such great amounts of energy.

The initial step of catabolism involves a hydrolysis to yield glycerol and three fatty acids. The glycerol molecule must then be phosphorylated before it can enter the metabolic system. A glycerol kinase enzyme does this at the expense of an ATP. The glycerol phosphate can then undergo a dehydrogenase reaction yielding dihydroxyacetone phosphate. A summary of reactions whereby glycerol enters the glycolytic pathway at the level of the triose phosphate is

$$\begin{array}{l} CH_2OH \\ | \\ CHOH \\ | \\ CH_2OH \end{array} \xrightarrow[\text{kinase}]{\text{Glycerol}} \begin{array}{l} CH_2OH \\ | \\ HCOH \\ | \\ CH_2O\,Ⓟ \end{array} \xrightleftharpoons[]{\text{Glycerol phosphate dehydrogenase}} \begin{array}{l} CH_2OH \\ | \\ C{=}O \\ | \\ CH_2O\,Ⓟ \end{array} + NADH + H^+$$

In this reaction series one ATP is utilized, but an NADH is generated. If the NADH could be fed into the electron-transport system with coupled oxidative phosphorylation, three ATP molecules could be generated. The other reactions of glycolysis and TCA cycle could be followed to ultimately yield three CO_2 molecules from the glycerol. The energy from glycerol catabolism could drive the phosphorylation process to yield 19 ATP molecules. However, it is the oxidative metabolism of the fatty acids of the triglycerides that holds the potential for the greatest energy yields.

Fatty Acid Metabolism

The fatty acids are metabolized within the mitochondria of liver cells by a **β-oxidation spiral**. In this system, two carbon atoms at a time are cleaved from the fatty acid. This occurs in repeated spiral-like sequences until nothing remains of the carbon chain except two carbon units. The two carbon units are in the form of the acetate portions of acetyl CoA. The ultimate fate of the acetate is the formation of CO_2 as catalyzed by the TCA cycle, which is found in the mitochondria of the cell. Recall that the TCA system can generate 12ATP's from the metabolism of each acetate. The β-oxidation spiral also drives the synthesis of ATP.

The first step in the β-oxidation system requires an input of energy in the form of an ATP. As was true in the case of glucose metabolism, the cell must provide an initial ATP to start the process. Throughout all nature it seems necessary to make an effort before results can be expected. An enzyme, acyl coenzyme A synthetase, brings about this activation:

$$RCH_2CH_2\overset{\overset{\displaystyle O}{\|}}{C}{-}O^- + CoA{-}S{-}H + ATP \rightleftharpoons RCH_2CH_2\overset{\overset{\displaystyle O}{\|}}{C}{-}S{-}CoA + AMP + P_iP_i$$

Note that in this reaction AMP and pyrophosphate are products.

This activation process sometimes can occur at other sites than the mitochondrion. When this extramitochondrial activation occurs, there is some question as to how the long-chain fatty acids move across the mitochondrial membrane. It has been found that a reversible set of transport reactions may be involved using carnitine in some cases:

$$(CH_3)_3\overset{+}{N}{-}CH_2{-}\underset{\displaystyle OH}{\underset{|}{C}H}{-}CH_2\overset{\displaystyle O}{\overset{\|}{C}}{-}O^-$$

Enzymes located on either side of the mitochondrial membrane exchange the fatty acid acyl CoA with the alcohol group of carnitine. The CoA part of active fatty acid does not transverse the membrane barrier in this mechanism.

The reaction continues within the mitochondria. The acyl CoA derivative is next subjected to dehydrogenation. The cofactor FAD is involved in the reaction catalyzed by acyl CoA dehydrogenase:

$$RCH_2CH_2\overset{\displaystyle O}{\overset{\|}{C}}{-}S{-}CoA + FAD \rightleftharpoons RCH{=}CH\overset{\displaystyle O}{\overset{\|}{C}}{-}S{-}CoA + FAD{\cdot}H_2$$

This reaction, involving an alkane being converted to an alkene, is reminiscent of one in the TCA cycle. The FAD is the electron acceptor of choice for this type of dehydrogenase. The $FAD{\cdot}H_2$ can be oxidized by the electron-transport system and drive the synthesis of two ATP molecules through coupled phosphorylation.

The following reactions are also similar to those found in the TCA cycle. Water is added to the alkene to form a secondary alcohol. This hydration reaction is catalyzed by the enzyme enoyl CoA hydrase:

$$RCH{=}CH\overset{\displaystyle O}{\overset{\|}{C}}{-}S{-}CoA + H_2O \rightleftharpoons R\overset{\displaystyle OH}{\overset{|}{C}}HCH_2\overset{\displaystyle O}{\overset{\|}{C}}{-}S{-}CoA$$

Some molecular specificity might be noted. The alkene linkage is a trans form, and a specific asymmetric carbon is formed here.

The reaction sequence continues with a dehydrogenation of the alcohol to a ketone. As is common in these cases, NAD is involved with the enzyme β-hydroxyacyl CoA dehydrogenase:

$$R\overset{\displaystyle OH}{\overset{|}{C}}HCH_2\overset{\displaystyle O}{\overset{\|}{C}}{-}S{-}CoA + NAD \rightleftharpoons R\overset{\displaystyle O}{\overset{\|}{C}}{-}CH_2\overset{\displaystyle O}{\overset{\|}{C}}{-}S{-}CoA + NADH + H^+$$

The NADH can generate three ATP by oxidative phosphorylation.

The final step in the spiral sequence involves a cleavage of the molecule adjacent to the ketone group. The cleavage involves a transfer of a portion of the molecule to another molecule of coenzyme A. The enzyme thiolase catalyzes the cleavage to form two activated molecules:

$$R{-}\overset{\displaystyle O}{\overset{\|}{C}}{-}CH_2\overset{\displaystyle O}{\overset{\|}{C}}{-}S{-}CoA + CoA{-}S{-}H \longrightarrow R{-}\overset{\displaystyle O}{\overset{\|}{C}}{-}S{-}CoA + CH_3\overset{\displaystyle O}{\overset{\|}{C}}{-}S{-}CoA$$

The acetyl CoA may then proceed through TCA intermediates to form CO_2 and water and to drive the synthesis of 12 ATP's. The remaining new acyl CoA derivative is already activated and can undergo the same series of reactions just considered. Each series of reactions yields an active acetate and an acyl CoA derivative containing two less carbon atoms. Thus we use the name *fatty acid spiral* to denote this special kind of recycling reaction series. This can be visualized as in Figure 56–1.

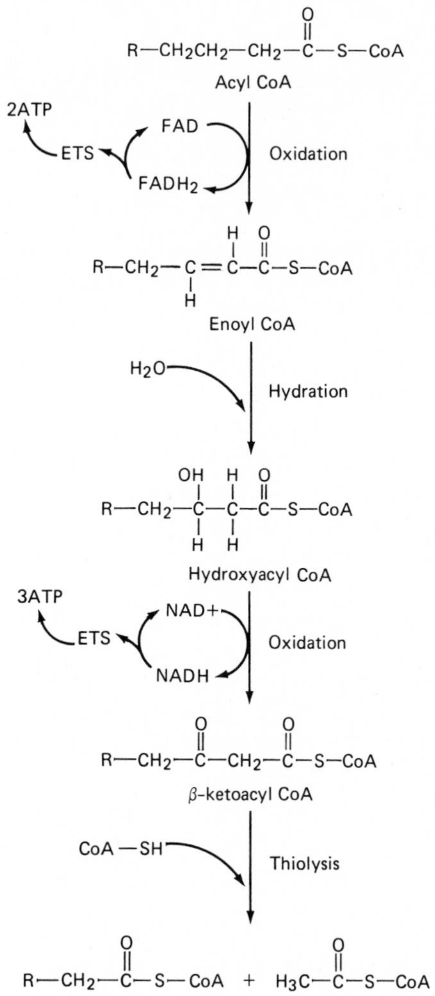
R—CH2CH2—CH2—C(=O)—S—CoA
Acyl CoA
2ATP
ETS
FAD
FADH2
Oxidation
R—CH2—C(H)=C(H)—C(=O)—S—CoA
Enoyl CoA
H2O
Hydration
R—CH2—C(OH)(H)—C(H)(H)—C(=O)—S—CoA
Hydroxyacyl CoA
3ATP
ETS
NAD+
NADH
Oxidation
R—CH2—C(=O)—CH2—C(=O)—S—CoA
β-ketoacyl CoA
CoA—SH
Thiolysis
R—CH2—C(=O)—S—CoA + H3C—C(=O)—S—CoA

Figure 56–1 The fatty acid spiral.

An acyl CoA derivative with four carbon atoms is the shortest chain length that can undergo this spiral. The products of such a derivative would be two acetyl CoA molecules.

The high energy yield from the oxidation of a fatty acid now becomes obvious. For example, stearic acid (18 carbons) requires only one ATP for activation. This long acyl derivative could undergo eight turns around the spiral to yield nine acetyl CoA molecules:

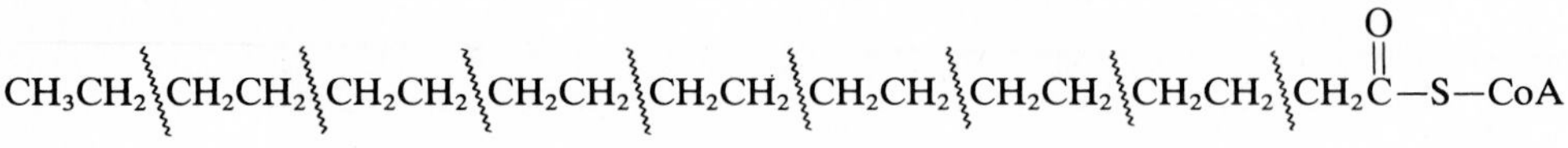

The total energy yield from stearic acid can be summarized as follows:

Activation	−1 ATP
8 turns of spiral (5 each)	40 ATP
9 acetyl CoA (12 each: TCA)	108 ATP
Net	147 ATP

If we assume that each ATP is equivalent to only 7 kcal, this would yield 1020 kcal for the catabolism of only one fatty acid! Now we can understand on a molecular level the high energy yields of fats.

Unsaturated fatty acids undergo just about the same series of reactions as the saturated acids. There are some problems to be surmounted of both placement and configuration of some of the double bonds before the catabolism can take place. The necessary enzymes are present to allow for the utilization of the unsaturated acids, with minimal modifications for the scheme just considered.

The acid chains containing an odd number of carbon atoms present a special problem. Although such chain lengths occur rarely, there is a special system to handle their catabolism. The spiral we have just considered is adequate until the final step. With an odd number of carbon atoms, the final turn of the spiral would yield one acetyl CoA and one propionic acid derivative of CoA. The propionyl CoA can undergo a series of special reactions, one of which involves vitamin B_{12} as a cofactor. The final products of catabolism would again be CO_2 and H_2O.

Under normal circumstances, the majority of acetyl CoA formed in the β-oxidation spiral enters the TCA cycle for conversion to CO_2 and H_2O. There are two other major fates of acetyl CoA, one being the resynthesis of fatty acids. The other possibility involves a condensation leading to the synthesis of steroids and *ketone bodies.*

Ketone Bodies

The **ketone bodies** are acetoacetic acid, β-hydroxybutyric acid, and acetone.

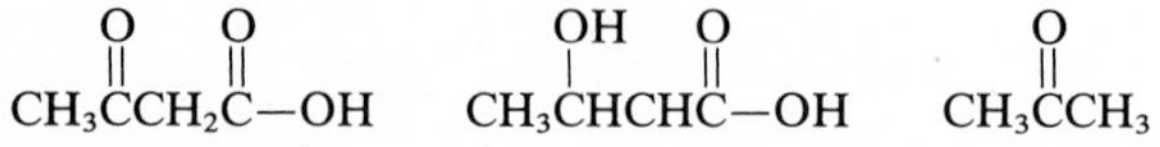

It appears that acetoacetic acid could be formed quite simply by loss of the CoA in the final step of the β-oxidation spiral. This does not usually happen, however. The end product of the β-oxidation scheme is indeed acetyl CoA. Acetyl CoA undergoes a condensation to yield a new molecule of acetoacetyl CoA. Within the mitochondria of the liver, two separate condensation reactions can take place:

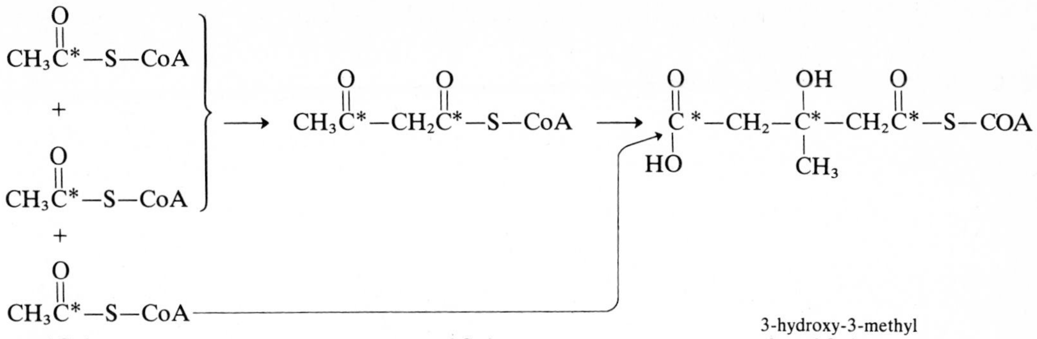

$$CH_3\overset{O}{\underset{\|}{C}}{}^*-S-CoA$$

$$+$$

$$CH_3\overset{O}{\underset{\|}{C}}{}^*-S-CoA \Bigg\}$$

$$CH_3\overset{O}{\underset{\|}{C}}{}^*-S-CoA$$

acetyl CoA

$$\longrightarrow \quad CH_3\overset{O}{\underset{\|}{C}}{}^*-CH_2\overset{O}{\underset{\|}{C}}{}^*-S-CoA \quad \longrightarrow$$

acetoacetyl CoA

$$\overset{O}{\underset{\underset{HO}{|}}{\overset{\|}{C}}}{}^*-CH_2-\overset{OH}{\underset{\underset{CH_3}{|}}{\overset{|}{C}}}{}^*-CH_2\overset{O}{\underset{\|}{C}}{}^*-S-COA$$

3-hydroxy-3-methyl glutaryl CoA

The carbonyl carbons are marked using radioactive labels.

This 3-hydroxy-3-methyl glutaryl CoA synthetic step is also the initial one for steroid formation. In the case of ketone-body formation, this molecule moves in a different pathway involving a cleavage:

$$HO-\overset{O}{\underset{\|}{C}}-CH_2\overset{OH}{\underset{\underset{CH_3}{|}}{\overset{|}{C}}}-CH_2\overset{O}{\underset{\|}{C}}-S-CoA \longrightarrow CH_3\overset{O}{\underset{\|}{C}}-CH_2\overset{O}{\underset{\underset{OH}{|}}{\overset{\|}{C}}} + CH_3\overset{O}{\underset{\|}{C}}-S-CoA$$

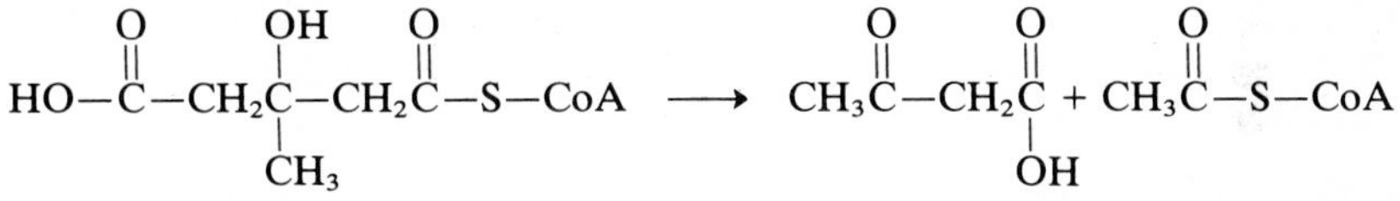

3-hydroxy-3-methyl glutaryl CoA · acetoacetic acid · active acetate

The other two ketone bodies are formed from acetoacetic acid by reduction and decarboxylation, respectively.

$$CH_3\overset{O}{\underset{\|}{C}}CH_2\overset{O}{\underset{\underset{OH}{|}}{\overset{\|}{C}}} \quad \begin{array}{c} \xrightarrow{\text{Reduction}} \\ \xrightarrow{\text{Decarboxylation}} \end{array} \quad \begin{array}{c} CH_3\overset{OH}{\underset{|}{C}}HCH_2\overset{O}{\underset{\underset{OH}{|}}{\overset{\|}{C}}} \\ \\ CH_3\overset{O}{\underset{\|}{C}}CH_3 \end{array}$$

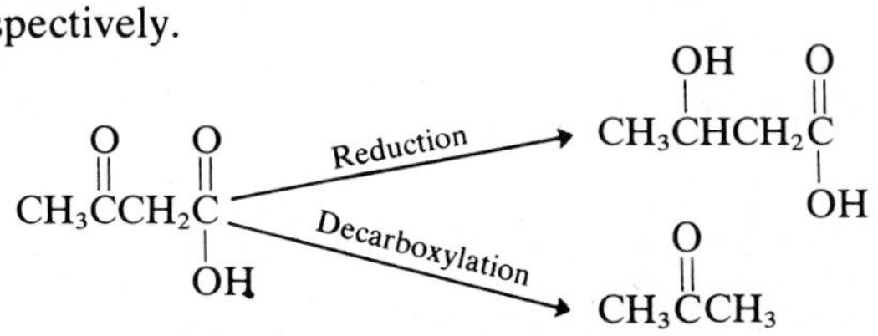

The production of the ketone bodies has considerable clinical significance. An abnormally high level of the ketone bodies in the blood is known as **ketonemia**. When abnormally high levels of acetone are reached, as we shall consider shortly, the acetone can actually be expelled by the lungs (acetone breath). Abnormally large amounts of acetoacetic acid and β-hydroxybutyric acid can be excreted by the kidneys into the urine. This is known as **ketonuria**. Since carboxyl groups are ionized, the loss of these anions also leads to the loss of cations like Na^+, which serve as counter ions. The depletion of these cations can influence the entire acid–base chemistry within the body, as well as cause considerable water loss. This fluid loss may lead to dehydration and fatal complications if uncorrected. A general term **ketosis** is used to describe this combination of ketonemia, ketonuria, and acetone breath.

One of the most common disease states associated with ketosis is diabetes mellitus. In this disease a hormonal deficiency (insulin) prevents glucose from being properly utilized. If the glucose cannot be utilized, the body turns to abnormally high fatty acid metabolism to supply its energy needs. The normal metabolism of acetyl CoA by the TCA cycle can only take place if the fat and carbohydrate degradation are properly balanced. This relationship is an excellent example of the need to understand living processes on a molecular level.

Biosynthesis of Lipids

To bring about the synthesis of the neutral lipids, an abundant supply of both glycerol and fatty acids must be available. Lipid biosynthesis also requires a tremendous supply of energy (ATP) and a biochemical reducing agent (NADPH). At first, it appears that the biosynthesis of fatty acids might be simply a reverse of the degradative reactions, but they are quite separate and distinct. Actually, there are several different processes by which the fatty acid chains are constructed. One of these, a **de novo** process, builds chains with up to 14 carbon atoms using a cytoplasmic enzyme complex. A secondary system, known as **elongation**, catalyzes the construction of the fatty acids within the mitochondria.

de novo Synthesis

The *de novo* synthesis of fatty acids in mammalian systems involves a tightly organized complex found in the cytoplasm. One of the important participants in this complex, an acyl carrier protein (ACP), provides the site for the synthesis. This protein is a single chain containing 77 amino acids with a phosphopantetheinic acid group attached to a serine. This yields another carrier molecule with a sulfhydryl group like Coenzyme A as the site of action:

$$(ACP)-Serine-CH_2O-\overset{\overset{O}{\|}}{\underset{\underset{O^-}{|}}{P}}-OCH_2\overset{\overset{CH_3}{|}}{\underset{\underset{CH_3}{|}}{C}}-\overset{\overset{}{|}}{\underset{\underset{OH}{|}}{CH}}-\overset{\overset{O}{\|}}{C}-\overset{}{\underset{\underset{H}{|}}{N}}-CH_2CH_2\overset{\overset{O}{\|}}{C}-\overset{}{\underset{\underset{H}{|}}{N}}-CH_2CH_2-S-H$$

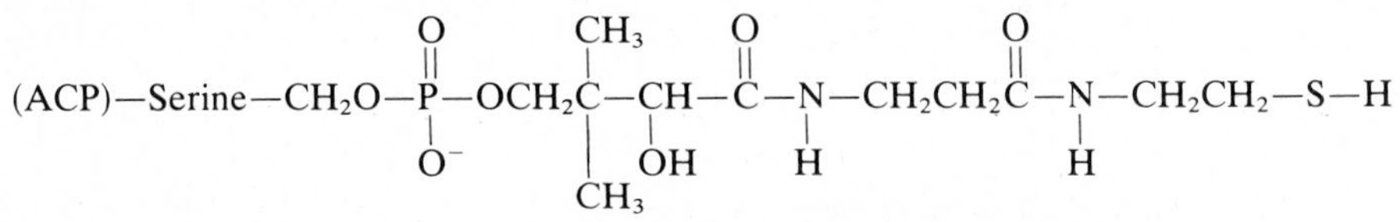

phosphopantetheine group attached to the serine

The biosynthetic system was found to involve CO_2 (or HCO_3^-), acetyl CoA, ATP, and NADPH. However, radioactively labeled $H^{14}CO_3^-$ did not become incorporated in the acids that were synthesized. Biochemists found that the first reaction was the formation of a malonic acid derivative catalyzed by an enzyme acetyl CoA carboxylase:

$$CH_3\overset{\overset{O}{\|}}{C}-S-CoA + HCO_3^- + ATP \longrightarrow \overset{\overset{O}{\diagdown}}{\underset{\underset{O^-}{\diagup}}{C}}-CH_2\overset{\overset{O}{\|}}{C}-S-CoA + ADP + P_i$$

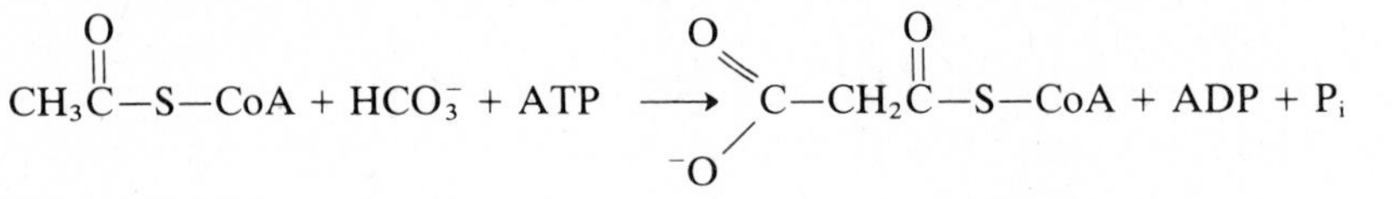

acetyl CoA   malonyl CoA

This reaction occurs under allosteric control in a two-step process involving biotin. Citrate activates this enzyme, whereas the long acyl CoA derivatives inhibit this carboxylase.

The next reaction in the process involves transferring the participants to the ACP (transacylases):

$$CH_3\overset{\overset{O}{\|}}{C}-S-CoA + ACP \rightleftharpoons CH_3\overset{\overset{O}{\|}}{C}-S-ACP + CoA-S-H$$

$$\overset{O}{\underset{^-O}{\diagdown}}C-CH_2\overset{\overset{O}{\|}}{C}-S-CoA + ACP \rightleftharpoons \overset{O}{\underset{^-O}{\diagdown}}C-CH_2\overset{\overset{O}{\|}}{C}-S-ACP + CoA-S-H$$

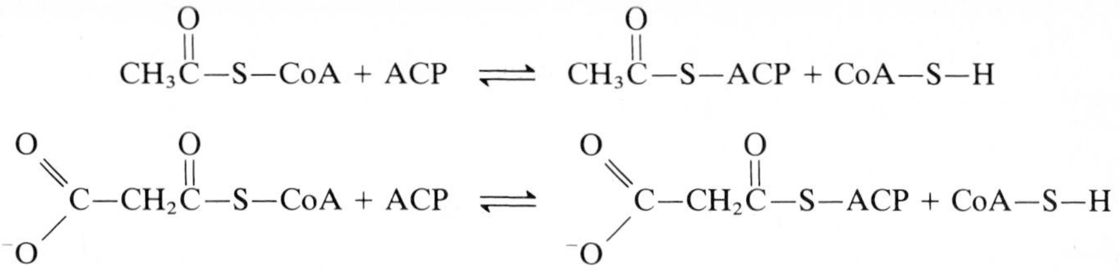

The fatty acid forms by condensation:

$$\overset{O}{\underset{^-O}{\diagdown}}C-CH_2-\overset{\overset{O}{\|}}{C}-S-ACP + CH_3\overset{\overset{O}{\|}}{C}-S-ACP \longrightarrow CH_3\overset{\overset{O}{\|}}{C}-CH_2-\overset{\overset{O}{\|}}{C}-S-ACP +$$
$$ACP + CO_2$$

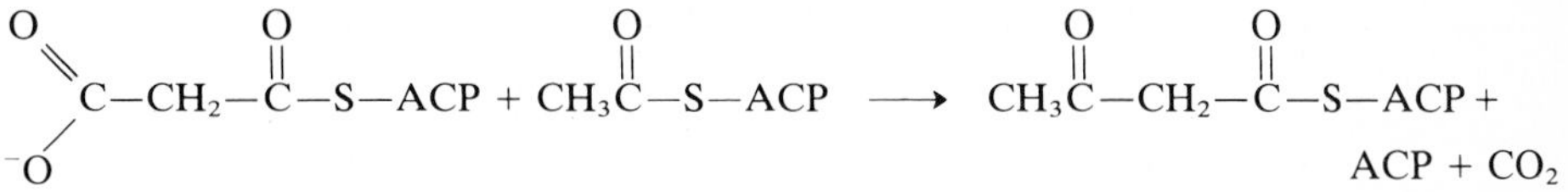

This releases the same CO_2 that was just added, which accounts for the fact that no labeled $^{14}CO_2$ is directly incorporated into the fatty acid chain.

The reductive process whereby the acetoacetyl-S-ACP is converted to a butyryl-S-ACP is outlined in Figure 56–2 which summarizes the total biosynthetic pathway. Similar chemistry as the reverse of the β-oxidative degradative is involved with several major differences:

1. The reduction of the ketone to the alcohol requires NADPH as a cofactor.
2. The alcohol formed is the D- rather than L-form found in the degradative process.
3. The reduction of the alkene to the alkane linkage requires NADPH rather than a flavin-containing coenzyme, as found in the degradative process.
4. The reactions take place as attached to ACP rather than coenzyme A. These important differences clearly differentiate the synthetic process from the degradative scheme.

The biosynthetic reaction cycle continues adding malonyl portions, increasing the chain by two carbon atoms each time. A variety of different fatty acids could be formed depending upon where the reaction was terminated. The last step moves the fatty acid to Coenzyme A for subsequent reactions.

Elongation

Some of the longer-chained fatty acids (C_{18} to C_{24}) may be synthesized by elongation within the mitochondrion. The elongation is a reversal of the degradation process and does not involve the malonyl derivatives.

Unsaturated Fatty Acids

The unsaturated fatty acids are synthesized by separate dehydrogenase systems. The essential polyunsaturated fatty acids cannot be synthesized within the human body but must be supplied in the diet.

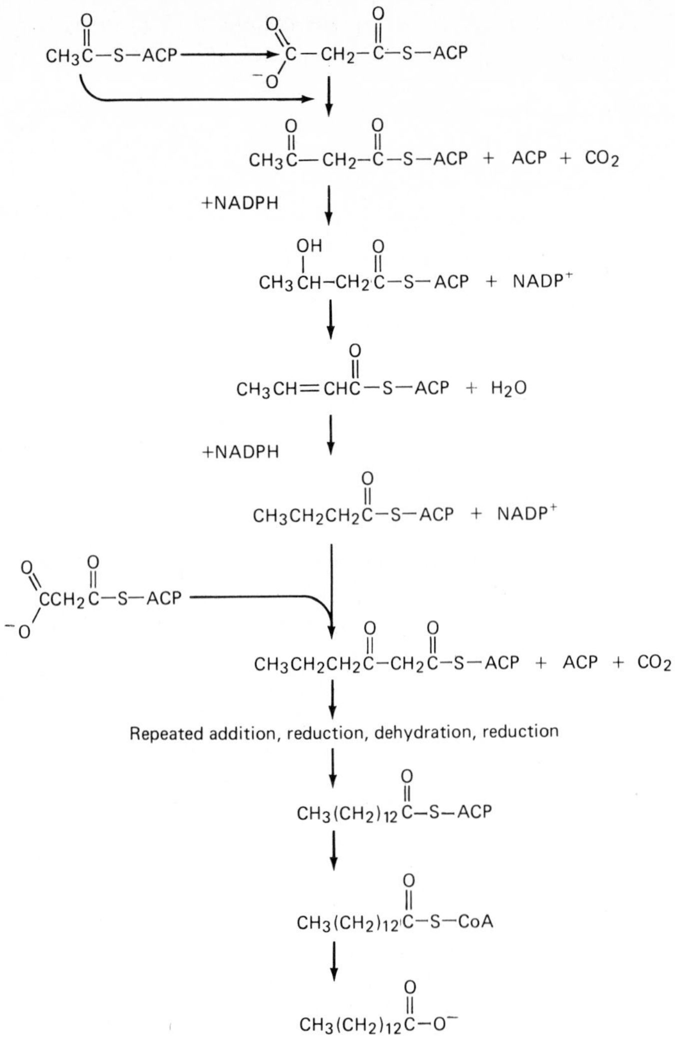
+NADPH
+NADPH
Repeated addition, reduction, dehydration, reduction

Figure 56–2 Outline of biosynthetic de novo pathway of fatty acids.

Triglyceride Synthesis

Fatty acids do not accumulate as such but form triglycerides. Glycerol-3-phosphate reacts with two molecules of acyl CoA to transform phosphatidic acid:

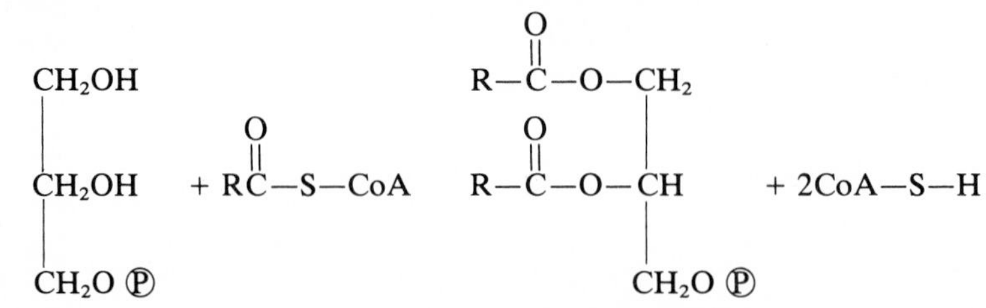

glycerol-phosphate phosphatidic acid

The phosphatidic acid loses a phosphate by hydrolysis. The diglyceride is then reacted upon by a final acyl CoA to form the neutral triglyceride.

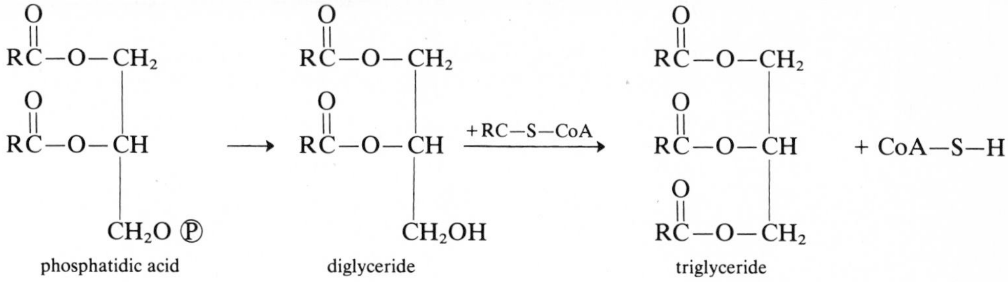
+RC—S—CoA
+ CoA—S—H
phosphatidic acid
diglyceride
triglyceride

The phosphatidic acid is a common intermediate for the synthesis of both the triglycerides and the phospholipids. In the case of phospholipid biosynthesis, the phosphate is already in place on the molecule. A variety of nitrogenous bases can be added to make complete phospholipids.

It is often assumed that the triglycerides represent a rather static energy-storage form. Actually, the lipid stores are in a more dynamic state than might be expected, especially in the liver. The liver shows quite rapid replacement of the fatty acids of triglycerides; the brain, for example, shows a much slower turnover. It has been estimated that in rats nearly 10% of the total fatty acids undergo a daily replacement. It is known that lipid deposits reflect the dietary intake of specific fatty acids to a certain extent. The overall regulation of lipid storage and mobilization are not clearly understood in many cases and will be the subject of future biochemical research.

UNIT 56 Problems and Exercises

56–1. In some disease states the body does not adequately make certain lipoproteins. Would it be better to use triglycerides made from long-chained fatty acids or short-chained fatty acids in the diets of these individuals?

56–2. Butyric acid is present in the triglycerides of butter. Outline the reactions involved as butyric acid is converted to two molecules of acetyl CoA.

56–3. Calculate the energy yield in ATP's formed from the following:

(a) The formation to two acetyl CoA molecules from one butyric acid molecule
(b) The formation of four CO_2 molecules from one butyric acid molecule by the β-oxidation scheme and TCA cycle
(c) The formation of 15 CO_2 molecules from one molecule of tributyrin

56–4. Dihydroxyacetone phosphate directly results from the metabolism of glycerol. What reaction would you expect this molecule to undergo next as it enters the glycolytic pathway?

56–5. What are the three major fates of the acetyl CoA molecules formed from the β-oxidation spiral?

56–6. Given a sample labeled as follows $^{14}CH_3\overset{O}{\overset{\|}{C}}—S—CoA$, which carbon atoms of acetone from the ketone-body-forming process would bear the ^{14}C label?

56–7. Show the form of the β-hydroxybutyric acid that you would expect to find eliminated in the urine at pH 7.

56–8. What is the difference between ketonemia and ketouria?

56–9. Explain the role of citrate in its influence on the acetyl CoA carboxylase system.

56–10. Account for the fact that, if $^{14}CO_2$ was introduced into a *de novo* synthesis system for fatty acids, the radioactive label would appear in malonate but not in a palmitate that was formed.

56–11. In the conversion alkane $\rightleftharpoons$ alkene, different coenzymes are involved depending upon the direction of the reaction.
(a) Which coenzymes are involved in each case?
(b) How is this selection of specific coenzymes consistent with the general laws of nature? (*Hint*: Consider the energetics.)

56–12. Using structures, outline the reactions whereby acetyl CoA can be added to hexanyl ACP (C_6) to form octonyl ACP (C_8).

56–13. Explain how the feeding of ^{14}C-labeled glucose could lead to the formation of triglycerides bearing the ^{14}C label. Use narrative rather than equations here.

CHAPTER 21

protein metabolism

The metabolism of proteins is really the metabolism of the amino acids of which the proteins are composed. This anabolic process of building the carefully sequenced proteins will be considered later as the culmination of nature's information-transfer system. The catabolic reactions of the amino acids start with the digestion of the proteins we consume in our diets. These metabolic reactions of amino acids are absolutely essential to maintain life and preserve our health.

As we might expect, nature operates in a cyclic fashion in these biochemical reactions. The nitrogen atom provides uniqueness to the chemistry of amino acids. With the discovery of the ^{15}N isotope, scientists were able to trace the flow of this atom from species to species and molecule to molecule. The dynamic reactions of nitrogen-containing compounds proved that there is a constant turnover of these molecules. Amino acid metabolism is the study of active biochemistry.

The reactions of amino acids do not stand alone, but must be considered as integrated with the other classes of compounds that make up our cells. The amino acids share many common metabolic pathways with the carbohydrates, for example. This is not obvious until we examine these reaction systems on a molecular basis. We somehow know that "our daily bread" can be converted into the proteins necessary to build the muscle of our bodies. By studying the metabolism of amino acids, we shall see how this is actually accomplished.

UNIT 57 GENERAL PROTEIN DIGESTION AND METABOLISM

Free amino acids are present in our diets in only very small amounts. Therefore, proteins are the major source of the amino acids required to meet our nutritional needs. The proteins themselves are not specifically required in our diets; the amino acids of which they are composed are required. Obviously, hydrolysis reactions must be involved in digestion. Normal recommended dietary allowances of proteins range from 25 g of proteins per day for young children to 65 g for fully grown men. As expected, these allowance levels are highly dependent upon the composition and availability of the amino acids that make up the proteins. As is true of any food, the initial problem is to transfer the ingested material to tissues in a form that is biochemically useful for metabolic reactions.

Digestion and Absorption

The first major digestive action on proteins occurs in the stomach, as shown in the outline of Figure 57–1. A proteolytic enzyme **pepsin** acts to initiate the hydrolysis of the large proteins. Pepsin is produced within specific cells of the walls of the stomach and secreted as a zymogen or proenzyme. This precursor is called **pepsinogen**. The active pepsinogen is composed of the same amino acid chain as pepsin plus an

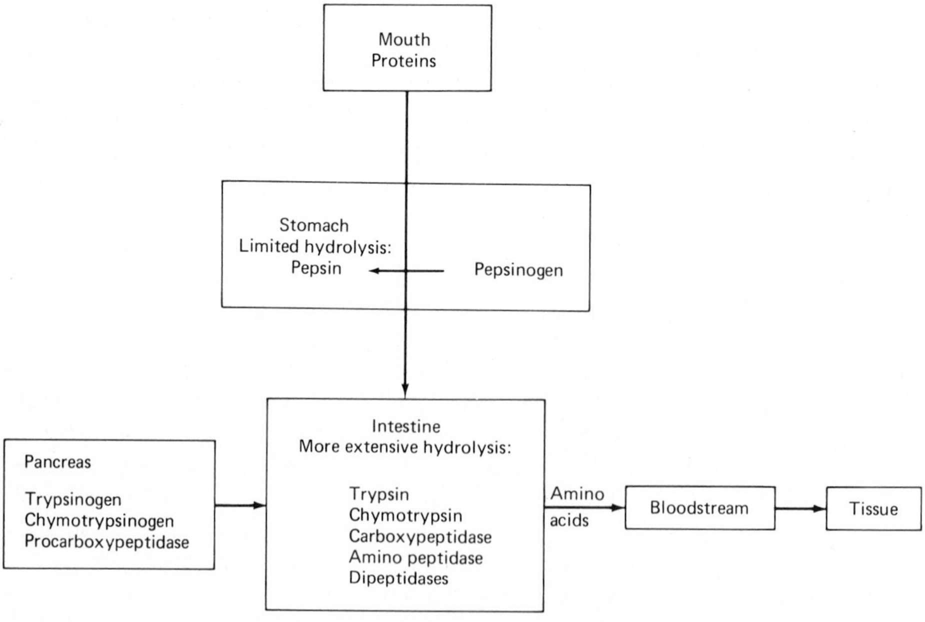

Mouth
Proteins
Stomach
Limited hydrolysis:
Pepsin
Pepsinogen
Intestine
More extensive hydrolysis:
Trypsin
Chymotrypsin
Carboxypeptidase
Amino peptidase
Dipeptidases
Pancreas
Trypsinogen
Chymotrypsinogen
Procarboxypeptidase
Amino
acids
Bloodstream
Tissue

Figure 57–1 Outline of protein digestion.

additional peptide at the N-terminal end. When this additional peptide is attached, the pepsinogen is inactive. When the pepsinogen enters the acidic environment of the stomach, it loses its terminal peptide and is converted to the active pepsin. This conversion is catalyzed by both the H^+ in the stomach as well as pepsin itself, once it is formed. This is an example of an autocatalytic process.

The pepsin starts the digestion by catalyzing the hydrolysis of a few of the peptide linkages of the large proteins in our foods. These partially digested proteins then pass into the intestinal tract.

The hydrolysis of the proteins continues within the intestine in a considerably different chemical environment. The pH is slightly alkaline. The pancreas secretes its important digestive juices into the intestine to continue the hydrolysis of the proteins. Once again, these potent proteolytic enzymes are formed and secreted as zymogens. The most important zymogens secreted at this point are trypsinogen, several different chymotrypsinogens, and procarboxypetidases.

As discussed previously, trypsinogen is a single peptide chain with the same composition as trypsin plus an additional hexapeptide at the N-terminal end of the molecule. The activation process of trypsinogen is especially important and can be accomplished in two ways. First there is a special intestinal enzyme, enterokinase, that is very active in this conversion. Once trypsin is formed, it is autocatalytic for its own conversion. Besides being one of the important protein-digesting enzymes in its own right, trypsin also is involved in the conversion of the other inactive zymogens in the intestine to their active forms.

The important proteolytic enzymes operating within the intestine besides **trypsin** are **chymotrypsin(s)**, **carboxypeptidase A** and **B**, and **aminopeptidase**. Each of these enzymes has a special hydrolytic mechanism by which it catalyzes the cleavage of the peptide bond. The carboxypeptidases are **exopeptidases**, since they catalyze the hydrolysis of peptide linkages on the exterior of the proteins, in this case, the carboxy terminal end. Trypsin, chymotrypsin, and pepsin (in the stomach) are **endopeptidases**, since they catalyze in the hydrolysis of peptide bonds within the interior of the protein molecule. These proteolytic enzymes show a specificity as to the amino acids of the peptide bond involved at the point of hydrolysis, as well as the position of the bond within the protein molecule. These preferences are worth considering as another example of the high degree of molecular specificity of biochemical systems.

Carboxypeptidase A reacts more rapidly to free the amino acids from the carboxyl end of proteins when the terminal amino acids have aromatic or aliphatic side chains. Carboxypeptidase B prefers to act on peptides when basic amino acids like lysine and arginine are at the carboxyl terminal end. Aminopeptidase, from mucosal cells lining the intestine, can free amino acids from the N-terminal end of proteins and peptides. The products of the exopeptidases are free amino acids.

The endopeptidases also catalyze peptide hydrolysis at differing rates. Trypsin acts most effectively when the carboxyl groups of basic amino acids like lysine and arginine are involved on peptide linkages. Chymotrypsin prefers to act on peptide bonds when the carboxyl groups of aromatic amino acids like phenylalanine, tyrosine, and tryptophan are involved. Pepsin also prefers to bring about hydrolysis on peptide bonds involving the carboxyl groups of these aromatic amino acids.

The combined action of these proteolytic enzymes leads to some free amino acids and some small peptides. Dipeptidases, from mucosal cells, are present to complete the job of converting the ingested proteins to amino acids.

The absorption of the amino acids from the small intestine is a complex process. As with the transport of other small molecules across cell walls, amino acid absorption is an active process requiring an input of energy. The amino acids absorbed into the plasma reach normal levels of about 35 to 65 mg of mixed amino acids per 100 mℓ of blood. The amino acids normally quickly pass from the bloodstream to the various tissues of the body, especially the liver. Upon intravenous injection, most amino acids are absorbed by the tissues nearly completely within a few minutes.

There is of course a constant gain and loss of nitrogenous compounds by our bodies under normal circumstances. As indicated in the earlier discussion of isotopic tracers, ^{15}N studies have shown the dynamic nature of nitrogen metabolism. Individuals may be in what is known as positive or negative nitrogen balance or a dynamic steady state. The proteins within our bodies are constantly in a process of both anabolism and catabolism. The rates of these two competing reaction schemes vary with the particular protein and the particular cell of the body. The "restless" nature of the metabolic systems and the tremendous degree of control that is exercised over these processes is truly remarkable.

Once amino acids enter the cells, they are utilized in one or more of the following metabolic possibilities: (1) synthesis of proteins and peptides, (2) synthesis of small molecules that contain nitrogen, and (3) as α-keto acids after the removal of the amino group.

Protein biosynthesis can be considered only after we understand some of the molecules that direct this process. Therefore, this part of protein metabolism will be considered following a discussion of nucleic acids. It is interesting that for protein synthesis to take place all the amino acids must be available at about the same time. For example, studies have shown that rats fail to grow normally if only one of the required amino acids is withheld daily for a period of about 3 hr after the other amino acids were fed. This indicates that mammals store amino acids to only a very limited extent.

Essential and Nonessential Amino Acids

A great variety of small nitrogen-containing molecules is normally synthesized from the amino acids that enter our cells. One interesting group of such molecules is the **nonessential** amino acids themselves. These are the amino acids whose carbon skeletons can be synthesized by a given organism. The **essential** amino acids must be supplied to the organism. Actually, the list of either of these categories must indicate the species involved, the state of the organism, and a balance of the amino acids present. For man, the 10 essential amino acids are arginine, histidine, isoleucine, leucine, threonine, lysine, methionine, phenylalanine, tryptophan, and valine. The remaining common amino acids are considered nonessential because of their possible synthesis within our cells. Actually, these nonessential amino acids are in great abundance within our dietary proteins and are utilized in one or another of the possible pathways of metabolism mentioned.

It should be pointed out that the amino acids are highly interrelated as far as requirements are concerned. The sulfur-containing amino acids, methionine and cysteine, are interdependent. Also, phenylalanine requirements depend to some extent upon the presence of tyrosine because of their metabolic interrelationships. The study of these metabolic systems on a molecular basis will show these relationships.

Amino acid requirements may vary somewhat depending upon the state of the organism. For example, studies with rats indicate that arginine needs to be supplied only (essential) while the rat is growing. The adult rat may not need this amino acid supplied in the diet under conditions of nitrogen balance. This means that the rate at which this amino acid can be synthesized is insufficient to meet all the needs of a growing animal.

It is interesting that there are no well-characterized diseases of man that can be shown to result from the deficiency of a single amino acid. This is different from the case of vitamins, even though both are required in the diet. Certain amino acid deficiencies may lead to nutritionally related conditions, however. For example, kwashiorkor, a disease with many adverse symptoms, arises from low protein intake in general. The proteins that are available to those who suffer from this disease are frequently especially low in lysine. The diets of these individuals are often limited to proteins from plant sources. Cereal grains, as dietary protein sources, need to be supplemented so as to supply the proper balance of all the essential amino acids. A biological solution to this would be to genetically alter plants so as to produce a better balance of essential amino acids within the grains. This is being done and is a splendid example of applying a basic understanding of scientific principles to a practical problem facing mankind.

Amino Acid Derivatives

A number of other small molecules are derived from amino acids by rather direct metabolic routes. It would be impractical to catalog all these reactions, but a few examples of importance should be considered. These examples also illustrate the necessity of examining molecules on a structural basis. Without a knowledge and appreciation of chemical structures, the important biochemical relationships that exist would be completely meaningless.

One common example is the formation of histamine. Histamine is formed by the rather direct decarboxylation of histidine:

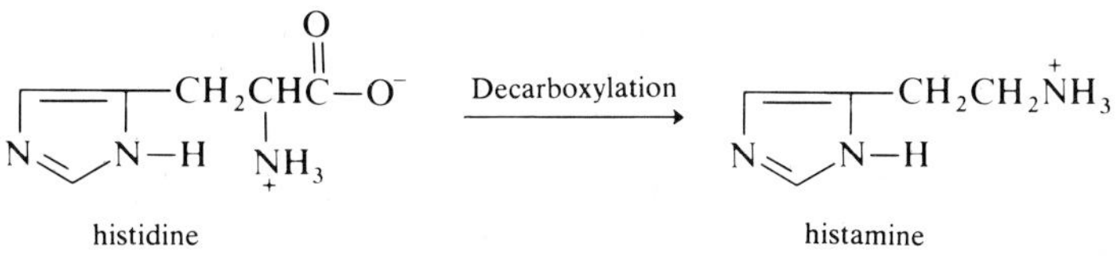
CH2CHC—O⁻
N—H
NH3
Decarboxylation
CH2CH2NH3
N—H
histidine
histamine

Histamine is synthesized within the cells and released in response to certain chemicals. The histamine can cause vascular changes and blood pressure changes along with other undesired effects. Antihistamines, which compete with histamines to inhibit their physiological effects, are widely used drugs. You might examine the antihistamines to see if they contain amine functional groups.

β-alanine is another good example of a nitrogenous compound derived from *α*-amino acids. Microbiological systems are especially rich in enzymes that can decarboxylate aspartic acid:

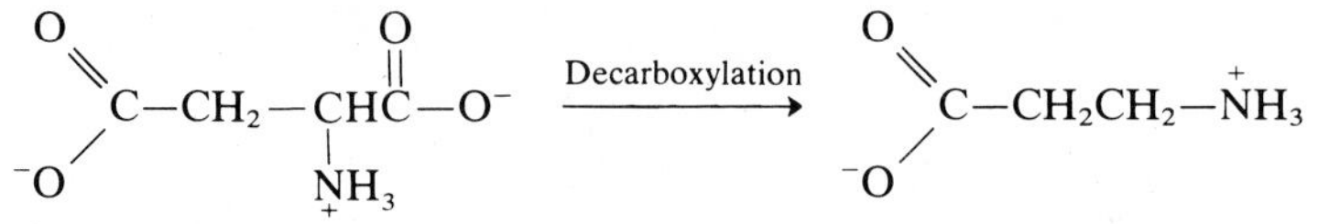

Aspartate β-alanine

The hormone thyroxine arises from tyrosine in a reaction sequence requiring iodination:

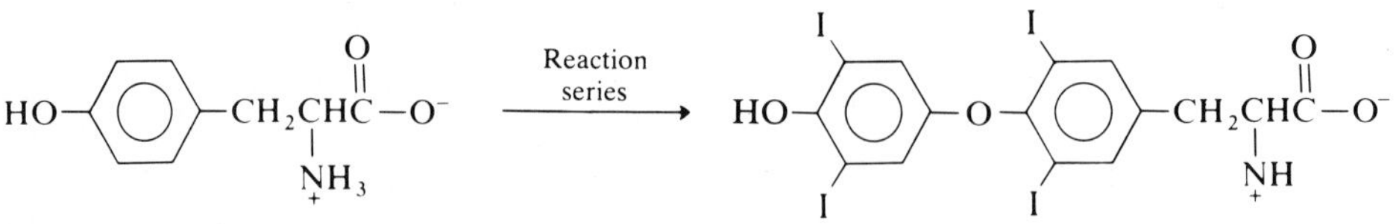

tyrosine thyroxine

Creatine, a nitrogenous compound found in muscle, is synthesized from components supplied by several different amino acids. The amino acids arginine, glycine, and methionine contribute components to the synthesis of creatine. This biosynthesis is sketched in Figure 57–2.

Transamination and *α*-Keto Acids

The third major fate of amino acids that enter our cells is a removal of the amino group to form an *α*-keto acid. The specific mechanism of this reaction will be considered in the following unit. There is one reaction, however, that should be considered here because of its important role. A reversible reaction is catalyzed by the enzyme glutamic dehydrogenase:

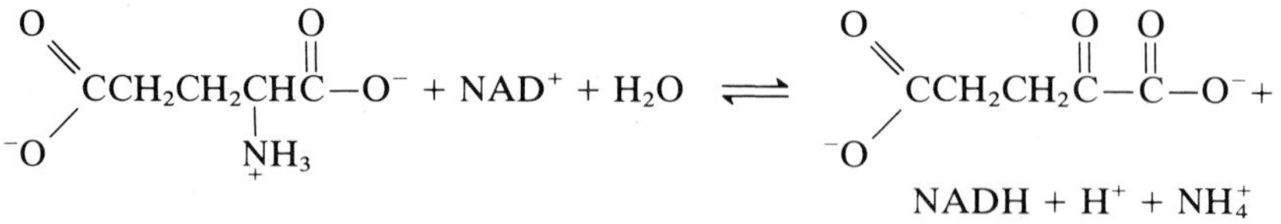

glutamate α-ketoglutarate

This reaction provides the principal source of ammonia, which must undergo further metabolism because of its toxic nature. The NADH formed could be channeled into the electron-transport system. It is the loss of the product NADH to the ets that causes this reaction to be favored in the direction written above. The *α*-ketoglutarate, being one of the components of the tricarboxylic acid cycle, can ultimately be degraded to CO_2.

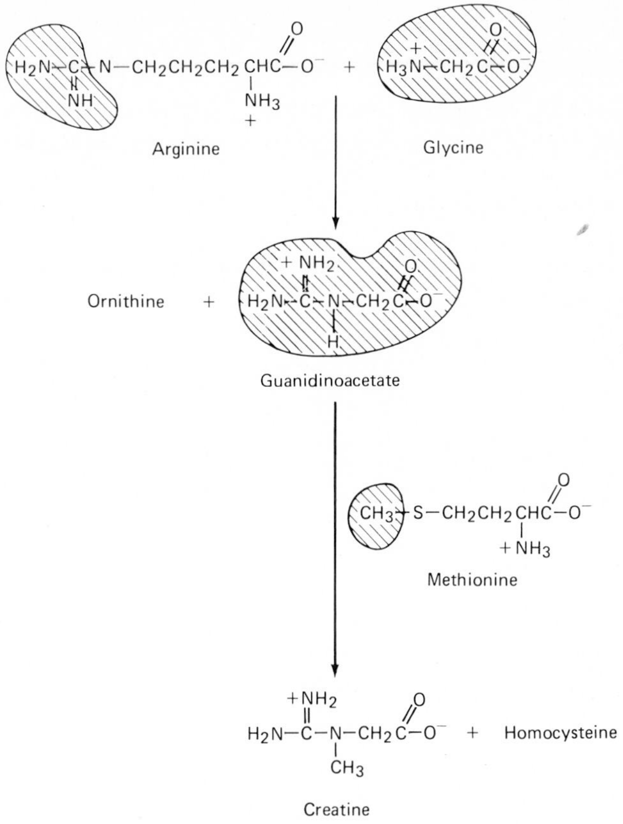
Arginine
Glycine
Ornithine
Guanidinoacetate
Methionine
Homocysteine
Creatine

Figure 57–2 Creatine biosynthesis.

The importance of this reaction is increased by the fact that many other amino acids can undergo an aminotransferase reaction (or transamination) with α-ketoglutarate. This exchange of an amino functional group for an oxygen functional group on two different carbon skeletons can be shown:

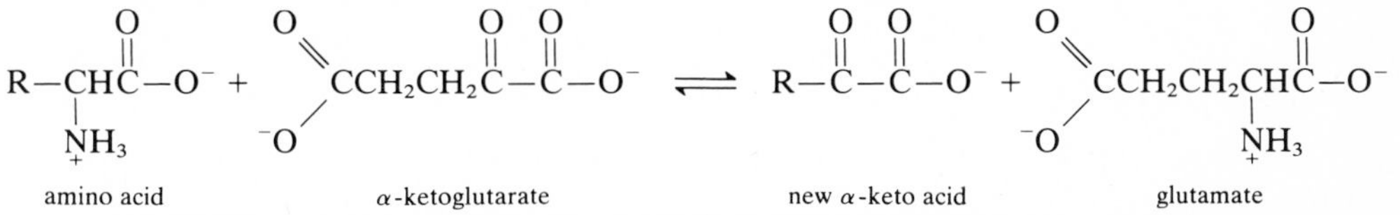
amino acid
α-ketoglutarate
new α-keto acid
glutamate

These two reactions can be coupled to provide a mode by which many amino acids can be degraded to α-keto acids. This can be shown as

$$\text{Amino acid} + \alpha\text{-ketoglutarate} \rightleftharpoons \text{glutamate} + \alpha\text{-keto acid}$$
$$\text{Glutamate} + NAD^+ + H_2O \rightleftharpoons \alpha\text{-ketoglutarate} + NH_4^+ + NADH$$

Net:

$$\text{Amino acid} + NAD^+ + H_2O \rightleftharpoons \alpha\text{-keto acid} + NH_4^+ + NADH$$

Generally, the most important initial step in amino acid metabolism within cells is the formation of an α-keto acid. The fates of the α-keto acids can generally be classified as leading to pathways involving mainly glucose metabolism or ketone-body formation (or both). Isotopically labeled amino acids can be administered to fasting test animals to determine the reaction systems involved. Some amino acids are metabolized by schemes involving glucose metabolism under these conditions (**glucogenic**). Obviously, amino acids leading directly to α-ketoglutarate, oxaloacetate, and pyruvate would be glucogenic. Many other amino acids follow more indirect pathways to ultimately enter the glucogenic scheme. Other amino acids, notably leucine, are metabolized to acetyl CoA and thus can yield ketone bodies under these conditions of study (**ketogenic**). Isoleucine, lysine, phenylalanine, and tyrosine have been shown to be involved in both **glucogenic** and **ketogenic** schemes in these studies.

It is beyond the scope of this presentation to consider the stepwise catabolism of each of the amino acids that enter our cells. More comprehensive biochemistry texts should be consulted for this material. Obviously, every amino acid must have both a specific biosynthetic and degradative pathway to complete the cyclic requirements of nature.

We have touched on a few of the types of reactions of amino acids; the next two units deal with these pathways in greater detail. The general reactions are considered in unit 58. Unit 59 covers the completion of nitrogen metabolism and the formation and elimination of the end products involved.

UNIT 57 Problems and Exercises

57–1. Given an interior section of a protein with the following sequence of amino acids:

· · · ala-lys-gly-gly-arg-ala-tyr-gly · · ·

indicate the specific point of action of the following enzymes on this section of the protein.

(a) Pepsin (b) Trypsin (c) Chymotrypsin

57–2. Would carboxypeptidase A be expected to act on the section of the protein above? Why?

57–3. Why is it an advantage that amino acids are transported across cell membranes rather than whole proteins?

57–4. Explain why the nonessential amino acids are found in the diet.

57–5. What are some of the common chemical features shared by the nonessential amino acids?

57–6. Trace the reactions whereby the nitrogen atom of valine is converted to ammonia using glutamic dehydrogenase coupled with transamination.

57–7. In a genetic-deficiency disease involving amino acid metabolism (maple syrup urine disease), α-ketoisovalerate is found in the urine in excessive amounts.

(a) Which amino acid would you expect as a source of this compound?

(b) If isobutyryl CoA is normally formed from this compound, what is the enzyme that is deficient in this disease?

57–8. A number of small molecules are known to be derived from amino acids by rather direct reactions. Name the amino acid that you might expect to lead to the following:

(a) $HOCH_2CH_2CHNH_2COOH$, homoserine

(b) $H_2NCH_2CH_2CH_2COOH$, γ-aminobutyric acid

(c) 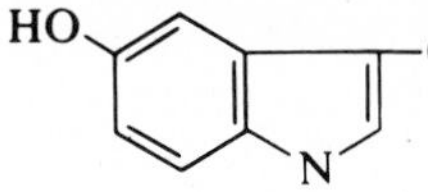

$-CH_2CH_2NH_2$, serotonin

(d) $H_2NCH_2CH_2SO_3H$, taurine

(e) 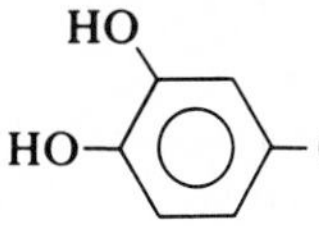

CH_2CHNH_2COOH, dihydroxyphenylalanine (dopa)

(f) 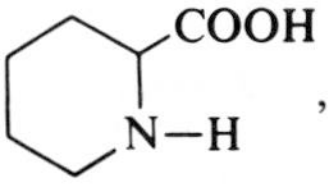

, pipecolic acid

UNIT 58 REACTIONS OF AMINO ACIDS

We have just considered the digestive processes whereby the proteins we ingest are broken down and enter our cells as mixtures of amino acids. Possibilities were indicated whereby the amino acids could be transformed into α-keto acids, which in turn had a variety of possible metabolic schemes. A few examples of amino acids converted to other small organic molecules were also shown. This unit will show how amino acids become integrated into the overall network of metabolic systems, including those previously considered for carbohydrates and lipids.

Transamination

One of the most frequently encountered families of enzymes is the **transaminases** or **aminotransferases**. An example of this reaction was used in the last unit. These enzymes are responsible for catalyzing reactions that bring about the exchange of a nitrogen (amino group) with an oxygen (keto group) on two different carbon skeletons:

$$R^1-\underset{\overset{+}{N}H_3}{\underset{|}{CH}}-\overset{O}{\overset{\|}{C}}-O^- + R^2-\underset{O}{\underset{\|}{C}}-\overset{O}{\overset{\|}{C}}-O^- \rightleftharpoons R^1-\underset{O}{\underset{\|}{C}}-\overset{O}{\overset{\|}{C}}-O^- + R^2-\underset{\overset{+}{N}H_3}{\underset{|}{CH}}-\overset{O}{\overset{\|}{C}}-O^-$$

There is a separate transaminase known for each α-amino acid, with the exception of threonine and lysine. Actually, all the enzymes may not have an absolute specificity, but some may be involved with more than one amino acid.

These enzymes bring about an active exchange of nitrogen among most of the amino acids and provide the variety necessary in the amino acid pool.

These enzymes are named to designate the two amino acids involved, for example, alanine-glutamate aminotransferase. Glutamate and aspartate are common as participants in the exchange because of their relationship to the tricarboxylic acid cycle. The importance of glutamic dehydrogenase, as previously discussed, also adds to this consideration in completing a metabolic system.

The aminotransferases catalyze reversible reactions. The direction of any given reaction system would be governed by what was required and removed for other reactions by the cell.

Either pyridoxal phosphate or pyridoxamine phosphate, with a metal ion, is involved in the aminotransferase reaction. An intermediate with an amino linkage or Schiff base is formed, $-\overset{\mathrm{H}}{\overset{|}{\mathrm{C}}}{=}\mathrm{N}-$. A reaction could be initiated between pyridoxal phosphate (aldehyde form) and an amino acid, or between pyridoxamine phosphate (amino form) and a keto acid. The key to the reversible nature of these reactions is the formation of an intermediate Schiff base, which can be rearranged in either of two directions as shown in Figure 58–1.

The aminotransferases have been widely used in clinical chemistry and are of diagnostic importance. Glutamic-aspartate aminotransferase, sometimes called glutamic-oxaloacetic transaminase (GOT), is present in large amounts in heart, liver, and skeletal muscle. The GOT activity present in the blood is elevated in over 90% of cases of myocardial infarction. The activity of this enzyme in the blood reaches a maximum within 24 hr of the infarction. Glutamic-alanine aminotransferase, sometimes called glutamic-pyruvic transaminase (GPT), is present in large amounts in the liver, but in much smaller amounts in other tissues. The blood GPT levels remain nearly constant during infarctions, but would show specific elevation in cases where liver damage was involved.

Decarboxylation

Many tissues have enzymes that catalyze the loss of the carboxyl carbon of amino acids to form CO_2. The amino acid **decarboxylases** generally require pyridoxal phosphate as a coenzyme in a reaction initially proceeding in a similar manner to the transamination just described. A Schiff base forms as shown in Figure 58–1. A hydrogen replaces the carboxyl carbon:

$$\mathrm{R{-}CH_2{-}\underset{\underset{+}{NH_3}}{\underset{|}{CH}}{-}\overset{\overset{O}{\|}}{C}{-}O^-} \xrightarrow{\text{Decarboxylation}} \mathrm{RCH_2\underset{\underset{+}{NH_3}}{\underset{|}{CH_2}}} + \mathrm{CO_2}$$

Since the CO_2 is lost, the reaction proceeds in the direction of decarboxylation.

The decarboxylation of diamino amino acids, such as lysine and ornithine, lead

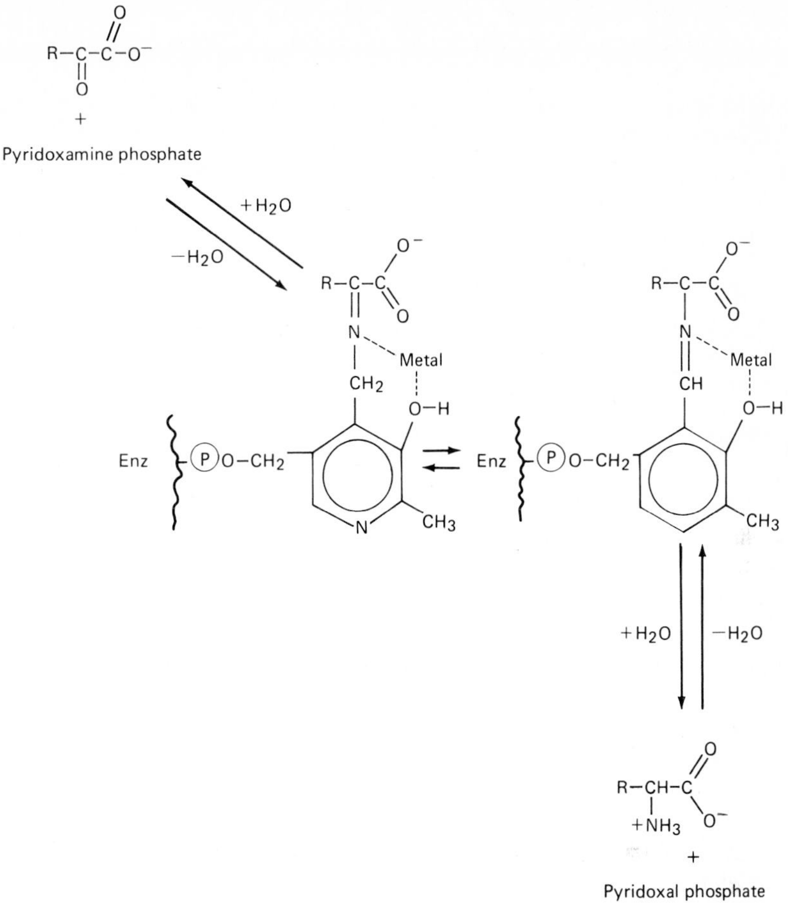
Pyridoxamine phosphate
+H2O
−H2O
Metal
O−H
Enz
CH2
CH
CH3
+H2O
−H2O
Pyridoxal phosphate

Figure 58–1 Aminotransferase system. The arrows show the specific bond involved with either of the possible Schiff bases. The phosphate of the coenzyme interacts with a specific site on the enzyme and the metal ion stabilizes the complex during the reaction.

to the formation of diamines. The decarboxylation of ornithine yields a compound with the common name putrescine:

$$\underset{\text{ornithine}}{H_3\overset{+}{N}{-}(CH_2)_3\underset{\underset{+}{NH_3}}{\underset{|}{C}}H{-}\overset{O}{\overset{\|}{C}}{-}O^-} \xrightarrow{\text{Decarboxylation}} \underset{\text{putrescine}}{H_3\overset{+}{N}{-}(CH_2)_4{-}\overset{+}{N}H_3}$$

The name given this compound is most descriptive of its odor. These diamines commonly arise from bacterial decay of proteins, leading to the undesirable odor associated with this degradative process.

Deamination

We have also considered the enzyme of greatest importance for this type of reaction. Glutamic dehydrogenase catalyzes **oxidative deamination** with either NAD or NADP as a cofactor:

$$\text{Glutamate} + \text{NAD}^+ \xrightarrow{\text{Glutamic dehydrogenase}} \alpha\text{-ketoglutarate} + \text{NADH} + \text{NH}_4^+$$

This reaction has its greatest significance in that the nitrogen is liberated from an organic molecule. When this reaction is coupled with transamination, nearly all the amino acids can lose their nitrogen in this manner.

Another group of enzymes, known as the amino acid oxidases, also catalyzes deamination. They bring about the loss of ammonia, but require a flavin cofactor, generally FMN. One role that has been assigned to these enzymes is the deamination of D-amino acids. Although D-amino acids are not common, they do enter our systems and have toxic effects. These enzymes could convert the D-amino acids to harmless keto acids, which of course would not be stereospecific. In this way a potentially harmful material could be detoxified and become part of the mainstream of catabolic reactions.

The fate of the ammonia formed in these reactions deserves further comment. Since free ammonia is toxic, some innocuous transport scheme must be available. One of these is the synthesis of amides using the enzyme glutamine synthetase:

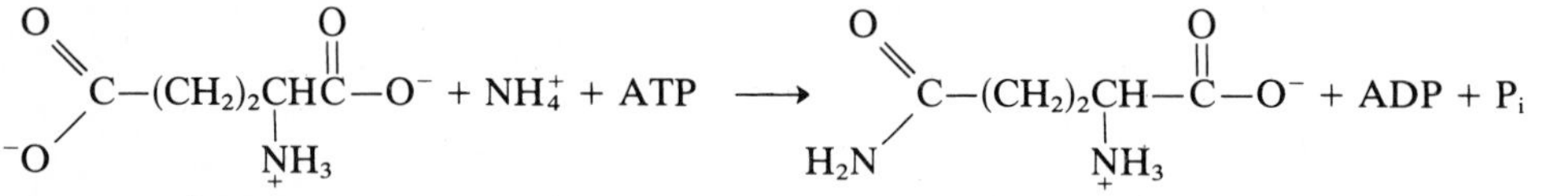

glutamate glutamine

The glutamine can become incorporated into proteins or act as a transport molecule. The normal range of glutamine in blood plasma is about 5 to 10 mg%; the normal range for glutamate is 0.4 to 4.0 mg%. Since this molecule carries two nitrogens per molecule, it is efficient for nitrogen transport.

It is also interesting that glutamine appears to have some different permeability characteristics from glutamate, especially for the brain, which has some special transport and permeability characteristics. For example, glutamine readily enters the brain from the blood, whereas glutamate does not. Asparagine (from aspartic acid) metabolism appears to be quite similar to that of glutamine.

Transmethylation

The metabolism of two carbon units involves acetyl CoA or active acetate. We shall now direct our attention to the metabolism of the one-carbon units, especially the methyl group. Methionine, an essential amino acid, is a ready source of methyl groups. The methionine assumes an active or coenzyme form before the methyl group can be readily transferred. An ATP is utilized to form the activated molecule

S-adenosylmethionine (SAM):

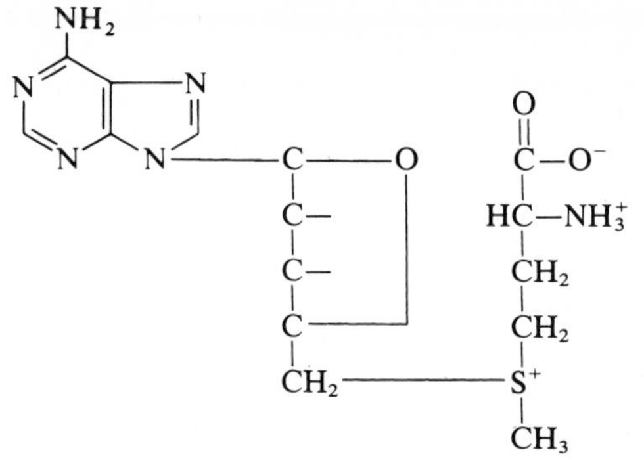

NH2
N
N
N
N
C
O
C—
C—
C
CH2
O
C—O⁻
HC—NH3+
CH2
CH2
S+
CH3

In this form the methyl group may be transferred to form creatine, considered earlier, as well as to form other molecules requiring a methyl source.

It is important to consider the molecule remaining from SAM after the methyl group is lost. The adenosyl group can also become detached from the sulfur after the loss of the methyl group, leaving a molecule homocysteine:

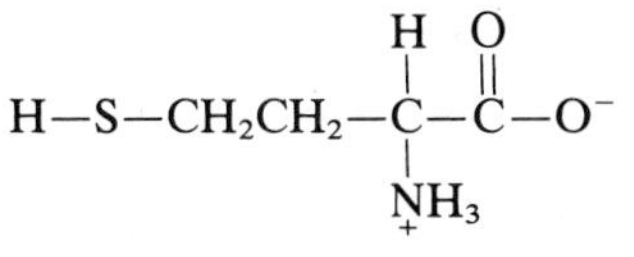

H O
H—S—CH2CH2—C—C—O⁻
NH3
+

This molecule is important because it is this portion of methionine that man is unable to synthesize. This accounts for the fact that methionine is essential. If homocysteine is fed along with a proper source of methyl groups (like choline), methionine is not necessary in our diets.

The homocysteine is also a key compound in interrelating sulfur-containing organic molecules. For example, cysteine biosynthesis arises from homocysteine and serine. The sulfur for cysteine can be derived from methionine, and the carbon chain can come from serine. The overall scheme including the source of serine is shown in Figure 58–2. The sulfur is transferred from a four-carbon chain to a three-carbon chain by two enzymes. One enzyme forms cystathione, and the other cleaves it to form cysteine. We can now see that the need for dietary methionine actually is a combination of needs for both the methionine itself and cysteine. If significant amounts of cysteine are included in the diet, the amount of the methionine required could be decreased. If enough methionine is in the diet, no cysteine needs to be ingested.

It may seem strange that although various complex intermediates can be synthesized, the relatively simple homocysteine cannot. The specificities of such enzyme systems are still beyond our present level of understanding.

Homocysteine can be converted to methionine if an adequate supply of methyl groups is available and two important coenzymes are present. This requires another coenzyme that acts as the methyl (one-carbon) carrier. The first coenzyme is the 5-methyltetrahydrofolate (5-methyl THF). The complete structure of folic acid was given previously. The tetrahydro form of folic acid is involved, and a methyl group is attached to a nitrogen at the number 5 position. The number 5 position on the

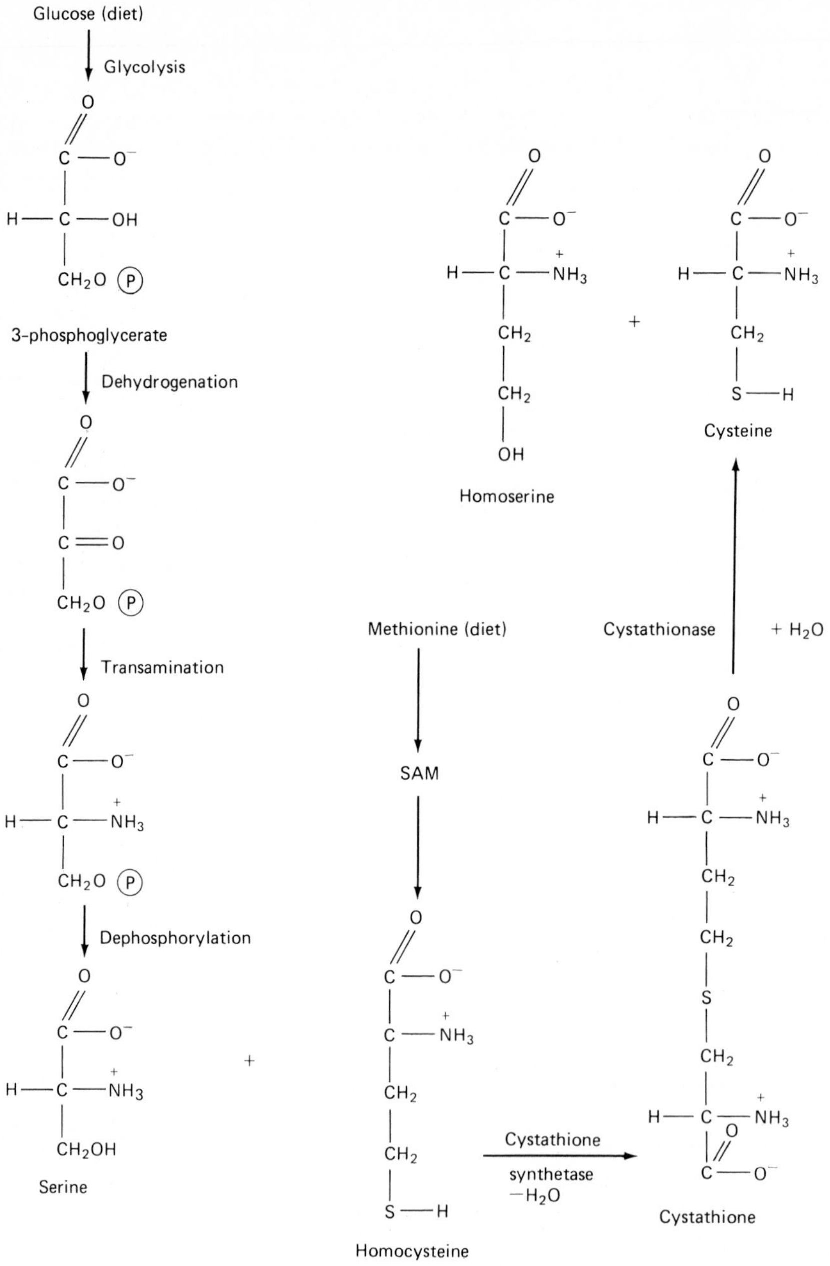
Glucose (diet)
Glycolysis
3-phosphoglycerate
Dehydrogenation
Transamination
Dephosphorylation
Serine
Homoserine
Cysteine
Methionine (diet)
Cystathionase
+ H2O
SAM
Homocysteine
Cystathione
synthetase
−H2O
Cystathione

Figure 58–2 The interrelationships of serine, cysteine, and methionine metabolism.

molecule is the nitrogen of the ring system, marked by an arrow on the structure shown in unit 50. This methyl transferase reaction proceeds to form methionine:

$$\text{Homocysteine} + \text{5-methyl THF} \longrightarrow \text{methionine} + \text{THF}$$

A special enzyme mediates this reaction, and a vitamin B_{12} derivative (aquocobalamin) is also required as a second cofactor. This may be one of the most important reactions of B_{12}. In spite of the possibility of other metabolic reactions, the methionine of the diet normally provides the major source of methyl groups in human systems.

Glycine

There is one more reaction system that should be considered, and that is the biosynthesis of the simplest of all our amino acids, glycine. Glycine is a nonessential amino acid and is formed from serine. A tetrahydrofolate coenzyme is involved in this reaction to accept another one-carbon transfer function.

$$\text{HOCH}_2\underset{\underset{+}{\text{NH}_3}}{\text{CH}}\overset{\text{O}}{\overset{\|}{\text{C}}}\text{—O}^- + \text{THF} \longrightarrow \text{H}_3\overset{+}{\text{N}}\text{CH}_2\overset{\text{O}}{\overset{\|}{\text{C}}}\text{—O}^- + \text{H}_2\text{O} + \text{5,10-methylene THF}$$

The number 10 nitrogen is adjacent to the phenyl ring, as shown previously in the complete structure of folic acid. The active portion of the coenzyme bearing the methylene derivative is sketched in Figure 58–3. The one carbon lost from the serine

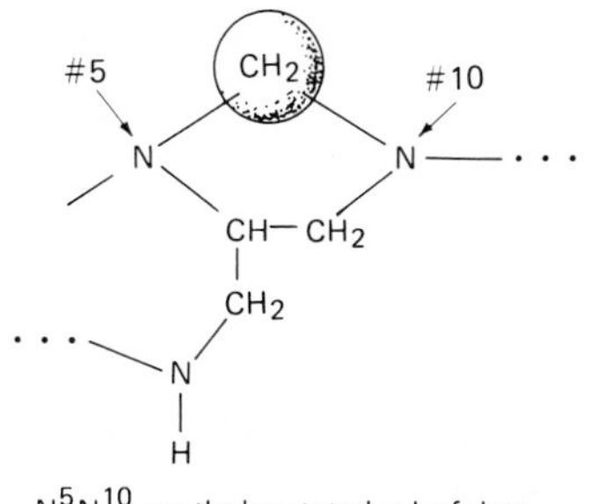
#5
CH2
#10
N
N
CH— CH2
CH2
N
H

N^5N^{10}-methylenetetrahydrofolate

Figure 58–3 Portion of folic acid derivative showing "active formaldehyde."

is shaded in the sketch. The 5,10-methylene THF can become oxidized and be a source for formaldehyde. It is sometimes given the name *active formaldehyde.* The folic acid derivatives, along with biotin considered previously, are to one-carbon metabolism what coenzyme A and the vitamin pantothenic acid are to two-carbon metabolism.

The serine, also a nonessential amino acid, was derived from glucose in the reaction scheme shown in Figure 58–2. Once again, glucose fulfills its metabolic responsibility of supplying both carbon skeletons and energy. We can now understand the necessity of examining the detailed metabolic schemes of glucose presented in the previous sections. It is only through such insight that molecular understanding can be developed. The real beauty and interwoven complexities of nature can only be realized following these levels of comprehension. Proper nutrition, better health, and biochemical understanding are inseparable.

UNIT 58 Problems and Exercises

58–1. Using structures, show the transamination reaction that takes place between alanine and oxaloacetate.

58–2. Draw the structure of the cofactor pyridoxal phosphate in a form you might expect as it reacted with alanine.

58–3. Why would you expect aminotransferase reactions to be readily reversible, whereas decarboxylase reactions are unidirectional?

58–4. Brain tissue is known to possess enzymes for the *GABA shunt.* In this reaction γ-aminobutyric acid (GABA) is formed by decarboxylation. The amino group of the GABA is then lost by a special transamination. The resulting aldehyde can be oxidized for further metabolism.
(a) Using structures, outline the GABA shunt.
(b) How would the GABA ultimately be metabolized?

58–5. The hydrolysis of SAM would yield a high negative free energy. How does this coincide with the role of the molecule?

58–6. Outline how the methionine-cysteine interrelationship is a good example of the sparing effect.

58–7. Why is methionine an essential amino acid and cysteine is not, even though the above relationship exists?

58–8. If ^{14}C-1-glucose was introduced in an animal diet,
(a) Would the methylene carbon of THF bear the ^{14}C label in the serine–glycine conversion?
(b) Would the ^{14}C label be found in the glycine?

UNIT 59 NITROGEN METABOLISM: COMPLETION OF THE CYCLE

We have considered some of the metabolic reactions of the amino acids resulting from the proteins that are ingested in our diets. All the catabolic reactions of amino acids lead to glucogenic or ketogenic ends, with the carbon atoms of amino acids ultimately lost as CO_2. The photosynthetic reactions of green plants and some microorganisms fix the CO_2 and return it to an organic form. This allows the carbon cycle to be completed. As expected, the nitrogen cycle must be completed also.

Reactions of Ammonia

We have seen that ammonia, or NH_4^+ in aqueous solution at biological pH, is an important form of nitrogen. The two major sources of ammonia considered were oxidative deamination by glutamic dehydrogenase and general deamination reactions of amino acids. Some ammonia is also released within the intestine during the digestive processes. Some of this ammonia can be taken up and passed to the liver in the portal blood. The normal ammonium level in peripheral blood must be kept very low by detoxification reactions. Ammonium ions are very toxic to our cells and must be eliminated in the urine or dealt with by other reactions.

There are three major reaction pathways that ammonia can take within a cell:

1. Reverse of the glutamate dehydrogenase, considered earlier.

   α-ketoglutarate + ammonia + NADH → glutamate + NAD

2. Glutamine formation by glutamine synthetase, considered earlier.

   Glutamate + ammonia + ATP → glutamine + ADP + P_i

3. Synthesis of carbamyl phosphate, $H_2N{-}\overset{\overset{\displaystyle O}{\|}}{C}{-}O{-}P$.

   Bicarbonate + ammonia + 2ATP → carbamyl phosphate + 2ADP + P_i

Carbamyl phosphate is a participant in urea synthesis and in the synthesis of pyrimidines, constituents of nucleic acids. It is the formation of carbamyl phosphate leading to urea biosynthesis that is of concern at this time. The formation of carbamyl phosphate in the process of urea biosynthesis utilizes ammonia as the nitrogen donor. An enzyme found in the liver, carbamyl phosphate synthetase, catalyzes the essentially irreversible reaction:

$$NH_4^+ + HCO_3^- + 2ATP \longrightarrow H_2N{-}\overset{\overset{\displaystyle O}{\|}}{C}{-}O{-}\underset{\underset{\displaystyle O^-}{|}}{\overset{\overset{\displaystyle O}{\|}}{P}}{-}O^- + 2ADP + P_i$$

This enzyme operates by a somewhat different mechanism from the enzyme that catalyzes the formation of carbamyl phosphate for pyrimidine biosynthesis. This is a situation in which two rather different and distinct enzyme systems form the same compound within the same organism. The two enzyme systems are located at different points within the cell, and are subject to quite different control mechanisms. One enzyme starts the process for the formation of an end product that will be eliminated (urea). The other enzyme starts the process toward the biosynthesis of an essential component (nucleic acids) of dividing cells. The differences in these systems might be expected.

Urea Biosynthesis

Urea biosynthesis proceeds by reacting the carbamyl phosphate with an amino acid, ornithine. The enzyme ornithine transcarbamylase brings about the formation of another amino acid, citrulline.

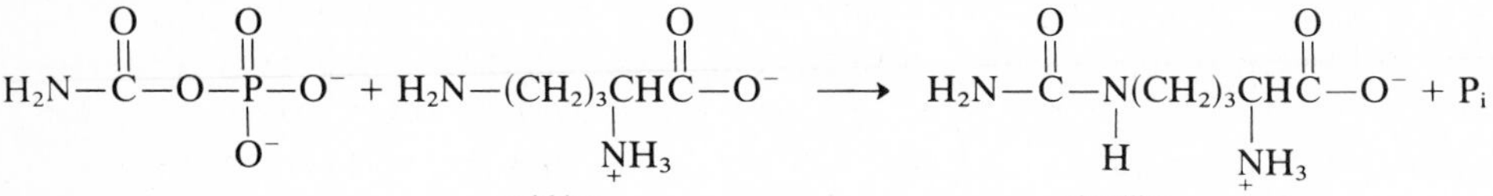
ornithine citrulline

Neither ornithine nor citrulline are found as normal constituents of proteins but have these special functions in urea biosynthesis. They can be found also at other points in nature. For example, citrulline was first isolated as a free amino acid in watermelon juice.

The next reaction is a condensation step that requires another ATP and another amino acid. Aspartate combines with the citrulline, leading to argininosuccinate synthesis:

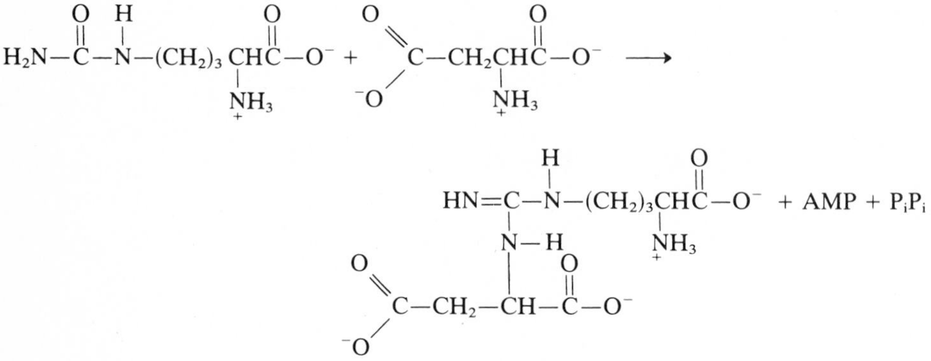
argininosuccinate

This large molecule next undergoes a cleavage, catalyzed by argininosuccinase:

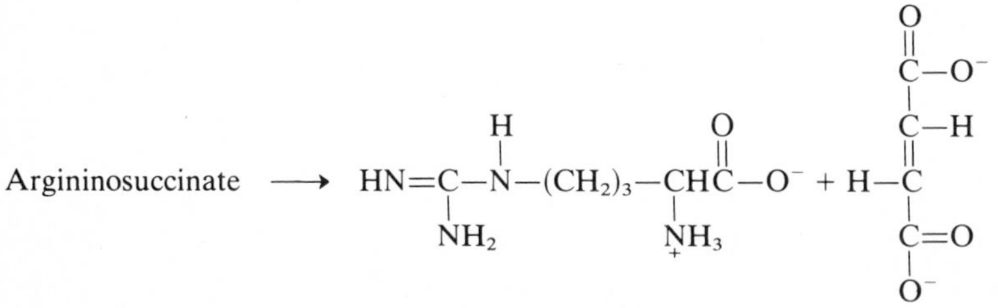
arginate fumarate

Both of the nitrogen atoms of urea are fixed in organic forms at this point. Arginate can be hydrolyzed by the enzyme arginase to form urea and re-form ornithine:

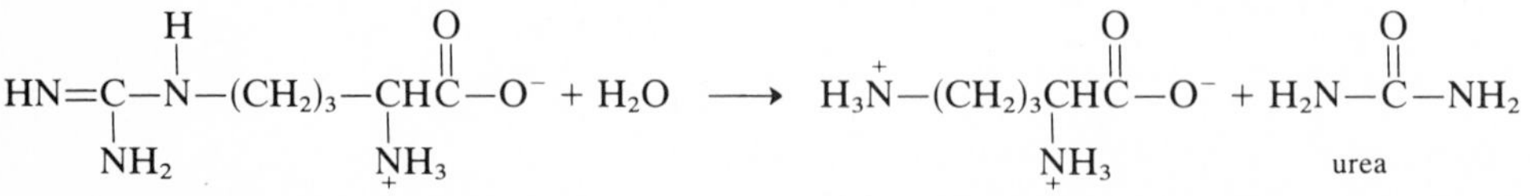
urea

The enzyme arginase is quite interesting because it is limited, almost exclusively, to the liver. This means that the entire urea biosynthesis process takes place within the liver. Because of its involvement with the elimination process, it was first thought that the kidney was the site of urea biosynthesis. Both biochemical and physiological studies have proven that the liver is the organ mainly responsible for the synthesis of urea, and the kidney is the organ responsible for its elimination from the bloodstream. Since urea is an innocuous compound, it can be tolerated at rather high levels within our systems. The normal level of urea in our blood ranges from 20 to 40 mg%. Since urea is nearly half nitrogen, blood urea nitrogen (BUN) levels are 10 to 20 mg% N.

The cyclic nature of urea biosynthesis is summarized in Figure 59–1. This shows the sources of the nitrogen atoms of urea and the interrelationships with other

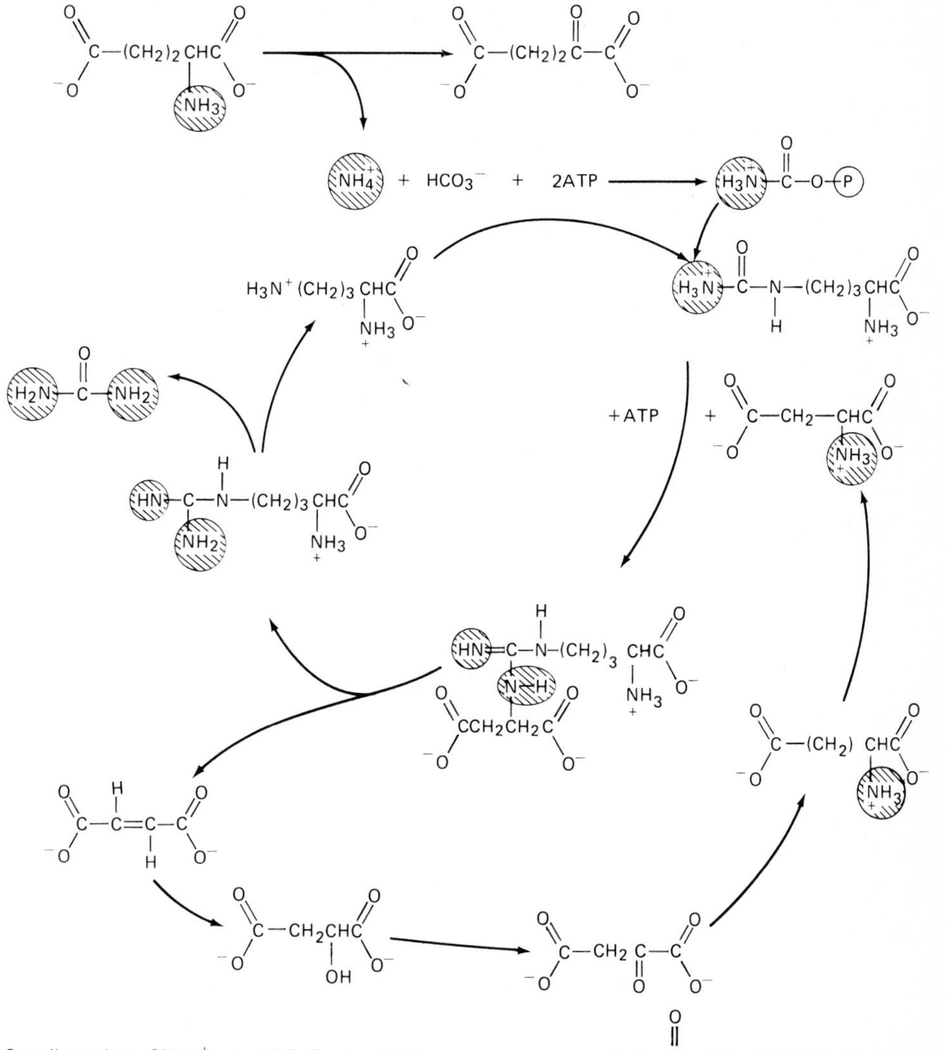

NH4 + HCO3 + 2ATP
+ATP

Overall reaction: $2NH_4^+ + HCO_3^- + 3ATP \longrightarrow H_2N{-}\overset{O}{\overset{\|}{C}}{-}NH_2 + 2ADP + 2P_1 + P_1P$

Figure 59–1 Urea biosynthesis. The nitrogen atoms of urea are shaded.

reactions, such as those mediated by glutamate dehydrogenase and transaminase. The close relation of this biosynthetic pathway with the tricarboxylic acid cycle is important. This part of the urea cycle also takes place within the mitochondria, where the energy necessary to drive the process is readily available.

Once the urea is eliminated from our bodies, nitrogen metabolism must continue in the cyclic relationships so important in nature. An enzyme is rather widespread in plants and microorganisms that breaks down the urea. Urease specifically catalyzes the hydrolytic reaction:

$$H_2N-\overset{\overset{\displaystyle O}{\|}}{C}-NH_2 + H_2O \longrightarrow 2NH_4^+ + CO_2$$

Nitrogen Fixation

The fixation of atmospheric nitrogen (N_2) into organic molecules does not take place in animal systems but is important in nature. The enzyme system whereby N_2 becomes reduced to ammonia is known as nitrogenase:

$$N_2 \xrightarrow{\text{Reduction}} 2NH_3$$

This is a complex reaction system, which is known to operate in both aerobic and anaerobic microorganisms. Higher plants, especially the legumes (soybeans, peas, clover), often operate in a symbiotic relationship with these nitrogen-fixing microorganisms (e.g., *Clostridium pasteuriarium*). This reductive process requires an energy source and a source of electrons. Carbohydrate metabolism almost always supplies the electrons necessary for the reduction and acts as the energy source.

Another important part of nitrogen utilization involves the formation of nitrates (NO_3^-) and nitrites (NO_2^-). This process is called **nitrification**. Many microorganisms, especially in the soil, derive their energy from the oxidation of ammonia to nitrites or nitrates. The *Nitrosomonas* are involved in the formation of nitrites; the *Nitrobacter* species continue the oxidation to form nitrates. The reduced form of nitrogen is an important energy source to these soil bacteria, just as the reduced form of carbon is to our metabolism.

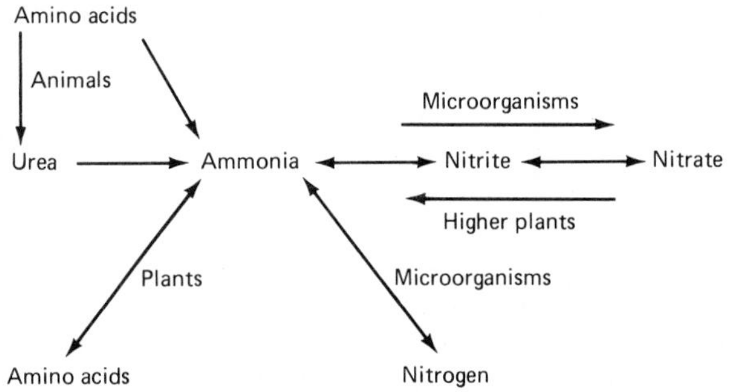
Amino acids
Animals
Microorganisms
Urea
Ammonia
Nitrite
Nitrate
Higher plants
Plants
Microorganisms
Amino acids
Nitrogen

Figure 59–2 The nitrogen cycle.

Higher plants use the nitrate of the soil as their principal nitrogen source. These plants reverse the process to reduce the oxidized forms of nitrogen to ammonia. The higher plants then convert the ammonia to amino acids and proteins. The animal kingdom then indulges itself upon the plant kingdom. Also, some nitrate can be converted to atmospheric nitrogen by certain other bacteria found in the soil. This is called **denitrification**. The scheme is outlined in Figure 59–2.

Urinary Constituents

It would not be complete to leave a discussion of nitrogen metabolism without considering some of the nitrogenous constituents of normal urine. Table 59–1 shows the amounts of these molecules normally present in urine. Ranges are given for each

Table 59–1 Nitrogen-containing compounds in a 24-hour sample of normal human urine

Material	Amount (as g of N)	Relative percentage of nitrogen excreted
Urea	6–18	85–90
Creatinine	0.3–0.8	4–5
Uric acid	0.08–0.2	1–2
Ammonia	0.4–1.0	2–4
Other molecules		3–5

value; there is some variation depending on such things as the diet, state of the individual, activity of the individual, and so on. A few of these molecules should be mentioned. Hippuric acid (benzoylglycine) appears in the urine and results from the presence of benzoic acid in the diet. Very small amounts of sodium benzoate are commonly used as food preservatives. As discussed previously, the benzene ring is not readily attacked by enzyme systems, but is eliminated following a detoxification by conjugation with glycine:

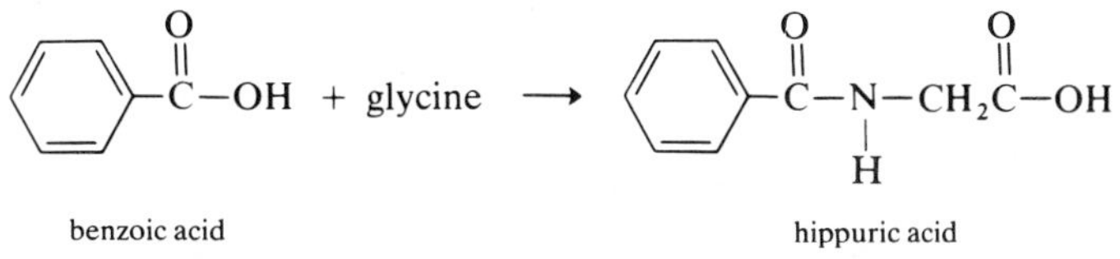
C—OH + glycine
C—N—CH₂C—OH
benzoic acid
hippuric acid

Other compounds, including peptides, water-soluble vitamins, and end products of bile pigment metabolism (urochrome or urobilinogen), are also present.

The processes whereby the metabolic systems of the various organs of the body control the fate of the nitrogen atom is truly remarkable. Man fits into only a small part of the overall nitrogen cycle that is so important in nature. The biochemistry of nitrogen metabolism is certainly one of the key features that contributes to the unity of all living systems.

UNIT 59 Problems and Exercises

59–1. If phenaceturic acid, 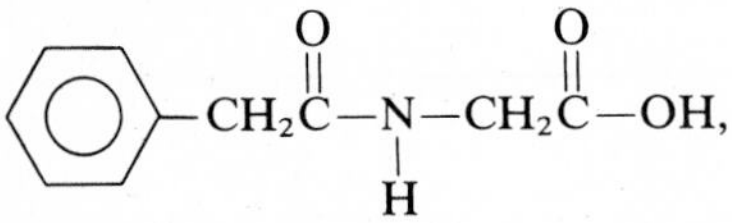

O
O
CH₂C—N—CH₂C—OH
H

, was found in the urine, what would you expect as sources of this material? Using equations, outline how this material might be formed within the body.

59–2. If 25 mg% of urea was eliminated in the urine, calculate the BUN levels from the loss of this constituent.

59–3. Would you expect that the amount of urea excreted would increase or decrease in situations that lead to a positive nitrogen balance?

59–4. Briefly explain why you might expect to find two different enzyme systems within the liver that both catalyze the formation of carbamyl phosphate.

59–5. Starting with ^{15}N-labeled aspartate, show the pathways whereby one of the nitrogens of urea might bear the ^{15}N label.

59–6. Starting with ^{15}N-labeled aspartate, show the pathway whereby both of the nitrogens of urea might bear the ^{15}N label.

59–7. If you had $HOOCCH_2CH_2CHNH_2{}^{14}COOH$, show by equations how the carbon atom of urea might bear the ^{14}C label.

59–8. Would you expect nitrification to be an endergonic or exergonic process?

CHAPTER 22

nucleic acids and protein biosynthesis

The nucleic acids have been studied intensely since it became established that they were the primary informational macromolecules of living systems. A central dogma of molecular biology interrelates the nucleic acids, deoxyribonucleic acid (DNA) and ribonucleic acid (RNA), with protein biosynthesis. This involves the transfer of genetic information not only from generation to generation but also from nucleic acid molecules to proteins. The major flow of information is as follows:

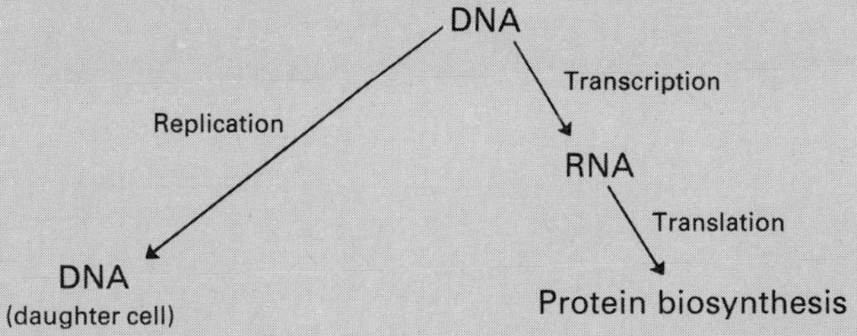
DNA
Transcription
Replication
RNA
Translation
DNA
(daughter cell)
Protein biosynthesis

Information is transferred in a highly regulated chemical fashion. The linear arrangement of repeating units must be exactly duplicated in the replication process to maintain the integrity of the information necessary to sustain life. This linear arrangement of repeating units in the nucleic acids must also be translated into the sequence of amino acids in proteins. At first the biochemical understanding of these processes almost seems to be beyond the hope of man. The wonder and beauty of living processes cannot be better visualized than in these reactions of nucleic acids and protein biosynthesis. To understand how these intricate processes occur, we must examine the chemical nature of each reactant, molecule by molecule.

UNIT 60 INTRODUCTION TO NUCLEIC ACIDS: DNA

In 1869, a German biochemist isolated a new class of acidic compounds from cell nuclei. At about this same time, Mendel worked out his basic rules of heredity. There was no firm connection made between these two important discoveries until 1944 when it was proved that **nucleic acids** were the carriers of genetic information. This gave a whole new impetus to the search for the information at the crossroads of chemistry and biology that would lead to a better understanding of the molecular basis of life itself.

The earliest chemical studies on nuclein, or nucleic acids as they were later named, involved the separation and identification of the component parts of DNA from genetic or chromosomal material. Both microorganisms and mammalian cell nuclei provide a good source of DNA. If, for example, thymus tissue is homogenized and extensively washed with isotonic saline (0.9% NaCl), a gelatinous mass of **nucleoprotein** remains. This deoxyribonucleoprotein can be broken into two major components if placed in a concentrated salt solution. A family of proteins known as histones dissolves, leaving the insoluble DNA. Since this separation can be brought about by using a high salt concentration, the linkage between the DNA and protein is assumed to be an ionic one. The DNA has many anionic sites (−) owing to the ionization of its phosphate groups. The histones contain the amino acids lysine and arginine, which possess a cationic (+) charge. In the nucleoprotein, the negatively charged DNA is neutralized by positively charged proteins.

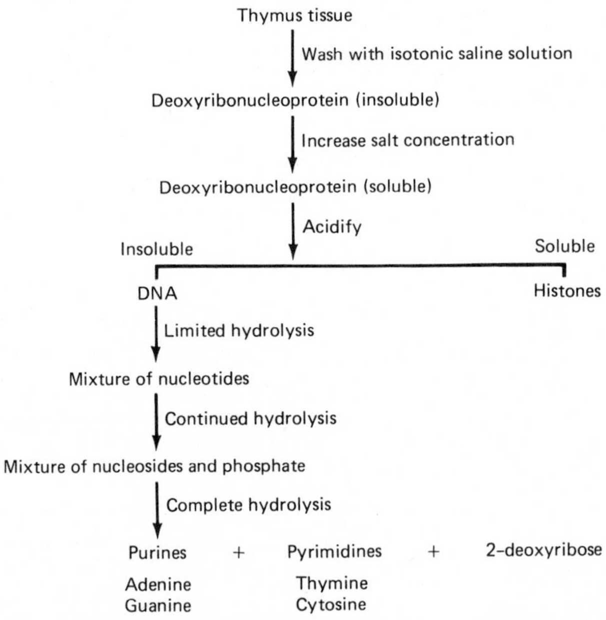
Thymus tissue
Wash with isotonic saline solution
Deoxyribonucleoprotein (insoluble)
Increase salt concentration
Deoxyribonucleoprotein (soluble)
Acidify
Insoluble
Soluble
DNA
Histones
Limited hydrolysis
Mixture of nucleotides
Continued hydrolysis
Mixture of nucleosides and phosphate
Complete hydrolysis
Purines + Pyrimidines + 2-deoxyribose
Adenine
Guanine
Thymine
Cytosine

Figure 60–1 The products of hydrolysis of deoxyribonucleoprotein. Each step represents a sequential hydrolysis procedure.

Protein-free DNA can be precipitated by acid and hydrolyzed to its components as shown in Figure 60–1. The first step in the hydrolysis of DNA yields a mixture of nucleotides, the second a mixture of nucleosides and phosphate. The final step yields nitrogen bases (purines and pyrimidines) and a carbohydrate, deoxyribose. Let us start with the simplest parts and build the nucleoprotein.

Nitrogen Bases

Nucleic acids have two different ring systems containing carbon and nitrogen. The simpler of the two heterocyclic ring systems has the parent molecule, **pyrimidine**. The general abbreviation Py will be used for an unspecified pyrimidine and the first letter will be used for a specific nitrogen base. The structures of thymine (T) and cytosine (C), which are present in DNA, and uracil (U), which is present in RNA, are as follows:

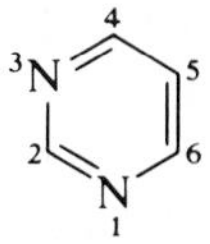

pyrimidine (Py)

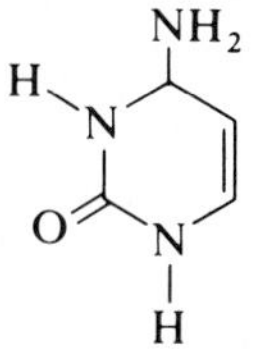

cytosine (C)

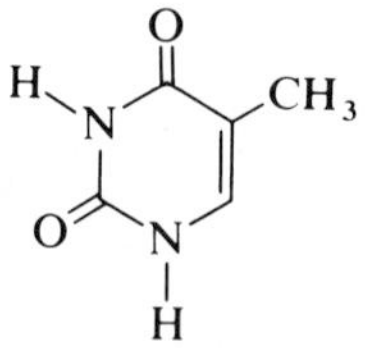

thymine (T)

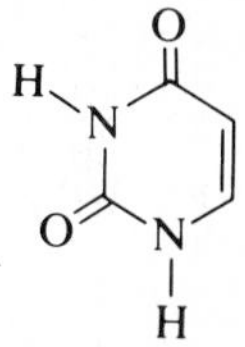

uracil (U)

There are other pyrimidines found in nature, which can be visualized using the number system as shown on the pyrimidine. Uracil (U) is not found in DNA, but is an important component of RNA.

The second family of heterocyclic ring systems found in nucleic acids is the **purines** (Pu). Adenine (A) and guanine (G) are part of both DNA and RNA molecules:

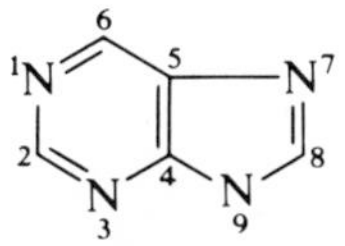

purine (Pu)

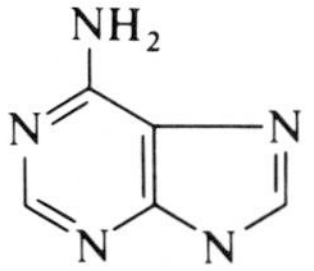

adenine (A)

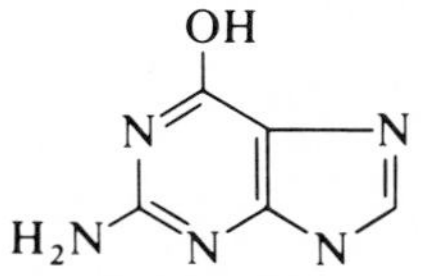

guanine (G)

We have used the purines before, as in the case of ATP and GTP.

Other purines of interest are not components of nucleic acids. One of these is caffeine. Another is uric acid, a compound formed in excessive amounts in the blood in the disease gout.

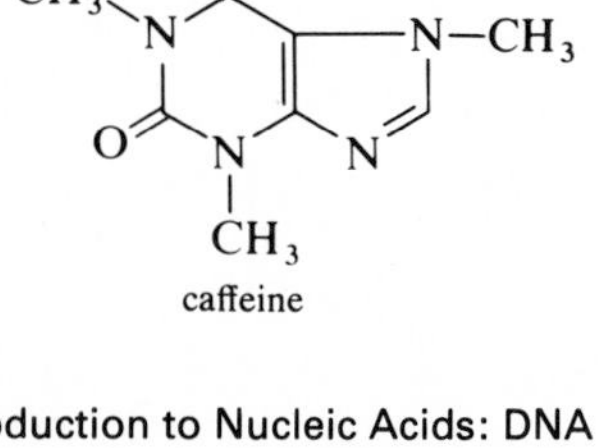

caffeine

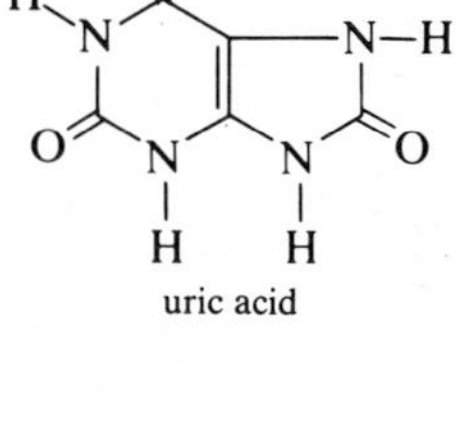

uric acid

Nucleosides and Nucleotides

The nitrogen bases combine with either deoxyribose or ribose to yield **deoxyribonucleosides** or **ribonucleosides**, respectively. The names assigned to the nucleic acid components are shown in Table 60–1. Notice that prime numbers are assigned to the carbons of the deoxyribose or ribose part of the molecules.

Table 60–1 Examples of nomenclature of components of nucleic acids

Base	Nucleoside	Nucleotide	Abbreviation of nucleotide
Adenine	Adenosine	Adenosine-2′-monophosphate or 2′-adenylic acid	2′-AMP
		Adenosine-3′-monophosphate or 3′-adenylic acid	3′-AMP or Ap
		Adenosine-5′-monophosphate or 5′-adenylic acid	5′-AMP or pA
	Deoxyadenosine	Deoxyadenosine-3′-monophosphate or 3′-deoxyadenylic acid	3′-dAMP or dAp
		Deoxyadenosine-5′-monophosphate or 5′-deoxyadenylic acid	5′-dAMP or pdA
Thymine	Thymidine	Thymidine-5′-monophosphate or thymidylic acid	dTMP
	Thymine riboside	Thymine riboside-5′-phosphate	TMP

The nucleotides are the phosphate esters of the nucleosides:

Nucleotides = nucleosides + phosphate

Table 60–1 shows two systems of nomenclature for both the nucleotides and deoxyribonucleotides of adenine. Recall the structure of the nucleotide AMP. The bases guanine, cytosine, and uracil would be similar to this system. Since thymine (T) only occurs in quantity in DNA as a deoxyribonucleotide, a special nomenclature system is shown for it. The conventional abbreviations are also shown. When the numerical prefix is not shown, it is assumed to be the 5′ derivative. A second abbreviation scheme using a lower case *p* for the phosphate group is also shown. In this system the 5′ terminal phosphate is shown to the left and the 3′ terminal to the right of the base involved. The notation for the ATP we have studied would be given as pppA; ADP would be ppA.

Nucleic acids are known to be nucleotides joined through **phosphodiester** (O—P—O—) linkages. There are different ways in which neighboring deoxyribonucleotides could be linked. Research has shown that the linkage is from the 5′ carbon of one deoxyribose to the 3′ carbon of its adjacent deoxyribose. A section of the linear arrangement of DNA can be represented as in Figure 60–2. The actual DNA molecule may have 3000 such deoxyribonucleotides for every 1 million molecular weight. DNA molecules range in molecular weight from about 1×10^6 daltons for a simple virus to 2×10^9 daltons for the single molecule of DNA found in

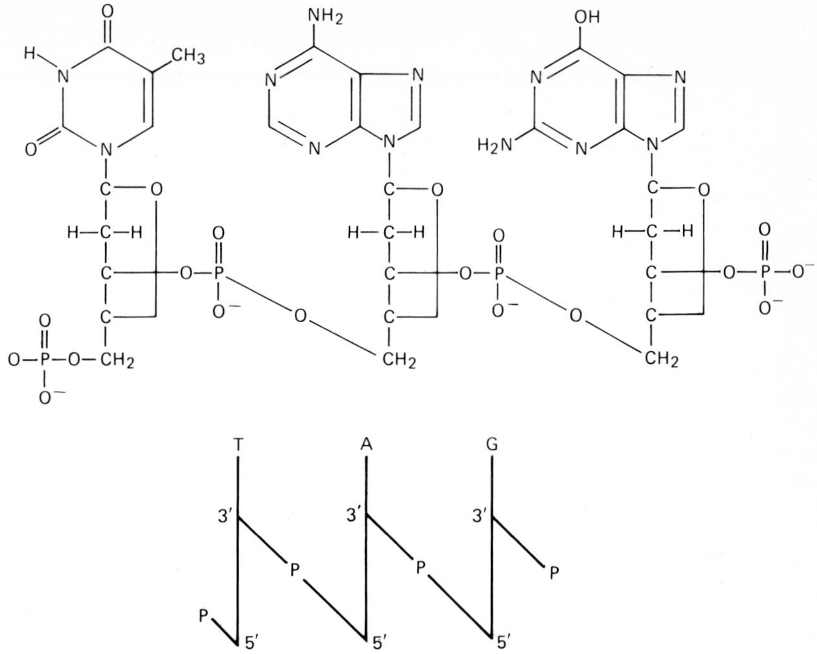
T
A
G
3′
5′
P

pT pdA pdGp

Figure 60–2 A section of a single chain of DNA. Two forms of shorthand notations are shown below the structural representation. Note that this chain is running in a 3′–5′ direction.

the bacterium *E. coli.* Most generally, the DNA molecules isolated are fragmented mixtures of the huge molecules that actually exist within the cell. Since DNA is such a long molecule, it is fragile and difficult to isolate in an intact form.

The Helical Structure

DNA is not simply a long single strand of polynucleotides. In 1953, James Watson and F. H. C. Crick suggested that DNA was actually in the form of a double **helix**. The helical model arose from work using X-ray methods of analysis. Although this observation is very important, it was preceded by other studies that need to be investigated before we can understand how DNA is formed or how it functions in nature.

Certain regularities are known to exist in the distribution of nitrogen bases of DNA. The **ratio** of the sum of the purines to the pyrimidines was found to be about 1. This could be stated as

$$\frac{\mathrm{Pu}}{\mathrm{Py}} = 1 = \frac{\mathrm{A+G}}{\mathrm{T+C}}$$

However, it was noted that the ratios A:G or T:C ranged over a great variety of values depending upon the source of the material under study. It was found that each species studied had a characteristic composition of nitrogen bases. Furthermore, the DNA from different organs and tissues of the same species of higher organisms had similar base compositions.

The DNA is depicted as two long polydeoxyribonucleotide strands intertwined about each other in a double helix. In this double helix there are matched pairs of bases interacting to provide part of the stability known to exist in this long molecule. The match-up in the double-stranded DNA corresponds to the analytical data: for every A there is a paired T, for every G there is a paired C. The chemical structures of the pairs are matched to allow **hydrogen bonding** between the nitrogen bases of the opposite strands, as shown in Figure 60–3. The paired nitrogen bases of the

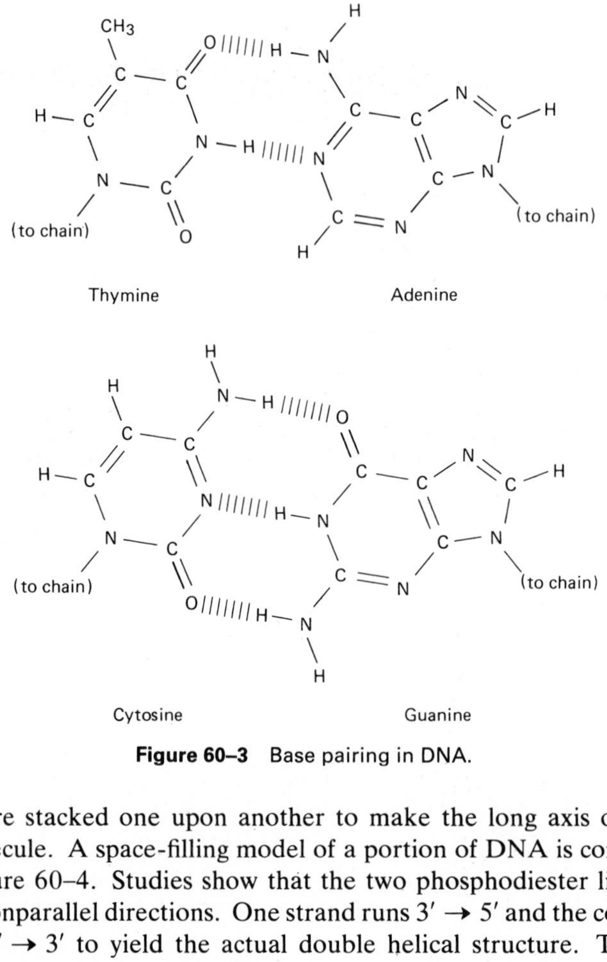

(to chain)
Thymine
Adenine
Cytosine
Guanine

Figure 60–3 Base pairing in DNA.

nucleotides are stacked one upon another to make the long axis of the double-stranded molecule. A space-filling model of a portion of DNA is compared with a sketch in Figure 60–4. Studies show that the two phosphodiester linkages run in opposite or nonparallel directions. One strand runs 3′ → 5′ and the complementary strand runs 5′ → 3′ to yield the actual double helical structure. The phosphate groups are on the outside of the helical molecule and give DNA a multitude of negative charges at physiological pH.

DNA exists as a coiled double helix that can be uncoiled. If a solution DNA is gently heated, the coil falls apart and the DNA is said to be **denatured.** Subsequent

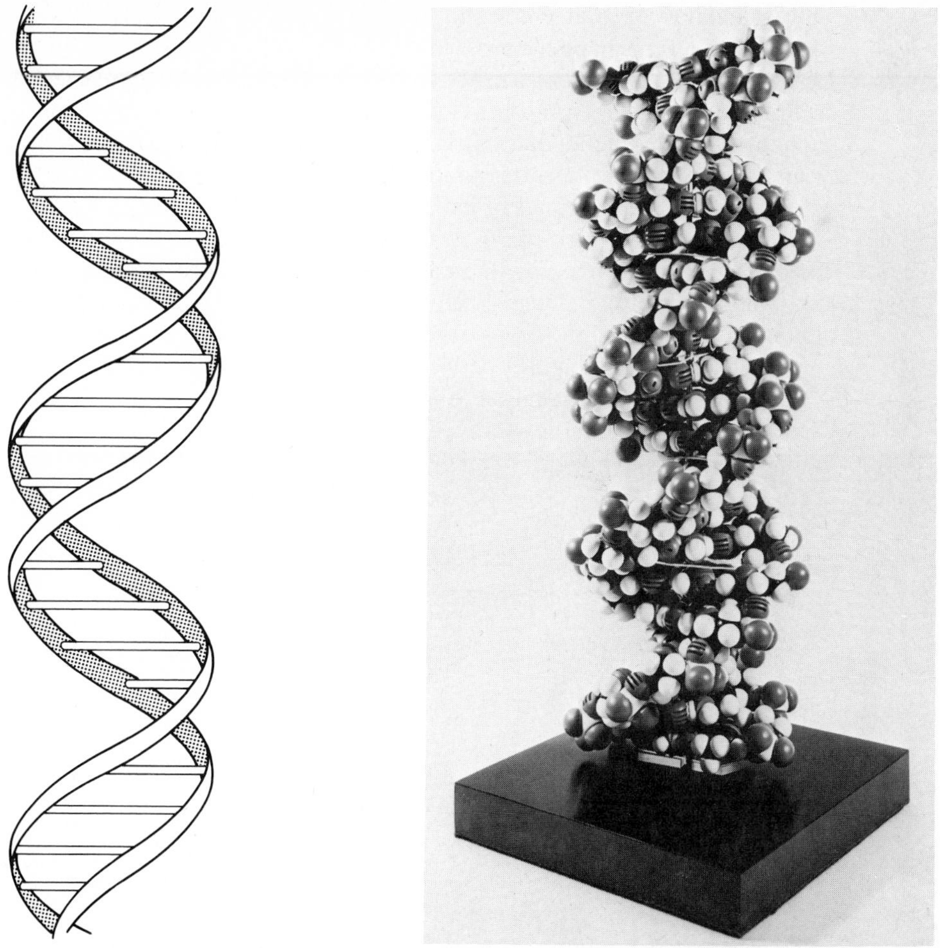

Figure 60–4 Sketch and space-filling model of DNA. The sketch was redrawn from the original paper by Watson and Crick which uses horizontal rods to show the pairs of hydrogen-bonded bases of the two separate chains. (Photograph courtesy of the Ealing Corporation, South Natick, Ma.)

cooling of the solution will re-form or "renature" the DNA molecule, but seldom to the extent that it was in its native state. This ordered $\rightleftharpoons$ disordered relationship is rather similar to what happens in protein molecules. Since the G–C pair is stabilized by three hydrogen bonds and the A–T pair by only two, samples that are rich in G–C are relatively more stable toward this denaturation than those with a large number of A–T pairing.

DNA Replication

The double-helix model suggests that DNA replication could be directed by base pairing, as the complementary chains unwind. This unwinding would provide two new **templates** or patterns upon which new double-stranded DNA could be synthesized. If this happens on each of the complementary strands, the result could be two new DNA molecules exactly like the parent. This is required if the genetic information is to be transmitted to the next generation. The experiment that demonstrated this model of duplication was reported by Meselson and Stahl about 5 years after the double helix was announced. They first grew *E. coli* on a ^{15}N source to allow it to label both strands of its DNA with this heavy isotope. Carefully washed cells (containing ^{15}N-labeled DNA) were allowed to undergo only one cell division to form daughter cells in a growth medium devoid of ^{15}N, but containing the naturally occurring ^{14}N isotope. During cell division, the total DNA content of the culture doubled. The total DNA was carefully analyzed using ultracentrifugation techniques that could distinguish between DNA containing ^{15}N and DNA containing ^{14}N. Each DNA of the daughter cells now was shown to have one ^{15}N-DNA and one ^{14}N-DNA. All the double-stranded DNA of the daughter cells was found to have both ^{15}N and ^{14}N. This was called the **semiconservative replication** scheme, since the

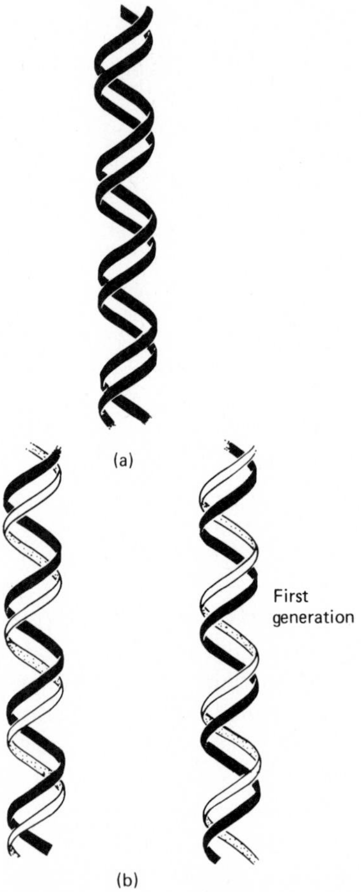
(a)
First generation
(b)

Figure 60–5 (a) Heavy lines indicating DNA containing heavy ^{15}N. (b) Replication in presence of light isotope containing ^{14}N (indicated by light lines).

new DNA was not either all new or all old but rather a hybrid of both. The resulting DNA of the daughter cells was formed in a manner consistent with the template hypothesis and could be as shown in Figure 60–5.

At about the same time, Arthur Kornberg and coworkers isolated an enzyme, DNA polymerase, that was capable of synthesizing DNA from simpler molecules in a cell-free system. In the presence of a small amount of preformed DNA to start or "prime" the reaction, this enzyme could incorporate all four deoxynucleotides into new DNA. Deoxynucleotides were required as triphosphates to bring about the reaction; Mg^{2+} was required as well. Kornberg received the Nobel prize for this demonstration of the *in vitro* synthesis of DNA.

Kornberg and coworkers went on to synthesize a DNA from a natural source and then to test this synthetic copy for biological activity. In this set of experiments, a single-stranded form of DNA from a bacterial virus, denoted $\phi\times$ 174, was used. This viral DNA is a small molecule by DNA standards (1.6×10^6 daltons) and is circular in form. This experiment is sketched in Figure 60-6. A single-stranded template of the DNA was directed by a polymerase to synthesize a complementary strand containing a special isotopic marker. The newly formed strand was then

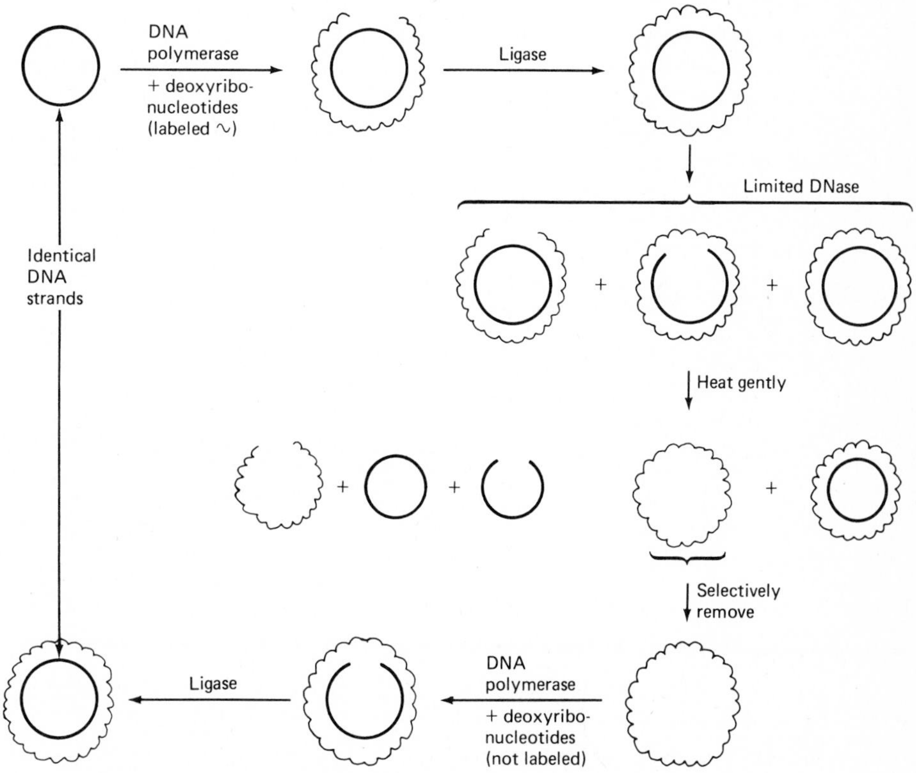
DNA
polymerase
+ deoxyribo-
nucleotides
(labeled ∿)
Ligase
Limited DNase
Identical
DNA
strands
+
+
Heat gently
+
+
+
Selectively
remove
DNA
polymerase
+ deoxyribo-
nucleotides
(not labeled)
Ligase

Figure 60–6 *In vitro* synthesis of a biologically active DNA using $\phi\times$ 174 virus. The solid line represents the original DNA strand. The wavy line represents a complementary strand.

closed into a circular form by the use of a joining enzyme known as ligase. This was the only other enzyme needed to bring about the DNA synthesis. The DNA was in a double-stranded form at this point in the experiment. This double-stranded form was subjected to a single break (DNase) on at most only one of the DNA strands. By using centrifugation techniques, the newly formed (marked) strand could be separated from all the other single- and double-stranded molecules. This synthetic circular DNA could be isolated and subjected to the same series of reactions that was used in its own synthesis. If it is a true complement of the original molecule, it should undergo the necessary reactions to yield a molecule identical with its parent. Radioactive tracers confirmed that this is exactly what happened during the course of this experiment. This then represents the first synthesis of a molecule that can direct its own replication.

One of the more recent studies involving nucleic acids involves research on recombinant DNA. In this work it has become possible to use microorganisms as "factories" for the preparation of segments of DNA from other organisms. Once a purified segment of DNA is made available to the microorganism, it can incorporate this foreign DNA into its own genetic machinery. This research effort has been very controversial because of the awesome potential of its applications. On one hand, this technique could be used to synthesize molecules like insulin in large quantities by incorporating pancreatic DNA into microbial systems. However, the hazards of forming new man-made pathogenic bacteria or viruses by this technique have caused great concern in the scientific community. The experiments, regardless of their application, remain landmarks of biochemistry. This series of experiments is the culmination of years of painstaking research, and represents an example of man's effort to unravel some of the mystery of living processes.

UNIT 60 Problems and Exercises

60–1. Some unusual or minor bases are known to occur in some nucleic acid molecules. Draw the structure for each of these.
(a) 5-methyl-cytosine (b) 4-thiouracil
(c) 2-methyl adenine (d) 7-methyl guanine

60–2. Draw the complete structure of a dinucleotide containing deoxyribose and the bases adenine and cytosine. What would be the abbreviated way of designating this molecule?

60–3. If you had a single strand of DNA with the following base sequence,

A–A–A–G–T–T–G–C–G–C

what would be the base sequence of the complementary strand in the double-helix form?

60–4. Using the model in Figure 60–4, estimate the number of base pairs in each complete turn of the helix.

60–5. Draw a rough sketch for a segment of a deoxyribonucleoprotein as you might expect to find it in a chromosome. Show the charged forms of the two components.

60–6. Assume that you isolated a DNA sample from a new source, and you determined that 27% of the nitrogen bases was guanine. Estimate the percentage composition of A, T, and C, respectively.

60–7. Would you expect the DNA isolated in problem (60–6) to have a higher or lower temperature of denaturation (uncoiling) than DNA from *E. coli*? (*Note*: the DNA of *E. coli* has been found to be 24% G and 25% A.)

60–8. Assume that you carried the Meselson–Stahl experiment through one more generation. You continued this experiment using the same ^{14}N-containing media as you used in the first generation. Draw the resulting DNA helix forms that you would expect to find at the end of this second generation. Use light lines for the ^{14}N strands.

60–9. There are several different DNase enzymes that are known to produce hydrolysis products from the same DNA sample. Some DNases yield 3′ phosphates and some yield 5′ phosphates. Sketch what this means, and comment on how these two different products might arise.

UNIT 61 THE MOLECULES BETWEEN: RNA

As studies on nucleic acids progressed, it became known that there were two main types. It was first thought that DNA was the nucleic acid found in animal tissues and RNA was the nucleic acid from plants. We now know that both DNA and RNA are found in both the animal and plant kingdoms. Except for viruses, most cells contain some of each of the nucleic acids. Furthermore, both DNA and RNA are distributed at several sites within the cell itself. Although most of the DNA may be found within the nucleus, an important DNA is also known to exist within the mitochondrion. RNA molecules can be found both within the nucleus and the cytoplasm of cells.

The hydrolysis products of RNA show that it is similar to DNA in many respects. There is a difference in the carbohydrate portion of the molecule, as the name suggests. The hydrolysis of RNA yields nucleotides and nucleosides with a phosphodiester linkage between the 3′ and the 5′ carbons on the ribose of adjacent nucleotides. This is true although in ribose there is a 2′ hydroxyl group that could be involved in such a linkage.

Another major difference between the two types of nucleic acids is in the pyrimidine bases. DNA and RNA both have the purines adenine and guanine, as well as the pyrimidine cytosine. In RNA, uracil is found in the place of thymine as one of the major bases. A number of *minor bases* are found in some types of RNA. An unusual nucleoside, pseudouridine, is also found in RNA but not DNA. However, RNA is still a polyribonucleotide of the same general form discussed previously.

Much of the RNA is single stranded. It is possible to have double-stranded portions stabilized by specific hydrogen bonding between the A=U and G≡C base pairs. These base pairs are important also in the recognition and direction of

RNA-dependent protein synthesis. Three major types of RNA are distributed within a cell:

1. Ribosomal RNA = r-RNA (70 to 80% of total RNA).
2. Messenger RNA = m-RNA (> 5% of total RNA).
3. Transfer RNA = t-RNA (10 to 20% of total RNA).

Ribosomal RNA

The r-RNA is found associated with proteins in a **ribonucleoprotein** complex (RNP). This RNP particle provides the site of protein biosynthesis within the cell. Three major types of RNP complexes exist. The three ribosomal components vary a great deal in size and molecular weights. They can be separated from each other by use of an ultracentrifuge and are designated by their sedimentation rates. The designation S, after Svedberg the inventor of the ultracentrifuge, is assigned to each particle. The smallest RNA component is designated as 5S and has a molecular weight of about 36,000 daltons. The medium-sized component has an RNA of 16 to 18S, with a molecular weight of slightly under 10^6 daltons. The biggest RNA of this family is designated 23 to 28S, corresponding to a molecular weight of over 10^6 daltons. There is some variation in these RNA molecules, depending upon the source. The form of the r-RNA from the bacterium *E. coli* has been characterized to the greatest extent. In this microorganism, the two larger RNA molecules are known to combine with a number of different proteins to yield RNP particles, which are also assigned sedimentation values. These two nucleoproteins are known as 30S and 50S particles, respectively. There is a small component of the 50S particle that contains the 5S particle. The 50S particle is now thought to contain 34 different proteins. These macromolecular complexes have a role in the process of protein biosynthesis considered in the following unit.

There various ribosomal particles can undergo an *in vitro* reversible association $\rightleftharpoons$ disassociation process, depending upon the Mg^{2+} concentration in which they are placed. For example, at $10^{-4}\,M$ Mg^{2+} or less, separate 30S and 50S particles are present. At Mg^{2+} concentrations greater than $10^{-4}\,M$, an association takes place between one of each of these particles to yield a 70S particle, a complete ribosome in bacterial systems (mammalian systems yield a heavier 80S particle):

$$30S + 50S \underset{\text{Lower } Mg^{2+}}{\overset{\text{Higher } Mg^{2+}}{\rightleftharpoons}} 70S$$

This reversible process will be seen as an essential part of protein biosynthesis.

Messenger RNA

Messenger RNA (m-RNA) is, as its name implies, the bearer of the information from DNA that is necessary to dictate the linear sequence of amino acids in protein biosynthesis. Messenger-RNA has been found to be a single stranded polynucleotide of variable length. There are from several hundred to several thousand nucleotides in an m-RNA molecule. The shorter ones may contain and carry the information from a single gene required for the synthesis of a single protein. The larger m-RNA molecules could hold the information carried in several adjacent genes. At any given time, there may be a different population of m-RNA molecules within a cell. This would depend upon the protein synthesis underway at that moment.

As you might expect, this long m-RNA molecule is very fragile. This is actually desirable, since the cell would have to get rid of a specific m-RNA once it had directed the synthesis of a sufficient quantity of its specific protein. This unstable nature made it very difficult to isolate and study. As a matter of fact, the existence of m-RNA was postulated before it was ever found to be present within the cell. Jacob and Monod correctly predicted that such a molecule should exist, if only the experimental techniques (and perseverance) were developed to detect it. m-RNA has subsequently been confirmed as an essential component of the protein biosynthesis system.

The point to remember is that the sequence of the nitrogen bases in the m-RNA molecule is precisely directed by the sequence of the nitrogen bases in a segment of a DNA molecule. Each protein would have its own m-RNA (or a segment of it). The r-RNA does not have this protein-directing ability.

Transfer RNA

In 1956, the third type of RNA was discovered in liver cells. It was first called soluble RNA because it was found to be suspended in the cytoplasm of the cell. It was later called **transfer**-RNA (t-RNA), which is more fitting to describe its role in the biosynthesis of proteins. Like m-RNA, t-RNA is not usually found to be associated with proteins in any nucleoprotein complex. There are many t-RNA molecules in any given cell. Some estimates are that there may be as many as 60 slightly different t-RNA's in a given cell. As we shall see, there must of necessity be at least one t-RNA for every amino acid that will eventually become a part of a protein molecule. Actually, there is often more than one t-RNA for each amino acid that might be incorporated into a protein. We can distinguish between the various t-RNA's by the amino acids that they are involved with in protein biosynthesis. For example, t-RNA^{ala} would be a transfer RNA involved with the incorporation of the amino acid alanine, t-RNA^{met} would be involved with the incorporation of the amino acid methionine and so on.

The t-RNA molecules are relatively short strands of polynucleotides. The nitrogen bases that make up a t-RNA molecule consist of the usual ones for RNA (A, G, U, C) and in addition contain a great variety of minor bases. These minor bases are very likely just modifications of the usual bases as they occur in polynucleotides. As many as 15 rather unusual nucleosides are incorporated in t-RNA molecules. As yet there is no clear-cut explanation as to why the t-RNA carries these unique molecules. Typical examples of these minor base nucleosides include:

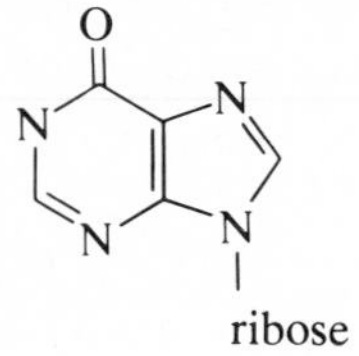

inosine

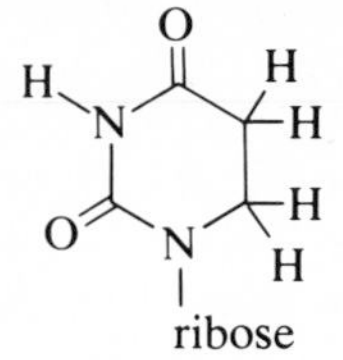

dihydrouridine

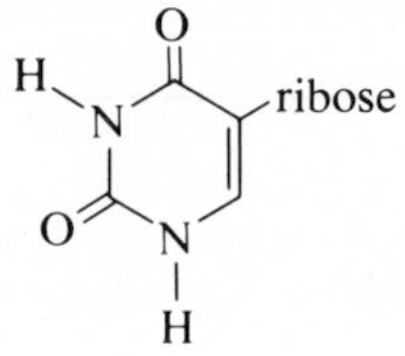

pseudouridine

The t-RNA's usually contain less than 90 nucleotides in a linear arrangement. The complete sequence of an entire t-RNA molecule was first reported in 1965 by R. W. Holley and coworkers. They started with 300 lb of yeast from which they finally isolated 1 g of the purified t-RNA^{ala}. By the use of a variety of techniques and great patience, the sequence of all 77 nucleotides of the t-RNA was determined. Other t-RNA sequences have been determined since that time, and special features were found to be common to all these molecules studied thus far.

The 3′-OH ends of all the t-RNA molecules have the same three nucleotides: pCpCpA. This means that this end of the polynucleotide chain has the 3′-hydroxyl of its terminal ribose free. As we shall see, this is now known to be the site of attachment of an amino acid when t-RNA acts as its name implies, as a transfer agent.

When the sequence of the nucleotides in t-RNA became known, several other bits of chemical information were fitted together to get a more complete picture of

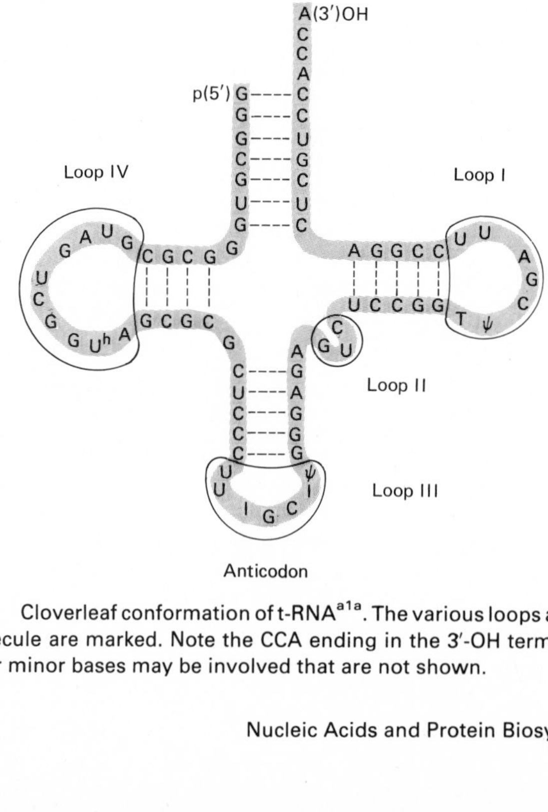
A(3′)OH
p(5′)
Loop IV
Loop I
Loop II
Loop III
Anticodon

Figure 61–1 Cloverleaf conformation of t-RNA^{ala}. The various loops and regions of the molecule are marked. Note the CCA ending in the 3′-OH terminal. Some modified or minor bases may be involved that are not shown.

these molecules. For example, it was known that t-RNA's could have a specific three-dimensional ordered conformation in their native state. These could be stabilized by rather short regions of hydrogen bonding of a type quite similar to that we saw earlier stabilizing the double helix of DNA. In the case of the RNA molecules, the pairs would be G≡C and A=U. All the data were put together to give the *cloverleaf* conformation, as shown for t-RNA^{ala} in Figure 61–1. Note that there are four loops in the open or extended cloverleaf representation. Each loop may be involved in a particular part of the function of t-RNA. For example, loop III contains the important anticodon region of three bases that must match with the tripled codon region of m-RNA in the process of protein biosynthesis. Further studies have shown that t-RNA has a special three-dimensional structure resulting in an L-shaped molecule. Each of the loops appears at a surface of the molecule where it can directly participate in a molecule-to-molecule interaction.

Synthesis of RNA

The synthesis of all these various types of RNA must be under careful control within the cell. According to our overall plan, the sequence of the bases within the DNA must somehow provide the pattern of the sequence of bases within the RNA molecules. This is called the **transcription** process. In 1959 an enzyme was discovered that would form an RNA polymer. This enzyme is now known as a DNA-dependent RNA polymerase. This means that this enzyme requires DNA as a template or guide for the linear arrangement of the bases within the forming RNA. Furthermore, studies have shown that the base sequences of the DNA template and the newly formed RNA are perfectly complementary. When DNA templates of a given base sequence are used, the RNA synthesized under the direction of this enzyme has a base sequence that would be expected from our usual hydrogen-bond base pairing. For example, if a synthetic polynucleotide composed of dAT were used as the template in this system, an RNA polymer would be formed that would be exclusively poly UA. It turns out that a double-stranded DNA is a primer or template of choice, but only one strand of the DNA is usually copied. It also has been found that the newly formed RNA strand has a polarity opposite from the template DNA. This is similar to the situation found in the double-helical conformation of DNA itself. This DNA-dependent RNA polymerase uses the various triphosphates as the source of the synthesis of RNA:

$$\left.\begin{matrix}\text{ATP}\\+\\\text{GTP}\\+\\\text{UTP}\\+\\\text{CTP}\end{matrix}\right\} \underset{\substack{\text{(1) Enzyme}\\\text{(2) Mg}^{2+}\\\text{(3) Some specific proteins}\\\text{(4) DNA}}}{\overset{\text{Requires}}{\rightleftharpoons}} (\text{ApGpUpC}) + 4(\text{P}_i\text{P}_i)$$

The reaction yields the RNA and pyrophosphate. It also requires specific proteins, which are involved in the initiation and regulation of this process. This enzyme appears to be the major one in the transcription process.

Another enzyme was found in bacteria, which at first appeared to be the one responsible for RNA biosynthesis. This enzyme, discovered in 1955, is known as polynucleotide phosphorylase and catalyzes a reversible reaction which could yield RNA from a mixture of diphosphates of ribonucleosides:

$$\left.\begin{matrix} \text{ADP} \\ + \\ \text{GDP} \\ + \\ \text{UDP} \\ + \\ \text{CDP} \end{matrix}\right\} \xrightleftharpoons[\substack{\text{(1) Enzyme} \\ \text{(2) } Mg^{2+} \\ \text{(3) RNA primer}}]{\text{Requires}} \text{(ApGpUpC)} + P_i$$

This reaction requires a primer rather than a true template. The primer merely acts as a site (3′-OH) upon which a growing polynucleotide chain may be formed. The newly formed RNA chain does not have a sequence directed by the primer, but is dependent upon the relative concentrations of the various diphosphates present in the reaction mixture. It is now thought that this is not a synthetic enzyme for RNA, but more likely functions to catalyze the degradation of RNA molecules within the cell (reverse of the reaction direction shown). Remember that we discussed the labile nature of m-RNA? This enzyme may be involved with this degradative process. It is interesting that nature attempts to conserve some energy in this degradative reaction. The products of the hydrolysis are preformed diphosphates. It would not place as much of an energy drain on the cell to convert these diphosphates to the triphosphates necessary for RNA synthesis. If the RNA were hydrolyzed to monophosphates, there would be an extra, and apparently unnecessary, energy requirement—another good example of nature's way of conservation on a molecular level.

The process of synthesizing the nucleoside triphosphates necessary for RNA formation is both complex and energy demanding. We apparently have all the necessary synthetic machinery within our bodies because we do not need preformed nucleic acids in our diets. It appears that mammals can use some of the preformed nucleic acids that they ingest, however. The details of the biosynthesis of the nucleotides of both the pyrimidines and purines are well worked out but are beyond our present interest. More advanced textbooks in biochemistry should be consulted for the step-by-step anabolism of these important molecules. As is true in the case of the synthesis of other large molecules, small precursor molecules are combined in specific ways to yield the nucleotides (or deoxyribonucleotides) from which the nucleic acids are formed.

Metabolism of RNA Components

The sugar moiety and the phosphate of the nucleotides come from 5-phosphoribosyl-1-pyrophosphate (PRPP). The ribose-5-phosphate was derived from metabolic schemes discussed previously in the pentose phosphate pathways of carbohydrate metabolism. The atoms of both heterocyclic ring systems are derived from the precursors shown in Figure 61–2. It is interesting to recall the structures of the precursor molecules and examine how they contribute a part or all of their components to these more complex compounds. There are at least six distinct enzyme systems involved in the pyrimidine biosynthesis and ten enzymes involved in the purine biosynthesis.

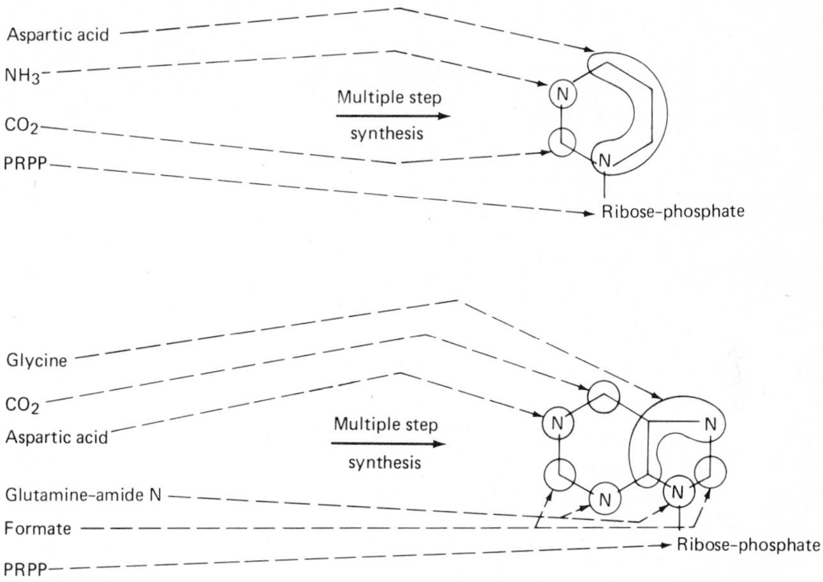

Aspartic acid
NH_3
CO_2
PRPP
Multiple step
synthesis
N
N
Ribose-phosphate
Glycine
CO_2
Aspartic acid
Glutamine-amide N
Formate
PRPP
Multiple step
synthesis
N
N
N
N
Ribose-phosphate

Figure 61–2 The origin of the atoms of the pyrimidine and purine portions of the nucleotides.

Studies have shown that the deoxyribonucleotides of both the purines and pyrimidines arise from a deoxygenation of the ribose at the nucleotide level. Therefore, both families of nucleic acids can be synthesized by common pathways. This is another beautiful choice of nature, since these biosynthetic pathways are so complex and critical to the living cell.

Important studies have been conducted on the manner in which humans catabolize or degrade the nucleic acids. There are several hydrolytic enzymes, deoxyribonuclease (DNase) and ribonuclease (RNase), that act to break down the phosphodiester linkages. Other enzymes can cause further degradation to yield free bases, carbohydrates, and phosphate. It should be mentioned that there are some salvage pathways to recover some of the nitrogen bases. In these schemes the nitrogen heterocycles can be reused by being incorporated into newly forming molecules. This is part of the dynamic turnover in the metabolism of nucleic acids.

Studies on the way in which the heterocylic rings are catabolized are health related. In man, the major end product of purine metabolism is uric acid, which may be found in the blood of normal adults in the range of 3 to 7 mg%. In the disease gout, high blood levels of uric acid are present, and in some cases actual deposits of uric acid may be found in the kidneys or in cartilage.

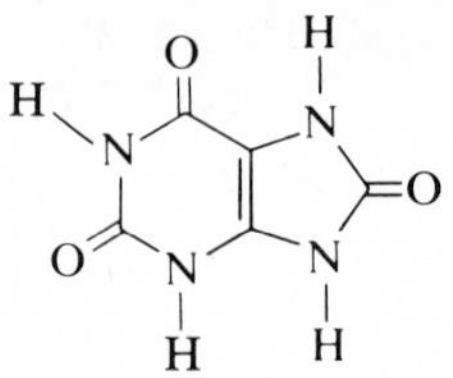

uric acid

It is now thought that primary gout is a metabolic disease that arises from the overproduction of uric acid. Some other animals are known to possess other enzyme systems that can bring about the breakdown of uric acid to urea or other compounds. On the other hand, some species like birds tend to excrete uric acid as their principal nitrogenous end product, because they cannot form urea to any significant extent.

The catabolism of the free bases cytosine and uracil are representative of the degradative pathways of the pyrimidines:

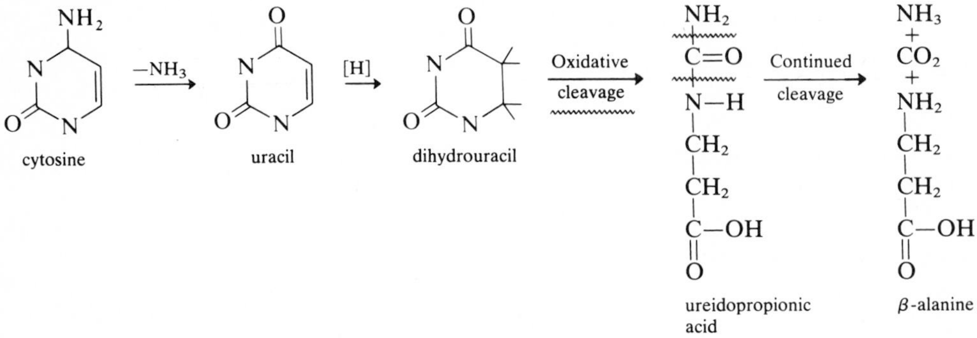

cytosine
$-NH_3$
uracil
[H]
dihydrouracil
Oxidative cleavage
Continued cleavage
ureidopropionic acid
β-alanine

In this case, the pyrimidine ring is hydrogenated and several subsequent cleavage reactions occur. The points of cleavage are shown in the reaction system. It is interesting to note that with the formation of the end products CO_2 and NH_3 the cycle is complete. The β-alanine may be incorporated in other molecules or reused in other ways that are beneficial to the body. As we have seen in so many other situations, the schemes of nature provide for a chemical and molecular continuity that is both amazing and practical.

UNIT 61 Problems and Exercises

61–1. If a sample of RNA and a sample of DNA were both subject to hydrolysis, list what components would be common and which would be different in each nucleic acid sample.

61–2. Suggest a reason why RNA may contain some of the minor bases whereas DNA usually does not.

61–3. Give a segment of DNA with the following base sequence:

A–A–A–G–T–T–G–C–G–C

What would be the sequence of bases in a corresponding m-RNA molecule?

61–4. Draw the specific base pairing of A–U, showing the hydrogen bonds involved.

61–5. Briefly explain the difference between the concept of a *template* and a *primer* as they influence enzymatic reaction involved with RNA synthesis.

61–6. Compare the similarities and contrast the differences between the enzymes DNA-dependent RNA polymerase and polynucleotide phosphorylase.

61–7. Briefly account for the fact that polynucleotide phosphorylase can act as a catabolic enzyme, but cannot exclusively be the anabolic enzyme responsible for the RNA synthesis.

61–8. Draw the structure of the precursor molecules that provide the N that is attached to the ribose of both purine and pyrimidine nucleosides. Circle the specific N atom involved in each of the molecules.

61–9. On the basis of just the carbon-to-nitrogen ratio, which end product, urea or uric acid, is the most efficient molecule for nitrogen elimination? Assume that you wish to conserve the carbon atoms while eliminating nitrogen atoms.

61–10. Assume that thymine is subject to the same catabolic scheme as outlined for pyrimidines. What would be the other end product in addition to NH_3 and CO_2?

UNIT 62 PROTEIN BIOSYNTHESIS

The studies on the molecules we have considered thus far reach a climax in the process of protein biosynthesis. Our earlier studies on the amino acids and proteins provided the first clue as to the wonder and complexity of this process. The tremendous variety in protein molecules became obvious. We noted the importance of the sequence to the unique structure of the proteins. The flow of information between molecules became part of our thinking as we established the central dogma of molecular biology whereby nucleic acids direct the synthesis of proteins. It now remains for us to fit all these individual pieces together to describe the process that lies at the very foundation of living systems.

Energy Requirements

One part of the process should be considered before we investigate the mechanisms of protein biosynthesis. This is the requirement of energy. The biosynthesis of protein is an anabolic process that has definite energy needs in order for it to proceed. There are several steps along the route where we can clearly identify the need for energy in the form of ATP and GTP. It may very well be that some of the energy-requiring steps in the process are yet to be identified. In addition to the more direct points of energy utilization, which we shall consider in detail, there are hidden energy costs to protein biosynthesis. There are a great many factors and other participants involved within the overall system that must be synthesized and maintained. It also takes energy to transport amino acids across cell membranes to bring them to a point where synthesis can take place. Since protein biosynthesis is so important to the sustenance of life, nature seems to have spared no expense in making this truly remarkable system operate.

The Code

Once you consider the process of protein synthesis, you are immediately overwhelmed by a serious dilemma. On the one hand, there are a great many different proteins, each with its own particular sequence of amino acids. On the other hand, each protein must be synthesized in almost a letter perfect copy, because there is little allowance for error in most cases. Consider writing a book with no errors. Consider taking an examination without making at least one rather foolish mistake.

Think of the case of a simple tripeptide. It could be composed of amino acids in arrangements ABC, ACB, BAC, BCA, CAB, and CBA. As the number of participants increases, the number of possible arrangements becomes staggering. Consider also that 20 amino acids are found in protein structure. One problem to be solved is the number of nucleotides that it takes to direct or control the placement of each amino acid. In other words, how many bases in DNA must be transcribed into bases of the m-RNA to specify the placement of each amino acid? There are four different major bases to direct the synthesis of 20 different amino acids. We shall assume that one base in DNA will direct the arrangement of one base in the m-RNA. We shall also assume that the usual hydrogen-bonding pairs will be involved as a stabilizing force in the selection process involving nucleic acid–nucleic acid interactions.

It is easy to see that one base cannot be assigned to direct the placement of each amino acid. Furthermore, two bases in a specific arrangement could only yield 16 different combinations and thus direct the placement of only 16 different amino acids. There must be three or more bases involved with each amino acid if we have 20 amino acids to use in the process. We shall examine several experiments of the many that have been used over the past few years to show first, that the m-RNA does indeed specify the sequence of the amino acids and second, that three bases (**triplet**) in the messenger are needed to direct the placement of each amino acid.

A key experiment was conducted by M. Nirenberg (Nobel laureate, 1968) and his coworkers at the National Institutes of Health. They developed a cell-free system that could synthesize protein and yet be carefully controlled by the addition or deletion of certain components. In one classic study, they used a system that contained all the necessary components for protein synthesis except the messenger RNA. For the m-RNA they substituted a synthetic polynucleotide containing a long chain of uridine, poly U. The system was allowed to synthesize a peptide under the direction of the poly U. A peptide containing just the amino acid phenylalanine was synthesized, although the synthetic system had all the amino acids available to it in the reaction mixture. It also was observed that the length of the phenylalanine chain was approximately one third the length of the poly U chain, suggesting a triplet relationship. The real significance, however, was the selection of one particular amino acid for incorporation into a peptide from a mixture of all the amino acids.

A second step along this path of research was conducted by Khorana (Nobel laureate, 1968) and coworkers at the University of Wisconsin. They used a polyribonucleotide as a messenger that contained a known alternating sequence:

UGUGUGUGUGUGUGUGUGUG

When this was added to the protein-synthesizing system containing all the amino acids, a peptide was synthesized with an alternating sequence of cysteine and valine.

cys–val–cys–val–cys–val

This result proved that three bases were necessary to code for each amino acid rather than two. You will note that, if only two bases were required for one amino acid, the peptide resulting from the use of this alternating polynucleotide would have contained only one kind of amino acid. Try it for yourself by starting anywhere you like along the polynucleotide chain.

If we now accept the theory that a triplet of bases is required to code for each amino acid, we can make an even more specific assignment for a given amino acid. We shall designate these three coding bases on the m-RNA as the **codon**. The only possible triplets in the above polynucleotide are UGU and GUG, regardless of the starting point that you choose. One of these codons must code for cysteine and the other for valine, but which one?

The specific codons for each amino acid were worked out as the result of collaborative studies by a number of workers. All 64 possible triplet codons of the four nitrogen bases were synthesized. The separate triplet codons of known sequence were placed in a protein-synthesizing system. The transfer RNA's with their specific amino acids attached were added. Radioactive labels were employed, since only very small amounts of materials could be used in this sort of study. The given triplet codon became attracted to the t-RNA bearing its corresponding amino acid. Only the specific t-RNA bearing the coded amino acid would become stabilized and involved in the interaction, regardless of how many of these t-RNA's were presented to the system. After this reaction was allowed to take place, all the unbound t-RNA (with their different amino acids) could be removed from the reaction mixture. The t-RNA bearing the proper amino acid would interact in its complementary **anticodon** region with the synthetic triplet codon.

This type of study was repeated with each codon. The complete genetic coding assignment of the codons for each amino acid was finally worked out; it is summarized in Table 62–1.

As you can see, there are often several codons that can direct the placement of a single amino acid. This is known as **degeneracy** of the code. Just as there is more than one t-RNA for a given amino acid, there is more than one codon for many of the amino acids. This undoubtedly is one of the built-in safety devices for this important system, perhaps a sort of backup. For example, if an error appeared in one of the nucleic acids influencing the system, it might be rectified by this degeneracy. It is important to note that the first two bases of the triplet seem to be the most specific for an amino acid. The observation that the third base has less influence on specifying

Table 62–1 Codon assignment for amino acids

First position (5′ end)	Second position U	C	A	G	Third position (3′ end)
U	Phe	Ser	Thy	Cys	U
	Phe	Ser	Tyr	Cys	C
	Leu	Ser	Term	Term	A
	Leu	Ser	Term	Trp	G
C	Leu	Pro	His	Arg	U
	Leu	Pro	His	Arg	C
	Leu	Pro	Gln	Arg	A
	Leu	Pro	Gln	Arg	G
A	Ile	Thr	Asn	Ser	U
	Ile	Thr	Asn	Ser	C
	Ile	Thr	Lys	Arg	A
	Met*	Thr	Lys	Arg	G
G	Val	Ala	Asp	Gly	U
	Val	Ala	Asp	Gly	C
	Val	Ala	Glu	Gly	A
	Val	Ala	Glu	Gly	G

NOTE: The initiator codon is marked *, the peptide chain terminating codons are marked Term.

the amino acid has led to what is sometimes described as a *wobble* at this position. For example, GU specifies valine regardless of the third base in the triplet. Note also that the degeneracy in this position frequently involves one purine substituting for the other purine or one pyrimidine substituting for another. This is a good example of our need to study living systems on a molecular basis. Without knowledge of the structures of these molecules, these sorts of interchanges would be meaningless.

Some of the codons in Table 62–1 are especially marked. These are either signals for the initiation point of synthesis or its termination. There must be a regulated way in which both ends of the protein are defined. We shall consider this in more detail under the specific mechanisms of protein biosynthesis.

This code assignment appears to be **universal** throughout all the systems that have been studied. This is of great importance to molecular biologists and points out to all of us the grand plan of nature that interrelates the species. It still remains for mankind to recognize this relationship in many of his dealings both within his species and beyond.

Upon careful examination of Table 62–1, we find some of the amino acids missing that are known to be present in proteins. Hydroxyproline is a good example. In this case, and in others, there can be a modification of some amino acids after they are placed within the protein chain. The proline is hydroxylated while it is part of a peptide. Nature seems to have chosen this method to provide yet another degree of specificity in proteins beyond that directed by the usual nucleic acid sequencing mechanism.

Direction of Biosynthesis

Before we consider the stepwise details of protein biosynthesis, let us consider the direction of peptide synthesis. Labeling studies on partly synthesized proteins have shown that the sequence of peptide synthesis starts at the amino terminal end of the chain and proceeds toward the carboxyl end. Luckily, this is the direction in which we have chosen to name peptides, as we discussed earlier. The codon direction as shown in Table 62–1 is from the 5′ end toward the 3′ end. This then would require that the corresponding base sequence in the anticodon region of the t-RNA be antiparallel and read from the 3′ end to the 5′ end:

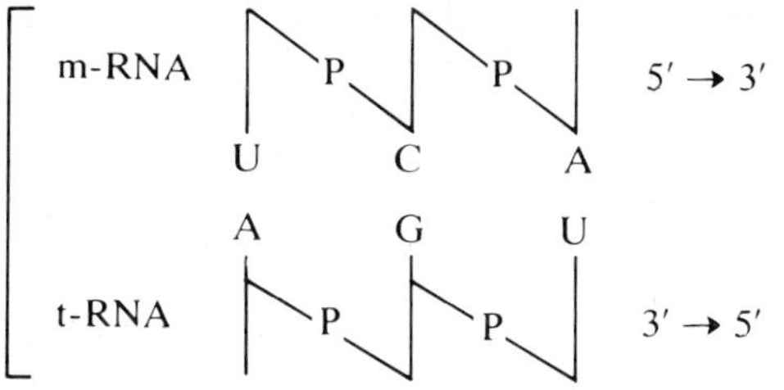

m-RNA
P
P
5′ → 3′
U
C
A
A
G
U
t-RNA
P
P
3′ → 5′

The Steps of the Process

For the purposes of our study, we shall consider the detailed mechanism of protein biosynthesis as consisting of four distinct steps. Since many of the studies are best worked out with bacteria, we shall use this as a general model for our system. Some differences exist among the species insofar as protein biosynthesis is concerned, but a general unity of mechanism appears to follow through all systems studied thus far. Naturally, all these processes may be going on simultaneously within the cell or system under study. The process of protein biosynthesis is a very dynamic one, with macromolecular systems interacting with each other and with considerable relative movements both within and among the participating entities. We shall consider the steps as follows:

STEP 1: **Activation** of an amino acid and its interaction with its specific t-RNA.

STEP 2: **Initiation** of peptide chain formation.

STEP 3: **Elongation** or growth of the peptide chain.

STEP 4: **Termination** of the peptide chain growth and its release as a free protein.

Activation

The first critical step of **recognition** of an amino acid and a nucleic acid takes place at this point. An enzyme aminoacyl-t-RNA synthetase is specific for both an amino acid and its proper t-RNA. There is at least one such enzyme for each amino acid.

As we discussed previously, there may be two or more specific t-RNA's that will react in this system. The t-RNA's that would react must have the triplet in the anticodon region that corresponds to the amino acid as shown in Table 62–1. The way in which these three different kinds of molecules, amino acid, nucleic acid, and enzyme, come together still remains a mystery and is the subject of considerable research. It is important to note that this first recognition must be correct or all the other processes of the synthetic systems cannot produce the desired product.

The enzyme reacts with Mg^{2+} and uses an ATP to activate the amino acid. This is a two-step process in which the amino acid first becomes attached to the AMP portion resulting from the hydrolysis of ATP. This hydrolysis provides the energy for

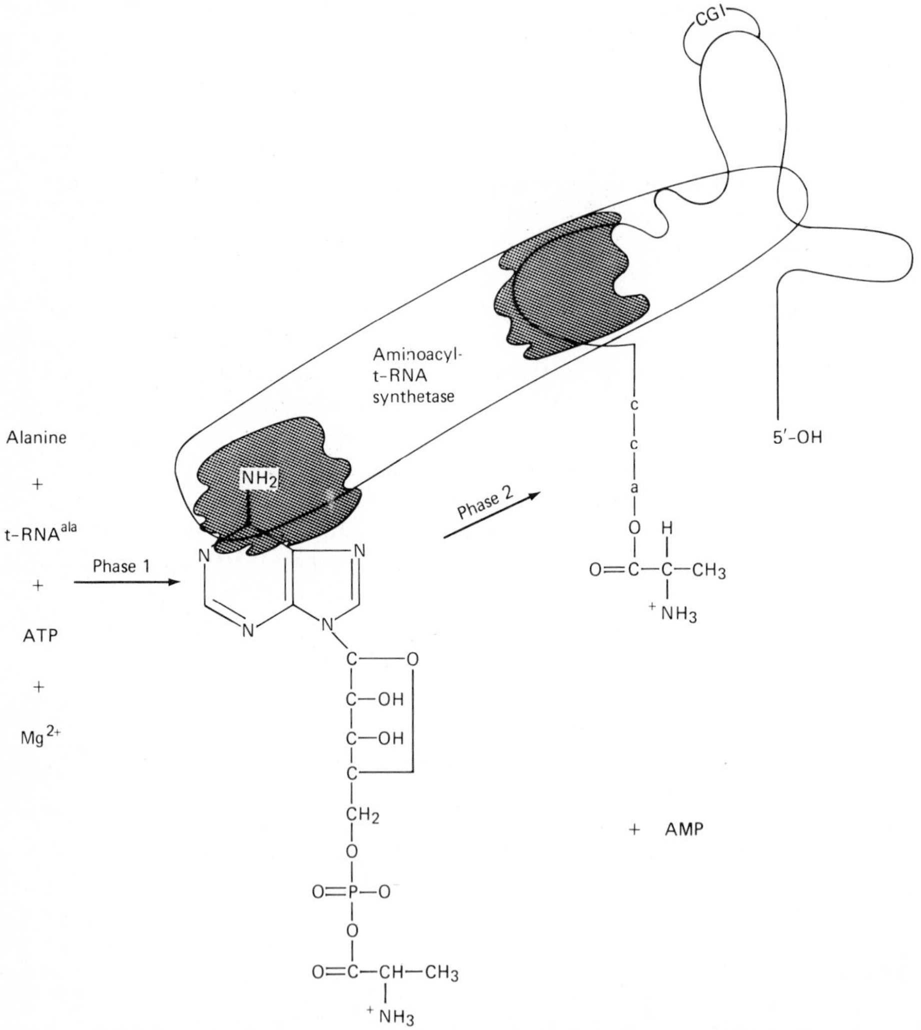

CGI
Aminoacyl-
t-RNA
synthetase
Alanine
+
t-RNA^ala
+
ATP
+
Mg 2+
Phase 1
NH2
N
N
N
N
C—O
C—OH
C—OH
C
CH2
O
O=P—O
O
O=C—CH—CH3
+NH3
Phase 2
c
c
a
O H
O=C—C—CH3
+NH3
5′-OH
+ AMP

Figure 62–1 The activation of an amino acid leading to the formation of a "charged" t-RNA. This two-phase reaction is mediated by aminoacyl-t-RNA synthetase.

the coupling and forms a high-energy anhydride that is capable of undergoing the second phase of this reaction. In this second phase, the carboxyl end of the amino acid disengages itself from the AMP and attaches itself by an ester linkage to the 3′-OH of the terminal ribose of its proper t-RNA. These phases are shown in Figure 62–1, which shows the formation of a **charged t-RNA** with its activated amino acid attached to its own specific t-RNA, here designated ala-t-RNA^{ala}. If this reaction took place on each of the 20 different enzymes, we would have all the necessary charged t-RNA molecules to continue with the synthetic process. This step may actually be thought of as getting the amino acids prepared for biosynthesis.

Initiation

The second step of the process, **initiation**, has been worked out in detail using bacterial systems as shown in Figure 62–2.

In the case shown, it was observed that the first or N-terminal amino acid was a derivative of methionine. This methionine had a formyl group attached to its amino group (fmet), which effectively blocked the N-terminal position to any other chemical reaction. The codon associated with this group was the initiator codon AUG. It has been found that there is a special t-RNA^{fmet} that directs this amino acid to its initiator position. The AUG codon, with t-RNA^{fmet}, could direct the proper positioning of methionine within a polypeptide chain. This initiation step takes place with a complex series of interactions involving the ribosomes, several specific protein factors (initiation factors F_1, F_2, and F_3), and energy in the form of GTP. The 30S and 50S ribosomes are known to undergo a reversible association during the initiation. This association may be under the control of one of the initiation factors, rather than the concentration of Mg^{2+} that we considered previously. The free 30S ribosome subunit reacts with the 5′-OH terminal end of the m-RNA bearing the AUG triplet initiator signal. GTP hydrolysis is required as well as several of the initiation factors.

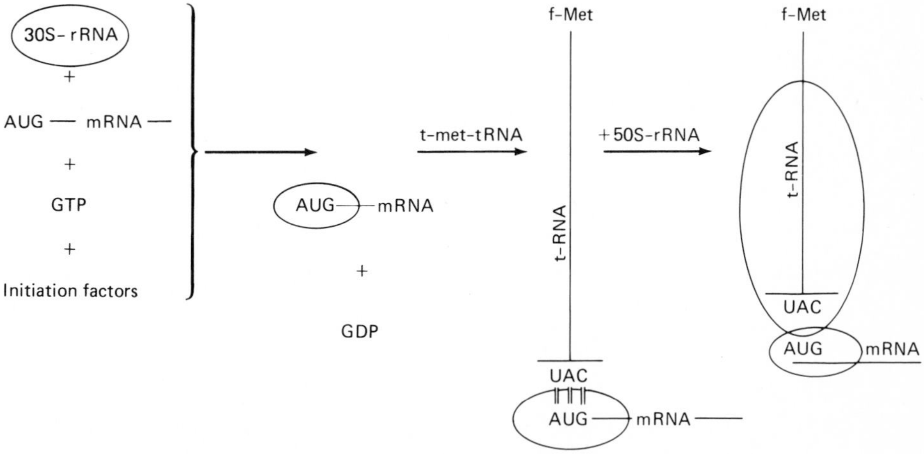
30S-rRNA
+
AUG — mRNA —
+
GTP
+
Initiation factors
AUG
mRNA
+
GDP
t-met-tRNA
f-Met
t-RNA
UAC
AUG
mRNA
+50S-rRNA
f-Met
t-RNA
UAC
AUG
mRNA

Figure 62–2 Initiation process leading to protein biosynthesis.

The f-met-t-RNA^{fmet} then becomes involved in this complex at what we shall designate as site I. A free 50S ribosome subunit then is added to complete the molecular complex, which is now ready for the next step in protein biosynthesis.

Elongation

The next step involves the growth or **elongation** of the peptide chain by adding properly charged t-RNA molecules to the ribosomal unit described above. The second charged t-RNA will interact at a second site (site II) on the ribosomal surface as directed by the next codon of the m-RNA. A peptide bond is formed between the two amino acids carried by the two t-RNA's. This enzyme is called **peptidyl transferase**, because it allows the carboxyl group on the amino acid at site I to react with the free amino group of the amino acid attached to the t-RNA at site II. This is depicted in Figure 62–3. Note that the t-RNA occupying site I is now free of any amino acid. This *uncharged* t-RNA can return to the pool of t-RNA molecules available to accept another activated amino acid from step 1. The next thing that happens in this dynamic process is a relative movement of the m-RNA and the ribosome. The t-RNA on site II moves to site I to make the new site II available to accept the next t-RNA as dictated by its own specific codon. This process is sometimes called **translocation**.

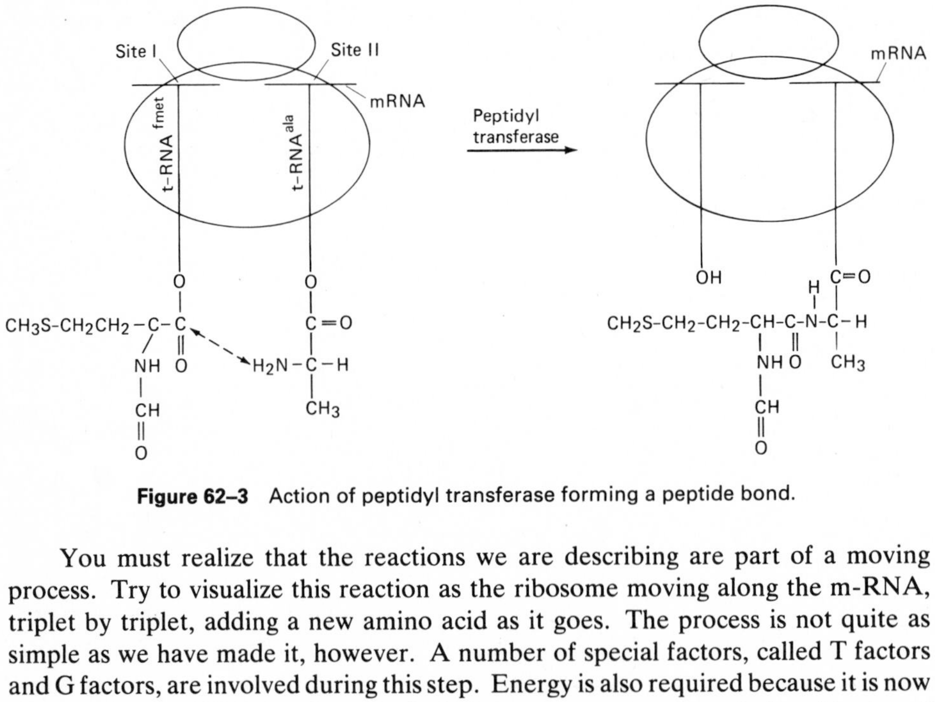

Site I
Site II
mRNA
t-RNA fmet
t-RNA ala
Peptidyl transferase
mRNA
CH3S-CH2CH2-C-C
NH O
CH
O
O
C=O
H2N-C-H
CH3
OH
C=O
H
CH2S-CH2-CH2-CH-C-N-C-H
NH O
CH3
CH
O

Figure 62–3 Action of peptidyl transferase forming a peptide bond.

You must realize that the reactions we are describing are part of a moving process. Try to visualize this reaction as the ribosome moving along the m-RNA, triplet by triplet, adding a new amino acid as it goes. The process is not quite as simple as we have made it, however. A number of special factors, called T factors and G factors, are involved during this step. Energy is also required because it is now thought that two GTP's undergo hydrolysis in separate reactions during this step of protein biosynthesis.

Termination

All good things must come to an end, and the synthesis of a protein is no exception. The **termination** step occurs when one of the three chain-terminating codons is reached: UAA, UAG, or UGA. The completed peptide chain in its proper conformation becomes free, and the ribosome dissociates into its subunits so that the entire cycle can repeat itself. Several specific factors known as R (release) factors, S factors, and others are thought to be essential to complete this step.

One other reaction remains. Since not all proteins have methionine or formylmethionine at their N-terminal positions, exopeptidases must come into play that can selectively remove this amino acid. This methionine seems to play a unique role in the initiation of protein biosynthesis in bacterial systems. Other protein-synthesizing systems may not use this specific initiator.

The active nature of this overall process cannot be overemphasized. Although it is a very complex and carefully controlled scheme, as we have seen, it takes place at a very great rate. It has been estimated that one complete chain of hemoglobin, containing over 140 amino acids can be synthesized in 3 min. Bacterial systems are even faster.

The precise manner in which this transcription and translation process takes place is remarkable. There are some opportunities to correct a limited number of errors through the degeneracy of the code and the wobble effect. Any alterations along the way could be detrimental or might even be fatal to the cell or system involved. A number of genetic disease conditions are now known to result from these kinds of errors. The classic example of this is **sickle cell anemia**. In this case the alteration of only one amino acid in a hemoglobin chain (a valine for a glutamic acid) of over 140 amino acids brings about profound physiological and chemical changes.

Insulin Biosynthesis

One interesting case of synthesis involves the formation of the protein hormone insulin as shown in Figure 62–4.

As we have mentioned, insulin in its final or active form has two chains, an A chain with 21 amino acids and a B chain with 30 amino acids. These chains are held together by two interchain disulfide bridges, and the A chain is stabilized by yet

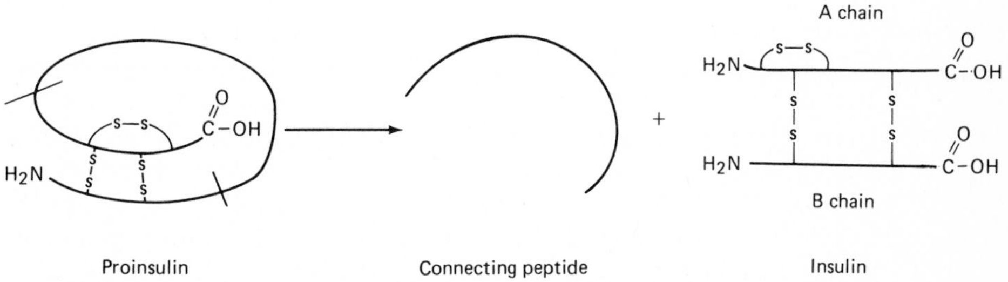
A chain
H2N
S—S
C—OH
O
S
S
H2N
B chain
Proinsulin
Connecting peptide
Insulin
+

Figure 62–4 Insulin biosynthesis.

another intrachain disulfide linkage. It has been found that the biosynthesis of this protein hormone actually results from the synthesis of one longer precursor molecule, proinsulin. The pancreas actually produces the inactive proinsulin as a single peptide chain. The hormone is activated by removing the connecting peptide to form the active molecule.

UNIT 62 Problems and Exercises

62–1. It has been suggested that, instead of a three-base codon, we actually should consider a two and one-half base codon. Comment on the meaning of this suggestion.

62–2. m-RNA is the least stable of all the RNA molecules. From a chemical point of view, why would you expect this to be the case? From a biological point of view, why is it desirable that m-RNA have the least stability?

62–3. Why is it chemically favorable that the initial amino acid have its amino group blocked?

62–4. What specific sequence of bases must exist on the DNA molecule to direct the location of the first three amino acids in proinsulin? (Neglect any initiator triplets.)

62–5. If a synthetic messenger of a repeating AU polynucleotide were used in a protein-synthesizing system, what amino acid sequence would result?

62–6. Some proteins are known to contain methyl lysine. Since this amino acid does not appear in the table of codons, account for the presence of this less common amino acid in a protein.

62–7. The antibiotic chloramphenicol seems to bind to the 50S ribosomes of many microbial systems, but does not interfere with the large ribosomes of systems in higher organisms. Which step in the protein-synthesis system would you expect to find blocked by this antibiotic?

62–8. Since active insulin contains 51 amino acids, how many nucleotides would you expect to be necessary in the m-RNA directing the synthesis of this hormone? It now is thought that there are at least 240 nucleotides in the gene region directing this synthesis. Account for these differences.

62–9. Some proteins that are essential to living systems have quite similar amino acid sequences throughout all the species that have been studied. Briefly comment on the significance of this observation.

CHAPTER 23

Biological control

We covered previously the various facets of metabolic reactions of the carbohydrates, lipids, proteins, and nucleic acids. Almost every reaction considered was under the **control** or **regulation** of a specially designed enzyme system. The complexity of these systems becomes even more dramatic upon the realization that the synthesis of each of these enzymes is in turn a product of the sequence of the enzyme-controlled DNA → RNA → protein chain of reactions. Each enzyme must be present in its proper form and amount with the necessary cofactors to meet the demands of the normal living process. Some controls were considered on a molecular level where the enzymes were specifically involved, for example, the influence of high levels of ATP on phosphofructokinase or high levels of ADP on isocitrate dehydrogenase. The presence of the proper oxidation states of both NAD and NADP were also examples of this kind of control.

Hormones function at an additional level of control of the metabolic activity. Hormonal activity was most obvious in the careful balance of the sex hormones considered earlier. The dramatic cascade effect of hormones like epinephrine on the control of carbohydrate metabolism and blood sugar levels is another excellent example of the great precision with which nature operates.

An additional level of control or influence upon the living process is exhibited by both the physical and chemical environment. Everything from the air that we breathe to the drugs that we ingest are examples of the factors that exert control on the biochemistry of cells and organisms.

Some of these levels of control will be considered in this summary chapter.

Control of Enzyme Synthesis

Basic to the entire concept of bioregulation is the concept that the cell must be capable of **synthesizing** the enzyme involved in the control process. Naturally, this in turn depends upon the nature of the DNA within the cell that can direct this synthesis. The DNA in a given cell carries coding information greatly in excess of what is needed for the unique function of that particular cell. Although all cells, and all DNA molecules within cells, may have a great many similarities, each will have its own particular role to play in the living process. The basic question of cell specificity is what makes a liver cell synthesize its own enzymes for its own special reactions. Most of the information in this regard has come from studies involving the regulation of bacteria enzyme synthesis. Higher organisms have similarities in regulations, with the added influence of the products of other tissues, hormones, and external stimulants.

Bacterial studies have been especially useful because certain of their enzymes are produced in response to the nutrients in their culture media. These studies then become models for extrapolation to more complex systems and organisms. Some enzymes are always present in fairly constant amounts and are called **constitutive** enzymes. Other enzymes are present in bacteria in only extremely small amounts when the substrate of that enzyme is not present in the growth medium. The quantity of these enzymes rises dramatically when this substance is added to the medium under controlled conditions. Such enzymes, called either **adaptive** or **inducible** enzymes, are formed by the cells to allow the substrate in question to be utilized in metabolic pathways. The substrate acts as an **inducer** to cause the biosynthesis of these adaptive enzymes. A good example of this is the enzymes responsible for the splitting of lactose, which can be used as a carbon source for microorganisms.

A very important observation in this series of events occurs when the substrate of interest is removed from the medium of the bacteria. The induced enzyme is then no longer formed by an effect known as **repression**. Therefore, the bacteria must possess the DNA and the capacity to synthesize all other intermediates necessary for the formation of the inducible enzyme under controlled conditions.

To explain these events, it has been suggested that enzyme repression is the basis of control and that enzyme induction is actually the release of repression. Accordingly, there are specific areas on the DNA involved with each step in this overall process. The gene or nucleotide sequence coding for the sequence of the amino acids in the enzyme is known as the **structural gene**. This portion of DNA directs the transcription of the appropriate m-RNA unless it is repressed by a specific repressor. This repressor becomes specifically bound to a portion of DNA known as the **operator** or O site. When the repressor is bound to the O site, the DNA cannot form the m-RNA necessary to provide for the synthesis of the inducible enzyme. The repressor itself is synthesized from information contained within the DNA.

In the absence of inducers, the repressor interacts with the operator and prevents the enzyme formation. When the inducer is present, it becomes involved with the repressor in a manner that makes it unable to combine with the operator DNA. The structural gene is then free to direct the synthesis of m-RNA, leading to the formation of the inducible enzyme. This can be visualized in the very simplified scheme shown in Figure 63–1.

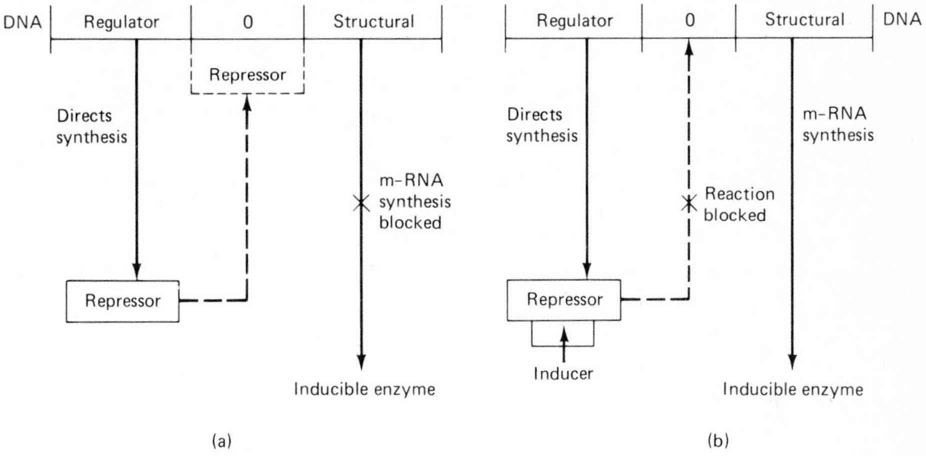
DNA
Regulator
0
Structural
Repressor
Directs synthesis
m-RNA synthesis blocked
Repressor
Inducible enzyme
(a)
Regulator
0
Structural
DNA
Directs synthesis
m-RNA synthesis
Reaction blocked
Repressor
Inducer
Inducible enzyme
(b)

Figure 63–1 Schematic of indirect enzyme formation.

Control by Hormones

In man and other higher animals, another type of regulation and control involves hormones. **Hormones** are chemicals synthesized and secreted by various glands and organs which act as messengers to influence the functions of another gland or cell.

A moderately well understood hormone system, discussed previously, dealt with the influence of epinephrine on the metabolism of glycogen. Only a tiny amount of hormone is necessary to elicit this response. It has been estimated that concentrations in the blood of no more than 10^{-9} M epinephrine is sufficient for activity. The hormonal activity is enhanced by a cascade effect. The epinephrine first interacts with a specific receptor site on the membrane of the muscle cell. This interaction, at the cell membrane, brings about the enzymatic formation of cyclic AMP.

Studies have shown that cyclic AMP may have a special role as a more or less universal intracellular hormone or **second messenger** in the action. The site of activity at the cell membrane solves many problems presented by permeability difficulties. Once again, nature seems to have chosen a common molecule to fulfill a number of important roles.

The hormones themselves are interrelated by a complex series of reactions and controls. The anterior pituitary of the brain secretes a series of **master hormones**, which in turn stimulate various other endocrine glands to secrete their own unique hormones. The pituitary gland (hypophysis) secretes at least nine different peptide hormones. These include the hormones that regulate the thyroid and adrenal glands, ovaries, and testes. For example, the same hormone, follicle-stimulating hormone (FSH), causes growth of the ovarian follicles in the female and stimulates spermatogenesis in the male. They are present in extremely small amounts in the blood, and must be determined clinically by a technique known as radioimmunoassay (RIA).

The secretion of the pituitary hormones is regulated in turn by specific polypeptides termed **releasing hormones** or releasing factors produced in the hypothalamus. These releasing factors are very small peptides, ranging from tripeptides to decapeptides. An interesting observation has been made that these releasing factors are not species specific; this promises to have considerable clinical significance and is another demonstration of the unity of living systems. The feedback control of the various participants in the elaboration of thyroid hormones is used as an example and is shown in Figure 63–2.

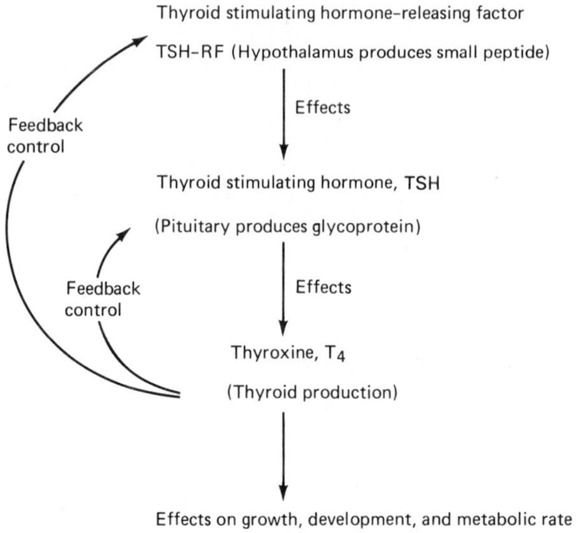
Thyroid stimulating hormone-releasing factor
TSH-RF (Hypothalamus produces small peptide)
Effects
Feedback control
Thyroid stimulating hormone, TSH
(Pituitary produces glycoprotein)
Feedback control
Effects
Thyroxine, T4
(Thyroid production)
Effects on growth, development, and metabolic rate

Figure 63–2 Hormonal reactions leading to thyroxine production and action.

Control by Administered Drugs

Besides the various factors discussed, which are **endogenous**, or produced within our own bodies, many chemicals are ingested (**exogenous**) that greatly influence our systems. The control exerted by these drugs may be to release or influence a naturally occurring function, treat a disease caused by an invading organism, alter our mood, or change our response to some stimulation.

A classic example of a **replacement** drug is insulin. Aqueous solutions derived from beef or pork pancreas can be suitably prepared for injection, and function to treat successfully people suffering from diabetes mellitus. In less severe cases of this hormonal insufficiency, drugs like tolbutamide (Orinase) can be administered orally to stimulate the increased secretion of insulin from the pancreas. Orinase is not chemically related to insulin, as seen by its structure:

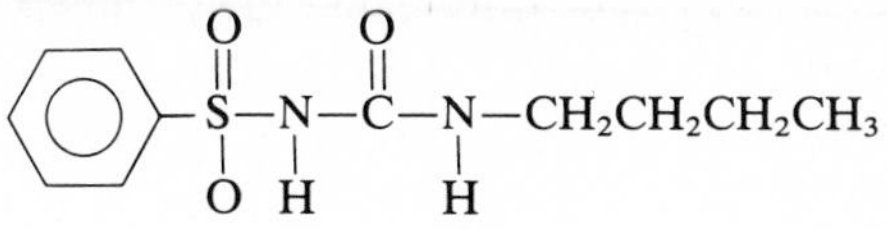

The treatment of infectious diseases by a variety of drugs has become very commonplace. Antibiotics are excellent examples of drugs that **control** invading organisms and free the host from their effects. The sulfa drugs were employed early in this role. Penicillin compounds are good examples of control agents that interfere with the synthesis of the cell walls of some invading bacteria. A member of the family of penicillins, which now includes both natural and synthetic compounds, is penicillin G:

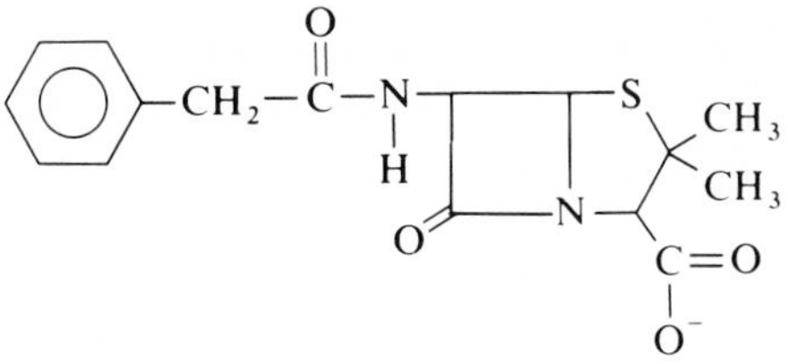

A number of different penicillins exist with different functional groups substituting for the aromatic ring. Two other widely used antibiotics include the tetracyclines and streptomycins. Tetracyclines are known to bind to the 30S ribosomal subunits and inhibit the charged t-RNA molecules from reacting properly. Streptomycin is known to cause a misreading of the m-RNA. Thus these compounds inhibit protein biosynthesis of bacteria and act to control the multiplication of invading organisms. Their structures are:

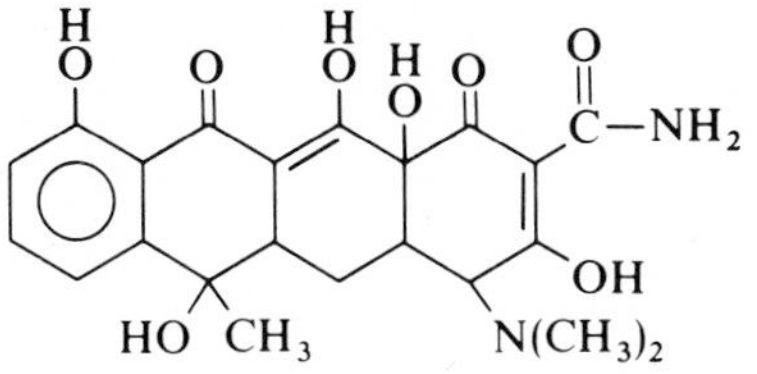

tetracycline

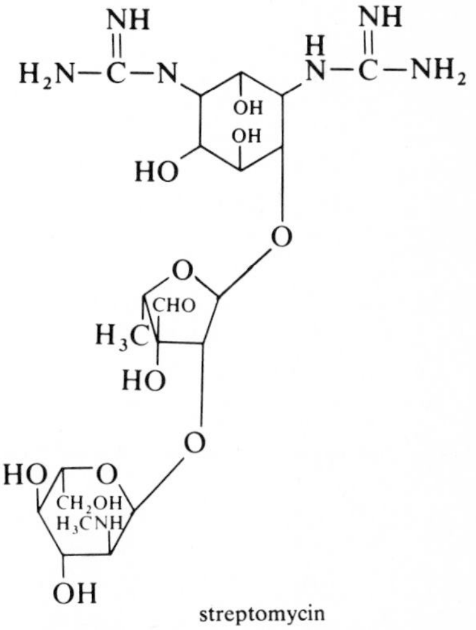

streptomycin

In the 1930's, the sulfa drugs launched the modern era of chemotherapy and antiobiotic therapy. Although penicillin was actually recognized in 1929 as an accidental contaminant of a bacterial culture, it did not have an impact on medicine until about 1939. As the result of such discoveries, the infectious diseases that killed about 500 of every 100,000 Americans in 1900 now account for only about 50 deaths in this same number of people. This is an excellent example of the blending of basic scientific research with the applications of the pharmaceutical industry.

A number of drugs have been employed in our search to provide relief from pain. **Morphine** is an alkaloid that has been used for years in this regard. Heroin, a highly addictive drug of abuse, is the diacetyl derivative of morphine. Morphine has the structure

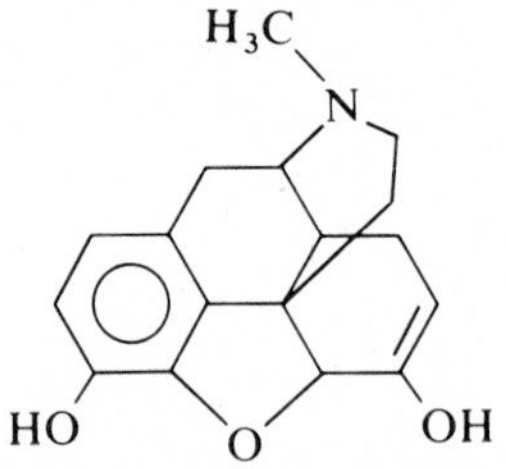

H3C
N
HO
O
OH

Other drugs of abuse include the more common ones like caffeine, nicotine, ethanol, and cannabinol (marijuana). These compounds exhibit a wide variety of chemical structures that function in as yet a largely unknown fashion.

Some of the common drugs ingested in huge quantities are those that alter the mood of individuals and act as **tranquilizers**. Drugs like Librium and Valium act to provide relatively safe control and relief of anxiety and tension. The similarity in the structures of these two common tranquilizers is shown:

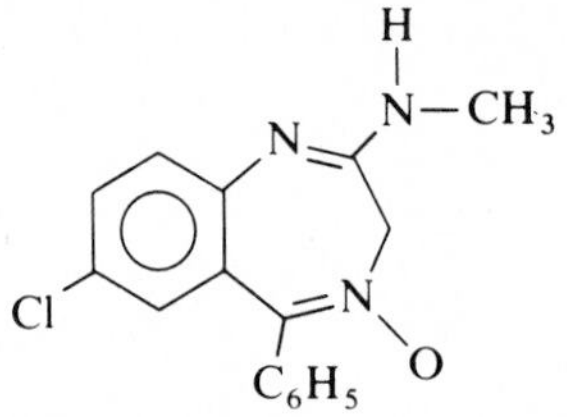

H
N—CH3
N
N
O
Cl
C6H5

Librium

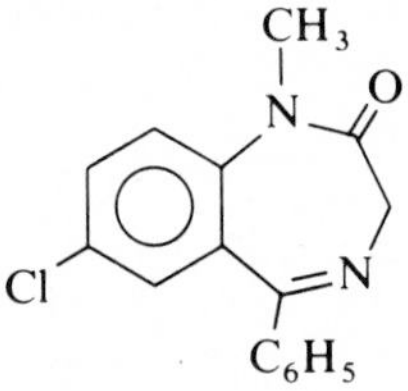

CH3
N
O
N
Cl
C6H5

Valium

The variety of chemical compounds that act to control one or another of our biochemical systems is enormous. The way in which each of the various proteins (enzymes) specifically interacts with these compounds adds a factor of complexity that makes our hope of understanding such processes and their controls on a molecular basis just short of impossible. The study of biochemistry is that endless search to apply the principles of chemistry and biology to both this level of understanding and its application toward improving the quality of life.

Answers to odd-numbered problems and exercises

Answers for Unit 1

1–1. (a) Milli. (b) Kilo. (c) Deca. (d) Micro. (e) Mega.

1–3. (a) 3 places to the left. (b) 3 places to the right.
(c) 3 places to the left. (d) 3 places to the right.
(e) 2 places to the left. (f) 2 places to the right.
(g) 1 place to the left. (h) 3 places to the left.

1–5. Experimental work.

1–7. Use Table 1–4 for this problem.

1–9. (a) 22.2°C. (b) 37°C. (c) −28.9°C. (d) −12.2°C.

1–11. (a) $k\ell \rightarrow \ell \rightarrow m\ell = 10^6\ m\ell$. (b) kg → g → lb = 2.2 lb.
(c) g → lb → ton = 1.1×10^{-6} ton. (d) ℓ → qt → gal = 1.32 gal.
(e) in. → ft → mi = 0.0785 mi. (f) gal → qt → cups = 80 cups.

1–13. \$4.86/kg.

1–15. (a) 7.5 grains. (b) 0.5 g.

1–17. \$0.18/ℓ.

1–19. 500 g = 271.1 mℓ; beaker is large enough.

1–21. 0.00576 g.

1–23. 934.2 mℓ.

1–25. (a) 9800 dyn/cm^2. (b) 98 dyn/cm^2. (c) 17,240 dyn/cm^2.

Answers for Unit 2

2–1. (a) Compound. (b) Compound. (c) Element. (d) Mixture. (e) Mixture.

2–3. Temp. change = 60°C. 60,000 cal.

2–5. 1 cup = 240 g; requires 16.8 kcal (Cal.).

2–7. 10^{-2} Å has greater energy than 10 Å has greater energy than 100 Å has greater energy than 1.

2–9. Energy releasing: a, e; energy absorbing: b, c, d, f.

2–11. (a) Thermal → chemical. (b) Chemical → kinetic → gravitational. (c) Solar → chemical. (d) Chemical → kinetic. (e) Gravitational → kinetic. (f) Solar → chemical.

Answers for Unit 3

3–1. Dalton thought an atom was as small as one could get; we know that electrons, protons, and neutrons are smaller. Dalton thought all atoms of one element were exactly identical, but we know that isotopes can exist.

3–3. They have the same mass. Protons have a positive charge, but neutrons have no charge.

3–5. See Table 3-3.

3–7. A molecule always contains more than one atom, so we cannot obtain one atom of a compound. Molecules are the smallest units of compounds.

3–9. (a) False; isotopes exist.
(b) False; protons, neutrons, and electrons are smaller than atoms.
(c) False; each atom of different elements contains different numbers of protons, neutrons, and electrons.
(d) True.
(e) False; an atom of copper weighs more than an atom of iron.
(f) True.
(g) True; isotopes exist.
(h) False; the atomic weight = sum of numbers of protons and neutrons = 19.

3–11. (a) 3 molecules on left, 3 molecules on right.
(b) 3 molecules on left, 3 molecules on right.
(c) 3 molecules on left, 1 molecule on right.

3–13. (a) Positive 2. (b) No charge. (c) Negative 3. (d) Negative 2.

3–15. (a) Ca. (b) Fe. (c) Hg. (d) K. (e) Ag. (f) Na. (g) C. (h) N. (i) Se.

3–17. Isotopes contain different numbers of neutrons, but they contain the same numbers of protons and electrons.

3–19. (a) Yes. (b) No. (c) No.

3–21. It showed that most of the mass of an atom was located in a very small portion of the atom. Had the atom been a dense solid throughout, all of the neutrons would have bounced back.

Answers for Unit 4

4–1. (a) 38 protons, 52 neutrons. (b) 55 protons, 82 neutrons. (c) 92 protons, 143 neutrons. (d) 53 protons, 78 neutrons. (e) 30 protons, 41 neutrons.

4–3. Gamma rays can penetrate the deepest into human tissue and so would cause most damage.

4–5. Strontium-85, carbon-14, and cobalt-60 all have half-lives that exceed 1 year. They would accumulate in the body.

4–7. (a) An alpha particle. (b) A negatron. (c) A negatron.

4–9. (a) ${}^{4}_{2}He$, ${}^{218}_{84}Po$. (b) ${}^{218}_{84}Po$. (c) ${}^{210}_{82}Pb$.
(d) ${}^{4}_{2}He$, ${}^{234}_{90}Th$. (e) ${}^{214}_{83}Bi$.

4–11. (a) Betatron emission. (b) Betatron emission. (c) Alpha particle emission. (d) Gamma-ray emission.

4–13. (a) ${}^{3}_{1}H - {}^{0}_{-1}\beta \rightarrow {}^{3}_{2}He$. (b) ${}^{14}_{6}C - {}^{0}_{-1}\beta \rightarrow {}^{14}_{7}N$.
(c) ${}^{226}_{88}Ra - {}^{4}_{2}He \rightarrow {}^{222}_{86}Rn$. (d) ${}^{51}_{24}Cr - {}^{0}_{0}\gamma \rightarrow {}^{51}_{24}Cr$.
(e) ${}^{131}_{53}I - {}^{0}_{-1}\beta \rightarrow {}^{131}_{54}Xe$.

4–15. 88% will decay in 3 half-lives. 3 half-lives = 366 min. 1 half-life = 122 min.

Answers for Unit 5

5–1. Atomic orbitals are subdivisions of the energy levels.
5–3. Four electron volts would be required.
5–5. An s-orbital is spherical. As you move from the first level to the second level, the radius of the sphere gets larger. The energy possessed by the s-orbital also increases.
5–7. Absorption of energy = a, c, e; loss of energy = b, d, f.
5–9. See Table 5–1.
5–11. (a) $1s^22s^22p^1$. (b) $1s^22s^22p^63s^23p^4$. (c) $1s^22s^22p^63s^23p^64s^2$. (d) $1s^22s^22p^63s^23p^64s^23d^4$.
5–13. There are no partially filled orbitals in the low-energy portions. All electrons are present in the lowest-energy orbitals available.
5–15. All the ions have the electron configuration of $1s^22s^22p^6$.

Answers for Unit 6

6–1. (a) False; it is based upon atomic numbers.
(b) False; it is based upon the number of protons, the atomic number.
(c) False; they all have the same numbers of electrons in the outermost occupied shell.
(d) True.
(e) False; the number increases by 1 from left to right.
(f) True; when a shell is filled, you jump down to a new period.
(g) True.
(h) False; all elements in a family have the same number of electrons in the outermost occupied shell.
(i) True. (j) True. (k) True.
(l) False; they have completely filled s + p orbitals; d + f may not be filled.
(m) True.
(n) False; they tend to gain electrons.

6–3. Similar physical and chemical properties seem to repeat at regular intervals as atomic number increases by one unit at a time.
6–5. (a) $1s^22s^22p^63s^1$; metal; period 3; group IA; lose one electron.
(b) $1s^22s^22p^63s^23p^64s^1$; metal; period 4; group IA; lose one electron.
(c) $1s^22s^22p^63s^23p^64s^23d^6$; metal; period 4; group VIIIB; cannot predict.
(d) $1s^22s^22p^63s^23p^4$; nonmetal; period 3; group VIA; gain two electrons.
(e) $1s^22s^22p^63s^23p^64s^23d^{10}4p^6$; nonmetal; period 4; group VIIIA; inert gas.
(f) $1s^22s^22p^63s^23p^5$; nonmetal; period 3; group VIIA; gain one electron.
(g) $1s^22s^22p^63s^2$; metal; period 3; group IIA; lose two electrons.
6–7. B = 8, C = 9, D = 15, F = 17, G = 33, H = 34.
6–9. Both react to obtain filled outer-occupied energy levels. Lithium does this by losing one electron and fluorine does it by gaining one electron.
6–11. (a) $Na + Cl \rightarrow Na^{1+} + Cl^{1-}$ (b) N; farther to the right. (c) $2K + O \rightarrow 2K^{1+} + O^{2-}$ (d) $Ca + O \rightarrow Ca^{2+} + O^{2-}$
6–13. As one goes down a group, the ionization energy decreases. The outer-level electrons that are removed are present in levels farther away from the nucleus; they are easier to remove. As one goes from left to right across a period, the ionization energy increases because the number of electrons in the outer-occupied level increases.
6–15. (a) Ca; farther to the right. (b) N; farther to the right. (c) S; higher in the same group. (d) F; farther to the right.
6–17. It should be low, they do not want more electrons.

Answers for Unit 7

7–1. An atom is a neutral species: the number of protons equals the number of electrons. An ion is a charged species: the number of protons does not equal the number of electrons.

7–3. (a) K^{1+} (b) Mg^{2+} (c) Sr^{2+} (d) Ra^{2+}
(e) Al^{3+} (f) S^{2-} (g) Fe^{2+} (h) Cu^{2+}
(i) Hg^{2+} (j) CO_3^{2-} (k) OH^{1-} (l) NO_3^{1-}
(m) PO_4^{3-} (n) MnO_4^{1-} (o) HCO_3^{1-} (p) NH_4^{1+}
(q) HSO_4^{1-} (r) SO_4^{2-}

7–5. (a) 6. (b) 5. (c) 11. (d) 29. (e) 9. (f) 7. (g) 17.

7–7. (a) $HgSO_4$ (b) $HgCl_2$ (c) K_2CO_3
(d) $Fe(OH)_3$ (e) $Al(HSO_4)_3$ (f) NH_4Br
(g) $Cu(CN)_2$ (h) NH_4Cl (i) $Ca(HCO_3)_2$
(j) Ag_2O (k) NaOH (l) Al_2O_3

7–9. Sodium has one electron in its outer energy level. Loss of that electron gives sodium a stable electron configuration equivalent to that of the inert gas neon. Loss of two electrons would give it an electron configuration equivalent to fluorine, an unstable configuration.

Answers for Unit 8

8–1. A hydrogen molecule is more stable because both of the atoms possess the required two electrons in the energy level.

8–3. Stable electron configurations are those which have s- and p-orbitals filled (eight electrons in the outer occupied level). When an atom forms a single covalent bond, it gains one electron. So the number of covalent bonds an atom will form depends upon the number of electrons it needs to fill its s- and p-orbitals. Nitrogen needs three electrons, so it forms three covalent bonds; oxygen forms two covalent bonds, and chlorine forms one covalent bond.

8–5. In a triple bond there are six electrons present to act as glue to hold the two carbon atoms together; in a single bond only two electrons are present to act as the glue. The greater the number of electrons shared between two atoms, the tighter are those two atoms held together.

8–7. (a) :B̤̈r:B̤̈r: (b) H:H
(c) See (g) (d) :N:::N: N≡N
(e)
```
  H
H:C:H
  H
```
(f) H:C̤̈l:
(g) H:Ö̤:H

8–9. Phosphorous has an electron configuration of $1s^2 2s^2 2p^6 3s^2 3p^3$ and needs three electrons, so it forms three covalent bonds. A probable structure is

```
H—P—H = PH3
  |
  H
```

8–11. Yes. Many different arrangements of atoms are possible. In this unit we listed nine different structures for the formula C_3H_6O.

8–13. Cl—Cl is the least polar because both of the atoms involved possess the same electronegativity.

8–15. (Most polar) B—F C—O H—Cl H—Br N—O N—C C—Br (Least polar)

8–17. Any molecule possessing a very polar covalent bond can hydrogen bond:

$CH_3—CH_2—O—H$, H—Br, $H_3C—NH—CH_3$, $CH_3—CH_2—NH_2$

8–19. (a) Covalent. (b) Ionic. (c) Covalent. (d) Covalent. (e) Ionic.

Answers for Unit 9

9–1. (a) 6×10^{23}. (b) 6×10^{23}. (c) 12×10^{23}. (d) 16. (e) 32.
9–3. (a) 56 g. (b) 84 g. (c) 174 g. (d) 267 g. (e) 106 g. (f) 150 g. (g) 132 g. (h) 294 g.
9–5. (a) 12×10^{23} molecules. (b) 6×10^{22} molecules. (c) 6 molecules. (d) 3.0×10^{26} molecules. (e) 4.8×10^{21} molecules.
9–7. 15,133 mol.
9–9. (a) 18 g. (b) 212 g. (c) 0.06 g. (d) 920 g. (e) 0.924 g.
9–11. (a) 360 g. (b) 1629.6 g. (c) 556.5 g. (d) 176 g.
9–13. (a) 0.121 g. (b) 7.2×10^4 g. (c) 2.36×10^{-3} g.
9–15. 96.25 mℓ.
9–17. (a) 6.15×10^{22} molecules. (b) 1.846×10^{23} atoms.
9–19. (a) 1.1×10^{23} molecules. (b) 6.6×10^{23} atoms of carbon.
9–21. 3×10^{12} yr.

Answers for Unit 10

10–1. (a) C_3H_6O (b) C_4H_4O (c) C_1H_1 (d) C_5H_7N
(e) $C_{12}H_{22}O_{11}$ (f) C_5H_4 (g) $C_9H_8O_4$
10–3. (a) % K = 56.5, % C = 8.7, % O = 34.8.
(b) % Al = 12.9, % H = 0.96, % C = 17.2, % O = 68.9.
(c) % Ca = 43.5, % C = 26.1, % N = 30.4.
(d) % K = 26.7, % Cr = 34.9, % O = 38.4.
10–5. $C_1H_2O_1$
10–7. $C_3H_6O_2$
10–9. (a) NO (b) NO_2 (c) $N_2 + O_2 \rightarrow 2NO$; $2NO + O_2 \rightarrow 2NO_2$
10–11. $C_6H_{12}O_6$
10–13. (a) 1 mol = 78 g. (b) 1 mol = 44 g.
10–15. 0.00086 mol = 31 g; 1 mol = 36046.5 g.

Answers for Unit 11

11–1. The particles in a solid are closely packed in orderly, tight lattices and cannot move past one another. In a liquid, the particles are closely packed but can move past one another.
11–3. (a) Any compound whose density is less than 1 g/mℓ will float on water. Ice, cedar, balsa, butter, and gasoline would float on water.
(b) All the materials listed have densities less than mercury and would float on mercury.
11–5. The volume of a gas means nothing unless the pressure and temperature at which the volume is measured are given.
11–7. $P_1/T_1 = P_2/T_2$. Pressure is directly proportional to temperature.
11–9. The molecules of a gas are widely separated and move freely about a container (the room).
11–11. Final pressure = 34.1 lb/in.^{2}
11–13. Final volume = 2.50 ℓ.
11–15. Final temperature = 177°C.
11–17. Number of moles = 1.38 mol.
11–19. Molecular weight = 49.8 g.
11–21. Density = 0.0022 g/mℓ.
11–23. Molecular formula = $C_4H_6O_2$

Answers for Unit 12

12–1. (a) Sublimation is the process by which a solid is converted directly to a gas.
(b) Evaporation is the process by which a liquid is converted to a gas.
(c) Boiling is the process by which a liquid at its boiling point is converted to a gas.
(d) Melting is the process by which a solid at its melting point is converted to a liquid.

12–3. No. The boiling temperature of liquids is proportional only to the nature of the liquid and to the external pressure.

12–5. (a) Oxygen is heavier than nitrogen so it would boil at a higher temperature.
(b) HF is much more polar than HCl, so HF has the higher boiling point.

12–7. CH_4 less than CF_4 less than CCl_4 less than CBr_4.

12–9. $CH_3CH_2CH_2CH_3$ boils at the lowest temperature; it is a nonpolar covalent compound.
$CH_3CH_2—CH_2—OH$ boils next lowest; one end can hydrogen bond.
$HO—CH_2CH_2—OH$ boils next; both ends can hydrogen bond.
NaCl boils at highest temperature; it is an ionic compound.

12–11. Heat of fusion.

12–13. The boiling point is proportional to the pressure. As the pressure decreases, the boiling point will also decrease.

12–15. As long as ice and water are in equilibrium, the temperature of the mixture must be equal to 0°C.

12–17. As the alcohol evaporates, it absorbs heat from the skin equal to the heat of vaporization of the alcohol. The skin loses heat and feels cooler.

12–19. (a) 800 cal. (b) 84 cal.

12–21. The change in temperature = 80°C. (a) 2.4 cal. (b) 37.6 cal. (c) 160 cal.
(d) 7409 cal.

Answers for Unit 13

13–1. A solution consists of two or more distinct compounds that can be separated from each other. The amounts of each compound present can vary. In a compound there is only one type of molecule, and the ratio of the atoms of each element present is a constant.

13–3. Colloids = b, c, f, h, i, j.

13–5.

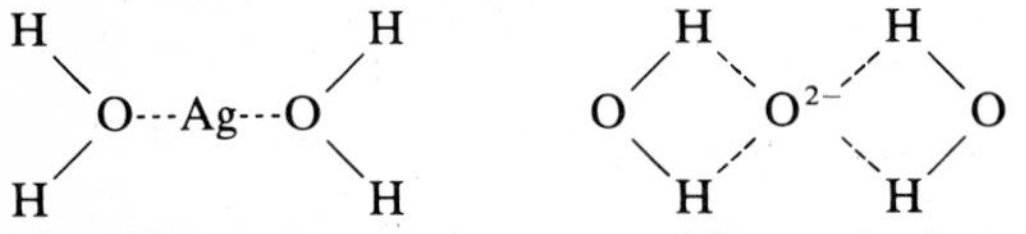
H
H
O---Ag---O
H
H
H
H
O
O^{2-}
O
H
H

13–7. No. Often water can be removed from a hydrate by heating it. Covalent bonds usually require more energy to break them.

13–9. (a) Solubility decreases. (b) Solubility decreases. (c) No change in solubility.
(d) Cannot predict.

13–11. (a) Not saturated: 35.7 g will dissolve in 100 mℓ.
(b) Not saturated: 3.12 g will dissolve in 1 mℓ of hot water.
(c) Saturated.

13–13. 39.75 g will dissolve in 25 mℓ of hot water.
18.6 g will dissolve in 25 mℓ of cold water.
21.15 g will precipitate if the solution were cooled.

13–15. (a) No. The rate of dissolving decreases as temperature decreases.
(b) Yes. Stirring brings fresh solvent into contact with the solid.
(c) Yes. The rate of dissolving increases with increasing temperature.
(d) Yes. There is more surface area for the solid and thus the solvent can come into contact with more solid.
(e) No. Dissolving of solids in liquids is independent of pressure.

13–17. Pure water boils lower than a solution.

13–19. Boil them; the liquid boiling at the lowest temperature is pure water.

13–21. (a) From the 2% solution to the 8% solution.
(b) The volume of the 8% solution would increase.

Answers for Unit 14

14–1. (a) 0.025 mol. (b) 0.171 mol. (c) 0.204 mol. (d) 0.174 mol. (e) 0.00067 mol.

14–3.

	GMW	Wt. solute	Moles solute	$M\ell$ soln.	ℓ soln.	M
NaOH	40 g	400 g	10	1×10^4	10	1
NaCl	58.5	11.7	0.2	20	0.02	10
$NaHCO_3$	84	21	0.25	2.5×10^3	2.5	0.1
$Ca(OH)_2$	74	1.85	0.025	250	0.25	0.1
$C_6H_{12}O_6$	180	1080	6	500	0.5	12

14–5. 0.005 *M*.

14–7. (a) 5.0 g. (b) 120 g. (c) 0.03 g. (d) 0.49 g. (e) 21.3 g. (f) 16 g. (g) 24.5 g. (h) 1.80 g.

14–9. 3.2 g of bromide ion are present.

14–11. (a) 1.7 *M* NaCl. (b) 0.97 *M* NaBr. (c) 0.66 *M* NaI.

14–13. (a) 50 ppm; 5×10^3 ppb. (b) 16200 mℓ.

14–15. (a) 3.0 *M*. (b) 0.05 *M*.

14–17. 0.23 *M*.

14–19. (a) 2 gram equivalents. (b) 0.5 gram equivalent. (c) 0.1 gram equivalent.

Answers for Unit 15

15–1. The first solid must have undergone an addition reaction with some gas in the air; it would gain weight. The second solid must have undergone a decomposition reaction and released some gas; it would lose weight. The third solid either did not undergo reaction or it underwent rearrangement and thus did not lose or gain weight.

15–3. Coal = ash, CO_2, and H_2O. The water and CO_2 escape; weight is lost.
Iron + oxygen = Fe_2O_3. The material gains weight; nothing escapes.

15–5. (a) $MgBr_2 + Cl_2 \rightarrow MgCl_2 + Br_2$
(b) $Zn + 2HCl \rightarrow ZnCl_2 + H_2$
(c) $Fe_2O_3 + 3CO \rightarrow 2Fe^0 + 3CO_2$
(d) $2CuCl + Fe^0 \rightarrow FeCl_2 + 2Cu^0$
(e) $2Na^0 + 2H_2O \rightarrow 2NaOH + H_2$
(f) $Mg + 2Fe^{3+} \rightarrow Mg^{2+} + 2Fe^{2+}$

15–7. (a) $4Na + O_2 \rightarrow 2Na_2O$
(b) $CaO + SO_3 \rightarrow CaSO_4$
(c) $2Al + 3Br_2 \rightarrow 2AlBr_3$
(d) $2Fe + 3S \rightarrow Fe_2S_3$

15–9. (a) $2AgNO_3 + H_2S \rightarrow Ag_2S + 2HNO_3$
(b) $Cu_2O + 2HCl \rightarrow 2CuCl + H_2O$
(c) $Zn(OH)_2 + 2HCl \rightarrow ZnCl_2 + 2H_2O$
(d) $FeCl_2 + 2NaOH \rightarrow Fe(OH)_2 + 2NaCl$

15–11. (a) $C_{25}H_{52} + 38O_2 = 25CO_2 + 26H_2O$.
(b) 190 mol of O_2. (c) 130 mol of H_2O.

15–13. 3.0 g of sulfur.

15–15. 5.33 g of rust were removed.

15–17. The formula is HgO.

15–19. 754.7 g of NaOH would be formed.

15–21. 92965 g of oxygen gas (O_2) would be consumed.

Answers for Unit 16

16–1. Exothermic = a, b, e, f; endothermic = c, d.

16–3. Spontaneous changes: water runs downhill, hot water cools, moving objects slow down. Nonspontaneous changes: moving objects can be made to speed up; food cooks when heat is applied; we toss objects into the air.

16–5. Positive = a, c, d, g, h; negative = b, e, f.

16–7. (a) Endothermic c. (b) Increases, $\Delta H = +$. (c) Increases $\Delta S = +$.

16–9. (a) ΔG can be either positive or negative, depending on the temperature.
(b) ΔG is positive; nonspontaneous.
(c) ΔG can be either positive or negative, depending on the temperature.
(d) ΔG is negative; spontaneous.
(e) ΔG is negative; spontaneous.

Answers for Unit 17

17–1.

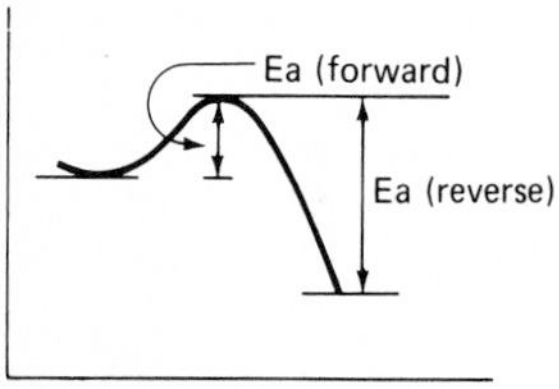

Ea (forward)
Ea (reverse)

Exothermic

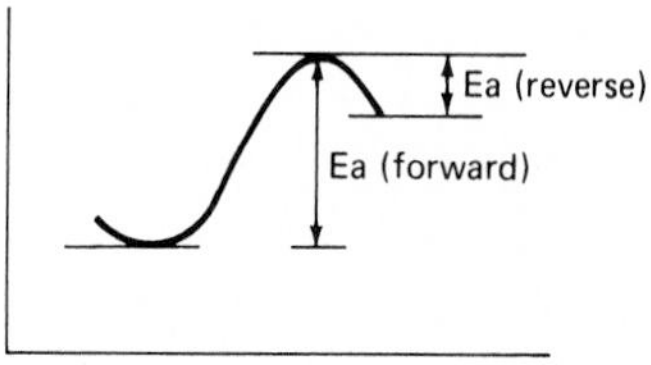

Ea (reverse)
Ea (forward)

Endothermic

17–3. The reactants must collide.
The reactants must collide in the proper orientation.
The collision must be forceful enough to overcome the activation energy barrier.

17–5. The rate of spoilage of milk depends upon the temperature. This reaction occurs more rapidly outside the refrigerator (the temperature is higher).

17–7. The reaction between hydrogen and chlorine has an activation energy barrier. Absorption of light energy provides sufficient energy to overcome the barrier.

17–9. (a) No effect; pressure changes do not affect solids.
(b) The rate would increase for the forward reaction.
(c) The rate would increase for the forward reaction.
(d) The rate of the forward reaction would decrease.
(e) The rates of both the forward and reverse reactions would increase.
(f) Hard to predict; the rate of one or both would increase.
(g) The rates of both the forward and reverse reactions would decrease.

Answers for Unit 18

18–1. They are acids so they would turn litmus paper red, conduct electricity, and react with basic solutions.

18–3. (a) Acids = HCl, H_2O; bases = NaOH, Cl^{1-}.
(b) Acids = HNO_3, H_2O; bases = KOH, NO_3^{1-}.
(c) Acids = H_2O, NH_4^+; bases = NH_3, OH^{1-}.
(d) Acids = HCl, H_3O^+; bases = H_2O, Cl^{1-}.

18–5. (a) HCO_3^{1-} (b) $H_2PO_4^{1-}$ (c) NO_3^{1-} (d) NH_2^{1-}
(e) OH^{1-}

18–7. (a) CH_3^{1-} (b) OH^{1-} (c) NH_2^{1-} (d) Cl^{1-}
(e) NO_3^{1-} (f) S^{2-} (g) SO_4^{2-} (h) HPO_4^{2-}

18–9. (a) NH_3 and NH_4^{1+}, H_3O^{1+} and H_2O; H_3O^{1+} is the stronger acid.
(B) (A) (A) (B)
(b) CH_3OH and CH_3O^{1-}; NH_2^{1-} and NH_3; CH_3OH is the stronger acid.
(A) (B) (B) (A)
(c) NH_2^{1-} and NH_3; H_2O and OH^{1-}; H_2O is the stronger acid.
(B) (A) (A) (B)

Answers for Unit 19

19–1. Some ingredient in tea must be an acid–base indicator that changes color when citric acid is added.

19–3. (a) True.
(b) False; the first solution has pH = 6; the second has pH = 4.
(c) False; any acid with a concentration greater than 1 *M* will have a negative pH.
(d) False; it allows you to determine pH. From pH you can calculate the concentration of acid.

19–5. (a) 4.456. (b) 12.097. (c) 3.222. (d) 0.0.
(e) 1.222. (f) 0.311. (g) −0.699.

19–7. (a) 10^{-6}. (b) 3.315×10^{-6}. (c) 10^{-7}. (d) 2.5×10^{-9}.
(e) 1.6×10^{-11}. (f) 10^{1}. (g) 1 *M*. (h) 6.3×10^{-8}.

19–9. 4.74×10^{-5} g.

19–11. (a) Methyl orange (red at pH = 1, pink at pH = 3).
(b) Litmus (red at pH = 5, bluish at pH = 7).
(c) Phenolphthalein (colorless at pH = 8, red at pH = 10).
(d) Litmus (red at pH = 6, blue at pH = 8).

19–13. 0.365 g of HCl.

19–15. Yes. If the volumes differ, the numbers of grams of acid present would differ.

Answers for Unit 20

20–1. They would be used to neutralize the strong base, sodium hydroxide. Vinegar contains acetic acid, and lemon juice and orange juice contain citric acid.

20–3. 1 g of calcium carbonate is present.

20–5. 238 mℓ of stomach acid will be neutralized.

20–7. 0.16 *M*. **20–9.** 0.2 *M*.

20–11. The solution will be acidic because there is 0.015 mol of H_2SO_4 in excess.

20–13. The solution is acidic; the pH will be 1.7.

20–15. 0.25 g of calcium carbonate is present. The rock was 2.5% $CaCo_3$.

Answers for Unit 21

21–1. The acetic acid–acetate solution is a buffer. Its pH should not change much when small amounts of HCl or NaOH are added. The 1 *M* HCl solution is not a buffer; addition of small amounts of HCl or NaOH will dramatically change its pH.

21–3. (a) $K_{eq} = \frac{[CH_3CO_2CH_3][H_2O]}{[CH_3OH][CH_3CO_2H]}$

(b) $4HCl + O_2 \rightleftharpoons 2H_2O + 2Cl_2$

$K_{eq} = \frac{[H_2O]^2[Cl_2]^2}{[HCl]^4[O_2]}$

(c) $2HBr \rightleftharpoons Br_2 + H_2$

$K_{eq} = \frac{[Br_2][H_2]}{[HBr]^2}$

(d) $N_2O_4 \rightleftharpoons 2NO_2$

$K_{eq} = \frac{[NO_2]^2}{[N_2O_4]}$

(e) $2C + O_2 \rightleftharpoons 2CO$

$K_{eq} = \frac{[CO]^2}{[C]^2[O_2]}$

21–5. $H_2CO_3 < CH_3CO_2H < HCO_2H < HF < H_2SO_4$.

21–7. C > D > A > B.

21–9. (a) Sodium acetate solution. (b) Sodium carbonate solution.

21–11. Solution (a) will have the highest pH.

21–13. The pH of the solution is 8.94.

21–15. (a) $(H^{1+}) = 3.5 \times 10^{-5}\,M$; $(OH^{1-}) = 2.86 \times 10^{-10}\,M$; pH = 4.46.
(b) $(H^{1+}) = 3.6 \times 10^{-4}\,M$; $(OH^{1-}) = 2.78 \times 10^{-11}\,M$; pH = 4.07.
(c) $(H^{1+}) = 3.53 \times 10^{-4}\,M$; $(OH^{1-}) = 2.83 \times 10^{-11}\,M$; pH = 3.17.

21–17. (a) For pH = 5, use acetic acid/acetane buffer HOAc/OAc = $\frac{0.56}{1}$.
(b) For pH = 10, use ammonia or hydrocyanic acid buffer $NH_4^{1+}/NH_3 = \frac{1.8}{1}$.
(c) For pH = 2, use HSO_4^{1-}/SO_4^{2-} buffer $HSO_4^{1-}/SO_4^{2-} = \frac{0.83}{1}$.

21–19. $(H^{1+}) = 1.34 \times 10^{-2}$; pH = 1.87.

21–21. (a) $CH_3CH_2—NH_3^{1+}$.
(b) Equal concentrations.
(c) $CH_3CH_2—NH_2$.

Answers for Unit 22

22–1. An atomic orbital is located only on a single atom, whereas a molecular orbital covers at least two atoms.

22–3. (a) sp^2. (b) sp^3. (c) sp^2. (d) sp. (e) sp. (f) sp^3.

22–5.

(a)	$—CH_2—OH \rightarrow CH_2—Br$	Substitution on sp^3 center
(b)	C(H)—C(OH)— → C=C	Elimination
(c)	C(OH) → C(=O)	Oxidation
(d)	C—Br → C—OH	Substitution of sp^3 center
(e)	$C(=O)—OCH_3 \rightarrow C(=O)—NH_2$	Substitution on sp^2 center
(f)	$CH=CH_2 \rightarrow CH_2—CH_3$	Reduction
(g)	—CH(OH)— → —C(=O)—	Oxidation
(h)	$—HC=CH_2 \rightarrow CH(OH)—CH_3$	Addition

Answers for Unit 23

23–1. (a) Hexane. (b) Pentane. (c) Heptane. (d) Octane. (e) Cyclohexane. (f) Pentane.

23–3. (a) 2-methylbutane. (b) 2,5-dimethylhexane.
(c) 3-methylhexane. (d) 4-methyl-4-isopropyloctane.
(e) 1-iodo-3-*t*-butylcyclopentane. (f) 3,3-dimethylhexane.
(g) 4-methyl-5-*n*-propylnonane. (h) 1-bromo-3-isobutylcyclohexane.
(i) 2-methylheptane. (j) 3,5-diethylheptane.

23–5. (a) *n*-pentane. (b) 2-methylpentane.
(c) 2,3,6-trimethylheptane. (d) 4-*n*-propylheptane.
(e) 3-bromo-6-ethyloctane. (f) 2-methylpentane.
(g) 1-bromo-3-isobutylcyclohexane. (h) Correct.
(i) 1,1-dimethyl-3-*sec*-pentylcyclohexane.

23–7. (a) 2-methylpentane. (b) 3,3-dimethylhexane.
(c) 3,5-diethylheptane. (d) 2,2-dimethyloctane.
(e) Nonane.

23–9. Same = a, b, e; different = c, d.

23–11. (a) $2C_4H_{10} + 13O_2 = 8CO_2 + 10H_2O$

(b) $C_6H_{12} + 9O_2 = 6CO_2 + 6H_2O$

(c) $C_4H_{10} + Cl_2 =$

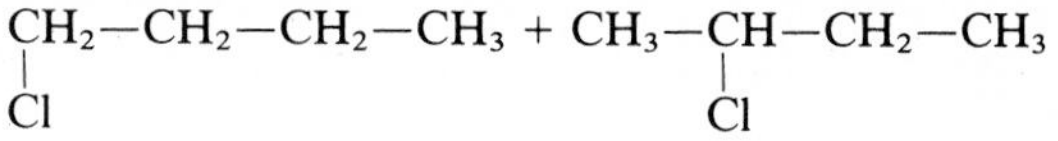

(d)

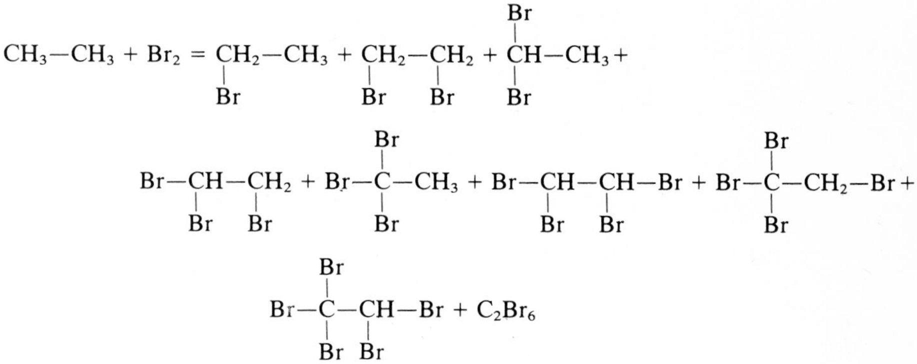

23–13. (a) 2,3-dimethylbutane. (b) *n*-hexane. (c) 2-methylpentane.

Answers for Unit 24

24–1. (a) 2-pentene. (b) 1-pentene. (c) 3-methyl-1-pentene.
(d) 3-ethyl-1-hexene. (e) Cyclohexene.
(f) 3,5-dimethyl-4-isopropyl-3-heptene. (g) 3-isobutylcyclohexene.
(h) 3,3-6-trimethyl-4-nonene. (i) 2-ethyl-1-pentene.
(j) 3-propyl-1,4-hexadiene.

24–3. (a)

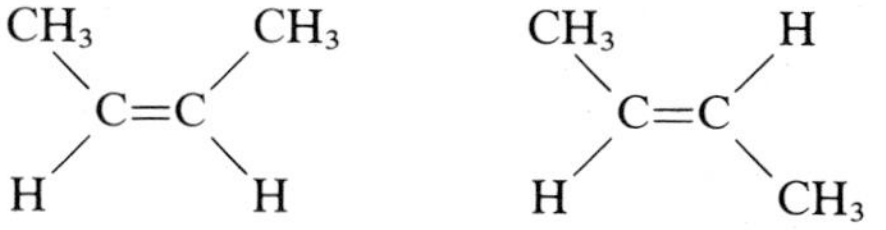

cis-2-butene *trans*-2-butene

(b) No.

(c)

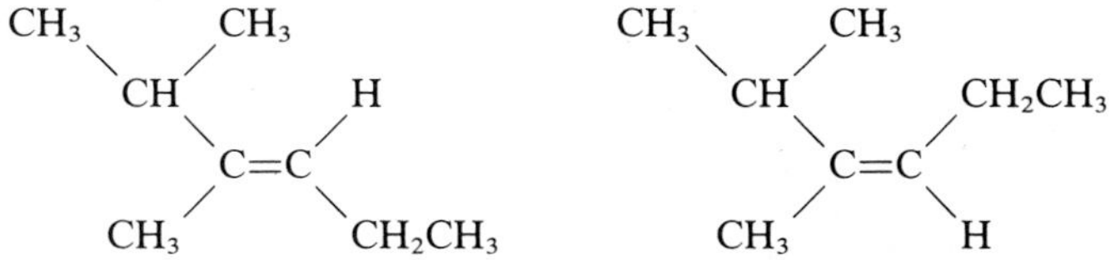

trans-2,3-dimethyl-3-hexene *cis*-2,3-dimethyl-3-hexene

(d)

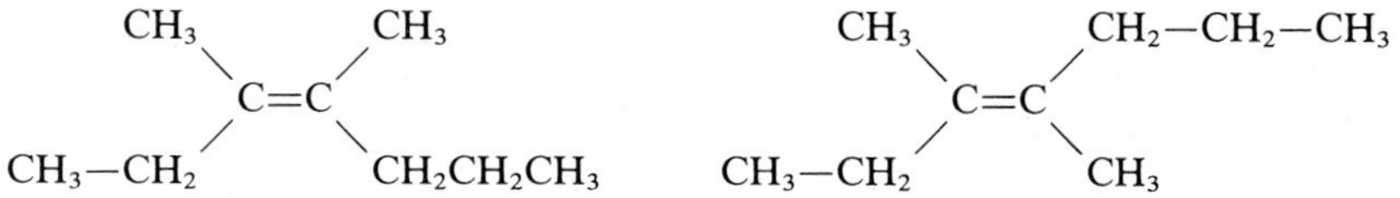

cis *trans*

(e) No. (f) No.

24–5. (a) C_6H_{10}. (b) C_6H_8. (c) C_7H_{12}. (d) $C_{10}H_{16}$.

24–7. (a) Hydration. (b) Oxidation-ozonolysis. (c) Dehydration. (d) Reduction. (e) Bromination.

Answers for Unit 25

25–1. (a) 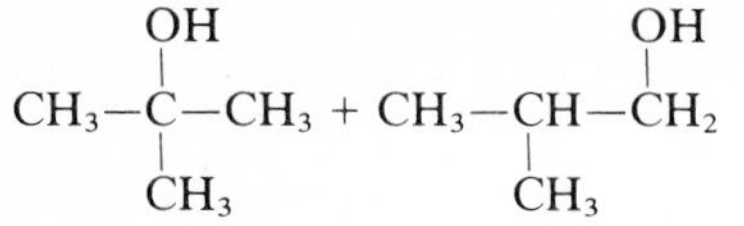

major

(b) 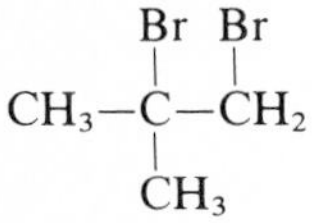

(c) 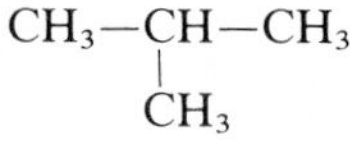

(d) 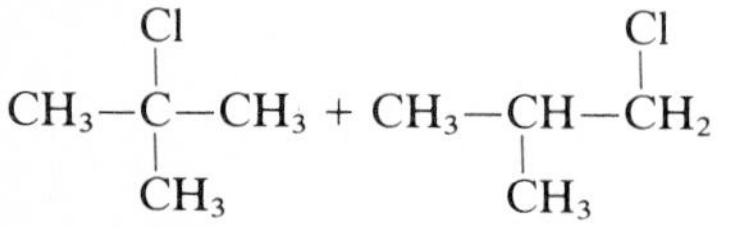

major

(e) 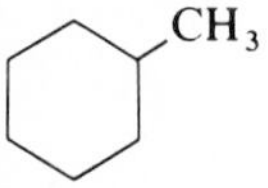

(f) 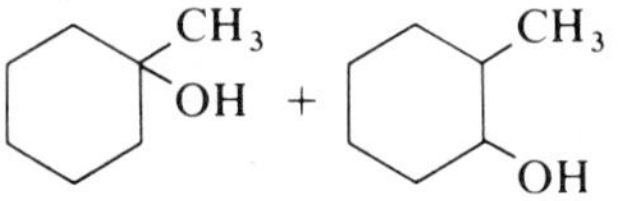

major

(g) $CH_3-CH=CH-CH_3 + CH_2=CH-CH_2CH_3$

major

(h) Same as (g).

(i) 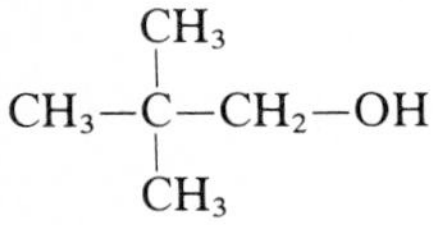

(j) 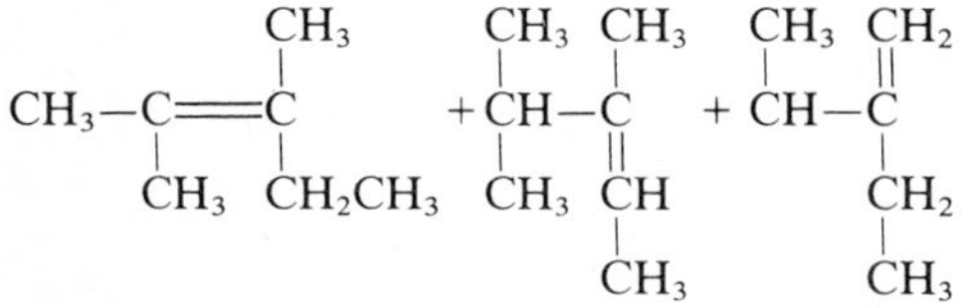

25–3. 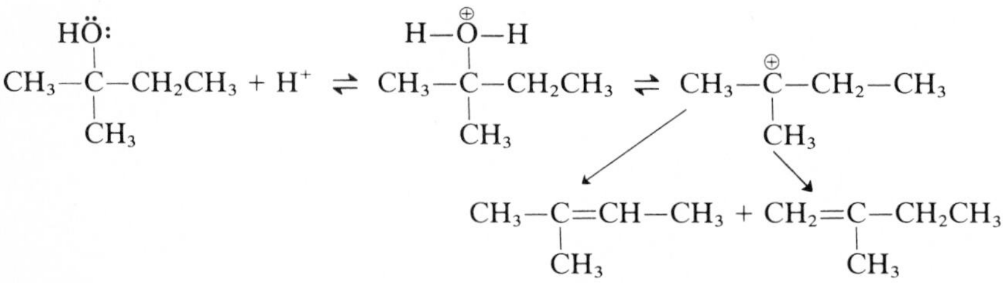

major
(most stable)

25–5. Mech.: $R{-}OH \overset{H+}{\rightleftharpoons} R{-}\overset{\oplus}{O}H_2 \rightleftharpoons R^{\oplus} \rightarrow$ Alkene.

The alcohol forming the most stable carbonium will react the most rapidly.

$$\underset{\text{forms 3° carbonium ion}}{CH_3{-}\overset{\overset{CH_3}{|}}{\underset{\underset{CH_3}{|}}{C}}{-}OH} \;>\; \underset{\text{forms 2° carbonium ion}}{CH_3{-}\underset{\underset{CH_3}{|}}{C}H{-}OH} \;>\; \underset{\text{forms 1° carbonium ion}}{CH_3CH_2{-}OH} \;>\; \underset{\text{forms methyl carbonium ion}}{CH_5OH}$$

25–7. (a) $CH_3{-}CH_2{-}CH_2{-}OH \xrightarrow[\Delta]{H^+} CH_3{-}CH{=}CH_2 \xrightarrow[H^+]{H_2O} CH_3{-}\overset{\overset{OH}{|}}{C}H{-}CH_3$

(b) $CH_3{-}CH_2{-}CH_2{-}OH \xrightarrow[\Delta]{H^+} CH_3{-}CH{=}CH_2 \xrightarrow[Pt]{H_2} CH_3{-}CH_2{-}CH_3$

(c) $CH_3{-}CH_2{-}CH_2{-}OH \xrightarrow[\Delta]{H^+} CH_3{-}CH{=}CH_2 \xrightarrow{HBr} CH_3{-}\overset{\overset{Br}{|}}{C}H{-}CH_3$

(d) $CH_3{-}CH_2{-}CH_2{-}Br \xrightarrow{OH^-} CH_3{-}CH{=}CH_2 \xrightarrow{HBr} CH_3{-}\overset{\overset{Br}{|}}{C}H{-}CH_3$

(e) $CH_3{-}CH_2{-}CH_2{-}Br \xrightarrow{OH^-} CH_3{-}CH{=}CH_2 \xrightarrow[H^+]{H_2O} CH_3{-}\overset{\overset{OH}{|}}{C}H{-}CH_3$

(f) $CH_3{-}\underset{\underset{Br}{|}}{C}H{-}CH_3 \xrightarrow{OH^-} CH_3{-}CH{=}CH_2 \xrightarrow[Pt]{H_2} CH_3CH_2CH_3$

(g) $CH_3{-}\underset{\underset{Br}{|}}{C}H{-}CH_3 \xrightarrow{OH^-} CH_3{-}CH{=}CH_2 \xrightarrow{HCl} CH_3{-}\overset{\overset{Cl}{|}}{C}H{-}CH_3$

Answers for Unit 26

26–1. (a) 1-propyne.
(b) 2-pentyne.
(c) 2-methyl-5-ethyl-3-octyne.
(d) 3-bromo-1-propyne.
(e) 2,2-dimethyl-6-isopropyl-3-nonyne.
(f) 6,6-dichloro-3-heptyne.

26–3. (a) 2,2,3,3-tetrabromobutane.
(b) Butane.

(c) $CH_3{-}CH_2{-}\overset{\overset{O}{\|}}{C}{-}CH_2$

(d) $\underset{\text{major}}{CH_3{-}\overset{\overset{Br}{|}}{\underset{\underset{Br}{|}}{C}}{-}CH_3} + CH_3{-}CH_2{-}CHBr_2$

(e) $\underset{\text{major}}{CH_3{-}\underset{\underset{O}{\|}}{C}{-}CH_3} + CH_3{-}CH_2{-}\underset{\underset{O}{\|}}{C}$

(f) $CH_3{-}\overset{\overset{Br}{|}}{\underset{\underset{Br}{|}}{C}}{-}CH\begin{matrix}\diagup Br\\ \diagdown Br\end{matrix}$

(g) No reaction.

(h) $CH_3{-}C{\equiv}C^{-}{:} + NH_3$

26–5. 1-pentyne; 2-pentyne; 3-methyl-1-butyne.

26–7. (a) 1-butyne will react with $NaNH_2$ since it is a terminal alkyne; 2-butyne should not react.
(b) Do a titration using bromine; 2-butyne will react with twice as much bromine solution as will 2-butene.

Answers for Unit 27

27–1. Aromatic hydrocarbons are much more inert than alkenes or alkynes. They have many fewer hydrogens than alkanes of the same number of carbon atoms. They have unusual odors.

27–3. (a)

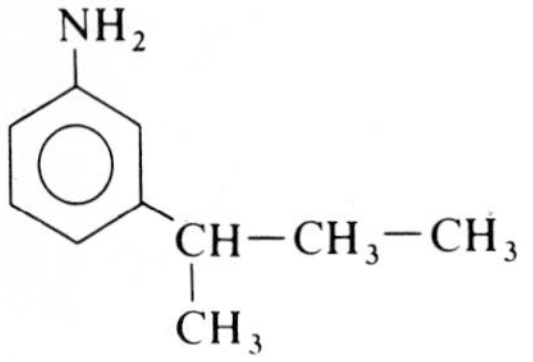

(b)

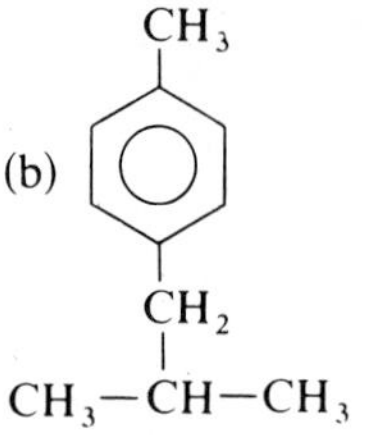

(c)

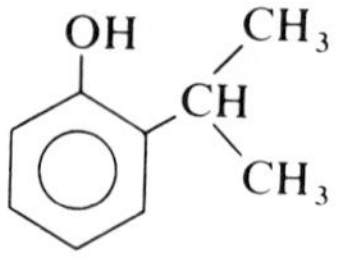

(d)

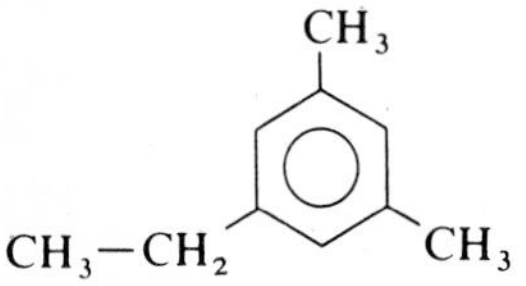

(e)

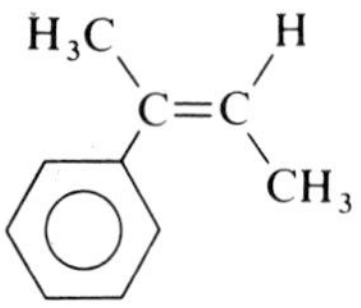

(f)

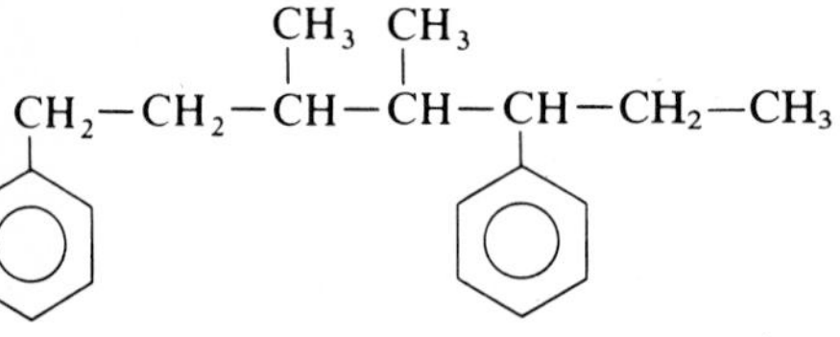

27–5. (a) C_7H_8 (b) $C_{10}H_{14}$ (c) $C_{10}H_{12}$
(d) $C_{10}H_{10}$ (e) $C_{10}H_8$ (f) $C_{14}H_{10}$
(g) $C_{12}H_{16}$

27–7. Only two different compounds are possible.

27–9. (a)

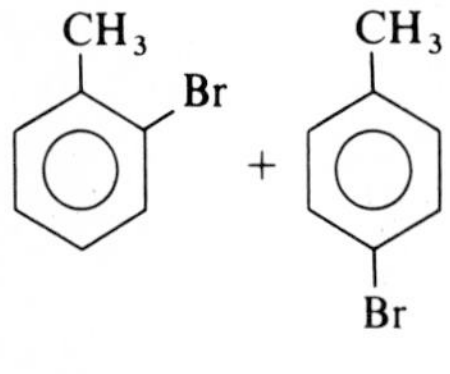

(b)

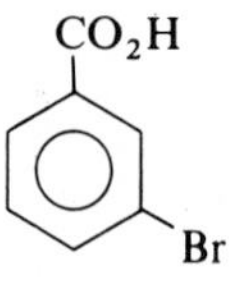

(c)

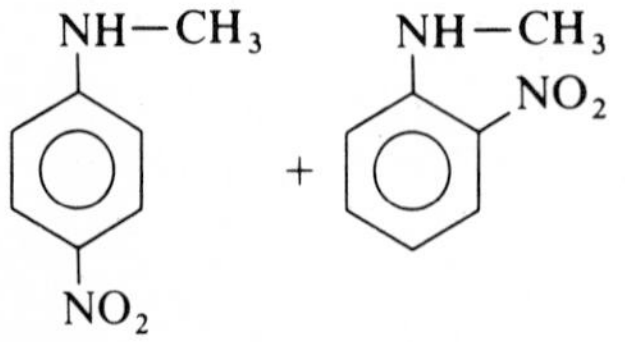

(d)

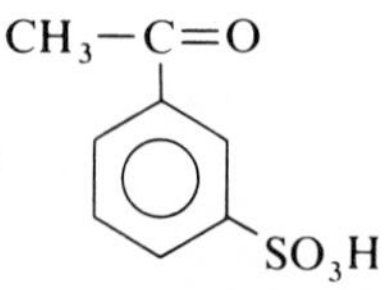

Answers for Unit 28

28–1. (a) 1-phenyl-2-butanol.
(b) 2,2-dimethyl-1-hexanol.
(c) 3-ethyl-1-cyclohexanol.
(d) 3-ethyl-4-phenyl-2-hexanol.
(e) 5-hepten-3-ol.

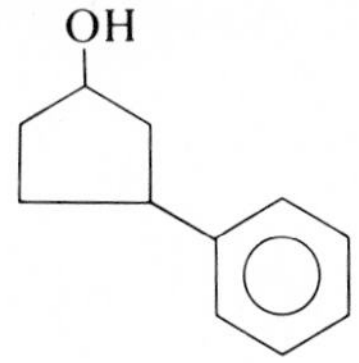

OH

(b) $CH_3-CH(OH)-CH_2-C(CH_3)_2-CH_2-CH_2CH_3$

(c) 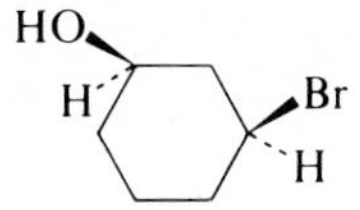

HO
H
Br
H

(d) $CH_2(OH)-C(CH_3)_2-CH_2-CH(CH-CH_3)-CH_2-CH_2-CH_2-CH_3$

(e) $CH_3-CH(OH)-CH=CH-CH_2-CH_3$

(f) $CH_2(OH)-CH(CH_3)-CH_2-CH(CH_3)-CH(OH)-CH_3$

28–5. (a) $CH_3-CH(OH)-CH_2-CH_3$
(b) $CH_2(OH)-CH_2-CH_2-CH_3$

(c) $CH_2(OH)-CH(CH_3)-CH_3$

(d) 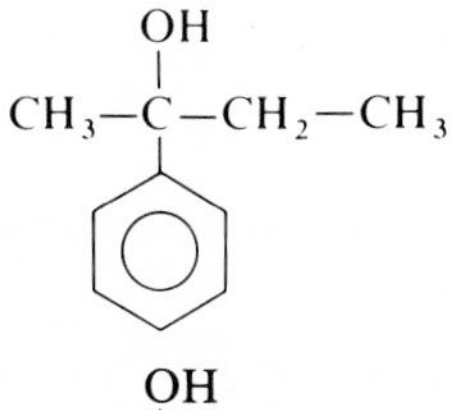

OH
CH3—C—CH2—CH3

(e) $CH_3-CH(OH)-CH=CH_2$

(f) $CH_2(OH)-CH_2-C\equiv C-CH_2-CH_3$

28–7. (a) $CH_3-CH=C-CH_3$ + $CH_2=CH-CH-CH_3$ 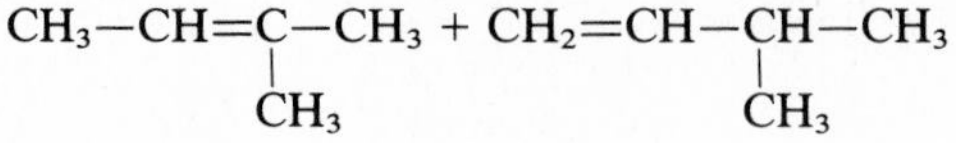
             $\quad\quad\quad\quad\;\;|$ $\quad\quad\quad\quad\quad\;|$
             $\quad\quad\quad\quad CH_3$ $\quad\quad\quad\quad\quad CH_3$
             $\quad\quad\quad$ major

(b) 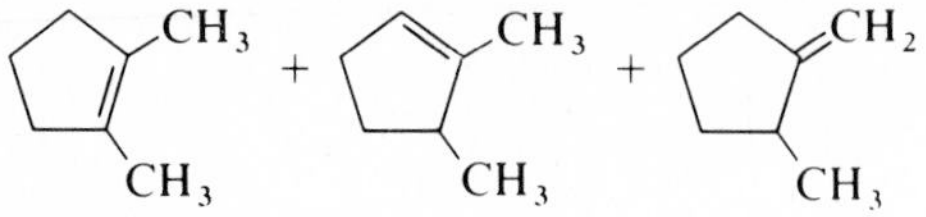 major

(c) No reaction.

(d) 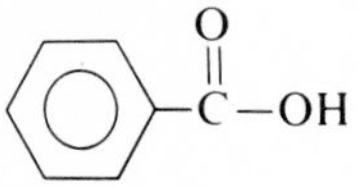

(e) $CH_3-\overset{\overset{\displaystyle O}{\|}}{C}-CH_3$ 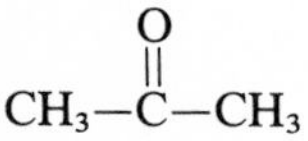

(f) $CH_3-CH_2-O^-Na^+ + H_2$

(g) 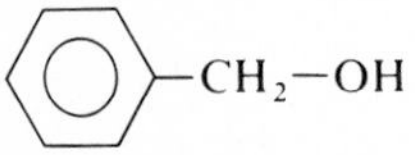

(h) No reaction.

28–9. A < C < D < B. Molecules with the larger numbers of OH groups will hydrogen bond the most, be the most polar, and have the highest boiling points.

28–11. $CH_3-CH=CH-CH_2-OH$

Answers for Unit 29

29–1. (a) Iodoethane (ethyl iodide).

(b) 2-bromopropane (isopropyl bromide).

(c) Chlorocyclohexane (cyclohexyl chloride).

(d) 2-bromobutane (*sec*-butyl bromide).

(e) 1-chloro-2-methylpropane (isobutyl chloride).

(f) 2-iodo-2-methylpropane (*t*-butyl iodide).

(g) Bromobenzene (phenyl bromide).

29–3. (a) CH_3-Cl

(b) $CH_3-\overset{\overset{\displaystyle Br}{|}}{\underset{\underset{\displaystyle CH_3}{|}}{C}}-CH_2-CH_3$ + $CH_3-\overset{\overset{\displaystyle Br}{|}}{CH}-\underset{\underset{\displaystyle CH_3}{|}}{CH}-CH_3$ 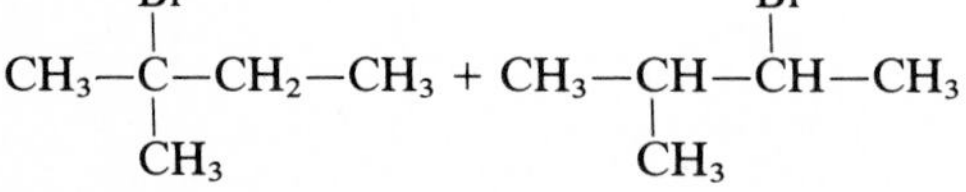
$\quad\quad\quad\quad\quad$ major

(c) 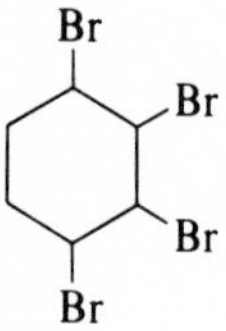

(d) 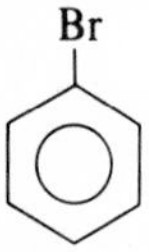

(e) $CH{-}C(CH_3){=}CH{-}CH_3 + CH_2{=}C(CH_3){-}CH_2{-}CH_3$
major (first product)

(f) 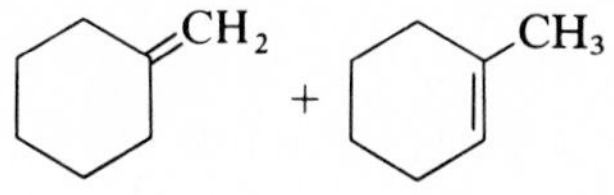
major (second product)

29–5. (a) $CH_2{=}CH{-}CH_3 + CH_3{-}CH(OH){-}CH_3$

(b) $CH_2{=}CH{-}CH_3 + CH_3{-}CH(O{-}CH_3){-}CH_3$

(c) $CH_2{=}CH{-}CH_3 + CH_3{-}CH(CH_3){-}O{-}CH(CH_3){-}CH_3$

(d) $CH_3{-}CH(CN){-}CH_3$

(e) $CH_3{-}CH(S{-}CH_3){-}CH_3$

(f) $CH_3{-}CH(NH_2){-}CH_3$

(g) $CH_3{-}CH(NH{-}CH_3){-}CH_3$

(h) 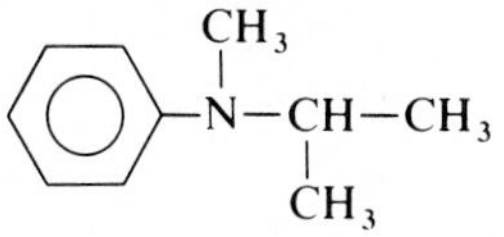

29–7. (a) $CH_3{-}CH{=}CH_2 \xrightarrow{HBr} CH_3{-}CH(Br){-}CH_3 \xrightarrow{CN^-} CH_3{-}CH(CN){-}CH_3$

(b) $CH_3{-}CH_2{-}CH_2{-}Br \xrightarrow{Base} CH_3{-}CH{=}CH_2 \xrightarrow[H^+]{H_2O} CH_3{-}CH(OH){-}CH_3$

(c) 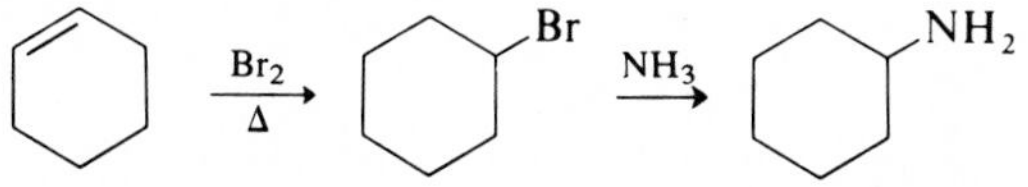

(d) $CH_3{-}CH_2{-}C({=}O){-}H \xrightarrow[Pt]{H_2} CH_3{-}CH_2{-}CH_2(OH) \xrightarrow[\Delta]{H^+} CH_3{-}CH{=}CH_2$

$\xrightarrow{HBr} CH_3{-}CH(Br){-}CH_3 \xrightarrow{CN^-} CH_3{-}CH(CN){-}CH_3$

29–9. 2,2,3,3-tetramethylbutane.

Answers for Unit 30

30–1. (a) Methyl ethyl ether. (b) Ethyl isopropyl ether.
(c) Diphenyl ether. (d) 1-methoxy-2-butene.
(e) *m*-methoxytoluene. (f) Propanethiol.
(g) Benzenethiol. (h) Methyl propyl thioether.
(i) Methyl propyl disulfide. (j) Isopropylphenyl thioether.

30–3. A < B < C.

30–5. (a) $CH_3{-}O{-}CH_3$

(b) $CH_3{-}\underset{CH_3}{\underset{|}{\overset{CH_3}{\overset{|}{C}}}}{-}O^-Na^+$

(c) 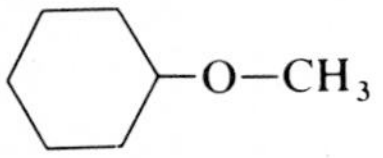

(d) $CH_3{-}\overset{CH_3}{\overset{|}{CH}}{-}O{-}CH_2{-}CH_3$

(e) $CH_3{-}CH_2{-}S{-}H$

(f) 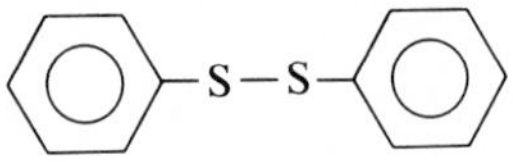

(g) 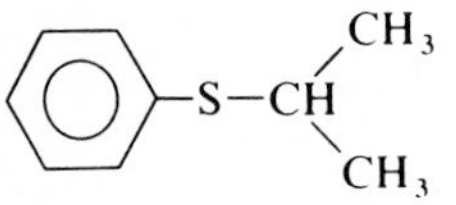

30–7. Treat both with elemental sodium; the alcohol will react and produce bubbles of hydrogen gas.

30–9. (a) $CH_3{-}CH_2{-}OH + CH_3{-}\overset{Br}{\overset{|}{CH}}{-}CH_3$ or $CH_3{-}CH_2{-}Br + CH_3{-}\overset{OH}{\overset{|}{CH}}{-}CH_3$

(b) 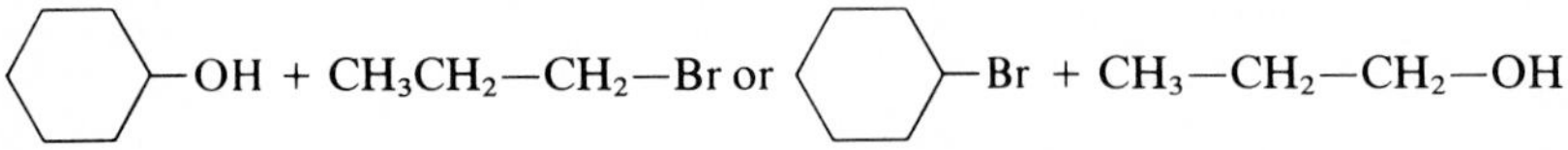

(c) $CH_3{-}CH_2{-}CH_2{-}CH_2{-}OH + CH_3CH_2{-}Br$ or $CH_3CH_2CH_2\underset{Br}{\underset{|}{CH_2}} + CH_3{-}\underset{OH}{\underset{|}{CH_2}}$

(d) $CH_3CH_2{-}\overset{OH}{\overset{|}{CH}}{-}CH_2CH_3 + CH_3{-}Br$ or $CH_3{-}CH_2{-}\underset{Br}{\underset{|}{CH}}{-}CH_2{-}CH_3 + CH_3{-}OH$

(e) $CH_3{-}\underset{CH_3}{\underset{|}{\overset{CH_3}{\overset{|}{C}}}}{-}OH + \underset{Br}{\underset{|}{\overset{CH_3}{\overset{|}{CH}}}}{-}CH_2CH_3$ or $CH_3{-}\underset{CH_3}{\underset{|}{\overset{CH_3}{\overset{|}{C}}}}{-}Br + \underset{OH}{\underset{|}{\overset{CH_3}{\overset{|}{CH}}}}{-}CH_2{-}CH_3$

30–11. $CH_3{-}OH + CH_3CH_2CH_2{-}OH \xrightarrow{H^+} \underset{(C_4H_{10}O)}{CH_3{-}O{-}CH_2{-}CH_2{-}CH_3}$

30–13. Compound 1: $C_4H_9{-}S{-}H$.
Compound 2: $C_4H_9{-}S{-}S{-}C_4H_9$.

Answers for Unit 31

31–1. (a) Diethylamine. (b) Methylethylamine.
(c) *N,N*-dimethylpropanamine. (d) *N*-methylaniline.
(e) 2-butanamine. (f) *N*-phenyl-3-methyl-2-butanamine.
(g) *N,N*-dimethyl-2-pentanamine. (h) Cyclopentylamine.
(i) *N*-ethyl-*p*-methylaniline. (j) *N*-methyl-*N*-ethyl-3-phenyl-2-butanamine.

31–3. From 31–1: primary amines = e, h; secondary amines = a, b, d, f, i; tertiary amines = c, g, j.
From 31–2: primary amines = a, c; secondary amines = b, d, e, h; tertiary amines = f, g.

31–5.

31–7. (a) $CH_3-CH_2-NH_2 + CH_3-Br$ or $CH_3CH_2-Br + CH_3-NH_2$

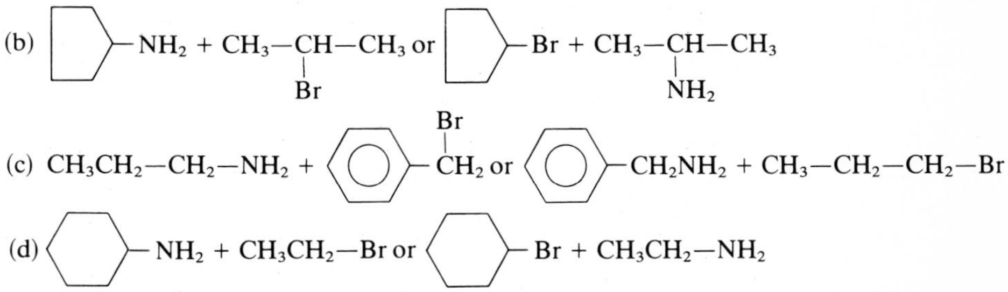

31–9. Treat with dilute HCl. Amines are soluble in dilute HCl; most other classes of compounds are not.

31–11. $CH_3-CH_2-NH-CH_3$

Answers for Unit 32

32–1. C < A < B < D. The mechanism involves the formation of carbonium ions. Compounds that form the most stable carbonium ions will react the most rapidly.

32–3. It is an SN-1 reaction.

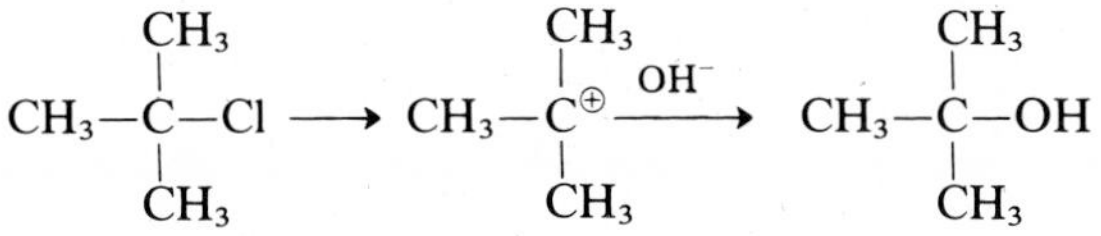

32–5. (a) CH_3-I, because I^- is a better leaving group than Br^-.
(b) *t*-butyl chloride, because Cl^- is a better leaving group than SH^-.

32–7. Compound (c) should react most rapidly; it has a primary center, a good leaving group, and a good nucleophile.

32–9. Measure the rate of the reaction using various concentrations of the nucleophile (OH^-). If the rate does not change much, the mechanism is probably SN-2. You could also start with optically active starting materials, and see if the products rotated polarized light or not. If the products did not rotate polarized light, the mechanism would probably be SN-1.

Answers for Unit 33

33–1. (a) 2-methylpropanal. (b) *p*-chlorobenzaldehyde. (c) 4-phenyl-2-pentanone. (d) Cyclohexanone. (e) 5-methyl-3-hexanone. (f) 2-methylcyclopentanone. (g) 2,4-dimethylpentanal.

33–3. (a) $CH_3-CH_2-CH_2-CH_2-C(=O)-H$

pentanal
(pentyl aldehyde)

$CH_3-CH_2-CH_2-C(=O)-CH_3$

2-pentanone
(methyl *n*-butyl ketone)

$CH_3-CH_2-C(=O)-CH_2CH_3$

3-pentanone
(diethyl ketone)

$CH_3-CH_2-CH(CH_3)-C(=O)-H$

2-methylbutanal

$CH_3-CH(CH_3)-CH_2-C(=O)-H$

3-methylbutanal

$CH_3-CH(CH_3)-C(=O)-CH_3$

3-methyl-2-butanone
(methyl isobutyl ketone)

$CH_3-C(CH_3)_2-C(=O)-H$

2,2-diethylpropanal

(b) 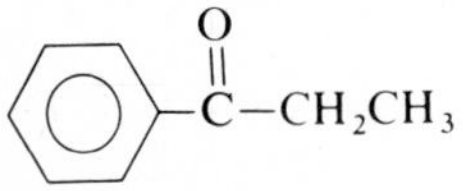

1-phenyl-1-propanone
(ethyl phenyl ketone)

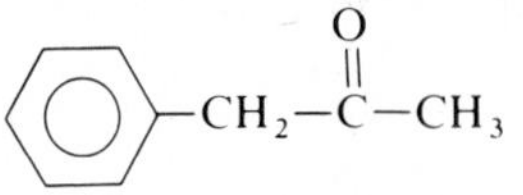

1-phenyl-2-propanone

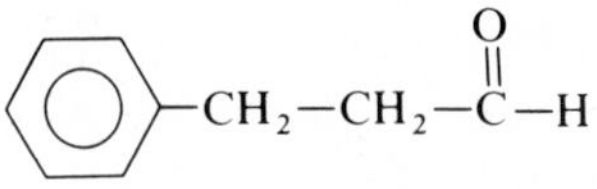

3-phenylpropanal

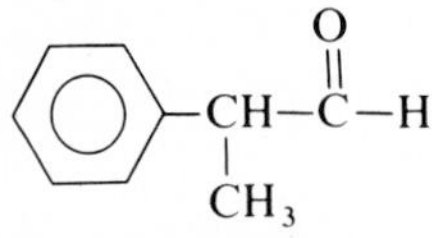

2-phenylpropanal

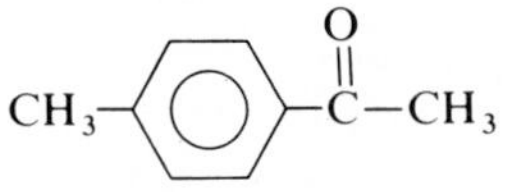

p-methylacetophenone

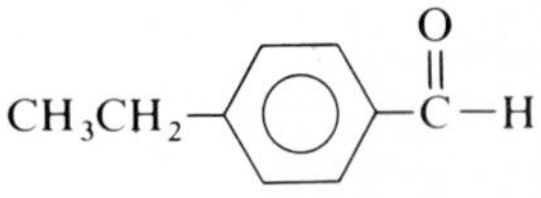

p-ethylbenzaldehyde

33–5. (a) $CH_3-C(=O)-CH_3$ (b) $CH_3CH_2CH_2-C(=O)-OH$ (c) No reaction.

(d) 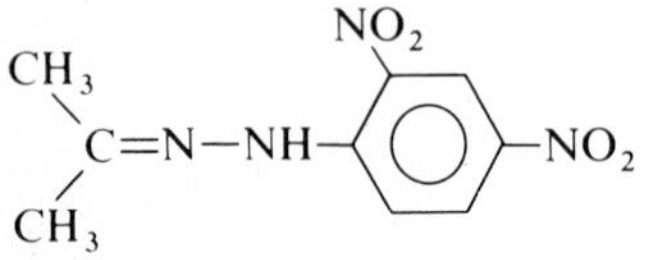

(e) 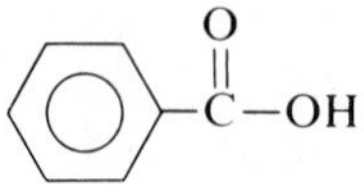

(f) No reaction.

(g) 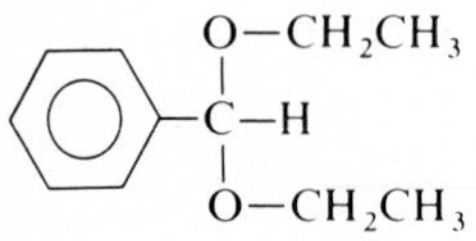

(h) $CH_3CH_2CH(OH)-CH_3$

(i) $CH_3CH_2-C(OH)(CN)-CH_3$

33–7. (a) $CH_3CH_2\overset{O}{\overset{\|}{C}}H + CH_3MgBr$ or $CH_3{-}\overset{O}{\overset{\|}{C}}{-}H + CH_3{-}CH_2MgBr$

(b) 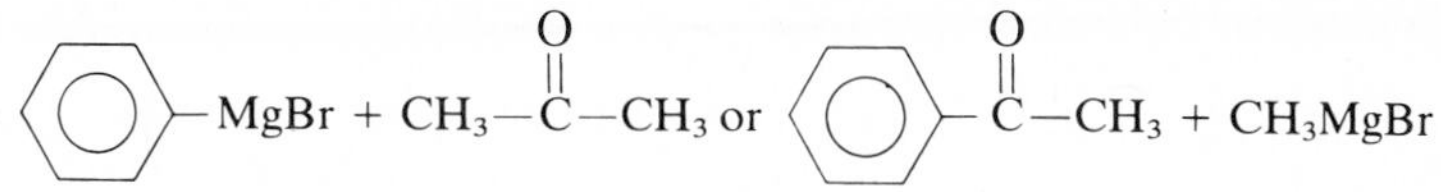

(c) $CH_3CH_2CH_2CH_2MgBr + H{-}\overset{O}{\overset{\|}{C}}{-}H$

33–9. (a) $CH_3CH_2\overset{O}{\overset{\|}{C}}H \xrightarrow[Pt]{H_2} CH_3CH_2\overset{OH}{\overset{|}{C}}H_2$

(b) (a) $\longrightarrow CH_3CH_2\overset{OH}{\overset{|}{C}}H_2 \xrightarrow[\Delta]{H^+} CH_3{-}CH{=}CH_2 \xrightarrow[H^+]{H_2O} CH_3{-}\overset{OH}{\overset{|}{C}}H{-}CH_3$

(c) See (b).

(d) $CH_3CH_2\overset{O}{\overset{\|}{C}}{-}H \xrightarrow{KMnO_4} CH_3CH_2\overset{O}{\overset{\|}{C}}{-}OH$

(e) (a) $\longrightarrow CH_3CH_2\overset{OH}{\overset{|}{C}}H_2 \xrightarrow{HBr} CH_3CH_2{-}CH_2{-}Br \xrightarrow{CN^-} CH_3CH_2CH_2{-}CN$

(f) (e) $\longrightarrow CH_3CH_2CH_2{-}CN \xrightarrow[Pt]{2H_2} CH_3CH_2CH_2CH_2{-}NH_2 \xrightarrow{CH_3Br}$

$CH_3{-}CH_2{-}CH_2{-}\underset{CH_3{-}NH}{\underset{|}{C}}H$

(g) Aldol: $2CH_3CH_2\overset{O}{\overset{\|}{C}}{-}H \xrightarrow{Base} CH_3\overset{H-C=O}{\overset{|}{C}}H{-}\overset{OH}{\overset{|}{C}}H{-}CH_2CH_3$

33–11. (a) $CH_3CH_2{-}\underset{OH}{\underset{|}{\overset{O-CH_2CH_3}{\overset{|}{C}}}}H \longrightarrow CH_3CH_2\underset{OCH_2CH_3}{\underset{|}{\overset{OCH_2CH_3}{\overset{|}{C}}}}H$

↑ hemiacetal ↓ ↑ acetal ↓

(b) 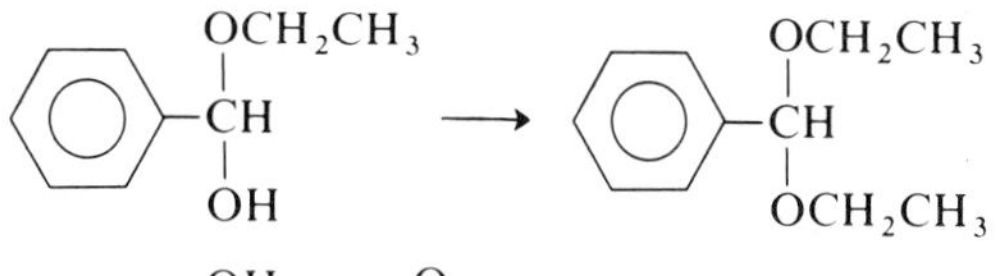

33–13. (a) 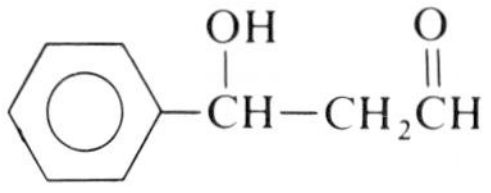

(b) $CH_3CH_2\overset{OH}{\overset{|}{C}}H{-}\underset{CH_3}{\underset{|}{C}}H{-}\overset{O}{\overset{\|}{C}}H$

(c) 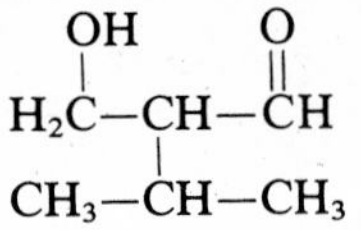

(d) 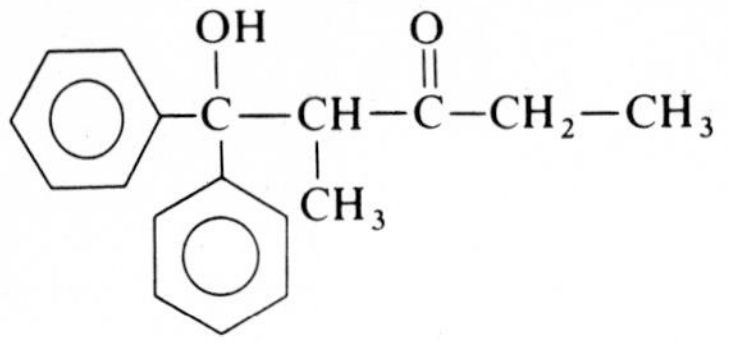

33–15. 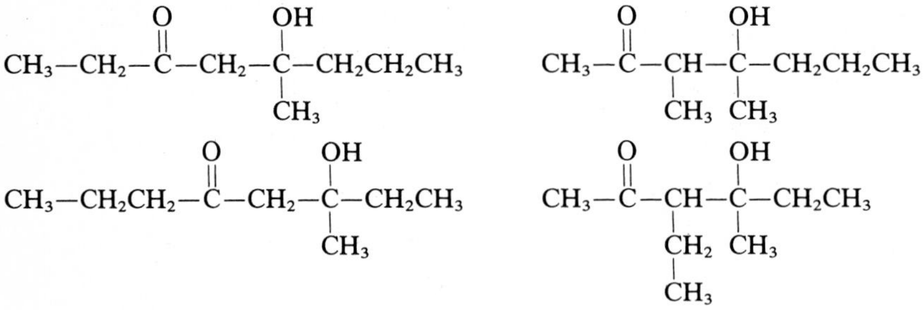

33–17. (a) 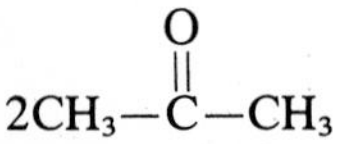

(b) 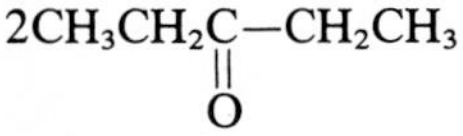

(c) 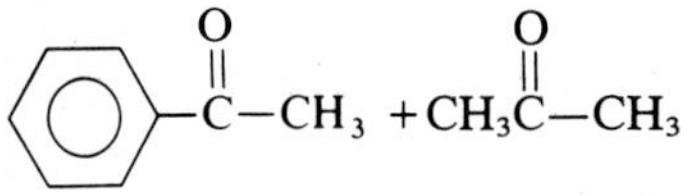

(d) 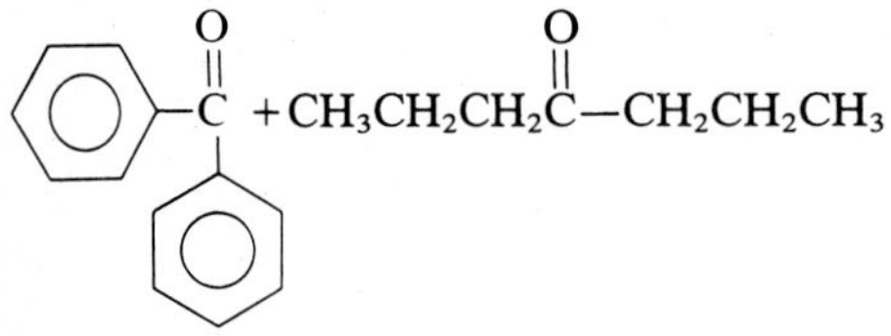

33–19. It must be a symmetrical ketone.

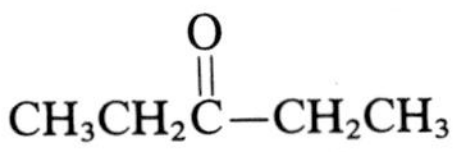

Answers for Unit 34

34–1. (a) *m*-bromobenzoic acid.
(b) Formic acid.
(c) Acetic acid.
(d) 2,2-dimethylbutanoic acid.
(e) 4-phenylpentanoic acid.
(f) 5-bromo-4-methylhexanoic acid.
(g) 2-butenoic acid.

34–3. (a) 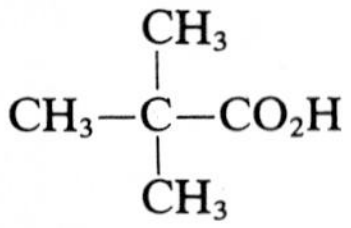

(b) 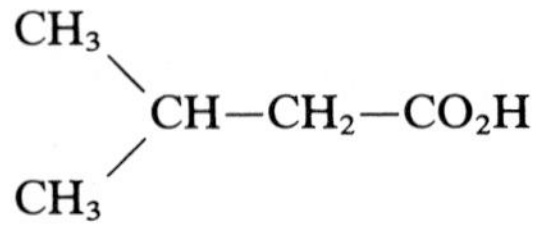

(c) 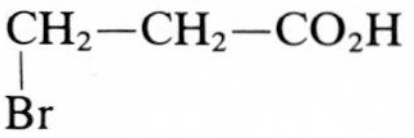

(d) 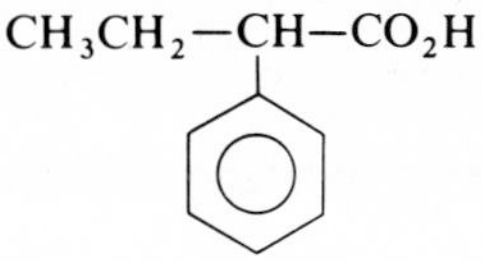

(e) 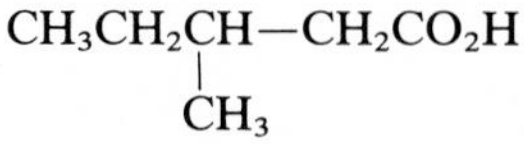

(f) 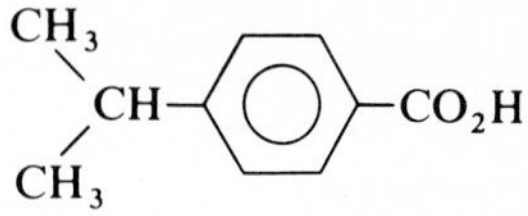

34–5. After loss of hydrogen ion, the carboxylate ion from carboxylic acid is resonance stabilized. Thus it forms easier than the alkoxide ion resulting from loss of hydrogen ion from an alcohol. The more stable (weaker) the base, the stronger is the conjugate acid.

34–7. (a) Propanoic acid and methyl alcohol.
(b) Benzoic acid and ethanol.
(c) Acetic acid and phenol.
(d) Acetic acid and 2-propanol.
(e) 2-methylpropanoic acid and ethanol.

34–9. Test the chemical with dilute sodium bicarbonate solution. If it is a carboxylic acid, it will dissolve and bubbles of carbon dioxide will evolve.

34–11. The molecular weight should be 88 g.

Answers for Unit 35

35–1. (a) Methyl propanate.
(b) *Sec*-butyl benzoate.
(c) 2-phenylbutanamide.
(d) 4-methylpentanoyl chloride.
(e) *p*-methylbenzoyl chloride.
(f) *N,N*-dimethylpentanamide.
(g) Phenyl 2-methylpropanate.
(h) *N*-methyl-*N*-phenyl-4,5-dimethylhexanamide.
(i) *n*-propyl 3-bromo-2-methylbutanate.

35–3. 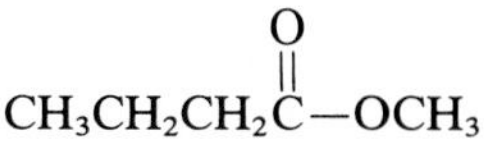

methyl butanate

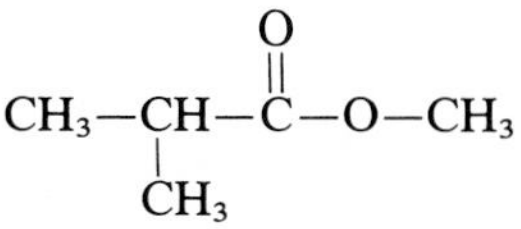

methyl 2-methylpropanate

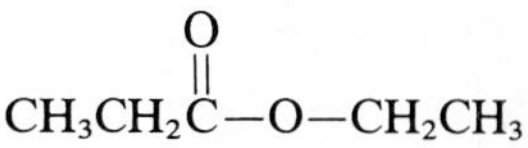

ethyl propanate

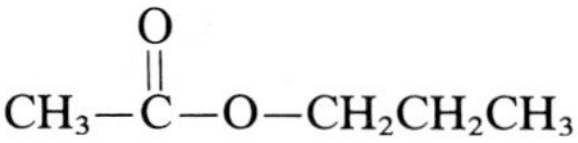

n-propyl acetate

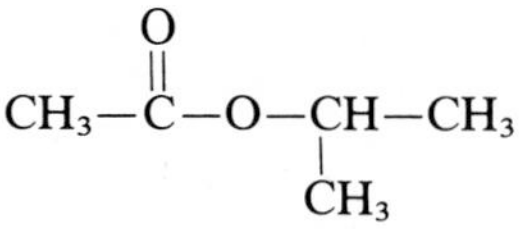

isopropyl acetate

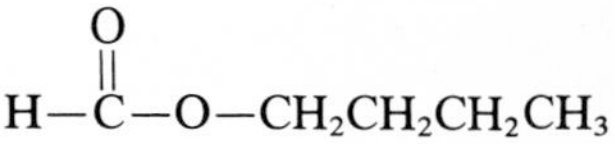

n-butyl formate

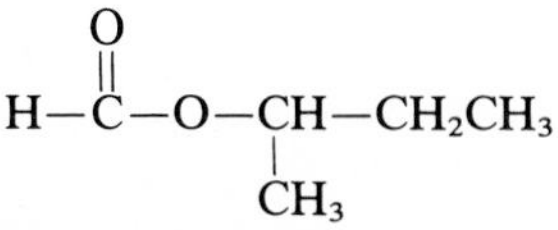

sec-butyl formate

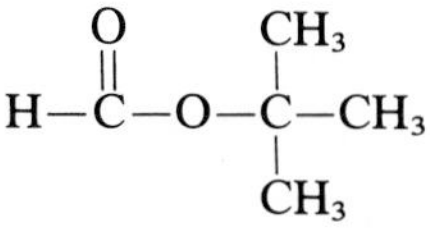

t-butyl formate

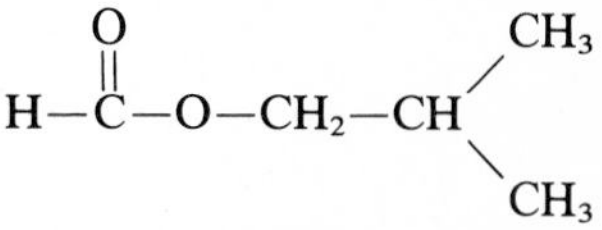

*iso*butyl formate

35–5. (a) 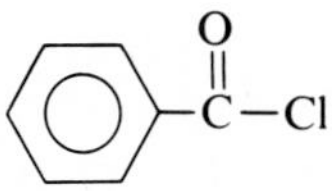

(b) 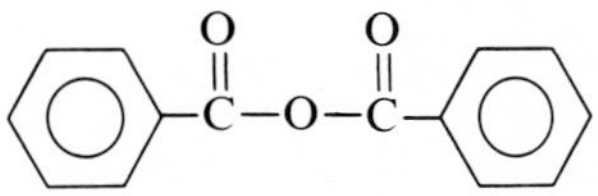

(c) $CH_3-\overset{O}{\overset{||}{C}}-NH-CH_2CH_3$

(d) $CH_3-\overset{O}{\overset{||}{C}}-O-\underset{CH_3}{\underset{|}{CH}}-CH_3$

(e) 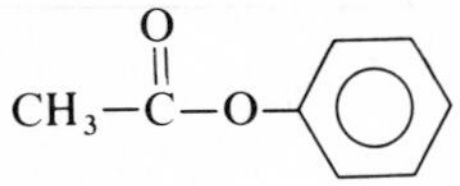

(f) $CH_3CH_2\overset{O}{\overset{||}{C}}-O-CH_2CH_2CH_3$

(g) 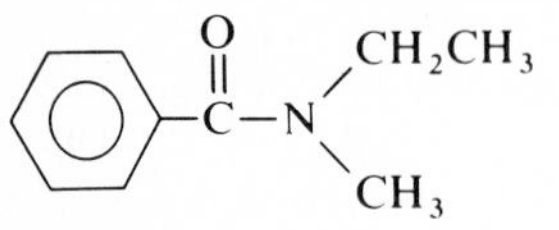

(h) 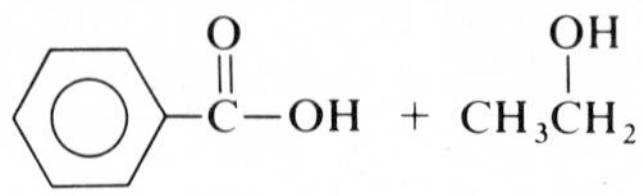

(i) $CH_3\underset{OH_3}{\underset{|}{CH}}-\overset{O}{\overset{||}{C}}-OH + \overset{+}{N}H_4$

35–7. The ester could be identified by its pleasant odor. Treat the other two compounds with dilute sodium bicarbonate; the compound that dissolved and gave off bubbles of carbon dioxide would be the acid.

35–9. (a) $CH_3\overset{O}{\overset{||}{C}}-O-CH_3$

(b) 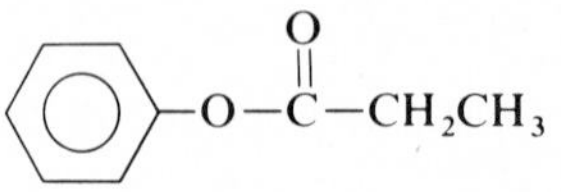

(c) $CH_3CH_2-O-\overset{O}{\overset{||}{C}}-CH_2CH_3$

(d) 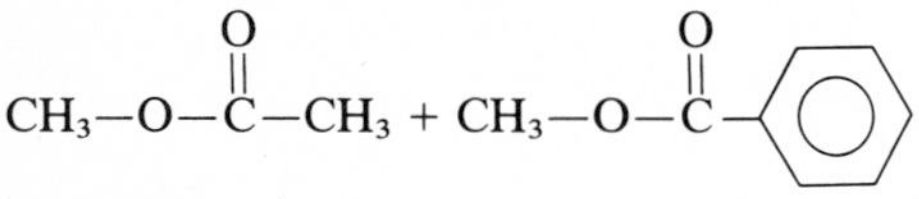

(e) 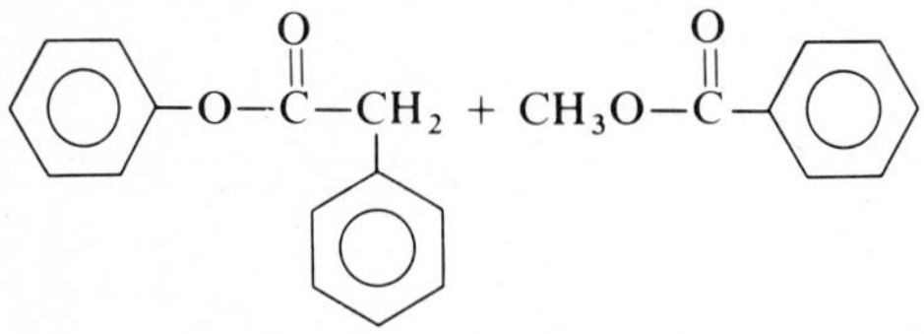

(f) $CH_3-O-\overset{O}{\overset{||}{C}}-CH_2CH_2CH_3 + CH_3-O-\overset{O}{\overset{||}{C}}-H$

35–11. (a) $CH_3-\overset{O}{\overset{||}{C}}-CH_3$

(b) $CH_3\overset{O}{\overset{||}{C}}-CH_2CH_3$

(c) $CH_3\overset{O}{\overset{||}{C}}-CH_2CH_3$

(d) $(CH_3)_2CH-\overset{O}{\overset{||}{C}}-CH(CH_3)_2$

Answers for Unit 36

36–1. In order to have substitution reactions in aldehydes and ketones, an H^- or R^- ion would have to be eliminated from the molecule. Both of these ions are strong bases and poor leaving groups. Thus *addition* reactions are the major mode of reaction.

36–3. Substitution reactions depend in part on the ability of a group to leave the reaction site. Table 36–1 lists the leaving group abilities, so acid chlorides are most reactive. Then come acids, followed by esters, amides, and finally ketones.

36–5.

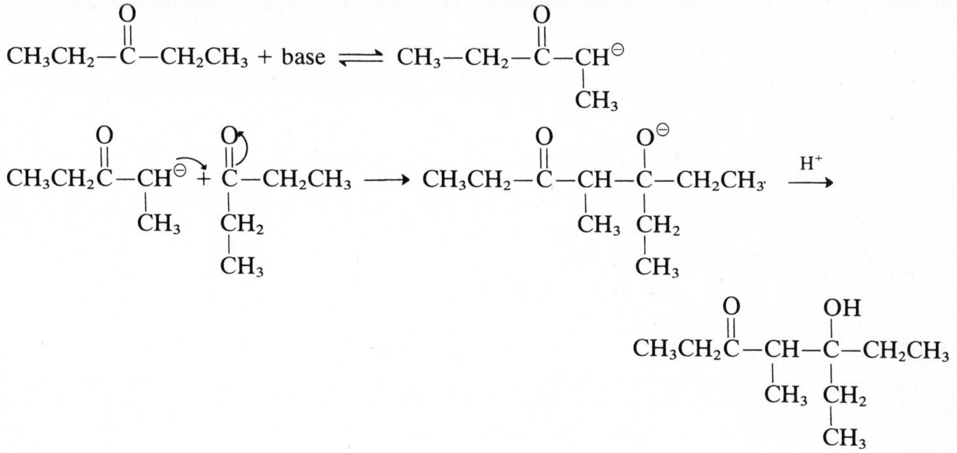

36–7. See problem (36–3).

36–9.

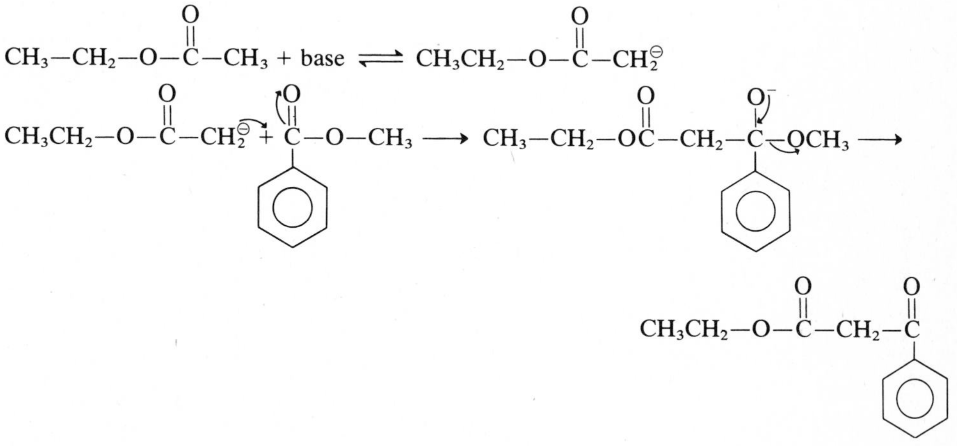

Answers for Unit 37

37–1.

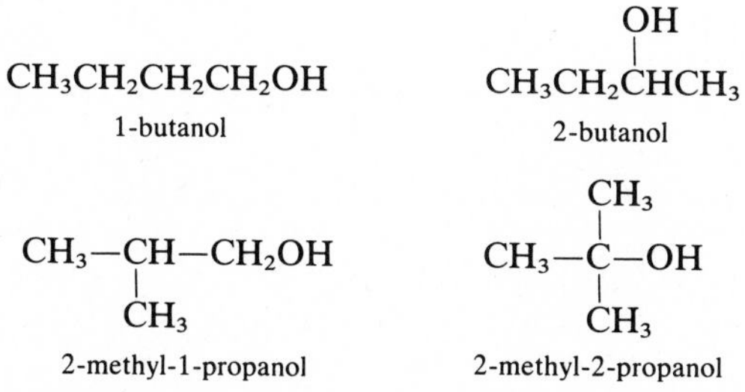

1-butanol
2-butanol
2-methyl-1-propanol
2-methyl-2-propanol

37–3.

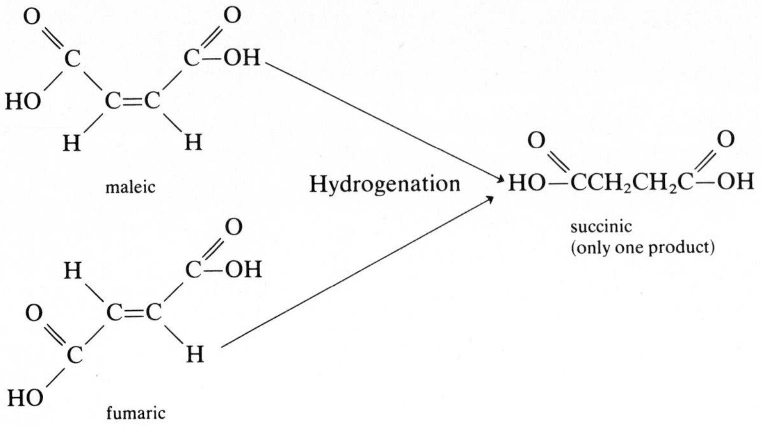

maleic
Hydrogenation
succinic
(only one product)
fumaric

37–5. Yes,

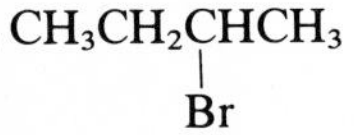

37–7. No, each would cancel the opposite rotation of the other.

Answers for Unit 38

38–1.

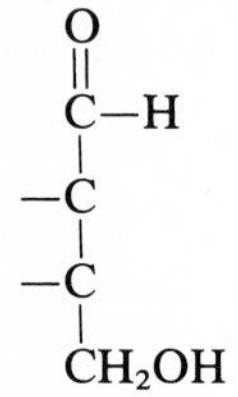

L-erythrose

38–3. No, neither an aldehyde nor a ketone.

38–5. No, the glucoside cannot form an aldehyde.

38–7.

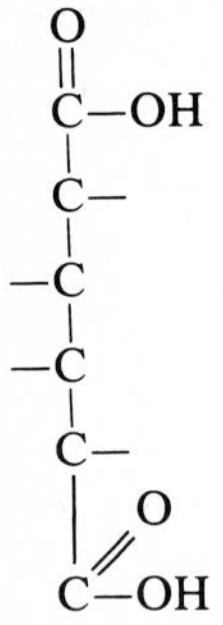

38–9.

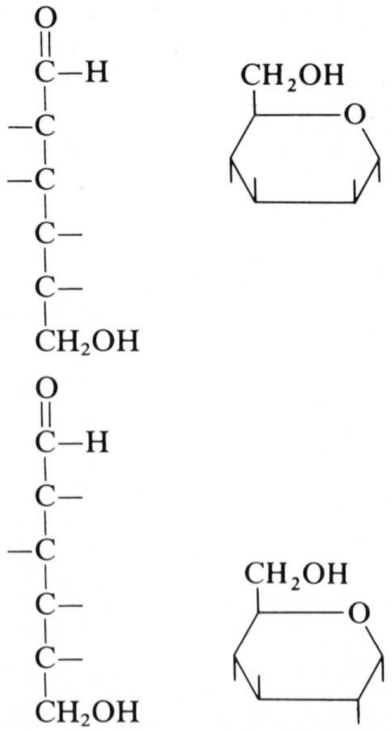

Answers for Unit 39

39–1.

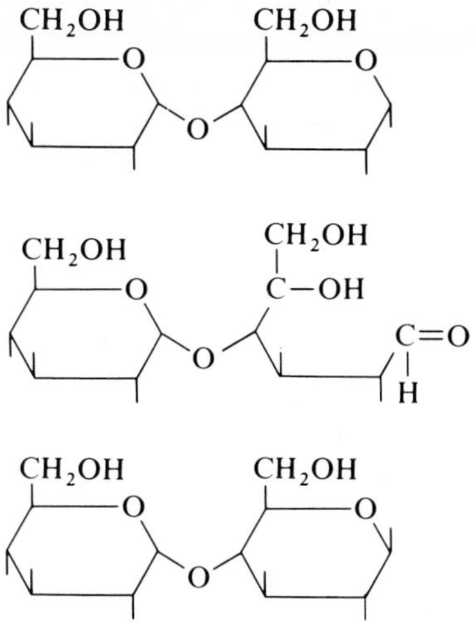

CH₂OH
O
O
CH₂OH
O
CH₂OH
O
O
CH₂OH
C—OH
C=O
H
CH₂OH
O
O
CH₂OH
O

39–3. There is one free aldehyde group that could be oxidized. There is only one end to a very large molecule, so the oxidation would very likely not be readily observed in most reactions.

39–5.

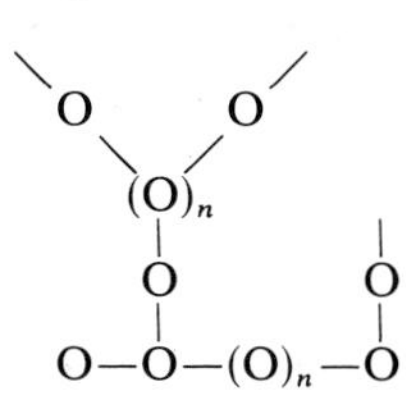

O
O
(O)ₙ
O
O
O—O—(O)ₙ—O

39–7.

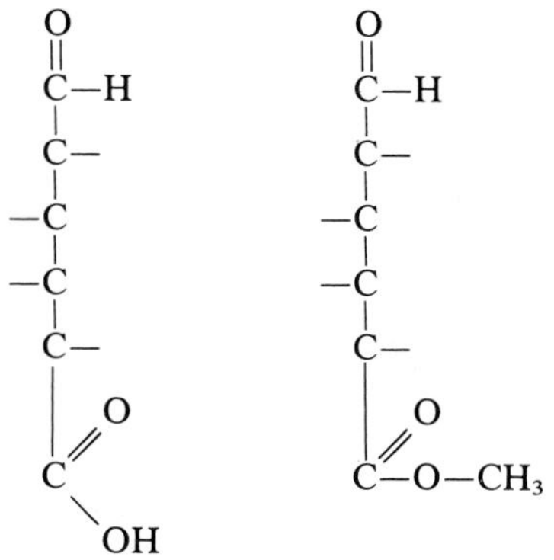

O
C—H
C—
—C
—C
C—
O
C
OH
O
C—H
C—
—C
—C
C—
O
C—O—CH₃

galacturonic acid

Answers for Unit 40

40–1. $CH_3(CH_2)_{14}\overset{\overset{\displaystyle O}{\|}}{C}—O—C_{16}H_{33}$

40–3.

$$CH_2O-\overset{\overset{\displaystyle O}{\|}}{C}-(CH_2)_{10}CH_3$$

$$CHO-\overset{\overset{\displaystyle O}{\|}}{C}-(CH_2)_{10}CH_3$$

$$CH_2O-\overset{\overset{\displaystyle O}{\|}}{C}-(CH_2)_{10}CH_3$$

$$CH_2O-\overset{\overset{\displaystyle O}{\|}}{C}-(CH_2)_7CH=CHCH_2CH=CH(CH_2)_4CH_3$$

$$CHO-\overset{\overset{\displaystyle O}{\|}}{C}-(CH)_2CH=CHCH_2CH=CH(CH_2)_4CH_3$$

$$CH_2O-\overset{\overset{\displaystyle O}{\|}}{C}-(CH)_2CH=CHCH_2CH=CH(CH_2)_4CH_3$$ 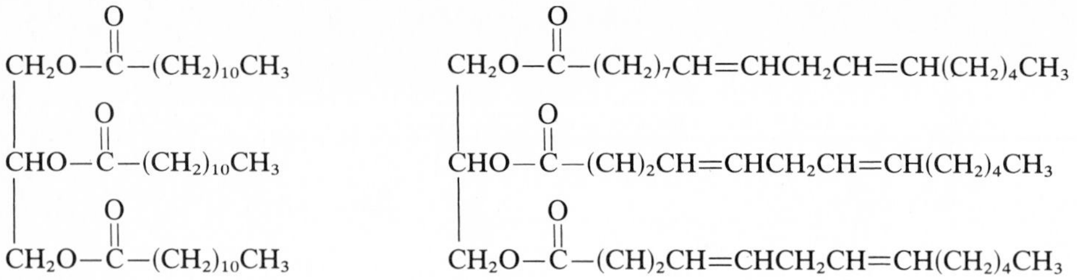

40–5. Peanut oil, because it contains a greater proportion of unsaturated fatty acids.

40–7. Higher, because it has a lower molecular weight and more molecules to saponify.

40–9. Must be capable of interacting with both aqueous and nonaqueous environments. Must be quite readily degradable, economical. Must not interact with a variety of ions to produce undesirable chemical reactions. Must not contain ions or portions with undesirable chemical consequences.

Answers for Unit 41

41–1. A phospholipid would be more water soluble because it contains an ionic portion.

41–3. Since membranes are rich in lipids and yet exposed on both sides to an aqueous environment, the double layer provides one good model.

41–5. Since the A ring of estradiol is aromatic, it would be impossible to have another constituent at position 10.

41–7. The body can synthesize cholesterol from simple precursor molecules in quite high amounts.

41–9. In the elderly the former predominate and sexually characteristic hormone levels may decrease. The balance shifts.

Answers for Unit 42

42–1. Note it does have an amine group.

42–3. Good sources: carrots, tomatoes.
Poor sources: potatoes.

42–5. Note the side chain at no. 17:

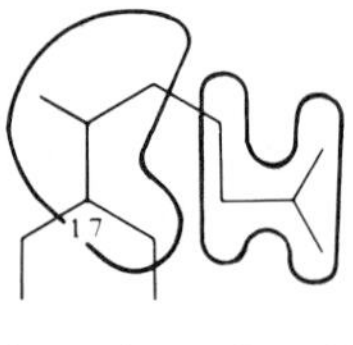

42–7. Overabundance of vitamin D may be harmful. The pigments may act to filter out the sunlight involved in the formation of some vitamins.

42–9. K_3 would have the greatest water solubility. The long aliphatic side chain of vitamin K_2 causes it to have limited water solubility.

Answers for Unit 43

43–1. Since insulin is a protein, it would be degraded in the digestive tract.

43–3. Alanine has an asymmetric carbon and thus can exist in both D and L forms. Glycine does not have an asymmetric carbon.

43–5. 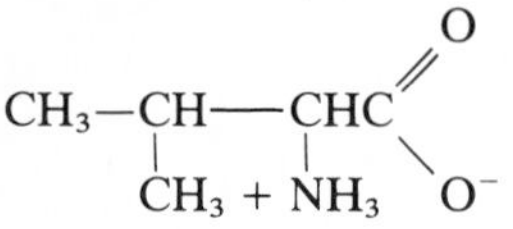

$$CH_3-CH-\underset{\underset{\displaystyle NH_3^+}{|}}{CHC}\overset{\overset{\displaystyle O}{/\!\!/}}{\underset{\displaystyle O^-}{\backslash}}$$
$$\quad\;\; \underset{\displaystyle CH_3}{|}$$

43–7. $\mathrm{pI} = \dfrac{8.9 + 10.5}{2} = 9.7.$

43–9. Creatinine levels are related to muscle mass. Many females have a lower muscle mass than males.

Answers for Unit 44

44–1. 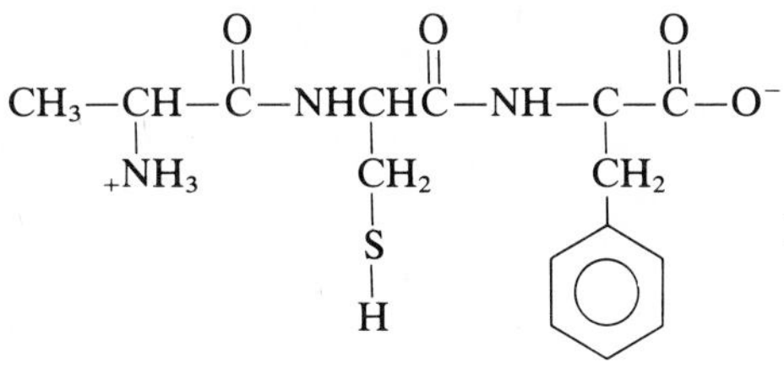

Reduced form

Under oxidizing conditions, two similar peptides would be joined by disulfide bond.

44–3. Alanine and cysteine would be isolated as amino acids along with

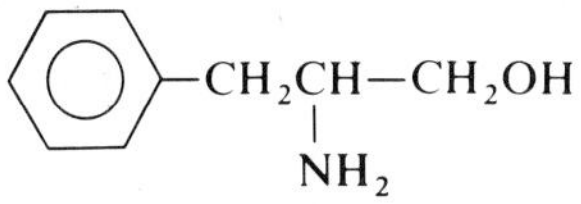

44–5. The oxidation would cause the disulfide bonds to re-form. There may be some mismatching in the formation of the disulfides. This would yield an altered molecule with decreased biological activity.

44–7.

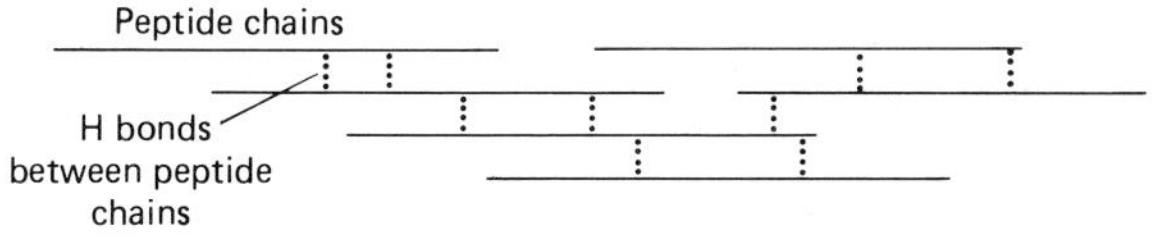
Peptide chains
H bonds between peptide chains

44–9. Interior: leu, ileu, phe.
Exterior: glu, asp, lys.

Answers for Unit 45

45–1. 31.8 g.

45–3. 68,000.

45–5. Protein concentration:

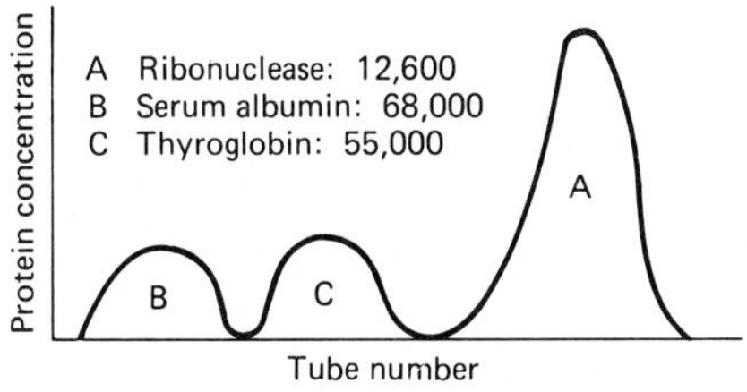
Protein concentration
A Ribonuclease: 12,600
B Serum albumin: 68,000
C Thyroglobin: 55,000
A
B
C
Tube number

45–7. Since the molecular weights are quite similar, you would not expect a great separation upon ultracentrifugation.

Answers for Unit 46

46–1. Nucleus: DNA synthesis.
Mitochondrion: oxidative metabolism.
Lysosomes: degradative reactions.
Ribosomes: protein biosynthesis.
Cytosol: metabolism of carbohydrates and amino acids.

46–3. Five percent of body weight is blood plasma.

46–5. ^3H is radioactive and can be detected more readily.

46–7. If the half-life were shortened, it would decay sooner and no longer pose a hazard to a living cell.

46–9. More nitrogen is being eliminated in a given period than taken into the body.

Answers for Unit 47

47–1. H, water; O, water; C, carbohydrates, lipids, proteins; N, proteins; Ca, bone structure.

47–3. Ca^{2+} and Mg^{2+} are closely related from a chemical point of view. Check your periodic table.

47–5. Fe is part of the cytochrome molecules.

47–7. B_{12}.

Answers for Unit 48

48–1. Coenzyme, it is much smaller in size.

48–3. Lysozyme could be influenced by oxidizing and reducing agents since it has disulfide bridges.

48–5. LDH.

48–7. Alcohol dehydrogenase.

48–9.

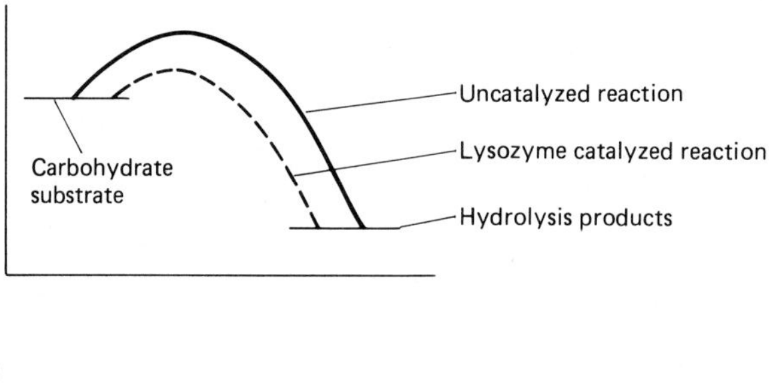
Uncatalyzed reaction
Lysozyme catalyzed reaction
Carbohydrate substrate
Hydrolysis products

Answers for Unit 49

49–1. dES/dt means the change (formation) of the enzyme substrate concentration with time. Since dP/dt would be the formation of the product with time, the ES complex would have to form first before any product would form.

49–3. In a closed system the substrate may be consumed, and thus its concentration may limit the rate of the reaction. There are a variety of other reasons, such as product inhibition, to contend with near the completion of a reacting system.

49–5. If you would eliminate an inhibitor that might have been present in the original crude preparation the specific activity would increase.

49–7. The ionization of the amino acid side chains of importance in the reaction would follow a rather symmetrical process. Heat denaturation might be rather abrupt and lead to a rather sharp decline in enzymatic activity.

49–9. (a)

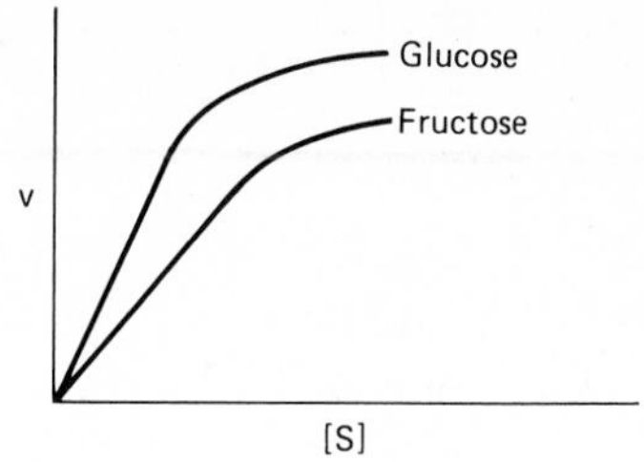

Glucose
Fructose
V
[S]

(b) Glucose would be considered the best substrate.

Answers for Unit 50

50–1. See text.
50–3. NADH + H^+ + FMN $\rightleftharpoons$ NAD + FMN·2H
50–5. The sulfur atom of the coenzyme is the active site to become chemically attached to the two-carbon unit.
50–7. Biotin.
50–9. THFA is heavily involved with one-carbon metabolism; coenzyme A is heavily involved with two-carbon metabolism.

Answers for Unit 51

51–1. (a) − (b) + (c) + (d) − (e) −
51–3. No. Yes.
51–5. O. Same kinds of bonds broken as bonds formed.
51–7. −10.3 kcal/mol.

Answers for Unit 52

52–1. (a) Reaction 1. (b) −3.8 kcal/mol; −11.8 kcal/mol.
52–3. See text.
52–5. No; the total energy yield is dependent only upon the initial and final states.
52–7. Stimulate electron flow.
52–9.

$$\Delta G^{0'} = -46{,}000 \frac{\text{cal}}{\text{V}} \times 0.04\ \text{V}$$

$$= -1840\ \text{cal} \quad \text{(not sufficient to drive the synthesis of an ATP)}$$

Answers for Unit 53

53–1.

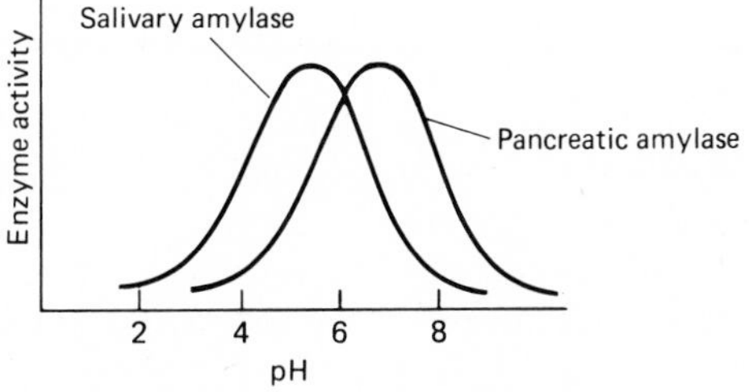

Salivary amylase
Pancreatic amylase
Enzyme activity
2
4
6
8
pH

53–3. High levels of citrate, a reactant nearer end of a sequence of enzymatic events, would be a negative inhibitor for phosphofructokinase, an enzyme near the beginning of this series of events.

53–5. (a) 6ATP. (b) 0ATP.

Answers for Unit 54

54–1. (a) Oxidative decarboxylation. (b) Succinic dehydrogenase. (c) Pyruvate carboxylase.

54–3. 38 net ATP's.

54–5. The energy release in the case of acetyl CoA is consumed in the synthesis of citrate. No synthetic reactions requiring energy are involved in the case of succinyl CoA.

54–7. (a)

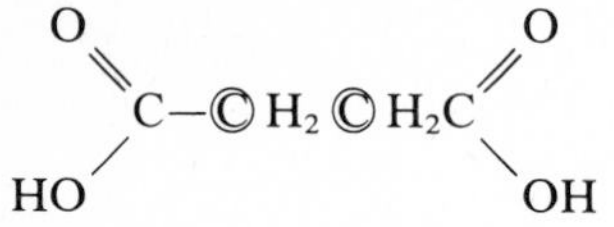

(b)

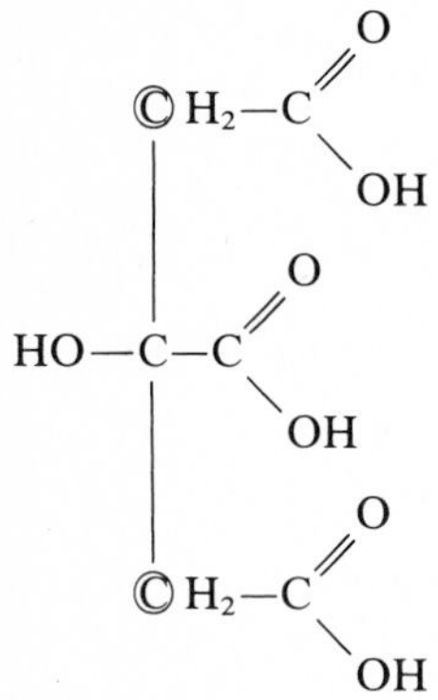

Circled atoms would bear ^{14}C label.

Answers for Unit 55

55–1. Same.

55–3. No galactose to lead to problems.

55–5.

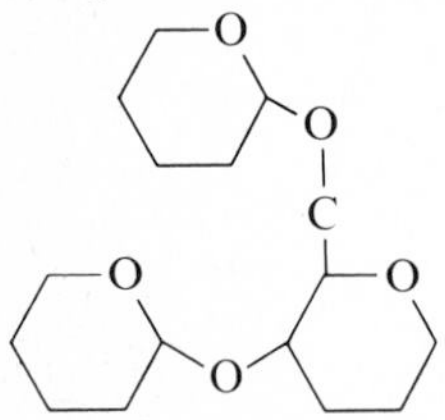

55–7. c-AMP influences the enzymes involved with both synthesis and degradation of glycogen.

55–9. Numbers 3 and 4.

Answers for Unit 56

56–1. Shorter-chained fatty acids, because their metabolism is not as dependent upon lipoproteins as the longer-chained fatty acids.

56–3. (a) 4 ATP. (b) 28 ATP. (c) 106 ATP.

56–5. Conversion to CO_2 via TCA; synthesis of new fatty acids; condensation to form steroids or ketone bodies.

56–7. $CH_3CH(OH)CH_2C(=O)O^-$

56–9. Acetyl CoA carboxylase is the first reaction leading to biosynthesis of fatty acids. When citrate is present in high concentrations, the cell is in a nutritional state of plenty. Fatty acid and lipid synthesis would be stimulated under such conditions.

56–11. (a) Alkane → alkene (flavin coenzyme involved generally).
Alkene → alkane (NADPH involved generally).
(b) $FADH_2$ generates two ATP's via oxidative phosphorylation. NADH generates three ATP's via oxidative phosphorylation.

56–13. The acetyl CoA formed from glucose can provide the crossover point leading to the synthesis of fatty acids. Glycerol also directly interrelated.

Answers for Unit 57

57–1. . . . ala-lys-gly-gly-arg-ala-tyr-gly . . .
(↑ (b) at lys; ↑ (b) at arg; ↑ (a, c) at tyr)

57–3. Besides immunological advantages, a few amino acids (20) are sufficient to build the great variety of different proteins needed within the cell.

57–5. They are often chemically quite similar to the structures of molecules found in other metabolic pathways. For example, glutamic acid and α-ketoglutaric acid.

57–7. (a) Valine. (b) Decarboxylase.

Answers for Unit 58

58–1. $CH_3CH(NH_3^+)C(=O)-O^- + {}^-O-C(=O)CH_2C(=O)C(=O)-O^- \rightleftharpoons CH_3C(=O)C(=O)-O^- + {}^-O(O=)CCH_2CH(NH_3^+)C(=O)-O^-$

58–3. Decarboxylation involves a loss of CO_2 from the system; aminotransferase involves only a transfer of groups between molecules within the system.

58–5. The reactions involving the transfer of the methyl group would be driven by the energy released in the overall reaction of this molecule with the high group transfer potential.

58–7. The reaction proceeds essentially in one direction from methionine to cysteine in the human system.

Answers for Unit 59

59–1. Phenylalanine and glycine.

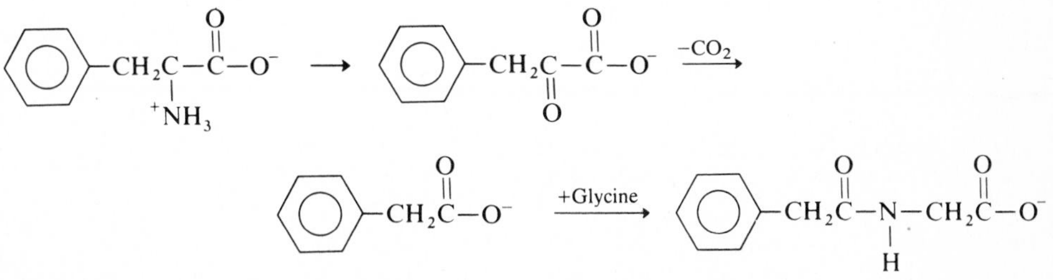

59–3. Decreases.

59–5. See Figure 59–1.

59–7. 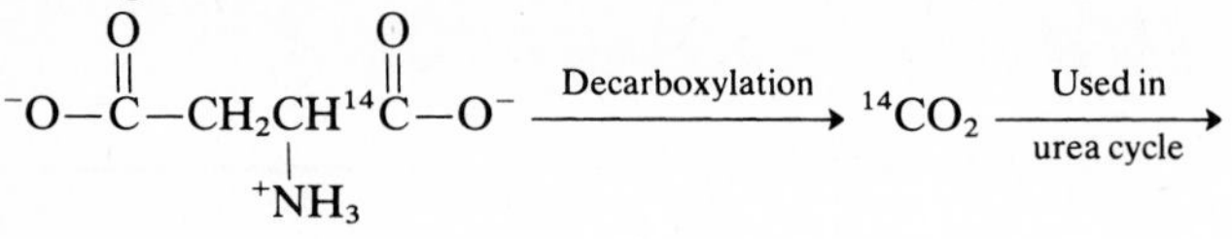

Decarboxylation
$^{14}CO_2$
Used in
urea cycle

Answers for Unit 60

(a) 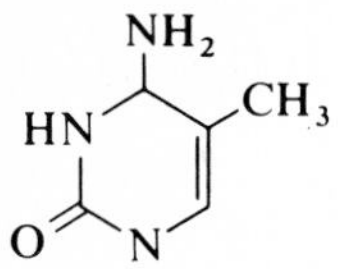

(b) 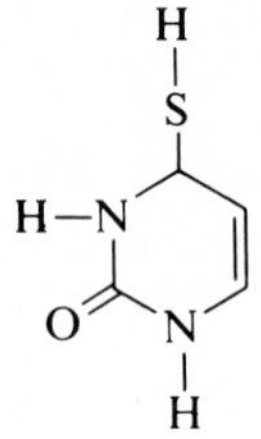

(c) 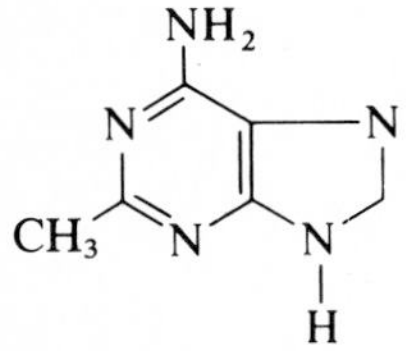

(d) 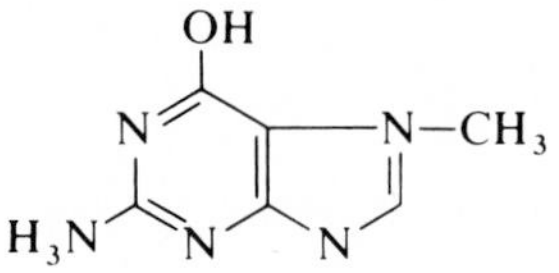

60–3. T–T–T–C–A–A–C–G–C–G.

60–5.

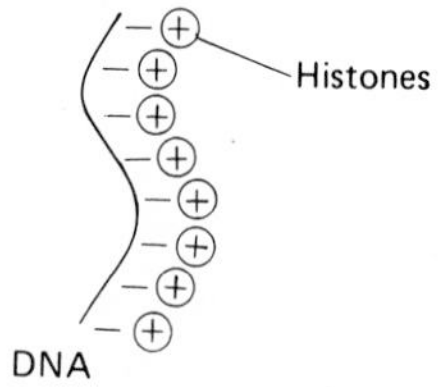

Histones
DNA

60–7. Higher.

60–9.

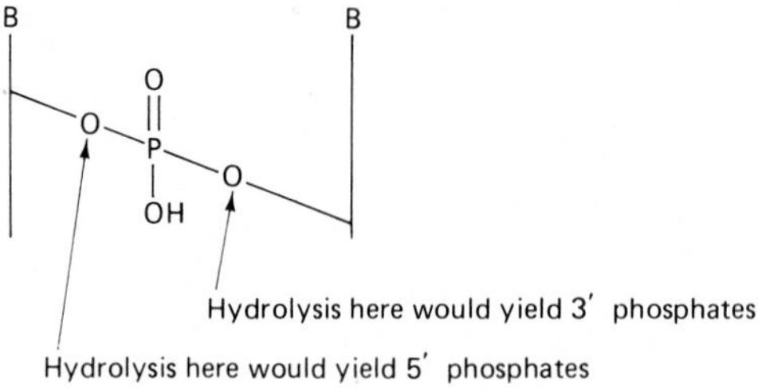

B
B
O
P
O
O
OH
Hydrolysis here would yield 3′ phosphates
Hydrolysis here would yield 5′ phosphates

Answers for Unit 61

61–1. Common: phos.; A, G, C.
Different: ribose and deoxyribose; U in RNA; T in DNA.

61–3. U–U–U–C–A–A–C–G–C–G.

61–5. A primer may act as a rather nonspecific strand upon which to build or lengthen a molecule. A template would be specifically involved in determining the sequences of bases in the molecule being synthesized.

61–7. Phosphorylase system would not yield sequence of bases in controlled fashion required in biosynthesis.

61–9. Urea: $\frac{C}{N} = \frac{1}{2}$, better efficiency.

Uric acid: $\frac{C}{N} = \frac{5}{4}$.

Answers for Unit 62

62–1. In many cases the third base in the triplet code can be one of several possibilities. (See wobble effect.)

62–3. There cannot be an addition of another group at this initial position.

62–5. AUAUA; Ile Tyr . . . .

62–7. In the initiation process, just after the addition of the charged t-RNA^{fmet}

62–9. This might indicate that there are some sequences in the DNA that are common. Perhaps these similar sequences arose from the same precursors.

index

A 2-PLACE LOGARITHM TABLE

	M	Decimal portion of number 0	1	2	3	4	5	6	7	8	9
Whole numbers	1	.000	.041	.079	.114	.146	.176	.204	.230	.255	.279
	2	.301	.322	.342	.362	.380	.398	.415	.431	.447	.462
	3	.477	.491	.505	.518	.531	.544	.556	.568	.580	.591
	4	.602	.613	.623	.634	.644	.653	.663	.672	.681	.690
	5	.699	.708	.716	.724	.732	.740	.748	.756	.763	.771
	6	.778	.785	.792	.799	.806	.813	.819	.826	.832	.839
	7	.845	.851	.857	.863	.869	.875	.881	.886	.892	.897
	8	.903	.908	.914	.919	.924	.929	.934	.939	.944	.949
	9	.954	.959	.963	.968	.973	.977	.982	.987	.991	.996

log 1.2 = 0.079 antilog .568 = 3.7
log 3.0 = 0.477 antilog .845 = 7.1
log 6.9 = 0.839 antilog .114 = 1.3
log 8.5 = 0.929 antilog .987 = 9.7